THEORY OF STOCHASTIC DIFFERENTIAL EQUATIONS WITH JUMPS AND APPLICATIONS

MATHEMATICAL AND ANALYTICAL TECHNIQUES WITH APPLICATIONS TO ENGINEERING

MATHEMATICAL AND ANALYTICAL TECHNIQUES WITH APPLICATIONS TO ENGINEERING

Alan Jeffrey, *Consulting Editor*

Published:

Inverse Problems
A. G. Ramm

Singular Perturbation Theory
R. S. Johnson

Methods for Constructing Exact Solutions of Partial Differential Equations with Applications
S. V. Meleshko

Stochastic Differential Equations with Applications
R. Situ

Forthcoming:

The Fast Solution of Boundary Integral Equations
S. Rjasanow and O. Steinbach

THEORY OF STOCHASTIC DIFFERENTIAL EQUATIONS WITH JUMPS AND APPLICATIONS

MATHEMATICAL AND ANALYTICAL TECHNIQUES WITH APPLICATIONS TO ENGINEERING

RONG SITU

Springer

Library of Congress Cataloging-in-Publication Data

Theory of Stochastic Differential Equations with Jumps and Applications:
Mathematical and Analytical Techniques with Applications to Engineering
By Rong Situ

ISBN-10: 0-387-25083-2 e-ISBN 10: 0-387-25175-8 Printed on acid-free paper.
ISBN-13: 978-0387-25083-0 e-ISBN-13: 978-0387-25175-2

Printed in the United States of America. (BS/DH)

9 8 7 6 5 4 3 2 1 SPIN 11399278

springeronline.com

Contents

Preface xi

Acknowledgement xvii

Abbreviations and Some Explanations xix

I Stochastic Differential Equations with Jumps in R^d **1**

1 Martingale Theory and the Stochastic Integral for Point Processes **3**

1.1 Concept of a Martingale. 3
1.2 Stopping Times. Predictable Process 5
1.3 Martingales with Discrete Time 8
1.4 Uniform Integrability and Martingales 12
1.5 Martingales with Continuous Time 17
1.6 Doob-Meyer Decomposition Theorem 19
1.7 Poisson Random Measure and Its Existence 26
1.8 Poisson Point Process and Its Existence 28
1.9 Stochastic Integral for Point Process. Square Integrable Martingales 32

2 Brownian Motion, Stochastic Integral and Ito's Formula **39**

2.1 Brownian Motion and Its Nowhere Differentiability 39
2.2 Spaces $\mathcal{L}^0$ and $\mathcal{L}^2$. 43
2.3 Ito's Integrals on $\mathcal{L}^2$ 44
2.4 Ito's Integrals on $\mathcal{L}^{2,loc}$ 47
2.5 Stochastic Integrals with respect to Martingales 49
2.6 Ito's Formula for Continuous Semi-Martingales 54
2.7 Ito's Formula for Semi-Martingales with Jumps 58
2.8 Ito's Formula for d-dimensional Semi-Martingales. Integration by Parts. 62
2.9 Independence of BM and Poisson Point Processes 64
2.10 Some Examples 65
2.11 Strong Markov Property of BM and Poisson Point Processes 67
2.12 Martingale Representation Theorem 68

3 Stochastic Differential Equations **75**
3.1 Strong Solutions to SDE with Jumps 75
3.1.1 Notation 75
3.1.2 A Priori Estimate and Uniqueness of Solutions . . . 76
3.1.3 Existence of Solutions for the Lipschitzian Case . . . 79
3.2 Exponential Solutions to Linear SDE with Jumps 84
3.3 Girsanov Transformation and Weak Solutions of SDE with Jumps 86
3.4 Examples of Weak Solutions 99

4 Some Useful Tools in Stochastic Differential Equations **103**
4.1 Yamada-Watanabe Type Theorem 103
4.2 Tanaka Type Formula and Some Applications 109
4.2.1 Localization Technique 109
4.2.2 Tanaka Type Formula in $d-$Dimensional Space . . . 110
4.2.3 Applications to Pathwise Uniqueness and Convergence of Solutions 112
4.2.4 Tanaka Type Formual in 1-Dimensional Space . . . 116
4.2.5 Tanaka Type Formula in The Component Form . . . 121
4.2.6 Pathwise Uniqueness of solutions 122
4.3 Local Time and Occupation Density Formula 124
4.4 Krylov Estimation 129
4.4.1 The case for $1-$dimensional space 129
4.4.2 The Case for $d-$dimensional space 130
4.4.3 Applications to Convergence of Solutions to SDE with Jumps 133

5 Stochastic Differential Equations with Non-Lipschitzian Coefficients **139**
5.1 Strong Solutions. Continuous Coefficients with $\rho-$ Conditions 140
5.2 The Skorohod Weak Convergence Technique 145

5.3 Weak Solutions. Continuous Coefficients 147
5.4 Existence of Strong Solutions and Applications to ODE . . 153
5.5 Weak Solutions. Measurable Coefficient Case 153

II Applications 161

6 How to Use the Stochastic Calculus to Solve SDE 163
6.1 The Foundation of Applications: Ito's Formula and Girsanov's Theorem 163
6.2 More Useful Examples 167

7 Linear and Non-linear Filtering 169
7.1 Solutions of SDE with Functional Coefficients and Girsanov Theorems 169
7.2 Martingale Representation Theorems (Functional Coefficient Case) 177
7.3 Non-linear Filtering Equation 180
7.4 Optimal Linear Filtering 191
7.5 Continuous Linear Filtering. Kalman-Bucy Equation 194
7.6 Kalman-Bucy Equation in Multi-Dimensional Case 196
7.7 More General Continuous Linear Filtering 197
7.8 Zakai Equation 201
7.9 Examples on Linear Filtering 203

8 Option Pricing in a Financial Market and BSDE 205
8.1 Introduction 205
8.2 A More Detailed Derivation of the BSDE for Option Pricing 208
8.3 Existence of Solutions with Bounded Stopping Times 209
8.3.1 The General Model and its Explanation 209
8.3.2 A Priori Estimate and Uniqueness of a Solution . . . 213
8.3.3 Existence of Solutions for the Lipschitzian Case . . . 215
8.4 Explanation of the Solution of BSDE to Option Pricing . . 219
8.4.1 Continuous Case 219
8.4.2 Discontinuous Case 220
8.5 Black-Scholes Formula for Option Pricing. Two Approaches 223
8.6 Black-Scholes Formula for Markets with Jumps 229
8.7 More General Wealth Processes and BSDEs 234
8.8 Existence of Solutions for Non-Lipschitzian Case 236
8.9 Convergence of Solutions 239
8.10 Explanation of Solutions of BSDEs to Financial Markets . . 241
8.11 Comparison Theorem for BSDE with Jumps. 243
8.12 Explanation of Comparison Theorem. Arbitrage-Free Market 250
8.13 Solutions for Unbounded (Terminal) Stopping Times 254
8.14 Minimal Solution for BSDE with Discontinuous Drift 258

8.15 Existence of Non-Lipschitzian Optimal Control. BSDE Case 262
8.16 Existence of Discontinuous Optimal Control. BSDEs in R^1 . 267
8.17 Application to PDE. Feynman-Kac Formula 271

9 Optimal Consumption by H-J-B Equation and Lagrange Method 277
9.1 Optimal Consumption 277
9.2 Optimization for a Financial Market with Jumps by the Lagrange Method 279
9.2.1 Introduction 280
9.2.2 Models 280
9.2.3 Main Theorem and Proof 282
9.2.4 Applications 286
9.2.5 Concluding Remarks 290

10 Comparison Theorem and Stochastic Pathwise Control 291
10.1 Comparison for Solutions of Stochastic Differential Equations 292
10.1.1 1−Dimensional Space Case 292
10.1.2 Component Comparison in d−Dimensional Space . . 293
10.1.3 Applications to Existence of Strong Solutions. Weaker Conditions. 294
10.2 Weak and Pathwise Uniqueness for 1-Dimensional SDE with Jumps 298
10.3 Strong Solutions for 1-Dimensional SDE with Jumps 300
10.3.1 Non-Degenerate Case 300
10.3.2 Degenerate and Partially-Degenerate Case 303
10.4 Stochastic Pathwise Bang-Bang Control for a Non-linear System 312
10.4.1 Non-Degenerate Case 312
10.4.2 Partially-Degenerate Case 316
10.5 Bang-Bang Control for d−Dimensional Non-linear Systems 319
10.5.1 Non-Degenerate Case 319
10.5.2 Partially-Degenerate Case 322

11 Stochastic Population Control and Reflecting SDE 329
11.1 Introduction 330
11.2 Notation 332
11.3 Skorohod's Problem and its Solutions 335
11.4 Moment Estimates and Uniqueness of Solutions to RSDE . 342
11.5 Solutions for RSDE with Jumps and with Continuous Coefficients 345
11.6 Solutions for RSDE with Jumps and with Discontinuous Coefficients 349
11.7 Solutions to Population SDE and Their Properties 352
11.8 Comparison of Solutions and Stochastic Population Control 363

11.9 Caculation of Solutions to Population RSDE 372

12 Maximum Principle for Stochastic Systems with Jumps 377
12.1 Introduction 377
12.2 Basic Assumption and Notation 378
12.3 Maximum Principle and Adjoint Equation as BSDE with Jumps 379
12.4 A Simple Example 380
12.5 Intuitive thinking on the Maximum Principle 381
12.6 Some Lemmas 383
12.7 Proof of Theorem 354 386

A A Short Review on Basic Probability Theory 389
A.1 Probability Space, Random Variable and Mathematical Expectation 389
A.2 Gaussian Vectors and Poisson Random Variables 392
A.3 Conditional Mathematical Expectation and its Properties . 395
A.4 Random Processes and the Kolmogorov Theorem 397

B Space D and Skorohod's Metric 401

C Monotone Class Theorems. Convergence of Random Processes 407
C.1 Monotone Class Theorems 407
C.2 Convergence of Random Variables 409
C.3 Convergence of Random Processes and Stochastic Integrals 411

References 415

Index 431

Preface

Stochastic differential equations (SDEs) were first initiated and developed by K. Ito (1942). Today they have become a very powerful tool applied to Mathematics, Physics, Chemistry, Biology, Medical science, and almost all sciences. Let us explain why we need SDEs, and how the contents in this book have been arranged.

In nature, physics, society, engineering and so on we always meet two kinds of functions with respect to time: one is determinstic, and another is random. For example, in a financial market we deposite money π_t in a bank. This can be seen as our having bought some units η_t^0 of a bond, where the bond's price P_t^0 satisfies the following ordinary differential equation

$dP_t^0 = P_t^0 r_t dt,\ P_0^0 = 1,\ t \in [0,T],$

where r_t is the rate of the bond, and the money that we deposite in the bank is $\pi_t = \eta_t^0 P_t^0 = \eta_t^0 \exp[\int_0^t r_s ds]$. Obviously, usually, $P_t^0 = \exp[\int_0^t r_s ds]$ is non-random, since the rate r_t is usually deterministic. However, if we want to buy some stocks from the market, each stock's price is random. For simplicity let us assume that in the financial market there is only one stock, and its price is P_t^1. Obviously, it will satisfy a differential equation as follows:

$dP_t^1 = P_t^1(b_t dt + d(\text{a stochastic perturbation})),\ P_0^1 = P_0^1,\ t \in [0,T],$

where all of the above processes are 1−dimensional. Here the stochastic perturbation is very important, because it influences the price of the stock, which will cause us to earn or lose money if we buy the stock. One important problem arises naturally. How can we model this stochastic perturbation? Can we make calculations to get the solution of the stock's price P_t^1, as we do in the case of the bond's price P_t^0? The answer is positive, usually a continuous stochastic perturbation will be modeled by a stochastic integral $\int_0^t \sigma_s dw_s$, where $w_t, t \geq 0$ is the so-called Brownian Motion process (BM), or the Wiener process. The 1−dimensional BM $w_t, t \geq 0$ has the following nice properties: 1) (Independent increment property). It has an independent increment property, that is, for any $0 < t_1 < \cdots < t_n$ the system $\{w_0, w_{t_1} - w_0, w_{t_2} - w_{t_1}, \cdots, w_{t_n} - w_{t_{n-1}}, \}$ is an independent system. Or say, the increments, which happen in disjoint time intervals, occured independently. 2) (Normal distribution property). Each increment is Normally distributed. That is, for any $0 \leq s < t$ the increment $w_t - w_s$ on this time interval is a normal random variable with mean m, and variance $\sigma^2(t-s)$. We write this as $w_t - w_s \sim N(m, \sigma^2(t-s))$. 3) (Stationary distribution property). The probability distribution of each increment only depends on the length of the time interval, and it does not depend on the starting point of the time interval. That is, the m and σ^2 appearing in property 3) are constants. 4) (Continuous trajectory property). Its trajectory is continuous. That is BM $w_t, t \geq 0$ is continuous in t.

Since the simplest or say, the most basic continuous stochastic perturbation, intuitively will have the above four properties, the modeling of the general continuous stochastic perturbation by a stochastic integral with respect to this basic BM $w_t, t \geq 0$ is quite natural. However, the 1−dimensional BM also has some strange property: Even though it is continuous in t, it is nowhere differentiable in t. So we cannot define the stochastic integral $\int_0^t \sigma_s(\omega) dw_s(\omega)$ for each given ω. That is why K. Ito (1942) invented a completely new way to define this stochastic integral.

Our first task in this book is to introduce the Ito stochastic integral and discuss its properties for later applications.

After we have understood the stochastic integral $\int_0^t \sigma_s(\omega) dw_s(\omega)$ we can study the following general stochastic differential equation (SDE):

$x_t = x_0 + \int_0^t \widetilde{b}(s, x_s) ds + \int_0^t \widetilde{\sigma}(s, x_s) dw_s,\ t \geq 0,$

or equivalently, we write

$$dx_t = \widetilde{b}(t, x_t) dt + \widetilde{\sigma}(t, x_t) dw_t, x_0 = x_0, t \geq 0. \tag{1}$$

Returning to the stock's price equation, we naturally consider it as the following SDE:

$$dP_t^1 = P_t^1(b_t dt + \sigma_t dw_t), P_0^1 = P_0^1, t \in [0, T]. \tag{2}$$

Comparing this to the solution of P_t^0, one naturally asks could the solution of this SDE be $P_t^1 = P_0^1 \exp[\int_0^t b_s ds + \int_0^t \sigma_s dw_s]$? To check this guess, obviously if we can have a differential rule to perform differentiation on $P_0^1 \exp x_t$, where $x_t = \int_0^t b_s ds + \int_0^t \sigma_s dw_s$, then we can make the check. Or more generally, if we have an $f(x) \in C^2(R)$ and $dx_t = b_t dt + \sigma_t dw_t$, can we have

$df'(x_t) = f'(x_t) dx_t = f'(x_t)\,(b_t dt + \sigma_t dw_t)$?

If as in the deterministic case, this differential rule holds true, then we immediately see that $P_t^1 = P_0^1 \exp[\int_0^t b_s ds + \int_0^t \sigma_s dw_s]$ satisfies (2). Unfortunately, such a differential rule is not true. K. Ito (1942) has found that it should obey another differential rule - the so-called Ito's formula:

$df'(x_t) = f'(x_t)dx_t + \frac{1}{2}f''(x_t)\left|\sigma_t\right|^2 dt.$

By this rule one easily checks that

$P_t^1 = P_0^1 \exp[\int_0^t b_s ds + \int_0^t \sigma_s dw_s - \frac{1}{2}\int_0^t \left|\sigma_s\right|^2 ds]$

is a solution of (2), and $\widetilde{P}_t^1 = P_0^1 \exp[\int_0^t b_s ds + \int_0^t \sigma_s dw_s]$ actually satisfies another SDE:

$d\widetilde{P}_t^1 = \widetilde{P}_t^1[\int_0^t b_s ds + \int_0^t \sigma_s dw_s + \frac{1}{2}\int_0^t \left|\sigma_s\right|^2 ds], \widetilde{P}_0^1 = P_0^1, \forall t \in [0,T].$

Our second task in this book is to establish the Ito formula and discuss its applications: solving SDE and solving other problems.

However, even if we have a powerful Ito formula (or say, Ito's differential rule) in our hand, we still need to discuss how to solve the general SDE, because usually, the form of the solution of SDE is not easy to guess. Moreover, for solutions of SDE, we actually meet a more complicated and hence also a more interesting case. Consider some physical quantity x_t determined by dynamics. If this dynamics is deterministic, that is, it is not disturbed by any random noises, say such that $dx_t = \widetilde{b}(t,x_t)dt, x_0 = x_0, t \geq 0$; then solving this ODE we immediately get this quantity x_t. However, if the dynamics are disturbed by some continuous random noise, say such that $dx_t = \widetilde{b}(t,x_t)dt + \widetilde{\sigma}(t,x_t)dw_t, x_0 = x_0, t \geq 0$, then for the amount x_t, or say, the solution of this SDE, two situations can arise. The first one is, if we think that the random noise - BM $w_s, s \leq t$, is an input, then after disturbing the dynamics we get an output x_t. This means that the solution x_t is a functional of the given noise - BM $w_s, s \leq t$ for each t. We will call such a solution a strong solution. Another situation is that for a given noise we cannot immediately find the solution. However, we can find a random process $\widetilde{x}_t, t \geq 0$, and maybe another random noise that is also a BM $\widetilde{w}_t, t \geq 0$, such that $(\widetilde{x}_t, \widetilde{w}_t), t \geq 0$ satisfy the SDE $d\widetilde{x}_t = \widetilde{b}(t,\widetilde{x}_t)dt + \widetilde{\sigma}(t,\widetilde{x}_t)d\widetilde{w}_t, \widetilde{x}_0 = x_0, t \geq 0$. In this case we will call $(\widetilde{x}_t, \widetilde{w}_t), t \geq 0$ or simply, $\widetilde{x}_t, t \geq 0$, a weak solution of the original SDE. Obviously, from the engineering point of view the strong solutions is more realistic and useful. However, since, if the strong solution $x_t, t \geq 0$ exists, then all finite dimensional probability ditributions of $(x_t, w_t)_{t\geq 0}$ are the same as that of $(\widetilde{x}_t, \widetilde{w}_t)_{t\geq 0}$. So, if we do not know the existence of a strong solution, but we do know the existence of a weak solution $(\widetilde{x}_t, \widetilde{w}_t), t \geq 0$, then from the probability point of view the weak solution can still help us in some sense. Therefore, for solutions of SDEs there are two kinds that need to be considered: strong solutions and weak solutions.

Our third task in this book is to introduce the concepts of solutions and to discuss their existence and uniquess and the related important theory. (For example, Girsanov's theorem and the matingale representation theorem, the first of which can help us find weak solutions, while the second one is necessary for finding the solutions of backward SDE and the filtering problem considered later).

Since, actually, in the realistic world we will always meet some jump type stochastic perturbation, in this book we also consider stochastic integrals with respect to a Poisson counting measure (which is generated by a Poisson point process), the Ito formula and SDE for this case. (To find the reason why we consider the Poisson point process and its related integral as a jump type stochastic perturbation see the subsection "The General Model and its Explanation" in Chapter 8 - "Option Pricing in a Financial Market and BSDE").

The first three Chapters are intended to solve the above three tasks. They are the basic foudation of the SDE theory and its applications.

However, interesting and important things for SDE do not only come from the above mentioned three chapters, where they exhibit the following facts: the definition of Ito's stochastic integrals and Ito's differential rule are completely different from the deterministic case, etc. The interesting and important things also come from the following facts:

1) For an ordinary differential equation (ODE) $dx_t = \widetilde{b}(t, x_t)dt, x_0 = x_0, t \geq 0$ if $\widetilde{b}(t, x)$ is only bounded and jointly continuous, then even though the solution exists, is not necessary unique. However, for the SDE (1) in $1-$dimensional case if $\widetilde{b}(t, x)$ and $\widetilde{\sigma}(t, x)$ are only bounded and jointly Borel-measurable, and $|\widetilde{\sigma}(t, x)|^{-1}$ is also bounded and $\widetilde{\sigma}(t, x)$ is Lipschitz continuous in x, then (1) will have a unique strong solution. (Here "strong" means that x_t is $\mathfrak{F}_t^w-$measurable). This means that adding a non-degenerate stochastic perturbation term into the differential equation, can even improve the nice property of the solution.

2) The stochastic perturbation term has an importaqnt practical meaning in some cases and it cannot be discarded. For example, in the investment problem and the option pricing problem from a Financial Market, the investment portfolio actually appears as the coefficient of the stochastic integral in an SDE, where the stochastic integral acts like a stochatic perturbation term.

3) The solutions of SDEs and backward SDEs can help us to explain the solutions of some deterministic partial differential equations (PDEs) with integral terms (the Feynman-Kac formula) and even to guess and find the solution of a PDE, for example, the soluition of the PDE for the price of a option can be solved by a solution of a BSDE - the Black-Scholes formula.

4) More and more.

So we have many reasons to study the SDE theory and its applications more deeply and carefully. That is why we have a Chapter that discusses useful tools for SDE, and a Chapter for the solutions of an SDE with non-Lipschitzian coefiicients. These are Chapter 4 and 5.

The above concerns the first part of our book, which represents the theory and general background of the SDE.

The second part of our book is about the Applications.

We first provide a short Chapter to help the reader to take a quick look at how to use Stochastic Analysis (the theory in the first part), to solve an SDE.

Then we discuss the estimation problem for a signal process : the so-called filtering problem, where the general linear and non-linear filtering problem for continous SDE systems and SDE systems with jumps, the Kalman-Bucy filtering equation for continuous systems, and the Zakai equation for non-linear filtering, etc. are also considered.

Since, now, research on mathematical finance, and in particular on the option pricing problem for the financial market has become very popular, we also provide a Chapter that discusses the option pricing problem and backward SDE, where the famous Black-Scholes formulas for a market with or without jumps are derived using a probability and a PDE approach, respectively; and the arbitrage-free market is also discussed. The interesting thing here is that we deal with the mathematical financial problem by using the backward stochastic differential equation (BSDE) technique, which now becomes very powerful when treating many problems in the finacial market, in mathematics and in other sciences.

Since deterministic population control has proved to be important and efficient, and the stochastic population control is more realistic, we also provide a Chapter that develops the stochastic population control problem by using the reflecting SDE approach, where the existence, the comparison and the calculation of the population solution and the optimal stochastic population control are established.

Besides, the stochastic Lagrange method for the stochastic optimal control, the non-linear pathwise stochastic optimal control, and the Maximum Principle (that is, the necessary conditions for a stochastic optimal control) are also formulated and developed in specific Chapters, respectively.

For the convenience of the readers three Appendixes are also provided: giving a short review on basic probability theory, space D and Skorohod's metric, and monotone class theorems and the convergence of random processes.

We suggest that the reader studies the book as follows:

For readers who are mainly interested in Applications, the following approach may be considered: Appendix A → Chapter 1 → Chapter 2 → Chapter 3 → Chapter 6 → Any Chapter in The second part "Applications" except Chapter 10, and at any time return to read the related sections in Chapters 4 and 5, or Appendixes B and C, when necessary. However, to read Chapter 10, knowledge of Chapters 4 and 5 and Appendixes B and C are necessary.

Acknowledgement

The author would like to express his sincere thanks to Professor Alan Jeffrey for kindly recommending publication of this book, and for his interest in the book from the very beginning to the very end, and for offering many valuable and important suggestions. Without his help this book would not have been possible

Abbreviations and Some Explanations

All important statements and results, like, Definitions, Lemmas, Propositions, Theorems, Corollaries, Remarks and Examples are numbered in sequential order throughout the whole book. So, it is easy to find where they are located. For example, Lemma 22 follows Definition 21, and Theorem 394 is just after Remark 393, etc. However, the numbers of equations are arranged independently in each Chapter and each Appendix. For example, (3.25) means equation 25 in Chapter 3, and (C.4) means the equation 4 in Appendix C.

The following abbreviations are frequently used in this book.

a.e. almost everywhere.
a.s. almost sure.
BM Brownian Motion.
BSDE backward stochastic differential equation.
FSDE forward stochastic differential equation.
H-J-B equation Hamilton-Jacobi-Bellman equation.
IDE integral-differential equation.
ODE ordinary differential equation.
PDE partial differential equation.
RCLL right continuous with left limit.
SDE stochastic differential equation.
$P - a.s$ almost sure in probability P.
a^+ $\max\{a, 0\}$.
a^- $\max\{-a, 0\}$.
$a \vee b$ $\max\{a, b\}$.
$a \wedge b$ $\min\{a, b\}$.

$\mu << \nu$ measure μ is absolutely continuous with respect to ν; that is, for any measurable set A, $\nu(A) = 0$ implies that $\mu(A) = 0$.
$\xi_n \to \xi$, *a.s.* ξ_n converges to ξ, almost surely; that is, $\xi_n(\omega) \to \xi(\omega)$ for all ω except at the points $\omega \in \Lambda$, where $P(\Lambda) = 0$.
$\xi_n \to \xi$, in P ξ_n converges to ξ in probability; that is, $\forall \varepsilon > 0$, $\lim_{n\to\infty} P(\omega : |\xi_n(\omega) - \xi(\omega)| > \varepsilon) = 0$.
$\#\{\cdot\}$ the numbers of $\cdot$ counted in the set $\{\cdot\}$.
$\sigma(x_s, s \le t)$ the smallest $\sigma-$field, which makes all $x_s, s \le t$ measurable.
$E[\xi|\eta]$ $E[\xi|\sigma(\eta)]$. It means the conditional expectation of ξ given $\sigma(\eta)$.

The following notations can be found on the corresponding pages. For example, $\mathfrak{F}, 4, 387$ means that notation $\mathfrak{F}$ can be found in page 4 and page 387.

Ω, 4, 387
$\mathfrak{F}$, 4, 387
$\mathfrak{F}_t$, 4
$\mathcal{F}_p$, 34
$\mathcal{F}_p^1$, 34
$\mathcal{F}_p^2$, 34
$\mathcal{F}_p^{2,loc}$, 34
$F_{\mathfrak{F}}^2(R^{d\otimes d_2})$, 212
$\mathcal{L}^0$, 43
$\mathcal{L}_T^0$, 43
$\mathcal{L}^2$, 43
$\mathcal{L}_T^2$, 43
$\mathcal{L}^{2,loc}$, 48
$\mathcal{L}_M^2$, 50
$\mathcal{L}_M^{2,loc}$, 51
$L_{\mathfrak{F}}^2(R^{d\otimes d_1})$, 212
$L_{\pi(\cdot)}^2(R^{d\otimes d_2})$, 212
$L^1(\Omega, \mathfrak{F}, P)$. 12
$\mathcal{M}^2$, 35
$\mathcal{M}^{2,c}$, 37
$\mathcal{M}^{2,loc}$, 35
$\mathcal{M}^{2,rcll}$, 37
$\mathcal{M}_T^2$, 35
$\mathcal{M}_T^{2,c}$, 37
$\mathcal{M}_T^{2,loc}$, 35
$\mathcal{M}_T^{2,rcll}$, 37
$\mathcal{O}$, 7
$\mathcal{P}$, 7
P, 4, 387
$S_{\mathfrak{F}}^2(R^d)$, 212

Part I

Stochastic Differential Equations with Jumps in R^d

1

Martingale Theory and the Stochastic Integral for Point Processes

A stochastic integral is a kind of integral quite different from the usual deterministic integral. However, its theory has broad and important applications in Science, Mathematics itself, Economic, Finance, and elsewhere. A stochastic integral can be completely charaterized by martingale theory. In this chapter we will discuss the elementary martingale theory, which forms the foundation of stochastic analysis and stochastic integral. As a first step we also introduce the stochastic integral with respect to a Point process.

1.1 Concept of a Martingale.

In some sense the martingale conception can be explained by a fair game. Let us interprete it as follows:

In a game suppose that a person at the present time s has wealth x_s for the game, and at the future time t he will have the wealth x_t. The expected money for this person at the future time t is naturally expressed as $E[x_t|\mathfrak{F}_s]$, where $E[\cdot]$ means the expectation value of $\cdot$, $\mathfrak{F}_s$ means the information up to time s, which is known by the gambler, and $E[\cdot|\mathfrak{F}_s]$ is the conditional expectation value of $\cdot$ under given $\mathfrak{F}_s$. Obviously, if the game is fair, then it should be

$E[x_t|\mathfrak{F}_s] = x_s, \forall t \geq s.$

This is exactly the definition of a martingale for a random process $x_t, t \geq 0$. Let us make it more explicit for later developement.

Let $(\Omega, \mathfrak{F}, P)$ be a probability space, $\{\mathfrak{F}_t\}_{t\geq 0}$ be an information family (in Mathematics, we call it a $\sigma-$algebra family or a $\sigma-$field family, see Appendix A), which satisfies the so-called "Usual Conditions":

(i) $\mathfrak{F}_s \subset \mathfrak{F}_t$, as $0 \leq s \leq t$; (ii) $\mathfrak{F}_{t+} =: \cap_{h>0}\mathfrak{F}_{t+h}$.

Here condition (i) means that the information increases with time, and condition (ii) that the information is right continuous, or say, $\mathfrak{F}_{t+h} \downarrow \mathfrak{F}_t$, as $h \downarrow 0$. In this case we call $\{\mathfrak{F}_t\}_{t\geq 0}$ a $\sigma-$field filtration.

Definition 1 *A real random process $\{x_t\}_{t\geq 0}$ is called a martingale (supermartingale, submartingale) with respect to $\{\mathfrak{F}_t\}_{t\geq 0}$, or $\{x_t, \mathfrak{F}_t\}_{t\geq 0}$ is a martingale (supermartingale, submartingale), if*
(i) x_t is integrable for each $t \geq 0$; that is, $E|x_t| < \infty, \forall t \geq 0$;
(ii) x_t is $\mathfrak{F}_t-$adapted; that is, for each $t \geq 0$, x_t is $\mathfrak{F}_t-$measurable;
(iii) $E(x_t|\mathfrak{F}_s] = x_s$, (respectively, $\leq$, $\geq$), a.s. $\forall 0 \leq s \leq t$.

For the random process $\{x_t\}_{t\in[0,T]}$ and the random process $\{x_n\}_{n=1}^{\infty}$ with discrete time similar definitions can be given.

Example 2 *If $\{x_t\}_{t\geq 0}$ is a random process with independent increments; that is, $\forall 0 < t_1 < t_2 < \cdots < t_n$, the family of random variables*
$\{x_0, x_{t_1} - x_0, x_{t_2} - x_{t_1}, \cdots, x_{t_n} - x_{t_{n-1}}\}$
is independent, and the increment $x_t - x_s, \forall t > s$, is integrable and with non-negative expectation, moreover, x_0 is also integrable, then $\{x_t\}_{t\geq 0}$ is a submartingale with respect to $\{\mathfrak{F}_t^x\}_{t\geq 0}$, where $\mathfrak{F}_t^x = \sigma(x_s, s \leq t)$, which is a $\sigma-$field generated by $\{x_s, s \leq t\}$ (that is, the smallest $\sigma-$field which makes all $x_s, s \leq t$ measurable) and makes a completion.

In fact, by independent and non-negative increments,
$0 \leq E(x_t - x_s) = E[(x_t - x_s)|\mathfrak{F}_s^x], \forall t \geq s.$
Hence the conclusion is reached.

Example 3 *If $\{x_t\}_{t\geq 0}$ is a submartingale, let $y_t := x_t \vee 0 = \max(x_t, 0)$, then $\{y_t\}_{t\geq 0}$ is still a submartingale.*

In fact, since $f(x) = x \vee 0$ is a convex function, hence by Jensen's inequality for the conditional expectation
$E[x_t \vee 0|\mathfrak{F}_s] \geq E[x_t|\mathfrak{F}_s] \vee E[0|\mathfrak{F}_s] \geq x_s \vee 0, \forall t \geq s.$
So the conclusion is true.

Example 4 *If $\{x_t\}_{t\geq 0}$ is a martingale, then $\{|x_t|\}_{t\geq 0}$ is a submartingale.*

In fact, by Jensen's inequality
$E[|x_t|\,|\mathfrak{F}_s] \geq |E[x_t|\mathfrak{F}_s]| = |x_s|\,, \forall t \geq s.$
Thus the conclusion is deduced.

Martingales, submartingales and supermartingales have many important and useful properties, which make them become powerful tools in dealing with many theoretical and practical problems in Science, Finance and elsewhere. Among them the martingale inequalites, the limit theorems, and the

Doob-Meyer decomposition theorem for submartingales and supermartingales are most helpful and are frequently encountered in Stochastic Analysis and its Applications, and in this book. So we will discuss them in this chapter. However, to show them clearly we need to introduce the concept called a stopping time, which will be important for us later. We proceed to the next section.

1.2 Stopping Times. Predictable Process

Definition 5 *A random variable $\tau(\omega) \in [0,\infty]$ is called a $\mathfrak{F}_t-$stopping time, or simply, a stoping time, if for any $(\infty >)t \geq 0$, $\{\tau(\omega) \leq t\} \in \mathfrak{F}_t$.*

The intuitive interpletation of a stopping time is as follows: If a gambler has a right to stop his gamble at any time $\tau(\omega)$, he would of course like to choose the best time to stop. Suppose he stops his game before time t, i.e. he likes to make $\tau(\omega) \leq t$, then the maximum information he can get about his decision is only the information up to t, i.e $\{\tau(\omega) \leq t\} \in \mathfrak{F}_t$. The trivial example for a stopping time is $\tau(\omega) \equiv t, \forall \omega \in \Omega$. That is to say, any constant time t actually is a stopping time.

For a discrete random variable $\tau(\omega) \in \{0,1,2,\cdots,\infty\}$ the definition can be reduced to that $\tau(\omega)$ is a stopping time, if for any $n \in \{0,1,2,\cdots\}$, $\{\tau(\omega) = n\} \in \mathfrak{F}_n$, since $\{\tau(\omega) = n\} = \{\tau(\omega) \leq n\} - \{\tau(\omega) \leq n-1\}$, and $\{\tau(\omega) \leq n\} = \cup_{k=1}^{n} \{\tau(\omega) = k\}$. The following examples of stopping time are useful later.

Example 6 *Let B be a Borel set in R^1 and $\{x_n\}_{n=1}^{\infty}$ be a sequence of real $\mathfrak{F}_t-$adapted random variables. Define the first hitting time $\tau_B(\omega)$ to the set B (i.e. the first time that $\{x_n\}_{n=1}^{\infty}$ hits B) by*
$\tau_B(\omega) = \inf\{n : x_n(\omega) \in B\}$.
Then $\tau_B(\omega)$ is a discrete stopping time.

In fact,
$\{\tau_B(\omega) = n\} = \cap_{k=1}^{n-1} \{x_k \in B^c\} \cap \{x_n \in B\} \in \mathfrak{F}_n$.
For a general random process with continuous time parameter t we have the following similar example.

Example 7 *Let x_t be a $d-$dimensional right continuous $\mathfrak{F}_t-$adapted process and let A be an open set in R^d. Denote the first hitting time $\sigma_A(\omega)$ to A by*
$\sigma_A(\omega) = \inf\{t > 0 : x_t(\omega) \in A\}$.
Then $\sigma_A(\omega)$is a stopping time.

In fact, by the open set property and the right continuity of x_t one has that
$\{\sigma_A(\omega) \leq t\} = \cap_{n=1}^{\infty} \{\sigma_A(\omega) < t + \frac{1}{n}\}$
$= \cap_{n=1}^{\infty} \cup_{r \in Q, r < t+1/n} \{x_r(\omega) \in A\} \in \mathfrak{F}_{t+0} = \mathfrak{F}_t,$

where Q is the set of all rational numbers.

The following properties of general stopping times will be useful later.

Lemma 8 *$\tau(\omega)$ is a stopping time, if and only if $\{\tau(\omega) < t\} \in \mathfrak{F}_t, \forall t$.*

Proof. $\Longrightarrow$: $\{\tau(\omega) < t\} = \cup_{n=1}^{\infty} \{\tau(\omega) \leq t - \frac{1}{n}\} \in \mathfrak{F}_t$.
$\Longleftarrow$: $\{\tau(\omega) \leq t\} = \cap_{n=1}^{\infty} \{\tau(\omega) < t + \frac{1}{n}\} \in \mathfrak{F}_{t+0} = \mathfrak{F}_t$. ∎

Lemma 9 *Let $\sigma, \tau, \sigma_n, n = 1, 2, \cdots$ be stopping times. Then*
(i) $\sigma \wedge \tau, \sigma \vee \tau$,
(ii) $\sigma = \lim_{n\to\infty} \sigma_n$, when $\sigma_n \uparrow$ or $\sigma_n \downarrow$,
are all stopping times.

Proof. (i): $\{\sigma \wedge \tau \leq t\} = \{\sigma \leq t\} \cup \{\tau \leq t\} \in \mathfrak{F}_t$,
$\{\sigma \vee \tau \leq t\} = \{\sigma \leq t\} \cap \{\tau \leq t\} \in \mathfrak{F}_t$.
(ii): If $\sigma_n \uparrow \sigma$, then
$\{\sigma \leq t\} = \cap_{n=1}^{\infty} \{\sigma_n \leq t\} \in \mathfrak{F}_t$.
If $\sigma_n \downarrow \sigma$, then
$\{\sigma < t\} = \cup_{n=1}^{\infty} \{\sigma_n < t\} \in \mathfrak{F}_t$.
By Lemma 8 σ is a stopping time. ∎

Now let us introduce a $\sigma-$ field which describes the information obtained up to stopping time τ. Set
$\mathfrak{F}_\tau = \{A \in \mathfrak{F}_\infty : \forall t \in [0, \infty), A \cap \{\tau(\omega) \leq t\} \in \mathfrak{F}_t\}$,
where we naturally define that $\mathfrak{F}_\infty = \vee_{t\geq 0}\mathfrak{F}_t$, i.e. the smallest $\sigma-$field including all $\mathfrak{F}_t, t \in [0, \infty)$. Obviously, $\mathfrak{F}_\tau$ is a $\sigma-$ algebra, and if $\tau(\omega) \equiv t$, then $\mathfrak{F}_\tau = \mathfrak{F}_t$.

Proposition 10 *Let $\sigma, \tau, \sigma_n, n = 1, 2, \cdots$ be stopping times.*
(1) If $\sigma(\omega) \leq \tau(\omega), \forall \omega$, then $\mathfrak{F}_\sigma \subset \mathfrak{F}_\tau$.
(2) If $\sigma_n(\omega) \downarrow \sigma(\omega), \forall \omega$, then $\cap_{n=1}^{\infty} \mathfrak{F}_{\sigma_n} = \mathfrak{F}_\sigma$.
(3) $\sigma \in \mathfrak{F}_\sigma$. (We use $f \in \mathfrak{F}_\sigma$ to mean that f is $\mathfrak{F}_\sigma-$measurable).

Proof. (1): $A \cap \{\tau \leq t\} = (A \cap \{\sigma \leq t\}) \cap \{\tau \leq t\} \in \mathfrak{F}_t$.
(2): By (1) $\mathfrak{F}_\sigma \subset \cap_{n=1}^{\infty} \mathfrak{F}_{\sigma_n}$. Conversely, if $A \in \cap_{n=1}^{\infty} \mathfrak{F}_{\sigma_n}$, then
$A \cap \{\sigma_n < t\} = \cup_{k=1}^{\infty} (A \cap \{\sigma_n \leq t - \frac{1}{k}\}) \in \mathfrak{F}_t, \forall t \geq 0, \forall n$.
Hence $A \cap \{\sigma < t\} = \cup_{n=1}^{\infty} (A \cap \{\sigma_n < t\}) \in \mathfrak{F}_t$, and
$A \cap \{\sigma \leq t\} = \cap_{k=1}^{\infty} (A \cap \{\sigma < t + \frac{1}{k}\}) \in \mathfrak{F}_{t+0} = \mathfrak{F}_t$, i.e. $A \in \mathfrak{F}_\sigma$.
(3): For any constant $0 \leq c < \infty$ one has that $\{\sigma \leq c\} \cap \{\sigma \leq t\} \in \mathfrak{F}_{t\wedge c} \subset \mathfrak{F}_t$, so $\{\sigma \leq c\} \in \mathfrak{F}_\sigma$. ∎

It is natural to ask that if $\{x_t\}_{t\geq 0}$ is $\mathfrak{F}_t-$adapted, and σ is a stopping time, is it true that $x_\sigma \in \mathfrak{F}_\sigma$? Generally speaking, it is not true. However, if $\{x_t\}_{t\geq 0}$ is a progressive measurable process, then it is correct. Let us introduce such a related concept.

Definition 11 *An R^d-valued process $\{x_t\}_{t\geq 0}$ is called measurable (respectively, progressive measurable), if the mapping*
$(t, \omega) \in [0, \infty) \times \Omega \to x_t(\omega) \in R^d$

(respectively, for each $t \geq 0, (s,\omega) \in [0,t] \times \Omega \to x_t(\omega) \in R^d$*)*
is $\mathfrak{B}([0,\infty)) \times \mathfrak{F} / \mathfrak{B}(R^d)$ *−measurable*
(respectively, $\mathfrak{B}([0,t]) \times \mathfrak{F}_t / \mathfrak{B}(R^d)$ *−measurable);*
that is, $\{(t,\omega) : x_t(\omega) \in B\} \in \mathfrak{B}([0,\infty)) \times \mathfrak{F}, \forall B \in \mathfrak{B}(R^d)$
(respectively, $\{(s,\omega) : s \in [0,t], x_s(\omega) \in B\} \in \mathfrak{B}([0,t]) \times \mathfrak{F}_t, \forall B \in \mathfrak{B}(R^d)$*).*

Let us introduce two useful $\sigma-$algebras as follows: Denote by $\mathcal{P}$ (respectively, $\mathcal{O}$) as the smallest $\sigma-$algebra on $[0,\infty) \times \Omega$ such that all left-continuous (respectively, right-continuous) $\mathfrak{F}_t-$adapted processes
$y_t(\omega) : [0,\infty) \times \Omega \to y_t(\omega) \in R^d$
are measurable. $\mathcal{P}$ (respectively, $\mathcal{O}$) is called the predictable (respectively, optional) $\sigma-$algebra. Thus, the following definition is natural.

Definition 12 *A process* $\{x_t\}_{t \geq 0}$ *is called predictable (optional), if the mapping*
$(t,\omega) \in [0,\infty) \times \Omega \to x_t(\omega) \in R^d$
is $\mathcal{P} / \mathfrak{B}(R^d)$ *−measurable (respectively* $\mathcal{O} / \mathfrak{B}(R^d)$ *−measurable).*

Let us use the notation $f \in \mathcal{P}$ to mean that f is $\mathcal{P}-$ measurable; etc. It is easily seen that the following relations hold:
$f \in \mathcal{P} \Rightarrow f \in \mathcal{O} \Rightarrow f$ is progressive measurable $\Rightarrow f$ is measurable and $\mathfrak{F}_t-$adapted.
We only need to show the first two implications. The last one is obvious.
Assume that $\{x_t\}_{t \geq 0}$ is left-continuous, let $x_t^n = x_{\frac{k}{2^n}}$, as $t \in [\frac{k}{2^n}, \frac{k+1}{2^n})$, $k = 0, 1, \cdots; n = 1, 2, \cdots$. Then obviously, x_t^n is right-continuous, and by the left-continuity of x_t, $x_t^n(\omega) \to x_t(\omega)$, as $n \to \infty, \forall t, \forall \omega$. So $\{x_t\}_{t \geq 0} \in \mathcal{O}$. From this one sees that $\mathcal{P} \subset \mathcal{O}$. Let us show that $\{x_t\}_{t \geq 0} \in \mathcal{O}$ implies that $\{x_t\}_{t \geq 0}$ is progressive measurable. For this for each given $t \geq 0$ we show that $\{x_s\}_{s \geq 0}$ restricted on $(s,\omega) \in [0,t] \times \Omega$ is $\mathfrak{B}([0,t]) \times \mathfrak{F}_t-$ measurable. In fact, without loss of generality we may assume that $\{x_t\}_{t \geq 0}$ is right-continuous. Now for each given $t \geq 0$, let $x_s^n = x_{\frac{k}{2^n}t}$, as $s \in [\frac{kt}{2^n}, \frac{(k+1)t}{2^n})$, $k = 0, 1, \cdots, 2^n - 1; n = 1, 2, \cdots$. Then obviously, $\{x_s^n\}_{s \in [0,t]}$ is $\mathfrak{B}([0,t]) \times \mathfrak{F}_t-$measurable, so is $\{x_s\}_{s \in [0,t]}$, since by the right continuity of x_t we have that as $n \to \infty, x_s^n(\omega) \to x_s(\omega)$ $, \forall s \in [0,t], \forall \omega$.
Let us show the following

Theorem 13 *If* $\{x_t\}_{t \geq 0}$ *is a* R^d-*valued progressive measurable process, then for each stopping time* σ, $Z_\sigma I_{\sigma < \infty}$ *is* $\mathfrak{F}_\sigma-$*measurable.*

We will use the compositon of measurable maps to show this theorem. For this we need the following lemma.

Lemma 14 *If* f_i *is a measurable mapping from* $(\Omega, \mathfrak{F})$ *to* $(\Omega_i', \mathfrak{F}_i'), i = 1, 2, \cdots$; *then*
$f(\omega) = (f_1(\omega), f_2(\omega), \cdots)$
is a measurable mapping from $(\Omega, \mathfrak{F})$ *to* $(\Omega_1' \times \Omega_2' \times \cdots, \mathfrak{F}_1' \times \mathfrak{F}_2' \times \cdots)$.

In fact, for any $B_i \in \mathfrak{F}'_i, i = 1, 2, \cdots, f^{-1}(B_1 \times B_2 \times \cdots) = \cap_{i=1}^{\infty} f_i^{-1}(B_i) \in \mathfrak{F}$. So $f^{-1}(\mathfrak{F}'_1 \times \mathfrak{F}'_2 \times \cdots) \subset \mathfrak{F}$.

Now let us prove Theorem 13.

Proof. Let $\Omega_\sigma = \{\sigma < \infty\}$. We need to show that x_σ is a measurable mapping from $(\Omega_\sigma, \mathfrak{F}_\sigma)$ to $(R^d, \mathfrak{B}(R^d))$. For any given $t \geq 0$ by Proposition 10 $\sigma \in \mathfrak{F}_\sigma$. So by the definition of $\mathfrak{F}_\sigma$, σ is a measurable mapping from $(\{\sigma \leq t\}, \mathfrak{F}_t)$ to $([0,t], \mathfrak{B}([0,t]))$. Hence by Lemma 14 $g_1(\omega) = (\sigma(\omega), \omega)$ is a measurable mapping from $(\{\sigma \leq t\}, \mathfrak{F}_t)$ to $([0,t] \times \Omega, \mathfrak{B}([0,t]) \times \mathfrak{F}_t)$. Note that by the progressive measurability of $\{x_t\}_{t \geq 0}$, $g_2(s,\omega) = x_s(\omega)$ is a measurable mapping from $([0,t] \times \Omega, \mathfrak{B}([0,t]) \times \mathfrak{F}_t)$ to $(R^d, \mathfrak{B}(R^d))$. Hence $x_{\sigma(\omega)}(\omega) I_{\sigma<\infty} = g_2 \circ g_1(\omega)$ is a measurable mapping from $(\{\sigma \leq t\}, \mathfrak{F}_t)$ to $(R^d, \mathfrak{B}(R^d))$ This shows that for any $B \in \mathfrak{B}(R^d)$, $\{x_\sigma I_{\sigma<\infty} \in B\} \cap \{\sigma \leq t\} \in \mathfrak{F}_t$.Since $t \geq 0$ is arbitrary by definition $\{x_\sigma I_{\sigma<\infty} \in B\} \in \mathfrak{F}_\sigma$. ∎

1.3 Martingales with Discrete Time

First we will show the Doob's stopping theorem (or called Doob's optional sampling theorem) for bounded stoping times.

Theorem 15 *Let $\{x_n\}_{n=0,1,2,\cdots}$ be a martingle (supermartingle, submartingale), $\sigma \leq \tau$ be two bounded stopping times. Then $\{x_n\}_{n=0,1,2,\cdots}$ is a strong martingle (respectively, strong supermartingle, strong submartingle), i.e.*

$E[x_\tau | \mathfrak{F}_\sigma] = x_\sigma$ (respectivly, $\leq$, $\geq$), a.s.

Proof. We only prove the conclusion for the case of submartingle. By assumption there exists a natural number $0 \leq n_0$ such that $\tau \leq n_0$. So $|x_\tau| \leq \max\{|x_n|, n = 0, 1, 2, \cdots, n_0\} \leq \sum_{n=0}^{n_0} |x_n|$. So $E|x_\tau| < \infty$. By the same manner $E|x_\sigma| < \infty$. Note that by the definition of a stopping time and $\mathfrak{F}_\sigma$ for $A \in \mathfrak{F}_\sigma$ and $0 \leq n \leq n_0$

$A \cap \{\sigma = n\} \cap \{\tau > n\} \in \mathfrak{F}_n$.

Now suppose $\tau - \sigma \leq 1$ in addition. Then by the definition of a submartingale

$\int_A (x_\sigma - x_\tau) dP = \sum_{n=0}^{n_0} \int_{A \cap \{\sigma=n\} \cap \{\tau>n\}} (x_n - x_{n+1}) dP \leq 0.$

In the general case set $T_n = \tau \wedge (\sigma + n), n = 1, 2, \cdots, n_0$. Then all T_n are stopping times, and

$\sigma \leq T_1 \leq T_2 \leq \cdots \leq T_{n_0} = \tau$, $T_1 - \sigma \leq 1$, $T_{n+1} - T_n \leq 1$,
$n = 1, 2, \cdots, n_0 - 1$.

Let $A \in \mathfrak{F}_\sigma \subset \mathfrak{F}_{T_{n_0}}$. Then by the above conclusion

$\int_A x_\sigma dP \leq \int_A x_{T_1} dP \leq \cdots \leq \int_A x_\tau dP.$

The proof is complete. ∎

Now we have the following martingale inequality:

Theorem 16 *Let $\{x_n\}_{n=0,1,\cdots}$ be a submartingale. Then for every $\lambda > 0$ and natural number N*

$\lambda P(\max_{0\leq n\leq N} x_n \geq \lambda) \leq E(x_N I_{\max_{0\leq n\leq N} x_n \geq \lambda}) \leq E(x_N^+) \leq E|x_N|,$
and

$\lambda P(\min_{0\leq n\leq N} x_n \leq -\lambda) \leq -Ex_0 + E(x_N I_{\min_{0\leq n\leq N} x_n > -\lambda})$
$\leq Ex_0^- + E(x_N^+) \leq E|x_0| + E|x_N|.$

Proof. Let us use the first hitting time technique and strong submartingale property to show this theorem. Set
$\sigma = \min\{n \leq N : x_n \geq \lambda\}$; $\sigma = N$, if $\{\cdot\} = \phi$.
Then σ is a bounded stopping time. By Theorem 15
$Ex_N \geq Ex_\sigma = Ex_\sigma I_{\max_{0\leq n\leq N} x_n \geq \lambda} + Ex_N I_{\max_{0\leq n\leq N} x_n < \lambda}$
$\geq \lambda P(\max_{0\leq n\leq N} x_n \geq \lambda) + Ex_N I_{\max_{0\leq n\leq N} x_n < \lambda}.$
Transferring the last term to the left hand side, we obtain the first inequality. Now set
$\tau = \min\{n \leq N : x_n \leq -\lambda\}$; $\tau = N$, if $\{\cdot\} = \phi$.
Then
$Ex_0 \leq Ex_\tau = Ex_\tau I_{\min_{0\leq n\leq N} x_n \leq -\lambda} + Ex_N I_{\min_{0\leq n\leq N} x_n > -\lambda}$
$\leq -\lambda P(\min_{0\leq n\leq N} x_n \leq -\lambda) + E(x_N I_{\min_{0\leq n\leq N} x_n > -\lambda}).$
Thus the second inequality is derived. ∎

Corollary 17 *1) Assume that* $\{x_n\}_{n=0,1,\cdots}$ *is a real submartingale such that* $E((x_n^+)^p) < \infty, n = 0,1,\cdots$, *for some* $p \geq 1$. *Then for every* N, *and* $\lambda > 0$,
$P(\max_{0\leq n\leq N} x_n^+ \geq \lambda) \leq E((x_N^+)^p)/\lambda^p,$
and if $p > 1$,
$E(\max_{0\leq n\leq N} (x_n^+)^p) \leq \left(\frac{p}{p-1}\right)^p E((x_N^+)^p).$
2) If $\{x_n\}_{n=0,1,\cdots}$ *is a real martingale such that* $E(|x_n|^p) < \infty, n = 0,1,\cdots$, *then the conclusions in 1) hold true for* x_n^+ *and* x_N^+ *replaced by* $|x_n|$ *and* $|x_N|$, *respectively.*

Proof. 1): By Example 3 $\{x_n^+\}_{n=0,1,\cdots}$ is a non-negative submartingale. Using Jensen's inequality again one has that $\{(x_n^+)^p\}_{n=0,1,\cdots}$ is still a non-negative submartingale. Hence the first inequality is obtained from Theorem 16. Now if $p > 1$, set $\xi = \max_{0\leq n\leq N}(x_n^+)$. then by Theorem 16 again one has that
$\lambda P(\xi \geq \lambda) \leq Ex_N^+ I_{\xi\geq\lambda}.$
Hence using Fubini's theorem and Hölder's inequality one derives that
$E(\xi^p) = E\int_0^\xi p\lambda^{p-1}d\lambda = E\int_0^\infty p\lambda^{p-1}I_{\lambda\leq\xi}d\lambda = p\int_0^\infty \lambda^{p-1}P(\xi \geq \lambda)d\lambda$
$\leq p\int_0^\infty \lambda^{p-2}E(x_N^+ I_{\xi\geq\lambda})d\lambda = \frac{p}{p-1}E[\xi^{p-1}x_N^+]$
$\leq \frac{p}{p-1}[E(x_N^+)^p]^{1/p}[E\xi^p]^{(p-1)/p}.$
Now if $E(\xi^p) = 0$, then the second inequality is trivial. If $E(\xi^p) > 0$, dividing both sides by $[E\xi^p]^{(p-1)/p}$, the second inequality is also obtained.

2): If $\{x_n\}_{n=0,1,\cdots}$ is a real martingale, then by Jensen's inequality we have that $\{|x_n|\}_{n=0,1,\cdots}$ is a submartingale, and $|x_n|^+ = |x_n|$. So by 1) the conclusions are derived in this case. ∎

In the following we will show the upcrossing inequality for a submartingale, which is the basis for proving the important limit property of a submartingale. First we introduce some notations.

For a real $\mathfrak{F}_t-$ adapted process $\{x_n\}_{n=0,1,\cdots}$ and an interval $[a,b]$, where $b>a$, let

$\tau_1 = \min\{n \geq 0 : x_n \leq a\}$,
$\tau_2 = \min\{n \geq \tau_1 : x_n \geq b\}$,
$\cdots\cdots$
$\tau_{2n+1} = \min\{n \geq \tau_{2n} : x_n \leq a\}$,
$\tau_{2n+2} = \min\{n \geq \tau_{2n+1} : x_n \geq b\}$,
$\cdots\cdots$;

where we recall that $\min \phi = +\infty$. Then $\{\tau_n\}$ is an increasing sequence of stopping times. In fact, $\forall k \geq 0$,

$\{\tau_1 = k\} = \{x_0 > a, x_1 > a, \cdots, x_{k-1} > a, x_k \leq a\} \in \mathfrak{F}_k$;
$\{\tau_2 = k\} = \cup_{j=0}^{k-1} \{\tau_1 = j, \tau_2 = k\}$
$= \cup_{j=0}^{k-1} \{\tau_1 = j, x_j \leq a, x_{j+1} < b, \cdots, x_{k-1} < b, x_k \geq b\} \in \mathfrak{F}_k$;
$\{\tau_3 = k\} = \cup_{j=1}^{k-1} \{\tau_2 = j, \tau_3 = k\}$
$= \cup_{j=1}^{k-1} \{\tau_2 = j, x_j \geq b, x_{j+1} > a, \cdots, x_{k-1} > a, x_k \leq a\} \in \mathfrak{F}_k$.

Hence τ_1, τ_2, and τ_3 are stopping times. The proofs for the rest are similar.

Now set

$U_a^b[x(.), N](\omega) = \max\{k \geq 1 : \tau_{2k}(\omega) \leq N\}$,
$D_a^b[x(.), N](\omega) = \max\{k \geq 1 : \tau_{2k-1}(\omega) \leq N\}$.

Obviously the first one is the number connected to the upcrossing of $\{x_n\}_{n=0}^N$ for the interval $[a,b]$, and the second one is the number connected to the downcrossing of $\{x_n\}_{n=0}^N$ for the interval $[a,b]$.

Theorem 18 *If $\{x_n\}_{n=0}^\infty$ is a submartingale, then for each $N \geq 1, n \geq 0$ and $a < b$*

$EU_a^b[x(.), N] \leq \frac{1}{b-a}(E[(x_N - a)^+ - (x_0 - a)^+)$,
$P(U_a^b[x(.), N] \geq n) \leq \frac{1}{b-a} E[(x_N - a)^+ I_{U_a^b[x(.),N]=n}]$,
$ED_a^b[x(.), N] \leq \frac{1}{(b-a)} E(x_N - b)^+$,
$P(D_a^b[x(.), N] \geq n+1) \leq \frac{1}{(b-a)} E(x_N - b)^+ I_{D_a^b[x(.),N]=n}$.

Proof. For a submartingale $\{x_n\}_{n=0}^\infty$ by Example 3 one sees that $\{y_n\}_{n=0}^\infty = \{(x_n - a)^+\}_{n=0}^\infty$ is a non-negative submartingale. Clearly $U_0^{b-a}[y(.), N](\omega) = U_a^b[x(.), N](\omega)$. Again define $\tau_1, \tau_2, \cdots$ as above, but with x, a, and b replaced by $y, 0$, and $b-a$ respectively. Then if $2k > N$

$E(y_N - y_0) = E\sum_{n=1}^{2k}(y_{\tau_n \wedge N} - y_{\tau_{n-1} \wedge N}) = E\sum_{n=1}^{k}(y_{\tau_{2n} \wedge N} - y_{\tau_{2n-1} \wedge N})$
$+ \sum_{n=0}^{k-1} E(y_{\tau_{2n+1} \wedge N} - y_{\tau_{2n} \wedge N}) \geq (b-a) EU_0^{b-a}[y(.), N]$,

where we have used the fact that $\{y_n\}_{n=0}^\infty$ is a submartingale, and hence a strong submartingale (Theorem 15), so $E(y_{\tau_{2n+1} \wedge N} - y_{\tau_{2n} \wedge N}) \geq 0$; and $y_n \geq 0, \forall n$. The first inequality is proved. Now observe that

$0 \geq E(y_{\tau_{2n} \wedge N} - y_{\tau_{2n+1} \wedge N})$
$= E[(y_{\tau_{2n} \wedge N} - y_{\tau_{2n+1} \wedge N})(I_{\tau_{2n} \leq N < \tau_{2n+1}} + I_{\tau_{2n+1} \leq N})]$

$= E[(b-a-y_N)I_{\tau_{2n}\le N<\tau_{2n+1}} + (b-a)I_{\tau_{2n+1}\le N}]$
$= E(b-a)I_{\tau_{2n}\le N} - Ey_N I_{\tau_{2n}\le N<\tau_{2n+1}}.$

Since $\{U_0^{b-a}[y(.),N]\ge n\} = \{N\ge\tau_{2n}\}$ and
$\{\tau_{2n}\le N<\tau_{2n}+1\}\subset\{\tau_{2n}\le N<\tau_{2n+2}\} = \{U_0^{b-a}[y(.),N]=n\}.$
Hence we find that $Ey_N I_{U_a^b[x(.),N]=n} \ge (b-a)P(U_0^{b-a}[y(.),N]\ge n).$

For the downcrossing inequality we have to discuss $\{x_n\}_{n=0}^\infty$ itself directly, since $\{x_n\wedge 0\}_{n=0}^\infty$ is not a submartingale. Let us set $y_n = x_n - b$. Then $\{y_n\}_{n=0}^\infty$ is still a submartingale, and
$D^0_{-(b-a)}[y(.),N](\omega) = D_a^b[x(.),N](\omega).$

Again define $\tau_1,\tau_2,\cdots$ as above but with x, a, and b replaced by y, $-(b-a)$, and 0 respectively. We will now use another method to show the last two inequalities. First, for the fourth inequality we have that as $n\ge 1$
$0\ge E(y_{\tau_{2n}\wedge N} - y_{\tau_{2n+1}\wedge N})$
$= E[(0-(x_N-b))I_{\tau_{2n}\le N<\tau_{2n+1}} + (b-a)I_{\tau_{2n+1}\le N}].$
Since
$\{D_a^b[x(.),N]\ge n+1\} = \left\{D^0_{-(b-a)}[y(.),N]\ge n+1\right\}$
$= \{N\ge\tau_{2n+2}\}\subset\{N\ge\tau_{2n+1}\}$
and $\{\tau_{2n}\le N<\tau_{2n+1}\}\subset\{\tau_{2n}\le N<\tau_{2n+2}\} = \{D_a^b[x(.),N]=n\}.$
Hence it follows that
$E(x_N-b)^+ I_{D_a^b[x(.),N]=n} \ge (b-a)P(D_a^b[x(.),N]\ge n+1).$
The fourth inequality holds. Now taking the summation for $n\ge 0$ it yields
$E(x_N-b)^+ \ge (b-a)\sum_{n=0}^\infty P(D_a^b[x(.),N]\ge n+1)$
$= (b-a)\sum_{n=0}^\infty nP(D_a^b[x(.),N]=n) = (b-a)ED_a^b[x(.),N].$
The third inequality is also established. ∎

Corollary 19 *If $\{x_n\}_{n=0}^\infty$ is a supermartingale, then for each $N\ge 1, n\ge 0$ and $a<b$*

$EU_a^b[x(.),N]\le \frac{1}{b-a}E[(x_N-a)^-],$
$P(U_a^b[x(.),N]\ge n+1)\le \frac{1}{b-a}E[(x_N-a)^- I_{U_a^b[x(.),N]=n}].$
$ED_a^b[x(.),N]\le \frac{1}{(b-a)}E[(x_N-b)^- - (x_0-b)^-],$
$P(D_a^b[x(.),N]\ge n)\le \frac{1}{(b-a)}E(x_N-b)^- I_{D_a^b[x(.),N]=n}.$

Proof. Let $y_n = -x_n$. Then $\{y_n\}_{n=0}^\infty$ is a submartingale. Hence
$U_a^b[x(.),N] = D_{-b}^{-a}[y(.),N],$
and
$D_a^b[x(.),N] = U_{-b}^{-a}[y(.),N].$
Applying Theorem 18 we arrive at the results. ∎

Theorem 18 and Corollary 19 are the classical crossing theorems on martingale. We can derive some other useful crossing results which are very useful in the mathematical finance.[16],[17],[180] Here we apply some of them to derive the important limit theorem on martingales.

Theorem 20 *If $\{x_n\}_{n=0}^{\infty}$ is a submartingale such that there exists a subsequence of $\{n\}$, denote it by $\{n_k\}$, such that*

$$\sup{}_k Ex_{n_k}^+ < \infty, \tag{1.1}$$

then $x_\infty = \lim_{n\to\infty} x_n$ exists a.s., and x_∞ is integrable. In particular, if $x_n \leq 0, \forall n$, then condition (1.1) is obviously satisfied, and in this case $\forall n$
$E[x_\infty|\mathfrak{F}_n] \geq x_n$, a.s.

Proof. First, clearly
condition (1.1) $\Longleftrightarrow \sup{}_n Ex_n^+ < \infty \Longleftrightarrow \sup{}_n E|x_n| < \infty$.
In fact, by the properties of submartingales one has that
$Ex_k^+ \leq Ex_{n_k}^+, \ \forall k.$
and
$E|x_n| = 2Ex_n^+ - Ex_n \leq 2Ex_n^+ - Ex_0.$
Hence the equivalent relations hold. Now let $U_a^b(x(.)) = \lim_{N\to\infty} U_a^b(x(.), N)$.
Then by Theorem 18
$EU_a^b(x(.)) \leq \frac{1}{b-a}\sup_N E(x_N - a)^+ < \infty.$
Hence $U_a^b(x(.)) < \infty$, *a.s.* Let
$W = \cup_{a,b\in Q,a<b} W_{a,b} = \cup_{a,b\in Q,a<b}\left\{\underline{\lim}_n x_n < a < b < \overline{\lim}_n x_n\right\}.$
Then
$P(W) \leq \sum_{a,b\in Q,a<b} P(W_{a,b}) \leq \sum_{a,b\in Q,a<b} P(U_a^b(x(.)) = \infty) = 0.$
Now we can let $x_\infty(\omega) = \lim_{n\to\infty} x_n(\omega)$, as $\omega \notin W$; and $x_\infty(\omega) = 0$, as $\omega \in W$. By Fatou's lemma
$E|x_\infty| \leq \sup_n E|x_n| < \infty.$
Hence x_∞ is integrable. In the case $x_n \leq 0, \forall n$, by the definition of a submartingale
$0 \geq E[x_m|\mathfrak{F}_n] \geq x_n$, a.s. $\forall m$.
Again by Fatou's lemma letting $m \to \infty$ one reaches the final conclusion.
■

1.4 Uniform Integrability and Martingales

It is well known in the theory of real analysis that if a sequence of measurable functions is dominated by an integrable function, then one can take the limit under the integral sign for the function sequence. That is the famous Lebesgue's dominated convergence theorem. However, sometimes it is difficult to find such a dominated function. In this case the uniform integrability of that function sequence can be a great help. Actually, in many cases it is a powerful tool .

Definition 21 *A family of functions $A \subset L^1(\Omega, \mathfrak{F}, P)$ is called uniformly integrable, if $\lim_{\lambda\to\infty}\sup_{f\in A} E(fI_{|f|>\lambda}) = 0$, where $L^1(\Omega, \mathfrak{F}, P)$ is the totality of random variables ξ, (that is, all ξ are $\mathfrak{F}-$measurable) such that $E|\xi| < \infty$.*

Lemma 22 *Suppose that* $\{x_n\}_{n=1}^{\infty} \subset L^1(\Omega, \mathfrak{F}, P)$ *is uniformly integrable, and as* $n \to \infty$,

$x_n \to x$, *in probability,*

i.e. $\forall \varepsilon > 0$, $\lim_{n\to\infty} P(|x_n - x| > \varepsilon) = 0$, *then*

$\lim_{n\to\infty} E|x_n - x| = 0$. *(i.e.* $x_n \to x$, *in* $L^1(\Omega, \mathfrak{F}, P)$*).*

In particular, $\lim_{n\to\infty} Ex_n = Ex$

Proof. In fact, $\forall \varepsilon > 0$,

$E|x_n - x| \leq E(|x_n - x| I_{|x_n - x| > \lambda}) + E(|x_n - x| I_{|x_n - x| \leq \lambda}) = I_1^{n,\lambda} + I_2^{n,\lambda}$.

Hence one can take a λ large enough such that $I_1^{n,\lambda} < \varepsilon/2$, since clearly $\{x_n - x\}_{n=1}^{\infty}$ is uniformly integrable. Then for this fixed λ by using Lebesgue's dominated convergence theorem one can have a sufficiently large N such that as $n \geq N$, $I_2^{n,\lambda} < \varepsilon/2$. ■

For the sufficient conditions of uniform integrability of a family A we have

Lemma 23 *Suppose that* $A \subset L^1(\Omega, \mathfrak{F}, P)$. *Any one of the following conditions makes* A *uniformly integrable:*

1) There exists an integrable $g \in L^1(\Omega, \mathfrak{F}, P)$ *such that*

$|x| \leq g$, $\forall x \in A$.

2) There exists a $p > 1$ *such that* $\sup_{x\in A} E|x(\omega)|^p < \infty$.

Proof. 1): Since as $\lambda \to \infty$

$\sup_{x\in A} P(|x(\omega)| > \lambda) \leq \frac{1}{\lambda} \sup_{x\in A} E|x(\omega)| \leq \frac{1}{\lambda} E|g(\omega)| \to 0$.

So by the integrability of g one has that as $\lambda \to \infty$

$E|x(\omega)| I_{|x(\omega)|>\lambda} \leq Eg(\omega) I_{|x(\omega)|>\lambda} \to 0$, uniformly w.r.t. $x \in A$.

2): Since $\sup_{x\in A} P(|x(\omega)| > \lambda) \leq \frac{1}{\lambda} \sup_{x\in A} E|x(\omega)| \to 0$, as $\lambda \to \infty$. So as $\lambda \to \infty$

$\sup_{x\in A} E|x(\omega)| I_{|x(\omega)|>\lambda}$

$\leq \sup_{x\in A}(E|x(\omega)|^p)^{1/p} \sup_{x\in A}[P(|x(\omega)| > \lambda)]^{(p-1)/p} \to 0$. ■

Now we know that the uniform integrability condition is weaker than the domination condition. Actually, it is also the neccessary condition for the L^1–convergence of the sequence of integrable random variables or, say, integrable functions.

Theorem 24 *Suppose that* $\{x_n\}_{n=1}^{\infty} \subset L^1(\Omega, \mathfrak{F}, P)$. *Then the following two statements are equivalent:*

1) $\{x_n\}_{n=1}^{\infty}$ *is uniformly integrable.*

2) $\sup_{n=1,2,\cdots} E|x_n| < \infty$; *and* $\forall \varepsilon > 0, \exists \delta > 0$ *such that* $\forall B \in \mathfrak{F}$, *as* $P(B) < \delta$

$\sup_{n=1,2,\cdots} E|x_n| I_B < \varepsilon$.

Furthermore, if there exists an $x \in L^1(\Omega, \mathfrak{F}, P)$ *such that as* $n \to \infty$, $x_n \to x$, *in probability; then the following statement is also equivalent to 1):*

3) $x_n \to x$, *in* $L^1(\Omega, \mathfrak{F}, P)$.

Proof. Since 1) $\Longrightarrow$ 3) is already proved in Lemma 22, we will show that 3) $\Longrightarrow$ 2) $\Longrightarrow$ 1) $\Longrightarrow$2).

1) $\Longrightarrow$ 2): Take a λ_0 large enough such that $\sup_{n=1,2,\cdots} E|x_n| I_{|x_n|>\lambda_0} < 1$. Then

$\sup_{n=1,2,\cdots} E|x_n| \leq \lambda_0 + 1.$

On the other hand, for any $B \in \mathfrak{F}$ since

$E|x_n| I_B = E|x_n| I_{\{|x_n|>\lambda\}\cap B} + E|x_n| I_{\{|x_n|\leq\lambda\}\cap B}$

$\leq \sup_{n=1,2,\cdots} E|x_n| I_{\{|x_n|>\lambda\}} + \lambda P(B) = I_1^\lambda + I_2^{\lambda,B}.$

Hence $\forall\varepsilon > 0$ one can take a $\lambda_\varepsilon > 0$ large enough such that $I_1^{\lambda_\varepsilon} < \frac{\varepsilon}{2}$, then let $\delta_\varepsilon = \frac{\varepsilon}{2\lambda_\varepsilon}$. For this $\delta_\varepsilon > 0$ one has that $\forall B \in \mathfrak{F}, P(B) < \delta_\varepsilon \Longrightarrow \sup_{n=1,2,\cdots} E|x_n| I_B < \varepsilon$.

2) $\Longrightarrow$ 1): $\forall\varepsilon > 0$ Take $\delta > 0$ such that 2) holds. Since

$P(|x_n| > \lambda) \leq \frac{1}{\lambda} \sup_{n=1,2,\cdots} E|x_n|.$

Hence one can take an N large enough such that as $\lambda > N$,

$P(|x_n| > \lambda) < \delta, \forall n = 1, 2, \cdots.$

Thus by 2) as $\lambda > N$,

$E|x_n| I_{|x_n|>\lambda} < \varepsilon, \forall n = 1, 2, \cdots.$

3) $\Longrightarrow$ 2): Take an N_0 large enough such that as $n > N_0$,

$E|x - x_n| < 1.$

Thus

$\sup_{n=1,2,\cdots} E|x_n| \leq \max\{1 + E|x|, E|x_1|, \cdots, E|x_{N_0}|\} < \infty.$

On the other hand, observe that

$E|x_n| I_B \leq E|x_n - x| + E|x| I_B.$

Hence $\forall\varepsilon > 0$, one can take an N_ε large enough such that as $n > N_\varepsilon$,

$E|x - x_n| < \frac{\varepsilon}{2}.$

Then take a $\delta > 0$ small enough such that $\forall B \in \mathfrak{F}$, as $P(B) < \delta$,

$\max_{n=1,\cdots,N_\varepsilon} \{E|x_n| I_B\} < \varepsilon$, and $E|x| I_B < \varepsilon/2$.

Thus as $P(B) < \delta, E|x_n| I_B < \varepsilon, \forall n = 1, 2, \cdots$. ∎

Now let us use uniform integrability as a tool to study the martingales.

Theorem 25 *If $\{x_n\}_{n=0}^\infty$ is a submartingale such that $\{x_n^+\}_{n=0}^\infty$ is uniformly integrable, then $x_\infty = \lim_{n\to\infty} x_n$ exists, a.s., and*

$E[x_\infty|\mathfrak{F}_n] \geq x_n, \forall n,$

i.e. $\{x_n\}_{n=0,1,2,\cdots,\infty}$ is also a submartingale, and we call it a right-closed submartingale.

This theorem actually tells us that a uniformly integrable submartingale is a right-closed submartingale.

Proof. By uniform integrability one has that

$\sup_{n=0,1,2,\cdots} Ex_n^+ < \infty.$

Hence applying Theorem 20 one has that $x_\infty = \lim_{n\to\infty} x_n$ exists, a.s. Now by the submartingale property $\{x_n^+\}_{n=0}^\infty$ is also a submartingale (Example 3). Hence for any $a > 0$, and $B \in \mathfrak{F}_n$, as $m \geq n$,

$\int_B [(-a) \vee x_n] dP \leq \int_B [(-a) \vee x_m] dP.$

Letting $m \to \infty$ by the uniform integrability of $\{x_n^+\}_{n=0}^\infty$ one has that

$\int_B [(-a) \vee x_n] dP \leq \int_B [(-a) \vee x_\infty] dP.$
Now letting $a \uparrow \infty$ by Fatou's lemma one obtains that
$\int_B x_n dP \leq \int_B x_\infty dP = \int_B E[x_\infty|\mathfrak{F}_n] dP, \forall B \in \mathfrak{F}_n.$
The conclusion is established. ■

We also have the following inverse theorem.

Theorem 26 *If $\{x_n\}_{n=0,1,2,\cdots,\infty}$ is a submartingale, where $x_\infty = \lim_{n\to\infty} x_n$ exists, a.s., then $\{x_n^+\}_{n=0,1,2,\cdots}$ is uniformly integrable.*

Proof. *By Jensen's inequality $\{x_n^+\}_{n=0,1,2,\cdots,\infty}$ is also a submartingale. Now $\forall \lambda > 0$, denote $B_\lambda^n = \{x_n^+ > \lambda\}$, then by the submartingale definition as $\lambda \to \infty$*

$P(B_\lambda^n) \leq \frac{1}{\lambda} E x_n^+ \leq \frac{1}{\lambda} E x_\infty^+ \to 0,$ *uniformly w.r.t. n.*

Therefore as $\lambda \to \infty$

$\int_{B_\lambda^n} x_n^+ dP \leq \int_{B_\lambda^n} x_\infty^+ dP \to 0,$ *uniformly w.r.t. n.* ■

Corollary 27 *1) If $\{x_n\}_{n=0,1,2,\cdots,\infty}$ is a right-closed martingale, then one has that $\{x_n\}_{n=0,1,2,\cdots}$ is uniformly integrable.*

2) For a sequence of random variables $\{y_n\}_{n=0,1,2,\cdots}$ if there exists a $z \in L^1(\Omega, \mathfrak{F}, P)$ such that $|y_n| \leq E(|z|\,|\mathfrak{F}_n)$, where $\mathfrak{F}_n \subset \mathfrak{F}_{n+1} \subset \mathfrak{F}$ are $\sigma-$fields, then $\{y_n\}_{n=0,1,2,\cdots}$ is uniformly integrable.

Proof. 1): It can be derived from Theorem 26, since $\{x_n\}_{n=0,1,2,\cdots,\infty}$ is both a submartingale and a supermartingale.

2): In fact, let $z_n = E(z|\mathfrak{F}_n)$. Then, obviously $\{z_n\}_{n=0,1,2,\cdots}$ is a martingale, and $\sup_{n=0,1,2,\cdots} E|z_n| \leq E|z| < \infty$. Hence by Theorem 20 $z_\infty = \lim_{n\to\infty} z_n$ exists, *a.s.*, $z_\infty \in L^1(\Omega, \mathfrak{F}, P)$, and $z_n = E(z_\infty|\mathfrak{F}_n), \forall n$. Therefore $\{z_n\}_{n=0,1,2,\cdots,\infty}$ is a right-closed martingale. Now by 1) it is uniformly integrable, so is $\{y_n\}_{n=0,1,2,\cdots}$. ■

Condition in 2) for the uniform integrability is weaker than the usual Lebesgue's dominated condition. Moreover, from the proof of 2) in Corollary 27 one also can obtain the following thoerem

Theorem 28 *(Levi's theorem) If $z \in L^1(\Omega, \mathfrak{F}, P)$, and $\mathfrak{F}_n \subset \mathfrak{F}_{n+1} \subset \mathfrak{F}$ are $\sigma-$fields, then as $n \uparrow \infty$, $E[z|\mathfrak{F}_n] \to E[z|\mathfrak{F}_\infty]$, where $\mathfrak{F}_\infty = \vee_{n=1,2,\cdots}\mathfrak{F}_n$, i.e. $\mathfrak{F}_\infty$ is the smallest $\sigma-$field including all $\mathfrak{F}_n, n = 1, 2, \cdots$.*

Proof. By the proof of 2) in Corollary 27 one already has that $z_\infty = \lim_{n\to\infty} z_n = \lim_{n\to\infty} E[z|\mathfrak{F}_n]$ exists, *a.s.*, $z_\infty \in L^1(\Omega, \mathfrak{F}, P)$, and $z_n = E(z_\infty|\mathfrak{F}_n), \forall n$. Let us show that $z_\infty = E[z|\mathfrak{F}_\infty], a.s.$ In fact, by limit one has $z_\infty \in \mathfrak{F}_\infty$, i.e. it is $\mathfrak{F}_\infty-$measurable. Moreover, $\forall n, \forall B \in \mathfrak{F}_n$,

$Ez_\infty I_B = Ez_n I_B = E(E[z|\mathfrak{F}_n] I_B) = E(E[zI_B|\mathfrak{F}_n]) = EzI_B$
$= EE[z|\mathfrak{F}_\infty] I_B.$

From this one also has that $\forall B \in \mathfrak{F}_\infty = \vee_{n=1,2,\cdots}\mathfrak{F}_n$,

$Ez_\infty I_B = EE[z|\mathfrak{F}_\infty] I_B.$

Since z_∞ and $E[z|\mathfrak{F}_\infty]$ both are $\mathfrak{F}_\infty$−measurable. Hence $z_\infty = E[z|\mathfrak{F}_\infty], a.s.$
∎

Now let us consider the discrete time $\{\cdots,-k,-k+1,\cdots,-2,-1,0\}$ with right-end point 0 but without the initial left–starting-time. We still call $\{x_n\}_{n\in\{\cdots,-k,-k+1,\cdots,-2,-1,0\}}$ a martingale (supermartingale, submartingale), if
(i) x_n is integrable for each $n = 0,-1,-2,\cdots$;
(ii) x_n is $\mathfrak{F}_n$−adapted, i.e. for each $n = 0,-1,-2,\cdots$, x_n is $\mathfrak{F}_n$−measurable, where $\mathfrak{F}_n, n = 0,-1,-2,\cdots$, are σ−fields still with an increasing property, i.e. $\mathfrak{F}_n \subset \mathfrak{F}_m$, as $n \leq m; \forall n,m \in \{\cdots,-k,-k+1,\cdots,-2,-1,0\}$;
(iii) $E[x_m|\mathfrak{F}_s] = x_n$, (respectively, $\leq$, $\geq$), $a.s.$ $\forall 0 \leq n \leq m; \forall n,m \in \{\cdots,-k,-k+1,\cdots,-2,-1,0\}$.
We have the following limit thoerem.

Theorem 29 *If* $\{x_n\}_{n\in\{\cdots,-k,-k+1,\cdots,-2,-1,0\}}$ *is a submartingale such that* $\inf_{n=0,-1,-2,\cdots} Ex_n > -\infty$,
then $\{x_n\}_{n\in\{\cdots,-k,-k+1,\cdots,-2,-1,0\}}$ *is uniformly integrable,*
$x_{-\infty} = \lim_{n\to-\infty} x_n$ *exists a.s.,*
and as $n \to -\infty$, $x_n \to x_{-\infty}$, *in* $L^1(\Omega,\mathfrak{F},P)$.

Proof. For each N consider the finite sequence of random variables $\{x_n\}_{n=-N,-N+1,\cdots,-1,0}$. Denot by $U_a^b[x(.),-N]$ the number of upcrossing of $\{x_n\}_{n=-N,-N+1,\cdots,-1,0}$ for the interval $[a,b]$. Then by Theorem 18 $EU_a^b[x(.),-N] \leq \frac{1}{b-a}E(x_0-a)^+$, and
$EU_a^b[x(.)] \leq \frac{1}{b-a}E(x_0-a)^+ < \infty$,
where $U_a^b[x(.)] = \lim_{N\to\infty} U_a^b[x(.),-N]$. By the proof of Theorem 18 one has that $x_{-\infty} = \lim_{n\to-\infty} x_n$ exists $a.s.$ However, it still remains to be proved that $x_{-\infty}$ is finite, $a.s.$. Let us show that $\{x_n\}_{n\in\{\cdots,-k,-k+1,\cdots,-2,-1,0\}}$ is uniformly integrable. If this can be done, then all conclusions will be derived immediately by the property of uniform integrability. Observe that $Ex_n \downarrow$, as $n \downarrow$, since $\{x_n\}_{n\in\{\cdots,-k,-k+1,\cdots,-2,-1,0\}}$ is a submartingale. Hence by assumption a finite limit exists:
$\lim_{n\to-\infty} Ex_n \geq \inf_{n=0,-1,-2,\cdots} Ex_n > -\infty$.
Note that $\{E[x_0|\mathfrak{F}_n]\}_{n\in\{\cdots,-k,-k+1,\cdots,-2,-1,0\}}$ is a martingale and uniformly integrable. Hence $\{x_n - E[x_0|\mathfrak{F}_n]\}_{n\in\{\cdots,-k,-k+1,\cdots,-2,-1,0\}}$ is a non-positive submartingale, and the uniform integrability of it is the same as that of $\{x_n\}_{n\in\{\cdots,-k,-k+1,\cdots,-2,-1,0\}}$. So we may assume that $x_n \leq 0$. Now $\forall\varepsilon > 0$, take a $-k$ large enough such that $Ex_k - \lim_{n\to-\infty} Ex_n < \varepsilon$. Then by the property of submartingales and the property that $Ex_n \downarrow$ as $n \downarrow$,
$P(x_n < -\lambda) \leq \frac{1}{\lambda}E|x_n| = \frac{1}{\lambda}(2Ex_n^+ - Ex_n)$
$\leq \frac{1}{\lambda}(2Ex_0^+ - \lim_{n\to-\infty} Ex_n) \to 0$, uniformly w.r.t. n, as $\lambda \to \infty$.
and if $n \leq k \leq 0$,
$0 \leq E[(-x_n)I_{x_n<-\lambda}] \leq -Ex_n - E[(-x_n)I_{x_n\geq-\lambda}]$
$\leq -\lim_{n\to-\infty} Ex_n + E[x_k I_{x_n\geq-\lambda}]$
$\leq -Ex_k + E[x_k I_{x_n\geq-\lambda}] + \varepsilon = E[(-x_k)I_{x_n<-\lambda}] + \varepsilon$.

From this one easily derives that $\{x_n\}_{n\in\{\cdots,-k,-k+1,\cdots,-2,-1,0\}}$ with $x_n \leq 0$ is uniformly integrable. ■

1.5 Martingales with Continuous Time

Now let us consider the martingale (submartingale, and supermartingale) $\{x_t\}_{t\in[0,\infty)}$ with continuous time. First, we will still introduce the upcrossing numbers of a random process for an interval. Let $\{x_t\}_{t\geq 0} = \{x_t\}_{t\in[0,\infty)}$be an adapted process, and $U = \{t_1, t_2, \cdots, t_n\}$ be a finite subset of $R_+ = [0, \infty)$. Denote its rearrangement to the natural order by $\{s_1, s_2, \cdots, s_n\}$, i.e. $s_1 < s_2 < \cdots < s_n$. Let $U_a^b[x(.), U]$ be the number of upcrossings of $\{x_{s_k}\}_{k=1}^n$ for interval $[a, b]$, and we also call it the number of upcrossings of $\{x_t\}_{t\in U}$ for interval $[a, b]$. For any subset D of R_+, define

$U_a^b[x(.), D] = \sup\left\{U_a^b[x(.), U] : U \text{ is a finite subset of } D\right\}$.

In case $D = \{t_1, t_2, \cdots, t_n, \cdots\}$, obviously

$U_a^b[x(.), D] = \lim_{n\to\infty} U_a^b[x(.), U_n]$,

where $U_n = \{t_1, t_2, \cdots, t_n\}$. By using the results on discrete time we have the following theorem.

Theorem 30 *If $\{x_t\}_{t\geq 0}$ is a submartingale, $D = \{t_1, t_2, \cdots, t_n, \cdots\}$, then for any $0 \leq r < s, a < b$ and $\lambda > 0$ one has that*

$\lambda P(\sup_{t\in D\cap[r,s]} |x_t| > -\lambda) \leq Ex_r^- + 2E(x_s^+)$,

$EU_a^b[x(.), D\cap[r, s]] \leq \frac{1}{b-a}(E[(x_s - a)^+ - (x_r - a)^+)$.

Proof. Set $\widehat{D} = \{r, s, t_1, t_2, \cdots, t_n, \cdots\}$ Notice that

$U_n = (\{r, s, t_1, t_2, \cdots, t_n\} \cap [r, s]) \uparrow (\widehat{D} \cap [r, s])$.

So the conlusions for the set $\widehat{D} \cup [r, s]$ are derived by applying Theorem 16 and 18 and taking the limit for $n \to \infty$. However, $(D \cap [r, s]) \subset (\widehat{D} \cap [r, s])$. So the two conclusions for $(D \cap [r, s])$ hold true. ■

Now we can generalize the limit theorem to submartingales with continuous time.

Theorem 31 *Let $\{x_t\}_{t\geq 0}$ be a submartingale. Then $\widehat{x}_t = \lim_{r\downarrow t, r\in Q} x_r$ exists and finite a.s. and $\{\widehat{x}_t\}_{t\geq 0}$ is still a submartingale such that $\widehat{x}_t$ is right continuous with left-hand limits a.s. Furthermore, $x_t \leq \widehat{x}_t$ a.s. for $\forall t \geq 0$.*

(Recall that $Q =$ the totality of real rational numbers). To establish the above theorem we divide it into two steps. The first step can be written as the following lemma.

Lemma 32 *If $\{x_t\}_{t\geq 0}$ is a submartingale, then $\forall t \geq 0$*

$\lim_{r\uparrow t, r\in Q} x_r$ *and* $\lim_{r\downarrow t, r\in Q} x_r$

exist and are finite a.s. (Here, we define $x_{0-} = x_0$, and $\mathfrak{F}_{0-} = \mathfrak{F}_0$).

Proof. Denote $Q = \{t_1, t_2, \cdots, t_n, \cdots\}$. Notice that for any given $T > 0$ if $\omega \in \Omega$ is such a point that there exists a $t \in [0,T]$ such that one of the limits: $\lim_{r\uparrow t, r\in Q} x_r(\omega)$ and $\lim_{r\downarrow t, r\in Q} x_r(\omega)$, does not exist, then there will exist $a < b$ such that

$U_a^b[x(.), Q \cap [0,T]](\omega) = \infty.$

On the other hand, if one of the above limits even exists but it is not finite, then

$\sup_{t\in Q\cap[0,T]} |x_t(\omega)| = \infty.$

However, by Theorem 30 one has that $\forall T > 0$, and for all a, b with $a < b$

$EH_{T,a,b} = 0,$

where $H_{T,a,b} = \left\{U_a^b[x(.), Q \cap [0,T]] = \infty\right\} \cup \left\{\sup_{t\in Q\cap[0,T]} |x_t| = \infty\right\}$. Hence the conclusion can be established. ■

The second step to show Theorem 31 is as follows:

Proof. By the above lemma $\widehat{x}_t = \lim_{r\downarrow t, r\in Q} x_r$ exists and are finite a.s. Obviously, it is right continuous with left-hand limit. Since $\widehat{x}_t$ is $\mathfrak{F}_t = \mathfrak{F}_{t+0}-$adapted and $\{x_t\}_{t\geq 0}$ is a submartingale. Hence $\forall \varepsilon_n \downarrow 0$, $\{x_{t+\varepsilon_n}\}_{n=1}^{\infty}$ is uniformly integrable by Theorem 29, and

$E\widehat{x}_t I_B = \lim_{n\to\infty} Ex_{t+\varepsilon_n} I_B \leq \lim_{n\to\infty} Ex_{s+\varepsilon_n} I_B = E\widehat{x}_s I_B,$

for $s > t$ and $B \in \mathfrak{F}_t$. This shows that $\{\widehat{x}_t\}_{t\geq 0}$ is a submartingale. Moreover, since $E\widehat{x}_t I_B = \lim_{n\to\infty} Ex_{t+\varepsilon_n} I_B \geq Ex_t I_B, \forall B \in \mathfrak{F}_t$. So $x_t \leq \widehat{x}_t$ a.s. for $\forall t \geq 0$. ■

Corollary 33 *Under the assumption of Theorem 31 one has the following conclusions:*

1) $P(x_t = \widehat{x}_t) = 1, \forall t \geq 0 \Longleftrightarrow E(x_t)$ *is right-continuous,* $\forall t \geq 0$.

2) If $\{x_t\}_{t\geq 0}$ *is a martingale, then* $P(x_t = \widehat{x}_t) = 1, \forall t \geq 0$. *So* $\{\widehat{x}_t\}_{t\geq 0}$ *may be called a right-continuous version of* $\{x_t\}_{t\geq 0}$.

Proof. 1): $\Longleftarrow$: By the uniform integrability of $\{x_{t+\varepsilon_n}\}_{n=1}^{\infty}$, where $\varepsilon_n \downarrow 0$, and the right-continuity of $E(x_t)$ one has

$E\widehat{x}_t = \lim_{n\to\infty} Ex_{t+\varepsilon_n} = E(x_t).$

Since $x_t \leq \widehat{x}_t$, a.s. Hence $x_t = \widehat{x}_t$, a.s.

$\Longrightarrow$: By assumption and the uniform integrability one has that $\forall \varepsilon_n \downarrow 0$,

$E(x_t) = E\widehat{x}_t = E\lim_{n\to\infty} x_{t+\varepsilon_n} = \lim_{n\to\infty} Ex_{t+\varepsilon_n}.$

So $E(x_t)$ is right-continuous.

2): Since in this case $E(x_t) = E(x_0), \forall t \geq 0$. The conclusion is established. ■

From now on we consider right-continuous martingales (supermartingales, submartingales) only. The martingale inequalities, Doob's stopping time theorem (Doob's optional sampling theorem) now are naturally generalized to the continuous time case.

Theorem 34 *1) If* $\{x_t\}_{t\geq 0}$ *is a real right-continuous submartingale such that* $E((x_t^+)^p) < \infty, t \geq 0$, *for some* $p \geq 1$, *then for every* $T > 0$, *and* $\lambda > 0$,

*$P(\sup_{t\in[0,T]} x_t^+ \geq \lambda) \leq E((x_T^+)^p)/\lambda^p$,
and if $p > 1$, then*
$E(\sup_{t\in[0,T]} (x_t^+)^p) \leq \left(\frac{p}{p-1}\right)^p E((x_T^+)^p)$.
2) If $\{x_t\}_{t\geq 0}$ is a real right-continuous martingale such that $E(|x_t|^p) < \infty, t \geq 0$, then the conclusions in 1) hold true for x_t^+ and x_T^+ replaced by $|x_t|$ and $|x_T|$, respectively.

Theorem 35 *(Doob's stopping time theorem). Let $\{x_t\}_{t\geq 0}$ be a real right-continuous martingle (supermartingale, submartingale) with respect to $\{\mathfrak{F}_t\}$, and $\{\sigma_t\}_{t\in[0,\infty)}$ be a family of bounded stopping times such that $P(\sigma_t \leq \sigma_s) = 1$, if $t < s$. Then $\{x_t\}_{t\geq 0}$ is a strong martingle (respectively, strong supermartingle, strong submartingle), i.e. as $t < s$,*
$E[x_{\sigma_s}|\mathfrak{F}_{\sigma_t}] = x_{\sigma_t}$ (respectivly, $\leq, \geq$), a.s.

In Theorem 34, 1) can be proved by using the same technique as the one in Theorem 30, and 2) is derived from 1). To show Theorem 35 let us use the standard discretization technique.

Proof. We show the conclusion for martingales only. For any bounded stopping time σ_t in the given family let
$\sigma_t^n = \frac{k}{2^n}$, as $\sigma_t \in [\frac{k-1}{2^n}, \frac{k}{2^n})$, $n, k = 1, 2, \cdots$.
Then for each n, σ_t^n is a discrete stopping time. In fact, $\{\sigma_t^n = \frac{k}{2^n}\} = \{\frac{k-1}{2^n} \leq \sigma_t < \frac{k}{2^n}\} \in \mathfrak{F}_{\frac{k}{2^n}}$. Moreover, as $n \uparrow \infty, \sigma_t^n \downarrow \sigma_t$. By Theorem 15 one has that $\forall B \in \mathfrak{F}_{\sigma_t} \subset \mathfrak{F}_{\sigma_t^n}$, as $t < s$,
$Ex_{\sigma_s^n} I_B = Ex_{\sigma_t^n} I_B$.
Moreover, since $\sigma_t^n \downarrow$ and is bounded, if we let $y^{-n} = x_{\sigma_t^n}$, $\mathfrak{F}_{-n} = \mathfrak{F}_{\sigma_t^n}$ then $\{y^n\}_{n=-1,-2,\cdots}$ with the $\sigma-$field family $\{\mathfrak{F}_n\}_{n=-1,-2,\cdots}$ obviously satisfies all conditions in Theorem 29. In fact, clearly it is a submartingale by Theorem 15, and
$\inf_n Ey^n \geq Ex_0 > -\infty$.
So $\{y^n\}_{n=-1,-2,\cdots}$ is uniformly integrable. Hence by the right-continuity of x_t one has that $\lim_{n\to\infty} Ex_{\sigma_t^n} I_B = Ex_{\sigma_t} I_B$. Similarly, $\lim_{n\to\infty} Ex_{\sigma_s^n} I_B = Ex_{\sigma_s} I_B$. Therefore, $\forall B \in \mathfrak{F}_{\sigma_t}$, as $t < s$,
$Ex_{\sigma_s} I_B = Ex_{\sigma_t} I_B$.
So $\{x_t\}_{t\geq 0}$ is a strong martingale. ■

1.6 Doob-Meyer Decomposition Theorem

In the incomplete financial market to price some option will involve the problem connected to the Doob-Meyer decomposition of some submartingales or supermartingales. Besides, this decomposition theorem is also a fundamental tool in stochastic analysis and its applications. First we will look at the discrete time case.

Theorem 36 *Let $\{x_n\}_{n=0,1,2,\cdots}$ be a submartingale. The there exists a unique decomposition such that*

$x_n = M_n + A_n, \ n = 0, 1, 2, \cdots,$

where $\{M_n\}_{n=0,1,2,\cdots}$ is a martingale, and $\{A_n\}_{n=0,1,2,\cdots}$ is an increasing process, both are $\{\mathfrak{F}_n\}_{n=0,1,2,\cdots}$ −adapted, and $\{A_n\}_{n=0,1,2,\cdots}$ is predictable, where predictable means that $A_n \in \mathfrak{F}_{n-1}, \forall n = 1, 2, \cdots$, and $A_0 = 0$.

Proof. Uniqueness. If x_n has the above decomposition, where $A_n \in \mathfrak{F}_{n-1}$, then it is easily checked that it will satisfy

$A_0 = 0,$

$A_n = A_{n-1} + E[x_n - x_{n-1}|\mathfrak{F}_{n-1}], \ n = 1, 2, \cdots,$

and

$M_n = x_n - A_n.$

So the decomposition must be unique.

Existence. Now let us define $\{A_n\}_{n=0,1,2,\cdots}$ and $\{M_n\}_{n=0,1,2,\cdots}$ as above. Then obviously $\{A_n\}_{n=0,1,2,\cdots}$ is increasing in n and predictable. Moreover, $\{M_n\}_{n=0,1,2,\cdots}$ is a martingale. In fact,

$E[M_n|\mathfrak{F}_{n-1}] = E[x_n - A_n|\mathfrak{F}_{n-1}] = E[x_n - (A_{n-1} + E[x_n - x_{n-1}|\mathfrak{F}_{n-1}])|\mathfrak{F}_{n-1}]$

$= x_{n-1} - A_{n-1} = M_{n-1}.$

Hence we have a decomposition for x_n. ■

In the continuous time case the situation is much more difficult. Let us introduce some neccesary related concepts first.

In the following we fix a probability space $(\Omega, \mathfrak{F}, P)$ with an increasing σ−field family $\{\mathfrak{F}_t\}_{t\geq 0}, \mathfrak{F}_t \subset \mathfrak{F}, \forall t \geq 0$. (Such a $\{\mathfrak{F}_t\}_{t\geq 0}$ is called a reference family). All martingales, adaptness, stopping times, etc. are relative to this reference family. Now let us introduce the first concept needed for Doob-Meyer decomposition theorem. For $0 < a \leq \infty$ let S_a be the set of all stopping times σ such that $\sigma < a$, a.s.

Definition 37 *A submartingale $\{x_t\}_{t\geq 0}$ is said to be of class (DL) (class (D)), if for every $0 < a < \infty$ (for $a = \infty$), the family of random variables $\{x_\sigma : \sigma \in S_a\}$ is uniformly integrable.*

Clearly, $(D) \subset (DL)$. Recall that we only consider the right continuous martingales (supermartingales, submartingales).

Theorem 38 *1) Any martingale $\{x_t\}_{t\geq 0}$ is of class (DL).*
2) Any uniformly integrable martingale $\{x_t\}_{t\geq 0}$ is of class (D).
3) Any non-negative submartingale $\{x_t\}_{t\geq 0}$ is of class (DL).

Proof. 1): By the optional sampling theorem (Theorem 35), for any $0 < a < \infty$, as $\lambda \to \infty$,

$\sup_{\sigma\in S_a} P(|x_\sigma| > \lambda) \leq \sup_{\sigma\in S_a} E|x_\sigma|/\lambda \leq E|x_a|/\lambda \to 0,$

$E|x_\sigma| I_{|x_\sigma|>\lambda} \leq E|x_a| I_{|x_\sigma|>\lambda} \to 0$, uniformly w.r.t. $\sigma \in S_a$,

where we have applied Jensen's inequality to show that $\{|x_t|\}_{t\geq 0}$ is a submartingale.

2): Since $\{x_t\}_{t\geq 0}$ is a uniformly integrable martingale, so by Theorem 25 one can derive that $x_\infty = \lim_{t\in Q, t\uparrow\infty} x_t$ exists, $x_\infty \in L^1(\Omega, \mathfrak{F}, P)$, and $x_t = E[x_\infty|\mathfrak{F}_t], \forall t \geq 0$. Hence by using the standard discretization technique as in the proof of Theorem 35 one also has that $x_{\sigma\wedge N} = E[x_\infty|\mathfrak{F}_{\sigma\wedge N}], \forall\sigma \in S_\infty, \forall N = 1, 2, \cdots$. Letting $N \uparrow \infty$ and applying Levi's Theorem one obtains

$x_\sigma = E[x_\infty|\mathfrak{F}_\sigma], \forall\sigma \in S_\infty$.

By this the uniform integrability of $\{x_\sigma : \sigma \in S_\infty\}$ is easily derived. (See Corollary 27).

3): Since $|x_t| = x_t \geq 0$, the proof in 1) goes through. ∎

The second important concept we need for the Doob-Meyer decomposion theorem is the natural increasing process.

Definition 39 *1) A right continuous random process $\{A_t\}_{t\geq 0}$ is called an integrable increasing process, if (i) $\{A_t\}_{t\geq 0}$ is $\{\mathfrak{F}_t\}_{t\geq 0}$ −asdapted, (ii) $A_0 = 0, A_s \leq A_t$, a.s., as $s \leq t$, (iii) $EA_t < \infty, \forall t \geq 0$.*

2) An integrable increasing process $\{A_t\}_{t\geq 0}$ is called natural, if for every bounded martingale $\{M_t\}_{t\geq 0}$

$E\int_0^t M_s dA_s = E\int_0^t M_{s-} dA_s, \quad \forall t \in [0, \infty)$.

Lemma 40 *The following statements are equivalent:*
1) The integrable increasing process $\{A_t\}_{t\geq 0}$ is natural.
2) For every bounded martingale $\{M_t\}_{t\geq 0}$

$EM_tA_t = E\int_0^t M_{s-} dA_s, \quad \forall t \in [0, \infty)$.

Proof. In fact, let $0 = t_0 < t_1 < t_2 < \cdots < t_n = t$. Then

$EM_tA_t = \sum_{k=1}^n E[M_t(A_{t_k} - A_{t_{k-1}})] = \sum_{k=1}^n E[(A_{t_k} - A_{t_{k-1}})E(M_t|\mathfrak{F}_{t_k})]$
$= \sum_{k=1}^n E[M_{t_k}(A_{t_k} - A_{t_{k-1}})] \to E\int_0^t M_s dA_s$,

as $max_{1\leq k\leq n}(t_k - t_{k-1}) \to 0$. ∎

The following are examples of natural processes.

Example 41 *1) Any continuous integrable increasing process $\{A_t\}_{t\geq 0}$ is natural.*
2) An integrable increasing process $\{A_t\}_{t\geq 0}$ is natural if and only if it is predictable.

Proof. 1): Any martingale has a right continuous version (actually, as we said before, we need only consider the right continuous martingales, supermartingales, and submartingales). However, for a right continuous function it can only have at most countable discontinuous points. Now A_t is continuous. So we have that $\int_0^t I_{M_s \neq M_{s-}} dA_s = 0$. Hence $\int_0^t M_s dA_s = \int_0^t M_{s-} dA_s$, *a.s.*

2): This proof needs more information, so we refer to Theorem 5.33 in [43] However, here we can give some intuitive view to see why it is possible that this is true. From 1) it is known that if A_t is left continuous, then for any bounded martingale $\{M_t\}_{t\geq 0}$,

$$\int_0^t M_s dA_s = \int_0^t M_{s-} dA_s, a.s. \tag{1.2}$$

In particular for any right continuous martingale $\{M_t\}_{t\geq 0}$ which is a simple process, i.e. there exists a sequence of times $0 = t_0 < t_1 < t_2 < \cdots < t_n < \cdots$ and a sequence of random variables $\{\xi_n\}_{n=1}^{\infty}, \xi_n \in \mathfrak{F}_{t_n}$ such that $M_t = \xi_n$, as $t \in [t_n, t_{n+1}), n = 0, 1, \cdots$; one has that (1.2) is true. Fix such an M_t temporarily. Notice that the predictable $\sigma-$field is generated by left continuous $\mathfrak{F}_t-$adapted process, and we have already known that (1.2) is true, if A_t is left continuous. So we can derive that (1.2) is still true, if A_t is $\mathfrak{F}_t-$predictable. Finally approximating the bounded right continuous martingale $\{M_t\}_{t\geq 0}$ by the simple processes we obtain that (1.2) is true when A_t is $\mathfrak{F}_t-$predictable. Conversely, suppose that A_t is $\mathfrak{F}_t-$predictable. We notice that $A_t \in \mathfrak{F}_{t-}$,because it is true when A_t is left continuous, hence this is so when A_t is $\mathfrak{F}_t-$predictable. From this we intuitively have $E\int_0^t M_s dA_s = E\int_0^t E[M_s dA_s|\mathfrak{F}_{s-}] = E\int_0^t M_{s-} dA_s$. Beware of the fact that the last equality is not rigorous. ∎

Now suppose that a submartingale $\{x_t\}_{t\geq 0}$ can be expressed as

$$x_t = M_t + A_t, \tag{1.3}$$

where $\{M_t\}_{t\geq 0}$ is a martingale, and $\{A_t\}_{t\geq 0}$ is an integrable increasing process. Then $\{x_t\}_{t\geq 0}$ is of class (DL). In fact, by Theorem 38 $\{M_t\}_{t\geq 0}$ is of class (DL), and obviously by $0 \leq A_\sigma \leq A_a$, as $\sigma \in S_a$, $\{A_t\}_{t\geq 0}$ is of class (DL) too, so is $\{x_t\}_{t\geq 0}$. Conversely, we have the following so-called Doob-Meyer decomposition theorem:

Theorem 42 *If $\{x_t\}_{t\geq 0}$ is a submartingale of class (DL), then there exists a unique expression (1.3) for $\{x_t\}_{t\geq 0}$, where $\{M_t\}_{t\geq 0}$ is a martingale, and $\{A_t\}_{t\geq 0}$ is a natural integrable increasing process. We call this decomposition a D-M decomposition.*

Proof. Uniqueness. If $x_t = M_t + A_t = M_t' + A_t'$ are two such decompositions, then $A_t - A_t' = M_t - M_t'$ is a martingale. Let $0 = t_0 < t_1 < t_2 < \cdots < t_n = t$, and $|\triangle| = \max_{k=1,\cdots,n}\{t_k - t_{k-1}\}$. Suppose that ξ is a bounded random variable. Then $m_t = E[\xi|\mathfrak{F}_t]$ is a bounded martingale. So

$E\int_0^t m_{s-} d(A_s - A_s')$

$= \lim_{|\triangle|\to 0} E[\sum_{k=0}^{n-1} m_{t_k}[(A_{t_{k+1}} - A'_{t_{k+1}}) - (A_{t_k} - A'_{t_k})]$

$= \lim_{|\triangle|\to 0} E\sum_{k=0}^{n-1} m_{t_k}[E(A_{t_{k+1}} - A'_{t_{k+1}}|\mathfrak{F}_{t_k}) - (A_{t_k} - A'_{t_k})] = 0,$

since $A_t - A_t'$ is a martingale. By the naturality property one has that $Em_tA_t = Em_tA_t'$. So $E\xi A_t = E\xi A_t'$. Hence for each t, $A_t = A_t', a.s.$ Now by the right continuity of A_t and A_t' one derives that $A_t = A_t'$, for all $t \geq 0, a.s.$

Existence. By uniqueness we only need to show the result holds on interval $[0,a]$, for every $a>0$. Set $y_t = x_t - E[x_a|\mathfrak{F}_t], t \in [0,a]$. Then $y_a = 0, a.s.$ and $\{y_t\}_{t\in[0,a]}$ is a non-positive submartingale, since $\{x_t\}_{t\in[0,a]}$ is a submartingale and hence $x_t \leq E[x_a|\mathfrak{F}_t], \forall t \in [0,a]$. Obviously,

x_t has a D-M decomposition $\Longleftrightarrow$ y_t has a D-M decomposition.

However, y_t has nicer properties. Now it is natural to discretize y_t and to use the D-M decomposition (Theorem 36) for the submartingale with discrete time, then to take the limit and get the general result. Let $\triangle_n$ be the partition $0 = t_0^n < t_1^n < \cdots < t_{2^n}^n = a$ of $[0,a]$ by $t_j^n = ja/2^n$. Then by Theorem 36 for each n, $\{y_{t_1^n}\}_{j=0,1,\cdots,2^n}$ has a D-M decomposition $y_{t_j^n} = M_{t_j^n}^n + A_{t_j^n}^n$, $j = 0,1,\cdots,2^n$, where

$$A_0^n = 0, A_{t_j^n}^n = A_{t_{j-1}^n}^n + E[y_{t_j^n} - y_{t_{j-1}^n}|\mathfrak{F}_{t_{j-1}^n}], M_{t_j^n}^n = y_{t_j^n} - A_{t_j^n}^n. \tag{1.4}$$

In what follows we are going to show that as $n \to \infty$, $\{A_a^n\}_{n=1,2,\cdots}$ will have a limit A_a. Noice that if $y_t = M_t + A_t$ holds, then by $0 = M_a + A_a$ one should have $A_t = y_t + E[A_a|\mathfrak{F}_t]$. This means that for given y_t from A_a one can define A_t, and then define M_t. So, possibly, the problem is solved.

Lemma 43 *$\{A_a^n\}_{n=1,2,\cdots}$ is uniformly integrable.*

Proof. Suppose that $t = t_{j-1}^n \in \triangle_n$. By (1.4)
$A_{t_j^n}^n = A_{t_{j-1}^n}^n + E[y_{t_j^n} - y_{t_{j-1}^n}|\mathfrak{F}_{t_{j-1}^n}]$,
take the summation $\sum_{k=j}^{2^n} A_{t_k^n}^n$ then take the conditional expectation $E[\cdot|\mathfrak{F}_t]$ on both sides and notice that $y_a = y_{t_{2^n}^n} = 0$ as a result one obtains that $\forall t \in \triangle_n$

$$E[A_a^n|\mathfrak{F}_t] = A_t^n - y_t. \tag{1.5}$$

Now let $\lambda > 0$ be fixed and set
$\sigma_\lambda^n = \inf\left\{t_{k-1}^n : A_{t_k^n}^n > \lambda\right\}$; $\sigma_\lambda^n = a$, if $\{\cdot\} = \phi$.
Then $\sigma_\lambda^n \in S_a$. In fact,
$\{\sigma_\lambda^n = t_{k-1}^n\} = \left\{A_{t_j^n}^n \leq \lambda, j = 0,1,\cdots,k-1; A_{t_k^n}^n > \lambda\right\} \in \mathfrak{F}_{t_{k-1}^n}$,
where we have used the result that by (1.4) $A_{t_k^n}^n \in \mathfrak{F}_{t_{k-1}^n}$. By Doob's stopping time theorem

$$E[A_a^n|\mathfrak{F}_{\sigma_\lambda^n}] = A_{\sigma_\lambda^n}^n - y_{\sigma_\lambda^n}. \tag{1.6}$$

Noting that $\{A_a^n > \lambda\} \subset \{\sigma_\lambda^n < a\}$. On the other hand, by the increasing property of A_t^n one also has that $\{A_a^n > \lambda\} \supset \{\sigma_\lambda^n < a\}$. So $\{A_a^n > \lambda\} = \{\sigma_\lambda^n < a\}$. Hence
$EA_a^n I_{A_a^n > \lambda} = -Ey_{\sigma_\lambda^n} I_{\sigma_\lambda^n < a} + EA_{\sigma_\lambda^n} I_{\sigma_\lambda^n < a} \leq -Ey_{\sigma_\lambda^n} I_{\sigma_\lambda^n < a} + \lambda P(\sigma_\lambda^n < a)$.
Notice also that by (1.6)
$-Ey_{\sigma_{\lambda/2}^n} I_{\sigma_{\lambda/2}^n < a} = E(A_a^n - A_{\sigma_{\lambda/2}^n}) I_{\sigma_{\lambda/2}^n < a} \geq E(A_a^n - A_{\sigma_{\lambda/2}^n}) I_{\sigma_\lambda^n < a}$
$\geq \frac{\lambda}{2} P(\sigma_\lambda^n < a)$,

where we have used the result that $A_a^n > \lambda$, as $\sigma_\lambda^n < a$; and $A_{\sigma_{\lambda/2}^n} \leq \frac{\lambda}{2}$. Therefore,

$EA_a^n I_{A_a^n > \lambda} \leq -Ey_{\sigma_\lambda^n} I_{\sigma_\lambda^n < a} - 2Ey_{\sigma_{\lambda/2}^n} I_{\sigma_{\lambda/2}^n < a}$.

Now we can show that $\{A_a^n\}_{n=1,2,\cdots}$ is uniformly integrable. In fact, the assumption that $\{x_t\}_{t \geq 0}$ is a submartingale of class (DL), means that $\{x_\sigma\}_{\sigma \in S_a}$ is uniformly integrable, and so is $\{y_\sigma\}_{\sigma \in S_a}$. However, as $\lambda \to \infty$,

$P(\sigma_\lambda^n < a) = P(A_a^n > \lambda) \leq EA_a^n/\lambda \leq -Ey_0/\lambda \to 0$,

uniformly w.r.t. n, where we have used the result that by (1.5) $EA_a^n = -Ey_0$. Similarly, one also has that as $\lambda \to \infty$,

$P(\sigma_{\lambda/2}^n < a) \to 0$, uniformly w.r.t. n.

So applying the fact that $\{y_\sigma\}_{\sigma \in S_a}$ is uniformly integrable, one finds that as $\lambda \to \infty$, $EA_a^n I_{A_a^n > \lambda} \to 0$. That is, $\{A_a^n\}_{n=1,2,\cdots}$ is uniformly integrable. ■

However, for the uniform integrability family we can quote the following lemma.

Lemma 44 [28], p292 *A subset K of $L^1(\Omega, \mathfrak{F}, P)$ is uniformly integrable if and only if it is weakly sequentially compact, i.e. for any sequence $\{f_n\} \subset K$, there exists a subsequence $\{f_{n_k}\} \subset \{f_n\}$ and a $f \in L^1(\Omega, \mathfrak{F}, P)$ such that as $k \to \infty$, $E(gf_{n_k}) \to E(gf), \forall g \in L^\infty(\Omega, \mathfrak{F}, P)$.*

Now let us return to the proof of Theorem 42.

Proof. By the above two lemmas there exists a subsequence $\{A_a^{n_k}\}_{k=1,2,\cdots}$ of $\{A_a^n\}_{n=1,2,\cdots}$ and a $A_a \in L^1(\Omega, \mathfrak{F}, P)$ such that as $k \to \infty$, $E(gA_a^{n_k}) \to E(gA_a), \forall g \in L^\infty(\Omega, \mathfrak{F}, P)$. Let $A_t = y_t + E[A_a|\mathfrak{F}_t]$, $t \in [0, a]$, where we always take the right continuous version of martingale $E[A_a|\mathfrak{F}_t]$. Recall that $y_t = x_t - E[x_a|\mathfrak{F}_t], t \in [0, a]$. Then $x_t = A_t + E[x_a - A_a|\mathfrak{F}_t]$. So we have a decomposition for x_t. Obviously, A_t is integrable, since y_t and A_a are integrable. However, to be a D-M decomposition we still need to prove: 1) A_t is increasing in t; 2) A_t is natural.

1): Since by (1.5) one has that $A_t^n = y_t + E[A_a^n|\mathfrak{F}_t]$, $t \in \triangle_m \subset \triangle_n$, as $n \geq m$. Hence one can show that as $k \to \infty$, $E(gA_t^{n_k}) \to E(gA_t), \forall g \in L^\infty(\Omega, \mathfrak{F}, P), t \in \triangle_m, \forall m$. In fact, as $k \to \infty$,

$E(gE[A_a^{n_k}|\mathfrak{F}_t]) = E(E[g|\mathfrak{F}_t]E[A_a^{n_k}|\mathfrak{F}_t]) = E(E[g|\mathfrak{F}_t]A_a^{n_k}])$

$\to E(E[g|\mathfrak{F}_t]A_a]) = E(gE[A_a|\mathfrak{F}_t])$.

Moreover, by (1.4) one sees that $A_t^n \leq A_s^n$, as $t \leq s$, and $t, s \in \triangle_m, n \geq m$. Therefore, one can easily show that $A_t \leq A_s$, as $t \leq s$, and $t, s \in \triangle_m, \forall m$. By the right continuity of A_t it follows that A_t is increasing in t.

2): For each $t \in [0, a]$ if m_t is a bounded martingale, then

$E\int_0^t m_{s-} dA_s = \lim_{n\to\infty} \sum_{k=0}^{2^n-1} E[m_{t_k^n \wedge t}(A_{t_{k+1}^n \wedge t} - A_{t_k^n \wedge t})]$

$= \lim_{n\to\infty} \sum_{k=0}^{2^n-1} E[m_{t_k^n \wedge t}(y_{t_{k+1}^n \wedge t}^n - y_{t_k^n \wedge t}^n)]$

$= \lim_{n\to\infty} \sum_{k=0}^{2^n-1} E[m_{t_k^n \wedge t}(A_{t_{k+1}^n \wedge t}^n - A_{t_k^n \wedge t}^n)]$

$= \lim_{n\to\infty} \sum_{k=0}^{2^n-1} E[m_t(A_{t_{k+1}^n \wedge t}^n - A_{t_k^n \wedge t}^n)] = \lim_{n\to\infty} Em_t A_t^n = Em_t A_t$.

Therefore A_t is natural. ■

Corollary 45 *If $\{x_t\}_{t\geq0}$ is a submartingale of class (D), then $\{M_t\}_{t\geq0}$ in the D-M decomposition is uniformly integrable.*

Proof. In fact, in this case $sup_{t\geq0}E\,|x_t| = k_0 < \infty$, since $\{x_\sigma\}_{\sigma\in S_\infty}$ is uniformly integrable. So we also have $sup_{t\geq0}E\,|y_t| \leq 2k_0 < \infty$, since $y_t = x_t - E[x_a|\mathfrak{F}_t], t \in [0,a], \forall a > 0$. Furthermore, by (1.5) one has that $0 \leq EA^n_a = -Ey_0 \leq 2k_0 < \infty, \forall a > 0$. Therefore $0 \leq EA_a \leq 2k_0 < \infty, \forall a > 0$. This shows that $sup_{t\geq0}E\,|M_t| \leq 3k_0 < \infty$, since $M_t = x_t - A_t, \forall t \geq 0$. Thus by Theorem 25 and the proof of Lemma 32 $M_\infty = \lim_{t\in Q, t\uparrow\infty} M_t \in L^1(\Omega, \mathfrak{F}, P)$ exists, and $M_t = E[M_\infty|\mathfrak{F}_t], \forall t \geq 0$. $\{M_t\}_{t\geq0}$ is a uniformly integrable martingale. ∎

Now let us discuss when A_t can be continuous.

Definition 46 *A submartingale is called regular, if for $\forall a > 0$ and $\sigma_n \in S_a$ such that $\sigma_n \uparrow \sigma$, one has $Ex_{\sigma_n} \to Ex_\sigma$.*

Theorem 47 *For a submartingale $\{x_t\}_{t\geq0}$ of class (DL) the natural increasing process $\{A_t\}_{t\geq0}$ in the D-M decomposition (1.3) is continuous, if and only if $\{x_t\}_{t\geq0}$ is regular.*

Proof. $\Longrightarrow$: Notice that any martingale $\{M_t\}_{t\geq0}$ is always regular, since $EM_\sigma = EM_0, \forall\sigma \in S_a$. Moreover, if $\{A_t\}_{t\geq0}$ is continuous, then as $\sigma_n \uparrow \sigma$, $0 \leq A_{\sigma_n} \uparrow A_\sigma$. Hence by Fatou's lemma $EA_{\sigma_n} \uparrow EA_\sigma$. So $\{x_t\}_{t\geq0}$ is regular, if $\{A_t\}_{t\geq0}$ is continuous.

$\Longleftarrow$: Suppose that $\{x_t\}_{t\geq0}$ is regular. Then $A_t = x_t - M_t$ is also regular. Moreover, as $\sigma_n \uparrow \sigma, \sigma_n \in S_a$, one has that

$$0 \leq A_{\sigma_n} \uparrow A_\sigma, a.s. \tag{1.7}$$

In fact, by regularity $\forall B \in \mathfrak{F}, 0 \leq E(A_\sigma - A_{\sigma_n})I_B \leq E(A_\sigma - A_{\sigma_n}) \downarrow 0$, as $n \uparrow \infty$. So if $P(B) = P(\lim_{n\to\infty} A_{\sigma_n} < A_\sigma) > 0$, this will be a contradiction. Now define the sequence $\triangle_n$ of the partition of $[0,a]$ as in the proof of Theorem 42. Let, for $c > 0$,

$$\widetilde{A}^n_t = E[A_{t^n_{k+1}} \wedge c|\mathfrak{F}_t], \tag{1.8}$$

as $t \in (t^n_k, t^n_{k+1}]$. Then $\widetilde{A}^n_t$ is a bounded martingale on $t \in (t^n_k, t^n_{k+1}]$, and by the naturality of A_t one has $E\int_{t^n_k}^t \widetilde{A}^n_s dA_s = E\int_{t^n_k}^t \widetilde{A}^n_{s-} dA_s, \forall t \in (t^n_k, t^n_{k+1}]$, and $E\int_{t^n_j}^{t^n_{j+1}} \widetilde{A}^n_s dA_s = E\int_{t^n_j}^{t^n_{j+1}} \widetilde{A}^n_{s-} dA_s$. Taking the summation for j from 0 to $k - 1$ one finds that $\forall t \in [0, a]$,

$$E\int_0^t \widetilde{A}^n_s dA_s = E]\sum_{j=0}^{k-1} \int_{t^n_j}^{t^n_{j+1}} \widetilde{A}^n_s dA_s + \int_{t^n_k}^t \widetilde{A}^n_s dA_s] = E\int_0^t \widetilde{A}^n_{s-} dA_s,$$

i.e.

$$E\int_0^t (\widetilde{A}^n_s - \widetilde{A}^n_{s-}) dA_s = 0. \tag{1.9}$$

From this one sees that if one can prove the following statement:

Statement (A): There exists a subsequence $\left\{\widetilde{A}_t^{n_k}\right\}$ such that $\lim_{k\to\infty}\sup_{t\in[0,a]}\left|\widetilde{A}_t^{n_k}-A_t\wedge c\right|=0$, a.s.
Then $A_t\wedge c$ is continuous in t, and so is A_t, since c can be taken arbitrarily. In fact, if Statement (A) is true, then by (1.9) one has that

$0=E\int_0^t(A_s\wedge c-A_{s-}\wedge c)dA_s\geq E\{\sum_{s\leq t}[(A_s\wedge c)-(A_{s-}\wedge c)]^2\}$.

So $A_t\wedge c$ is continuous in t. Now we are going to establish Statement (A). For this we only need to show that as $n\to\infty$, $\sup_{t\in[0,a]}\left|\widetilde{A}_t^n-A_t\wedge c\right|\to 0$, in probability; i.e. for any $\varepsilon>0$, as $n\to\infty$, $P[\sup_{t\in[0,a]}\left|\widetilde{A}_t^n-A_t\wedge c\right|>\varepsilon]\to 0$. Now define

$\sigma_\varepsilon^n=\inf\{t\in[0,a]:\widetilde{A}_t^n-A_t\wedge c>\varepsilon\};\sigma_\varepsilon^n=a$, if $\{\cdot\}=\phi$.

Then σ_ε^n is a stopping time by Example 7. Moreover, it has the following properties 1) for every n, $\sigma_\varepsilon^n=a$ implies that $\widetilde{A}_t^n-A_t\wedge c\leq\varepsilon,\forall t\in[0,a]$. 2) $\sigma_\varepsilon^n\uparrow$, as $n\uparrow$. In fact, A_s is increasing in s, hence by definition $\widetilde{A}_t^n$ is decreasing in n. So $\sigma_\varepsilon^n\uparrow$, as $n\uparrow$. Thus $\sigma_\varepsilon=\lim_{n\to\infty}\sigma_\varepsilon^n\leq a$ exists. Notice that $\widetilde{A}_t^n$ is a martingale on $t\in(t_k^n,t_{k+1}^n]$. Hence by (1.8) one has that

$EI_{t\in(t_k^n,t_{k+1}^n]}\widetilde{A}_t^n=EI_{t\in(t_k^n,t_{k+1}^n]}(A_{t_{k+1}^n}\wedge c)$.

Now let us generalize this equality for t to that for the stopping time σ_ε^n by means of Doob's stopping time theorem. For this we introduce a function $\varphi^n(t)$ by defining $\varphi^n(t)=t_{k+1}^n$, as $t\in(t_k^n,t_{k+1}^n]$. Then $\varphi^n(\sigma_\varepsilon^n)$ is obviously a discrete stopping time, since $\{\varphi^n(\sigma_\varepsilon^n)=t_{k+1}^n\}=\{t_k^n<\sigma_\varepsilon^n\leq t_{k+1}^n\}\in\mathfrak{F}_{t_{k+1}^n}$. Moreover, $\varphi^n(\sigma_\varepsilon^n)\leq a$, since $\sigma_\varepsilon^n\leq a$. Furthermore, $\lim_{n\to\infty}\varphi^n(\sigma_\varepsilon^n)=\sigma_\varepsilon$, since $\lim_{n\to\infty}\sigma_\varepsilon^n=\sigma_\varepsilon$. Now applying Doob's stopping time theorem one finds that $EI_{\sigma_\varepsilon^n\in(t_k^n,t_{k+1}^n]}\widetilde{A}_{\sigma_\varepsilon^n}^n=EI_{\sigma_\varepsilon^n\in(t_k^n,t_{k+1}^n]}(A_{\varphi^n(\sigma_\varepsilon^n)}\wedge c)$. Take the summation for $k=0$ up to $k=2^n-1$. One finds that $E\widetilde{A}_{\sigma_\varepsilon^n}^n=E(A_{\varphi^n(\sigma_\varepsilon^n)}\wedge c)$. Hence by this and by the property 1) of σ_ε^n, and after applying (1.7) one finds that

$0=\lim_{n\to\infty}E[A_{\varphi^n(\sigma_\varepsilon^n)}\wedge c-A_{\sigma_\varepsilon^n}\wedge c]=\lim_{n\to\infty}E[\widetilde{A}_{\sigma_\varepsilon^n}^n-A_{\sigma_\varepsilon^n}\wedge c]$
$\geq\lim_{n\to\infty}\varepsilon P(\sigma_\varepsilon^n<a)\geq\lim_{n\to\infty}\varepsilon P(\sup_{t\in[0,a]}\left|\widetilde{A}_t^n-A_t\wedge c\right|>\varepsilon)$.

The proof is complete. ■

1.7 Poisson Random Measure and Its Existence

A dynamical system will always encounter some jump stochastic purterbations. The simplest type comes from a stochastic point process. To understand it properly requires some preparation. Let $(Z,\mathfrak{B}_Z)$ be a measurable space.

Definition 48 *A map $\mu(B,\omega):\mathfrak{B}_Z\times\Omega\to R_+\cup\{\infty\}$ is called a random measure on $\mathfrak{B}_Z\times\Omega$, if 1) for any fixed $B\in\mathfrak{B}_Z$, $\mu(B,\cdot)$ is a random variable but with values in $R_+\cup\{\infty\}$; 2) for any fixed $\omega\in\Omega$, $\mu(\cdot,\omega)$ is a*

$\sigma-$finite measure. (Here, $\sigma-$finite means that there exists $\{U_n\}_{n=1}^{\infty} \subset \mathfrak{B}_Z$ such that $Z = \cup_{n=1}^{\infty} U_n$ and $\mu(U_n, \omega) < \infty, \forall n$).

Here the definition of a random variable taking values in $R_+ \cup \{\infty\}$, is the same as that taking values in R. Let us introduce the Poisson random measure as follows:

Definition 49 *A random measure $\mu(B, \omega)$ is called a Poisson random measure on $\mathfrak{B}_Z \times \Omega$, if it is non-negative integer valued (possibly ∞) such that*
1) for each $B \in \mathfrak{B}_Z$, $\mu(B, \cdot)$ is Poisson distributed; i.e.
$p(\{\omega : \mu(B, \omega) = n\}) = e^{-\lambda(B)} \frac{\lambda(B)^n}{n!}, n = 0, 1, \cdots;$
where $\lambda(B) = E\mu(B, \omega), \forall B \in \mathfrak{B}_Z$, is usually called the mean measure, or the intensity measure of μ;
2) if $\mathfrak{B}_Z \supset \{B_j\}_{j=1}^{m}$ are disjoint, then $\{\mu(B_j, \cdot)\}_{j=1}^{m}$ are independent.

Here as in the real analysis we still define $0 \cdot \infty = 0$. Thus if $\lambda(B) = \infty$, then all $p(\{\omega : \mu(B, \omega) = n\}) = 0$, $n = 0, 1, \cdots$, hence $\mu(B, \omega) = \infty, P-a.s.$ The existence of a Poisson random measure is given by the following theorem.

Theorem 50 *For any $\sigma-$finite measure λ on $(Z, \mathfrak{B}_Z)$ there exists a Poisson random measure μ with $\lambda(B) = E\mu(B), \forall B \in \mathfrak{B}_Z$.*

Proof. Since λ is a $\sigma-$finite measure on $(Z, \mathfrak{B}_Z)$, there exists a $\{U_n\}_{n=1}^{\infty} \subset \mathfrak{B}_Z$ such that all $U_n, n = 1, \cdots$, are disjoint, i.e. $U_i \cap U_j = \phi, i \neq j, \forall i, j$; and $Z = \cup_{n=1}^{\infty} U_n$ and $0 < \lambda(U_n) < \infty, \forall n$. Let us construct a probability space and construct random variables on it as follows:

(i) for each $n = 1, 2, \cdots$, and $i = 1, 2, \cdots$, construct a U_n- valued random variable ξ_i^n such that $P(\xi_i^n \in du) = \lambda(du)/\lambda(U_n)$;

(ii) $p_n, n = 1, 2, \cdots$, is an integral-valued random variable with $P(p_n = k) = \lambda(U_k)^k \exp[-\lambda(U_k)]/k!, k = 0, 1, 2, \cdots;$

(iii) $\xi_i^n, p_n, n = 1, 2, \cdots, i = 1, 2, \cdots$ are multually independent.

Now set $\mu(B) = \sum_{n=1}^{\infty} \sum_{i=1}^{p_n} I_{B \cap U_n}(\xi_i^n) I_{p_n \geq 1}, \forall B \in \mathfrak{B}_Z$. Let us show that for every disjoint $B_1, \cdots, B_n \in \mathfrak{B}_Z, \forall \alpha_i > 0, i = 1, 2, \cdots, m$

$$E \exp[-\sum_{i=1}^{m} \alpha_i \mu(B_i)] = \exp[\sum_{i=1}^{m} (e^{-\alpha_i} - 1)\lambda(B_i)]. \tag{1.10}$$

If this is true, then μ is a Poisson random measure with the intensity measure λ. Let us simply show the case for $m = 2$. Note that by the independence

$E \exp(-\sum_{j=1}^{2} \alpha_j \mu(B_j)) = \prod_{n=1}^{\infty} E \exp(-\alpha_1 \sum_{i=1}^{p_n} I_{U_n B_1}(\xi_i^n) I_{p_n \geq 1}$
$-\alpha_2 \sum_{i=1}^{p_n} I_{U_n B_2}(\xi_i^n) I_{p_n \geq 1}) = \prod_{n=1}^{\infty} J_n.$

However, by the complete probability formula one can derive that

$J_n = 1 \cdot P(p_n = 0) + \sum_{k=1}^{\infty} E[\exp(-\alpha_1 \sum_{i=1}^{k} I_{U_n B_1}(\xi_i^n)$

$-\alpha_2 \sum_{i=1}^k I_{U_nB_2}(\xi_i^n)) \,|p_n = k] \cdot P(p_n = k) = P(p_n = 0) + \sum_{k=1}^\infty J_{nk}.$
Note that by the independence of $\{\xi_i^n, p_n, \forall n, \forall i\}$

$P(p_n = 0) = e^{-\lambda(U_n)},$

$J_{n1} = e^{-\lambda(U_n)} \cdot \frac{\lambda(U_n)}{1!} E \exp(-\alpha_1 I_{U_nB_1}(\xi_1^n) - \alpha_2 I_{U_nB_2}(\xi_1^n))$

$= e^{-\lambda(U_n)} \cdot \frac{\lambda(U_n)}{1!} [e^{-\alpha_1} \frac{\lambda(U_nB_1)}{\lambda(U_n)} + e^{-\alpha_2} \frac{\lambda(U_nB_2)}{\lambda(U_n)}$

$+1 \cdot (1 - \frac{\lambda(U_nB_1)}{\lambda(U_n)} - \frac{\lambda(U_nB_2)}{\lambda(U_n)})]$

$= e^{-\lambda(U_n)} \cdot \frac{\lambda(U_n)}{1!} [(e^{-\alpha_1} - 1) \frac{\lambda(U_nB_1)}{\lambda(U_n)} + (e^{-\alpha_2} - 1)\lambda(U_nB_2) + 1]$

$= e^{-\lambda(U_n)} \cdot \frac{\lambda(U_n)}{1!} [1 + \sum_{j=1}^2 (e^{-\alpha_i} - 1) \frac{\lambda(U_nB_i)}{\lambda(U_n)}],$

where we have applied $P(\xi_i^n \in U_nB_1) = \frac{\lambda(U_nB_1)}{\lambda(U_n)}$, etc. Hence,

$J_{nk} = e^{-\lambda(U_n)} \cdot \frac{\lambda(U_n)^k}{k!} \prod_{i=1}^k E \exp(-\alpha_1 I_{U_nB_1}(\xi_i^n) - \alpha_2 I_{U_nB_2}(\xi_i^n))$

$= e^{-\lambda(U_n)} \cdot \frac{\lambda(U_n)^k}{k!} [1 + \sum_{j=1}^2 (e^{-\alpha_j} - 1) \frac{\lambda(U_nB_j)}{\lambda(U_n)}]^k$

$= e^{-\lambda(U_n)} \cdot [\lambda(U_n) + \sum_{j=1}^2 (e^{-\alpha_j} - 1)\lambda(U_nB_j)]^k / k!.$

Therefore,

$E \exp(-\sum_{j=1}^2 \alpha_j \mu(B_j)) = \exp[\sum_{j=1}^2 (e^{-\alpha_j} - 1)\lambda(U_nB_j)].$

The proof is complete. ∎

1.8 Poisson Point Process and Its Existence

Now let us introduce the concept of random point processes. Assume that $(Z, \mathfrak{B}_Z)$ is a measurable space. Suppose that $D_p \subset (0, \infty)$ is a countable set, then a mapping $p : D_p \to Z$, is called a point function (valued) on Z. Endow $(0, \infty) \times Z$ with the product $\sigma-$field $\mathfrak{B}((0, \infty)) \times \mathfrak{B}_Z$, and define a counting measure through p as follows:

$N_p((0, t] \times U) = \#\{s \in D_p : s \leq t, p(s) \in U\}, \forall t > 0, U \in \mathfrak{B}_Z,$

where $\#$ means the numbers of $\cdot$ counting in the set $\{\cdot\}$.

Now let us consider a function of two variables $p(t, \omega)$ such that for each $\omega \in \Omega$, $p(\cdot, \omega)$ is a point function on Z, i.e. $p(\cdot, \omega) : D_{p(\cdot,\omega)} \to Z$, where $D_{p(\cdot,\omega)} \subset (0, \infty)$ is a countable set. Naturally, its counting measure is defined by

$N_p((0, t] \times U, \omega) = N_{p(\omega)}((0, t] \times U)$

$= \#\{s \in D_p : s \leq t, p(s, \omega) \in U\}, \forall t > 0, U \in \mathfrak{B}_Z,$

and we introduce the definition as follows:

Definition 51 *1) If $N_p((0, t] \times U, \omega)$ is a random measure on $(\mathfrak{B}((0, \infty)) \times \mathfrak{B}_Z) \times \Omega$, then p is called a (random) point process.*

2) If $N_p((0, t] \times U, \omega)$ is a Poisson random measure on $(\mathfrak{B}((0, \infty)) \times \mathfrak{B}_Z) \times \Omega$, then p is called a Poisson point process.

3) For a Poisson point process p if its intesity measure $n_p(dtdx) = E(N_p((dtdx))$ satisfies that

$n_p(dtdx) = \pi(dx)dt,$

where $\pi(dx)$ is some measure on $(Z, \mathfrak{B}_Z)$, then p is called a stationary Poisson point process. $\pi(dx)$ is called the characteristic measure of p.

Because for a Poisson random measure on $\mathfrak{B}_Z \times \Omega$ (1.10) is its sufficient and necessary condition. Now we consider the Poisson random measure defined on $(\mathfrak{B}((0,\infty)) \times \mathfrak{B}_Z) \times \Omega$. Then it is easily seen that p is a stationary Poisson point process with the characteristic measure $dt\pi(dx)$, if and only if $\forall t > s \geq 0$, disjoint $\{U_i\}_{i=1}^m \subset \mathfrak{B}_Z$ and $\lambda_i > 0, P-a.s.$

$$E[\exp(-\sum_{i=1}^m \lambda_i N_p((s,t] \times U_i)|\widetilde{\mathfrak{F}}_s] = \exp[(t-s)\sum_{i=1}^m (e^{-\lambda_i} - 1)\pi(U_i)], \quad (1.11)$$

where $\widetilde{\mathfrak{F}}_s = \sigma[N_p((0,s'] \times U); s' \leq s, U \in \mathfrak{B}_Z]$. Now let us use this fact to show the existence of a stationary Poisson point process. Firstly, we will show a lemma on Poisson random process.

Lemma 52 *If $\{N_t\}_{t\geq 0}$ is a Poisson process with intensity $t\mu$, i.e. $\{N_t\}_{t\geq 0}$ is a random process such that $P(N(t) = k) = e^{-\mu t}\frac{(t\mu)^k}{k!}$, $EN(t) = t\mu$, and it has stationary independent incremenrs; then*
1) $E[e^{-\lambda(N_t - N_s)}|\widetilde{\mathfrak{F}}_s] = e^{(t-s)\mu(e^{i\lambda}-1)}$, $\forall t > s, \lambda > 0$;
2) $E[e^{-\lambda(N_{t+\sigma} - N_\sigma)}|\widetilde{\mathfrak{F}}_\sigma] = e^{t\mu(e^{i\lambda}-1)}$, $\forall t > 0, \lambda > 0$,
and $\forall \sigma-$bounded $\widetilde{\mathfrak{F}}_t-$stopping time.

Proof. 1): Since $\{N_t\}_{t\geq 0}$ has independent increments,
$E[e^{-\lambda(N_t-N_s)}|\widetilde{\mathfrak{F}}_s] = E[e^{-\lambda(N_t-N_s)}] = e^{(t-s)\mu(e^{i\lambda}-1)}$.
The last equality follows from the elementary probability theory. 2): Let us make a standard approximation of σ, i.e for any bounded stopping time σ with $0 \leq \sigma \leq T$, where T is a constant, let
$\sigma^n = \frac{kT}{2^n}$, as $\sigma \in [\frac{k-1}{2^n}T, \frac{k}{2^n}T)$, $n, k = 1, 2, \cdots$.
Then for each n, σ^n is a bounded discrete stopping time. Moreover, as $n \uparrow \infty, \sigma^n \downarrow \sigma$. By 1) $\forall B \in \widetilde{\mathfrak{F}}_{\sigma_n}$
$\int_{\{\frac{(k-1)T}{2^n} \leq \sigma < \frac{kT}{2^n}\} \cap B} e^{e^{-\lambda(N_{t+\sigma_n} - N_{\sigma_n})}} dP = \int_{\{\sigma_n = \frac{kT}{2^n}\} \cap B} e^{e^{-\lambda(N_{t+\frac{kT}{2^n}} - N_{\frac{kT}{2^n}})}} dP$
$= \int_{\{\sigma_n = \frac{kT}{2^n}\} \cap B} e^{t\mu(e^{i\lambda}-1)} dP.$
So
$\int_B e^{e^{-\lambda(N_{t+\sigma_n} - N_{\sigma_n})}} dP = \int_B e^{t\mu(e^{i\lambda}-1)} dP.$
In particular, $\forall B \in \widetilde{\mathfrak{F}}_\sigma \subset \widetilde{\mathfrak{F}}_{\sigma_n}$,
$\int_B e^{e^{-\lambda(N_{t+\sigma_n} - N_{\sigma_n})}} dP = \int_B e^{t\mu(e^{i\lambda}-1)} dP.$
Letting $n \to \infty$ one obtains that $\forall B \in \widetilde{\mathfrak{F}}_\sigma$.
$\int_B e^{e^{-\lambda(N_{t+\sigma} - N_\sigma)}} dP = \int_B e^{t\mu(e^{i\lambda}-1)} dP.$
Thus 2) is proved. ■

By means of this lemma one immediately obtains the following lemma.

Lemma 53 *If $\{N(t)\}_{t\geq 0}$ is a Poisson process with an intensity function μt, denote $\tau_1 = \inf\{t > 0 : N(t) = 1\}, \cdots$, and*
$\tau_k = \inf\{t - \tau_{k-1} > 0 : N(t) - N(\tau_{k-1}) = 1\}$,
then
1) $\{\tau_k\}_{k=1}^{\infty}$ is an independent family of random variables, and
$P(\tau_k > t) = e^{-\lambda t}, \forall k = 1, 2, \cdots$;
2) $P(\sum_{j=1}^{k-1} \tau_j \leq t < \sum_{j=1}^{k} \tau_j) = e^{-\mu t} \frac{(\mu t)^{k-1}}{(k-1)!}$.

Proof. 1): In fact, $P(\tau_1 > t) = P(N(t) - N(0) = 0) = e^{-\mu t}$. Hence by Lemma 52
$P(\tau_k > t) = P(N(t + \tau_{k-1}) - N(\tau_{k-1}) = 0)$
$= P(N(t) - N(0) = 0) = e^{-\mu t}, \forall k = 1, 2, \cdots$.
Since $\{N(t)\}_{t\geq 0}$ has independent increments, $\forall t_1, \forall t_2, \cdots, \forall t_m$
$P(\cap_{k=1}^{m} \{\tau_k > t\}) = P(\cap_{k=1}^{m} \{N(t + \tau_{k-1}) - N(\tau_{k-1}) = 0\})$
$= \prod_{k=1}^{m} P(N(t + \tau_{k-1}) - N(\tau_{k-1}) = 0) = \prod_{k=1}^{m} P(\tau_k > t)$.
Hence $\{\tau_k\}_{k=1}^{\infty}$ is an independent family.
2): $P(\tau_1 + \tau_2 > t \geq \tau_1) = P(N(t) = 1) = e^{-\mu t} \frac{(\mu t)}{1!}$.
$P(\sum_{j=1}^{k-1} \tau_j \leq t < \sum_{j=1}^{k} \tau_j) = P(N(t) = k - 1) = e^{-\mu t} \frac{(\mu t)^{k-1}}{(k-1)!}$. ∎
Now we are in a position to show the existence of a Poisson point process.

Theorem 54 *Given a $\sigma-$finite measure π on $(Z, \mathfrak{B}_Z)$ there exists a sationary Poisson point process on Z with the charteristic measure π.*

Proof. Since π is a $\sigma-$finite measure on $(Z, \mathfrak{B}_Z)$, there exists a disjoint $\{U_k\}_{k=1}^{\infty} \subset \mathfrak{B}_Z$ such that $\pi(U_k) < \infty$, and $Z = \cup_{k=1}^{\infty} U_k$. Let us construct probability spaces and the random variables defined on them as follows:

(i) On a probability space $(\Omega_0, \mathfrak{F}_0, P_0)$, $\forall k, i = 1, 2, \cdots$; ξ_i^k is a U_k-valued random variable with $P(\xi_i^k \in dx) = \pi(dx)/\pi(U_k)$ defined on it such that $\left\{\xi_i^k, \forall k, i = 1, 2, \cdots\right\}$ is an indepent random variable system.

(ii) For each $k = 1, 2, \cdots$ on a probability space $(\Omega_k, \mathfrak{F}_k, P_k)$, $\{N_t^k\}_{t\geq 0}$ is a Poisson random process with intensity $t\pi(U_k)$ defined on it, set $\forall i = 1, 2, \cdots$,
$\tau_i^k = \inf\{t - \tau_{i-1} > 0 : N^k(t) - N^k(\tau_{i-1}) = 1\}$;
then by Lemma 53 $\{\tau_i^k\}_{i=1}^{\infty}$ is an independent variable system such that $P(\tau_i^k > t) = \exp[-t\pi(U_k)]$, for $t \geq 0$.
Now let $\Omega = \times_{k=0}^{\infty} \Omega_k, \mathfrak{F} = \times_{k=0}^{\infty} \mathfrak{F}_k, P = \times_{k=0}^{\infty} P_k$. Then ξ_i^k, N_t^k are naturally extended to be defined on probability space $(\Omega, \mathfrak{F}, P)$, i.e. for $\omega = (\omega_0, \omega_1, \cdots, \omega_k, \cdots) \in \Omega$ let $\xi_i^k(\omega) = \xi_i^k(\omega_0)$, $N_t^k(\omega) = N_t^k(\omega_k)$. Then $\left\{\xi_i^k, \tau_i^k, \forall k, i = 1, 2, \cdots\right\}$ is an indepent random variable system on $(\Omega, \mathfrak{F}, P)$. Moreover, $\left\{N_\cdot^k, \xi_i^k, \forall k, i = 1, 2, \cdots\right\}$ is a independent system of random maps. Now set
$D_p = \cup_{k=1}^{\infty} \{\tau_1^k, \tau_1^k + \tau_2^k, \cdots, \tau_1^k + \tau_2^k + \cdots + \tau_m^k, \cdots\}$,
and

$p(\tau_1^k + \tau_2^k + \cdots + \tau_m^k) = \xi_m^k, \forall k, m = 1, 2, \cdots$.
Introduce a counting measure by p as follows:
$N_p((s,t] \times (U_k \cap B)) = \# \{r \in D_p : r \in (s,t], p(s) \in U_k \cap B\}$.
Then we have that

$$N_p((s,t] \times B) = \sum_{k=1}^{\infty} \sum_{m=1}^{\infty} I_{p(\tau_1^k+\tau_2^k+\cdots+\tau_m^k) \in U_k \cap B} I_{s<\tau_1^k+\tau_2^k+\cdots+\tau_m^k \leq t} \quad (1.12)$$

Note that if $\omega \in \Omega$ is such that $s < \sum_{j=1}^m \tau_j^k(\omega) \leq t$, then there exists an $\widetilde{s} \in (s,t]$ (actually, $\widetilde{s} = \sum_{j=1}^m \tau_j^k(\omega)$) such that $N^k(\widetilde{s}, \omega) = m$, and $N^k(u,\omega) = m-1, \forall u < \widetilde{s}$. Conversely, if there exists an $\widetilde{s} \in (s,t]$ such that $N^k(\widetilde{s}, \omega) = m$, and $N^k(u,\omega) = m-1, \forall u < \widetilde{s}$, then $\widetilde{s} \in (s,t]$ (and $\widetilde{s} = \sum_{j=1}^m \tau_j^k(\omega)$), because $\{\tau_j^k(\omega)\}_{j=1}^{\infty}$ is just the set of all jump times which have happened for the Poisson process $N^k(u,\omega)$. Thus by (1.12) and by the independence one has that
$N_p((s,t] \times B) = \sum_{k=1}^{\infty} \sum_{m=1}^{N^k(\widetilde{s})} I_{U_k \cap B}(\xi_m^k) I_{(s,t]}(\widetilde{s}) I_{N^k(\widetilde{s}) \geq 1}$.
Now the proof of (1.11) for this $N_p((s,t] \times B)$ can be completed by the complete probability formula, as in the proof of Theorem 50. ■
A special case is the following corollary.

Corollary 55 *Given a finite measure π on $(Z, \mathfrak{B}_Z)$ there exists a finite sationary Poisson point process on Z with the charteristic measure π. (Here, a finite measure π means that $\pi(Z) < \infty$; and, a finite point process p means that the counting measure $N_p((0,t], Z)$ generated by p, is always finite, $\forall 0 \leq t < \infty$).*

In the case of Corollary 55 by construction the domain of a finite Poisson point process p is
$D_p = \{\tau_1, \tau_1 + \tau_2, \cdots, \tau_1 + \tau_2 + \cdots + \tau_m, \cdots\}$,
and $N_p((0,t], Z), t \geq 0$ is a Poisson process, where its first jump happened at the random time τ_1, and its m−th jump happened at the random time $\tau_1 + \tau_2 + \cdots + \tau_m$. However, if one concerns with the counting measure $N_p((0,t], U)$, where $U \in \mathfrak{B}_Z$, then even $N_p((0,t], U), t \geq 0$ is still a Poisson process, but τ_1 is not necessary its first jump time, because now $p(\tau_1) \in U$ is not necessary true. Actually, Poisson processes $N_p((0,t], Z)$ and $N_p((0,t], U)$ have different intensity functions $t\pi(Z)$ and $t\pi(U)$, respectively, if $\pi(U) < \pi(Z)$. Therefore, one sees that the concept of a Poisson point process is finer than a Poisson process, because it also considers where jumps occur, as well as the jumps themselves. Sometimes to such situations more attention should be paid. For example, in many cases to count how many times the degree of an earthquake exceeds some level (that is, the point process drops in some area), where the earthquake happened in some area, is more important than counting all of the times it has happened. Actually, the forecast of an earthquake is only that its power is stronger than some degree. When the earthquake is very very small, usually, it is not necessary to forecast it. So the point process is more realistic. For the reason

why we consider the Poisson point process as a basic jump type stochastic perturbation see the subsection "The General Model and its Explanation" in Chapter 8 - "Option Pricing in a Financial Market and BSDE".

1.9 Stochastic Integral for Point Process. Square Integrable Martingales

In a dynamical system the stochastic jump perturbation usually can be modeled as a stochastic integral with respect to some point process (i.e. its counting measure), or its martingale measure. In this section we will discuss how to define such stochastic integral. The idea is first to define it in the simple case by Lebesgue-Stieltjes integral for each or almost all $\omega \in \Omega$ (is said to define it pathwise). Then consider it in the general case through some limits. For this now let us consider a probability space $(\Omega, \mathfrak{F}, P)$ with an increasing $\sigma-$ field family $\{\mathfrak{F}_t\}_{t\geq 0}$, which satisfies the usual condition explained in the first section of this chapter. From now on all random variables and random processes are defined on it if without further explanation.

Definition 56 *Suppose that p is a point process on Z, and* $N_p(t,U) = \sum_{s\in D_p, s\leq t} I_U(p(s))$ *is its counting measure.*
1) p is called $\mathfrak{F}_t-$*adapted, if its counting measure is* $\mathfrak{F}_t-$*measurable for each* $t \geq 0$ *and each* $U \in \mathfrak{B}_Z$.
2) p is called $\sigma-$ *finite, if* $\exists \{U_n\}_{n=1}^{\infty} \subset \mathfrak{B}_Z$, *such that* $EN_p(t,U_n) < \infty, \forall t > 0, \forall n$, *and* $Z = \cup_{n=1}^{\infty} U_n$.

From now on we only discuss the $\mathfrak{F}_t-$adapted and $\sigma-$ finite point process p. Denote $\Gamma_p = \{U \in \mathfrak{B}_Z : EN_p(t,U) < \infty, \forall t > 0\}$. Obviously, for any $U \in \Gamma_p$, $N_p(t,U)$ is a submartingale and is of class (DL), since it is non-negative and increasing in t. Hence by Doob-Meyer's decomposition theorem (Theorem 42) there exists a unique $\mathfrak{F}_t-$adapted martingale $\widetilde{N}_p(t,U)$ and a unique $\mathfrak{F}_t-$adapted natural increasing process $\widehat{N}_p(t,U)$ such that

$$N_p(t,U) = \widetilde{N}_p(t,U) + \widehat{N}_p(t,U). \tag{1.13}$$

Notice that the equality only holds true $P - a.s.$ for the given U. Hence $\widehat{N}_p(t,U)$ may not be a measure for $U \in \mathfrak{B}_Z$, a.s. Moreover, it also may not be continuous in t. However, in most practical case we need $\widehat{N}_p(t,U)$ to have such properties.

Definition 57 *A point process p is said to be of class (QL) (meaning Quasi Left-continuous) if in the D-M decomposition expression (1.13)*
(i) $\widehat{N}_p(t,U)$ *is continuous in t for any* $U \in \Gamma_p$;
(ii) $\widehat{N}_p(t,U)$ *is a* $\sigma-$ *finite measure on* $(Z, \mathfrak{B}_Z)$ *for any given* $t \geq 0$, *P-a.s.*

We will call $\widehat{N}_p(t,U)$ the compensator of $N_p(t,U)$ (or p). We now introduce the following definition for the $\mathfrak{F}_t-$ Poisson point process.

Definition 58 *A point process p is called an $\mathfrak{F}_t-$ Poisson point process, if it is a Poisson point process, $\mathfrak{F}_t-$adapted, and $\sigma-$ finite, such that $N_p(t+h,U) - N_p(t,U)$ is independent of $\mathfrak{F}_t$ for each $h>0$ and each $U \in \Gamma_p$.*

Notice that $\widetilde{\mathfrak{F}}_t = \sigma[N_p((0,s] \times U); s \leq t, U \in \mathfrak{B}_Z] \subset \mathfrak{F}_t$, and in general these may not equal to each other. This is why we have to assume that for a Poisson point process $N_p(t+h,U) - N_p(t,U)$ is independent of $\mathfrak{F}_t$. From now on we only discuss point processes which belong to class (QL). By definition one can consider that the following proposition holds true.

Proposition 59 *A ($\mathfrak{F}_t-$) point processes p is a stationary $\mathfrak{F}_t-$ Poisson point process of class (QL), if and only if its compensator has the form: $\widehat{N}_p(t,U) = t\pi(U), \forall t > 0, U \in \Gamma_p$, where $\pi(\cdot)$ is a $\sigma-$finite measure on $\mathfrak{B}_Z$.*

In fact, the "only if part" of the Proposition can be seen from the definition: If p is a stationary $\mathfrak{F}_t-$ Poisson point process, then its counting measure $N_p(t,U)$ is a Poisson random measure with the intensity measure $EN_p(t,U) = t\pi(U)$, where $\pi(\cdot)$ is a $\sigma-$finite measure on $\mathfrak{B}_Z$. From this, one sees that $\forall t \geq 0, \forall h > 0$,

$E[(N_p(t+h,U) - (t+h)\pi(U)) - (N_p(t,U) - t\pi(U))|\mathfrak{F}_t]$
$= E[N_p(h,U) - h\pi(U)] = 0.$

Hence $\{N_p(t,U) - t\pi(U)\}_{t\geq 0}$ is a $\mathfrak{F}_t-$martingale, i.e.

$N_p(t,U) - t\pi(U) = M_t,$

where M_t is $\mathfrak{F}_t-$martingale. However, by the uniqueness of decomposition of the submartingale $N_p(t,U)$ (Theorem 42) one should have $M_t = \widetilde{N}_p(t,U)$, and $\widehat{N}_p(t,U) = t\pi(U)$. The "if part" of the above Proposition will be proved in the next chapter by using Ito's formula.

Now let us discuss the integral with respect to the point process. In the simple case it can be defined by the Lebesgue-Stieltjse integral. First, we have the following Lemma.

Lemma 60 *For any given $U \in \mathfrak{B}_Z$ and any bounded $\mathfrak{F}_t-$predictable process $f(t,\omega)$ let*

$x_t(\omega) = \int_0^t f(s,\omega)d\widetilde{N}_p(s,U) =: \int_0^t f(s,\omega)dN_p(s,U) - \int_0^t f(s,\omega)d\widehat{N}_p(s,U)$
$= \sum_{s\leq t, s\in D_{p(\omega)}} f(s,\omega)I_U(p(s,\omega)) - \int_0^t f(s,\omega)d\widehat{N}_p(s,U).$

Then x_t is a $\mathfrak{F}_t-$martingale.

Proof. Assume that $f(t,\omega)$ is a left-continuous bounded $\mathfrak{F}_t-$predictable process. Let $f_n(t) = f(0)I_{\{s=0\}}(s) + \sum_{k=0}^{\infty} f(\frac{k}{2^n})I_{(\frac{k}{2^n},\frac{k+1}{2^n}]}(s)$. Then by definition one easily sees that

$\int_0^t f_n(s)d\widetilde{N}_p(s,U) = \sum_{k=0}^{\infty} f(\frac{k}{2^n})[\widetilde{N}_p(\frac{k+1}{2^n} \wedge t, U) - \widetilde{N}_p(\frac{k}{2^n} \wedge t, U)].$

So the left hand side is obviously an $\mathfrak{F}_t-$martingale. Now since $f(t)$ is left-continuous and bounded, applying Lebesgue's dominated convergence theorem and using

$E[\int_0^t f_n(s)d\widetilde{N}_p(s,U)|\mathfrak{F}_s] = \int_0^s f_n(s)d\widetilde{N}_p(s,U),\ \forall s \leq t.$

one obtains that as $n \to \infty$

$E[\int_0^t f(s)d\widetilde{N}_p(s,U)|\mathfrak{F}_s] = \int_0^s f(s)d\widetilde{N}_p(s,U),\ \forall s \leq t.$

Now by the Monotone-class Theorem (Theorem 392) it is easily seen that it also holds true for all bounded $\mathfrak{F}_t-$predictable process. ■

The integral defined in the above lemma motivates us to define the stochastic integrals with respect to the counting measure and martingale measure generated by a point process of the class (QL) for some class of stochastic processes as the integrands thru Lebesgue-Stieltjes integral. First, let us generalize the notion of predictable processes to functions $f(t,z,\omega)$ with three variables.

Definition 61 *1) By $\mathcal{P}$ we denote the smallest $\sigma-$field on $[0,\infty) \times Z \times \Omega$ such that it makes all g having the following properties $\mathcal{P}/\mathfrak{B}(R^1)-$ measurable:*
(i) for each $t > 0$, $Z \times \Omega \ni (z,\omega) \longrightarrow g(t,z,\omega) \in R^1$ is $\mathfrak{B}_Z \times \mathfrak{F}_t-$measurable;
(ii) for each (z,ω), $g(t,z,\omega)$ is left-continuous in t.

2) If a real function $g(t,z,\omega)$ is $\mathcal{P}/\mathfrak{B}(R^1)-$ measurable, then we call it $\mathfrak{F}_t-$ predictable, and denote $g \in \mathcal{P}$.

Now for any given $\mathfrak{F}_t-$point process of the class (QL) let us introduce four classes of random processes as follows:

$\mathcal{F}_p = \{f(t,x,\omega) : f$ is $\mathfrak{F}_t-$ predictable such that $\forall t > 0,$
$\int_0^{t+} \int_Z |f(s,x,\omega)|\, N_p(ds,dz) < \infty,$ a.s.$\}$,

$\mathcal{F}_p^1 = \{f(t,x,\omega) : f$ is $\mathfrak{F}_t-$ predictable such that $\forall t > 0,$
$E\int_0^{t+} \int_Z |f(s,x,\omega)|\, \widehat{N}_p(ds,dz) < \infty\}$,

$\mathcal{F}_p^2 = \{f(t,x,\omega) : f$ is $\mathfrak{F}_t-$ predictable such that $\forall t > 0,$
$E\int_0^{t+} \int_Z |f(s,x,\omega)|^2\, \widehat{N}_p(ds,dz) < \infty\}$,

$\mathcal{F}_p^{2,loc} = \{f(t,x,\omega) : f$ is $\mathfrak{F}_t-$ predictable such that $\exists \sigma_n \uparrow \infty$, a.s.,
σ_n is a stopping time, and $I_{[0,\sigma_n]}(t)\ f(t,x,\omega) \in \mathcal{F}_p^2, \forall n = 1,2,\cdots\}$.

It is natural that we define the stochastic integral for $f \in \mathcal{F}_p$ with respect to the counting measure by

$\int_0^{t+} \int_Z f(s,z,\omega) N_p(ds,dz) = \sum_{s \leq t, s \in D_p} f(s,p(s),\omega)$
$= \sum_{s \leq t, s \in D_{p(\omega)}} f(s,p(s,\omega),\omega),$

since the last series absolutely converges for a.s. $\omega \in \Omega$. Note that

$E\int_0^{t+} \int_Z |f(s,x,\omega)|\, N_p(ds,dz) = E\int_0^{t+} \int_Z |f(s,x,\omega)|\, \widehat{N}_p(ds,dz).$

Actually, the above equality holds for f being an $\mathfrak{F}_t-$simple process. Applying the monotone class theorem (Theorem 391) one easily sees that it is also true for f being an $\mathfrak{F}_t-$ predictable process. So $\mathcal{F}_p^1 \subset \mathcal{F}_p$, and for $f \in \mathcal{F}_p^1$ we can define the stochastic integral for $f \in \mathcal{F}_p^1$ with respect to the martingale measure by

$$\int_0^{t+}\int_Z f(s,z,\omega)\widetilde{N}_p(ds,dz) = \int_0^{t+}\int_Z f(s,z,\omega)N_p(ds,dz)$$
$$-\int_0^t\int_Z f(s,z,\omega)\widehat{N}_p(ds,dz). \tag{1.14}$$

As the proof of Lemma 60 one can show that it is a martingale. In fact, it is true for f being an $\mathfrak{F}_t$−simple process. Applying the monotone class theorem (Theorem 391) again one easily sees that it also holds true for f being an $\mathfrak{F}_t$− predictable process. Thus for $f \in \mathcal{F}_p^1$ the stochastic integral with respect to $\widetilde{N}_p(ds,dz)$ can be defined by (1.14), and it is a $\mathfrak{F}_t$− martingale.

However, for $f \in \mathcal{F}_p^2$ we cannot define the stochastic integral by (1.14), since each term on the right hand side of (1.14) may not have meaning in this case. So we have to define it through limit. Let us introduce the following notion:

$\mathcal{M}^2 = \{\{m_t\}_{t\geq 0} : \{m_t\}_{t\geq 0}$ is a square integrable martingale, i.e. $\{m_t\}_{t\geq 0}$

is a martingale, and for each $t > 0$, $E\left|m_t\right|^2 < \infty$. Moreover, $m_0 = 0\}$,

$\mathcal{M}^{2,loc} = \{\{m_t\}_{t\geq 0} : \{m_t\}_{t\geq 0}$ is a locally square integrable martingale, i.e.

$\exists\sigma_n \uparrow \infty$, each σ_n is a stopping time, such that for each n, $\{m_{t\wedge\sigma_n}\}_{t\geq 0} \in \mathcal{M}^2\}$,

$\mathcal{M}_T^2 = \left\{\{m_t\}_{t\in[0,T]} : \{m_t\}_{t\geq 0} \in \mathcal{M}^2\right\}$,

$\mathcal{M}_T^{2,loc} = \{\{m_t\}_{t\in[0,T]} : \{m_t\}_{t\geq 0} \in \mathcal{M}^{2,loc}\}$.

For each $\{m_t\}_{t\geq 0} \in \mathcal{M}^2$ by Jensen's inequality $\left\{\left|m_t\right|^2\right\}_{t\geq 0}$ is a non-negative submartingale. So it is of class (DL). In fact, for any constant $a > 0$ one has that $\sup_{\sigma\in S_a} E\left|m_\sigma\right|^2 \leq E\left|m_a\right|^2$. So $\{m_\sigma\}_{\sigma\in S_a}$ is uniformly integrable. Now by the D-M decomposition theorem $\left|m_t\right|^2$ has a unique decomposition, and we denote it by

$\left|m_t\right|^2 =$ a martingale $+ \langle m\rangle_t$,

i.e. $\langle m\rangle_t$ is the natural increasing process for the decomposition of submartingale $\left|m_t\right|^2$. Usually, $\langle m\rangle_t$ is called the characteristic process of m_t.

Let us show the following lemma.

Lemma 62 *If $f \in \mathcal{F}_p^1 \cap \mathcal{F}_p^2$, then*
$\left\{\int_0^{t+}\int_Z f(s,z,\omega)\widetilde{N}_p(ds,dz)\right\}_{t\geq 0} \in \mathcal{M}^2$; *and*
$\left\langle\int_0^{\cdot+}\int_Z f(s,z,\omega)\widetilde{N}_p(ds,dz)\right\rangle_t = \int_0^t\int_Z \left|f(s,z,\omega)\right|^2 \widehat{N}_p(ds,dz).$

Proof. Let us consider a special case first. Assume that $f(s,z,\omega) = I_U(z), U \in \Gamma$. We are going to show that $\left\langle \widetilde{N}_p(\cdot,U)\right\rangle_t = \widehat{N}_p(t,U)$. In fact, let $\forall m = 1,2,\cdots,$

$\sigma_m = \inf\left\{t \geq 0 : \left|\widetilde{N}_p(\cdot,U)\right| > m, \text{ or } \widehat{N}_p(t,U) > m\right\}.$

Then σ_m is a stopping time, since $\forall t > 0$,

$\{\sigma_m > t\} = \left\{\left|\widetilde{N}_p(\cdot,U)\right| \leq m, \text{ and } \widehat{N}_p(t,U) \leq m\right\} \in \mathfrak{F}_t.$

Denote $g(t) = \widetilde{N}_p^m(t,U) = \widetilde{N}_p(t \wedge \sigma_m, U)$, and $\widehat{N}_p^m(t,U) = \widehat{N}_p(t \wedge \sigma_m, U)$. Then both of them are bounded in t, and as $n \to \infty$,

$g(t)^2 = \sum_{k=1}^n [g(\frac{kt}{2^n})^2 - g(\frac{(k-1)t}{2^n})^2] = \sum_{k=1}^n g(\frac{(k-1)t}{2^n})[g(\frac{kt}{2^n}) - g(\frac{(k-1)t}{2^n})]$
$+ \sum_{k=1}^n g(\frac{kt}{2^n})[g(\frac{kt}{2^n}) - g(\frac{(k-1)t}{2^n})] \to \int_0^t g(s-)dg(s) + \int_0^t g(s)dg(s)$
$= \int_0^t \widetilde{N}_p^m(s-,U)\widetilde{N}_p^m(ds,U) + \int_0^t \widetilde{N}_p^m(s,U)\widetilde{N}_p^m(ds,U)$
$= 2\int_0^t \widetilde{N}_p^m(s-,U)\widetilde{N}_p^m(ds,U) + \int_0^t [\widetilde{N}_p^m(s,U) - \widetilde{N}_p^m(s-,U)]\widetilde{N}_p^m(ds,U)$
$=$ a martingale $+ \int_0^t [N_p^m(s,U) - N_p^m(s-,U)]N_p^m(ds,U)$
$=$ a martingale $+ N_p^m(t,U) =$ a martingale $+ \widehat{N}_p^m(t,U)$,

where we have applied Lemma 60 to obtain the result that the integral $\int_0^t \widetilde{N}_p^m(s-,U)\widetilde{N}_p^m(ds,U)$ is a martingale, and used the fact that $\widehat{N}_p^m(t,U)$ is continuous. Hence

$$\left\langle \widetilde{N}_p(\cdot \wedge \sigma_m, U)\right\rangle_t = \widehat{N}_p(t \wedge \sigma_m, U), \forall m. \tag{1.15}$$

So by Fatou's lemma

$E\left|\widetilde{N}_p(t,U)\right|^2 = E\lim_{m\to\infty}\left|\widetilde{N}_p(t\wedge\sigma_m,U)\right|^2 \leq \underline{\lim}_{m\to\infty}E\left|\widetilde{N}_p(t\wedge\sigma_m,U)\right|^2$
$= \underline{\lim}_{m\to\infty}E\widehat{N}_p(t\wedge\sigma_m,U) = E\widehat{N}_p(t,U) < \infty.$

That is to say, $\widetilde{N}_p(t,U)$ is a square integrable martingale, so $\left\langle \widetilde{N}_p(\cdot,U)\right\rangle_t$ exists and has meaning. However, $\left\langle \widetilde{N}_p(\cdot,U)\right\rangle_{t\wedge\sigma_m} = \left\langle \widetilde{N}_p(\cdot \wedge \sigma_m,U)\right\rangle_t$. So by (1.15) letting $m \to \infty$ one obtains that $\left\langle \widetilde{N}_p(\cdot,U)\right\rangle_t = \widehat{N}_p(t,U)$.

Now let us return the proof of Lemma 62. However, the conclusions are true for f being a simple process (just proved). So applying the monotone class theorem (Theorem 391) one sees that the conclusions also hold true for f being a predictable process belonging to $\mathcal{F}_p^1 \cap \mathcal{F}_p^2$. The proof is complete. ∎

Now assume that $f \in \mathcal{F}_p^2$. We are going to define the stochstic integral for f with respect to the martingale measure $\widetilde{N}_p(ds,dz)$. Since we only consider the point process being $\mathfrak{F}_t$-adapted, σ-finite and of class (QL), we have $\exists U_n \in \mathfrak{B}_Z, U_n \uparrow Z$, such that $EN_p(t,U_n) < \infty, \forall t > 0, \forall n$. Let

$f_n(s,z,\omega) = I_{[-n,n]}(f(s,z,\omega))I_{U_n}(z)f(s,z,\omega).$

Then obviously, f_n is bounded and $f_n \in \mathcal{F}_p^1 \cap \mathcal{F}_p^2, \forall n$. By (1.14) and Lemma 62 the integral $I(f_n)(t+) = \int_0^{t+}\int_Z f_n(s,z,\omega)\widetilde{N}_p(ds,dz)$ is defined and belongs to $\mathcal{M}^2$. Moreover, by Lemma 62 for each fixed $T > 0$, as $0 \leq t \leq T$,

$E[|I(f_n)(t+) - I(f_m)(t+)|^2] = E\int_0^t \int_Z |(f_n - f_m)(s,z,\omega)|^2 \widehat{N}_p(ds,dz)$
$\leq E\int_0^T \int_Z |(f_n - f_m)(s,z,\omega)|^2 \widehat{N}_p(ds,dz).$

So $\{I(f_n)(t+)\}_{t\in[0,T]}$ is a Cauchy sequence in $\mathcal{M}_T^2$ with the norm $\|m\|_T = \sqrt{E[|m_T|^2]} = \sqrt{E\langle m\rangle_T}$. By the completeness of $\mathcal{M}_T^2$ with such norm (Lemma 63 below) there exists a unique limit denoted by

$\{I(f)(t+)\}_{t\in[0,T]} = \left\{\int_0^{t+} \int_Z f(s,z,\omega)\widetilde{N}_p(ds,dz)\right\}_{t\in[0,T]} \in \mathcal{M}_T^2.$

Again by the uniqueness of the limit one can define $\forall T > 0$

$\int_0^{t+} \int_Z f(s,z,\omega)\widetilde{N}_p(ds,dz) = \int_0^{t+} \int_Z f(s,z,\omega)\widetilde{N}_p(ds,dz)$, as $t \in [0,T]$.

Then obviously, $\left\{\int_0^{t+} \int_Z f(s,z,\omega)\widetilde{N}_p(ds,dz)\right\}_{t\geq 0} \in \mathcal{M}^2$. The integral

$\int_0^{t+} \int_Z f(s,z,\omega)\widehat{N}_p(ds,dz)$

is usually called a "compensated sum".

Finally, for $f \in \mathcal{F}_p^{2,loc}$ by assumption $\exists \sigma_n \uparrow \infty$, a.s., σ_n is a stopping time, and $I_{[0,\sigma_n]}(t)\ f(t,x,\omega) \in \mathcal{F}_p^2, \forall n = 1,2,\cdots$. Let $\{x_t\}_{t\geq 0} \in \mathcal{M}^{2,loc}$ such that

$x_{t\wedge\sigma_n} = \int_0^{t+} \int_Z f(s,z,\omega)I_{[0,\sigma_n]}(s)\widetilde{N}_p(ds,dz).$

Then the integral is uniquely defined.

Lemma 63 *1) For $\{m_t\}_{t\in[0.T]} \in \mathcal{M}_T^2$ let $\|m\|_T = \sqrt{E[|m_T|^2]} = \sqrt{E\langle m\rangle_T}$, then $(\mathcal{M}_T^2, \|\cdot\|_T)$ is a Banach space.*

2) Let $\mathcal{M}^{2,c} = \left\{\{m_t\}_{t\geq 0} \in \mathcal{M}^2 : \{m_t\}_{t\geq 0} \text{ is continuous in } t\right\}$,

$$\mathcal{M}^{2,rcll} = \left\{ \begin{array}{c} \{m_t\}_{t\geq 0} \in \mathcal{M}^2 : \{m_t\}_{t\geq 0} \textit{ is right continuous} \\ \textit{with left limit in } t \end{array} \right\},$$

$$\mathcal{M}_T^{2,c} = \left\{\{m_t\}_{t\in[0,T]} : \{m_t\}_{t\geq 0} \in \mathcal{M}^{2,c}\right\},$$

and

$$\mathcal{M}_T^{2,rcll} = \left\{\{m_t\}_{t\in[0,T]} : \{m_t\}_{t\geq 0} \in \mathcal{M}^{2,llrc}\right\}.$$

Then $\mathcal{M}_T^{2,c} \subset \mathcal{M}_T^{2,rllc}$ both are closed subspaces of $\mathcal{M}_T^2$.

3) For $\{m_t\}_{t\geq 0} \in \mathcal{M}^2$ let $\|m\| = \sum_{n=1}^{\infty} \frac{1}{2^n}(\|m\|_n \wedge 1)$, and $d(m,n) = \|m-n\|$, then $(\mathcal{M}^2, d)$ is a complete metric space. Moreover, $\mathcal{M}^{2,c} \subset \mathcal{M}^{2,llrc}$ both are closed subspaces of $\mathcal{M}^2$.

Proof. 1): By functional analysis all others are true except that the completeness under this norm still needs to prove. Now suppose that $\{m_t^n\}_{t\in[0.T]} \in \mathcal{M}_T^2$ is a Cauchy sequence under this norm, i.e. as $n,k \to \infty$

$E[|m_T^n - m_T^k|^2 \to 0.$

By L^2- theory there exists an $m_T \in \mathfrak{F}_T$ such that $E[|m_T|^2] < \infty$ and as $n \to \infty, E[|m_T^n - m_T|^2] \to 0$. Let $m_t = E[m_T|\mathfrak{F}_t]$. Then by Jensen's inequality etc. $\{m_t\}_{t\in[0.T]} \in \mathcal{M}_T^2$. So the proof of 1) is complete.

2): Assume that $\{m_t^n\}_{t\in[0.T]}, \{m_t\}_{t\in[0.T]} \in \mathcal{M}_T^2$ such that

$\lim_{n\to\infty} \|m^n - m\|_T = \lim_{n\to\infty} E[|m_T^n - m_T|^2] = 0$. By the martingale inequality (Theorem 34) one has that as $n \to \infty$,

$E[\sup_{t\in[0.T]} |m_t^n - m_t|^2] \leq 4E[|m_T^n - m_T|^2] \to 0$.

So $\exists \{n_k\}_{k=1}^{\infty} \subset \{n\}_{n=1}^{\infty}$ such that as $k \to \infty$,

$\sup_{t\in[0.T]} |m_t^{n_k} - m_t|^2 \to 0,\ P - a.s.$

Since the convergence of $\{m_t^{n_k}\}_{t\in[0,T]}$ to $\{m_t\}_{t\in[0,T]}$ is uniform in t, we have that if for each k, $\{m_t^{n_k}\}_{t\in[0,T]}$ is continuous in t (right continuous with left limit in t), so is $\{m_t\}_{t\in[0,T]}$ (respectively, so also is $\{m_t\}_{t\in[0,T]}$).

3): We only need to note that $\|m - n\| = 0 \iff \|m - n\|_n = 0, \forall n$; and that $\|m^n - m\| \to 0 \iff \|m - n\|_n \to 0, \forall n$; provided that $\{m_t^n\}_{t\geq 0}$, $\{m_t\}_{t\geq 0} \in \mathcal{M}^2$. Then all of the conclusions are easily derived by means of 1) and 2). ∎

2

Brownian Motion, Stochastic Integral and Ito's Formula

For a dynamic system the simplest continuous stochastic perturbation is naturally considered to be a Brownian motion (BM), since it is a Normal process (or say, a Guassian process) with independent increments which are also normally distributed. In general, a continuous stochastic perturbation will be modeled as some stochastic integral with respect to the BM. However, the BM has the strange property that even though its trajectory is continuous in t, it is not differentiable for all t. So for a stochastic integral with respect to BM one has to use a different approach - the martingale approach is used to define it. In this chapter we will discuss such problems.

2.1 Brownian Motion and Its Nowhere Differentiability

Definition 64 *A $d-$dimensional random process $\{x_t\}_{t\geq 0}$ is called a Brownian Motion (BM) or a Weiner process, if*

1) its initial probability law is given by some probability measure μ, i.e. $\forall \Gamma \in \mathfrak{B}(R^d), P(x_0 \in \Gamma) = \mu(\Gamma)$;

2) it has independent increments, i.e. $\forall 0 = t_0 < t_1 < \cdots < t_m$, the increments $x_{t_0}, x_{t_1} - x_{t_0}, x_{t_2} - x_{t_1}, \cdots, x_{t_m} - x_{t_{m-1}}$ are independent;

3) $\forall 0 \leq s < t$, $x_t^i - x_s^i \sim N(0, (t-s))$, $i = 1, 2, \cdots, d$, i.e. each real component increment $x_t^i - x_s^i$ is Normally distributed with the mean $E(x_t - x_s) = 0$ and the variance $V(x_t - x_s) = (t-s)$; where $x_t = (x_t^1, x_t^2, \cdots, x_t^d)$, and $\{x_t^i\}_{i=1}^d$ is an independent random variable family for each $t > 0$;

4) it is continuous in t*, a.s., that is, for almost all* $\omega \in \Omega$ *the trajectory* $x_t(\omega)$ *is continuous in* t*.*

Now for $t > 0, x \in R^d$ let
$p(t,x) = (2\pi t)^{-d/2} \exp[-|x|^2/2t]$.
Then we have the following proposition.

Proposition 65 *If* $\{x_t\}_{t\geq 0}$ *is a continuous d-dimensional random process, the following statements are equivalent:*
(i) $\{x_t\}_{t\geq 0}$ *is a BM with some initial law* μ*.*
(ii) $\forall 0 = t_0 < t_1 < \cdots < t_m$*, and* $\Gamma_i \in \mathfrak{B}(R^d), i = 0,1,2,\cdots,m,$

$$P(x_{t_0} \in \Gamma_0, x_{t_1} \in \Gamma_1, \cdots, x_{t_m} \in \Gamma_m)$$
$$= \int_{\Gamma_0} \mu(dx) \int_{\Gamma_1} p(t_1, x_1 - x) dx_1 \int_{\Gamma_2} p(t_2 - t_1, x_2 - x_1) dx_2$$
$$\cdots \int_{\Gamma_m} p(t_m - t_{m-1}, x_m - x_{m-1}) dx_m. \tag{2.1}$$

(iii) $E[\exp(i\langle\lambda, x_t - x_s\rangle)|\mathfrak{F}_t^x] = \exp[-(t-s)|\lambda|^2/2]$*, a.s.,*
$\forall \lambda \in R^d, 0 \leq s < t,$
where $\mathfrak{F}_t^x = \cap_{n=1}^\infty \sigma(x_s : s \leq t + \frac{1}{n})$*, and* $\sigma(x_s; s \leq t + \frac{1}{n})$ *is the smallest* $\sigma-$*field such that it makes all* x_s*, for* $s \leq t + \frac{1}{n}$*, measurable.*

Proof. *(ii)* $\Longrightarrow$ *(i)* : Consider the probability law of $P(x_{t_0} \in \Gamma_0, x_{t_1} - x_{t_0} \in \Gamma_1, \cdots, x_{t_m} - x_{t_{m-1}} \in \Gamma_m)$. If we let $y_0 = x, y_1 = x_1 - x, y_2 = x_2 - x_1, \cdots, y_m = x_m - x_{m-1}$, then one can easily finds that the absolute value of the Jacobi determinant $\left|\frac{\partial(x_0,x_1,\cdots,x_m)}{\partial(y_0,y_1,\cdots,y_m)}\right| = 1$. Thus by (ii) one has that

$$P(x_{t_0} \in \Gamma_0, x_{t_1} - x_{t_0} \in \Gamma_1, x_{t_2} - x_{t_1} \in \Gamma_2, \cdots, x_{t_m} - x_{t_{m-1}} \in \Gamma_m) \tag{2.2}$$
$$= \int_{\Gamma_0} \mu(dx) \int_{\Gamma_1} p(t_1, y_1) dy_1 \int_{\Gamma_2} p(t_2 - t_1, y_2) dy_2 \cdots \int_{\Gamma_m} p(t_m - t_{m-1}, y_m) dy_m,$$

which is equivalent to saying that $\{x_t\}_{t\geq 0}$ is a BM with the initial law μ.
(i) $\Longrightarrow$ *(ii)* : Similarly, now consider the probability law of $P(x_{t_0} \in \Gamma_0, x_{t_1} \in \Gamma_1, \cdots, x_{t_m} \in \Gamma_m)$. If we let $x = y_0, x_1 - x = y_1, x_2 - x_1 = y_2, \cdots, x_m - x_{m-1} = y_m$, then one still finds that the absolute value of the Jacobi determinant $\left|\frac{\partial(y_0,y_1,\cdots,y_m)}{\partial(x_0,x_1,\cdots,x_m)}\right| = 1$. Thus by (2.2) we find that (ii) holds true.
(iii) $\Longrightarrow$ *(i)* : By (iii) $\forall 0 = t_0 < t_1 < \cdots < t_m, \lambda_k \in R^d, k = 0,1,2,\cdots,m,$
$E[\exp\{i(\lambda_0 x_0 + \sum_{k=1}^m \langle\lambda_k, x_{t_k} - x_{t_{k-1}}\rangle)\}]$
$= E[e^{i\lambda_0 x_0} E(\exp(i\sum_{k=1}^m \langle\lambda_k, x_{t_m} - x_{t_{k-1}}\rangle)|\mathfrak{F}_0^x)]$
$= E[e^{i\lambda_0 x_0} E(e^{i\lambda_1(x_{t_1} - x_0)} E[e^{i\lambda_2(x_{t_2} - x_{t_1})}$
$\cdots E(e^{i\lambda_m(x_{t_m} - x_{t_{m-1}})}|\mathfrak{F}_{t_{m-1}}) \cdots |\mathfrak{F}_{t_1}]|\mathfrak{F}_0)]$

$= (Ee^{i\lambda_0 x_0}) \cdot \prod_{j=1}^m \exp[-(t_j - t_{j-1}) |\lambda_j|^2 /2]$
$= (Ee^{i\lambda_0 x_0}) \cdot \prod_{j=1}^m Ee^{i\lambda_j (x_{t_j} - x_{t_{j-1}})}.$
Thus the increments $x_{t_0}, x_{t_1} - x_{t_0}, x_{t_2} - x_{t_1}, \cdots, x_{t_m} - x_{t_{m-1}}$ are independent, and the increment $x_{t_j} - x_{t_{j-1}}$ is Normally distributed with mean 0 and with the variance matrix such that all elements on the diagonal are equal to $t_j - t_{j-1}$, and all other elements are equal 0. Thus $\{x_t\}_{t\geq 0}$ is a ($d-$dimensional) BM with some initial law.

$(i) \Longrightarrow (iii)$: If $\{x_t\}_{t\geq 0}$ is a BM, then by the independent increment property, and because its increments are Normally distributed, one finds that

$E[\exp(i \langle \lambda, x_t - x_s \rangle)|\mathfrak{F}_t^x] = E[\exp(i \langle \lambda, x_t - x_s \rangle)] = \exp[-(t-s) |\lambda|^2 /2].$

■

Now let us set

$W^d =$ the set of all continuous $d-$dimensional functions $w(t)$ defined for $t \geq 0$.

$\mathfrak{B}(W^d) =$ the smallest $\sigma-$field including all Borel cylinder sets in W^d, where a Borel cylinder set means a set $B \subset W^d$ of the following form

$B = \{w : (w(t_1), \cdots w(t_n)) \in A\}$,

for some finite sequence $0 \leq t_1 < t_2 < \cdots < t_n$ and $A \in \mathfrak{B}(R^{nd})$.

From above one sees that given a Brownian motion $\{x_t\}_{t\geq 0}$, this will lead to the generation of a probability measure P defined on $\mathfrak{B}(W^d)$ such that its measure of the Borel cylinder set is given by (ii) in Proposition 65. Such a probability measure is called a Wiener measure with the initial measure (or say, the initial law) μ. Conversely, if we have a Wiener measure P with initial measure μ on $\mathfrak{B}(W^d)$, let $(\Omega, \mathfrak{F}, P) = (W^d, \mathfrak{B}(W^d), P), x(t,w) = w(t), \forall t \geq 0, w \in W^d$, then we obtain a BM $\{x_t\}_{t\geq 0}$ defined on the probability space $(\Omega, \mathfrak{F}, P)$. So the BM is in one to one correspondence with the Wiener measure. Now a natural question arises: does the BM, that is, the Wiener measure exist?

The existence of the Brownian motion is established by the following theorem.

Theorem 66 *For any probability measure μ on $(R^d, \mathfrak{B}(\mathfrak{R}^{\mathfrak{d}}))$ the Wiener measure P_μ with the initial law μ exists uniquely.*

Proof. *Uniqueness. Since the totality of Borel cylinder sets generates the $\sigma-$field $\mathfrak{B}(W^d)$, any two measures that coincide on Borel cylinder sets still coincide on $\mathfrak{B}(W^d)$.*

Existence. We shall show the result by using the Kolmogorov extension theorem (Theorem 379 in Appendix A). In fact, by this theorem there exists a random process $\{x_t\}_{t\geq 0}$ such that it has the finite probability distribution defined as (2.1). Moreover, since $(x_t^i - x_s^i) \sim N(0, (t-s)), i = 1, 2, \cdots, d$. Hence, by calculation, $E\left|x_t^i - x_s^i\right|^4 = 3\left|t-s\right|^2$. So by the Kolmogorov continuous version theorem[78] *$\{x_t^i\}_{t\geq 0}$ has a continuous version, $i = 1, 2, \cdots, d$; and so does $\{x_t\}_{t\geq 0}$. Let us again denote this continuous* *version by* $\{x_t\}_{t\geq 0}$. *Thus such a* $d-$ *dimensional random process is a BM with the initial law* μ. ∎

Definition 67 *If a d-dimensional BM* $\{x_t\}_{t\geq 0}$ *is such that* $P(x_0 = 0) = 1$, *that is,* $\mu = \delta_0-$ *the probability measure concentrated at the single point* $\{0\}$, *then it is called the standard BM and denoted by* $\{w_t\}_{t\geq 0}$.

From now on we always discuss the standard Brownian motion, and it is simply denoted by BM. Brownian motion has some nice properties. For example, its trajectory is continuous, i.e. $\{x_t\}_{t\geq 0}$ is continuous. Moreover, it can be a square integrable martingale.

Corollary 68 *If* $\{x_t\}_{t\geq 0}$ *is a* $d-$*dimensional* $\mathfrak{F}_t-$*BM, and* $E\left|x_0\right|^2 < \infty$, *then 1)* $\{x_t\}_{t\geq 0}$ *is a square integrable* $\mathfrak{F}_t-$*martingale; 2)* $x_t^i x_t^j - \delta_{ij} t$ *is a* $\mathfrak{F}_t-$*martingale. (Here "*$\mathfrak{F}_t -$*" means "*$\mathfrak{F}_t-$ *adapted").*

Proof. By the definition of a BM we have that as $0 \leq s < t$, $E[x_t^i - x_s^i|\mathfrak{F}_s] = 0$ and $E[(x_t^i - x_s^i)\left(x_t^j - x_s^j\right)|\mathfrak{F}_s] = \delta_{ij}(t-s)$. Since the left hand side of the above equality equals $E[x_t^i x_t^j|\mathfrak{F}_s] - x_s^i x_s^j$. So the proof is complete. ∎

However, a BM also has the following strange properties.

Theorem 69 *Suppose that* $\{x_t\}_{t\geq 0}$ *is a* $1-$*dimensional BM, then* $P-a.s.$, *for any given* $\alpha > \frac{1}{2}$, $\{x_t\}_{t\geq 0}$ *is not Hölder-continuous with index* α *for each* $t \geq 0$.

Definition 70 *We say that* $\{x_t\}_{t\geq 0}$ *is Hölder-continuous with index* α *at* $t_0 > 0$, *if* $\forall \varepsilon > 0, \exists \delta > 0$ *such that as* $|t - t_0| < \delta, |x_t - x_{t_0}| < \varepsilon\left|t - t_0\right|^\alpha$.

Theorem 69 actually tells us that the trajectory of BM is not Lipschitzian continuous at each point t, so it is nowhere differentiable for $t \geq 0$. Hence it is also not finite variational on any finite interval of t, since each finite variational function of t should be almost everywhere differentiable for t. Thus we arrive at the following corollary.

Corollary 71 *1) The trajectory of a BM is nowhere differentiable for* $t \geq 0$, $P - a.s.$

2) The trajectory of a BM is not finite variational on any finite interval of $t, P - a.s.$

Now let us show Theorem 69.

Proof. Take a positive integer N such that $N(\alpha - \frac{1}{2}) > 1$. For any positive integer $T > 0$ denote

$$A_n^\varepsilon = \left\{ \begin{array}{c} \omega : \text{ there exists a } s \in [0,T] \text{ such that } \forall t \in [0,T], \\ |t-s| < \frac{N}{n} \Longrightarrow |x_t(\omega) - x_s(\omega)| < \varepsilon\left|t-s\right|^\alpha \end{array} \right\}.$$

Obviously, $A_n^\varepsilon \uparrow$, as $n \uparrow$. Let $A^\varepsilon = \cup_{n=1}^\infty A_n^\varepsilon$. If one can show that $\forall \varepsilon > 0$, $P(A_n^\varepsilon) = 0$, then one finds that the conclusion of Theorem 69 holds true on the interval $[0,T]$. Now set

$Z_k = \max_{1\leq i\leq N}\left|x(\frac{k+i}{n}) - x(\frac{k+i-1}{n})\right|,\ k = 0, 1, \cdots, nT;$
$B_n^\varepsilon = \{\omega : \exists k \text{ such that } Z_k(\omega) \leq 2\varepsilon(N/n)^\alpha\}.$

Let us show that $A_n^\varepsilon \subset B_n^\varepsilon$. In fact, if $\omega \in A_n^\varepsilon$, then $\exists s \in [0, T]$ such that $\forall t \in [0, T]$,

$|t - s| < \frac{N}{n} \Longrightarrow |x_t(\omega) - x_s(\omega)| < \varepsilon\,|t - s|^\alpha.$

Set $k_0 = \max\{k : k/n \leq s\}$. Then $Z_{k_0}(\omega) \leq 2\varepsilon(N/n)^\alpha$, since in this case $s \in [\frac{k_0}{n}, \frac{k_0+N}{n}]$, and each point in this interval has a distance from s less than $\frac{N}{n}$. Hence $\omega \in B_n^\varepsilon$. Thus $p(A^\varepsilon) = \lim_{n\to\infty} P(A_n^\varepsilon) \leq \underline{\lim}_{n\to\infty} P(B_n^\varepsilon)$. However,

$P(B_n^\varepsilon) = P(\cup_{k=0}^{nT}\{Z_k \leq 2\varepsilon(N/n)^\alpha\}) \leq \sum_{k=0}^{nT} P(Z_k \leq 2\varepsilon(N/n)^\alpha)$
$= nTP(Z_0 \leq 2\varepsilon(N/n)^\alpha) \leq nT[P(\left|x(\frac{1}{n})\right| \leq 2\varepsilon(N/n)^\alpha)]^N$
$= nT(\sqrt{\frac{n}{2\pi}}\int_{-r/n^\alpha}^{r/n^\alpha} e^{-nx^2/2}dx)^N,\ (r = 2\varepsilon N^\alpha),$

where we have used the fact that since $\{x_t\}_{t\geq 0}$ is a BM, it has independent increments with the same probability law such that $x(\frac{k+i}{n}) - x(\frac{k+i-1}{n}) \sim N(0, \frac{1}{n}), \forall i, \forall k$. Let $y = n^\alpha x$. One finds that as $n \to \infty$,

$P(A_n^\varepsilon) \leq P(B_n^\varepsilon) \leq \frac{T}{(2\pi)^{N/2} n^{N(\alpha-(1/2))-1}}(\int_{-r}^{r} e^{-y^2/(2n^{2\alpha-1})}dy)^N \to 0.$

Thus $p(A^\varepsilon) = 0$. The proof is complete. ∎

The trajectory of a standard BM $\{w_t(\omega)\}_{t\geq 0}$ is nowhere differentiable on $t \geq 0$, $P - a.s.$ This means that $\exists\Lambda \in \mathfrak{F}$ such that $P(\Lambda) = 0$ and as $\omega_0 \notin \Lambda$, the function $w_t(\omega_0)$ cannot be differentiated at $\forall t \geq 0$. This means that we cannot simply define the stochastic integral $\int_0^T f(t, \omega)dw_t(\omega)$ for each $\omega \in \Lambda$ in terms of the usual integral. That is why Ito had to invent a new way to define this completely different integral which now is known as Ito's integral.[50],[51]

2.2 Spaces $\mathcal{L}^0$ and $\mathcal{L}^2$.

To discuss Ito's integral we first need to consider its integrand processes.

Definition 72 *Let*

$$\mathcal{L}^2 = \left\{\begin{array}{c} \{f(t,\omega)\}_{t\geq 0} : \text{ it is } \mathfrak{F}_t - \text{adapted, real such that } \forall T > 0 \\ \|f\|_{2,T}^2 = E\int_0^T f^2(t,\omega)dt < \infty. \end{array}\right\},$$

$$\mathcal{L}^0 = \left\{\begin{array}{c} \{f(t,\omega)\}_{t\geq 0} : \text{ it is } \mathfrak{F}_t - \text{adapted, real and} \\ \exists{:}0 = t_0 < t_1 < \cdots < t_n < \cdots \to \infty, \text{ and} \\ \exists\varphi_i(\omega) \in \mathfrak{F}_{t_i}, \sup_i \|\varphi_i\|_\infty < \infty \text{ such that} \\ f(t,\omega) = \varphi_0(\omega)I_{\{t=0\}}(t) + \sum_{i=0}^\infty \varphi_i(\omega)I_{(t_i,t_{i+1}]}(t). \end{array}\right\},$$

$$\mathcal{L}_T^2 = \left\{\{f(t,\omega)\}_{t\in[0,T]} : \{f(t,\omega)\}_{t\geq 0} \in \mathcal{L}^2\right\},$$

$$\mathcal{L}_T^0 = \left\{\{f(t,\omega)\}_{t\in[0,T]} : \{f(t,\omega)\}_{t\geq 0} \in \mathcal{L}^0\right\}$$

Here $\|\varphi_i\|_\infty = ess\sup|\varphi_i(t,\omega)|$. Now let us discuss the relationship between $\mathcal{L}^2$ and $\mathcal{L}^0$.

Lemma 73 *For* $f = \{f(t,\omega)\}_{t\geq 0} \in \mathcal{L}^2$ *let*

$\|f\|_2 = \sum_{n=0}^{\infty} \frac{1}{2^n}(\|f\|_{2,n} \wedge 1)$

Then

1) $\|\cdot\|_2$ *is a metric, and* $\mathcal{L}^2$ *is complete under this metric, if we make the idenfication* $f = f', \forall f, f' \in \mathcal{L}^2$, *as* $\|f - f'\|_{2,n} = 0, \forall n$.

2) $\mathcal{L}^0$ *is dense in* $\mathcal{L}^2$ *with respect to the metric* $\|\cdot\|_2$.

Proof. 1) is obvious. We only need to show 2). Suppose that $f = \{f_t\}_{t\geq 0}$ is a bounded left-continuous $\mathfrak{F}_t-$adapted process. Let

$f_n(0,\omega) = f(0,\omega)$, $f_n(t,\omega) = f(\frac{k}{2^n},\omega)$, as $t \in (\frac{k}{2^n}, \frac{k+1}{2^n}]$, $k = 0, 1, \cdots$.

Then $f_n \in \mathcal{L}_0$ and $\|f_n - f\|_2 \to 0$, as $n \to \infty$, by the left-continuity of f and the Lebesgue's dominated convergence theorem. Now collect all $f \in \mathcal{L}^2$ to form a family $\mathcal{H}$ such that each $f \in \mathcal{H}$ can be approximated by some $\{f_n\}_{n=1}^{\infty} \subset \mathcal{L}^0$ under the metric $\|\cdot\|_2$. Then $\mathcal{H}$ contains all bounded left-continuous $\mathfrak{F}_t-$ process, and obviously it is closed for the non-negative increasing limit under the metric $\|\cdot\|_2$. So, by Theorem 392, $\mathcal{H}$ contains all bounded $\mathfrak{F}_t-$ predictable process. However, for each bounded $f \in \mathcal{L}_2$ one can let $f_n(t,\omega) = n\int_{t-\frac{1}{n}}^{t} f(s,\omega)ds$ to form a bounded sequence $\{f_n\}_{n=1}^{\infty}$ such that each f_n is a bounded $\mathfrak{F}_t-$ continuous process, and by real function theory for each $\omega \in \Omega$, $f_n(t,\omega) \to f(t,\omega)$ for a.e. t. Hence one sees that for each k, as $n \to \infty$, $\|f_n - f\|_{2,k} = E\int_0^k |f_n(t,\omega) - f(t,\omega)|^2 \to 0$, by Lebesgue's dominated convergence theorem. Hence $\mathcal{H}$ contains all bounded $f \in \mathcal{L}_2$.Finally, for any $f \in \mathcal{L}_2$ let $f_n(t,\omega) = f(t,\omega)I_{|f|\leq n}$. Then $|f_n| \leq n$, and for each k, as $n \to \infty$, $\|f_n - f\|_{2,k} \to 0$. So we can show that $\mathcal{H} \supset \mathcal{L}^2$.
■

2.3 Ito's Integrals on $\mathcal{L}^2$

First, we will define the Ito integral for $\mathcal{L}^0$. Suppose that a $\mathfrak{F}_t-$Brownian motion $\{w_t\}_{t\geq 0}$ (Wiener process) is given on $(\Omega, \mathfrak{F}, P)$.

Definition 74 *For every* $f = \{f(t,\omega)\}_{t\geq 0} \in \mathcal{L}^0$:

$f(t,\omega) = \varphi_0(\omega)I_{\{t=0\}}(t) + \sum_{i=0}^{\infty}\varphi_i(\omega)I_{(t_i,t_{i+1}]}(t),$

define for $t_n \leq t < t_{n+1}, n = 0, 1, 2, \cdots$,

$I(f)(t,\omega) = \int_0^t f(s,\omega)dw(s,\omega) = \sum_{i=0}^{n}\varphi_i(\omega)(w(t_{i+1},\omega) - w(t_i,\omega)).$

Firstly, it is easily seen that the stochastic integral also has an expression, which is actually a finite sum for each $0 < t < \infty$,

$I(f)(t) = \sum_{i=0}^{\infty}\varphi_i(w(t_{i+1}\wedge t) - w(t_i \wedge t)),$

moreover, $I(f)(t)$ is continuous in t.

Secondly, it has the following property.

Proposition 75 *1)* $I(f)(0) = 0$, *a.s. and for any* $\alpha, \beta \in R$; $f, g \in \mathcal{L}^0$

$I(\alpha f + \beta g) = \alpha I(f) + \beta I(g).$

2) $\forall f \in \mathcal{L}^0$, $I(f) = \{I(f)(t)\} \in \mathcal{M}^{2,c}$, *and for each* $t > s \geq 0$,
$E[(I(f)(t) - I(f)(s))^2|\mathfrak{F}_s] = E[\int_s^t f^2(u,\omega)du|\mathfrak{F}_s]$.

Proof. 1) is obvious. 2): Since BM $w(t)$ is a square integrable martingale, $\forall s \leq t$,
$E[\varphi_i(w(t_{i+1} \wedge t) - w(t_i \wedge t))|\mathfrak{F}_s] = \varphi_i(w(t_{i+1} \wedge s) - w(t_i \wedge s))$.
So $E[I(f)(t)|\mathfrak{F}_s] = I(f)(s)$. This means that $\{I(f)(t)\}_{t \geq 0}$ is a $\mathfrak{F}_t$ - martingale. Note that
$E[\varphi_i^2(w_{t_{i+1}} - w_{t_i})^2 \,|\mathfrak{F}_s]$
$= E[\varphi_i^2 E[(w_{t_{i+1}} - w_{t_i})^2 \,|\mathfrak{F}_{t_i}] \,|\mathfrak{F}_s] = E[\varphi_i^2(t_{i+1} - t_i) \,|\mathfrak{F}_s]$,
and as $i < j$,
$E[\varphi_i(w_{t_{i+1}} - w_{t_i}) \cdot \varphi_j(w_{t_{j+1}} - w_{t_j}) \,|\mathfrak{F}_s]$
$= E[\varphi_i(w_{t_{i+1}} - w_{t_i}) E[\varphi_j E((w_{t_{j+1}} - w_{t_j}) \,|\mathfrak{F}_{t_j})) \,|\mathfrak{F}_s] = 0$.
Now assume that $0 \leq s < t$, if $\exists t_n < t_{n+1}$ such that $t_n \leq s < t < t_{n+1}$, then obviously,
$E[(I(f)(t) - I(f)(s))^2|\mathfrak{F}_s] = E[\varphi_n^2(t-s)|\mathfrak{F}_s] = E[\int_s^t f^2(u,\omega)du|\mathfrak{F}_s]$.
Suppose that $\exists t_n < t_{n+1} \leq t_m < t_{m+1}$ such that $t_n \leq s < t_{n+1}$ and $t_m \leq t < t_{m+1}$. Then
$E[(I(f)(t) - I(f)(s))^2|\mathfrak{F}_s] =$
$E([\sum_{i=n+1}^{m-1} \varphi_i(w_{t_{i+1}} - w_{t_i}) + \varphi_n(w_{t_{n+1}} - w_s) + \varphi_m(w_t - w_{t_m})]^2 \,|\mathfrak{F}_s)$
$= E[(\sum_{i=n+1}^{m-1} \varphi_i^2(t_{i+1} - t_i) + \varphi_n^2(t_{n+1} - s) + \varphi_m^2(t - t_m) \,|\mathfrak{F}_s]$
$= E[\int_s^t f^2(u,\omega)du \,|\mathfrak{F}_s]$. ■
From this one sees that, in particular, $\forall f \in \mathcal{L}^0$,

$$E[I(f)(t)^2] = E[\left|\int_0^t f(u,\omega)dw_u\right|^2] = E\int_0^t f^2(u,\omega)du. \tag{2.3}$$

Hence $I(f)(t)$ is uniquely determined by f and is independent of the particular choice of $\{t_i\}$. Now suppose that $f \in \mathcal{L}^2$. Since $\{w_t\}_{t \geq 0}$ is nowhere differentiable, we cannot define $\int_0^t f(u,\omega)dw_u(\omega)$ pathwise for each or a.s. $\omega \in \Omega$. However, by (2.3) one sees that for $f \in \mathcal{L}^0$, and for each given $T > 0$, $\{f_t\}_{t \in [0,T]}$ has a norm $\|f\|_{2,T}^2 = E\int_0^t f^2(u,\omega)du = E[I(f)(T)^2]$. So $\{f(t)\}_{t \in [0,T]} \in \mathcal{L}_T^0$ is in 1 to 1 correspondence with $\{I(f)(t)^2\}_{t \in [0,T]} \in \mathcal{M}_T^{2,c}$, and both with the same norm. This motivates us to define $\{I(f)(t)^2\}_{t \in [0,T]} \in \mathcal{M}_T^{2,c}$ for $f \in \mathcal{L}^2$ through the limit of $\{I(f_n)(t)^2\}_{t \in [0,T]} \in \mathcal{M}_T^{2,c}$, where $\{f_n\}_{t \in [0,T]} \in \mathcal{L}_T^0$, if $\|f_n - f\|_{2,T}^2 \to 0$. That is exactly what Ito's integral defines. Let us make it more precise. For any $f \in \mathcal{L}^2$ since $\mathcal{L}^0$ is dense in $\mathcal{L}^2$ with metric $\|\cdot\|_2$ (Lemma 73), $\exists f_n \in \mathcal{L}^0$ such that $\|f_n - f\|_2 \to 0$, as $n \to \infty$. So $\|I(f_n) - I(f_m)\|_2 = \|f_n - f_m\|_2 \to 0$, as $n, m \to \infty$. This means that $\{I(f_n)\}_{n=1}^{\infty} \subset \mathcal{M}^{2,c}$ is a Cauchy sequence. However, $\mathcal{M}^{2,c}$ is complete under the metric $\|\cdot\|_2$ (Lemma 63). Therefore there exists a unique limit, denoted by $I(f)$, belonging to $\mathcal{M}^{2,c}$. Let us show that $I(f)$ is uniquely determined by f and is independent of the particular

choice of $\{f_n\}_{n=1}^{\infty} \subset \mathcal{L}^0$. In fact, let there be two $\{f_n\}_{n=1}^{\infty}, \left\{\widetilde{f}_n\right\}_{n=1}^{\infty} \subset \mathcal{L}^0$ such that both $\|f_n - f\|_2, \left\|\widetilde{f}_n - f\right\|_2 \to 0$, as $n \to \infty$. Construct a new sequence $\{g_n\}_{n=1}^{\infty} \subset \mathcal{L}^0$ such that $g_{2n} = f_n, g_{2n+1} = \widetilde{f}_n$. Then one still has $\|I(g_n) - I(g_m)\|_2 = \|g_n - g_m\|_2 \to 0$, $\|g_n - f\|_2 \to 0$ as $n, m \to \infty$. So the limits should satisfy the conditions that
$\lim_{n\to\infty} I(f_n) = \lim_{n\to\infty} I(\widetilde{f}_n) = \lim_{n\to\infty} I(g_n) = I(f)$.

Definition 76 *$I(f) \in \mathcal{M}^{2,c}$ defined above is called the stochastic integral or the Ito integral of $f \in \mathcal{L}^2$ with respect to a BM $\{w(t)\}_{t\geq 0}$, and it is denoted by $I(f)(t) = \int_0^t f(s)dw(s) = \int_0^t f(s,\omega)dw(s,\omega)$.*

Beware of the fact that the integral is not defined pathwise. So, actually, $I(f)(t)(\omega_0) = (\int_0^t f(s)dw(s))(\omega_0) = (\int_0^t f(s,\omega)dw(s,\omega))(\omega_0)$, $P-a.s.\omega_0$.

Proposition 77 *(i) All conclusions in Proposition 75 still hold for $\forall f \in \mathcal{L}^2$.*

(ii) More generally, if $\tau \geq \sigma$ are both stopping times, then $\forall t > 0$, $\forall f \in \mathcal{L}^2$,
$E[(I(f)(t\wedge\tau) - I(f)(t\wedge\sigma))|\mathfrak{F}_\sigma] = 0, \quad a.s.$
$E[(I(f)(t\wedge\tau) - I(f)(t\wedge\sigma))^2|\mathfrak{F}_\sigma] = E[\int_{t\wedge\sigma}^{t\wedge\tau} f^2(u,\omega)du|\mathfrak{F}_\sigma], \quad a.s.$
(iii) Furthermore, $\forall f, g \in \mathcal{L}^2; \forall t > s \geq 0$; for all stopping times $\tau \geq \sigma$,
$E[(I(f)(t) - I(f)(s))(I(g)(t) - I(g)(s))|\mathfrak{F}_s]$
$= E[\int_s^t (f\cdot g)(u,\omega)du|\mathfrak{F}_s], \quad a.s.$
$E[(I(f)(t\wedge\tau) - I(f)(t\wedge\sigma))(I(g)(t\wedge\tau) - I(g)(t\wedge\sigma))|\mathfrak{F}_\sigma]$
$= E[\int_{t\wedge\sigma}^{t\wedge\tau} (f\cdot g)(u,\omega)du|\mathfrak{F}_\sigma], \quad a.s.$
(iv) For any stopping time $\sigma, \forall f \in \mathcal{L}^2$,
$I(f)(t\wedge\sigma) = I(f')(t), \quad \forall t \geq 0,$
where $f'(t,\omega) = I_{t\leq\sigma(\omega)} f(t,\omega)$.

Proof. (i) is true for $\forall f \in \mathcal{L}^0$, and so is true for $\forall f \in \mathcal{L}^2$ through the limits. (ii): Since $\{I(f)(t)\}_{t\geq 0}$ is a $\mathfrak{F}_t-$martingale, and so by Doob's stopping time theorem, the first conclusion of (ii) holds true. On the other hand, by (i) for each $t > s \geq 0, \forall f \in \mathcal{L}^2$,

$$\begin{aligned} E[I(f)(t)^2 - I(f)(s)^2|\mathfrak{F}_s] &= E[(I(f)(t) - I(f)(s))^2|\mathfrak{F}_s] \\ &= E[\int_s^t f^2(u,\omega)du|\mathfrak{F}_s]. \end{aligned} \tag{2.4}$$

Thus $\left\{I(f)(t)^2 - \int_0^t f^2(u,\omega)du\right\}_{t\geq 0}$ is also a $\mathfrak{F}_t-$martingale, and by Doob's stopping time theorem, (2.4) still holds true when t and s are substituted by the stopping times $t\wedge\tau$ and $t\wedge\sigma$, respectively. So (ii) is proved. (iii): The first conclusion is true for $f, g \in \mathcal{L}^0$, and so is for $f, g \in \mathcal{L}^2$ through the limits. The second conclusion follows from the first one through Doob's stopping time theorem, as in the proof of (ii). Finally, let us establish (iv).

We still show it first for $f \in \mathcal{L}^0$. This can be done by evaluation using the standard discretization of the approximation to the stopping time σ, followed by taking the limit. In fact, suppose that

$f(t,\omega) = \varphi_0(\omega)I_{\{t=0\}}(t) + \sum_{i=0}^\infty \varphi_i(\omega)I_{(t_i,t_{i+1}]}(t)$.

Introduce $\{s_i^n\}_{i=0}^\infty$, which is the refinement of subdivisions $\{t_i\}_0^\infty$ and $\{\frac{i}{2^n}\}_{i=0}^\infty$. Now on this partition we re-express f as

$f(t,\omega) = \varphi_0(\omega)I_{\{t=0\}}(t) + \sum_{i=0}^\infty \varphi_i^n(\omega)I_{(s_i^n,s_{i+1}^n]}(t)$,

where $\varphi_i^n(\omega) = \varphi_j(\omega)$, as $t_j < s_i^n \leq t_{j+1}$, and make a standard discretization of the approximation to the stopping time σ as follows: Let

$\sigma^n(\omega) = s_{i+1}^n$, if $\sigma(\omega) \in (s_i^n, s_{i+1}^n]$.

Then as in the proof of Theorem 35, for each n σ^n is a discrete $\mathfrak{F}_t-$stopping time only valued on $\{s_i^n\}_{i=0}^\infty$, and $\sigma^n \downarrow \sigma$, as $n \uparrow \infty$. So, if we let $f_n'(s,\omega) = I_{s\leq\sigma^n(\omega)}f(s,\omega)$, then $f_n' \in \mathcal{L}^0$. In fact, $\exists \{s_i^n\}_{i=0}^\infty$ such that

$f_n'(s,\omega) = \varphi_0(\omega)I_{\{s=0\}}(s) + \sum_{i=0}^\infty \varphi_i^{n\prime}(\omega)I_{(s_i^n,s_{i+1}^n]}(s)$,

where $\varphi_i^{n\prime}(\omega) = \varphi_i^n(\omega)$, if $s \in (s_i^n, s_{i+1}^n], s \leq \sigma^n(\omega)$ ($\iff$ $s \in (s_i^n, s_{i+1}^n]$, $\sigma^n(\omega) \geq s_{i+1}^n$); $\varphi_i^{n\prime}(\omega) = 0$, if $s \in (s_i^n, s_{i+1}^n], s > \sigma^n(\omega)$ ($\iff$ $s \in (s_i^n, s_{i+1}^n]$, $\sigma^n(\omega) \leq s_i^n$). Obviously, $\varphi_i^{n\prime}(\omega) \in \mathfrak{F}_{s_i^n}$, since $\{\sigma^n(\omega) \leq s_i^n\} \in \mathfrak{F}_{s_i^n}$, and $\varphi_i^n(\omega) \in \mathfrak{F}_{s_i^n}$. Now by evaluation we are going to show that

1) $I(f_n')(t) = I(f)(t \wedge \sigma^n)$,

2) $\|I(f_n') - I(f')\|_{2,t} \to 0$, as $n \to \infty$.

If these results can be established, then (iv) is proved for $f \in \mathcal{L}^0$. However, as $n \to \infty$,

$\|I(f_n') - I(f)\|_{2,t}^2 = \|f_n' - f'\|_{2,t}^2 = E\int_0^t f^2(s,\omega)I_{(\sigma(\omega),\sigma^n(\omega)]}(s)ds \to 0$.

Thus, 2) is proved. Notice that if $s \in (s_i^n, s_{i+1}^n]$, then $I_{s\leq\sigma^n(\omega)} = I_{s_i^n\leq\sigma(\omega)}$.

$$\begin{aligned} I(f_n')(t) &= \sum_{i=0}^\infty \varphi_i^{n\prime}(\omega)(w_{t\wedge s_{i+1}^n} - w_{t\wedge s_i^n}) \\ &= \sum_{i=0}^\infty \varphi_i^n(\omega)I_{s\leq\sigma^n(\omega)}(w_{t\wedge s_{i+1}^n} - w_{t\wedge s_i^n}) \\ &= \sum_{i=0}^\infty \varphi_i^n(\omega)I_{s_i^n\leq\sigma(\omega)}(w_{t\wedge s_{i+1}^n} - w_{t\wedge s_i^n}) \\ &= \sum_{i=0}^\infty \varphi_i^n(\omega)I_{s_i^n\leq\sigma(\omega)}(w_{t\wedge\sigma^n\wedge s_{i+1}^n} - w_{t\wedge\sigma^n\wedge s_i^n}) \\ &= \sum_{i=0}^\infty \varphi_i^n(\omega)(w_{t\wedge\sigma^n\wedge s_{i+1}^n} - w_{t\wedge\sigma^n\wedge s_i^n}) = \int_0^{t\wedge\sigma^n} f(s,\omega)dw_s. \end{aligned}$$

Thus 1) is also proved. Hence (iv) is true for $\forall f \in \mathcal{L}^0$, and so is also true for $\forall f \in \mathcal{L}^2$ by taking limits. ■

2.4 Ito's Integrals on $\mathcal{L}^{2,loc}$

First, let us introduce the concept of local martingales.

Definition 78 *1) An $\mathfrak{F}_t-$adapted real random process $\{x_t\}_{t\geq0}$ is called a local $\mathfrak{F}_t-$martingale and denoted by $\{x_t\}_{t\geq0} \in \mathcal{M}^{loc}$, if $\exists\sigma_n \uparrow \infty, \sigma_n < \infty$ is a $\mathfrak{F}_t-$stopping time for each n, such that $\{x_{t\wedge\sigma_n}\}_{t\geq0}$ is a $\mathfrak{F}_t-$martingale for each n. 2) In addition, if $\{x_{t\wedge\sigma_n}\}_{t\geq0}$ is a square integrable $\mathfrak{F}_t-$martingale for each n, then $\{x_t\}_{t\geq0}$ is called a locally square integrable $\mathfrak{F}_t-$martingale, and it is denoted by $\{x_t\}_{t\geq0} \in \mathcal{M}^{2,loc}$. 3) Write*

$$\mathcal{M}^{2,loc,c} = \left\{ \{x_t\}_{t\geq 0} \in \mathcal{M}^{2,loc} : \{x_t\}_{t\geq 0} \text{ is continuous in } t \text{ with } x_0 = 0 \right\}.$$

Now consider the definition of Ito's integral on $\mathcal{L}^{2,loc}$. For each $f \in \mathcal{L}^{2,loc}$, that is $\int_0^t |f(s,\omega)|^2 ds < \infty$, a.s., let $\sigma_n = \inf\left\{ t \geq 0 : \int_0^t |f(s,\omega)|^2 ds > n \right\} \wedge n$. Then $\sigma_n \uparrow \infty$, $\sigma_n < \infty$ is a stopping time for each n. Obviously, $\{f(t,\omega) I_{t\leq\sigma_n}\}_{t\geq 0} \in \mathcal{L}^2$ for each n, since

$E\int_0^T |f(t,\omega)I_{t\leq\sigma_n}|^2 dt = E\int_0^{T\wedge\sigma_n} f^2(t,\omega)dt \leq n < \infty, \forall T > 0.$

Define in a natural way

$I(f)(t\wedge\sigma_n) = \int_0^{t\wedge\sigma_n} f(s,\omega)dw_s = \int_0^t f(s,\omega)I_{s\leq\sigma_n}dw_s$, $\forall n = 1, 2, \cdots$.

This stochastic integral is well defined, since for $m > n$, by (iv) of Proposition 77

$\int_0^{t\wedge\sigma_n} f(s,\omega)I_{s\leq\sigma_m}dw_s = \int_0^t f(s,\omega)I_{s\leq\sigma_m}I_{s\leq\sigma_n}dw_s = \int_0^t f(s,\omega)I_{s\leq\sigma_n}dw_s.$

Moreover, by definition $\{I(f)(t)\}_{t\geq 0} \in \mathcal{M}^{2,loc,c}$.

Definition 79 *$\forall f \in \mathcal{L}^{2,loc}$, define $\{I(f)(t)\}_{t\geq 0} \in \mathcal{M}^{2,loc,c}$ as above, then it is called the stochastic intergral or the Ito integral of f with respect to the BM $\{w(t)\}_{t\geq 0}$, and it is always denoted by $I(f)(t) = \int_0^t f(s,\omega)dw_s(\omega) = \int_0^t f(s)dw_s$.*

All of these integrals are called stochastic intergrals.

Finally, let us consider an $r-$dimensional $\mathfrak{F}_t-$ BM

$\{w(t)\}_{t\geq 0} = \{(w^1(t), \cdots, w^r(t)\}_{t\geq 0}$.

Suppose that $f_i \in \mathcal{L}^{2,loc}, i = 1, \cdots, r$. Then the stochastic integral

$\left\{ \int_0^t f_i(s,\omega)dw_s^i \right\}_{t\geq 0} \in \mathcal{M}^{2,loc,c}$

is defined for each $i = 1, \cdots, r$. We have the following proposition.

Proposition 80 *There exist $\sigma_n \uparrow \infty$, $\sigma_n < \infty$ is a $\mathfrak{F}_t-$ stopping time such that for each n and $\forall t > s \geq 0$*

$E[\int_{s\wedge\sigma_n}^{t\wedge\sigma_n} f_i(u,\omega)dw_u^i \int_{s\wedge\sigma_n}^{t\wedge\sigma_n} f_j(u,\omega)dw_u^j|\mathfrak{F}_s] = \delta_{ij}E[\int_{s\wedge\sigma_n}^{t\wedge\sigma_n}(f_i f_j)(u,\omega)ds|\mathfrak{F}_s].$

Proof. In fact, since $\left\{ \int_0^t f_i(s,\omega)dw_s^i \right\}_{t\geq 0}$ is a locally square integrable continuous martingale for each $i = 1, \cdots, r$, so $\exists\sigma_n \uparrow \infty$, $\sigma_n < \infty$ is a $\mathfrak{F}_t-$ stopping time for each n such that all $\left\{ \int_0^{t\wedge\sigma_n} f_i(s,\omega)dw_s^i \right\}_{t\geq 0} \in \mathcal{M}^{2,c}, \forall 1 \leq i \leq r$. So to establish the last equality is equivalent to showing that $\forall f_i \in \mathcal{L}^2, i = 1, \cdots, r$,

$E[\int_s^t f_i(u,\omega)dw_u^i \int_s^t f_j(u,\omega)dw_u^j|\mathfrak{F}_s] = \delta_{ij}E[\int_s^t (f_i f_j)(u,\omega)ds|\mathfrak{F}_s].$

Now the proof can be completed first for $f_i \in \mathcal{L}^0, \forall 1 \leq i \leq r$, and then for $f_i \in \mathcal{L}^2, \forall 1 \leq i \leq r$ by taking limits. ■

2.5 Stochastic Integrals with respect to Martingales

In this section we are going to discuss the stochastic integral $\int_0^t f(s,\omega)dM_s$, where $\{M_t\}_{t\geq 0} \in \mathcal{M}^2$. Recall that the procedure for defining Ito's integral $\int_0^t f(s,\omega)dw_s$ is as follows: First we define it for $f \in \mathcal{L}^0$ which is a simple process, and find that $\int_0^t f(s,\omega)dw_s \in \mathcal{M}^{2,c}$. Then, after establishing the 1 to 1 correspondence between space $\mathcal{L}^2$ and $\mathcal{M}^{2,c}$ with the same metric, for each $f \in \mathcal{L}^2$, we can take a sequence of $\{f_n\}_{n=1}^\infty \subset \mathcal{L}^0$ which tends to f in $\mathcal{L}^2$. So the corresponding sequence of integrals $\left\{\int_0^t f_n(s,\omega)dw_s\right\}_{n=1}^\infty \in \mathcal{M}^{2,c}$ will also tend to a limit in $\mathcal{M}^{2,c}$, which we denote by $\int_0^t f(s,\omega)dw_s$, and define it to be the stochastic integral for f.

Note that for a BM $\{w_t\}_{t\geq 0}$ we have that

$w_t^2 =$ a martingale $+\ t, \ \forall t \geq 0,$

and we establish a one to one correspondence as follows: for each $T > 0$,

$\mathcal{L}_T^2 \ni f \ \leftrightarrow \left\{\int_0^t f(s,\omega)dw_s\right\}_{t\in[0,T]} \in \mathcal{M}_T^{2,c},$

with the same norm $\|f\|_{2,T} = [\int_0^T f^2(s,\omega)ds]^{1/2}$. Now for a $\{M_t\}_{t\geq 0} \in \mathcal{M}^2$ we want to do the same thing. So first we need a D-M decomposition for its square. For simplicity we discuss the 1$-$dimensional processes.

Proposition 81 *1) If $\{M_t\}_{t\geq 0} \in \mathcal{M}^2$, then $\{M_t^2\}_{t\geq 0}$ has a unique D-M decomposition as follows:*

$M_t^2 =$ *a martingale* $+ \langle M \rangle_t,$

where $\langle M \rangle_t$ is a natural (predictable) integrable incresing process, and it is called the (predictable) characteristic process for M_t.

2) If $\{M_t\}_{t\geq 0}, \{N_t\}_{t\geq 0} \in \mathcal{M}^2$, then $\{M_tN_t\}_{t\geq 0}$ has a unique decomposition (it may be still called the D-M decomposition) as follows:

$M_tN_t =$ *a martingale* $+ \langle M, N \rangle_t,$

where $\langle M, N \rangle_t$ is a natural (predictable) integrable finite variational process, i.e. it is the difference of two natural (predictable) integrable incresing processes, and it is called the cross (predictable) characteristic process (or (predictable) quadratic variational $\mathfrak{F}_t-$adapted process) for M_t.and N_t.

3) If $\{M_t\}_{t\geq 0}, \{N_t\}_{t\geq 0} \in \mathcal{M}^{2,loc}$, then (i) $\exists \sigma_n \uparrow \infty, \sigma_n < \infty$ is a stopping time for each n such that $\{M_{t\wedge\sigma_n}\}_{t\geq 0}, \{N_{t\wedge\sigma_n}\}_{t\geq 0} \in \mathcal{M}^2$ for each n; (ii) there exist a unique predictable process $\{\langle M, N \rangle_t\}_{t\geq 0}$ such that

$\langle M, N \rangle_{t\wedge\sigma_n} = \langle M^{\sigma_n}, N^{\sigma_n} \rangle_t, \ \forall n$ *and* $\forall t > 0,$

where we write $M_t^{\sigma_n} = M_{t\wedge\sigma_n}$, and $N_t^{\sigma_n} = N_{t\wedge\sigma_n}$.

Proof. 1): By Jensen's inequality $\{M_t^2\}_{t\geq 0}$ is a submartingale. Since it is also non-negative, it is of class (DL). So by the D-M decomposition theorem we arrive at 1).

2): Note that in this case $\{M_1(t)\}_{t\geq 0}, \{M_2(t)\}_{t\geq 0} \in \mathcal{M}^2$, where $M_1(t) = \frac{M_t+N_t}{2}, M_2(t) = \frac{M_t-N_t}{2}$. Hence by the D-M decomposition theorem

$M_1(t)^2 =$ a martingale $+ A_1(t)$, $M_2(t)^2 =$ a martingale $+ A_2(t)$,

where $A_1(t)$ and $A_2(t)$ both are natural (predictable) integrable incresing processes. So

$M_tN_t = M_1(t)^2 - M_2(t)^2 =$ a martingale $+ A_1(t) - A_2(t)$.

Let us show the uniqueness of this decomposition, In fact, if there are two D-M decompositions for M_tN_t :

$M_tN_t = \widetilde{m}_t + \widetilde{A}_t, \quad M_tN_t = \widetilde{m}'_t + \widetilde{A}'_t,$

where $\widetilde{m}_t$ and $\widetilde{m}'_t$ are martingales, and $\widetilde{A}_t = A_{1t} - A_{2t}$, $\widetilde{A}'_t = A'_{1t} - A'_{2t}$ such that all $A_{1t}, A_{2t}, A'_{1t}, A'_{2t}$ are natural (predictable) integrable incresing processes, then

$\widetilde{m}_t - \widetilde{m}'_t + A_{1t} + A'_{2t} = A'_{1t} + A_{2t}.$

However, by the D-M decomposition theorem we must have

$\widetilde{m}_t - \widetilde{m}'_t = 0, \ A_{1t} + A'_{2t} = A'_{1t} + A_{2t},$

since $A'_{1t}+A_{2t}$ is also a submartingale of class (DL). So $\widetilde{m}_t = \widetilde{m}'_t, \widetilde{A}_t = \widetilde{A}'_t$.

3): (i) is true by the definition of $\mathcal{M}^{2,loc}$. (ii) We only need to show that the equality in (ii) is well definied. In fact, if $m > n$, then

$\langle M^{\sigma_n}, N^{\sigma_n}\rangle_t = \langle M^{\sigma_m}, N^{\sigma_m}\rangle_{t\wedge\sigma_n}.$

So it is true. ∎

For the continuity of $\langle M, N\rangle_t$ we have the following proposition.

Proposition 82 *Any one of the following conditions makes $\langle M, N\rangle_t$ continuous in t :*

(i) $\{\mathfrak{F}_t\}_{t\geq 0}$ is continuious in time, i.e. if $\sigma_n \uparrow \sigma$ and they are all stopping times, then $\mathfrak{F}_\sigma = \vee_n \mathfrak{F}_{\sigma_n}$;

(ii) $M, N \in \mathcal{M}^{2,c}$.

Proof. (i): In this case by Levi's theorem $M_{t\wedge\sigma_n} = E[M_t|\mathfrak{F}_{t\wedge\sigma_n}] \longrightarrow E[M_t|\mathfrak{F}_{t\wedge\sigma}] = M_{t\wedge\sigma}$, as $\sigma_n \uparrow \sigma$. However, since $0 \leq M^2_{t\wedge\sigma_n} \leq E[M_t^2|\mathfrak{F}_{t\wedge\sigma_n}]$, by Corollary 27 $\{M^2_{t\wedge\sigma_n}\}_{n=1}^\infty$ is uniformly integrable. Hence $\lim_{n\to\infty} EM^2_{t\wedge\sigma_n} = EM^2_{t\wedge\sigma}$, i.e. $\{M_t^2\}_{t\geq 0}$ is regular. Similarly, $\{N_t^2\}_{t\geq 0}$ is regular. Moreover, by the same token, one also has that $M_{t\wedge\sigma_n} + N_{t\wedge\sigma_n} \to M_{t\wedge\sigma} + N_{t\wedge\sigma}$, as $\sigma_n \uparrow \sigma$. Hence $\{(M_t + N_t)^2\}_{t\geq 0}$ is also regular. However, one easily sees that

$\langle M + N\rangle_t = \langle M\rangle_t + \langle N\rangle_t + 2\langle M, N\rangle_t.$

Thus the continuity of $\langle M, N\rangle_t$ follows from the continuity of the other three, because they are all regular. (ii) is similarly proved. ∎

For stochastic integrals with respect to the martingale $\{M_t\}_{t\geq 0} \in \mathcal{M}^2$, we need to introduce the space of integrand processes as in the case with respect to BM $\{w_t\}_{t\geq 0} \in \mathcal{M}^{2,c}$.

Definition 83 *1) Write*

$$\mathcal{L}^2_M = \{\{f(t,\omega\}_{t\geq 0} : \textit{it is } \mathfrak{F}_t\textit{-predictable such that } \forall T > 0 \quad (\|f\|^M_{2,T})^2 = E\int_0^T f^2(t,\omega)d\langle M\rangle_t < \infty.\}$$

For $f = \{f(t,\omega)\}_{t\geq 0} \in \mathcal{L}_M^2$ *set*
$\|f\|_2^M = \sum_{n=1}^{\infty} \frac{1}{2^n}(\|f\|_{2,n}^M \wedge 1).$
2) $\mathcal{L}_M^{2,loc} = \{\{f(t,\omega)\}_{t\geq 0}$: *it is* $\mathfrak{F}_t-$*predictable such that if* $\exists \sigma_n \uparrow \infty, \sigma_n$ *is a* $\mathfrak{F}_t-$*stopping time for each* n, *and*
$E\int_0^{T\wedge\sigma_n} f^2(t,\omega)d\langle M\rangle_t < \infty,\ \forall T > 0, \forall n\}.$
3) $\mathcal{L}^0$ *is defined the same as in Definition 72.*

Note that if $E\int_0^{T\wedge\sigma_n} f^2(t,\omega)d\langle M\rangle_t < \infty,\ \forall T > 0, \forall n$, then $P-a.s.$ $\int_0^{N\wedge\sigma_n} f^2(t,\omega)d\langle M\rangle_t < \infty, \forall n, \forall N = 1,2,\cdots$. Therefore, $\int_0^T f^2(t,\omega)d\langle M\rangle_t < \infty, \forall T > 0$, $P-a.s.$ In general, the inverse is not necessary true. However, if $\langle M\rangle_t$ is continuous in t, then the inverse is also true.

Reasoning almost completely in the same way as in Lemma 73, one arrives at the following lemma.

Lemma 84 $\mathcal{L}^0$ *is dense in* $\mathcal{L}_M^2$ *with respect to the metric* $\|\cdot\|_2^M$.

Now we can define the stochastic integral $\int_0^t f(s,\omega)dM_s$ with respect to $\{M_t\}_{t\geq 0}$, first for $f \in \mathcal{L}^0$, and then for $f \in \mathcal{L}_M^2$, and finally for $f \in \mathcal{L}_M^{2,loc}$ in completely the same way as when defining $\int_0^t f(s,\omega)dw_s$. However, we would like to define it in another way, even if it is more abstract and different, because it is then faster and easier to show all of its properties.

Definition 85 *For* $M \in \mathcal{M}^{2,loc}$ *and* $f \in \mathcal{L}_M^{2,loc}$ *(or* $M \in \mathcal{M}^2$ *and* $f \in \mathcal{L}_M^2$*) if* $X = \{x_t\}_{t\geq 0} \in \mathcal{M}^{2,loc}$ $(X = \{x_t\}_{t\geq 0} \in \mathcal{M}^2)$ *satisfies that*

$$\langle X, N\rangle(t) = \int_0^t f(u)d\langle M, N\rangle(u), \tag{2.5}$$

$\forall N \in \mathcal{M}^{2,loc}$ $(N \in \mathcal{M}^2), \forall t \geq 0$,
then set $x_t = I^M(f)(t)$, *and call it the stochastic integral of* f *with respect to martingale* M.

In the rest of this section we always assume that $M \in \mathcal{M}^{2,loc}$. First let us show the uniqueness of $X \in \mathcal{M}^{2,loc}$ in Definition 85. In fact, if there is another $X' \in \mathcal{M}^{2,loc}$ such that (2.5) holds, then $\langle X - X', N\rangle = 0, \forall N \in \mathcal{M}^{2,loc}$. Hence by taking $N = X - X'$ one finds that $X = X'$.

Secondly, we need to show that such a definition is equivalent to the usual one, which was explained before this definition.

Proposition 86 *If* $f \in \mathcal{L}_M^2$ *is a stochastic step function, i.e.* $f \in \mathcal{L}_M^2$, *and* $\exists \sigma_n \uparrow, \sigma_0 = 0$, σ_n *is a* $\mathfrak{F}_t-$*stopping time for each* n *such that*
$f(t,\omega) = f_0(\omega)I_{t=0} + \sum_{n=0}^{\infty} f_n(\omega)I_{(\sigma_n,\sigma_{n+1}]}(t)$,
where $f_n \in \mathfrak{F}_{\sigma_n}$, *then*

$$I^M(f)(t) = \sum_{n=0}^{\infty} f_n(\omega)(M_{t\wedge\sigma_{n+1}} - M_{t\wedge\sigma_n}) = \int_0^t f(u)dM(u). \tag{2.6}$$

Proof. In fact, $\forall N \in \mathcal{M}^{2,loc}$,
$\langle I^M(f), N\rangle_t = \langle \sum_{n=0}^{\infty} f_n(\omega)(M^{\sigma_{n+1}} - M^{\sigma_n}), N\rangle_t$
$= \sum_{n=0}^{\infty} f_n(\omega)(\langle M^{\sigma_{n+1}}, N\rangle_t - \langle M^{\sigma_n}, N\rangle_t)$
$= \sum_{n=0}^{\infty} f_n(\omega)(\langle M^{\sigma_{n+1}}, N^{\sigma_{n+1}}\rangle_t - \langle M^{\sigma_n}, N^{\sigma_n}\rangle_t) = \int_0^t f(s,\omega)d\langle M, N\rangle_s.$
■

By this lemma if $f \in \mathcal{L}_M^0$, $f(t) = f_0(\omega)I_{t=0} + \sum_{n=0}^{\infty} f_n(\omega)I_{(t_n,t_{n+1}]}(t)$, where $0 = t_0 < t_1 < \cdots < t_n \to \infty$, and $f_n(\omega) \in \mathfrak{F}_{t_n}$, and all f_n are bounded, then
$I^M(f)(t) = \sum_{n=0}^{\infty} f_n(\omega)(M_{t_{n+1}} - M_{t_n}).$
That is just the usual way to define the stochastic integral $\int_0^t f(s,\omega)dM_s \in \mathcal{M}^2$, for $f \in \mathcal{L}_M^0$. Following this usual way for $f \in \mathcal{L}_M^2$ one can take a sequence $\{f^n\} \in \mathcal{L}_M^0$ such that $\|f^n - f\|_2^M \to 0$. So there exists a limit $\int_0^\cdot f(s,\omega)dM_s \in \mathcal{M}^2$ such that $\left\|\int_0^\cdot f(s,\omega)dM_s - \int_0^\cdot f^n(s,\omega)dM_s\right\|_2^M \to 0$. Therefore the stochastic integral $\int_0^t f(s,\omega)dM_s \in \mathcal{M}^2$ is also defined for $f \in \mathcal{L}_M^2$. Let us show that it is equal to the stochastic integral $I^M(f)(t)$ defined in Definition 85. To show this we need to use the Kunita-Watanabe's inequality (see the lemma below). In fact, by this inequality $\forall N \in \mathcal{M}^2$
$\left|\langle I^M(f^n) - I^M(f), N\rangle_t\right| \leq \int_0^t |(f^n - f)(s,\omega)|\, d\,|\langle M, N\rangle|_s$
$\leq (\int_0^t |(f^n - f)(s,\omega)|^2\, d\langle M\rangle_s)^{1/2} \cdot (\langle N\rangle_t)^{1/2}.$
Hence $E\left|\langle I^M(f^n)(t) - I^M(f)(t), N\rangle_t\right| \leq (\|1\|_2^N)^2(\|f^n - f\|_2^M)^2 \to 0$. On the other hand, one also has that
$\langle \int_0^\cdot f(s,\omega)dM_s - \int_0^\cdot f^n(s,\omega)dM_s, N\rangle_t = \langle \int_0^\cdot (f(s,\omega) - f^n(s,\omega))dM_s, N\rangle_t$
$= \int_0^t (f(s,\omega) - f^n(s,\omega))d\langle M, N\rangle_s.$
So in the same way we can show that the left hand side of the above result tends to zero as $n \to \infty$. Since by Proposition 86 $\langle I^M(f^n), N\rangle_t = \langle \int_0^\cdot f^n(s,\omega)dM_s, N\rangle_t$. We have $\langle I^M(f), N\rangle_t = \langle \int_0^\cdot f(s,\omega)dM_s, N\rangle_t, \forall N \in \mathcal{M}^2$. Therefore $I^M(f)(t) = \int_0^t f(s,\omega)dM_s, \forall f \in \mathcal{L}_M^2$. Furthermore, we can also easily obtain the same equality for all $f \in \mathcal{L}_M^{2,loc}$.

Lemma 87 *(Kunita-Watanabe's inequality). If $M, N \in \mathcal{M}^2$, $f \in \mathcal{L}_M^2$, $g \in \mathcal{L}_N^2$, then*
$$\int_0^t |f \cdot g|\,(u)d\,|\langle M, N\rangle|_u \leq \left|\int_0^t f^2(u)d\langle M\rangle_u\right|^{1/2} \left|\int_0^t g^2(u)d\langle N\rangle_u\right|^{1/2},$$
where $|\langle M\rangle|_u$ is the total variation of $M_t, t \in [0,u]$. Note that we always write $d\,|\langle M, N\rangle|_u = |d\langle M, N\rangle|_u$.

Proof. First, we see that $\forall \lambda \in R$,
$0 \leq \langle M\lambda - N\rangle_t - \langle M\lambda - N\rangle_s$
$= (\langle M\rangle_t - \langle M\rangle_s)\lambda^2 + (\langle N\rangle_t - \langle N\rangle_s) - 2(\langle M, N\rangle_t - \langle M, N\rangle_s)\lambda.$
So $|\langle M, N\rangle_t - \langle M, N\rangle_s| \leq \sqrt{\langle M\rangle_t - \langle M\rangle_s} \cdot \sqrt{\langle N\rangle_t - \langle N\rangle_s}$. This intuitively tells us that $d\,|\langle M, N\rangle_u| \leq \sqrt{d\langle M\rangle_u} \cdot \sqrt{d\langle N\rangle_u}$. However, to make it rigorous we need to introduce a measure as follows: Let $\mu(t) = \int_0^t d\,|\langle M, N\rangle|_u +$

$\langle M\rangle_t + \langle N\rangle_t$. For any given rational number λ let $\nu(t) = \lambda^2 \langle M\rangle_t + 2\lambda \langle M,N\rangle_t + \langle N\rangle_t$, then $\forall t \geq s$,

$\nu(t) - \nu(s) = \lambda^2(\langle M\rangle_t - \langle M\rangle_s) + 2\lambda(\langle M,N\rangle_t - \langle M,N\rangle_s) + (\langle N\rangle_t - \langle N\rangle_s)$.

Notice that $|\langle M,N\rangle_t - \langle M,N\rangle_s| \leq \sqrt{\langle M\rangle_t - \langle M\rangle_s}\sqrt{\langle N\rangle_t - \langle N\rangle_s}$, we have $\nu(t) - \nu(s) \geq 0$, i.e. $\nu(t)$ is increasing in t. Hence

$\frac{d\nu}{d\mu} = \lambda^2 \frac{d\langle M\rangle}{d\mu} + 2\lambda \frac{d\langle M,N\rangle}{d\mu} + \frac{d\langle N\rangle}{d\mu} \geq 0, \ d\mu - a.e.$

The above inequality holds for all $\lambda \in Q$, since Q is countable. Hence one has that

$\left|\frac{d\langle M,N\rangle}{d\mu}\right| \leq \sqrt{\frac{d\langle M\rangle}{d\mu}}\sqrt{\frac{d\langle N\rangle}{d\mu}}, \ d\mu - a.e.$

Notice that now the above inequality is rigorous, so we can use it to obtain the final result. Actually, by Schwarz inequality

$\int_0^t |f(u)g(u)|\,|d\langle M,N\rangle|_u = \int_0^t |f(u)g(u)| \left|\frac{d\langle M,N\rangle}{d\mu}\right| d\mu$

$\leq \int_0^t |f(u)g(u)| \sqrt{\frac{d\langle M\rangle}{d\mu}}\sqrt{\frac{d\langle N\rangle}{d\mu}} d\mu \leq \sqrt{\int_0^t |f(u)|^2 d\langle M\rangle_u}\sqrt{\int_0^t |g(u)|^2 \ d\langle N\rangle_u}.$

The proof is complete. ■

Now, for the convenience to the reader, we write out the properties of stochastic integral with respect to martingales $M \in \mathcal{M}^2$ as follows:

Theorem 88 *For $M, N \in \mathcal{M}^2$, $f, g \in \mathcal{L}^2_M$, $h \in \mathcal{L}^2_N$ the stochastic integral $I^M(f)$ has the following properties: $\forall \tau \geq \sigma$ a.s., τ, σ are stopping times, $\forall t > 0$*

(i) $I^M(0) = 0$, *a.s.*

(ii) $E[I^M(f)(t\wedge\tau) - I^M(f)(t\wedge\sigma)|\mathfrak{F}_\sigma] = 0$, *a.s.*

(iii) $E[(I^M(f)(t\wedge\tau) - I^M(f)(t\wedge\sigma))(I^M(g)(t\wedge\tau) - I^M(g)(t\wedge\sigma))|\mathfrak{F}_\sigma]$
$= E[\int_{t\wedge\sigma}^{t\wedge\tau}(f\cdot g)(u)d\langle M\rangle_u |\mathfrak{F}_\sigma]$, *a.s.*

(iv) $I^M(f)(t\wedge\sigma) = I^M(f')(t)$,
where $f'(t) = I_{t\leq\sigma}f(t)$;

(v) $E[(I^M(f)(t) - I^M(f)(s))(I^N(h)(t) - I^N(h)(s))|\mathfrak{F}_s]$
$= E[\int_s^t (f\cdot h)(u)d\langle M,N\rangle_u |\mathfrak{F}_s]$, *a.s.*
so, $\langle I^M(f), I^N(h)\rangle_t = \int_0^t (f\cdot h)(u)d\langle M,N\rangle_u$;

(vi) $I^M(f+g)(t) = I^M(f)(t) + I^M(g)(t)$;

(vii) $\forall\phi \in \mathcal{L}^2_M \cap \mathcal{L}^2_N$, *then* $\phi \in \mathcal{L}^2_{M+N}$, *and* $I^{M+N}(\phi)(t) = I^M(\phi)(t) + I^N(\phi)(t)$;

(viii) set $N = I^M(f)$, $\forall h \in \mathcal{L}^2_N$, *then* $h\cdot f \in \mathcal{L}^2_M$, *and* $I^M(h\cdot f)(t) = I^N(h)(t)$.

Proof. (i) is obvious. Since $I^M(f) \in \mathcal{M}^2$, so (ii) is true by Doob's stopping time theorem. (iii) and (v) can be checked first for $f, g, h \in \mathcal{L}^0$, and then for $f, g \in \mathcal{L}^2_M$, $h \in \mathcal{L}^2_N$ by taking a limit. (vi), (vii) and (viii) are easily proved by Definition 85. Finally, let us show (iv) by checking that definition 85 is satisfied. In fact, writing $X^\sigma(t) = X(t\wedge\sigma)$, by Doob's stopping time theorem $\forall N \in \mathcal{M}^2$,

$\langle I^M(f)^\sigma, N\rangle_t = \langle I^M(f)^\sigma, N^\sigma\rangle_t = \langle I^M(f), N\rangle_{t\wedge\sigma} = \int_0^{t\wedge\sigma} f(u)d\langle M,N\rangle_u$
$= \int_0^t f(u)I_{u\leq\sigma}d\langle M,N\rangle_u$. ■

For the properties of stochastic integral with respect to martingales $M \in \mathcal{M}^2$ we have the following result:

Theorem 89 *For $M, N \in \mathcal{M}^{2,loc}$, $f, g \in \mathcal{L}^2_M$, $h \in \mathcal{L}^2_N$ the stochastic integral $I^M(f)$ has the following properties:*

(1) Properties (i), (iv), (vi), (vii) and (viii) in Theorem 88 still hold true.

(2) There exists a sequence of stopping times $\sigma_n \uparrow \infty$ such that (ii), (iii) and (v) hold true for t and s substituted by $t \wedge \sigma_n$ and $s \wedge \sigma_n$, respectively.

2.6 Ito's Formula for Continuous Semi-Martingales

In calculus if $f(x), x(t) \in C^1$ and both are non-random, then
$df(x(t)) = f'(x(t))dx(t)$.
However, this formula for a random process $x(t,\omega)$ is not necessary true. For example, one can show that

$$|w(t)|^2 = 2\int_0^t w(s)dw(s) + t, \tag{2.7}$$

or symbolically, we can denote it by
$d|w(t)|^2 = 2w(t)dw(t) + dt$,
where $w(t)$ is the 1-dimensional standard BM. Actually, this is a special case of the famous Ito formula. So one can understand how important and useful Ito' formula is in stochastic analysis and stochastic calculus. Let us see how (2.7) holds true. In fact, let
$0 = t_0 < t_1 < \cdots < t_{n-1} < t_n = t$.
Then by Taylor's formula
$|w(t)|^2 = \sum_{i=1}^n (|w(t_i)|^2 - |w(t_{i-1})|^2)$
$= \sum_{i=1}^n 2w(t_{i-1}) \triangle w(t_i) + \sum_{i=1}^n |\triangle w(t_i)|^2 = I_1^n(t) + I_2^n(t)$,
where $\triangle w(t_i) = w(t_i) - w(t_{i-1})$. However, by the property of the stochastic integral
$E\left|2\int_0^t w(s)dw(s) - I_1^n(t)\right|^2 = E\left|2\sum_{i=1}^n \int_{t_{i-1}}^{t_i} (w(s) - w(t_{i-1}))dw(s)\right|^2$
$= 2E\sum_{i=1}^n \int_{t_{i-1}}^{t_i} |w(s) - w(t_{i-1})|^2 ds$
$= 2E\int_0^t \left|w(s) - \sum_{i=1}^n w(t_{i-1})I_{s\in(t_{i-1},t_i]}\right|^2 ds \to 0$,
as $|\triangle^n| = \max_{1\le i\le n} \triangle t_i = \max_{1\le i\le n}(t_i - t_{i-1}) \to 0$ by Lebesgue's dominated convergence theorem, where $|w(s)| \le \sup_{s\in[0,t]} |w(s)|, \forall s \in [0,t]$, and by the martingale inequality $E\sup_{s\in[0,t]} |w(s)|^2 \le 4E|w(t)|^2 = 4t < \infty$. On the other hand, as $|\triangle^n| \to 0$,

$$E\left|t - I_2^n(t)\right|^2 = E\left|\sum_{i=1}^n(\triangle t_i - |\triangle w(t_i)|^2)\right|^2 = E[\sum_{i=1}^n(\triangle t_i - |\triangle w(t_i)|^2)^2]$$
$$= \sum_{i=1}^n(-(\triangle t_i)^2 + E\left|\triangle w(t_i)\right|^4) = 2\sum_{i=1}^n(\triangle t_i)^2 \leq 2\left|\triangle^n\right| t \to 0, \qquad (2.8)$$

where for the second, third, and fourth equality sign above we have used the result that for the cross term as $i < j$,
$E[(\triangle t_i - |\triangle w(t_i)|^2)(\triangle t_j - |\triangle w(t_j)|^2)]$
$= E[(\triangle t_i - |\triangle w(t_i)|^2)E((\triangle t_j - |\triangle w(t_j)|^2)|\mathfrak{F}_{t_i})] = 0,$
and $\triangle w(t_i) \sim N(0, \triangle t_i)$, so $E\left|\triangle w(t_i)\right|^2 = \triangle t_i$, and $E\left|\triangle w(t_i)\right|^4 = 3\left(\triangle t_i\right)^2$.
So (2.7) holds.

By the above proof we see that approximately, by Taylor's formula
$\triangle(|w(t)|^2) \approx 2w(t) \triangle w(t) + (\triangle w(t))^2.$
However,
$E[(\triangle w(t))^2 - \triangle t] = 0.$
So letting $\triangle t \to 0$ one can guess that in the $L^2(\Omega)-$limit it is possible to obtain the result that
$d\left|w(t)\right|^2 = 2w(t)dw(t) + dt.$
However, this result would not happen in the determinate case. Because if $x(t)$ is non-random, then $o(\triangle x(t)) = o(x'(t) \triangle t + o(\triangle t)) = o(\triangle t)$ is a higher degree of infinitesimal relative to $\triangle t$. So by the definition of the first degree of a differential one has that
$\triangle(\left|x(t)\right|^2) = 2x(t) \triangle x(t) + o(\triangle x(t)) = 2x(t)dx(t) + o(\triangle t)$
$= d(\left|x(t)\right|^2) + o(\triangle t).$
Therefore,
$d(\left|x(t)\right|^2) = 2x(t)dx(t).$

The idea of showing (2.7) now motivates us to show the general Ito formula for continuous locally integrable martingales. Suppose now that $M = \{M_t\}_{t \geq 0} \in \mathcal{M}^{2,c}$. Naturally, comparing to (2.7) we want to show that

$$\left|M(t)\right|^2 = 2\int_0^t M(s)dM(s) + \langle M \rangle(t). \qquad (2.9)$$

Proceeding as above, only substituting $w(s)$ and ds by $M(s)$ and $d\langle M \rangle(s)$, respectively, one sees that similarly,
$\left|M(t)\right|^2 = \sum_{i=1}^n(\left|M(t_i)\right|^2 - \left|M(t_{i-1})\right|^2)$
$= \sum_{i=1}^n 2M(t_{i-1}) \triangle M(t_i) + \sum_{i=1}^n \left|\triangle M(t_i)\right|^2 = I_1^n(t) + I_2^n(t),$
where $\triangle M(t_i) = M(t_i) - M(t_{i-1})$. However, to use the Lebesgue's dominated convergence theorem we need to introduce the stopping time: for each N let

$$\tau_N = \inf\left\{t \geq 0 : |M_t| + \langle M \rangle_t > N\right\}. \qquad (2.10)$$

Then as $|\triangle^n| \to 0$,

$E\left|2\int_0^{t\wedge\tau_N} M(s)dM(s) - I_1^n(t\wedge\tau_N)\right|^2$

$= 2E\int_0^{t\wedge\tau_N} \left|M(s) - \sum_{i=1}^n M(t_{i-1})I_{s\in(t_{i-1},t_i]}\right|^2 d\langle M\rangle(s) \to 0.$

However, to show $E\left|\langle M\rangle(t\wedge\tau_N) - I_2^n(t\wedge\tau_N)\right|^2 \to 0$, as $|\triangle^n| \to 0$, requires more discussion. One cannot simply copy the proof of (2.8), because now the property of Normal random variables cannot be applied here. Notice that

$E\left|\langle M\rangle(t) - I_2^n(t)\right|^2 = E[\sum_{i=1}^n(|\triangle M(t_i)|^2 - \triangle\langle M\rangle(t_i))]^2$

$= E[\sum_{i=1}^n(|\triangle M(t_i)|^2 - \triangle\langle M\rangle(t_i))^2)]$

$\leq 2E[\sum_{i=1}^n(|\triangle M(t_i)|^4 + (\triangle\langle M\rangle(t_i))^2]$

$\leq 2E\max_{1\leq i\leq n}|\triangle M(t_i)|^2\sum_{i=1}^n|\triangle M(t_i)|^2$

$+2E\max_{1\leq i\leq n}\triangle\langle M\rangle(t_i)M(t) = 2I_{21}^n(t) + 2I_{22}^n(t).$

Since $\langle M\rangle(s)$ is increasing in s, so $\triangle\langle M\rangle(t_i) \leq \langle M\rangle(t), \forall i = 1,\cdots,n$. Hence applying Lebesgue's dominated convergence theorem, one easily sees that as $|\triangle^n| \to 0$, $I_{22}^n(t\wedge\tau_N) \to 0$. For the discussion to $I_{21}^n(t\wedge\tau_N)$ we introduce the following lemma.

Lemma 90 *Assume that $M \in \mathcal{M}^{2,loc,c}$ such that $|M(s)| \leq C, \forall s \in [0,t]$, where C is a constant. Let*

$V_i^\triangle(t) = \sum_{k=1}^i|\triangle M(t_k)|^2 = \sum_{k=1}^i[M(t_k) - M(t_{k-1})]^2$, $i = 1,\cdots,n.$

Then $E[(V_n^\triangle(t))^2] \leq 6C^4$.

Proof. Note that

$(V_n^\triangle(t))^2 = \sum_{k=1}^n|\triangle M(t_k)|^4 + 2\sum_{k=1}^n(V_n^\triangle - V_k^\triangle)|\triangle M(t_k)|^2 = I^1 + 2I^2,$

and

$E[(V_n^\triangle(t) - V_k^\triangle(t))|\mathfrak{F}_{t_k}] \leq E[V_n^\triangle(t)|\mathfrak{F}_{t_k}] = E[\sum_{i=1}^n|\triangle M(t_i)|^2|\mathfrak{F}_{t_k}]$

$= E[\sum_{i=1}^n(|M(t_i)|^2 - |M(t_{i-1})|^2)|\mathfrak{F}_{t_k}]$

$= E[(|M(t)|^2)|\mathfrak{F}_{t_k}] \leq C^2.$

Hence

$EI^2 \leq C^2E(V_n^\triangle(t)) \leq C^4.$

Moreover,

$EI^1 \leq (2C)^2E(V_n^\triangle) \leq 4C^4.$

Therefore, $E[(V_n^\triangle(t))^2] \leq 6C^4$. ∎

Now let us return to the discussion of $I_{21}^n(t\wedge\tau_N)$. Applying Lemma 90 we have that as $|\triangle^n| \to 0$,

$I_{21}^n(t\wedge\tau_N) \leq \sqrt{E\max_{1\leq i\leq n}|\triangle M(t_i\wedge\tau_N)|^4}\sqrt{E(V_n^\triangle(t\wedge\tau_N))^2}$

$\leq \sqrt{6}N^2\sqrt{E\max_{1\leq i\leq n}|\triangle M(t_i\wedge\tau_N)|^4} \to 0.$

Therefore, as $|\triangle^n| \to 0$,

$$\begin{aligned} & E\left|\langle M\rangle(t\wedge\tau_N) - I_2^n(t\wedge\tau_N)\right|^2 \\ = \ & E[\sum_{i=1}^n(|\triangle M(t_i\wedge\tau_N)|^2 - \triangle\langle M\rangle(t_i\wedge\tau_N))]^2 \to 0. \quad (2.11) \end{aligned}$$

Thus we have proved that for each N

$|M(t \wedge \tau_N)|^2 = 2\int_0^{t\wedge\tau_N} M(s)dM(s) + \langle M\rangle (t \wedge \tau_N).$

Letting $N \uparrow \infty$, one obtains (2.9). Furthermore, by taking a sequence of stopping times $\sigma_n \uparrow \infty$, one immediately obtains that for $M \in \mathcal{M}^{2,loc,c}$, (2.9) still holds.

Keeping the proof and result of (2.9) in mind, one easily deduces the following proposition.

Proposition 91 *(Ito's formula). If $f(x) \in C^2(R)$, and $M \in \mathcal{M}^{2,loc,c}$, then*

$$f(M_t) - f(M_0) = \int_0^t f'(M_s)dM_s + \frac{1}{2}\int_0^t f''(M_s)d\langle M\rangle_s. \tag{2.12}$$

Proof. Note that if we put $f(x) = |x|^2$, then Ito's formula (2.12) becomes (2.9). So we can proceed almost completely as in the proof above. We only show the result for $M \in \mathcal{M}^{2,c}$. In fact, by Taylor's formula,

$f(M(t)) - f(M_0) = \sum_{i=1}^n (f(M(t_i)) - f(M(t_{i-1})))$

$= \sum_{i=1}^n f'(M(t_{i-1})) \triangle M(t_i)$

$+\frac{1}{2}\sum_{i=1}^n f''(\xi_{i-1}) |\triangle M(t_i)|^2 = I_1^n(t) + \frac{1}{2}I_2^n(t),$

where $\xi_{i-1} = M(t_{i-1}) + \theta_{i-1} \triangle M(t_i), 0 \leq \theta_{i-1} \leq 1$. Hence as $|\triangle^n| \to 0$,

$E\left|\int_0^{t\wedge\tau_N} f'(M_s)dM_s - I_1^n(t \wedge \tau_N)\right|^2$

$= E\int_0^{t\wedge\tau_N} \left|f'(M_s) - \sum_{i=1}^n f'(M(t_{i-1}))I_{s\in(t_{i-1},t_i]}\right|^2 d\langle M\rangle (s) \to 0,$

where τ_N is defined by (2.10). Furthermore,

$E\left|\int_0^t f''(M_s)d\langle M\rangle_s - I_2^n(t)\right|^2$

$\leq 3[E\sum_{i=1}^n \left|f''(\xi_{i-1}) - f''(M_{t_{i-1}})\right| |\triangle M(t_i)|^2$

$+E[\sum_{i=1}^n f''(M_{t_{i-1}})(|\triangle M(t_i)|^2 - \triangle \langle M\rangle (t_i))]^2$

$+E\left|\int_0^t (f''(M_s) - \sum_{i=1}^n f''(M_{i-1})I_{s\in(t_{i-1},t_i]})d\langle M\rangle_s\right|^2] = 3\sum_{i=1}^3 I_{2i}^n(t).$

Obviously, by Lebesgue's dominated convergence theorem as $|\triangle^n| \to 0$,

$I_{23}^n(t \wedge \tau_N) \to 0.$

Moreover, as the above as $|\triangle^n| \to 0$,

$I_{21}^n(t \wedge \tau_N)$

$\leq \sqrt{E\sup_{1\leq i\leq n} \left|f''(\xi_{i-1} \wedge \tau_N) - f''(M_{t_{i-1}} \wedge \tau_N)\right|}\sqrt{EV_n^{\triangle}(t \wedge \tau_N)}$

$\leq \sqrt{6}N^2\sqrt{E\sup_{1\leq i\leq n} \left|f''(\xi_{i-1} \wedge \tau_N) - f''(M_{t_{i-1}} \wedge \tau_N)\right|} \to 0.$

Furthermore, by (2.11)

$I_{22}^n(t \wedge \tau_N)$

$\leq \sup_{|x|\leq N} |f(x)| E[\sum_{i=1}^n (|\triangle M(t_i \wedge \tau_N)|^2 - \triangle \langle M\rangle (t_i \wedge \tau_N))^2)] \to 0.$

So (2.12) is true for $\forall t \wedge \tau_N$. Letting $N \uparrow \infty$, the proof is complete. ■

Now let us consider Ito's formula for the continuous semi-martingale. Suppose that

$x_t = x_0 + A_t + M_t,$

where $x_0 \in \mathfrak{F}_0$, $\{A_t\}_{t\geq 0}$ is a continuous finite variational ($\mathfrak{F}_t$-adapted) process with $A_0 = 0$, $\{M_t\}_{t\geq 0} \in \mathcal{M}^{2,loc,c}$. We will call such an x_t a continuous semi-martingale. The same proof will show the following result.

Theorem 92 *If $f(x) \in C^2(R)$, then*

$$f(x_t) - f(x_0) = \int_0^t f'(x_s)dA_s + \int_0^t f'(x_s)dM_s + \frac{1}{2}\int_0^t f''(x_s)d\langle M\rangle_s . \tag{2.13}$$

Proof. In fact, as above by Taylor's formula,
$f(x(t)) - f(x(0)) = \sum_{i=1}^n (f(x(t_i)) - f(x(t_{i-1})))$
$= \sum_{i=1}^n f'(x(t_{i-1})) \triangle A(t_i) + \sum_{i=1}^n f'(x(t_{i-1})) \triangle M(t_i)$
$+\frac{1}{2}\sum_{i=1}^n f''(\xi_{i-1}) |\triangle M(t_i)|^2 = I_1^n(t) + I_2^n(t) + \frac{1}{2}I_3^n(t)$,
where $\xi_{i-1} = M(t_{i-1}) + \theta_{i-1} \triangle M(t_i), 0 \leq \theta_{i-1} \leq 1$. Let

$$\tau_N = \inf\{t \geq 0 : |x_0| + |A|_t + |M_t| + \langle M\rangle_t > N\}, \tag{2.14}$$

where $|A|_t$ is the total variation of $A_s, s \in [0, t]$. Then one easily sees that by Lebesgue's dominated convergence theorem as $|\triangle^n| \to 0$,
$E\left|\int_0^{t\wedge\tau_N} f'(M_s)dA_s - I_1^n(t\wedge\tau_N)\right|^2$
$= E\int_0^{t\wedge\tau_N} \left|f'(M_s) - \sum_{i=1}^n f'(M(t_{i-1}))I_{s\in(t_{i-1},t_i]}\right|^2 d|A|(s) \to 0.$
The rest of the proof is completely the same. So the proof is complete. ∎

2.7 Ito's Formula for Semi-Martingales with Jumps

In the practical case we will always encounter some stochastic perturbation with jumps. So we need an Ito formula for a semi-martingale with jumps. Consider
$x_t = x_0 + A_t + M_t$
$+\int_0^{t+}\int_Z f(s,z,\omega)N_p(ds,dz) + \int_0^{t+}\int_Z g(s,z,\omega)\widetilde{N}_p(ds,dz)$,
where x_0, $\{A_t\}_{t\geq 0}$ and $\{M_t\}_{t\geq 0}$ are the same as in Theorem 92, and p is a $\mathfrak{F}_t$-point process of the class (QL), $f \in \mathcal{F}_p$, $g \in \mathcal{F}_p^{2,loc}$ such that $f(s,z,\omega)g(s,z,\omega) = 0$. We will call such a x_t a semi-martingale with jumps. Here the last two terms are called jump terms. All jumps of x_t are caused by them, and the condition $f(s,z,\omega)g(s,z,\omega) = 0$ means that the last two terms do not have the same jump times. We have the following general Ito formula:

Theorem 93 *If $F(x) \in C^2(R)$, then*

$$F(x_t) - F(x_0) = \int_0^t F'(x_s)dA_s + \int_0^t F'(x_s)dM_s + \frac{1}{2}\int_0^t F''(x_s)d\langle M\rangle_s$$
$$+ \int_0^{t+}\int_Z [F(x_{s-} + f(s,z,\omega)) - F(x_{s-})]N_p(ds,dz)$$
$$+ \int_0^{t+}\int_Z [F(x_{s-} + g(s,z,\omega)) - F(x_{s-})]\widetilde{N}_p(ds,dz)$$
$$+ \int_0^{t+}\int_Z [F(x_s + g(s,z,\omega)) - F(x_s) - F'(x_s)g(s,z,\omega)]\widehat{N}_p(ds,dz). \tag{2.15}$$

holds. Moreover, (2.15) also can be rewritten as
$F(x_t) - F(x_0) = \int_0^t F'(x_s)dA_s + \int_0^t F'(x_s)dM_s + \frac{1}{2}\int_0^t F''(x_s)d\langle M\rangle_s$
$+\int_0^{t+}\int_Z F'(x_{s-})g(s,z,\omega)\widetilde{N}_p(ds,dz) + \int_0^{t+}\int_Z F'(x_{s-})f(s,z,\omega)N_p(ds,dz)$
$+\int_0^{t+}\int_Z [F(x_{s-} + f(s,z,\omega)) - F(x_{s-}) - F'(x_{s-})f(s,z,\omega)]N_p(ds,dz)$
$+\int_0^{t+}\int_Z [F(x_{s-} + g(s,z,\omega)) - F(x_{s-}) - F'(x_{s-})g(s,z,\omega)]N_p(ds,dz),$
or, more simply,

$$F(x_t) - F(x_0) = \int_0^t F'(x_{s-})dx_s + \frac{1}{2}\int_0^t F''(x_s)d\langle M\rangle_s$$
$$+ \int_0^{t+}\int_Z [F(x_{s-} + f(s,z,\omega)) - F(x_{s-}) - F'(x_{s-})f(s,z,\omega)]N_p(ds,dz)$$
$$+ \int_0^{t+}\int_Z [F(x_{s-} + g(s,z,\omega)) - F(x_{s-}) - F'(x_{s-})g(s,z,\omega)]N_p(ds,dz). \tag{2.16}$$

Proof. To show the result our scheme is as follows: We first rewrite x_t as the sum of two parts: a continuous semi-martingale part and a pure jump part, then apply Ito's formula to the continuous semi-martingale part, and after that add the jump part at all jump times to obtain the general Ito formula.

1) Assume that $F(x) \in C_b^2(R)$, i.e. F has up to second continuous derivatives and all derivatives are bounded. To simplify the notation we may first assume that $g \in \mathcal{F}_p^2$. Let $U_n \uparrow Z, U_n \in \mathfrak{B}_Z$ such that $EN_p((0,t],U_n) < \infty, \forall t, \forall n$. For each n let
$x_t^n = x_0 + M_t + A_t + \int_0^{t+}\int_Z f^n(s,z,\omega)N_p(ds,dz)$
$+\int_0^{t+}\int_Z g^n(s,z,\omega)\widetilde{N}_p(ds,dz) = x_0 + M_t + A_t$
$-\int_0^{t+}\int_Z g^n(s,z,\omega)\widehat{N}_p(ds,dz)$
$+\int_0^{t+}\int_Z (f+g)(s,z,\omega)I_{U_n}(z)N_p(ds,dz),$
where $f^n(s,z,\omega) = f(s,z,\omega)I_{U_n}(z)$, and g^n is similarly defined. For the pure jump term (the last term) the counting measure only counts the numbers of the point $p(s) = z$ falling in U_n. Since

$EN_p((0,t],U_n) = E(\#\{0 < s \leq t : p(s) \in U_n\}) < \infty,$
where $\#$ means the numbers of $\cdot$ counted in the set $\{\cdot\}$, which we introduced in Chapter 1. So $\#\{0 < s \leq t : p(s) \in U_n\} < \infty$, a.s. That is to say, for a.s. ω, there are only finite points of $s \in (0,t]$ such that $p(s,\omega) \in U_n$. So we can denote these points by $0 < \sigma_1(\omega) < \cdots < \sigma_m(\omega) < \cdots$. Let us show that for each m, σ_m is a stopping time. In fact,

$$\{\sigma_m(\omega) \leq t\} = \{N_p((0,t],U_n) \geq m\} \in \mathfrak{F}_t.$$

For convenience we also write $\sigma_0 = 0$. Then

$$x_t^n = x_0 + M_t + A_t$$
$$-\int_0^{t+}\int_Z g^n(s,z,\omega)\widehat{N}_p(ds,dz)$$
$$+\sum_{\sigma_m \leq t}(f+g)(\sigma_m(\omega),p(\sigma_m(\omega),\omega),\omega),$$
$$F(x_t^n) - F(x_0^n) = \sum_m [F(x^n_{\sigma_m \wedge t}) - F(x^n_{\sigma_m \wedge t-})]$$
$$+\sum_m [F(x^n_{\sigma_m \wedge t-}) - F(x^n_{\sigma_{m-1} \wedge t})] = I_1(t) + I_2(t).$$

Note that $F(x^n_{\sigma_m \wedge t-}) = F(x^n_{\sigma_m -})$, as $\sigma_m \leq t$; and $F(x^n_{\sigma_m \wedge t-}) = F(x^n_{t-}) = F(x^n_t)$, as $\sigma_{m-1} < t < \sigma_m$. Since as $s \in (\sigma_{m-1}\wedge t, \sigma_m \wedge t)$, x_s^n has no jumps. So we can apply Ito's formula for the continuous semi-martingale to obtain

$$F(x^n_{\sigma_m \wedge t-}) - F(x^n_{\sigma_{m-1} \wedge t}) = \int_{\sigma_{m-1}\wedge t}^{\sigma_m \wedge t} F'(x_s^n)dA_s + \int_{\sigma_{m-1}\wedge t}^{\sigma_m \wedge t} F'(x_s^n)dM_s$$
$$+\frac{1}{2}\int_{\sigma_{m-1}\wedge t}^{\sigma_m \wedge t} F''(x_s^n)d\langle M\rangle_s - \int_{\sigma_{m-1}\wedge t}^{\sigma_m \wedge t} F'(x_s^n)dA_s^n$$

where $A_t^n = \int_0^t \int_Z g^n(s,z,\omega)\widehat{N}_p(ds,dz)$, and we have used the fact that by the continuity of A_s, $\int_{\sigma_{m-1}\wedge t}^{\sigma_m \wedge t-} F'(x_s^n)dA_s = \int_{\sigma_{m-1}\wedge t}^{\sigma_m \wedge t} F'(x_s^n)dA_s$, etc. Taking the summation we find that

$$I_2(t) = \int_0^t F'(x_s^n)dA_s + \int_0^t F'(x_s^n)dM_s$$
$$+\frac{1}{2}\int_0^t F''(x_s^n)d\langle M\rangle_s - \int_0^t F'(x_s^n)dA_s^n.$$

Now let us evaluate $I_1(t)$. By assumption $fg = 0$, so

$$I_1(t) = \sum_m [F(x^n_{\sigma_m \wedge t}) - F(x^n_{\sigma_m \wedge t-})] I_{\sigma_m \leq t, f(\sigma_m, p(\sigma_m),\omega)\neq 0}$$
$$+\sum_m [F(x^n_{\sigma_m \wedge t}) - F(x^n_{\sigma_m \wedge t-})] I_{\sigma_m \leq t, g(\sigma_m, p(\sigma_m),\omega)\neq 0}$$
$$= \int_0^{t+}\int_Z [F(x^n_{s-} + f^n(s,z,\omega)) - F(x^n_{s-})]N_p(ds,dz)$$
$$+\int_0^{t+}\int_Z [F(x^n_{s-} + g^n(s,z,\omega)) - F(x^n_{s-})]\widetilde{N}_p(ds,dz)$$
$$+\int_0^{t+}\int_Z [F(x^n_s + g^n(s,z,\omega)) - F(x^n_s)]\widehat{N}_p(ds,dz).$$

Hence $F(x_t^n) - F(x_0^n) = I_1(t) + I_2(t) =$ the right hand side of (2.15) with x_{s-}, $f(s,z,\omega)$, and $g(s,z,\omega)$ substituted by x^n_{s-}, $f^n(s,z,\omega)$, and $g^n(s,z,\omega)$, respectively. Now let us discuss the limit. Since as $n \to \infty$,

$$E\left|\int_0^{T+}\int_Z (g^n(s,z,\omega) - g(s,z,\omega))\widetilde{N}_p(ds,dz)\right|^2$$
$$= E\int_0^T \int_Z |g^n(s,z,\omega) - g(s,z,\omega)|^2 \widehat{N}_p(ds,dz) \to 0.$$

Hence, if necessary, by taking a subsequence we may assume that a.s.

$$\int_0^{t+}\int_Z g^n(s,z,\omega)\widetilde{N}_p(ds,dz) \rightrightarrows \int_0^{t+}\int_Z g(s,z,\omega)\widetilde{N}_p(ds,dz), \text{ on any } [0,T],$$

where " $\rightrightarrows$ " means uniformly convergence. Similarly,

$$\int_0^{t+}\int_Z f^n(s,z,\omega)N_p(ds,dz) \rightrightarrows \int_0^{t+}\int_Z f(s,z,\omega)N_p(ds,dz), \text{ on any } [0,T].$$

Hence x^n_{t-} and x_t^n converge uniformly to x_{t-} and x_t on any $[0,T]$, respectively. From this, letting $n \to \infty$, one easily obtains that (2.15) holds true.

For example, let us show that as $n \to \infty$,

$$I_1^n = |\int_0^{t+}\int_Z h^n(s,z,\omega)\widehat{N}_p(ds,dz) - \int_0^{t+}\int_Z h(s,z,\omega)\widehat{N}_p(ds,dz)| \to 0, a.s. \quad (2.17)$$

where $h^n(s,z,\omega) = F(x_s^n + g^n(s,z,\omega)) - F(x_s^n) - F'(x_s^n)g^n(s,z,\omega)$, and $h(s,z,\omega)$ is similarly defined; and

$$I_2^n = E|\int_0^{t+}\int_Z \widetilde{h}^n(s,z,\omega)\widetilde{N}_p(ds,dz) - \int_0^{t+}\int_Z \widetilde{h}(s,z,\omega)\widetilde{N}_p(ds,dz)|^2 \to 0, \quad (2.18)$$

where $\widetilde{h}^n(s,z,\omega) = F(x_{s-}^n + g^n(s,z,\omega)) - F(x_{s-}^n)$, and $\widetilde{h}(s,z,\omega)$ is also similarly defined. In fact, $\forall s,z,$and a.s. ω, $h^n(s,z,\omega) \to h(s,z,\omega)$. Moreover, since $F \in C_b^2$,

$|h^n(s,z,\omega)| \leq \frac{1}{2}\sup_x |F''(x)| \, |g(s,z,\omega)|^2 \leq \widetilde{k}_0 \, |g(s,z,\omega)|^2 \in \mathcal{F}_p^1$.

Hence applying Lebesgue's dominated convergence theorem, one obtains (2.17). Notice that

$I_2^n = E\int_0^{t+}\int_Z \left|\widetilde{h}^n(s,z,\omega) - \widetilde{h}(s,z,\omega)\right|^2 \widehat{N}_p(ds,dz),$

and

$\left|\widetilde{h}^n(s,z,\omega) - \widetilde{h}(s,z,\omega)\right|^2 \leq \sup_x |F'(x)|^2 \, |g(s,z,\omega)|^2$.

So (2.18) is also similarly obtained. Thus (2.15) is true under the assumption that $g \in \mathcal{F}^2$.

However, for $g \in \mathcal{F}^{2,loc}$, $\exists \sigma_n \uparrow \infty$, σ_n is a stopping time for each n such that $I_{[0,\sigma_n]}(t)g(t,z,\omega) \in \mathcal{F}^2$. Hence one easily sees that (2.15) is true for t substituted by $t \wedge \sigma_n, \forall t, \forall n$. Letting $n \uparrow \infty$, one derives (2.15), $\forall t \geq 0$.

2) Now assume that $F(x) \in C^2(R)$. First let us show that each term in the right hand side of (2.16) has meaning. In fact, since $\{x_t\}_{t\geq 0}$ is a right continuous with left limit (RCLL) process, $\{x_{t-}\}_{t\geq 0}$ is a locally bounded process, i.e. $\exists \sigma_n' \uparrow \infty$, σ_n' is a stopping time for each n such that $\{x_{(t\wedge\sigma_n')-}\}_{t\geq 0}$ is bounded, for each n. Indeed, let

$\sigma_n' = \inf\{t \geq 0 : |x_t| > n\}$.

Then $|x_{(t\wedge\sigma_n)-}| \leq n, \forall t \geq 0$. So $\int_0^{t\wedge\sigma_n'} F'(x_s)dM_s = \int_0^{t\wedge\sigma_n'} F'(x_{s-})dM_s$ makes sence, so also does $\int_0^t F'(x_s)dM_s$. On the other hand, since $g \in \mathcal{F}^{2,loc}$,

$\sum_{s\in D_p, s\leq t} |g(s,p(s),\omega)|^2 = \int_0^{t+}\int_Z |g(s,z,\omega)|^2 N_p(ds,dz) < \infty$, a.s.

because by 1)

$E\int_0^{t\wedge\sigma_n}\int_Z |g(s,z,\omega)|^2 N_p(ds,dz)$

$= E\int_0^{t\wedge\sigma_n}\int_Z |g(s,z,\omega)|^2 \widehat{N}_p(ds,dz) < \infty, \forall n.$

Notice that for each $\omega \in \Omega$, $\{x_s(\omega)\}_{s\in[0,t]}$ is RCLL, so it is bounded; that is, $\exists k_0(\omega)$, such that $|x_s(\omega)| \leq k_0(\omega), \forall s \in [0,t]$. Hence $\exists \widetilde{k}_0(\omega)$ is such that $|F'(x)| + |F''(x)| \leq \widetilde{k}_0(\omega)$, as $x \in [-k_0(\omega), k_0(\omega)]$. Thus,

$\sum_{s\in D_p, s\leq t}[F(x_{s-} + g(s,p(s),\omega)) - F(x_{s-}) - F'(x_{s-})g(s,p(s),\omega)]$

$\leq \frac{1}{2}\widetilde{k}_0(\omega) \sum_{s\in D_p, s\leq t} |g(s,p(s),\omega)|^2 < \infty.$
Similarly, by
$\sum_{s\in D_p, s\leq t} |f(s,p(s),\omega)| = \int_0^{t+} \int_Z |f(s,z,\omega)| N_p(ds,dz) < \infty$, a.s.
one has that
$\sum_{s\in D_p, s\leq t}[F(x_{s-} + f(s,p(s),\omega)) - F(x_{s-})]$
$\leq \widetilde{k}_0(\omega) \sum_{s\in D_p, s\leq t} |f(s,p(s),\omega)| < \infty.$
Therefore one easily deduces that the right hand side of (2.16) makes sense, and so also does (2.15). Now let $W_N(x) \in C^\infty(R)$ such that $W_N(x) = 1$, as $|x| \leq N$; $W_N(x) = 0$, as $|x| > N+2$; and $|W_N'(x)| \leq 1, |W_N''(x)| \leq 1, \forall x \in R$, where W_N' and W_N'' are the first and second derivatives, respectively. Set $F_N(x) = F(x)W_N(x)$. Then $|F_N(x)| \leq |F(x)|, |F_N'(x)| \leq |F'(x)| + |F(x)|, |F_N''(x)| \leq |F''(x)| + 2|F'(x)| + |F(x)|$. Moreover, $F_N(x) = F_N'(x) = F_N''(x) = 0$, as $|x| > N+2$. Hence for each N, $F_N(x) \in C_b^2(R)$. Thus by 1) Ito's formula (2.16) holds true for F_N. Letting $N \uparrow \infty$, by Lebesgue's dominated convergence theorem one easily obtains that (2.16) still holds for F. ∎

2.8 Ito's Formula for d-dimensional Semi-Martingales. Integration by Parts.

The above Ito's formula for one-dimensional semi-martingales with jumps is easily generalized to that for the n-dimensional case. Consider a $d-$dimensional semi-martingales with jumps as follows: $x_t = (x_t^1, \cdots, x_t^d)$, where for $i = 1, 2, \cdots, d$
$x_t^i = x_0^i + A_t^i + M_t^i$
$+ \int_0^{t+} \int_Z f^i(s,z,\omega) N_p(ds,dz) + \int_0^{t+} \int_Z g^i(s,z,\omega) \widetilde{N}_p(ds,dz),$
where $x_0 \in \mathfrak{F}_0$, $\{A_t\}_{t\geq 0}$ is a finite variational ($\mathfrak{F}_t-$ adapted) process, and $\{M_t\}_{t\geq 0} \in \mathcal{M}^{2,loc,c}$, all are $d-$ dimensional, and p is a $\mathfrak{F}_t-$point process of the class (QL), $f \in \mathcal{F}_p$, $g \in \mathcal{F}_p^{2,loc}$ such that $f^i(s,z,\omega)g^j(s,z,\omega) = 0, \forall i,j = 1,2,\cdots,d$. Then we have the following theorem, but as the proof is completely the same we omit it.

Theorem 94 *(Ito's formula). If a real function $F(x) \in C^2(R^d)$, then*

$$F(x_t) - F(x_0) = \sum_{i=1}^d \int_0^t F'_{x^i}(x_{s-}) dx_s^i + \frac{1}{2} \sum_{i,j=1}^d \int_0^t F''_{x^i x^j}(x_s) d\left\langle M^i, M^j \right\rangle_s$$

$$+ \int_0^{t+} \int_Z [F(x_{s-} + f(s,z,\omega)) - F(x_{s-}) - \sum_{i=1}^d F'_{x^i}(x_{s-}) f^i(s,z,\omega)] N_p(ds,dz)$$

$$+ \int_0^{t+} \int_Z [F(x_{s-} + g(s,z,\omega)) - F(x_{s-}) - \sum_{i=1}^d F'_{x^i}(x_{s-}) g^i(s,z,\omega)] N_p(ds,dz), \tag{2.19}$$

or,

$F(x_t) - F(x_0) = \sum_{i=1}^d \int_0^t F'_{x^i}(x_{s-})dA_s^i + \sum_{i=1}^d \int_0^t F'_{x^i}(x_{s-})dM_s^i$
$+\frac{1}{2}\sum_{i,j=1}^d \int_0^t F''_{x^ix^j}(x_s)d\left\langle M^i, M^j\right\rangle_s$
$+\int_0^{t+}\int_Z [F(x_{s-} + g(s,z,\omega)) - F(x_{s-})]\widetilde{N}_p(ds,dz)$
$+\int_0^{t+}\int_Z [F(x_{s-} + f(s,z,\omega)) - F(x_{s-})]N_p(ds,dz)$
$+\int_0^{t+}\int_Z [F(x_{s-}+g(s,z,\omega))-F(x_{s-})-\sum_{i=1}^d F'_{x^i}(x_{s-})g^i(s,z,\omega)]\widehat{N}_p(ds,dz).$

Remark 95 *1) If we denote*
$[x^i, x^j]_t = \left\langle x^{ic}, x^{jc}\right\rangle_t + \sum_{s\leq t}(\triangle x^i \triangle x^j) = \left\langle M^i, M^j\right\rangle_t + \sum_{s\leq t}(\triangle x^i \triangle x^j),$
which is called the cross quadractic variational process (cross characteristics) of semi-martingales $\{x_t^i\}_{t\geq 0}$ *and* $\{x_t^i\}_{t\geq 0}$, *then (2.19) can be rewritten as*

$F(x_t) - F(x_0) = \sum_{i=1}^d \int_0^t F'_{x^i}(x_{s-})dx_s^i + \frac{1}{2}\sum_{i,j=1}^d \int_0^t F''_{x^ix^j}(x_{s-})d[x^i, x^j]_s$
$+\int_0^{t+}\int_Z [F(x_{s-} + f(s,z,\omega)) - F(x_{s-}) - \sum_{i=1}^d F'_{x^i}(x_{s-})f^i(s,z,\omega)$
$-\frac{1}{2}\sum_{i,j=1}^d F''_{x^ix^j}(x_{s-})f^i(s,z,\omega)f^j(s,z,\omega)]N_p(ds,dz)$
$+\int_0^{t+}\int_Z [F(x_{s-} + g(s,z,\omega)) - F(x_{s-}) - \sum_{i=1}^d F'_{x^i}(x_{s-})g^i(s,z,\omega)$
$-\frac{1}{2}\sum_{i,j=1}^d F''_{x^ix^j}(x_{s-})g^i(s,z,\omega)g^j(s,z,\omega)]N_p(ds,dz)].$

2) Ito's formula (2.19) can also be written symbolically in differential form as

$dF(x_t) = \sum_{i=1}^d F'_{x^i}(x_{t-})dx_t^i + \frac{1}{2}\sum_{i,j=1}^d F''_{x^ix^j}(x_t)d\left\langle M^i, M^j\right\rangle_t$
$+\int_Z [F(x_{t-} + f(t,z,\omega)) - F(x_{t-}) - \sum_{i=1}^d F'_{x^i}(x_{t-})f^i(t,z,\omega)]N_p(dt,dz)$
$+\int_Z [F(x_{t-} + g(t,z,\omega)) - F(x_{t-}) - \sum_{i=1}^d F'_{x^i}(x_{t-})g^i(t,z,\omega)]N_p(dt,dz),$

or,

$$dF(x_t) = \sum_{i=1}^d F'_{x^i}(x_{t-})dx_t^i + \frac{1}{2}\sum_{i,j=1}^d F''_{x^ix^j}(x_{t-})d[x^i, x^j]_t$$
$$+d\eta_t(F,g) + d\eta_t(F,f) \tag{2.20}$$

where

$d\eta_t(F,g) = \int_Z [F(x_{t-} + g(t,z,\omega)) - F(x_{t-}) - \sum_{i=1}^d F'_{x^i}(x_{t-})g^i(t,z,\omega)$
$-\frac{1}{2}\sum_{i,j=1}^d F''_{x^ix^j}(x_{t-})g^i(t,z,\omega)g^j(t,z,\omega)]N_p(dt,dz),$

and $d\eta_t(F,f)$ *is similarly defined.*

By Ito's formula one easily derives the formula of integration by parts for semi-martingales with jumps. Suppose the semimartingales $\{x_t^i\}_{t\geq 0}$ are given as above, $i = 1, 2, \cdots, d$. Then we have the following thoerm.

Theorem 96 *(Integration by parts).* $dx_t^i x_t^j = x_t^i dx_t^j + x_t^j dx_t^i + d[x^i, x^j]_t$, *or equivalently,*

$x_t^i x_t^j - x_0^i x_0^j = \int_0^t x_s^i dx_s^j + \int_0^t x_s^j dx_s^i + [x^i, x^j]_t.$

Proof. Let $F(x) = x^i x^j$, for $x \in R^d$. Applying (2.20) one arrives at the conclusion of the theorem. ■

2.9 Independence of BM and Poisson Point Processes

As an application of Ito's formula we can prove the independence of BM and Poisson point processes, which is very important in stochastic analysis. For simplicity let us first discuss the independence of a 1-dimensional Brownian motion and a Poisson point process.

Theorem 97 *Assume that* $\{x_t\}_{t\geq 0}$ *is a 1-dimensional* $\mathfrak{F}_t-$*semimartingale, and* p *is a* $\mathfrak{F}_t-$*point process of class (QL). If*

1) $M_t = x_t - x_0 \in \mathcal{M}^{2,loc,c}$, $\langle M \rangle_t = t$;

2) the compensator $\widehat{N}_p(dt,dz)$ *of* p *is a non-random* $\sigma-$ *finite measure on* $[0,\infty) \times Z$;

then $\{x_t\}_{t\geq 0}$ *is a 1-dimensional* $\mathfrak{F}_t-$*BM, and* p *is a* $\mathfrak{F}_t-$*Poisson point process such that they are independent.*

Proof. It is enough to prove that $\forall t > s > 0$

$$
\begin{aligned}
&E[\exp(i\lambda \cdot (x_t - x_s)) \exp(-\sum_{i=1}^m \lambda_i N_p((s,t] \times U_i)|\mathfrak{F}_s] \\
&= \exp[-(t-s)\left|\lambda\right|^2/2] \exp[\sum_{i=1}^m (e^{-\lambda_i} - 1)\widehat{N}_p((s,t],U_i)], \quad (2.21)
\end{aligned}
$$

where $\lambda \in R, \lambda_i > 0, i = 1, \cdots, m$; and $U_i \in \mathfrak{B}_Z$, $i = 1,2,\cdots,m$ such that $\widehat{N}_p([0,t],U_i) < \infty, \forall i$, and $U_i \cap U_y = \phi$, as $i \neq j$.

Let $F(x,y^1,\cdots,y^m) = e^{i\lambda x} e^{-\sum_{i=1}^m \lambda_i y^i}$. Apply Ito's formula to the $1+m-$dimensional semi-martingale $(x_t, y_t^1, \cdots, y_t^m)$, where

$x_t = x_t$,

$y_t^i = N_p((0,t],U_i) = \int_0^{t+} \int_Z I_{U_i}(z) N_p(dt,dz)$, $i = 1,\cdots,m$,

and denote $N_t = (N_p((0,t],U_1),\cdots,N_p((0,t],U_m))$, then

$\triangle F = F(x_t,y_t^1,\cdots,y_t^m) - F(x_s,y_s^1,\cdots,y_s^m)$

$= i\lambda \int_s^t e^{i\lambda x_u} e^{-\sum_{i=1}^m \lambda_i y_u^i} dM_u - \frac{\lambda^2}{2}\int_s^t e^{i\lambda x_u} e^{-\sum_{i=1}^m \lambda_i y_u^i} du$

$+ \int_{s+}^{t+} \int_Z [F(x_u, N_{u-} + I_{U_i}(z)) - F(x_u, N_{u-})] \widetilde{N}_p(du,dz)$

$+ \int_{s+}^{t+} \int_Z [F(x_u, N_{u-} + I_{U_i}(z)) - F(x_u, N_{u-})] \widehat{N}_p(du,dz)$.

However,

$F(x_u, N_{u-} + I_{U_i}(z)) - F(x_u, N_{u-}) = e^{i\lambda x_u} e^{-\sum_{i=1}^m \lambda_i N_p((0,u],U_i)}\cdot$

$\cdot(e^{-\sum_{i=1}^m \lambda_i I_{U_i}(z)} - 1)$.

Therefore, $\forall A \in \mathfrak{F}_s$, multiplying all terms in the expression of $\triangle F$ by $e^{-i\lambda x_u} e^{\sum_{i=1}^m \lambda_i N_p((0,u],U_i)} I_A$ and taking the expectation, one finds that

$Ee^{i\lambda(x_t - x_s)} e^{-\sum_{i=1}^m \lambda_i N_p((s,t],U_i)} V^s(t) - P(A) = -\frac{|\lambda|^2}{2} \int_s^t EI_A V^s(u) du$

$+ \int_s^t EI_A V^s(u) \int_Z (e^{-\sum_{i=1}^m \lambda_i I_{U_i}(z)} - 1) \widehat{N}_p(du,dz)$

$= -\frac{\lambda^2}{2} \int_s^t EI_A V^s(u) du + \int_s^t EI_A V^s(u) \sum_{i=1}^m (e^{-\lambda_i} - 1) \widehat{N}_p(du,U_i)$,

where $V^s(t) = e^{i\lambda(x_t - x_s)} e^{-\sum_{i=1}^m \lambda_i N_p((s,t],U_i)}$. Note that the ordinary differential equation

$Y_t = \alpha + \beta \int_s^t Y_u du + \gamma \int_s^t Y_u dG(u),$
where $G(u)$ is an increasing function, has a unique solution
$Y_t = \alpha e^{\beta(t-s)} e^{\gamma(G(t)-G(s))}.$
So
$Ee^{i\lambda(x_t-x_s)}e^{-\sum_{i=1}^m \lambda_i N_p((s,t],U_i)} = P(A)e^{-\frac{\lambda^2}{2}(t-s)}e^{\sum_{i=1}^m (e^{-\lambda_i}-1)\widehat{N}_p((s,t],U_i)}.$
The proof is complete. ■

The above theorem is easily generalized to the $d-$dimensional case.

Theorem 98 *Assume that $\{x_t\}_{t\geq 0}$ is a d-dimensional $\mathfrak{F}_t-$semimartingale, where $x_t = (x_t^1, \cdots, x_t^d)$, and $p_i, i = 1, 2, \cdots, n$, are $\mathfrak{F}_t-$point processes of class (QL) on state spaces $Z_i, i = 1, 2, \cdots, n$, respectively. If*
1) $M_t^i = x_t^i - x_0^i \in \mathcal{M}^{2,loc,c}$, $\langle M^i, M^j \rangle_t = \delta_{ij} t$; $i, j = 1, 2, \cdots, d$,
2) the compensator $\widehat{N}_{p_i}(dt, dz)$ of p_i is a non-random $\sigma-$ finite measure on $[0, \infty) \times Z$, $i = 1, 2, \cdots, n$; and the domains $D_{p_i(\omega)}, i = 1, 2, \cdots, n$, are mutually disjoint, a.s.
Then $\{x_t\}_{t\geq 0}$ is a d-dimensional $\mathfrak{F}_t-$BM, and p_i $(i = 1, 2, \cdots, n)$ is a $\mathfrak{F}_t-$Poisson point process such that they are mutually independent.

Proof. The only thing we need to do is to combine all point process p_i, $i = 1, 2, \cdots, n$, into one new point process p on a new space Z. Then we can prove (2.21) in exactly the same way as for x_t and p, only with $\lambda \in R$ substituted by $\widetilde{\lambda} = (\widetilde{\lambda}^1, \cdots, \widetilde{\lambda}^d) \in R^d$, and denote the inner product of two $d-$dimensional vectors $\widetilde{\lambda}$ and x by $\widetilde{\lambda} \cdot x$; that is, $\widetilde{\lambda} \cdot x = \sum_{i=1}^d \widetilde{\lambda}^i x^i$. For this we introduce a space Z such that $Z = \cup_{i=1}^n H_i$, where $H_i, i = 1, \cdots, n$, are disjoint such that for each i, H_i is in 1 to 1 correspondence with Z_i. For simplicity let us identify H_i and Z_i. So we can set $Z = \cup_{i=1}^n Z_{.i}$. Note that $\mathfrak{B}_Z = \cup_{i=1}^n \mathfrak{B}_{Z_i}$, and all $\mathfrak{B}_{Z_i}, i = 1, 2, \cdots, n$ are mutually disjoint. Now let $D_p = \cup_{i=1}^n D_{p_i}$, and set $p(t) = p_i(t)$, as $t \in D_{p_i}$. Then we have a point process p on Z, and it is easy to see that p is a point process of the class (QL). Moreover, $\widehat{N}_p(dt, dz) = \sum_{i=1}^n I_{Z_i}(z)\widehat{N}_{p_i}(dt, dz)$. Obviously, $\widehat{N}_p(dt, dz)$ is a non-random $\sigma-$ finite measure on $[0, \infty) \times Z$. Hence by the proof of Theorem 97 $\{x_t\}_{t\geq 0}$ is a d-dimensional $\mathfrak{F}_t-$BM, and p is a $\mathfrak{F}_t-$Poisson point process such that they are independent. Note that for each i, p_i is also a $\mathfrak{F}_t-$Poisson point process by Theorem 97. However, we still need to prove that all p_i, $i = 1, 2, \cdots, n$ are mutualy independent. In fact, if $A_i \in \mathfrak{B}_{Z_i}, i = 1, \cdots, n$, then they are mutually disjoint. Moreover, $\{N_{p_i}((0, t_i] \times A_i)\}_{i=1}^n = \{N_p((0, t_i] \times A_i)\}_{i=1}^n$ is an independent system of random variables, since p is a Poisson point process. The proof is complete.
■

2.10 Some Examples

In Calculus if $x(t) = f(y(t)) = e^{y(t)}$, $y(t) \in C^1$, $y(t)$ is non-random, then

$dx(t) = e^{y(t)}dy(t)$.
However, by Ito's formula we will see that some extra terms will occur in the expression for $dx(t)$, when $y(t,\omega)$ is some random process.

Example 99 *Assume that $\{w_t\}_{t\geq 0}$ is a d_1- dimensional BM, $\left\{\widetilde{N}_t\right\}_{t\geq 0}$ is a stationary $1-$ dimensional centralized Poisson process with the compensator λt; that is, $\widetilde{N}_t = N_t - \lambda t$, where $\lambda > 0$ is a constant, and N_t is a Poisson process such that $EN_t = \lambda t$. Suppose that $b(t,\omega)$ is a $\mathfrak{F}_t-$adapted $R^{1\otimes d_1}$ -valued process and $c(t,\omega)$ is a $\mathfrak{F}_t-$predictable R^1- valued process such that for each $T < \infty$*

$$E\int_0^T [|b(t,\omega)|^2 + |c(t,\omega)|^2]dt < \infty.$$

Then the following assertions hold true:

1) Let $x_t = k_0 \exp[\int_0^t b_s dw_s + \int_0^t c_s d\widetilde{N}_s] = k_0 e^{y_t}$. Then

$$\begin{aligned} dx_t &= x_t b_t dw_t + x_{t-}[e^{c_t} - 1]d\widetilde{N}_t + \tfrac{1}{2}|b_t|^2 x_t dt + \lambda x_t[e^{c_t} - 1 - c_t]dt \\ &= x_t b_t dw_t + x_{t-}c_t d\widetilde{N}_t + \tfrac{1}{2}|b_t|^2 x_t dt + x_{t-}[e^{c_t} - 1 - c_t]dN_t \\ &= x_{t-}dy_t + \tfrac{1}{2}|b_t|^2 x_t dt + x_{t-}[e^{c_t} - 1 - c_t]dN_t, \end{aligned}$$

$$x_0 = k_0,$$

that is, x_t satisfies the following linear stochastic differential equation:

$$\begin{aligned} x_t = {} & k_0 + \int_0^t x_s b_s dw_s + \int_0^t x_{s-}[e^{c_s} - 1]d\widetilde{N}_s + \tfrac{1}{2}\int_0^t |b_s|^2 x_s ds \\ & + \lambda \int_0^t x_s[e^{c_s} - 1 - c_s]ds, \ \forall t \geq 0; \end{aligned}$$

or,

$$\begin{aligned} x_t = {} & k_0 + \int_0^t x_s b_s dw_s + \int_0^t x_{s-}c_s d\widetilde{N}_s + \tfrac{1}{2}\int_0^t |b_s|^2 x_s ds \\ & + \int_0^t x_{s-}[e^{c_s} - 1 - c_s]dN_s, \ \forall t \geq 0. \end{aligned}$$

2) Let $x_t = k_0 \exp[\int_0^t b_s dw_s - \int_0^t \frac{1}{2}|b_s|^2 ds]$. Then

$$dx_t = x_t b_t dw_t, \ x_0 = k_0.$$

3) Let $x_t = \int_0^t b_s dw_s + \int_0^t c_s d\widetilde{N}_s$. Assume that $E\int_0^t [|b_s|^4 + \lambda |c_s|^4]ds < \infty$. Then

$$x_t^3 = 3\int_0^t x_s^2 dx_s + 3\int_0^t x_{s-}[|b_s|^2 ds + |c_s|^2 dN_s] + \int_0^t |c_s|^3 dN_s,$$

$$x_t^4 = 4\int_0^t x_s^3 dx_s + 6\int_0^t x_{s-}^2[|b_s|^2 ds + |c_s|^2 dN_s] + \int_0^t [4x_{s-}|c_s|^3 + |c_s|^4]dN_s,$$

and

$$Ex_t^3 = \lambda E\int_0^t |c_s|^3 ds,$$

$$Ex_t^4 = 6E\int_0^t [|b_s|^2 + \lambda |c_s|^2][\int_0^s (|b_r|^2 + \lambda |c_r|^2)dr ds + \lambda E\int_0^t |c_s|^4 ds.$$

Furthermore,

$$Ex_t^4 \leq 6t\int_0^t [|b_s|^2 + \lambda |c_s|^2]^2 ds + \lambda E\int_0^t |c_s|^4 ds.$$

Proof. 1): Let $y_t = \int_0^t b_s dw_s + \int_0^t c_s d\widetilde{N}_s$, $f(y) = e^y$. Applying Ito's formula to $f(y_t)$, we obtain the result.

2): Let $y_t = \int_0^t b_s dw_s - \int_0^t \frac{1}{2}|b_s|^2 ds$. From Ito's formula applied to e^{y_t} result 2) then follows.

3): Let $x_t = \int_0^t b_s dw_s + \int_0^t c_s d\widetilde{N}_s$. Applying the Ito formula:

$$F(x_t) - F(x_0) = \int_0^t F'(x_{s-})dx_s + \frac{1}{2}\int_0^t F''(x_s)d[x,x]_s$$
$$+ \int_0^t [F(x_{s-} + c_s) - F(x_{s-}) - F'(x_{s-})c_s - \frac{1}{2}|c_s|^2]dN_s$$

to $F(x_t) = x_t^3$ and $F(x_t) = x_t^4$, one obtains the expressions for x_t^3 and x_t^4, respectively. Notice that x_t is a martingale with $x_0 = 0$ and by Ito's formula again one has that

$Ex_t^2 = E\left|\int_0^t b_s dw_s\right|^2 + E\left|\int_0^t c_s d\widetilde{N}_s\right|^2 = E\int_0^t [|b_s|^2 + \lambda |c_s|^2]ds.$

Moreover, $\widetilde{N}_t = N_t - \lambda t$, and $E\widetilde{N}_t = 0, E\widetilde{N}_t^2 = \lambda t$. From these results one easily deduces the expressions for Ex_t^3 and Ex_t^3. Finally, applying the Schwartz inequality one obtains the last inequality. ■

2.11 Strong Markov Property of BM and Poisson Point Processes

The martingale characterization of BM and Poisson point processes (Theorem 98) can be used to show that a BM is still a BM if it starts again from any stopping time, and this property is also true for a stationary Poisson point process.

Theorem 100 *If $x_t = (x_t^1, \cdots, x_t^d)$ is a $d-$dimensional $\mathfrak{F}_t-$BM, and σ is a $\mathfrak{F}_t-$stopping time with $P(\sigma < \infty) = 1$, then $\{x_t^*\}_{t\geq 0} = \{x_{t+\sigma}\}_{t\geq 0}$ is a $d-$dimensional $\mathfrak{F}_t^* = \mathfrak{F}_{t+\sigma}-$BM. In particular, $\{w_t^*\}_{t\geq 0} = \{x_{t+\sigma} - x_\sigma\}_{t\geq 0}$ is a standard BM independent of $\mathfrak{F}_0^* = \mathfrak{F}_\sigma$.*

Proof. Let $\sigma_n = \sigma \wedge n$. Then $\sigma_n \uparrow \sigma$, and for each n, σ_n is a bounded stopping time. By Doob's stopping time theorem $\{M_t^n\}_{t\geq 0} = \{x_{t+\sigma_n} - x_{\sigma_n}\}_{t\geq 0} \in \mathcal{M}^{2,c}$ with respect to $\{\mathfrak{F}_{t+\sigma_n}\}_{t\geq 0}$, and $M_t^{ni}M_t^{nj} - \langle M^{ni}, M^{nj}\rangle_t$ is a $\mathfrak{F}_{t+\sigma_n}-$martingale. However, $\langle M^{ni}, M^{nj}\rangle_t = \delta_{ij}(t + \sigma_n - \sigma_n) = \delta_{ij}t$. So for each n, by Theorem 98 $\{M_t^n\}_{t\geq 0}$ is a $\{\mathfrak{F}_{t+\sigma_n}\}_{t\geq 0}-$BM. Furthermore, since $\mathfrak{F}_{t+\sigma_n} \uparrow \mathfrak{F}_t^*$. By Levi's theorem one also has that $\forall t > s$,

$E[w_t^* | \mathfrak{F}_s^*] = E[x_{t+\sigma} - x_\sigma | \mathfrak{F}_s^*] = x_{s+\sigma} - x_\sigma = w_s^*.$

In fact, $\sup_n E[|M_t^{ni}|^2] = t < \infty, \forall i = 1, 2, \cdots, d$. Hence $\{M_t^{ni}\}_{n=1}^\infty$ is uniformly integrable for each i. Therefore, as $n \to \infty$,

$E\left|E[M_t^{ni} - w_t^* | \mathfrak{F}_{t+\sigma_n}]\right| \leq E\left|M_t^{ni} - w_t^*\right| \to 0.$

On the other hand, by Levi's thoerem as $n \to \infty$,

$E[w_t^* | \mathfrak{F}_{t+\sigma_n}] \to E[w_t^* | \mathfrak{F}_{t+\sigma}] = E[w_t^* | \mathfrak{F}_t^*]$, a.s.

Hence as $n \to \infty$, the equality $E[M_t^{ni} | \mathfrak{F}_{t+\sigma_n}] = M_s^{ni}$ will tend to $E[w_t^{*i} | \mathfrak{F}_t^*] = w_s^{*i}$ in $L^1(\Omega, P)$, where $t > s; \forall i$. Similarly, one sees that

$\langle w^{*i}, w^{*j} \rangle_t = \delta_{ij}(t+\sigma-\sigma) = \delta_{ij}t$ holds. Hence, by Theorem 98, $\{w_t^*\}_{t\geq 0} = \{x_{t+\sigma} - x_\sigma\}_{t\geq 0}$ is a standard BM independent of $\mathfrak{F}_0^* = \mathfrak{F}_\sigma$. Moreover, $\{x_t^*\}_{t\geq 0} = \{x_{t+\sigma}\}_{t\geq 0} = \{w_t^* + x_\sigma\}_{t\geq 0}$ is a $d-$dimensional $\mathfrak{F}_t^* = \mathfrak{F}_{t+\sigma}-$BM with the initial random variable $x_\sigma \in \mathfrak{F}_0^*$. The proof is complete. ∎

Theorem 101 *If p is a stationary $\mathfrak{F}_t-$ Poisson point process on some space Z with the characterictic measure $\pi(dz)$, and σ is a $\{\mathfrak{F}_t\}_{t\geq 0}-$stopping time with $P(\sigma < \infty) = 1$, then $p^* = \{p^*(t)\}_{t\in D_{p^*}} = \{p(t+\sigma)\}_{t+\sigma\in D_p}$ is a stationary $\{\mathfrak{F}_t^*\}_{t\geq 0} = \{\mathfrak{F}_{t+\sigma}\}_{t\geq 0}-$Poisson point process with the same characterictic measure $\pi(dz)$.*

Proof. The proof is similar to that of Theorem 100. Let σ_n be the bounded stopping time defined in Theorem 100. By Doob's stopping time theorem

$\{N_p((\sigma_n, t+\sigma_n] \times U) - t\pi(U)\}_{t\geq 0}$ is a $\{\mathfrak{F}_{t+\sigma_n}\}_{t\geq 0}-$martingale for each n, where $U \in \Gamma_p$; that is, $EN_p((0,t] \times U) < \infty, \forall t$. Furthermore, by Levi's theorem and taking the limit as in the proof of Theorem 100 one easily sees that $\{N_p((\sigma, t+\sigma] \times U) - t\pi(U)\}_{t\geq 0}$ is a $\{\mathfrak{F}_{t+\sigma}\}_{t\geq 0}-$martingale. So the proof is complete. ∎

By the previous two theorems one immediately sees that a standard BM and a stationary Poisson point processes are both stationary strong Markov processes. That is to say, if $\{w_t\}_{t\geq 0}$ is a $d-$dimensional $\mathfrak{F}_t-$standard BM, then for any $\mathfrak{F}_t-$stopping time σ with $P(\sigma < \infty) = 1$, it satisfies $\forall A \in \mathfrak{B}(R^d), \forall t > 0$,

$P(w_{t+\sigma} \in A|\mathfrak{F}_\sigma) = P(w_t \in A|\mathfrak{F}_0)$, a.s.

In fact, by Theorem 100 one has that it is equivalent to

$P(w_t^* \in A|\mathfrak{F}_0^*) = P(w_t \in A|\mathfrak{F}_0) \iff P(w_t^* \in A) = P(w_t \in A)$

where $\{w_t^*\}_{t\geq 0}$ is a $\mathfrak{F}_t^*-$BM. The last equality is obviously true.

It is natural to define the strong Markov property of a stationary $\mathfrak{F}_t-$point process p as follows: If p satisfies that $\forall t > 0, \forall k = 1, 2, \cdots$,

$P(N_p((0, t+\sigma] \times U) = k|\mathfrak{F}_\sigma) = P(N_p((0,t] \times U) = k|\mathfrak{F}_0)$, a.s.,

then p is called a strong Markov $\mathfrak{F}_t-$point process. Since the proof is the same we omit it. Thus we arrive at the following corollary.

Corollary 102 *The $\mathfrak{F}_t-$ standard BM and $\mathfrak{F}_t-$ stationary Poisson point processes are both stationary strong Markov processes.*

2.12 Martingale Representation Theorem

In this section we are going to show that any square integrable $\mathfrak{F}_t^{w,p}-$ Martingales can be represented as a sum of an integral with respect to the BM $\{w_t\}_{t\geq 0}$ and an integral with respect to the martingale measure $\widetilde{N}_p(dt, dz)$ generated by the Poisson point process p. Here $\mathfrak{F}_t^{w,p}$ is the smallest $\sigma-$ field such that makes all $w_s, s \leq t$; and all $\widetilde{N}_p((0,s], U), s \leq t, U \in \mathfrak{B}_Z$ measurable. Such a representation theorem is very useful in the mathematical financial market and in the filtering problems. More precisely, we have the following theorem.

Theorem 103 . *(Martingale representation). Let $m(t)$ be a square integrable R^d-valued $\mathfrak{F}_t^{w,k}-$martingale, where $\mathfrak{F}_t^{w,k}$ is the $\sigma-$algebra generated (and completed) by $\{w_s, k_s, s \leqslant t\}$, and $\{w_t\}_{t\geq o}$ is a d_1- dimensional BM, $\{k_t\}_{t\geq o}$ is a stationary d_2- dimensional Poisson point process of the class (QL) such that the components $\{k_t^1\}_{t\geq o}, \cdots, \left\{k_t^{d_2}\right\}_{t\geq o}$ have disjoint domains and disjoint ranges. Then there exists a unique $(q_t, p_t) \in L^2_{\mathfrak{F}^{w,k}}(R^{d\otimes d_1}) \times F^2_{\mathfrak{F}^{w,k}}(R^{d\otimes d_2})$ such that*
$m(t) = m(0) + \int_0^t q_s dw_s + \int_0^t \int_Z p_s(z) \widetilde{N}_k(ds, dz).$

Here we write
$L^2_{\mathfrak{F}}(R^{d\otimes d_1}) = \{f(t,\omega) : f(t,\omega)$ is $\mathfrak{F}_t-$adapted, $R^{d\otimes d_1}-$valued such that
$E\int_0^T |f(t,\omega)|^2 dt < \infty$, for any $T < \infty$ $\}$
and
$F^2_{\mathfrak{F}}(R^{d\otimes d_2}) = \{f(t,z,\omega) : f(t,z,\omega)$ is $R^{d\otimes d_2}-$valued, $\mathfrak{F}_t$-predictable such that
$E\int_0^T \int_Z |f(t,z,\omega)|^2 \pi(dz)dt < \infty, \forall T < \infty.\}$
The following two lemmas are useful and interesting.

Lemma 104 . $\mathfrak{F}_{t+0}^{w,k} = \mathfrak{F}_t^{w,k}, \forall t \geqslant 0.$

Proof. Let
$H_n(t_1, t_2, \cdots, t_n; f_1, f_2, \cdots, f_n)$
$= H_{n-1}(t_1, t_2, \cdots, t_{n-1}; f_1, f_2, \cdots, f_{n-2}, f_{n-1}H_{t_n - t_{n-1}} f_n),$
$H_1(t; f)(x) = H_t(f)(x) = \int_{R^d} p(t, x-y) f(y) dy, \forall f \in C_0(R^d),$
$p(t,x) = (2\pi t)^{-d/2} \exp[-|x|^2/2t],$
and
$\widetilde{H}^j_n(t_1, t_2, \cdots, t_n; f_1, f_2, \cdots, f_n)$
$= \widetilde{H}^j_{n-1}(t_1, t_2, \cdots, t_{n-1}; f_1, f_2, \cdots, f_{n-2}, f_{n-1}H_{t_n - t_{n-1}} f_n),$
$\widetilde{H}^j_1(t; f) = \widetilde{H}^j_t(f), \forall f \in C_0(R^d),$
$\left(\widetilde{H}^j_t(g)\right)(m) = \sum_{n=0}^{\infty} g(n+m) e^{-t\pi(U_j)} (t\pi(U_j))^n / n!$
$= \sum_{n=m}^{\infty} g(n) e^{-t\pi(U_j)} (t\pi(U_j))^{n-m} / (n-m)!,$
where $0 = t_0 < t_1 < t_2 < \cdots < t_n$, $f_1, f_2, \cdots, f_n \in C_0(R^d)$,
$U_i \cap U_j = \phi, i \neq j, \pi(U_i) < \infty, i = 1, \cdots m; \cup_{i=1}^m U_i = Z$, and
$C_0(R^d) = \{f : f$ is a continuous function defined on R^d such that
$\lim_{|x|\to\infty} |f(x)| = 0$ $\}$.
Hence, if $t_{k-1} \leqslant t < t_k$, by the stationary Markov property and the independent property of BM and Poisson process
$E[f_1(w(t_1)) f_2(w(t_2)) \cdots f_n(w(t_n)) \prod_{j=1}^m g_1(N_k((0,t_1], U_j))$

$\cdot g_2(N_k((0,t_2],U_j))\cdots g_n(N_k((0,t_n],U_j))]\Big|\mathfrak{F}_t^{w,k}]$
$=\prod_{i=1}^{k-1} f_i(w(t_i))\prod_{j=1}^m g_i(N_k((0,t_i],U_j))$
$\cdot E^{(w_t,k_t)}\left[\prod_{i=k}^n f_i(w(t_i-t))\prod_{j=1}^m g_i(N_k((0,t_i-t],U_j))\right]$
$=\prod_{i=1}^{k-1} f_i(w(t_i))\prod_{j=1}^m g_i(N_k((0,t_i],U_j))E^{(w_t,k_t)}[\prod_{i=k}^n f_i(w(t_i-t))]$
$\cdot E^{(w_t,k_t)}\prod_{i=k}^n\prod_{j=1}^m g_i(N_k((0,t_i-t],U_j))$
$=\prod_{i=1}^{k-1} f_i(w(t_i))\prod_{j=1}^m g_i(N_k((0,t_i],U_j))$
$\cdot\prod_{j=1}^m \widetilde{H}_{n-k+1}^j(t_k-t,t_{k+1}-t,\cdots,t_n-t;g_k,g_{k+1},\cdots,g_n)(N_k((0,t],U_j))$
$\cdot H_{n-k+1}^j(t_k-t,t_{k+1}-t,\cdots,t_n-t;f_k,f_{k+1},\cdots,f_n)(w(t)).$
Therefore,
$E[\prod_{i=1}^n f_i(w(t_i))\prod_{j=1}^m g_i(N_k((0,t_i],U_j))\Big|\mathfrak{F}_{t+0}^{w,k}]$
$=\lim_{h\downarrow 0} E[\prod_{i=1}^n f_i(w(t_i))\prod_{j=1}^m g_i(N_k((0,t_i],U_j))\Big|\mathfrak{F}_{t+h}^{w,k}]$
$=E[\prod_{i=1}^n f_i(w(t_i))\prod_{j=1}^m g_i(N_k((0,t_i],U_j))\Big|\mathfrak{F}_t^{w,k}]\,.$
This proves that $\mathfrak{F}_{t+0}^{w,k}=\mathfrak{F}_t^{w,k}$. ■

Lemma 105 . *For any real-valued increasing sequence σ_n of $\mathfrak{F}_t^{w,k}$ – stopping times,*

$$\vee_n\mathfrak{F}_{\sigma_n}^{w,k}=\mathfrak{F}_\sigma^{w,k},$$

where $\sigma=\lim_{n\to\infty}\sigma_n$.

Proof. By the strong Markov property,
$E[\prod_{i=1}^n f_i(w(t_i))\prod_{j=1}^m g_i(N_k((0,t_i],U_j))\,|\mathfrak{F}_\tau^{w,k}]$
$=\sum_{k=1}^n I_{(t_{k-1}\leqslant\tau<t_k)}\prod_{i=1}^{k-1}(f_i(w(t_i))\prod_{j=1}^m g_i(N_k((0,t_i],U_j)))$
$\cdot H_{n-k+1}(t_k-\tau,t_{k+1}-\tau,\cdots,t_n-\tau;f_k,f_{k+1},\cdots,f_n)\,(w_\tau)$
$\cdot\prod_{j=1}^m\widetilde{H}_{n-k+1}^j(t_k-\tau,t_{k+1}-\tau,\cdots,t_n-\tau;g_k,g_{k+1},\cdots,g_n)$
$(N_k((0,\tau],U_j))+\prod_{i=1}^n(f_i(w(t_i))\prod_{j=1}^m g_i(N_k((0,t_i],U_j)))I_{(t_n\leqslant\tau)}.$
However, $N_k((0,t],U_j)$ is quasi-left-continuous, $P-a.s.$, i.e.

$$\lim_{n'\to\infty} N_k((0,\sigma_{n'}]U_j)=N_k((0,\sigma],U_j),P-a.s. \tag{2.22}$$

Indeed, if not, then as $n'\uparrow\infty$
$N_k((0,\sigma_{n'}]U_j)\uparrow Y\leqslant N_k((0,\sigma],U_j),$
and
$P(\lim_{n'\to\infty}(N_k((0,\sigma],U_j)-N_k((0,\sigma_{n'}],U_j))>0)$
$\geqslant P(N_k((0,\sigma],U_j)-Y>0)>0,$
Set
$\Gamma=\left\{\omega:\lim_{n'\to\infty}(N_k((0,\sigma],U_j)-N_k((0,\sigma_{n'}],U_j))>0\right\}.$
Thus by Fatou's lemma
$0<E\,(N_k((0,\sigma],U_j)-Y)\,I_\Gamma=\lim_{n'\to\infty}E[N_k((0,\sigma],U_j)-N_k((0,\sigma_{n'}],U_j)]I_\Gamma$

$= \pi(U_j) \lim_{n' \to \infty} E[\sigma - \sigma_{n'}] I_\Gamma = 0.$

This is a contradiction. Therefore (2.22) holds. Now it is easily shown that $P - a.s.$

$\lim_{n' \to \infty} E[\prod_{i=1}^n f_i(w(t_i)) \prod_{j=1}^m g_i(N_k((0,t_i],U_j)) \left| \mathfrak{F}_{\sigma_{n'}}^{w,k} \right.]$
$= E[\prod_{i=1}^n f_i(w(t_i)) \prod_{j=1}^m g_i(N_k((0,t_i],U_j)) \left| \mathfrak{F}_{\sigma}^{w,k} \right.] .$

For this, for any given $\omega \in \Omega$, let us show a special case: $\sigma(\omega) = t_n, \sigma_{n'}(\omega) < \sigma(\omega), \sigma_{n'}(\omega) \uparrow \sigma(\omega)$, as $n' \uparrow \infty$. All the other cases are similar or even much easier to handle. For simplicity let us omit the ω. For n' large enough, we have $t_{n-1} < \sigma_{n'} < \sigma = t_n$, and

$E[\prod_{i=1}^n f_i(w(t_i)) \prod_{j=1}^m g_i(N_k((0,t_i],U_j)) \left| \mathfrak{F}_{\sigma_{n'}}^{w,k} \right.]$
$= \prod_{i=1}^{n-1} f_i(w(t_i)) \prod_{j=1}^m g_i(N_k((0,t_i],U_j))$
$\cdot \left(H_{\sigma - \sigma_{n'}}(f_n)\right)(w(\sigma_{n'})) \left(\widetilde{H}^j_{\sigma - \sigma_{n'}}(g_n)\right)(N_k((0,\sigma_{n'}],U_j)).$

However, as $n' \to \infty$

$\left(\widetilde{H}^j_{\sigma - \sigma_{n'}}(g_n)\right)(N_k((0,\sigma_{n'}],U_j)) = \sum_{m=0}^{\infty} g_n(m + N_k((0,\sigma_{n'}],U_j))$
$\cdot e^{-(\sigma - \sigma_{n'})\pi(U_j)} ((\sigma - \sigma_{n'})\pi(U_j))^m / m! \to g_n(N_k((0,\sigma],U_j)).$

Similarly, one has that as $n' \to \infty$

$$\left(H_{\sigma - \sigma_{n'}}(f_n)\right)(w_{\sigma_{n'}}) \to f_n(w_\sigma). \tag{2.23}$$

In fact, one notes that

$(2\pi(t - s_{n'}))^{-d/2} \int_{R^d} e^{-|x-y|^2/2(t-s_{n'})} f(y) dy = I(R^d)$
$= I(|x-y| < \delta) + I(|x-y| \geqslant \delta) = I_1 + I_2.$

However, by the continuity of f at point x one has that for arbitrary given $\varepsilon > 0$ there exists a $\delta > 0$ such that for $|y - x| < \delta$

$|f(y) - f(x)| < \varepsilon/3.$

Hence as $|y - x| < \delta$, $\forall n'$

$\left| I_1 - (2\pi(t - s_{n'}))^{-d/2} \int_{(|x-y|<\delta)} f(x) e^{-|x-y|^2/2(t-s_{n'})} dy \right| < \varepsilon/3.$

On the other hand, one can also choose a large enough N such that $\forall n'$

$(2\pi(t - s_{n'}))^{-d/2} \int_{|y-x|>N} e^{-|x-y|^2/2(t-s_{n'})} |f(y)| dy < \varepsilon/3,$

because f is bounded. Now by Lebesgue's dominated convergence theorem as $s_{n'} \to t$

$\int_{(N \geqslant |x-y| \geqslant \delta)} (2\pi(t - s_{n'}))^{-d/2} e^{-\delta^2/2(t-s_{n'})} f(y) dy \to 0.$

Hence there exists an N_0 such that for $n' \geqslant N_0$

$I_2 < \varepsilon/3.$

Thus (2.23) is proved. Therefore in this case as $n' \to \infty$

$E[\prod_{i=1}^n f_i(w(t_i)) \prod_{j=1}^m g_i(N_k((0,t_i],U_j)) \left| \mathfrak{F}_{\sigma_{n'}}^{w,k} \right.]$
$\to E[\prod_{i=1}^n f_i(w(t_i)) \prod_{j=1}^m g_i(N_k((0,t_i],U_j)) \left| \mathfrak{F}_{\sigma}^{w,k} \right.] .$ ■

Now denote the totality of square integrable $\mathfrak{F}_t^{w,k}-$ martingales M with $M(0) = 0$ by $\mathfrak{M}_2$, and let $\mathfrak{M}_2^* \subset \mathfrak{M}_2$ be defined by

$\mathfrak{M}_2^* = \{M(t) = \int_0^t \varphi(s) dw_s + \int_0^t \int_Z \psi(s,z) \widetilde{N}_k(ds,dz),$

$(\varphi, \psi) \in L^2_{\mathfrak{F}_t}(R^{d\otimes d_1}) \times F^2_{\mathfrak{F}_t}(R^{d\otimes d_2}).\}$

We have the following lemmas.

Lemma 106 *Every $M \in \mathfrak{M}_2$ can be expressed as*

$M(t) = M_1(t) + M_2(t),$

where $M_1 \in \mathfrak{M}_2^*$ and $M_2 \in \mathfrak{M}_2$ satisfies the condition that $\langle M_2, N\rangle = 0, \forall N \in \mathfrak{M}_2^*$.

Proof. We only need to prove this result on $t \in [0,T]$ for any given $T < \infty$. We will use the following scheme: 1) We construct a (closed) subspace

$\mathcal{H} = \{M_1(T) : M_1 \in \mathfrak{M}_2^*\} \subset L^2(\Omega, \mathfrak{F}, P),$

where $L^2(\Omega, \mathfrak{F}, P) = \left\{\xi(\omega) : \xi(\omega) \in \mathfrak{F}, E\left|\xi(\omega)\right|^2 < \infty\right\}$. Then we obtain its orthogonal (closed) complement subspace $\mathcal{H}^{\perp} \subset L^2(\Omega, \mathfrak{F}, P)$. Any $M \in \mathfrak{M}_2$ has a decomposition for its terminal (random) value $M(T) = M_1(T) + H_2$, where $M_1(T) \in \mathcal{H}, H_2 \in \mathcal{H}^{\perp}$. 2) Since M and M_1 are square integrable $\mathfrak{F}_t^{w,k}-$ martingales,

$M(t) = M_1(t) + E(H_2|\mathfrak{F}_t^{w,k}) = M_1(t) + M_2(t)$. Obviously, $M_2 \in \mathfrak{M}_2$. Finally, the last step is 3): We prove the last equality. Obviously, steps 1) and 2) are easily seen. Let us show 3). By Lemma 107 below we only need to show that for any $N(t) = \int_0^t \varphi(s)dw_s + \int_0^t \int_Z \psi(s,z)\widetilde{N}_k(ds,dz) \in \mathfrak{M}_2^*$, and any $\mathfrak{F}_t^{w,k}-$stopping time $\sigma \le T$, $E[M_2(\sigma)N(\sigma)] = 0$. In fact,

$E[M_2(\sigma)N(\sigma)] = E[N(\sigma)E(M_2(T)|\mathfrak{F}_\sigma^{w,k})] = E[N(\sigma)M_2(T)]$

$= E[H_2(\int_0^T \varphi(s)I_{s\le\sigma}dw_s + \int_0^T \int_Z \psi(s,z)I_{s\le\sigma}\widetilde{N}_k(ds,dz))] = 0.$

The proof is complete. ∎

Lemma 107 *Assume that $\{X_t\}_{t\ge0}, \{Y_t\}_{t\ge0}$ with $X_0 = Y_0 = 0$ are two $\mathfrak{F}_t-$adapted random process such that $E|X_t| < \infty, E|Y_t| < \infty, \forall t \ge 0$. Then the following assertions hold true:*

1) $\{X_t\}_{t\ge0}$ is a $\mathfrak{F}_t-$martingale $\Longleftrightarrow$ For any bounded $\mathfrak{F}_t-$stopping time σ, $EX_\sigma = 0$.

2) If, in addition, $E|X_t|^2 < \infty, E|Y_t|^2 < \infty, \forall t \ge 0$, then $E\langle X, Y\rangle_\sigma = 0, \forall\sigma$, which is a $\mathfrak{F}_t-$stopping time $\Longleftrightarrow$ $\{X_tY_t\}_{t\ge0}$ is a $\mathfrak{F}_t-$martingale $\Longleftrightarrow$ $E[X,Y]_\sigma = 0, \forall\sigma$, which is a $\mathfrak{F}_t-$stopping time.

Proof. 1): The one way implication " $\Longrightarrow$ " is true by Doob's stopping time theorem. Now for any $s < t \le T$ and $A \in \mathfrak{F}_s$ let

$\sigma_s(\omega) = s$, as $\omega \in A$; and $\sigma_s(\omega) = T$, as $\omega \notin A$. Similarly, define $\sigma_t(\omega)$. Then σ_s is a $\mathfrak{F}_t-$stopping time, and so is σ_t. In fact, $\forall u \in [0,T]$

$\{\omega : \sigma_s(\omega) \le u\} = A \in \mathfrak{F}_s \subset \mathfrak{F}_u$, as $T > u \ge s$; and $\{\sigma_s \le u\} = \phi \in \mathfrak{F}_u$, as $u < s$; and $\{\sigma_s \le u\} = \Omega \in \mathfrak{F}_u$, as $u = T$. Hence by definition σ_s is a $\mathfrak{F}_t-$stopping time. Therefore, by assumption, $EX_{\sigma_s} = EX_{\sigma_t}$, that is,

$EXI_A + EX_TI_{A^c} = EX_tI_A + EX_TI_{A^c}.$

Thus $\{X_t\}_{t\ge0}$ is a $\mathfrak{F}_t-$martingale.

2): By the D-M decomposition theorem

$X_tY_t = \text{a martingale} + \langle X, Y\rangle_t$.

So the first equivalence is easily seen. (Actually, the sufficiency is derived by Doob's stopping time theorem, and the necessity is concluded from 1)). On the other hand by Ito's formula one has that

$X_tY_t = \int_0^t X_s dY_s + \int_0^t Y_s dX_s + [X,Y]_t = \text{a martingale} + [X,Y]_t$.

So the second equivalence is also easily seen. ∎

Lemma 108 . *Let* $(M_t, \mathfrak{F}_t^{w,k})_{t\geq 0} \in \mathfrak{M}_2$. *If* M *satisfies that* $EM_tN_t = 0, \forall N \in \mathfrak{M}_2^*, \forall t \geq 0$, *then* $M = 0$.

We will prove this lemma by showing the following two propositions.

Proposition 109 *Under assumptions of Lemma 108 for any* $0 \leq s < t$,

$\forall \lambda_j \in R, \forall U_j \in \mathfrak{B}(Z), \pi(U_j) < \infty$,

$U_i \cap U_j = \phi, i \neq j, i,j = 1,2,\cdots,m; \forall \widetilde{\lambda} \in R$

$E[\prod_{p=1}^n f_t g_t M_T | \mathfrak{F}_s^{w,k}] = M_s e^{-\frac{|\widetilde{\lambda}|^2}{2}(t-s)} e^{(t-s)\sum_{j=1}^m [(e^{i\lambda_j}-1)-\lambda_j]\pi(U_j)}$,

where

$f_t = e^{i\widetilde{\lambda}(w(t)-w(s))}, \; g_t = e^{i\sum_{j=1}^m \lambda_j \widetilde{N}_k((s,t],U_{,j})}$.

Proof. Applying Ito's formula to $f_r\widetilde{g}_r$ on $r \in [s,t]$, where

$\widetilde{g}_r = e^{i\sum_{j=1}^m \lambda_j N_k((s,r],U_{,j})}$,

one finds that

$f_t\widetilde{g}_t - 1 = i\widetilde{\lambda}\int_s^t f_r\widetilde{g}_r dw_r - \frac{|\widetilde{\lambda}|^2}{2}\int_s^t f_r\widetilde{g}_r dr$

$+ \int_{(s,t]}\int_Z f_r\widetilde{g}_r(e^{i\sum_{j=1}^n \lambda_j I_{U_j}(z)} - 1)\pi(dz)dr$

$+ \int_{(s,t]}\int_Z f_r\widetilde{g}_{r-}(e^{i\sum_{j=1}^n \lambda_j I_{U_j}(z)} - 1)\widetilde{N}_k(dr,dz)$.

Now for arbitrary $A \in \mathfrak{F}_s^{w,k}$ multiplying both sides of the above equality by M_tI_A, and taking the mathematical expectation, one finds that

$y_t = EM_tf_t\widetilde{g}_tI_A = E[M_sI_A] - \frac{|\widetilde{\lambda}|^2}{2}\int_s^t y_r dr$

$+ \int_{(s,t]}\int_Z y_r(e^{i\sum_{j=1}^n \lambda_j I_{U_j}(z)} - 1)\pi(dz)dr$

$= E[M_sI_A] - \frac{|\widetilde{\lambda}|^2}{2}\int_s^t y_r dr + \int_s^t y_r \sum_{j=1}^n (e^{i\lambda_j} - 1)\pi(U_j)dr$,

where we have applied the result that by assumption for any $N \in \mathfrak{M}_2^*$, $EM_t(N_t - N_s) = 0 + EM_tN_s = E[E(M_t|\mathfrak{F}_s^{w,k})N_s] = EM_sN_s = 0$. Solving the above linear ODE, one obtains that

$y_t = E[M_sI_A]e^{-\frac{|\widetilde{\lambda}|^2}{2}(t-s)}e^{(t-s)\sum_{j=1}^n (e^{i\lambda_j}-1)\pi(U_j)}$.

That is,

$E[M_te^{i\sum_{j=1}^n \lambda_j N_k((s,t],U_i)}|\mathfrak{F}_s^{w,k}] = M_se^{-\frac{|\widetilde{\lambda}|^2}{2}(t-s)}e^{(t-s)\sum_{j=1}^n (e^{i\lambda_j}-1)\pi(U_j)}$.

Mmultiplying both sides of the above equality by $e^{-i(t-s)\sum_{j=1}^n \lambda_j \pi(U_i)}$ the conclusion of this proposition follows. ∎

Proposition 110 *Under assumptions of Lemma 108 for any* $0 \leq t_1 < t_2 < \cdots < t_n \leq T, \forall p = 1,2,\cdots,n$;

$\forall \lambda_{p,j} \in R, \forall U_{p,j} \in \mathfrak{B}(Z), \pi(U_{p,j}) < \infty$,

$U_{p,i} \cap U_{p,j} = \phi, i \neq j, i,j = 1,2,\cdots,m_p; \forall \widetilde{\lambda}_p \in R$

$E[\prod_{p=1}^n f_p g_p M_T] = 0,$
where for $p = 1, 2, \cdots, n$
$f_p = e^{i\widetilde{\lambda}_p(w(t_p)-w(t_{p-1})}, \ g_p = e^{i\sum_{j=1}^{m_p}\lambda_{p,j}\widetilde{N}_k((t_{p-1},t_p],U_{p,j})}.$

Proof. The proof is a repeated application of Propositions 109. In fact,
$E[\prod_{p=1}^n f_p g_p M_T] = E[E(\prod_{p=1}^n f_p g_p M_T|\mathfrak{F}_0^{w,k})]$
$= E[\prod_{p=1}^{n-1} f_p g_p E(f_n g_n M_T|\mathfrak{F}_{t_{n-1}}^{w,k})|\mathfrak{F}_0^{w,k}]$
$= E[\prod_{p=1}^{n-1} f_p g_p M_{n-1}|\mathfrak{F}_0^{w,k}]e^{(t_n-t_{n-1})[-\frac{|\widetilde{\lambda}_n|^2}{2}+\sum_{j=1}^{m_n}[e^{i\lambda_{n,j}}-1-\lambda_{n,j}]\pi(U_{n,j})]}$
$= \cdots\cdots\cdots$
$= E[M_0 e^{\sum_{p=1}^n(t_p-t_{p-1})[-\frac{|\widetilde{\lambda}_p|^2}{2}+\sum_{j=1}^{m_p}[e^{i\lambda_{p,j}}-1-\lambda_{p,j}]\pi(U_{p,j})]}] = 0,$
because $M_0 = 0$. ■

Now one easily completes the proof of Lemma 108 by using Proposition 110. In fact, since BM $\{w_t\}_{t\geq 0}$ and Poisson point process $\{k_t\}_{t\geq 0}$ have independent increments, by Proposition 110 one finds that for all bounded $\eta \in \mathfrak{F}_t^{w,k}$ $E[\eta M_T] = 0$. Thus $M_T = 0, a.s.$ for any $T \geq 0$. The proof of of Lemma 108 is complete.

Now we are in a position to prove Theorem 103.

Proof. Combining Lemma 106 and 108 the result follows. ■

Corollary 111 *In Theorem 103 if $m(t)$ is a locally square integrable R^d- valued $\mathfrak{F}_t^{w,k}-$martingale, then the conclusion is that there exists a unique $(q_t, p_t) \in \mathfrak{F}_t^{w,k}$ with*
$P(\int_0^T |q_s|^2 ds < \infty) = 1, P(\int_0^T \int_Z |p_s(z)|^2 \pi(dz)ds < \infty) = 1,$
$\forall T \in [0, \infty)$, *such that* $\forall t \geq 0$,
$m(t) = m(0) + \int_0^t q_s dw_s + \int_0^t \int_Z p_s(z)\widetilde{N}_k(ds, dz).$

Proof. By the definition of a locally square integrable martingale there exists a $\sigma_n \uparrow \infty$, with $\sigma_n < \infty$ a stopping time for each n such that $\{m_{t\wedge\sigma_n}\}_{t\geq 0}$ is a square integrable $\mathfrak{F}_t-$ martingale for each n. Applying Theorem 103, we have that for each n there exists a unique $(q_t^n, p_t^n) \in L^2_{\mathfrak{F}^{w,k}}(R^{d\otimes d_1}) \times F^2_{\mathfrak{F}^{w,k}}(R^{d\otimes d_2})$ such that $\forall t \geq 0$,
$m(t \wedge \sigma_n) = m(0) + \int_0^t q_s^n dw_s + \int_0^t \int_Z p_s^n(z)\widetilde{N}_k(ds, dz).$
By the uniqueness property we can, without confusion, define
$q_t = q_t^n$ and $p_t(z) = p_t^n(z)$, as $t \in [0, \sigma_n)$.
For any given $0 < T < \infty$ since $\forall n, E\int_0^T [|q_s^n|^2 + \int_Z |p_s^n(z)|^2 \pi(dz)]ds < \infty$. Hence $P-a.s.$ $\forall n, \int_0^T [|q_s^n|^2 + \int_Z |p_s^n(z)|^2 \pi(dz)]ds < \infty$. Therefore, $P-a.s.$
$\int_0^T [|q_s|^2 + \int_Z |p_s(z)|^2 \pi(dz)]ds$
$= \sum_{n=1}^\infty \int_{T\wedge\sigma_{n-1}}^{T\wedge\sigma_n} [|q_s^n|^2 + \int_Z |p_s^n(z)|^2 \pi(dz)]ds < \infty,$
because for each $\omega \in \Omega$ the middle term is only a finite sum, where we define $\sigma_0 = 0$. The proof is complete. ■

3

Stochastic Differential Equations

In the practical case a dynamical system will always be disturbed by some stochastic perturbation, one type of which is continuous, and it can be modeled by some stochastic integral with respect to the BM, and the other is of the jump type, which is usually modeled by some stochastic integral with respect to the martingale measure generated by a point process. In this chapter we will discuss such kinds of stochastic differential equations (SDE) with jumps and some of their applications.

3.1 Strong Solutions to SDE with Jumps

3.1.1 Notation

Suppose we are given a probability space $(\Omega, \mathfrak{F}, P)$ with a $\sigma-$field filtration $\{\mathfrak{F}_t\}_{t\geq 0}$. Consider the following SDE with jumps in $d-$dimensional space :

$$\begin{aligned} x_t &= x_0 + \int_0^t b(s, x_s, \omega)ds + \int_0^t \sigma(s, x_s, \omega)dw_s \\ &\quad + \int_0^t \int_Z c(s, x_{s-}, z, \omega)\tilde{N}_k(ds, dz), \ t \geq 0 \end{aligned} \tag{3.1}$$

where $w_t^T = (w_t^1, \cdots, w_t^{d_1}), 0 \leqslant t$, is a d_1-dimensional $\mathfrak{F}_t-$adapted standard Brownian motion (BM), w_t^T is the transpose of w_t^T; $k^T = (k_1, \cdots, k_{d_2})$ is a d_2-dimensional $\mathfrak{F}_t-$adapted stationary Poisson point process with independent components, and $\tilde{N}_{k_i}(ds,dz)$ is the Poisson martingale measure generated by k_i satisfying

$$\tilde{N}_{k_i}(ds,dz) = N_{k_i}(ds,dz) - \pi(dz)ds, \quad i = 1, \cdots, d_2.$$

Here $\pi(.)$ is a σ-finite measure on a measurable space $(Z, \mathfrak{B}(Z))$, and $N_{k_i}(ds,dz)$ is the Poisson counting measure generated by k_i. From now on, for simplicity, we will always denote the integral $\int_0^t = \int_{(0,t]} = \int_0^{t+}$.

Definition 112 . *$\{x_t\}_{t\geq 0}$ (or, simply, x_t) is said to be a ($\mathfrak{F}_t-$) solution of* (3.1), *iff $\{x_t\}_{t\geq 0}$ satisfies* (3.1). *In the case that $x_t \in \mathfrak{F}_t^{w,k}$, $\forall t \geq 0$, where $\mathfrak{F}_t^{w,k}$ is the $\sigma-$algebra generated (and completed) by $w_s, k_s, s \leqslant t$, and then it is called a strong solution.*

From Definition 112 it is seen that for discussing the solution of (3.1) we always need to assume that the coefficients satisfy the following assumption
$(A)_1$ b and $\sigma : [0,\infty) \times R^d \times \Omega \to R^d$,
$c : [0,\infty) \times R^d \times Z \times \Omega \to R^d$
are jointly measurable and $\mathfrak{F}_t-$adapted where, furthermore, c is $\mathfrak{F}_t-$ predictable.

Moreover, to simplify the discussion of (3.1), we will also suppose that all $N_{k_i}(ds, Z), 1 \leqslant i \leqslant d_2$, have no common jump time;
i.e. we always make the following assumption
$(A)_2$ $N_{k_i}(\{t\}, U) N_{k_j}(\{t\}, U) = 0$, as $i \neq j$, for all $U \in \mathfrak{B}(Z)$ such that $\pi(U) < \infty$.

Now for the uniqueness of solutions to (3.1) we have the following definition.

Definition 113 *We say that the pathwise uniqueness of solutions to (3.1) holds, if for any two solutions $\{x_t^i\}_{t\geq 0}, i = 1, 2$ satisfying (3.1) on the same probability space with the same BM $\{w_t\}_{t\geq 0}$ and Poisson martingale measure $\tilde{N}_k(dt,dz)$, $P(\sup_{t\geq 0} |x_t^1 - x_t^2| = 0) = 1$.*

3.1.2 A Priori Estimate and Uniqueness of Solutions

Now let us introduce some notation which is useful later.

$$S_{\mathfrak{F}}^{2,loc}(R^d) = \left\{ \begin{array}{c} f(t,\omega) : f(t,\omega) \text{ is } \mathfrak{F}_t - \text{adapted}, R^d - \text{valued such that} \\ E \sup_{t\in[0,T]} |f(t,\omega)|^2 < \infty, \ \forall T < \infty, \end{array} \right\}.$$

Lemma 114 *Assume that x_t is a solution of (3.1), and assume that*
1° $E|x_0|^2 < \infty$, *and*
$2\langle x \cdot b(t,x,\omega)\rangle \leqslant c(t)(1+|x|^2)$,
$|\sigma(t,x,\omega)|^2 + \int_Z |c(t,x,z,\omega)|^2 \pi(dz) \leqslant c(t)(1+|x|^2)$,

where $0 \leqslant c(t)$ *is non-random, such that*

$C_T = \int_0^T c(t)dt < \infty,$

for any $0 < T < \infty$.

Then

$$E(\sup_{t \in [0,T]} |x_t|^2) \leqslant k_T < \infty,$$

where $k_T \geqslant 0$ *is a constant only depending on* C_T *and* $E|x_0|^2$. *Hence one has that under the assumption of this lemma the solution of (3.1) always satisfies* $\{x_t\}_{t \geq 0} \in S_{\mathfrak{F}}^{2,loc}(R^d)$.

Proof. Let $\tau_N = \inf\{t \geq 0 : |x_t| > N\}$. By Ito's formula

$$|x_{t\wedge\tau_N}|^2 = |x_0|^2 + 2\int_0^{t\wedge\tau_N} x_s \cdot b(s, x_s, \omega)ds + 2\int_0^{t\wedge\tau_N} x_s \cdot \sigma(s, x_s, \omega)dw_s$$
$$+ \int_0^{t\wedge\tau_N} |\sigma(s, x_s, \omega)|^2 ds + 2\int_0^{t\wedge\tau_N}\int_Z x_s \cdot c(s, x_{s-}, z, \omega)\tilde{N}_k(ds, dz)$$
$$+ \int_0^{t\wedge\tau_N}\int_Z |c(s, x_{s-}, z, \omega)|^2 N_k(ds, dz). \tag{3.2}$$

Since for any $T < \infty$, as $t \in [0, T]$,

$$E\int_0^{t\wedge\tau_N} |x_s \cdot \sigma(s, x_s, \omega)|^2 ds = E\int_0^{t\wedge\tau_N -} |x_s \cdot \sigma(s, x_s, \omega)|^2 ds$$
$$\leqslant N^2 \int_0^t c(s)(1 + |N|^2)ds < \infty.$$

Hence $\left\{\int_0^{t\wedge\tau_N} x_s \cdot \sigma(s, x_s, \omega)dw_s\right\}_{t\in[0,T]}$ is a martingale. A similar conclusion holds for $\int_0^{t\wedge\tau_N}\int_Z x_s \cdot c(s, x_{s-}, z, \omega)\tilde{N}_k(ds, dz)$. Hence by the martingale inequality

$$E\|x\|^2_{t\wedge\tau_N} \leqslant E|x_0|^2 + k_0 E\int_0^{t\wedge\tau_N} c(s)(1 + \|x\|_s^2)ds$$
$$+\frac{1}{2}E\|x\|^2_{t\wedge\tau_N} \leq E|x_0|^2 + \frac{1}{2}E\|x\|^2_{t\wedge\tau_N} + k_0\int_0^t c(s)(1 + E\|x_{s\wedge\tau_N}\|^2)ds, \tag{3.3}$$

where we write $\|x\|_t^2 = \sup_{s\leq t}|x_s|^2$, and we have used the fact that

$2E\sup_{s\leq t}\left|\int_0^{s\wedge\tau_N} x_s \cdot \sigma(s, x_s, \omega)dw_s\right| \leq k_0' E\sqrt{\int_0^{t\wedge\tau_N} |x_s \cdot \sigma(s, x_s, \omega)|^2 ds}$

$\leq \frac{1}{4}E\|x\|^2_{t\wedge\tau_N} + k_0'^2 E\int_0^{t\wedge\tau_N} |\sigma(s, x_s, \omega)|^2 ds$

$\leq \frac{1}{4}E\|x\|^2_{t\wedge\tau_N} + k_0'^2 E\int_0^{t\wedge\tau_N} c(s)(1 + \|x\|_s^2)ds,$

and a similar inequality also holds for

$2E\sup_{s\leq t}\left|\int_0^{s\wedge\tau_N}\int_Z x_s\cdot c(s,x_{s-},z,\omega)\tilde{N}_k(ds,dz)\right|.$

Therefore,

$\frac{1}{2}E\left\|x\right\|^2_{t\wedge\tau_N}\leqslant E\left|x_0\right|^2+k_0\int_0^t c(s)(1+E\left\|x_{s\wedge\tau_N}\right\|^2)ds.$

By Gronwall's inequality one finds that when $t\in[0,T]$

$$E\left\|x\right\|^2_{t\wedge\tau_N}\leqslant k'_T e^{2k_0\int_0^T c(s)ds}=k_T<\infty, \tag{3.4}$$

where $k'_T=2E\left|x_0\right|^2+2\int_0^T c(s)ds$. Letting $N\uparrow\infty$, by Fatou's lemma one finds that

$$E\left\|x\right\|^2_T\leqslant k_T,$$

where $k_T=(2E\left|x_0\right|^2+2\int_0^T c(s)ds)e^{2k_0\int_0^T c(s)ds}$ depends on $E\left|x_0\right|^2$ and $\int_0^T c(s)ds$ only. ∎

Lemma 115 *Assume that $b(t,x,\omega)$ and $\sigma(t,x,\omega)$ are uniformly locally bounded in x, that is, for each $0<r<\infty$,*

$|b(t,x,\omega)|+|\sigma(t,x,\omega)|\leq k_r$, *as* $|x|\leq r$,

where $k_r\geq 0$ is a constant depending only on r; and assume that for each $N=1,2,\cdots,T<\infty$ there exist non-random functions $c_T^N(t)$ and $\rho_T^N(u)$ such that as $|x_1|,|x_2|\leq N$; and $t\in[0,T]$

$$2(x_1-x_2)\cdot(b(t,x_1,\omega)-b(t,x_2,\omega))+|\sigma(t,x_1,\omega)-\sigma(t,x_2,\omega)|^2$$
$$+\int_Z|c(t,x_1,z,\omega)-c(t,x_2,z,\omega)|^2\pi(dz)\leqslant c_T^N(t)\rho_T^N(|x_1-x_2|^2),$$

where $c_T^N(t)$ is non-negative such that $\int_0^T c_T^N(t)dt<\infty$, and $\rho_T^N(u)$ defined on $u\geqslant 0$, is non-negative, increasing, continuous and concave such that $\int_{0+}du/\rho_T^N(u)=\infty$.
Then the solution of (3.1) is pathwise unique.

Proof. Asume $\{x_t^i\}_{t\geq 0}, i=1,2$ are two solutions of (3.1) with the same BM $\{w_t\}_{t\geq 0}$ and Poisson martingale measure $\widetilde{N}_k(dt,dz)$. Let

$$X_t=x_t^1-x_t^2,\ \widehat{b}(s,x_s^1,x_s^2,\omega)=b(s,x_s^1,\omega)-b(t,x_s^2,\omega),$$

etc. and

$\tau_N=\inf\left\{t\geq 0:\left|x_t^1\right|+\left|x_t^2\right|>N\right\}.$

Then by Ito's formula as in (3.2) one sees that

$Z_{t\wedge\tau_N}=E[|X_{t\wedge\tau_N}|^2=E\int_0^{t\wedge\tau_N}[2X_s\cdot\widehat{b}(s,x_s^1,x_s^2,\omega)+\left|\widehat{\sigma}(s,x_s^1,x_s^2,\omega)\right|^2$
$+\int_Z\left|\widehat{c}(s,x_s^1,x_s^2,z,\omega)\right|^2\pi(dz)]ds\leqslant E\int_0^{t\wedge\tau_N-}c_T^N(s)\rho_T^N\left(|X_s|^2\right)ds$
$\leq\int_0^t c_T^N(s)\rho_T^N\left(Z_{s\wedge\tau_N}\right)ds$, as $t\in[0,T]$.

Hence by the following Lemma 116 for any $T<\infty$, $P-a.s.$ $Z_{t\wedge\tau_N}=0,\forall t\in[0,T]$. Letting $N\to\infty$ one finds that $P-a.s.$

$Z_t=0,\forall t\in[0,T].$

By the RCLL (right continuous with left limit) property of $\{x_t^i\}_{t\geq 0}, i = 1, 2$ the conclusion now follows. ■

Lemma 116 *If* $\forall t \geq 0$ *a real non-random function* y_t *satisfies*
$0 \leq y_t \leq \int_0^t \rho(y_s)ds < \infty,$
where $\rho(u)$ *defined on* $u \geq 0$, *is non-negative, increasing such that* $\rho(0) = 0, \rho(u) > 0$, *as* $u > 0$; *and*
$\int_{0+} du/\rho(u) = \infty,$
then
$y_t = 0, \forall t \geq 0.$

Proof. Let
$z_t = \int_0^t \rho(y_s)ds (\geq y_t \geq 0).$
Obviously, one only needs to show that $\forall t \geq 0$
$z_t = 0.$
Indeed, z_t is absolutely continuous, increasing and *a.e.*

$$\dot{z}_t = \rho(y_t) \leq \rho(z_t). \tag{3.5}$$

Set
$t_0 = \sup\{t \geq 0 : z_s = 0, \forall s \in [0, t]\}.$
If $t_0 < \infty$, then $z_t > 0$, as $t > t_0$. Hence by assumption and from (3.5) for any $\delta > 0$
$\infty = \int_{(0,z(t_0+\delta))} du/\rho(u) = \int_{(t_0,t_0+\delta)} dz_t/\rho(z_t) \leq \int_{(t_0,t_0+\delta)} dt \leq \delta.$
This is a contradiction. Therefore $t_0 = \infty$. ■

3.1.3 Existence of Solutions for the Lipschitzian Case

In this section we are going to discuss the existence and uniqueness of solution to SDE (3.1). First, we introduce a notation which will be used later.

$$L^2_{\mathfrak{F}}(R^d) = \left\{ \begin{array}{c} f(t,\omega) : f(t,\omega) \text{ is } \mathfrak{F}_t - \text{adapted, } R^d - \text{valued} \\ \text{such that } E\int_0^T |f(t,\omega)|^2 dt < \infty \end{array} \right\}$$

Theorem 117 *Assume that*
1° b *and* $\sigma : [0, \infty) \times R^d \times \Omega \to R^d$,
$c : [0, \infty) \times R^d \times Z \times \Omega \to R^d$
are jointly measurable and $\mathfrak{F}_t$*−adapted, where furthermore,* c *is* $\mathfrak{F}_t$*− predictable such that* $P - a.s.$
$|b(t, x, \omega)| \leqslant c(t)(1 + |x|),$
$|\sigma(t, x, \omega)|^2 + \int_Z |c(t, x, z, \omega)|^2 \pi(dz) \leqslant c(t)(1 + |x|^2),$
where $c(t)$ *is non-negative and non-random such that*
$\int_0^T c(t)dt < \infty;$
2° $|b(t, x_1, \omega) - b(t, x_2, \omega)| \leqslant c(t) |x_1 - x_2|,$

$$|\sigma(t,x_1,\omega)-\sigma(t,x_2,\omega)|^2+\int_Z|c(t,x_1,z,\omega)-c(t,x_2,z,\omega)|^2\,\pi(dz)$$
$$\leqslant c(t)\,|x_1-x_2|^2,$$
where $c(t)$ satisfies the same conditions as in 1°;
3°

$$x_0\in\mathfrak{F}_0, E\,|x_0|^2<\infty.$$

Then (3.1) has a pathwise unique $\mathfrak{F}_t-$ adapted solution $\{x_t\}_{t\geq 0}\in S^{2,loc}_{\mathfrak{F}}(R^d)$. In the case that $b(t,x,\omega)$ and $\sigma(t,x,\omega)$ are $\mathfrak{F}^{w,\tilde{N}_k}_t-$ adapted, and $c(t,x,z,\omega)$ is $\mathfrak{F}^{w,\tilde{N}_k}_t-$ predictable, then the solution is also $\mathfrak{F}^{w,\tilde{N}_k}_t-$ adapted, i.e. it is a strong solution.

Proof. Let us use the contraction mapping principle to prove this result. Introduce a new norm as follows: For any given $T<\infty$ and for $(x.)\in\widetilde{B}=L^2_{\mathfrak{F}}(R^d)$ let

$$\|(x.)\|^2_M=\sup_{t\in[0,T]}e^{-b_0A(t)}E\,\|x\|^2_t\,,$$

where $\|x\|_t=\sup_{s\leq t}|x_s|$, and $b_0\geqslant 0$ is a constant, which will be determined later, and

$$A(t)=\int_0^t c(s)ds.$$

Write

$$H=\left\{(x.)\in\widetilde{B}:\|(x.)\|_M<\infty\right\}.$$

Then H is a Banach space. For any $(\overline{x}^i.)\in H, i=1,2$, denote $(x^i.)\,,i=1,2$, by the following SDE

$$x^i_t=x_0+\int_0^t b(s,\overline{x}^i_s,\omega)ds+\int_0^t\sigma(s,\overline{x}^i_s,\omega)dw_s$$
$$+\int_0^t\int_Z c(s,\overline{x}^i_{s-},z,\omega)\tilde{N}_k(ds,dz),0\leqslant t,\ i=1,2.$$

By assumption 1° one easily sees that $(x^i.)\in H, i=1,2$. Let

$$X_t=x^1_t-x^2_t.$$

Similarly, define $\overline{X}_t$. Then by Ito's formula as in (3.2), and discussing similarly as in (3.3), one has that

$$E[\|X\|^2_t\leqslant\gamma^{-1}\int_0^t k_0c(s)E\left\|\overline{X}\right\|^2_s ds+\gamma\int_0^t k_0c(s)E\,\|X\|^2_s\,ds$$
$$=\gamma^{-1}I^1_t+\gamma\int_0^t k_0c(s)E\left\|\overline{X}\right\|^2_s ds$$
$$\leqslant\gamma^{-1}I^1_t+\int_0^t\exp(\gamma k_0(A(t)-A(s))k_0c(s)I^1_s ds,$$

where $k_0 \geq 1$ is a fixed constant, and we have applied Lemma 118 below. Note that $0 \leqslant A(t)$ is increasing, so that

$$e^{-b_0 A(t)} I_t^1 = e^{-b_0 A(t)} \int_0^t k_0 c(s) E \left\| \overline{X} \right\|_s^2 ds$$
$$\leqslant \sup_{s \leqslant t} e^{-b_0 A(s)} E \left\| \overline{X} \right\|_s^2 \int_0^t e^{-b_0(A(t)-A(s))} k_0 c(s) ds$$
$$\leq \sup_{s \leqslant t} e^{-b_0 A(s)} E \left\| \overline{X} \right\|_s^2 \cdot k_0' \int_0^t c(s) ds$$
$$\leq k_T'' \sup_{s \leqslant T} e^{-b_0 A(s)} E \left\| \overline{X} \right\|_s^2,$$

where $k_T'' > 0$ is a constant depending on $\int_0^T c(s)ds$ only. Thus, if we write $u(s) = E \left| \overline{X}_s \right|^2$, then

$$e^{-b_0 A(t)} \int_0^t \exp(\gamma k_0((A(t) - A(s)) k_0 c(s) I_s^1 ds$$
$$\leqslant \sup_{s \leqslant t} e^{-b_0 A(s)} I_s^1 \int_0^t e^{-k_0(\frac{b_0}{k_0} - \gamma)(A(t)-A(s))} k_0 c(s) ds$$
$$\leqslant k_T'' \sup_{s \leqslant t} e^{-b_0 A(s)} u(s) (\frac{b_0}{k_0} - \gamma)^{-1} \leqslant k_T'' \sup_{s \leqslant T} e^{-b_0 A(s)} u(s) (\frac{b_0}{k_0} - \gamma)^{-1}.$$

Hence

$$\left\| (X.) \right\|_M^2 \leqslant \max(k_T'' \gamma^{-1}, k_T'' (\frac{b_0}{k_0} - \gamma)^{-1}) \left\| (\overline{X}.) \right\|_M^2.$$

After appropriately choosing γ and b_0 to make
$\max(k_T'' \gamma^{-1}, k_T'' (\frac{b_0}{k_0} - \gamma)^{-1}) < 1$
by the contraction mapping principle one finds that there exisits a unique solution $\{\overline{x}_t\}_{t \geq 0} \in L_{\mathfrak{F}}^2(R^d)$ satisfying (3.1). Let us show the following result:

There exists a version $\{x_t\}_{t \geq 0}$ of $\{\overline{x}_t\}_{t \geq 0}$, that is, for each $t \in [0,T]$ $P(x_t \neq \overline{x}_t) = 0$, such that $\{x_t\}_{t \geq 0}$ is RCLL (right continuou and with left limit) and $\{x_t\}_{t \geq 0}$ is a solution of (3.1).

In fact, write
$x_t = x_0 + \int_0^t b(s, \overline{x}_s, \omega) ds + \int_0^t \sigma(s, \overline{x}_s, \omega) dw_s$
$+ \int_0^t \int_Z c(s, \overline{x}_{s-}, z, \omega) \tilde{N}_k(ds, dz),\ t \in [0,T];$
$y_t = x_0 + \int_0^t b(s, x_s, \omega) ds + \int_0^t \sigma(s, x_s, \omega) dw_s$
$+ \int_0^t \int_Z c(s, x_{s-}, z, \omega) \tilde{N}_k(ds, dz),\ t \in [0,T].$
Then $E \int_0^T |x_t - \overline{x}_t|^2 dt = 0$. So, there exists a set $\Lambda_1 \times \Lambda_2 \in \mathfrak{B}([0,T]) \times \mathfrak{F}$ such that $E \int_0^T I_{\Lambda_1 \times \Lambda_2}(t, \omega) dt = 0$, and $x_t(\omega) = \overline{x}_t(\omega)$, as $(t, \omega) \notin \Lambda_1 \times \Lambda_2$. Hence, for each $t \in [0,T]$,

$E |x_t - y_t|^2 \leq 3[E \left| \int_0^t |b(s, \overline{x}_s) - b(s, x_s)| ds \right|^2 + E \int_0^t |\sigma(s, \overline{x}_s) - \sigma(s, x_s)|^2 ds$

$+E\int_0^t \int_Z |c(s,\overline{x}_{s-},z) - c(s,x_{s-},z)|^2 \pi(dz)ds] = 0.$
So, the above fact holds true.

Now, by Lemma 114 and 115 the solution is also pathwise unique such that $E(\sup_{t\in[0,T]} |x_t|^2) \leqslant k_T < \infty$, for each $T < \infty$, where k_T is a constant depending on T and $\int_0^T c(s)ds$ only. So we have show that there is a unique solution $\{x_t\}_{t\in[0,T]}$ for each given $T < \infty$. By the uniqueness of the solution we immediately obtain a solution $\{x_t\}_{t\geq 0}$, which is also unique. When all coefficients are $\mathfrak{F}_t^{w,\widetilde{N}_k}-$ adapted, etc., then by construction one easily sees that $\{x_t\}_{t\geq 0}$ is also $\mathfrak{F}_t^{w,\widetilde{N}_k}-$ adapted, i.e. it is a strong solution. ■

Lemma 118 . *(Gronwall's inequality). If* $0 \leqslant y_t \leqslant \gamma v_t + \int_0^t c(s)y_s ds, \forall t \geqslant 0$, *where* $\gamma > 0$ *is a constant, and* $c(s) \geqslant 0$, *then* $\forall t \geq 0$

$$y_t \leqslant \gamma v_t + \gamma \int_0^t \exp\left(\int_s^t c(r)dr\right) c(s)v_s ds.$$

Proof. Let $0 \leqslant z_t = \int_0^t c(s)y_s ds$, $g_t = y_t - \gamma v_t - z_t \leqslant 0$. Then

$$\dot{z}_t = c(t)y_t = c(t)\left(g_t + \gamma v_t + z_t\right).$$

From this one sees that

$$z_t = \int_0^t \exp\left(\int_s^t c(r)dr\right) c(s)\left(g_s + \gamma v_s\right) ds$$
$$\leqslant \gamma \int_0^t \exp\left(\int_s^t c(r)dr\right) c(s)v_s ds.$$

Hence

$$y_t \leqslant \gamma v_t + z_t \leqslant \gamma v_t + \gamma \int_0^t \exp\left(\int_s^t c(r)dr\right) c(s)v_s ds. \ ■$$

The above Theorem 117 is easily generalized to the locally Lipschitzian case.

Theorem 119 *If the condition 2° in Theorem 117 is weakened to*

2°′ for each $N = 1, 2, \cdots$ *there exists a non-random function* $c^N(t)$ *such that as* $|x_1|$ *and* $|x_2| \leq N$,

$$|b(t,x_1,\omega) - b(t,x_2,\omega)| \leqslant c^N(t)\,|x_1 - x_2|\,,$$

$$|\sigma(t,x_1,\omega) - \sigma(t,x_2,\omega)|^2 + \int_Z |c(t,x_1,z,\omega) - c(t,x_2,z,\omega)|^2 \pi(dz)$$
$$\leqslant c^N(t)\,|x_1 - x_2|^2\,,$$

where $c^N(t) \geq 0$ *satisfies the condition that* $\int_0^T c^N(t)dt < \infty$ *for each* $T < \infty$; *and all other conditions remain true, then the conclusion of Theorem 117 still holds.*

Proof. Let

$$b^N(t,x,\omega) = \begin{cases} b(t,x,\omega), & \text{as } |x| \leq N, \\ b(t, N\frac{x}{|x|}, \omega), & \text{as } |x| > N. \end{cases}$$

$\sigma^N(t,x,\omega)$ and $c^N(t,x,z,\omega)$ are similarly defined. Then by Theorem 117 there exists a unique $\mathfrak{F}_t$-solution $\{x_t^N\}_{t\geq 0} \in S_{\mathfrak{F}}^{2,loc}(R^d)$ solving the following SDE

$$\begin{aligned} x_t^N &= x_0 + \int_0^t b^N(s,x_s^N,\omega)ds + \int_0^t \sigma^N(s,x_s^N,\omega)dw_s \\ &\quad + \int_0^t \int_Z c^N(s,x_{s-}^N,z,\omega)\tilde{N}_k(ds,dz), \ t \geq 0. \end{aligned}$$

Set
$\tau^{N,m} = \inf\{t \geq 0 : |x_t^N| > m\}$.
Then we have that

$$\begin{aligned} x_t^N &= x_0 + \int_0^t b(s,x_s^N,\omega)ds + \int_0^t \sigma(s,x_s^N,\omega)dw_s \\ &\quad + \int_0^t \int_Z c(s,x_{s-}^N,z,\omega)\tilde{N}_k(ds,dz), \ t \in [0,\tau^{N,N}). \end{aligned}$$

By the uniqueness theorem $x_t^{N+m} = x_t^N$, as $t \in [0,\tau^{N,N}), \forall m = 0,1,2,\cdots$. Hence, $\tau^{N,N} \uparrow$, as $N \uparrow$. Let us show that

$$\lim_{N\to\infty} \tau^{N,N} = \infty, P-a.s. \tag{3.6}$$

In fact, if this is not true, then there exists a $T_0 < \infty$ such that $P(A) > 0$, where $A = \{\tau \leq T_0\}$, and $\tau = \lim_{N\to\infty} \tau^{N,N}$. Hence,
$E\underline{\lim}_{N\to\infty} \left|x_{\tau^{N,N}}^N\right|^2 I_A \geq \lim_{N\to\infty} NP(A) = \infty$
On the other hand, by Fatou's lemma and by the a priori estimate
$E\underline{\lim}_{N\to\infty} \left|x_{\tau^{N,N}}^N\right|^2 I_A \leq \underline{\lim}_{N\to\infty} E\left|x_{\tau^{N,N}}^N\right|^2 I_A \leq \sup_N E\sup_{t\leq T_0} \left|x_t^N\right|^2$
$\leq k_{T_0} < \infty$.

This is a contradiction. Therefore, (3.6) holds. Now let
$x_t = x_t^N$, as $t \in [0,\tau^{N,N})$,
then by uniqueness it is well defined. Moreover, since $\tau^{N,N} \uparrow \infty$, as $N \uparrow \infty$, $\{x_t\}_{t\geq 0}$ is a unique solution of (3.1). ■

The above Theorem 117 is also easily generalized to a more general case, which is useful in the non-linear filtering problem.

Theorem 120 *If the condition 3° in Theorem 117 is weakened to be*

3°′ there is a given process $\xi_t \in \mathfrak{F}_t^{w,\tilde{N}_k}$ *such that* $E|\xi_t|^2 < \infty$;

and assume that $b(t,x,\omega)$ *and* $\sigma(t,x,\omega)$ *are* $\mathfrak{F}_t^{w,\tilde{N}_k}-$ *adapted, and* $c(t,x,z,\omega)$ *is* $\mathfrak{F}_t^{w,\tilde{N}_k}-$ *predictable, then there exists a pathwise unique strong solution* $x_t, t \geq 0$, *satisfying the following SDE:*

$$\begin{aligned} x_t &= \xi_t + \int_0^t b(s,x_s,\omega)ds + \int_0^t \sigma(s,x_s,\omega)dw_s \\ &\quad + \int_0^t \int_Z c(s,x_{s-},z,\omega)\tilde{N}_k(ds,dz), \ t \geq 0 \end{aligned} \tag{3.7}$$

The proof of Theorem 120 can be completed in exactly the same way as that of Theorem 117.

Now let us give an example to show that the condition on $c(t)$ in Theorem 117 cannot be weakened.

Example 121 . *(Condition $\int_0^T c(t)dt < \infty$ cannot be weakened). Consider the following BSDE in 1-dimensional space*

$$x_t = 1 + \int_0^t I_{s\neq 0} s^{-\alpha} x_s ds + \int_0^t x_s dw_s + \int_0^t \int_Z x_s I_U(z) \tilde{N}_k(ds, dz),$$
$$0 \leqslant t \leqslant T,$$

where x_t, w_t, and k_t are all 1-dimensional, and $U \in \mathfrak{B}_Z$ such that $\pi(U) < \infty$. Obviously, if $\alpha < 1$, then by Theorem 117 it has a unique solution $\{x_t\}_{t\geq 0}$. However, if $\alpha \geqslant 1$, (in this case $\int_0^T I_{s\neq 0} s^{-\alpha} ds = \infty$), then it has no solution. Otherwise for the solution $\{x_t\}_{t\geq 0}$ one has that

$$Ex_t = 1 + \int_0^t I_{s\neq 0} s^{-\alpha} Ex_s ds = e^{\int_0^t I_{s\neq 0} s^{-\alpha} ds}.$$

Hence

$$Ex_t = \infty, \ \forall t > 0, \ \text{as } \alpha \geqslant 1.$$

This is a contradiction.

3.2 Exponential Solutions to Linear SDE with Jumps

In Calculus it is well konwn that a 1−dimensional linear ordinary differential equation (ODE)

$dx_t = a_t x_t dt,\ x_0 = c_0;\ t \geq 0,$

has a unique solution $x_t = c_0 e^{\int_0^t a_s ds}, t \geq 0$. However, for a 1−dimensional simple linear SDE

$d\widetilde{x}_t = a_t \widetilde{x}_t dw_t,\ x_0 = c_0;\ t \geq 0,$

even if $\{w_t\}_{t\geq 0}$ is a 1−dimensional BM (Brownian Motion process), write

$x_t = c_0 e^{\int_0^t a_s dw_s}, t \geq 0,$

x_t does not satisfies the above linear SDE. Actually, it satisfiess the following SDE:

$dx_t = \frac{1}{2} a_t^2 x_t dt + a_t x_t dw_t,\ x_0 = c_0;\ t \geq 0.$

In fact, let $f(y) = e^y$, and $y_t = \int_0^t a_s dw_s$. Then by Ito's formula

$dx_t = df(y_t) = f'(y_t)dy_t + \frac{1}{2} f''(y_t) d\langle y\rangle_t = c_0 a_t e^{y_t} dw_t + \frac{1}{2} c_0 e^{y_t} a_t^2 dt$

$= \frac{1}{2} a_t^2 x_t dt + a_t x_t dw_t.$

So x_t satisfies another linear SDE. What is the solution of the above original given SDE? In this section we will discuss this kind of problems.

First, we can use Ito's formula to verify that

$$z_t^w = \exp[\int_0^t \theta_s^w \cdot dw_s - \frac{1}{2}\int_0^t |\theta_s^w|^2 \, ds] \tag{3.8}$$

solves the following SDE

$$dz_t^w = \theta_t^w \cdot z_t^w dw_t, z_0^w = 1, \tag{3.9}$$

where w_t is a $d-$dimensional BM, and we assume that θ_t^w is a $d-$dimensional $\mathfrak{F}_t-$adapted process such that

$$P(\int_0^t |\theta_s^w|^2 \, ds < \infty) = 1. \tag{3.10}$$

In fact, $f(x) = e^x \in C^2(R^1)$. So if we let
$x_t = \int_0^t \theta_s^w \cdot dw_s - \frac{1}{2}\int_0^t |\theta_s^w|^2 \, ds,$
then by Ito's formula we have that
$e^{x_t} = 1 + \int_0^t e^{x_s}\theta_s^w \cdot dw_s - \frac{1}{2}\int_0^t e^{x_s} |\theta_s^w|^2 \, ds + \frac{1}{2}\int_0^t e^{x_s} |\theta_s^w|^2 \, ds$
$= 1 + \int_0^t e^{x_s}\theta_s^w \cdot dw_s.$
Therefore (3.9) holds. Furthermore, let us use Ito's formula to verify that

$$z_t^N = \exp[\int_0^t \int_Z \theta_s^N(z) \cdot \tilde{N}_k(ds, dz)] \cdot \prod_{0<s\leqslant t} (1 + \int_Z \theta_s^N(z) N_k(\{s\}, dz))$$

$$\cdot e^{-\int_Z \theta_s^N(z) N(\{s\},dz))} = \prod_{0<s\leqslant t} (1 + \int_Z \theta_s^N(z) N_k(\{s\}, dz)) e^{-\int_0^t \int_Z \theta_s^N(z)\pi(dz)ds} \tag{3.11}$$

solves the following SDE

$$dz_t^N = z_{t-}^N \int_Z \theta_t^N(z) \widetilde{N}_k(dt, dz), z_0^N = 1, \tag{3.12}$$

where for simplicity we assume that $\widetilde{N}_k(dt, dz)$ is a $1-$dimensional Poisson martingale measure, (for the $n-$dimensional case the discussion is similar), and θ_t^N is a $1-$dimensional $\mathfrak{F}_t-$predictable process such that

$$\theta_t^N \in \mathcal{F}_k^{2,loc}(R^1) \cap \mathcal{F}_k^1(R^1). \tag{3.13}$$

For this we first show that the right hand side of (3.11) makes sense. In fact, by assumption $\exists \sigma_n \uparrow \infty$, σ_n is a stoping time for each n such that $\theta_t^N I_{t \leq \sigma_n} \in \mathcal{F}_k^{2,loc}(R^1)$. So $\int_0^{t\wedge\sigma_n} \int_Z \theta_s^N \cdot \tilde{N}_k(ds, dz)$ makes sense. On the other hand, since $\int_0^{t\wedge\sigma_n} \int_Z \theta_s^N \cdot N_k(ds, dz)$ is RCLL in t, it has only a finite number of discontinuous points in t for each ω. So as the product is a finite product, it also makes sense. Therefore $z_{t\wedge\sigma_n}^N$ has meaning for each n, and so also does $z_{t\wedge\sigma}^N$. Now let

$$x_t = \prod_{0<s\leqslant t}(1+\int_Z \theta_s^N(z)N_k(\{s\},dz)),$$
$$y_t = e^{-\int_0^t \int_Z \theta_s^N(z)\pi(dz)ds}.$$

By the formula for integration by parts (Theorem 96)
$dx_t y_t = x_{t-}dy_t + y_{t-}dx_t + d[x,y]_t,$
where $[x,y]_t = \langle x^c, y^c\rangle_t + \sum_{0<s\leq t}\triangle x_s \triangle y_s$, x_t^c is the continuous martingale part of x_t. However, we easily see that $[x,y]_t = 0$, since y_t^c and $\triangle y_t$ are zero. Hence
$x_t y_t - 1 = \int_0^t x_{s-}dy_s + \sum_{0<s\leq t} y_{s-} \triangle x_s = \int_0^t x_{s-}dy_s$
$+\sum_{0<s\leq t} y_{s-}(x_s - x_{s-}) = -\int_0^t x_{s-}y_s \int_Z \theta_s^N(z)\pi(dz)ds$
$+\sum_{0<s\leq t} x_{s-}y_{s-}[(1+\int_Z \theta_s^N(z)N_k(\{s\},dz)) - 1]$
$= \int_0^t x_{s-}y_s \int_Z \theta_s^N(z)\widetilde{N}_k(ds,dz)).$
(3.12) is proved. Combining the two results and applying the formula for integration by parts we immediately obtain the following thoerem. (The uniqueness of the solution comes from the linearity of the coefficients).

Theorem 122 *Under assumption that* $P(\int_0^t |\theta_s^w|^2 ds < \infty) = 1$, *and* $\theta_t^N \in \mathcal{F}_k^{2,loc}(R^1) \cap \mathcal{F}_k^1(R^1)$, *let*
$z_t = z_t^w z_t^N,$
where z_t^w *and* z_t^N *are defined by (3.8) and (3.11), respectively, then* z_t *uniquely solves the following SDE with jumps:*

$$dz_t = z_t\theta_t^w dw_t + z_{t-}\int_Z \theta_t^N(z)\widetilde{N}_k(dt,dz), z_0 = 1. \tag{3.14}$$

3.3 Girsanov Transformation and Weak Solutions of SDE with Jumps

In the previous section we have already known the solution of a linear SDE:
$dz_t = z_t b_t dw_t, z_0 = 1.$
This motivates us to think about the following question: Could we use this simpler known result helping us to solve a more complicated SDE? For example, if we have such a SDE:
$dz_t = z_t b_t d\widetilde{w}_t + z_t b_t \widetilde{b}_t dt, z_0 = 1.$
then we could write it as
$dz_t = z_t b_t dw_t, z_0 = 1,$
where $dw_t = d\widetilde{w}_t + \widetilde{b}_t dt$. Could we solve the second SDE first, then let $d\widetilde{w}_t + \widetilde{b}_t dt = dw_t$, or equivalently, let $d\widetilde{w}_t = dw_t - \widetilde{b}_t dt$, to get back to the first SDE and get the solution of the first SDE? Here, obviously, some problems arise: 1) If w_t is a BM, can we see $d\widetilde{w}_t = dw_t - \widetilde{b}_t dt$ or equivalently, see

$\widetilde{w}_t = w_t - \int_0^t \widetilde{b}_s ds$ as a new BM maybe under some new probability measure? 2) If it does, then the same solution z_t from the second SDE will also satisfy the first SDE but obviously under a new probability (measure), or say, in a new probasbility space. What is the meaning of such solution which exists in a new but not the original probability space? So, the solution of this kind of problem actually leads us to study two direction of problems: 1) Find a new probability and appropriate conditions such that $\widetilde{w}_t = w_t - \int_0^t \widetilde{b}_s ds$ becomes a new BM under this new probability. This is exactly the content concerned by the Girsanov type theorem. 2) Under a new probability space find the solution of SDE. This is the discussion of the scope of a weak solution to a SDE.

In this section we will concern with such kind of problems.

First, by using the exponential solutions of the linear SDE we can establish some Girsanov type theorems, which are useful when discussing the (weak) solutions to SDE (3.1), where the coefficients may not be Lipschitz continuous. Moreover, they are also helpful in estasblishing some martingale representation theorems, which are the basic tools when deriving the non-linear filtering SDE in estimation problems involving stochastic processes.

Now in Theorem 122 let us make an additional assumption that

$$\theta_t^N > -1.$$

Then the exponential solution

$$z_t = \exp[\int_0^t \theta_s^w \cdot dw_s - \frac{1}{2}\int_0^t |\theta_s^w|^2 ds] \cdot \prod_{0<s\leqslant t} (1 + \int_Z \theta_s^N(z) N_k(\{s\}, dz)) e^{-\int_0^t \int_Z \theta_s^N(z)\pi(dz)ds}$$

of the linear SDE (3.14) is non-negative, i.e.

$$0 \leqslant z_t, \forall t \geqslant 0.$$

(Recall that w_t is a $d-$dimensional BM, and $N_k(dt, dz)$ is a $1-$dimensional Poisson counting measure). We have

Lemma 123 *Let* $\forall 0 \leqslant r \leqslant t$

$$z_t^0 = z_t = z_t^w z_t^N,$$

$$z_t^r = \exp[\int_r^t \theta_s^w \cdot dw_s - \frac{1}{2}\int_r^t |\theta_s^w|^2 ds] \cdot \prod_{r<s\leqslant t} (1 + \int_Z \theta_s^N(z) N_k(\{s\}, dz)) e^{-\int_r^t \int_Z \theta_s^N(z)\pi(dz)ds}.$$

Then

1) z_t^r *has the following properties:*

(i) $E(z_t^r | \mathfrak{F}_r) \leqslant 1, \forall 0 \leqslant r \leqslant t,$

(ii) z_t *is a non-negative* $P-$*supermartingale.*

(iii) If $E(z_t) = 1$, *then*

$$E(z_t^r | \mathfrak{F}_r) = 1, \forall 0 \leqslant r \leqslant t,$$

hence z_t *is a* $P-$*martingale.*

2) Set

$$d\widetilde{P}_t = z_t dP = z_t^w z_t^N dP. \tag{3.15}$$

If

$$|\theta_t^w|^2 + \int_Z \left|\theta_t^N(z)\right|^2 \pi(dz) \leqslant c(t), \forall 0 \leqslant t \leqslant T, \tag{3.16}$$

where $c(t) \geqslant 0$ *is non-random such that* $\int_0^T c(s)ds < \infty$, *then*

$$E|z_t|^2 \leqslant e^{2\int_0^t c(s)ds}, \forall t \in [0,T],$$

$\widetilde{P}_T$ *is a probability measure,* $\forall 0 \leqslant T < \infty$; *and* z_t *is a* $P-$*martingale.*

Proof. 1): By Ito's formula

$$dz_t = z_{t-}^w dz_t^N + z_t^N dz_t^w + 0 = z_t^N \theta_t^w \cdot z_t^w dw_t$$

$$+ z_t \int_Z \theta_t^N(z) \widetilde{N}_k(dt,dz) = \theta_t^w \cdot z_t dw_t + z_{t-} \int_Z \theta_t^N(z)\widetilde{N}_k(dt,dz). \tag{3.17}$$

Set

$$\widetilde{\tau}_N^r = \inf\left\{ s \geqslant r : \int_r^s (|\theta_t^w z_t|^2 + \int_Z \left|\theta_t^N(z) z_t\right|^2 \pi(dz))dt > N \right\}.$$

Then $\forall 0 \leqslant r \leqslant t$

$$E(z_{t\wedge\widetilde{\tau}_N^r}^r |\mathfrak{F}_r) = 1.$$

Hence by Fatou's lemma letting $N \to \infty$ one obtains (i). Notice that

$$z_r z_t^r = z_t.$$

Hence by (i)

$$E(z_t |\mathfrak{F}_r) \leqslant z_r.$$

(ii) is obtained. Now notice that for $0 \leqslant r \leqslant t$

$$E(z_t) = E(z_r E(z_t^r |\mathfrak{F}_r)).$$

Hence $E(z_t^r |\mathfrak{F}_r) < 1$ implies that $E(z_t) < 1$. Therefore (iii) is derived.

2): By Ito's formula

$$|z_{t\wedge\tau_n}|^2 = 1 + 2\int_0^{t\wedge\tau_n} z_{s-}dz_s + [z]_s$$

$$= 1 + 2\int_0^{t\wedge\tau_n} \int_Z \theta_s^N(z)(z_{s-})^2 \widetilde{N}_k(ds,dz) + 2\int_0^{t\wedge\tau_n} \theta_s^w \cdot (z_s)^2 dw_s$$

$$+ \int_0^{t\wedge\tau_n} |\theta_s^w z_s|^2 ds + \int_0^{t\wedge\tau_n} \int_Z \left|\theta_s^N(z) z_{s-}\right|^2 N_k(ds,dz),$$

where

$$\tau_n = \inf\{t \in [0,T] : |z_t| > n\}; \tau_n = T, \text{ for } inf\,\{\phi\}.$$

Thus

$$E|z_{t\wedge\tau_n}|^2 = 1 + E\int_0^{t\wedge\tau_n} |\theta_s^w z_s|^2 ds + E\int_0^{t\wedge\tau_n}\int_Z \left|\theta_s^N(z) z_{s-}\right|^2 N_k(ds,dz)$$

$$\leqslant 1 + 2\int_0^t c(s) E|z_{s\wedge\tau_n}|^2 ds.$$

By Gronwall's inequality and Fatou's lemma $\forall t \geqslant 0$

$$E|z_t|^2 \leqslant e^{2\int_0^t c(s)ds}.$$

Recall that

$$dz_t = \theta_t^w \cdot z_t dw_t + z_{t-}\int_Z \theta_t^N(z)\hat{N}(dt,dz), z_0 = 1.$$

Hence z_t is a $P-$martingale. It yields that

$$Ez_t = Ez_t^w z_t^N = 1.$$

Therefore $\widetilde{P}_T$ is a probability measure. ■

Now we can give the following interesting and useful Girsanov transformation theorem.

Theorem 124 ***(A Girsanov type Theorem).*** *Assume that $(\Omega, \mathfrak{F})$ is a standard measurable space. (For standard measurable space see the definition below).*

1) If for each $0 < T < \infty$,

$$|\theta_t^w|^2 + \int_Z \left|\theta_t^N(z)\right|^2 \pi(dz) \leqslant c(t),$$

where $c(t) \geqslant 0$ is non-random such that $\int_0^T c(t)dt < \infty$, for each $T < \infty$, then $\hat{P}_T$, defined by (3.15), for each $0 \leqslant T < \infty$ is a probability measure, and there exists a probability measure $\widetilde{P}$ defined on $(\Omega, \mathfrak{F})$ such that $\widetilde{P}\mid_{\mathfrak{F}_T} = \hat{P}_T$, for each $0 \leqslant T < \infty$. Furthermore,

$$w_t' = w_t - \int_0^t \theta_s^w ds, 0 \leqslant t,$$

is a BM under probability $\hat{P}$;

$$\begin{aligned} \widetilde{N'}(dt, dz) &= \widetilde{N}(dt, dz) - \theta_t^N(z)\pi(dz)dt \\ &= N(dt, dz) - (1 + \theta_t^N(z))\pi(dz)dt, 0 \leqslant t, \end{aligned}$$

is a $\hat{P}-$martingale measure with the new compensator $(1 + \theta_t^N(z))\pi(dz)dt$.

In particular, if $\theta_t^N(z) \equiv 0$, then under the above condition we have that $\widetilde{N'}(dt, dz) = \widetilde{N}(dt, dz) = N(dt, dz) - \pi(dz)dt$,
that is, $\widetilde{N}(dt, dz)$ is still a Poisson martingale measure with the same compensator $\pi(dz)dt$ under the new probability measure $\widetilde{P}$.

2) Assume that $\widetilde{P}_t$ defined by (3.15) with $\theta_t^N(z) \equiv 0$, i.e. $z_t^N = 1$ and for each $0 \leqslant t < \infty$, $d\widetilde{P}_t = z_t^w dP$ is a probability measure, i.e.

$$E\widetilde{P}_T = 1, \forall 0 \leqslant T < \infty,$$

and there exists a probability measure $\widetilde{P}$ defined on $(\Omega, \mathfrak{F})$ such that $\widetilde{P}\mid_{\mathfrak{F}_T} = \widetilde{P}_T$, for each $0 \leqslant T < \infty$.

If for any given $0 < T < \infty$,

$$\int_0^T |\theta_t^w|^2 \, dt < \infty, P - a.s.,$$

then $w_t', t \geqslant 0$, defined in 1), is still a $\widetilde{P}-BM$; and the original $\widetilde{N}(dt, dz), t \geqslant 0, z \in Z$, is still a $\widetilde{P}-$Poisson martingale with the same compensator $\pi(dz)dt$.

Definition 125 *A measurable space is called a standard measurable space, if it can be Borel isomorphic to a measurable subset of a complete separable metric space with the induced topological $\sigma-$field, that is, there is a bijection map between these two sets such that both the map itself and its inverse map are measurable.*

Many measurable spaces are standard measurable spaces. For example, R^d with its Borel sets is a standard measurable space. $C([0,\infty),R^1)$, the totality of all continuous R^1-valued functions defined on $[0,\infty)$ with the metric $\|X\| = \sum_{i=1}^{\infty} \frac{\|X\|_i \wedge 1}{2^i}$, where $\|X\|_T = \sup_{t\leq T}|X(t)|$, $\forall T<\infty, \forall X \in C([0,\infty),R^1)$ such that the Borel field is generated by all open sets of the topology from this metric, is also a standard measurable space. Furthermore, the space $D([0,\infty),R^1)$, the totality of all RCLL (right continuous with left limit) R^1-valued functions defined on $[0,\infty)$ with the Skorohod metric, (see Lemma 388 in the Appendix), such that the Borel field is generated by all open sets of the topology from this metric, is also a standard measurable space.

The standard measurable space has many nice properties. For example, we have the following lemma about the existence of an extension probability measure for a family of consitent probability measures. (See 4.1 in Chapter 4 of [49]).

Lemma 126 *Suppose that $(\Omega,\mathfrak{F})$ is a standard measurable space, $\{\mathfrak{F}_t\}_{t\geq 0} \subset \mathfrak{F}$ is an increasing $\sigma-$ field fitration such that $\mathfrak{F} = \vee_{t\geq 0}\mathfrak{F}_t$. If $\{\mu_t, \mathfrak{F}_t\}_{t\geq 0}$ is a family of consistent probability measures, that is, as $s<t$, $\mu_t|_{\mathfrak{F}_s} = \mu_s$, and they are absolutely continuous with respect to a probability measure P, and P is defined on $\mathfrak{F}$; then there exists a unique probability measure μ such that $\mu|_{\mathfrak{F}_t} = \mu_t, \forall t \geq 0$.*

Now let us prove Theorem 124.

Proof. 1): Under assumption 2) of Lemma 123 one has that $\hat{P}_T$, defined by (3.15), for each $0 \leqslant T < \infty$, is a probability measure. Since $\left\{\widetilde{P}_T, 0 \leqslant T < \infty\right\}$ is a system of consistent probability measures, which is absolutely continuous with respect to the original probability measure P, that is, as $t<T$, $\widetilde{P}_T|_{\mathfrak{F}_t} = \widetilde{P}_t$, $\widetilde{P}_T \ll P$, and $(\Omega,\mathfrak{F})$ is a standard measurable space, by Lemma 126 a unique extension probability measure $\widetilde{P}$ defined on $(\Omega,\mathfrak{F})$ exists. Now by Ito's formula for any $A \in \mathfrak{F}_s, \eta \in R^d$

$$d(e^{i\eta\cdot(w'_t-w'_s)}z_tI_A) = I_Ae^{i\eta\cdot(w'_t-w'_s)}[z_t^N\theta_t^w \cdot z_t^w dw_t$$
$$+z_t^w \int_Z \theta_t^N(z) \cdot z_{t-}^N \widetilde{N}_k(dt,dz) + i\eta\cdot(dw_t - \theta_t^w dt)z_t^N z_t^w$$
$$-\tfrac{1}{2}|\eta|^2 z_t^N z_t^w dt + i\eta\cdot\theta_t^w z_t^N z_t^w dt]$$
$$= dM_t - \tfrac{1}{2}|\eta|^2 z_t I_A e^{i\eta(w'_t-w'_s)}dt,$$

where M_t is a local $P-$martingale, and we have used the fact that

$$\sum_{r\in dt}[e^{i\eta\cdot(w'_{r-}-w'_s)}z_{r-}^w(z_{r-}^N + \triangle z_r^N) - e^{i\eta\cdot(w'_{r-}-w'_s)}z_{r-}^w z_{r-}^N$$
$$-e^{i\eta\cdot(w'_{r-}-w'_s)}z_{r-}^w \triangle z_{r-}^N] = 0.$$

1): If (3.16) holds, then by Lemma 123 M_t is a $P-$martingale. $EM_t = 0$. Hence after solving an ordinary differential equation one finds that

$$E[e^{i\eta\cdot(w'_t-w'_s)}z_tI_A] = \widetilde{P}(A)e^{-\frac{1}{2}|\eta|^2(t-s)}.$$

Or,

$$E_{\widetilde{P}}[e^{i\eta\cdot(w'_t-w'_s)} \mid \mathfrak{F}_s] = e^{-\frac{1}{2}|\eta|^2(t-s)}.$$

Therefore $w'_t, 0 \leqslant t$, is a BM under probability $\hat{P}$. Now for any given $U \in \mathfrak{B}(Z)$, with $\pi(U) < \infty$, and $A \in \mathfrak{F}_s$ write
$\widetilde{N_t} = \widetilde{N}((0,t],U), \widetilde{N'_t} = \widetilde{N'}((0,t],U), N_t = N((0,t],U)$.
and let
$x_t = z_t(\widetilde{N'_t} - \widetilde{N'_s})I_A$.
Then by Ito's formula

$$\begin{aligned} dx_t &= I_A z_{t-}(d\tilde{N}_t - \int_U \theta_t^N(z)\pi(dz)dt) + x_t\theta_t^w \cdot dw_t \\ &+x_{t-}\int_Z \theta_t^N(z)\widetilde{N}(dt,dz) + I_A z_{t-}\int_U \theta_t^N(z)N(dt,dz) \\ &= x_t\theta_t^w \cdot dw_t + \int_Z (I_A z_{t-}I_U(z) + x_{t-}\theta_t^N(z) \\ &+I_A\theta_t^N(z)z_{t-}I_U(z))\widetilde{N}(dt,dz). \end{aligned}$$

Again by Ito's formula

$$\begin{aligned} |x_t|^2 = 2\int_s^t \int_Z x_{r-}(I_A z_{r-}I_U(z) + x_{r-}\theta_r^N(z) \\ +I_A\theta_r^N(z)z_{r-}I_U(z))\widetilde{N}(dr,dz) + 2\int_s^t |x_r|^2\theta_r^w \cdot dw_r + \int_s^t |x_r\theta_r^w|^2 dr \\ +\int_s^t\int_Z \left|I_A z_{r-}I_U(z) + x_{r-}\theta_r^N(z) + I_A\theta_r^N(z)z_{r-}I_U(z)\right|^2 N(dr,dz). \end{aligned}$$

Set
$\overline{\tau}_n = \inf\{t \geqslant s : |x_t| > n\}$.
By 2) of Lemma 123 as $t \in [0,T]$

$$\begin{aligned} E|x_{t\wedge\overline{\tau}_n}|^2 &\leqslant E\int_s^{t\wedge\overline{\tau}_n} c(r)|x_r|^2 dr \\ +3E\int_s^{t\wedge\overline{\tau}_n} &(|z_r|^2\pi(U) + c(r)|x_r|^2 + c(r)|z_r|^2)dr \\ &\leqslant k'_0\int_s^t (E|x_{r\wedge\overline{\tau}_n}|^2 + E|z_{r\wedge\overline{\tau}_n}|^2)c'(r)dr, \end{aligned}$$

where $c'(r) = c(r)+1$. Hence by Lemma 123, Gronwall's inequality and Fatou's lemma $\forall T \geqslant t \geqslant s$
$E|x_t|^2 \leqslant k''_T e^{k'_0\int_s^t c'(r)dr}$,
where $0 \leqslant k''_T$ is a constant depending on $T, \pi(U)$ and k_0 only. Therefore $x_t, s \leqslant t \leqslant T$, is a $P-$martingale. This shows that $\forall A \in \mathfrak{F}_s$,
$E[(\widetilde{N'_t} - \widetilde{N'_s})z_t I_A] = 0$.

Therefore $\widetilde{N}'_t$ is a $\widetilde{P}-$martingale. Since $\widetilde{P}-a.s.$

$\widetilde{N}'(dt,dz) = \widetilde{N}(dt,dz) - \theta_t^N(z)\pi(dz)dt$

$= N(dt,dz) - (1+\theta_t^N(z))\pi(dz)dt, 0 \leqslant t.$

Hence $\widetilde{N}'(dt,dz)$ is a $\widetilde{P}-$martingale measure with the compensator $(1+\theta_t^N)\pi(dz)dt$. In the case that $\theta_t^N(z) \equiv 0$ we immediately find that $\widetilde{P}-a.s.$

$\widetilde{N}'(dt,dz) = \widetilde{N}(dt,dz) = N(dt,dz) - \pi(dz)dt, 0 \leqslant t.$

Hence in this case $\widetilde{N}(dt,dz)$ is a $\widetilde{P}-$Poisson martingale measure with the same compensator $\pi(dz)dt$.

2): Now suppose that

$\int_0^T |\theta_t^w|^2 dt < \infty, \forall 0 \leqslant t \leqslant T, P-a.s.$

$\theta_t^N(z) \equiv 0, \forall 0 \leqslant t, P-a.s.$

Let

$\theta_{n,t}^w = \theta_t^w I_{|\theta_t^w|<n}, z_{n,t}^w = \exp[\int_0^t \theta_{n,s}^w \cdot dw_s - \frac{1}{2}\int_0^t |\theta_{n,s}^w|^2 ds], z_{n,t} = z_{n,t}^w.$

Obviously, as $n \to \infty$

$\int_0^T |\theta_t^w - \theta_{n,t}^w|^2 dt \to 0, P-a.s.$

Hence, as $n \to \infty$

$\int_0^T |\theta_t^w - \theta_{n,t}^w|^2 dt \to 0,$ in probability P.

Write

$d\widetilde{P}_n|_{\mathfrak{F}_t} = z_{n,t}dP = z_{n,t}^w dP,$

$w'_{n,t} = w_t - \int_0^t \theta_{n,s}^w ds.$

Then by 1) $w'_{n,t}, t \in [0,T]$, is a BM under probability measure $\widetilde{P}_n$, and $\widetilde{N}((dt,dz)$ is a $\widetilde{P}_n-$ martingale with the same compensator $\pi(dz)dt$. Moreover, $\forall t \in [0,T]$, as $n \to \infty$

$w'_{n,t} \to w'_t,$ in probability P.

Now let us show that as $n \to \infty$

$E[e^{i\eta\cdot(w'_{n,t}-w'_{n,s})} z_{n,t} I_A] \to E[e^{i\eta\cdot(w'_t-w'_s)} z_t I_A],$

for $\forall s < t, A \in \mathfrak{F}_s$. First, since $w'_{n,t}$ converges to w'_t in probability,

$E[e^{i\eta\cdot(w'_{n,t}-w'_{n,s})} z_t I_A] \to E[e^{i\eta\cdot(w'_t-w'_s)} z_t I_A],$ as $n \to \infty$.

Secondly, since by assumption $Ez_t = Ez_{n,t} = 1, \forall n, \forall 0 \leqslant t \leqslant T,$

$E|z_t - z_{n,t}| = E(|z_t - z_{n,t}| + z_t - z_{n,t}).$

However,

$0 \leqslant |z_t - z_{n,t}| + z_t - z_{n,t} \leqslant 2z_t,$

and it is not difficult to see that as $n \to \infty$ for a fixed $t \in [s,T]$ (take a subsequence, if necessary)

$|z_t - z_{n,t}| + z_t - z_{n,t} \to 0,$ in probability.

Applying Lebesgue's dominated convergence theorem one finds that for a fixed $t \in [s,T]$ as $n \to \infty$

$E|z_t - z_{n,t}| = E(|z_t - z_{n,t}| + z_t - z_{n,t}) \to 0.$

Therefore, $\forall A \in \mathfrak{F}_s$, for a fixed $t \in [s,T]$, by Lebesgue's dominated convergence theorem

$$\left|Ee^{i\eta\cdot(w'_{n,t}-w'_{n,s})}z_{n,t}I_A - Ee^{i\eta\cdot(w'_t-w'_s)}z_tI_A\right| \leqslant E\left|z_{n,t}-z_t\right|$$
$$+E(\left|e^{i\eta\cdot(w'_{n,t}-w'_{n,s})} - e^{i\eta\cdot(w'_t-w'_s)}\right| z_t) \to 0, \quad \text{as } n \to \infty$$

Recall that
$Ee^{i\eta\cdot(w'_{n,t}-w'_{n,s})}z_{n,t}I_A = e^{-\frac{1}{2}|\eta|^2(t-s)}Ez_{n,t}I_A.$
By letting $n \to \infty$ one finds that
$Ee^{i\eta\cdot(w'_t-w'_s)}z_tI_A = e^{-\frac{1}{2}|\eta|^2(t-s)}Ez_tI_A = \widetilde{P}_T(A)e^{-\frac{1}{2}|\eta|^2(t-s)},$
where we have used the result that
$Ez_tI_A = E(z_tI_AE(z_T^t\,|\mathfrak{F}_t)) = Ez_TI_A = \widetilde{P}_T(A).$
Therefore, $w'_t, t \in [0,T]$, is a $\widetilde{P}_T-$BM. Similarly, by
$E[e^{i\sum_{j=1}^m \lambda_j N((s,t],U_j)}z_{n,t}I_A] = E[e^{(t-s)\sum_{j=1}^m (e^{i\lambda_j}-1)\pi(U_j)}z_{n,t}I_A]$
and letting $n \to \infty$ one has that
$E[e^{i\sum_{j=1}^m \lambda_j N((s,t],U_j)}z_tI_A] = E[e^{(t-s)\sum_{j=1}^m (e^{i\lambda_j}-1)\pi(U_j)}z_tI_A],$
where $\forall s < t, A \in \mathfrak{F}_s, U_j \in \mathfrak{B}_Z, i = 1,2,\cdots,m$ such that $U_i \cap U_j = \phi$, as $i \neq j$, and $\pi(U_j) < \infty, \forall j$, i.e. $U_j \in \Gamma, \forall j$. Therefore, $N((0,t],U), t \in [0,T], U \in \Gamma$ is a $\widetilde{P}_T-$Poisson counting measure with the compensator $\pi(U)t$. That is to say $\widetilde{N}((0,t],U) = N((0,t],U) - \pi(U)t$ is a $\widetilde{P}_T-$Poisson martingale measure with the compensator $\pi(U)t$. Since the results obtained here are true for any $0 < T < \infty$. By assumption $\widetilde{P}$ is a probability measure such that $\widetilde{P}\,|_{\mathfrak{F}_T} = \widetilde{P}_T, \forall T > 0$. Hence the final conclusion is derived. ■

Girsanov type theorems are powerful tools for obtaining the existence of weak solutions to SDE (3.1) under very weak conditions. Let us first introduce the concep of a weak solution for a SDE. For this we rewrite the $d-$dimensional SDE as follows:

$$dx_t = b(t,x_t)dt + \sigma(t,x_t)dw_t + \int_Z c(t,x_{t-},z)\tilde{N}_k(dt,dz), \qquad (3.18)$$

and suppose that x_0 has a probability distribution law μ, i.e. $\mu(A) = P(x_0 \in A), \forall A \in \mathfrak{B}(R^d)$.

Definition 127 *If there exist a probability space (with a $\sigma-$field filtration) $(\widetilde{\Omega}, \widetilde{\mathfrak{F}}, \left(\widetilde{\mathfrak{F}}_t\right)_{t\geqslant 0}, \widetilde{P})$ and a $\widetilde{\mathfrak{F}}_t-$adapted BM $\overline{w}_t, t \geqslant 0$, a Poisson martingale measure $\overline{\widetilde{N}}(dt,dz)$ (also $\widetilde{\mathfrak{F}}_t-$adapted) with the given beforehand compensator $\pi(dz)dt$, and a $\widetilde{\mathfrak{F}}_t-$adapted processes $\{x_t\}_{t\geq 0}$ define on this probability space such that* (3.18) *holds, $\overline{P}-a.s.$, and $\overline{P}(x_0 \in A) = \mu(A), \forall A \in \mathfrak{B}(R^d)$, then we say that*
$(\widetilde{\Omega}, \widetilde{\mathfrak{F}}, \left(\widetilde{\mathfrak{F}}_t\right)_{t\geqslant 0}, \widetilde{P}; \{w_t\}_{t\geq 0}, \overline{\widetilde{N}}(dt,dz), \{x_t\}_{t\geq 0})$ *(or simply, $\{x_t\}_{t\geq 0}$) is a weak solution of (*3.18*).*

From the definition the "weak" sense is that it may happen that $x_t \notin \widetilde{\mathfrak{F}}_t^{\widetilde{w},\widetilde{\widetilde{N}}}$.

Remark 128 *Given a Poisson martingale measure $\widetilde{\widetilde{N}}(dt,dz)$ with the compensator $\pi(dz)dt$ on a probability space $(\widetilde{\Omega},\widetilde{\mathfrak{F}},\left(\widetilde{\mathfrak{F}}_t\right)_{t\geqslant 0},\widetilde{P})$ means that there exists a Poisson random counting measure $\overline{N}(dt,dz)$ with the compensator $\pi(dz)dt$ such that*

$\widetilde{\overline{N}}(dt,dz) = \overline{N}(dt,dz) - \pi(dz)dt,$

where $\widetilde{\overline{N}}((0,t],U)$ and $\overline{N}((0,t],U), t>0, U \in \mathfrak{B}_Z$, are $\widetilde{\mathfrak{F}}_t-$ adapted RCLL (right continuous with left limit). However, from this one easily constructs a $\widetilde{\mathfrak{F}}_t-$ Poisson point process $\overline{p}$ that generates this Poisson counting measure. In fact, let

$D_{\overline{p}(\omega)} = \left\{t>0 : \int_Z \overline{N}(\{t\},dz,\omega)=1\right\}$

$= \{t>0 :$ *there exists a unique point, denote it by* $z(t,\omega)\in Z$,

such that $\overline{N}(\{t\},\{z(t,\omega)\},\omega)=1\}$.

Define $\overline{p}(t,\omega) = z(t,\omega), \forall t \in D_{\overline{p}(\omega)},$. Then we get a $\widetilde{\mathfrak{F}}_t-$ Poisson point process $\overline{p}$ such that

$\overline{N}(dt,dz) = N_{\overline{p}}(dt,dz).$

For the weak solution of (3.18) we sometimes also need to discuss its weak uniqueness.

Definition 129 *If for any two weak solutions*

$(\widetilde{\Omega}^i,\widetilde{\mathfrak{F}}^i,\left(\widetilde{\mathfrak{F}}_t^i\right)_{t\geqslant 0},\widetilde{P}^i;\{w_t^i\}_{t\geq 0},\widetilde{\overline{N}}^i(dt,dz),\{x_t^i\}_{t\geq 0}), i=1,2,$

of (3.18) having the same initial law μ, i.e. $\widetilde{P}^1(x_0^1\in A) = \widetilde{P}^2(x_0^1 \in A) = \mu(A), \forall A \in \mathfrak{B}(R^d)^{\otimes n}$, implies that they have the same probability distribution, that is, $\forall t_1<t_2<\cdots<t_n, \forall A \in \mathfrak{B}(R^d)^{\otimes n}$,

$\widetilde{P}^1((x_{t_1}^1,x_{t_2}^1,\cdots,x_{t_n}^1)\in A) = \widetilde{P}^2((x_{t_1}^2,x_{t_2}^2,\cdots,x_{t_n}^2)\in A),$

then we say that the weak uniqueness of weak solutions to (3.18) holds, where

$\mathfrak{B}(R^d)^{\otimes n} = \mathfrak{B}(R^d)\times\cdots\times\mathfrak{B}(R^d)$ *(n-times).*

Now we easily answer the question put out at the beginning of this section.

Proposition 130 *Assume that for each $0<T<\infty$,*

$$|\theta_t^w|^2 + \int_Z \left|\theta_t^N(z)\right|^2 \pi(dz) \leqslant c(t),$$

where $\theta_t^w(\omega)$ is a $d-$dimensional $\mathfrak{F}_t-$adapted process, $\theta_t^N(z,\omega)$ is a 1 - dimensional $\mathfrak{F}_t-$predictable process, and $c(t)\geqslant 0$ is non-random such that $\int_0^T c(t)dt<\infty$, for each $T<\infty$. If $b_t(\omega)$ is a $d-$dimensional $\mathfrak{F}_t-$adapted *process, such that* $|b_t(\omega)| \leq k_0$*, where* $k_0 \geq 0$ *is a constant, then the following SDE has a weak unique weak solution:* $\forall t \geq 0$,

$dz_t = \theta_t^w \cdot z_t b_t(\omega)dt + \theta_t^w \cdot z_t dw_t + z_{t-} \int_Z \theta_t^N(z)\widetilde{N}_k(dt, dz),$

$z_0 = c_0,$

where $c_0 \geq 0$ *is a constant,* w_t *is a* $d-$*dimensional BM, and* $\widetilde{N}_k(dt, dz)$ *is a* $1-$*dimensional Poisson martingale measure.*

Proof. Let

$$z_t = c_0 \exp[\int_0^t \theta_s^w \cdot dw_s - \frac{1}{2}\int_0^t |\theta_s^w|^2 ds] \cdot$$

$$\cdot \prod_{0<s\leqslant t} (1 + \int_Z \theta_s^N(z) N_k(\{s\}, dz))e^{-\int_0^t \int_Z \theta_s^N(z)\pi(dz)ds}. \quad (3.19)$$

Then by Lemma 123 $P-a.s.$

$$dz_t = \theta_t^w \cdot z_t dw_t + z_{t-} \int_Z \theta_t^N(z)\widetilde{N}_k(dt, dz). \quad (3.20)$$

Moreover, z_t is the unique solution of the above SDE. Now since $|b_t(\omega)| \leq k_0$, by Theorem 124 $\widetilde{N}_k(dt, dz)$ and

$\widetilde{w}_t = w_t - \int_0^t b_s ds, 0 \leqslant t,$

are a Poisson martingale measure (the same original one) and a BM under the new probability (measure) $\widetilde{P}$ with

$\frac{d\widetilde{P}}{dP}|_{\mathfrak{F}_T} = \exp[\int_0^T b_s \cdot dw_s - \frac{1}{2}\int_0^T |b_s|^2 ds],$

respectively. Hence, $\widetilde{P}-a.s.$

$dz_t = \theta_t^w \cdot z_t b_t(\omega)dt + \theta_t^w \cdot z_t d\widetilde{w}_t + z_{t-} \int_Z \theta_t^N(z)\widetilde{N}_k(dt, dz).$

That is, z_t is a weak solution of the above SDE. The weak unqueness is derived from the fact that any weak solution of the above SDE can always be returned to the expression (3.19) with some BM and the same Poisson counting measure under some probability measure. So the probability distribution of the weak solution should be the same. ∎

Furthermore, from Theorem 124 it is immediately seen that the following theorem holds.

Theorem 131 *Assume that* $b^0(t, x)$ *is* $R^{d\otimes 1}-$*valued, and* $|b^0(t, x)|^2 \leqslant k_T$, $\forall 0 \leqslant t \leqslant T$*, for any given* T*, where* k_T *is a constant depending on* T *only, and* $\sigma^{-1}(t, x)$ *exists such that,* $|\sigma^{-1}(t, x)| \leq c_1(t)$*, where* $c_1(t) \geq 0$ *is non-random and* $\int_0^T c_1(t)dt < \infty$*, for each* $T < \infty$*. Then the following statements are equivalent:*

1) (3.18) has a weak solution $\{x_t\}_{t\geq 0}$ *with the initial value* $x_0 = x$*, where* $x \in R^d$ *is a constant vector.*

2) SDE

$$\begin{aligned} dx_t &= (b(t, x_t) + b^0(t, x_t))dt + \sigma(t, x_t)dw_t \\ &\quad + \int_Z c(t, x_{t-}, z)\tilde{N}_k(dt, dz), \end{aligned} \quad (3.21)$$

has a weak solution $\{x_t\}_{t\geq 0}$ *with the initial value* $x_0 = x_0$, *where* $x_0 \in R^d$ *is a constant vector.*

In fact, (3.21) can be rewritten as

$dx_t = b(t,x_t)dt + \sigma(t,x_t)d\widetilde{w}_t + \int_Z c(t,x_{t-},z)\tilde{N}_k(dt,dz),$

where $d\widetilde{w}_t = \sigma^{-1}(t,x_t)b^0(t,x_t)dt + dw_t$, and the rest involves applying the Girsanov type theorem.

Beware of the fact that the coefficient $b(s,x) + b^0(t,x)$ in (3.21) now can be very discontinuous in x. Moreover, if we let $b \equiv 0$, then under the condition of Theorem 131, (3.21) always has a weak solution.

To relax the bounded condition on $b^0(t,x)$ we only need to weaken the sufficient conditions for Girsanov theorems to be satisfied. The following Theorem is useful.

Theorem 132 ***(Girsanov type Theorem).*** *Assume that* x_t *solves the following* $d-$ *dimensional SDE*

$dx_t = b(t,x_t)dt + \sigma(t,x_t)dw_t + \int_Z c(t,x_{t-},z)\tilde{N}_k(dt,dz),$

with the initial value $x_0 = x_0$, *where* $x_0 \in R^d$ *is a constant vector and, moreover, assume that*

$\left|b^0(t,x)\right|^2 + |b(t,x)|^2 + |\sigma(t,x)|^2 + \int_Z |c(t,x,z)|^2 \pi(dz)$
$\leq k_0(1+|x|^2),$

and σ^{-1} *exists such that*

$\left|\sigma^{-1}(t,x)\right| \leq c_1(t)$, *where* $c_1(t) \geq 0$ *is non-random and* $\int_0^T c_1(t)dt < \infty$, *for each* $T < \infty$.

Write

$z_t(\sigma^{-1}b^0) = \exp[\int_0^t (\sigma^{-1}b^0)(s,x_s)\cdot dw_s - \frac{1}{2}\int_0^t \left|(\sigma^{-1}b^0)(s,x_s)\right|^2 ds],$

$d\widetilde{P}_t = z_t dp,$

$\widetilde{w}_t = \int_0^t (\sigma^{-1}b^0)(s,x_s)ds + w_t,$

Then $\widetilde{P}_t$ *is a probability measure, and for each* $0 < T < \infty$, $\widetilde{w}_t, t \in [0,T]$, *is a BM under the new probability measure* $\widetilde{P}_T$, *and* $\tilde{N}_k((0,t],dz)$, $t \in [0,T]$, *is a Poisson martingale measure under the new probability measure* $\widetilde{P}_T$ *with the same compensator* $\pi(dz)t$. *Notice that here the space* Ω *and the* $\sigma-$*filtration* $\{\mathfrak{F}_t\}_{t\geqslant 0}$ *do not change. Furthermore, for each* $T < \infty$,

$$(\Omega, \mathfrak{F}, \{\mathfrak{F}_t\}_{t\in[0,T]}, \widetilde{P}_T; \{x_t\}_{t\in[0,T]}, \{\widetilde{w}_t\}_{t\in[0,T]}, \left\{\tilde{N}_k((0,t],dz)\right\}_{t\in[0,T],dz\in\mathfrak{B}_Z})$$

is a weak solution of the following SDE: $\widetilde{P}_T - a.s.$ $\forall t \in [0,T]$,

$dx_t = (b^0(t,x_t) + b(t,x_t))dt + \sigma(t,x_t)d\widetilde{w}_t + \int_Z c(t,x_{t-},z)\tilde{N}_k(dt,dz),$

with the initial value $x_0 = x_0$, *where* $x_0 \in R^d$ *is a constant vector. Furthermore, if* $(\Omega,\mathfrak{F})$ *is a standard measurable space, then a probability* $\widetilde{P}$ *exists such that* $\widetilde{P}|_{\mathfrak{F}_T} = \widetilde{P}_T, \forall T < \infty$. *Thus the result is generalized to all* $t \geq 0$. *Or, simply speaking,* $\{x_t\}_{t\geq 0}$ *is a weak solution of the above SDE,* $\forall t \geq 0, \widetilde{P} - a.s.$

Proof. Let

$$b^N(t,x) = \begin{cases} b^0(t,x), & \text{as } |x| \le N, \\ 0, & \text{otherwise.} \end{cases}$$

Then by Lemma 123

$$d\widetilde{P}_t^N = z_t(\sigma^{-1}b^N)dp$$

is a probability measure. Applying Theorem 124 for any given $0 \le T < \infty$ as $0 \le t \le T$

1) $w_t^N = w_t - \int_0^t (\sigma^{-1}b^N)(s,x_s)ds$,
is a BM under the probability $\hat{P}_T^N$;

2) $\tilde{N}_k(dt,dz) = N_k(dt,dz) - \pi(dz)dt$
is still a Poisson random martingale measure with the same compensator $\pi(dz)dt$ under the probability $\hat{P}_T^N$, where $N_k(dt,dz)$ is a Poisson counting measure with the compensator $\pi(dz)dt$ under the original probability P. Let

$$\tau_N = \inf(t \in [0,T] : |x_t| > N), \text{ and } \tau_N = T, \text{ for } \inf\{\phi\}$$

Then x_t solves the following SDE: as $0 \le t \le T$

$$x_{t\wedge\tau_N} = x_0 + \int_0^{t\wedge\tau_N}(b^0(s,x_s) + b(s,x_s))ds + \int_0^{t\wedge\tau_N}\sigma(s,x_s)dw_s^N$$
$$+ \int_0^{t\wedge\tau_N}\int_Z c(s,x_{s-},z)\tilde{N}_k(ds,dz),\ \hat{P}_T^N - a.s.$$

By Lemma 114

$$E^{\tilde{P}_T^N}(\sup_{t\le T}|x_{t\wedge\tau_N}|^2) \le k_T.$$

Hence

$$N^2\hat{P}_T^N(\tau_N < T) \le E^{\tilde{P}_T^N}(|x_{T\wedge\tau_N}|^2 I_{\tau_N<T}) \le k_T,$$

Therefore as $N \to \infty$

$$\tilde{P}_T^N(\tau_N < T) \to 0.$$

This yields

$$Ez_T(\sigma^{-1}b^0) \ge Ez_T(\sigma^{-1}b^N)I_{\tau_N\ge T} = P^N(\tau_N \ge T) \to 1, \text{ as } N \to \infty.$$

On the other hand we always have that

$$Ez_T(\sigma^{-1}b^0) \le 1.$$

Hence $\widetilde{P}_T = Ez_T(\sigma^{-1}b^0) = 1$. From now on the rest is derived by 2) of Theorem 124 ■

Furthermore, we can obtain the following more general theorem.

Theorem 133 ***(Girsanov type Theorem).*** *Assume that x_t solves the following $d-$ dimensional SDE*

$$dx_t = b(t,x_t)dt + \sigma(t,x_t)dw_t + \int_Z c(t,x_{t-},z)\tilde{N}_k(dt,dz),$$

with the initial value $x_0 = x_0$, where $x_0 \in R^d$ is a constant vector and, moreover, assume that

$$\int_Z |c(t,x,z)|^2 \pi(dz) \le c_1(t),$$
$$[(x \cdot b^0(t,x)) \vee (x \cdot b(t,x)) \vee |\sigma(t,x)|^2]$$
$$\le c_1(t)(1 + |x|^2 \textstyle\prod_{k=1}^m g_k(x)),$$

σ^{-1} exists such that $|\sigma^{-1}(t,x)| \le c_1(t)$,

where

$$g_k(x) = 1 + \underbrace{\ln(1 + \ln(1 + \cdots \ln}_{k-times}(1 + |x|^{2n}))),$$

(n is some natural number), and $c_1(t) \geq 0$ *is non-random such that for each* $T < \infty$, $\int_0^T c_1(t)dt < \infty$; *furthermore,* $b^0(t,x)$ *is locally bounded for* x, *that is, for each* $r > 0$, *as* $|x| \leq r$,

$$|b^0(t,x)| \leq k_r,$$

where $k_r > 0$ *is a constant only depending on* r.

Let

$$z_t(\sigma^{-1}b^0) = \exp[\int_0^t (\sigma^{-1}b^0)(s,x_s) \cdot dw_s - \frac{1}{2}\int_0^t \left|(\sigma^{-1}b^0)(s,x_s)\right|^2 ds],$$

$$d\widetilde{P}_t = z_t dp,$$

$$\widetilde{w}_t = \int_0^t (\sigma^{-1}b^0)(s,x_s)ds + w_t,$$

Then all conclusions in Theorem 132 still hold true.

Proof. Let

$$b^N(t,x) = \begin{cases} b^0(t,x), & \text{as } |x| \leq N, \\ 0, & \text{otherwise.} \end{cases}$$

By using the notation in Theorem 132 we still have that x_t solves the following SDE: as $0 \leq t \leq T$

$$x_{t\wedge\tau_N} = x_0 + \int_0^{t\wedge\tau_N} (b^0(s,x_s) + b(s,x_s))ds + \int_0^{t\wedge\tau_N} \sigma(s,x_s)dw_s^N$$
$$+ \int_0^{t\wedge\tau_N} \int_Z c(s,x_{s-},z)\tilde{N}_k(ds,dz),\ \hat{P}_T^N - a.s.$$

Notice that by evaluation

$$\frac{\partial}{\partial x_i} g_{m+1}(x) = \prod_{k=1}^m g_k^{-1}(x) \frac{2nx_i|x|^{2n-2}}{1+|x|^{2n}},$$

$$\frac{\partial^2}{\partial x_i \partial x_j} g_{m+1}(x) = \prod_{k=1}^m g_k^{-1}(x)[\frac{2n\delta_{ij}|x|^{2n-2}+4n(n-1)x_ix_j|x|^{2n-4}}{1+|x|^{2n}}$$
$$- \frac{4n^2x_ix_j|x|^{4n-4}}{(1+|x|^{2n})^2}] - \prod_{k=1}^m g_k^{-1}(x)\frac{4n^2x_ix_j|x|^{4n-4}}{(1+|x|^{2n})^2} \sum_{k=0}^m \prod_{l=1}^m g_l^{-1}(x),$$

where we write $g_0(x) = 1$. Denote $g'_{m+1}(x) = \text{grad} g_{m+1}(x)$, and $g''_{m+1}(x) = \left[\frac{\partial^2}{\partial x_i \partial x_j} g_{m+1}(x)\right]_{i,j=1}^d$. Then by assumption we find that

$$g'_{m+1}(x) \cdot b^0(t,x) + g'_{m+1}(x) \cdot b(t,x) \leq k_0 c_1(t),$$

$$\left|g''_{m+1}(x)\right| |\sigma(t,x)|^2 \leq k_0 c_1(t),$$

$$\left|g''_{m+1}(x)\right| \leq k_0,$$

where $k_0 > 0$ is a constant. Now applying Ito's formula to $g(x_t) = g_{m+1}(x_t)$ we have that as $0 \leq t \leq T$

$$0 \leq g(x_{t\wedge\tau_N}) = g(x_0) + \int_0^{t\wedge\tau_N} g'(x_s)(b^0(s,x_s) + b(s,x_s))ds$$
$$+ \int_0^{t\wedge\tau_N} g'(x_s)\sigma(s,x_s)dw_s^N + \int_0^{t\wedge\tau_N} \int_Z g'(x_{s-})c(s,x_{s-},z)\tilde{N}_k(ds,dz)$$
$$+ \int_0^{t\wedge\tau_N} \int_Z [g(x_{s-} + c(s,x_{s-},z)) - g(x_{s-}) - g'(x_{s-})c(s,x_{s-},z)]N_k(ds,dz)$$
$$= \sum_{i=1}^5 I^i,\ \hat{P}_T^N - a.s.$$

Taking the mathematical expectation with respect to the probability measure $\tilde{P}_T^N$ in both sides we see that

$$0 \leq E^{\tilde{P}_T^N}(g(x_{t\wedge\tau_N})) \leq k_T,$$

where we have applied the facts that

$$\left|EI^5\right| \leq \frac{1}{2}E\int_0^{t\wedge\tau_N} \int_Z |g''(x_{s-} + \theta c(s,x_{s-},z))| \left|c(s,x_{s-},z)\right|^2 \pi(dz)ds \leq \frac{1}{2}k_0 \int_0^T c_1(s)ds.$$

Hence

$$g(N)\hat{P}_T^N(\tau_N < T) \leq E^{\tilde{P}_T^N}(g(x_{T\wedge\tau_N})I_{\tau_N<T}) \leq k_T,$$

Therefore, as $N \to \infty$,
$\tilde{P}_T^N(\tau_N < T) \to 0.$
Thus all of the conclusions in Theorem 132 now follow. ■

3.4 Examples of Weak Solutions

For the existence and weak uniqueness of a weak solution of SDE with jumps we have the following examples.

Example 134 *Assume that*
$|b(t,x)| \leq k_0,$
$|\sigma(t,x)|^2 + \int_Z |c(t,x,z)|^2 \pi(dz) \leq k_0(1+|x|^2),$
and for each $N = 1,2,\cdots$ *there exist a non-random function* $c^N(t)$ *such that as* $|x_1|$ *and* $|x_2| \leq N$,
$|\sigma(t,x_1,\omega) - \sigma(t,x_2,\omega)|^2 + \int_Z |c(t,x_1,z,\omega) - c(t,x_2,z,\omega)|^2 \pi(dz)$
$\leqslant c^N(t)\,|x_1 - x_2|^2,$
where $c^N(t) \geq 0$ *satisfies that* $\int_0^T c^N(t)dt < \infty$ *for each* $T < \infty$;
and $\sigma^{-1}(t,x)$ *exists and is bounded* $|\sigma^{-1}(t,x)| \leq k_0$, *where* $x, b, c \in R^d, \sigma \in R^{d\otimes d}$.
Then the following SDE with jumps in $d-$*dimensional space on* $t \in [0,T]$:

$$x_t = x_0 + \int_0^t b(s,x_s)ds + \int_0^t \sigma(s,x_s)dw_s + \int_0^t \int_Z c(s,x_{s-},z)\tilde{N}_k(ds,dz), \tag{3.22}$$

where $x_0 \in R^d$ *is a constant vector, has a weak unique weak solution.*

Proof. Since the coefficients $\sigma(t,x)$ and $c(t,x,z)$ are less than linear in growth and are locally Lipschitzian continuous, by Theorem 119 there exists a pathwise unique strong solution $\{x_t\}_{t\in[0,T]}$ satisfying the following SDE: $P - a.s.$

$$x_t = x_0 + \int_0^t \sigma(s,x_s)dw_s + \int_0^t \int_Z c(s,x_{s-},z)\tilde{N}_k(ds,dz), t \in [0,T], \tag{3.23}$$

on the original given probability space $(\Omega, \mathfrak{F}, \{\mathfrak{F}_t\}_{t\in[0,T]}, P)$. However, by the Girsanov type theorem (Theorem 132) if we let

$$d\widetilde{P} = \exp[\int_0^T \sigma^{-1}(t,x_s)b(s,x_s)dw_s - \frac{1}{2}\int_0^T |\sigma^{-1}(t,x_s)b(s,x_s)|^2 ds]dP, \tag{3.24}$$

then $\widetilde{P}$ is a probability measure, and
$\widetilde{w}_t = w_t - \int_0^t \sigma^{-1}(t,x_s)b(s,x_s)ds, t \in [0,T],$
is a BM under the new probability measure $\widetilde{P}$, and $\tilde{N}_k(ds,dz)$ is still a Poisson martingale measure under $\widetilde{P}$ with the same compensator $\pi(dz)ds$

such that $\tilde{N}_k(ds,dz) = N_k(ds,dz) - \pi(dz)ds$. Hence we can rewrite (3.23) as follows: $\widetilde{P} - a.s.\ t \in [0,T]$,

$$x_t = x_0 + \int_0^t b(s,x_s)ds + \int_0^t \sigma(s,x_s)d\widetilde{w}_s + \int_0^t \int_Z c(s,x_{s-},z)\tilde{N}_k(ds,dz). \tag{3.25}$$

Thus we have obtained a weak solution
$(\Omega, \mathfrak{F}, \{\mathfrak{F}_t\}_{t\in[0,T]}, \widetilde{P}; \widetilde{w}_t, \tilde{N}_k(dt,dz), \{x_t\}_{t\in[0,T]})$.
Now let us prove the weak uniqueness. Suppose we have another weak solution $(\Omega', \mathfrak{F}', \{\mathfrak{F}'_t\}_{t\in[0,T]}, \widetilde{P}'; \widetilde{w}'_t, \tilde{N}_{k'}(dt,dz), \{x_t\}_{t\in[0,T]})$, where $k'(\cdot)$ is a $\mathfrak{F}'_t-$Poisson Point Process, which generates the Poisson martingale measure but with the same compensator $\pi(dz)dt$ such that $\tilde{N}_{k'}(dt,dz) = N_{k'}(dt,dz) - \pi(dz)dt$. Then we have $\widetilde{P}' - a.s.\ t \in [0,T]$,

$$x'_t = x_0 + \int_0^t b(s,x'_s)ds + \int_0^t \sigma(s,x'_s)d\widetilde{w}'_s + \int_0^t \int_Z c(s,x'_{s-},z)\tilde{N}_{k'}(ds,dz). \tag{3.26}$$

So we see again that by the Girsanov type theorem (Theorem 132) if we let

$$dP' = \exp[-\int_0^T \sigma^{-1}(t,x'_s)b(s,x'_s)d\widetilde{w}'_s - \frac{1}{2}\int_0^T \left|\sigma^{-1}(t,x'_s)b(s,x'_s)\right|^2 ds]d\widetilde{P}', \tag{3.27}$$

then P' is a probability measure, and
$w'_t = \widetilde{w}'_t + \int_0^t \sigma^{-1}(t,x'_s)b(s,x'_s)ds, t \in [0,T]$,
is a BM under the new probability measure P', and $\tilde{N}_{k'}(dt,dz)$ is still a Poisson martingale measure under P' with the same compensator $\pi(dz)dt$ such that $\tilde{N}_{k'}(dt,dz) = N_{k'}(dt,dz) - \pi(dz)dt$. Hence we can rewrite (3.26) as follows: $P' - a.s.$

$$x'_t = x_0 + \int_0^t \sigma(s,x'_s)dw'_s + \int_0^t \int_Z c(s,x'_{s-},z)\tilde{N}_{k'}(ds,dz), t \in [0,T]. \tag{3.28}$$

Notice that the solution for SDE (3.23) is a pathwise unique strong solution. Hence there exists a measurable function F such that
$x_t = F(w_s, \tilde{N}_k((0,s],U); \forall s \leq t, \forall U \in \mathfrak{B}_Z), P - a.s.$
This means that for (3.28) we must also have
$x'_t = F(w'_s, \tilde{N}_{k'}((0,s],U); \forall s \leq t, \forall U \in \mathfrak{B}_Z), P' - a.s.$
So $\forall A \in (R^d)^{\otimes n}$, and $\forall t_1, \cdots, t_n \in [0,T]$,
$P((x_{t_1}, \cdots, x_{t_n}) \in A) = P'((x'_{t_1}, \cdots, x'_{t_n}) \in A)$.
By (3.24) and (3.27) one immediately sees that
$\widetilde{P}((x_{t_1}, \cdots, x_{t_n}) \in A) = \widetilde{P}'((x'_{t_1}, \cdots, x'_{t_n}) \in A)$.
Hence the weak uniqueness for the weak solution holds. ■

In the case that the SDE (3.22) has no jump term, that is, $c = 0$, if $\sigma = \sigma(x)$ does not depend on t, then we can weaken the condition on σ to get a weak solution. However, in this case the weak uniqueness is not necessarily true.

Example 135 *Assume that* $\sigma = \sigma(x)$ *does not depend on* t, *and*
$|b(t,x)| \leq k_0,$
$|\sigma(x)|^2 \leq k_0(1+|x|^2),$
and $\sigma^{-1}(x)$ *exists and is bounded* $|\sigma^{-1}(x)| \leq k_0$, *where* $x, b, \in R^d, \sigma \in R^{d\otimes d}$.

Then for any $T < \infty$
$x_t = x_0 + \int_0^t b(s,x_s)ds + \int_0^t \sigma(x_s)dw_s,\ t \in [0,T],$
has a weak solution, where $x_0 \in R^d$ *is a constant vector.*

Proof. Let $(w_t, \mathfrak{F}_t)_{t\geq 0}$ be a BM given on a probability space $(\Omega, \mathfrak{F}, \{\mathfrak{F}_t\}_{t\geq 0}, P)$. Write
$\widetilde{w}_t = \int_0^{\tau_t} \sigma^{-1}(w_s)dw_s,\ \forall t \geq 0,$
where τ_t is the inverse function of the strictly increasing non-random continuous function $A(t) = E\int_0^t |\sigma^{-1}(w_s)|^2 ds$, that is, $A(\tau_t) = t$. Set
$\widetilde{\mathfrak{F}}_t = \mathfrak{F}_{\tau_t},\ \forall t \geq 0.$
Notice that $\forall t \geq s$,
$E[(\widetilde{w}_t - \widetilde{w}_s)^2 | \widetilde{\mathfrak{F}}_s] = E\int_{\tau_s}^{\tau_t} |\sigma^{-1}(w_u)|^2 du = t - s.$
So that by Theorem 97 $\left\{\widetilde{w}_t, \widetilde{\mathfrak{F}}_t\right\}_{t\geq 0}$ is a BM on the probability space $(\Omega, \mathfrak{F}, \left\{\widetilde{\mathfrak{F}}_t\right\}_{t\geq 0}, P)$. Let $x_t = x_0 + w_{\tau_t}$. Then x_t satisfies the SDE: $\forall t \geq 0$,
$dx_t = \sigma(x_t)d\widetilde{w}_t, x_0 = x_0,\ \ P-a.s.$
Now for any given $T < \infty$ write
$z_t(\sigma^{-1}b) = \exp[\int_0^t (\sigma^{-1}(x_s)b(s,x_s)) \cdot d\widetilde{w}_s - \frac{1}{2}\int_0^t |\sigma^{-1}(x_s)b(s,x_s)|^2 ds],$
$d\widetilde{P}_t = z_t dp,$
$\widetilde{w}'_t = \int_0^t \sigma^{-1}(x_s)b(s,x_s)ds + \widetilde{w}_t,$
Then by Theorem 132 $\widetilde{P}_t$ is a probability measure, and for each $0 < T < \infty$, $\widetilde{w}'_t, t \in [0,T]$, is a BM under the new probability measure $\widetilde{P}_T$. Furthermore, for each $T < \infty$,
$(\Omega, \mathfrak{F}, \left\{\widetilde{\mathfrak{F}}_t\right\}_{t\in[0,T]}, \widetilde{P}_T; \{x_t\}_{t\in[0,T]}, \{\widetilde{w}'_t\}_{t\in[0,T]})$
is a weak solution of the following SDE: $\widetilde{P}_T - a.s.$ $\forall t \in [0,T]$,
$dx_t = b(t,x_t)dt + \sigma(x_t)d\widetilde{w}'_t,\ x_0 = x_0.$ ∎

4

Some Useful Tools in Stochastic Differential Equations

In this chapter we will discuss some powerful tools in the theory and applications of stochatic differential equations (SDE). We will establish a Yamada-Watanabe type theorem, which is very helpful in getting a strong solution for the SDE, when we have already got a weak solution and we know that the pathwise uniqueness holds for this SDE. We will derive a Tanaka type formula, which is very useful when discussing the uniqueness, and the convergence of solutions for SDEs. In particular it can help us to get comparison results for solutions of SDE, which are important for the discussion of the properties of solutions and the pathwise stochastic optimal control in a later chapter. We will introduce the local time technique which can help us to simplify and weaken the conditions for a 1−dimensional SDE when disccusing the uniqueness and existence of solutions. Finally, we will establish the Krylov type estimate which is very powerful when discussing the weak solution of an SDE when the coefficients are only measurable.

4.1 Yamada-Watanabe Type Theorem

Consider a d-dimensional stochastic differential equation with Poisson jumps as follows:

$$\begin{aligned} dx_t &= b(t, x_t, \omega)dt + \sigma(t, x_t, \omega)dw_t + \int_Z c(t, x_{t-}, z, \omega)q(dt, dz), \\ x_0 &= x. \end{aligned} \tag{4.1}$$

where w_t is a d-dimensional standard Brownian motion process (BM), $q(dt, dz)$ is a Poisson martingale measure with a compensator $\pi(dz)dt$, $\pi(.)$ is a $\sigma-$finite measure on a measurable space $(Z, \mathfrak{B}(Z))$ such that

$q(dt, dz) = p(dt, dz) - \pi(dz)dt,$

where $p(dt, dz)$ is a Poisson random measure with the intensity measure $Ep(dt, dz) = \pi(dz)dt$.

Remark 136 *By Remark 128 we know that given a Poisson martingale measure $q(dt, dz)$ there always exists a $\mathfrak{F}_t-$Poisson point process $k(.)$ such that*

$q(dt, dz) = \widetilde{N}_k(dt, dz),$

where $\widetilde{N}_k(dt, dz)$ is the Poisson martingale measure generated by $k(.)$.

Theorem 137 *(Yamada-Watanabe type theorem). Assume that the coefficients $b(t, x), \sigma(t, x)$, and $c(t, x, z)$ in (4.1) do not depend on ω, and make the assumption (H):*

$$Z = R^d - \{0\}, \text{ and } \int_{R^d-\{0\}} \frac{|z|^2}{1+|z|^2}\pi(dz) < \infty. \tag{4.2}$$

Then we have the following two assertions.

1) If (4.1) has a weak solution and the pathwise uniqueness holds for (4.1), then (4.1) has a pathwise unique strong solution, $t \geq 0$.

2) If (4.1) has two weak solutions $(x_t^i, w_t^i, q^i(dt, dz)), i = 1, 2$, defined on two different probability spaces $(\Omega^i, \mathfrak{F}^i, (\mathfrak{F}_t^i), P^i), i = 1, 2$, respectively, where w_t^i is a P^i-BM, $q^i(dt, dz)$ is a P^i-Poisson martingale measure with the same compensator $\pi(dz)dt, i = 1, 2$, then there exist a probability space $(\Omega, \mathfrak{F}, (\mathfrak{F}_t), P)$ and two $\mathfrak{F}_t-$adapted cadlag (right continuous with left limit) processes $(\widetilde{x}_t^i), i = 1, 2$, with a BM $\widetilde{w}_t$ and a Poisson martingale measure $\widetilde{q}(dt, dz)$ having the same compensator $\pi(dz)dt$ as the given one defined on it such that $(\widetilde{x}_t^i), i = 1, 2$, are adapted to $\mathfrak{F}_t$, and the probability law of $(\widetilde{x}_t^i, \widetilde{w}_t, \widetilde{q}(dt, dz))$ coincides with that of $(x_t^i, w_t^i, q^i(dt, dz)), i = 1, 2$. Moreover, $(\widetilde{x}_t^i, \widetilde{w}_t, \widetilde{q}(dt, dz))$ satisfy (4.1) on the same probability space $(\Omega, \mathfrak{F}, (\mathfrak{F}_t), P), i = 1, 2$.

Before we establish this theorem let us give an example for the measure $\pi(.)$ such that (4.2) holds.

Example 138 *Let $\pi(dz) = \frac{dz}{|z|^{d+1}}$. Then (4.2) holds.*

Proof. In fact, in this case

$\int_{R^d-\{0\}} \frac{|z|^2}{1+|z|^2}\pi(dz) = \int_{R^d-\{0\}} \frac{|z|^2}{1+|z|^2} \frac{dz}{|z|^{d+1}}$

$\leq \int_{|z|<1} |z|^2 \frac{dz}{|z|^{d+1}} + \int_{|z|\geq 1} \frac{dz}{|z|^{d+1}} < \infty.$ ■

To show Theorem 137 let us first establish a Lemma.

Lemma 139 *Suppose that x_t is a solution of (4.1) with the BM w_t, and the Poisson martingale measure $q(dt,dz)$, which has the compensator $\pi(dz)dt$ such that*

$q(dt,dz) = p(ds,dz) - \pi(dz)dt,$

where $\pi(dz)$ satisfies (4.2). Denote

$\zeta_t = \int_0^t \int_{|z|<1} zq(ds,dz) + \int_0^t \int_{|z|\geq 1} zp(ds,dz).$

Suppose that three $\mathfrak{F}_t$−adapted cadlag processes $(\widetilde{x}_t, \widetilde{w}_t, \widetilde{\zeta}_t)$ have the same finite-dimensional probability distributions as that of (x_t, w_t, ζ_t). Write

$\widetilde{p}((0,t],U) = \sum_{0<s\leq t} I_{0\neq \triangle\widetilde{\zeta}_s \in U}$, *for* $t \geq 0, U \in \mathfrak{B}(Z)$,

$\widetilde{q}(dt,dz) = \widetilde{p}(dt,dz) - \pi(dz)dt.$

Then $\widetilde{x}_t$ is also a solution of (4.1) with the BM $\widetilde{w}_t$, and the Poisson martingale measure $\widetilde{q}(dt,dz)$, which has the compensator $\pi(dz)dt$.

Proof. First let us show that such a ζ_t is finite $P-a.s.$ In fact,

$\frac{1}{2}\int_0^t \int_Z |z|^2 I_{|z|<1}\pi(dz)ds \leq t\int_Z \frac{|z|^2}{1+|z|^2} I_{|z|<1}\pi(dz) < \infty.$

Hence $\int_0^t \int_{|z|<1} zq(ds,dz)$ is a square integrable $\mathfrak{F}_t$− martingale. On the other hand,

$E\frac{1}{2}\int_0^t \int_Z I_{|z|\geq 1}p(ds,dz) = \frac{1}{2}\int_0^t \int_Z I_{|z|\geq 1}\pi(dz)ds = t\int_{|z|\geq 1} \frac{|z|^2}{2|z|^2}\pi(dz)$

$\leq t\int_{|z|\geq 1} \frac{|z|^2}{1+|z|^2}\pi(dz) < \infty.$

Hence $p((0,t],\{|z|\geq 1\}) < \infty, P-a.s.$ Denote the number of jumps, the jump times, and the values of jumps of $p((0,t],\{|z|\geq 1\})$ by $\nu(\omega)$,

$\tau_1(\omega) < \tau_2(\omega) < \cdots < \tau_{\nu(\omega)}(\omega),$

and $\{z_1(\omega), \cdots, z_{\nu(\omega)}(\omega)\}$, respectively. Then $P-a.s.$

$\int_0^t \int_{|z|\geq 1} |z|\, p(ds,dz) = \sum_{k=1}^{\nu(\omega)} |z_k(\omega)| < \infty,$

because it is a finite sum for $P-a.s.\ \omega$. Therefore, ζ_t is finite $P-a.s.$ Now notice that p is the counting measure induced by counting the jumps of ζ_t with compensator $\pi(dz)dt$. Hence it is a Poisson counting measure. Since $\widetilde{\zeta}_t, t\geq 0$, and $\zeta_t, t\geq 0$, have the same finite-dimensional probability distributions, $\widetilde{p}$ is also a Poisson counting measure with the same compensator $\pi(dz)dt$. Hence $\widetilde{q}(dt,dz) = \widetilde{p}(ds,dz) - \pi(dz)dt$ is also a Poisson martingale measure. By the coincidence of finite-dimensional probability distributions of $(\widetilde{x}_t, \widetilde{w}_t, \widetilde{\zeta}_t), t\geq 0$, and $(x_t, w_t, \zeta_t), t\geq 0$, applying the monotone class theorem (Theorem 390), it is easily seen that the probability laws of $(\widetilde{x}_t, \widetilde{w}_t, \widetilde{\zeta}_t), t\geq 0$, and $(x_t, w_t, \zeta_t), t\geq 0$, also coincide. Notice that there exists a Baire function F such that

$F(x_s, w_s, \zeta_s; s\leq t) = x_t - x - \int_0^t b(s,x_s)ds - \int_0^t \sigma(s,x_s)dw_s$

$-\int_0^t \int_Z c(s,x_{s-},z)q(ds,dz).$

Hence one has that

$P(F(\widetilde{x}_s, \widetilde{w}_s, \widetilde{\zeta}_s; s\leq t) = 0) = P(F(x_s, w_s, \zeta_s; s\leq t) = 0) = 1.$ ∎

Now let us show Theorem 137.

Proof. The proof is similar to that of Ikeda & Watanabe (1989), which is for the case of an SDE without jumps. Let us show 2) first. Suppose that the weak solutions $(x_t^i, w_t^i, q^i(dt, dz)), i = 1, 2$, are defined on probability spaces $(\Omega^i, \mathfrak{F}^i, (\mathfrak{F}_t^i), P^i)$, $i = 1, 2$, respectively. By the assumption of 2) in Theorem 4.1 we can let

$\zeta_t^i = \int_0^t \int_{|z|\leq 1} zq^i(ds, dz) + \int_0^t \int_{|z|>1} zp^i(ds, dz), i = 1, 2.$

Construct space $\Omega = D^d \times D^d \times W_0^d \times D^d$, where $D^d = D([0, \infty); R^d)$ (the totality of R^d-valued right continuous with left limit functions defined on $[0, \infty)$), and W_0^d is the set of continuous R^d-valued functions $w(t)$ defined on $t \geq 0$ with $w(0) = 0$.

Map $(x^1(., \omega), w^1(., \omega), \varsigma^1(., \omega))$ and $(x^2(., \omega), w^2(., \omega), \varsigma^2(., \omega))$ into the $\Omega^1 =$ $1st \times 3rd \times 4th$ component space of Ω and $\Omega^2 = 2nd \times 3rd \times 4th$ component space of Ω, respectively. From these two maps we get the probability laws P_x^1 and P_x^2 on Ω^1 and Ω^2, respectively. Then both marginal distributions $\widetilde{\pi}(P_x^1)$ and $\widetilde{\pi}(P_x^2)$ coincide with $P^w \times P^\varsigma$, where P^w is the Wiener measure on W_0^d, P^ζ is the measure on D^d induced by ζ^1 or ζ^2 (both of them have the same probability distribution), and $\widetilde{\pi} : D^d \times W_0^d \times D^d \to W_0^d \times$ D^d is the projection. Let $Q_1^{(w_3,w_4)}(dw_1)$ and $Q_2^{(w_3,w_4)}(dw_2)$ be the regular conditional distributions of (w_1) given (w_3, w_4) and of (w_2) given (w_3, w_4), respectively. Define a probability measure Q on the space Ω by

$Q(dw_1 dw_2) = Q_1^{(w_3,w_4)}(dw_1) Q_2^{(w_3,w_4)}(dw_2) P^w(dw_3) P^\zeta(dw_4).$

Let $\mathfrak{F}$ be the completion of the $\sigma-$field $\mathfrak{B}(\Omega)$ by Q, and $\mathfrak{F}_t = \cap_{\varepsilon>0}(\mathfrak{B}_{t+\varepsilon} \vee$ $\mathcal{N})$, where $\mathfrak{B}_t = \mathfrak{B}_t(D^d) \times \mathfrak{B}_t(W_0^d) \times \mathfrak{B}_t(D^d)$ and $\mathcal{N}$ is the set of all $Q-$null sets. Then clearly (w_1, w_3, w_4) and (x^1, w^1, ζ^1) have the same distribution, and so do (w_2, w_3, w_4) and (x^2, w^2, ζ^2). Now let us show the following facts.

Fact A. For $A \in \mathfrak{B}_t(D^d)$,

$(w_3, w_4) \in W_0^d \times D^d \longmapsto Q_1^{(w_3,w_4)}(A)$

and

$(w_3, w_4) \in W_0^d \times D^d \longmapsto Q_2^{(w_3,w_4)}(A)$

are $\mathfrak{B}_t(W_0^d \times D^d)^{P^{w,\zeta}}-$ measurable, where $P^{w,\zeta} = P^w \times P^\zeta$, and $\mathfrak{B}_t(W_0^d \times$ $D^d)^{P^{w,\zeta}}$ is the completion of $\mathfrak{B}_t(W_0^d \times D^d)$ by $P^{w,\zeta}$.

Indeed, for fixed $t > 0$ and $A \in \mathfrak{B}_t(D^d \times D^d)$, there exists a conditional probability $Q_{1t}^{(w_3,w_4)}(A)$ such that $(w_3, w_4) \in W_0^d \times D^d \to Q_{1t}^{(w_3,w_4)}(A)$ is $\mathfrak{B}_t(W_0^d \times D^d)^{P^{w,\zeta}} -$ measurable, and

$P_x^1(A \times C) = \int_C Q_{1T}^{(w_3,w_4)}(A) P^w(dw_3) P^\zeta(dw_4)$, for any $C \in \mathfrak{B}_t(W_0^d \times D^d)$.

Now let

$C = \{(w_3, w_4) \in W_0^d \times D^d : \rho_t(w_3, w_4) \in A_1, \theta_t(w_3, w_4) \in A_2\},$
$A_1, A_2 \in \mathfrak{B}_t(W_0^d \times D^d),$

where θ_t is defined by $\theta_t(w_3, w_4)(s) = (w_3(t+s) - w_3(t), w_4(t+s) - w_4(t))$, $s \geq 0$,and ρ_t is defined by $\rho_t(w_3, w_4)(s) = (w_3(t \wedge s), w_4(t \wedge s))$. Since $\theta_t(w_3, w_4)$ is independent of $\mathfrak{B}_t(W_0^d \times D^d)$ with respect to $P^{w,\zeta}$ we have

$\int_C Q_{1t}^{(w_3,w_4)}(A) P^w(dw_3) P^\zeta(dw_4)$
$= \int_{\rho_t(w_3,w_4)\in A_1} Q_{1t}^{(w_3,w_4)}(A) P^w(dw_3) P^\zeta(dw_4) P^{w,\zeta}(\theta_t(w_3, w_4) \in A_2)$
$= P_x^1(A \times \{\rho_t(w_3, w_4) \in A_1\}) P^{w,\zeta}(\theta_t(w_3, w_4) \in A_2)$

$= P^1(x^1(.) \in A, \rho_t(w,\zeta) \in A_1)P^1(\theta_t(w,\zeta) \in A_2)$
$= P^1(x^1(.) \in A, \rho_t(w,\zeta) \in A_1, \theta_t(w,\zeta) \in A_2)$
$= P^1(x^1(.) \in A, (w,\zeta) \in C) = P^1_x(A \times C),$
where we have used the result that $\{x^1(.) \in A, \rho_t(w,\zeta) \in A_1\} \in \mathfrak{F}_t$, moreover, $\theta_t(w,\zeta)$ and $\mathfrak{F}_t$ are independent. Hence it is easily shown that

$Q_{1t}^{(w_3,w_4)}(A) = Q_1^{(w_3,w_4)}(A), a.a.(w_3,w_4)\ (P^{w,\zeta}).$

(Here "$a.a.$" means "almost all"). Fact A is proved.

Fact B. $w_3(t)$ is an d-dimensional $\mathfrak{F}_t$−BM on $(\Omega, \mathfrak{F}, Q)$, and $p(dt,dz)$ is a Poisson counting measure with the same compensator $\pi(dz)dt$ on $(\Omega, \mathfrak{F}, Q)$, where

$p((0,t],U) = \sum_{0<s\leq t} I_{0\neq\triangle\zeta_s\in U}$, for $t \geq 0, U \in \mathfrak{B}(Z)$.

Indeed, by using Fact A we have that for $\lambda \in R^d, A_i \in \mathfrak{B}_s(D^d), i = 1,2,4; A_3 \in \mathfrak{B}_s(W_0^d), 0 \leq s \leq t$

$E^Q[e^{i(\lambda,w_3(t)-w_3(s))}I_{A_1\times A_2\times A_3\times A_4}]$
$= \int_{A_3\times A_4} e^{i(\lambda,w_3(t)-w_3(s))}Q_1^{(w_3,w_4)}(A_1)Q_2^{(w_3,w_4)}(A_2)P^w(dw_3)P^\zeta(dw_4)$
$= e^{-(|\lambda|^2/2)(t-s)}\int_{A_3\times A_4} Q_1^{(w_3,w_4)}(A_1)Q_2^{(w_3,w_4)}(A_2)P^w(dw_3)P^\zeta(dw_4)$
$= e^{-(|\lambda|^2/2)(t-s)}Q(A_1 \times A_2 \times A_3 \times A_4).$

Therefore the first conclusion of Fact B is true. Now for $t > s \geq 0$, disjoint $U_1, \cdots, U_m \in \mathfrak{B}(Z)$ such that $p((0,t],U_i) < \infty, \forall i = 1, \cdots, m$ and $\lambda_i > 0, i = 1, \cdots, m$; $A_i \in \mathfrak{B}_s(D^d), i = 1,2,4; A_3 \in \mathfrak{B}_s(W_0^d), 0 \leq s \leq t$

$E^Q[\exp[-\sum_{i=1}^m \lambda_i p((s,t],U_i)]I_{A_1\times A_2\times A_3\times A_4}]$
$= \int_{A_5\times A_6} \exp[-\sum_{i=1}^m \lambda_i p((s,t],U_i)]Q_1^{(w_3,w_4)}(A_1)$
$\cdot Q_2^{(w_3,w_4)}(A_2)P^w(dw_3)P^\zeta(dw_4) = \exp[(t-s)\sum_{i=1}^m(e^{-\lambda_i}-1)\pi(U_i)]$
$\cdot \int_{A_5\times A_6} Q_1^{(w_3,w_4)}(A_1)Q_2^{(w_3,w_4)}(A_2)P^w(dw_3)P^\zeta(dw_4)$
$= \exp[(t-s)\sum_{i=1}^m(e^{-\lambda_i}-1)\pi(U_i)]Q(A_1 \times A_2 \times A_3 \times A_4)$

Therefore the second conclusion of Fact B is also true.

Now let us return to the proof of Theorem 4.1. From Fact B it is not difficult to show that $(w_1, w_3, q(\cdot,\cdot))$ and $(w_2, w_3, q(\cdot,\cdot))$ are solutions of (4.1) on the same space $(\Omega, \mathfrak{F}, (\mathfrak{F}_t), Q)$, where

$q(dt,dz) = p(dt,dz) - \pi(dz)dt$

is a Poisson martingale measure with the compensator $\pi(dz)dt$. Hence 2) of Theorem 4.1 is proved. Now assume that the condition in 1) holds. Then the pathwise uniqueness implies that $w_1 = w_2, Q - a.s.$ This implies that

$Q_1^{(w_3,w_4)} \times Q_2^{(w_3,w_4)}(w_1 = w_2) = 1,\ P^{w,\zeta} - a.s.$

Now it is easy to see that there exists a function $(w_3,w_4) \in W_0^d \times D^d \to F_x(w_3,w_4) \in D^d \times D^d$ such that $Q_1^{(w_3,w_4)} = Q_2^{(w_3,w_4)} = \delta_{\{F_x(w_3,w_4)\}}, P^{w,\zeta} - a.s.$ By Fact A this function is $\mathfrak{B}_t(W_0^d \times D^d)^{P^{w,\zeta}}/\mathfrak{B}_t(D^d \times D^d)$−measurable. Clearly, $F_x(w_3,w_4)$ is uniquely determined up to $P^{w,\zeta}$−measure 0. Now by the assumption of 1) (4.1) has a weak solution. Denote it by $(x_t, w_t, q(dt,dz))$, where $q(dt,dz)$ is a Poisson martingale measure with the compensator $\pi(dz)dt$. By above $F_x(w_s,\zeta_s, s \leq t)$, where $\zeta_t = \int_0^t\int_{|z|\leq 1} zq(ds,dz) + \int_0^t\int_{|z|>1} zp(ds,dz)$, is also a weak solution on the same probability space, since it has the same probability distributions as that of $F_x(w_3(s), w_4(s), s \leq t)$. Then by the pathwise uniqueness $x_t = F_x(w_s, \varsigma_s, s \leq t)$. Hence (x_t) is a strong solution of (4.1) . ∎

Corollary 140 *The pathwise uniqueness of solutions implies the weak uniqueness of weak solutions. In general, The pathwise uniqueness of solutions implies the uniqueness of probability laws of $(x_t^i, w_t^i, q^i(dt, dz))$ from the weak solutions to the SDE (4.1).*

Proof. Suppose that two weak solutions $(x_t^i, w_t^i, q^i(dt, dz)), i = 1, 2$, are defined on probability spaces $(\Omega^i, \mathfrak{F}^i, (\mathfrak{F}_t^i), P^i), i = 1, 2$, respectively. As the proof in Theorem 137 construct a space $\Omega = D^d \times D^d \times W_0^d \times D^d$. Map $(x^1(., \omega), w^1(., \omega), \varsigma^1(., \omega))$ and $(x^2(., \omega), w^2(., \omega), \varsigma^2(., \omega))$ into the $\Omega^1 = 1st \times 3rd \times 4th$ component space of Ω and $\Omega^2 = 2nd \times 3rd \times 4th$ component space of Ω, respectively. From these two maps we get the probability laws P_x^1 and P_x^2 on Ω^1 and Ω^2, respectively. By the proof of Theorem 137 one sees that the pathwise uniqueness deduces that
$(w_1, w_3, q(\cdot, \cdot)) = (w_2, w_3, q(\cdot, \cdot))$. This implies that the probability law P_x^1 coincides with the probability law P_x^2. In particular, the probaility law of $x^1(., \omega)$ coincides with the probability law of $x^2(., \omega)$. That is, the weak uniqueness holds. In general, the probability laws of
$(x^1(., \omega), w^1(., \omega), \varsigma^1(., \omega))$ and $(x^2(., \omega), w^2(., \omega), \varsigma^2(., \omega))$
coincide, because $P_x^1 = P_x^2$. That is, $\forall A_1 \in \mathfrak{B}(D^d), \forall A_2 \in \mathfrak{B}(W_0^d), \forall A_3 \in \mathfrak{B}(D^d)$,
$P_x^1((x^1(.), w^1(.), \varsigma^1(.)) \in A_1 \times A_2 \times A_3)$
$= P_x^2((x^2(.), w^2(.), \varsigma^2(.)) \in A_1 \times A_2 \times A_3)$. ∎

By using the Yamada-Watanabe Theorem and the pathwise uniqueness theorem (Lemma 115) one easily obtains the existence of a pathwise unique strong solution for a SDE with discontinuous coefficients.

Example 141 *Assume that (4.2) (that is, Assumption (H)) holds and assume that*
$|\sigma(t, x)|^2 + \int_Z |c(t, x, z)|^2 \pi(dz) \leq k_0(1 + |x|^2)$,
$|\sigma(t, x) - \sigma(t, y)|^2 \leq c_N^T(t) |x - y|^2$, *as* $|x|, |y| \leq N, t \in [0, T]$,
$\int_Z |c(t, x, z) - c(t, y, z)|^2 \pi(dz) \leq c_N^T(t) |x - y|^2$,
as $|x|, |y| \leq N, t \in [0, T]$,
where $c_N^T(t) \geq 0$ is non-random only depending on N and T such that $\int_0^T c_N^T(t) dt < \infty$, and $\sigma^{-1}(t, x)$ exists and is bounded $|\sigma^{-1}(t, x)| \leq k_0$, where $x, b, c \in R^d, \sigma \in R^{d \otimes d}$. If
$b(t, x) = -\frac{x}{|x|} I_{x \neq 0}$,
then for each $T < \infty$ the following SDE with jumps in $d-$dimensional space on $t \in [0, T]$:

$$x_t = x_0 + \int_0^t b(s, x_s) ds + \int_0^t \sigma(s, x_s) dw_s + \int_0^t \int_Z c(s, x_{s-}, z) \tilde{N}_k(ds, dz), \tag{4.3}$$

where $x_0 \in R^d$ is a constant vector, has a pathwise unique strong solution. Therefore, (4.3) on $t \geq 0$ also has a pathwise unique solution.

Proof. For each $T < \infty$ by Example 134 (4.3) has a weak solution. If we can show that the pathwise uniqueness holds for (4.3), then by Theorem 137 we will obtain a pathwise unique strong solution for (4.3) on $t \in [0, T]$. In fact,

$\left\langle x - y, -\frac{x}{|x|} + \frac{y}{|y|} \right\rangle = -|x| - |y| + \frac{x \cdot y}{|x|} + \frac{x \cdot y}{|y|}$
$\leq -|x| - |y| + |x| + |y| = 0.$

So Lemma 115 applies, the pathwise uniqueness holds for (4.3). ■

4.2 Tanaka Type Formula and Some Applications

4.2.1 Localization Technique

Consider a real semi-martingale x_t as follows:

$$x_t = x_0 + A_t + M_t + \int_0^t \int_Z c(s, z, \omega) \widetilde{N}_k(ds, dz), t \geq 0 \qquad (4.4)$$

where A_t is a real continuous, finite variational process with $A_0 = 0$, M_t is a real continuous, local square integrable martingale with $M_0 = 0$, and $\widetilde{N}_k(dt, dz)$ is a Poisson random martingale measure such that

$\widetilde{N}_k(dt, dz) = N_k(dt, dz) - \pi(dz)dt,$

where $N_k(dt, dz)$ is a Poisson counting process generated by a $\mathfrak{F}_t-$Poisson point process $k(\cdot)$ with the compensator $\pi(dz)dt$, $\pi(\cdot)$ is a non-random $\sigma-$finite measure in some measurable space $(Z, \Re(Z))$. Obviously, we may assume that x_t is a right continuous with a left limit process (denote it by a cadlag process). To make sense of (4.18) let us assume that

$$c \in \mathcal{F}_k^{2,loc} = \left\{ \begin{array}{c} f(t, z, \omega) : f \text{ is } \mathfrak{F}_t - \text{predictable such that} \\ \int_0^t \int_Z |f(s, z, \omega)|^2 \pi(dz)ds < \infty, P - a.s. \text{ for all } t \geq 0. \end{array} \right\} \qquad (4.5)$$

For simplicity, from now on, we will always use the phrase "by localization" to simplify our proofs or statements. It means that we let

$$\tau_n = \inf \left\{ t \geq 0 : |x_t|^2 + |A|_t^2 + |M_t|^2 + |N_t|^2 + \langle M \rangle_t + \langle N \rangle_t > n \right\}, \qquad (4.6)$$

where

$|A|_t =$ the total variation of A on $[0, t]$,
$N_t = \int_0^t \int_Z c(s, z, \omega) \widetilde{N}_k(ds, dz)$,
$\langle N \rangle_t = \int_0^t \int_Z |c(s, z, \omega)|^2 \pi(dz)ds$,

and discuss our problem for processes $x_{t \wedge \tau_n}, A_{t \wedge \tau_n}, M_{t \wedge \tau_n}$, and $N_{t \wedge \tau_n}$. After obtaining the result we then let $n \to \infty$. The advantage of such a

localization technique is that $|A|^2_{t\wedge\tau_n}, |M|^2_{t\wedge\tau_n}, \langle M\rangle_{t\wedge\tau_n}$, and $\langle N\rangle_{t\wedge\tau_n}$ are bounded by n for each n, and
$E\left|N_{t\wedge\tau_n}\right|^2 \leq n,\ E\left|x_{t\wedge\tau_n}\right|^2 \leq 4E\left|x_0\right|^2 + 12n.$

4.2.2 Tanaka Type Formula in d–Dimensional Space

Now let us consider some more concrete SDEs with jumps in d–dimensional space

$$x_t^i = x_0^i + \int_0^t b^i(s, x_s^i, \omega)dA_s + \int_0^t \sigma(s, x_s^i, \omega)dM_s$$
$$+ \int_0^t \int_Z c^i(s, x_{s-}^i, z, \omega)\widetilde{N}_k(ds, dz), \forall t \geq 0, \quad (4.7)$$

$i = 1, 2$, where $A_t = (A_{1t}, \cdots, A_{mt}), M_t = (M_{1t}, \cdots, M_{rt})$; A_{it} is a continuous increasing process, $1 \leq i \leq m$; M_{it} is a continuous local square integrable martingale, $1 \leq i \leq r$; such that $\langle M_i, M_j\rangle_t = 0$, as $i \neq j$; $\widetilde{N}_k(dt, dz)$ is a Poisson martingale measure, which has the same meaning as that in (4.4) c is a R^d–valued vector, and b and σ are $d \times m$ and $d \times r$ matrices, respectively. Recall that (4.11) can also be written as
$dx_t^i = x_0^i + b^i(t, x_t^i, \omega)dA_t + \sigma(t, x_t^i, \omega)dM_t$
$+ \int_Z c^i(t, x_{t-}^i, z, \omega)\widetilde{N}_k(dt, dz), x_t^i = x_0^i, \forall t \geq 0,\ i = 1, 2.$
We have

Theorem 142 *(Tanaka type formula) Assume that*
1° $\left|b^i(t, x, \omega)\right|^2 + \left|\sigma(t, x, \omega)\right|^2 + \int_Z \left|c^i(t, x, z, \omega)\right|^j \pi(dz) \leq g(|x|)$,
$i, j = 1, 2$, *where* $g(u)$ *is a continuous non-random function on* $u \geq 0$;
2° $\left|\sigma(t, x, \omega) - \sigma(t, y, \omega)\right|^2 \leq k_N^T(t)\rho_N^T(|x - y|)\left|x - y\right|$,
as $|x|, |y| \leq N$, *and* $t \in [0, T]$;
where $0 \leq k_N^T(t)$ *is such that* $\sum_{k=1}^r E\int_0^T k_N^T(t)d\langle M_k\rangle_t < \infty$, *for each* $T < \infty$; $0 \leq \rho_N^T(u)$ *is non-random, strictly increasing, and continuous in* $u \geq 0$ *with* $\rho_N^T(0) = 0$, *and*
$\int_{0+} du/\rho_N^T(u) = \infty.$
If x_t^i *satisfies (4.7) with coefficients* (b^i, σ, c^i) *and the initial value* $x_0^i, i = 1, 2$, *respectively, but defined on the same probability space and with the same* M_t, A_t *and* $\widetilde{N}_k(dt, dz)$, *then* $\forall t \geq 0$
$\left|x_t^1 - x_t^2\right| = \left|x_0^1 - x_0^2\right|$
$+ \int_0^t sgn(x_{s-}^1 - x_{s-}^2) \cdot d(x_s^1 - x_s^2)$
$+\frac{1}{2}\sum_{i,j,k=1}^{d,d,r} \int_0^t I_{(x_s^1 \neq x_s^2)} \frac{\left|x_s^1 - x_s^2\right|^2 \delta_{ij} - (x_{is}^1 - x_{is}^2)(x_{js}^1 - x_{js}^2)}{\left|x_s^1 - x_s^2\right|^3}$
$\cdot(\sigma_{ik}(s, x_s^1, \omega) - \sigma_{ik}(s, x_s^2, \omega))(\sigma_{jk}(s, x_s^1, \omega) - \sigma_{jk}(s, x_s^2, \omega))d\langle M_k\rangle_s$
$+ \int_0^t \int_Z [\left|x_{s-}^1 - x_{s-}^2 + c^1(s, x_{s-}^1, z, \omega) - c^2(s, x_{s-}^2, z, \omega)\right| - \left|x_{s-}^1 - x_{s-}^2\right|$
$-sgn(x_{s-}^1 - x_{s-}^2) \cdot (c^1(s, x_{s-}^1, z, \omega) - c^2(s, x_{s-}^2, z, \omega))]N_k(ds, dz),$
where

$$sgnx = \begin{cases} (\frac{x_1}{|x|}, \cdots, \frac{x_d}{|x|}), \ as\ x \neq 0, \\ 0, \ as\ x = 0. \end{cases}$$

Proof. To show the formula for $t \geq 0$ we only need to prove that it holds on $t \in [0,T]$ for each given $T < \infty$. For simplicity, in the following for an arbitrary given $T < \infty$, let us write $k_N(t)$ and $\rho_N(u)$ for $k_N^T(t)$ and $\rho_N^T(u)$, respectively. Write the equality which we want to prove as

$\left|x_t^1 - x_t^2\right| = \left|x_0^1 - x_0^2\right| + \sum_{i=2}^6 I_i.$

Let

$\tau_N = \inf\left\{t \geq 0 : \left|x_t^1\right| + \left|x_t^2\right| > N\right\}.$

Now for any fixed N take $1 \geq a_n \downarrow 0$ such that

$\int_{a_n}^{a_{n-1}} \rho_N^{-1}(u)du = n,$

and for each n take a continuous function $f_n(u)$ such that

$$f_n(u) = \begin{cases} 0, \text{ as } 0 \leq u \leq a_n, \text{ or } u \geq a_{n-1}, \\ \text{any number less than } 2(n\rho_N(u))^{-1}, \text{ otherwise;} \end{cases}$$

with $\int_{a_n}^{a_{n-1}} f_n(u)du = 1$. Set for $x \in R^d$

$\varphi_n(x) = \int_0^{|x|} dy \int_0^y f_n(u)du, n = 1, 2, \cdots.$

Then $\varphi_n \in C^2(R^d)$, and

$(\partial/\partial x_i)\varphi_n(x) = I_{(a_n<|x|)} \frac{x_i}{|x|} \int_0^{|x|} f_n(u)du,$

$(\partial^2/\partial x_i \partial x_j)\varphi_n(x) = I_{(a_n<|x|)}(\frac{|x|^2\delta_{ij} - x_i x_j}{|x|^3} \int_0^{|x|} f_n(u)du + f_n(|x|) \frac{x_i x_j}{|x|^2}).$

Obviously,

$|(\partial/\partial x_i)\varphi_n(x)| \leq 1,$

and as $n \to \infty$

$\varphi_n(x) \to |x|, \ (\partial/\partial x_i)\varphi_n(x) \to I_{(x\neq 0)} \frac{x_i}{|x|}.$

By Ito's formula

$$\begin{aligned}
&\varphi_n(x_{t\wedge\tau_N}^1 - x_{t\wedge\tau_N}^2) - \varphi_n(x_0^1 - x_0^2) \\
&= \textstyle\sum_{i,k=1}^{d,m} \int_0^{t\wedge\tau_N} (\partial/\partial x_i)\varphi_n(x_s^1 - x_s^2) \cdot (b_{i,k}^1(s, x_s^1, \omega) - b_{i,k}^2(s, x_s^2, \omega))dA_{ks} \\
&+ \textstyle\sum_{i,k=1}^{d,r} \int_0^{t\wedge\tau_N} (\partial/\partial x_i)\varphi_n(x_s^1 - x_s^2) \cdot (\sigma_{ik}(s, x_s^1, \omega) - \sigma_{ik}(s, x_s^2, \omega))dM_{ks} \\
&+ \tfrac{1}{2} \textstyle\sum_{i,j,k=1}^{d,d,r} \int_0^{t\wedge\tau_N} (\partial^2/\partial x_i \partial x_j)\varphi_n(x_s^1 - x_s^2) \\
&\cdot(\sigma_{ik}(s, x_s^1, \omega) - \sigma_{ik}(s, x_s^2, \omega))(\sigma_{jk}(s, x_s^1, \omega) - \sigma_{jk}(s, x_s^2, \omega))d\left\langle M_k\right\rangle_s \\
&+ \textstyle\sum_{i=1}^{d} \int_0^{t\wedge\tau_N} \int_Z (\partial/\partial x_i)\varphi_n(x_{s-}^1 - x_{s-}^2) \cdot (c_i^1(s, x_{s-}^1, z, \omega) \\
&- c_i^2(s, x_{s-}^2, z, \omega))\widetilde{N}_k(ds, dz) \\
&+ \textstyle\int_0^{t\wedge\tau_N} \int_Z [\varphi_n(x_{s-}^1 - x_{s-}^2 + c^1(s, x_{s-}^1, z, \omega) - c^2(s, x_{s-}^2, z, \omega)) \\
&- \varphi_n(x_{s-}^1 - x_{s-}^2) - \textstyle\sum_{i=1}^d (\partial/\partial x_i)\varphi_n(x_{s-}^1 - x_{s-}^2) \\
&\cdot(c_i^1(s, x_{s-}^1, z, \omega) - c_i^2(s, x_{s-}^2, z, \omega))]N_k(ds, dz) = \textstyle\sum_{i=1}^6 I_i^n.
\end{aligned}$$

By Lebesgue's dominated convergence theorem it is easily seen that as $n \to \infty$

$I_1^n \to \int_0^{t\wedge\tau_N} sgn(x_s^1 - x_s^2) \cdot (b^1(s, x_s^1, \omega) - b^2(s, x_s^2, \omega))dA_s,$

$$\begin{aligned}
&E\left| \textstyle\sum_{i,k=1}^{d,r} \int_0^{t\wedge\tau_N} (\partial/\partial x_i)\varphi_n(x_s^1 - x_s^2) \cdot (\sigma_{ik}(s, x_s^1, \omega) - \sigma_{ik}(s, x_s^2, \omega))dM_{ks} \right. \\
&\left. - \textstyle\int_0^{t\wedge\tau_N} sgn(x_s^1 - x_s^2) \cdot (\sigma(s, x_s^1, \omega) - \sigma(s, x_s^2, \omega))dM_s \right|^2 \to 0.
\end{aligned}$$

Note that as $n \to \infty, \forall t \in [0,T], \forall N$

$\sum_{i,j,k=1}^{d,d,r}\int_0^{t\wedge\tau_N} f_n\left(\left|x_s^1 - x_s^2\right|\right) \frac{(x_{is}^1 - x_{is}^2)(x_{js}^1 - x_{js}^2)}{|x_s^1 - x_s^2|^2} \cdot (\sigma_{ik}(s,x_s^1,\omega) - \sigma_{ik}(s,x_s^2,\omega))$
$\cdot(\sigma_{jk}(s,x_s^1,\omega) - \sigma_{jk}(s,x_s^2,\omega))d\langle M_k\rangle_s$
$\leq 2Nk_0'\sum_{k=1}^r\int_0^T k_N(s)d\langle M_k\rangle_s/n \to 0.$
Moreover, as $0 \leq s \leq t\wedge\tau_N$
$\left|I_{(a_n < |x_s^1 - x_s^2|)}\left(\frac{|x_s^1 - x_s^2|^2\delta_{ij} - (x_{is}^1 - x_{is}^2)(x_{js}^1 - x_{js}^2)}{|x_s^1 - x_s^2|^3}\int_0^{|x_s^1 - x_s^2|} f_n(u)du\right|\right.$
$\cdot\left|(\sigma_{ik}(s,x_s^1,\omega) - \sigma_{ik}(s,x_s^2,\omega))(\sigma_{jk}(s,x_s^1,\omega) - \sigma_{jk}(s,x_s^2,\omega))\right|$
$\leq \rho_N(2N)k_N(s).$
Hence by Lebesgue's dominated convergence theorem it is seen that as $n \to \infty$
$I_4^n \to I_4$ (with $t\wedge\tau_N$ substituting t).
Similarly, it is seen that as $n \to \infty$
$I_6^n \to I_6$ (with $t\wedge\tau_N$ substituting t).
Moreover, by Lebesgue's dominated convergence theorem as $n \to \infty$
$E\left|\sum_{i=1}^d\int_0^{t\wedge\tau_N}\int_Z(\partial/\partial x_i)\varphi_n(x_{s-}^1 - x_{s-}^2)\cdot(c_i^1(s,x_{s-}^1,z,\omega)\right.$
$-c_i^2(s,x_{s-}^2,z,\omega))\widetilde{N}_k(ds,dz) - \int_0^t\int_Z sgn(x_{s-}^1 - x_{s-}^2)$
$\left.\cdot(c^1(s,x_{s-}^1,z,\omega) - c^2(s,x_{s-}^2,z,\omega))\widetilde{N}_k(ds,dz)\right|^2 \to 0.$
Finally, notice that
$\lim_{N\to\infty}\tau_N = \infty.$
The desired conclusion now follows. ∎

4.2.3 Applications to Pathwise Uniqueness and Convergence of Solutions

The Tanaka type formula is a powerful tool when discussing the convergence of solutions for a sequence of SDEs and discussing the pathwise uniqueness of solutions to a SDE. Actually, we can give the following theorems.

Theorem 143 *Suppose that in (4.7)*
$b = b^1 = b^2,\ c = c^1 = c^2,\ x_0 = x_0^1 = x_0^2,$
that is, we have only one SDE. Assume that conditions 1° and 2° in Theorem 142 hold, moreover,
3° $\rho_N(u)$ *in 2° is concave,* $\forall N = 1, 2, \cdots;$
4° $\sum_{i=1}^d(sgn(x-y))_i\cdot(b(t,x,\omega) - b(t,y,\omega))_{ik} \leq c_N^T(t)\overline{G}_N^T(|x-y|),$
as $|x|, |y| \leq N, t \in [0,T]$, $\forall k = 1,\cdots,m;$
where $c_N^T(t)$ *and* $\overline{G}_N^T(u)$ *have the same property as that of* $k_N^T(t)$ *and* $\rho_N(u)$,
5° $\int_Z|c(t,x,z,\omega) - c(t,x,z,\omega)|\,\pi(dz) \leq \overline{c}_N^T(t)\widetilde{G}_N^T(|x-y|),$
as $|x|, |y| \leq N, t \in [0,T];$
where $\overline{c}_N^T(t)$ *and* $\widetilde{G}_N^T(u)$ *have the same property as that of* $k_N^T(t)$ *and* $\rho_N(u)$.
Then the pathwise uniqueness of solutions for (4.7) holds.

Proof. Assume that $x_t^i, i = 1,2$, are two solutions of (4.7) on the same probability space with the same $A_t, M_t, \widetilde{N}_k(dt,dz)$. Then by the above Tanaka type formula as $t \in [0,T]$

$E\left|x^1_{t\wedge\tau_N} - x^2_{t\wedge\tau_N}\right| \leq \sum_{k=1}^m E\int_0^{t\wedge\tau_N} c_N^T(s)\overline{G}_N^T(\left|x_s^1 - x_s^2\right|)dA_{ks}$
$+k_0'\sum_{k=1}^m E\int_0^{t\wedge\tau_N} k_N^T(s)\rho_N^T(\left|x_s^1 - x_s^2\right|)d\left\langle M\right\rangle_{ks}$
$+2\sum_{k=1}^m E\int_0^{t\wedge\tau_N} \overline{c}_N^T(s)\widetilde{G}_N^T(\left|x_s^1 - x_s^2\right|)ds$
$\leq E\int_0^{t\wedge\tau_N} \widetilde{\overline{G}}_N^T(\left|x_s^1 - x_s^2\right|)d\widetilde{A}_s,$
where $\tau_N = \inf\left\{t \geq 0 : \left|x_t^1\right| + \left|x_t^2\right| > N\right\}$,
$\widetilde{A}_t = \sum_{k=1}^m \int_0^t c_N^T(s)A_{ks} + \sum_{k=1}^m \int_0^t k_N(s)d\left\langle M\right\rangle_{ks} + \sum_{k=1}^m \int_0^t \overline{c}_N^T(s)ds + t,$
$\widetilde{\overline{G}}_N^T(u) = \overline{G}_N^T(u) + \rho_N^T(u) + \widetilde{G}_N^T(u),$
and $\widetilde{\overline{G}}_N^T(u)$ still has the property as that of $\rho_N(u)$ by Lemma 144 below. Let
$\widetilde{T}_t = \widetilde{A}_t^{-1}.$
Notice that $\widetilde{A}_t > t$. Hence $\widetilde{T}_t = \widetilde{A}_t^{-1} < t$. Thus
$E\left|x^1_{\widetilde{T}_{t\wedge\tau_N}} - x^2_{\widetilde{T}_{t\wedge\tau_N}}\right| \leq E\int_0^{t\wedge\tau_N} \widetilde{G}_N(\left|x^1_{\widetilde{T}_s} - x^2_{\widetilde{T}_s}\right|)ds$
$\leq \int_0^t \widetilde{G}_N(E\left|x^1_{\widetilde{T}_{s\wedge\tau_N}} - x^2_{\widetilde{T}_{s\wedge\tau_N}}\right|)ds.$
From this it easily found that $P-a.s.\ \forall t \geq 0$
$x_t^1 = x_t^2$. ■

Lemma 144 *If $\rho_i(u)$ is strictly increasing, concave, and continuous in $u \geq 0$ with $\rho_i(0) = 0$ such that*
$\int_{0+} du/\rho_i(u) = \infty,\ i = 1,2,$
then $\rho_1(u) + \rho_2(u)$ is still strictly increasing, concave, and continuous in $u \geq 0$ with $\rho_1(0) + \rho_2(0) = 0$ and satisfies

$$\int_{0+} du/\left(\rho_1(u) + \rho_2(u)\right) = \infty. \tag{4.8}$$

Proof. One only needs to prove (4.8). The rest of the conclusions in Lemma 144 are obviously true. Let
$\overline{u} = \inf\left\{u > 0 : \rho_1(u) = \rho_2(u)\right\}.$
1) Suppose that $\overline{u} > 0$. Since $\rho_1(u) - \rho_2(u)$ is a continuous function. It must be true that $\rho_1(u) - \rho_2(u) > 0$, for all $u \in (0,\overline{u})$, or $\rho_1(u) - \rho_2(u) < 0$, for all $u \in (0,\overline{u})$. Thus, in the first case, $\int_{0+} du/\left(\rho_1(u) + \rho_2(u)\right) \geq \int_{0+} du/\left(2\rho_1(u)\right) = \infty$. For the second case the proof is just the same.
2) Assume that $\overline{u} = 0$. Again by the continuity of $\rho_1(u) - \rho_2(u)$ it should be $\rho_1(u) - \rho_2(u) > 0$, for all $u > 0$ or $\rho_1(u) - \rho_2(u) < 0$, for all $u > 0$. The proof can still be completed as in 1). ■

We can also use Theorem 142 to discuss the convergence of solutions to (4.7). Suppose that $x_t^n, n = 0,1,2,\cdots$ satisfy the following SDE's, respectively,

$$x_t^n = x_0 + \int_0^t b^n(s, x_s^n, \omega)dA_s + \int_0^t \sigma(s, x_s^n, \omega)dM_s$$
$$+ \int_0^t \int_Z c^n(s, x_{s-}^n, z, \omega)\widetilde{N}_k(ds, dz), \forall t \geq 0, n = 0, 1, 2, \cdots \tag{4.9}$$

We also denote $x_t^0 = x_t, b^0 = b$.

Theorem 145 *Assume that*
1° $|b^n(s,x,\omega)| + |\sigma(s,x,\omega)| + \sum_{i=1}^2 \int_Z |c^n(s,x,z,\omega)|^i \pi(dz) \leq g(s,\omega)$,
$\forall n = 0, 1, 2, \cdots$, *where* $g(s,\omega)$ *is continuous,*
2° $\sum_{i=1}^d (sgn(x-y))_i \cdot (b(s,x,\omega) - b(s,y,\omega))_{ik} \leq F_k(s,\omega)|x-y|$,
where $0 \leq F_k(s,\omega), k = 1, \cdots, m$, *satisfy* $\forall t \geq 0$
$\int_0^t F_k(s,\omega)dA_{ks} < \infty$,
$\|b^n(t,x,\omega) - b(t,x,\omega)\| \leq F^n(t,\omega)$,
$\lim_{n\to\infty} \sum_{k=1}^m E\int_0^t F^n(s,\omega)dA_{ks} = 0$,
3° $\|\sigma(t,x,\omega) - \sigma(t,y,\omega)\|^2 \leq K(s,\omega)|x-y|^2$,
where $0 \leq K(s,\omega)$ *satisfies* $\forall t \geq 0$
$\sum_{k=1}^r \int_0^t K(s,\omega)^2 d\langle M_k\rangle_s < \infty$,
4° $|c^n(t,x,z,\omega) - c(t,x,z,\omega)| \leq G^n(t,z,\omega)$,
$\lim_{n\to\infty} E\int_0^t \int_Z G^n(s,z,\omega)\pi(dz)ds = 0$,
$|c(t,x,z,\omega) - c(t,y,z,\omega)| \leq G(t,z,\omega)|x-y|$,
$E\int_0^t \int_Z G(s,z,\omega)\pi(dz)ds < \infty$, *etc.*
5° $\lim_{n\to\infty} E|x_0^n - x_0| = 0$.
Then we have that $\forall t \geq 0$

$$\lim_{n\to\infty} E[\exp(-\sum_{k=1}^m \int_0^t F_k(s,\omega)dA_{ks}$$
$$-2\int_0^t \int_Z G(s,z,\omega)\pi(dz)ds) \cdot |x_t^n - x_t|] = 0. \tag{4.10}$$

Proof. For arbitrary given $0 \leq T < \infty$ let us show that (4.10) holds as $t \in [0,T]$. Applying Theorem 142 we find
$|x_{t\wedge\tau_N}^n - x_{t\wedge\tau_N}| = |x_0^n - x_0|$
$+ \int_0^{t\wedge\tau_N} sgn(x_s^n - x_s) \cdot (b^n(s,x_s^n,\omega) - b(s,x_s,\omega))dA_s$
$+ \int_0^{t\wedge\tau_N} sgn(x_s^n - x_s) \cdot (\sigma(s,x_s^n,\omega) - \sigma(s,x_s,\omega))dM_s$
$+\frac{1}{2}\sum_{i,j,k=1}^{d,d,r} \int_0^{t\wedge\tau_N} I_{(x_s^1 \neq x_s^2)} \frac{|x_s^n - x_s|^2\delta_{ij} - (x_{is}^n - x_{is})(x_{js}^n - x_{js})}{|x_s^n - x_s|^3}$
$\cdot(\sigma_{ik}(s,x_s^n,\omega) - \sigma_{ik}(s,x_s,\omega))(\sigma_{jk}(s,x_s^n,\omega) - \sigma_{jk}(s,x_s,\omega))d\langle M_k\rangle_s$
$+ \int_0^{t\wedge\tau_N} \int_Z [|x_{s-}^n - x_{s-} + c^n(s,x_{s-}^n,z,\omega) - c(s,x_{s-},z,\omega)|$
$- |x_{s-}^n - x_{s-}|]\widetilde{N}_k(ds,dz)$
$+ \int_0^{t\wedge\tau_N} \int_Z [|x_s^n - x_s + c^n(s,x_s^n,z,\omega) - c(s,x_s,z,\omega)|$
$- |x_s^n - x_s| - sgn(x_s^n - x_s)(c^n(s,x_s^n,z,\omega) - c(s,x_s,z,\omega))]\pi(dz)ds,$

where

$\tau_N = \inf\left\{t \in [0,T] : \int_0^t g(s,\omega)d(\langle M\rangle_s + s) > N\right\}.$

Write

$X_t^n = x_t^n - x_t,\ \sigma_n^*(t,\omega) = \sigma(t, x_t^n, \omega) - \sigma(t, x_t, \omega),$

$c_n^*(t,z,\omega) = c^n(t, x_{t-}^n, z, \omega) - c(t, x_{t-}, z, \omega).$

Then by the above

$\left|X_{t\wedge\tau_N}^n\right| \leq |X_0^n| + \int_0^{t\wedge\tau_N} F(s,\omega)\,|X_s^n|\,dA_s + \int_0^{t\wedge\tau_N} F^n(s,\omega)dA_s$

$+2\int_0^{t\wedge\tau_N}\int_Z G^n(s,z,\omega)\pi(dz)ds + 2\int_0^{t\wedge\tau_N}\int_Z G(s,z,\omega)\,|X_s^n|\,\pi(dz)ds$

$+k_0\sum_{k=1}^r \int_0^{t\wedge\tau_N} K(s,\omega)\,|X_s^n|\,d\langle M_k\rangle_s + N_{t\wedge\tau},$

where

$N_t = \int_0^t sgn(X_s^n)\cdot\sigma_n^*(s,\omega)dM_s + \int_0^t\int_Z[\left|X_{s-}^n + c_n^*(s,z,\omega)\right| - \left|X_{s-}^n\right|]\widetilde{N}_k(ds,dz).$

Applying Lemma 146 below we find

$\exp(-\int_0^{t\wedge\tau_r} F(s,\omega)dA_s - 2\int_0^{t\wedge\tau_r}\int_Z G(s,z,\omega)\pi(dz)ds)) \cdot \left|X_{t\wedge\tau_r}^n\right|$

$\leq |X_0^n| + \int_0^{t\wedge\tau_r} \exp(-\int_0^s F(u,\omega)dA_u - 2\int_0^s\int_Z G(u,z,\omega)\pi(dz)du))$

$\cdot(F^n(s,\omega)dA_s + 2\int_Z G^n(s,z,\omega)\pi(dz)ds + dN_s).$

Hence

$EH_{t\wedge\tau_r}\left|X_{t\wedge\tau_r}^n\right| \leq E\,|X_0^n|$

$+E\int_0^{t\wedge\tau_r} H_s(F^n(s,\omega)dA_s + 2\int_Z G^n(s,z,\omega)\pi(dz))ds$

$\leq E\,|X_0^n| + E\int_0^t H_s(F^n(s,\omega)dA_s + 2\int_Z G^n(s,z,\omega)\pi(dz))ds,$

where

$H_s = \exp(-\int_0^s F(u,\omega)dA_u - 2\int_0^s\int_Z G(u,z,\omega)\pi(dz)du)) \leq 1.$

Note that by the assumption for each ω, when r is large enough

$\tau_r(\omega) = \infty.$

Therefore by the above result letting $r \to \infty$, by Fatou's lemma we have

$EH_t\,|X_t^n| \leq E\,|X_0^n| + E\int_0^t F^n(s,\omega)dA_s + 2\int_Z G^n(s,z,\omega)\pi(dz)ds.$

Now letting $n \to \infty$, we obtain

$\lim_{n\to\infty} EH_t\,|X_t^n| = 0.$ ■

Notice that the condition 2° here is weaker than the usual Lipschitzian condition. Because now $b(t,x,\omega)$ can be discontinuous . For example, $b(t,x) = A_0 - A_1 sgn(x)$, where A_0, A_1 are constants.

Lemma 146 *(Stochastic Gronwall's inequality). Assume that V_t, N_t are RCLL (right continuous with left limit) processes, where N_t is a semi-martingale, and assume that B_t is a continuous increasing process. If*

$V_t \leq N_t + \int_0^t V_s dB_s, \forall t \geq 0;\ V_0 = 0,$

then

$e^{-B_t}V_t \leq N_0 + \int_0^t e^{-B_s}dN_s,\ \forall t \geq 0.$

Proof. Notice that by Ito's formula

$d(N_tB_t) = B_t dN_t + N_t dB_t,$

$d_s(N_s(B_t - B_s)^2/2) = [(B_t - B_s)^2/2]dN_s - (B_t - B_s)N_s dB_s,$

etc. Hence by assumption it can be seen that

$V_t \leq N_t + \int_0^t V_s dB_s \leq N_t + \int_0^t N_s dB_s + \int_0^t dB_s \int_0^s V_u dB_u$

$= N_0 + N_0 B_t + \int_0^t (B_t - B_s)dN_s + \int_0^t dN_s + \int_0^t V_s(B_t - B_s)dB_s$

$\leq \cdots \leq N_0(\sum_{n=0}^k (B_t)^n/n!) + \int_0^t (\sum_{n=0}^k (B_t - B_s)^n/n!) dN_s$
$+ \int_0^t V_s(B_t - B_s)^k/k! dB_s.$
Since V_t is RCLL, it must be bounded locally. (Its bound can depend on ω). Hence
$\left| \int_0^t V_s(B_t - B_s)^k/k! dB_s \right| \leq k_0(\omega) B_t^{k+1}/(k+1)! \to 0$, as $n \to \infty$.
Therefore
$V_t \leq N_0 e^{B_t} + \int_0^t e^{B_t - B_s} dN_s, \forall t \geq 0.$ ■
Conditions 3°, 4° in Theorem 143 can be some non-Lipschitzian condition. Let us give an example as follows.

Example 147 *Let*
$b_i(t,x) = -k_0 \frac{x_i}{|x|} I_{x \neq 0} + k_1 x_i + k_2 |x_i| + k_3 |x|.$
Then $b(t,x)$ *satisfies the condition 3°.*

In fact, as $x, y \neq 0$,
$\sum_{i=1}^d \frac{x_i - y_i}{|x-y|} \left(-\frac{x_i}{|x|} + \frac{y_i}{|y|} \right) = \sum_{i=1}^d \frac{1}{|x-y|} \left(-\frac{x_i^2}{|x|} - \frac{y_i^2}{|y|} + \frac{x_i y_i}{|x|} + \frac{x_i y_i}{|y|} \right)$
$\leq \frac{1}{|x-y|} \left(-|x| - |y| + \frac{|x||y|}{|x|} + \frac{|x||y|}{|y|} \right) = 0,$
where we have applied the Schwarz's inequality to get that
$\sum_{i=1}^d x_i y_i \leq |x| |y|.$
Note also that as $x \neq 0, y = 0$,
$\sum_{i=1}^d \frac{x_i}{|x|} \left(-\frac{x_i}{|x|} \right) \leq 0.$
Hence $b(t,x)$ satisfies the condition 3°.

4.2.4 Tanaka Type Formual in 1-Dimensional Space

Now consider SDEs (4.7) in 1 - dimensional space: $i = 1, 2$,

$$x_t^i = x_0^i + \int_0^t b^i(s, x_s^i, \omega) dA_s + \int_0^t \sigma(s, x_s, \omega) dM_s + \int_0^t \int_Z c^i(s, x_{s-}^i, z, \omega) \widetilde{N}_k(ds, dz), \forall t \geq 0, \tag{4.11}$$

where $d = 1$, but $A_t = (A_{1t}, \cdots, A_{mt})$ and $M_t = (M_{1t}, \cdots, M_{rt})$ are still $m-$dimensional and $r-$dimensional such that $\langle M_i, M_j \rangle_t = 0$, as $i \neq j$; respectively; so $b \in R^{1 \otimes m}$, and $\sigma \in R^{1 \otimes r}$. In this case we can prove that the Tanaka type formula will be simpler and the conditions to be satisfied are also weaker. Actually, we have

Theorem 148 *(Tanaka type formula) Assume that*
1° $\left| b^i(t,x,\omega) \right|^2 + |\sigma(t,x,\omega)|^2 + \int_Z \left| c^i(t,x,z,\omega) \right|^j \pi(dz) \leq g(|x|)$,
$i, j = 1, 2$, *where* $g(u)$ *is a continuous non-random function on* $u \geq 0$;
2° $|\sigma(t,x,\omega) - \sigma(t,y,\omega)|^2 \leq k_N^T(t) \rho_N^T(|x-y|)$, *as* $|x|, |y| \leq N, t \in [0,T]$;
where $0 \leq k_N^T(t)$, $\sum_{k=1}^r E \int_0^T k_N^T(t) d\langle M_k \rangle_t < \infty$, *for each* $T < \infty; 0 \leq \rho_N^T(u)$ *is non-random, strictly increasing in* $u \geq 0$ *with* $\rho_N^T(0) = 0$, *and*

$\int_{0+} du/\rho_N(u) = \infty.$

If x_t^i satisfies (4.11) with coefficients (b^i, σ, c^i) and the initial value $x_0^i, i = 1, 2$, respectively, and they are defined on the same probability space and with the same M_t, A_t and $\widetilde{N}_k(dt, dz)$, then $\forall t \geq 0$

$$\left|x_t^1 - x_t^2\right| = \left|x_0^1 - x_0^2\right|$$
$$+ \int_0^t sgn(x_{s-}^1 - x_{s-}^2) d(x_s^1 - x_s^2)$$
$$+ \int_0^t \int_Z [\left|x_{s-}^1 - x_{s-}^2 + c^1(s, x_{s-}^1, z, \omega) - c^2(s, x_{s-}^2, z, \omega)\right| - \left|x_{s-}^1 - x_{s-}^2\right|$$
$$- sgn(x_{s-}^1 - x_{s-}^2) \cdot (c^1(s, x_{s-}^1, z, \omega) - c^2(s, x_{s-}^2, z, \omega))] N_k(ds, dz),$$

where

$$sgn x = \begin{cases} \frac{x}{|x|}, & as\ x \neq 0, \\ 0, & as\ x = 0. \end{cases}$$

Proof. We only need to point out the difference of the proofs between here and Theorem 142. Checking the proof of Theorem 142 we find that here we have $\varphi_n \in C^2(R^1)$. So

$$\varphi_n'(x) = I_{(a_n < |x|)} \frac{x}{|x|} \int_0^{|x|} f_n(u) du,$$
$$\varphi_n''(x) = I_{(a_n < |x|)} f_n(|x|),$$

and as $n \to \infty$

$$\varphi_n(x) \to |x|, \ \varphi_n'(x) \to sgn\ x.$$

Moreover here, after applying Ito's formula to $\varphi_n(x_t^1 - x_t^2)$, we see that as $n \to \infty, \forall t \in [0, T], \forall N$

$$0 \leq \frac{1}{2} \sum_{k=1}^r \int_0^{t \wedge \tau_N} \varphi_n''(x_s^1 - x_s^2) \cdot (\sigma_k(s, x_s^1, \omega) - \sigma_k(s, x_s^2, \omega))^2 d \langle M_k \rangle_s$$
$$= \frac{1}{2} \sum_{k=1}^r \int_0^{t \wedge \tau_N} f_n\left(\left|x_s^1 - x_s^2\right|\right) (\sigma_k(s, x_s^1, \omega) - \sigma_k(s, x_s^2, \omega))^2 d \langle M_k \rangle_s$$
$$\leq 2k_0' \sum_{k=1}^r \int_0^T k_N(s) d \langle M_k \rangle_s / n \to 0,$$

and the rest of the discussion remain true. So we arrive at the conclusion. ∎

Conditions for σ in Theorem 148 are weaker than that in Theorem 142. Moreover, here σ even can be non-Lipschitz continuous. In fact, we have the following example.

Example 149 *Set*

$$\sigma(t, x) = k_1(t) \sqrt{|x|} + k_2(t) x,$$

where $0 \leq k_i(t)$ is non-random such that $\int_0^T k_i(t)^2 dt < \infty, i = 1, 2$, for each $T < \infty$. Then condition 2° in Theorem 148 is satisfied.

To see this we use the elementary inequalities

$$\left|\sqrt{|x|} - \sqrt{|y|}\right| \leq \sqrt{|x - y|},$$

and

$$|x - y|^2 \leq \sqrt{|x - y|} \sqrt{|x + y|}$$

to get

$$|\sigma(t, x) - \sigma(t, y)|^2 \leq 2k_1(t)^2 |x - y| + 4N k_2(t)^2 |x - y|, \text{ as } |x|, |y| \leq N.$$

So, for this example, condition 2° in Theorem 148 is satisfied.

Furthermore, let us show that, actually, condition

$$\int_Z |c(t, x, z, \omega)| \pi(dz) \leq g(|x|)$$

in 1° of Theorem 148 also can be dropped. For this, let us first prove the following theorem.

Theorem 150 *Assume that*
$\int_Z \left|c^i(t,x,z,\omega)\right|^2 \pi(dz) \le g(|x|),\ i=1,2,$
where $g(u)$ *is a continuous non-random function on* $u \ge 0$.
If $x_t^i, i=1,2$ *are two solutions of (4.11) with the initial value* $x_0^i, i=1,2,$ *respectively, then* $\forall t \ge 0$

$$\begin{aligned}
&1)\ \int_0^t \int_Z \left|(\widehat{x}_{s-} + \widehat{c}(s,z,\omega))^+ - (\widehat{x}_{s-})^+ - I_{(\widehat{x}_{s-}>0)}\widehat{c}(s,z,\omega)\right| N_k(ds,dz) \\
&= \int_0^t \int_Z [(\widehat{x}_{s-} + \widehat{c}(s,z,\omega))^+ - (\widehat{x}_{s-})^+ - I_{(\widehat{x}_{s-}>0)}\widehat{c}(s,z,\omega)] N_k(ds,dz) \\
&= \int_0^t \int_Z [I_{(\widehat{x}_{s-}>0)}(\widehat{x}_{s-}+\widehat{c}(s,z,\omega))^- + I_{(\widehat{x}_{s-}\le 0)}(\widehat{x}_{s-}+\widehat{c}(s,z,\omega))^+] N_k(ds,dz) \\
&= \sum\nolimits_{0<s\le t} [I_{(\widehat{x}_{s-}>0)}(\widehat{x}_s)^- + I_{(\widehat{x}_{s-}\le 0)}(\widehat{x}_s)^+] \\
&\le \left|\widehat{x}_t^+ - \widehat{x}_0^+\right| + \sum\nolimits_{p=1}^m \int_0^t \left|\widehat{b}_p(s,\omega)\right| dA_p(s) \\
&+ \sum\nolimits_{k=1}^r \left|\int_0^t I_{(\widehat{x}_{s-}>0)}\widehat{\sigma}_k(s,\omega) dM_k(s)\right| \\
&+ \left|\int_0^t \int_Z I_{(\widehat{x}_{s-}>0)}\widehat{c}(s,z,\omega)\widetilde{N}_k(ds,dz)\right| < \infty,\ P-a.s.,
\end{aligned}$$

and similarly,

$$\begin{aligned}
&\int_0^t \int_Z \left|(\widehat{x}_{s-} + \widehat{c}(s,z,\omega))^- - (\widehat{x}_{s-})^- + I_{(\widehat{x}_{s-}\le 0)}\widehat{c}(s,z,\omega)\right| N_k(ds,dz) \\
&= \sum\nolimits_{0<s\le t} [I_{(\widehat{x}_{s-}>0)}(\widehat{x}_s)^- + I_{(\widehat{x}_{s-}\le 0)}(\widehat{x}_s)^+] \\
&\le \left|\widehat{x}_t^+ - \widehat{x}_0^+\right| + \sum\nolimits_{p=1}^m \int_0^t \left|\widehat{b}_p(s,\omega)\right| dA_p(s) \\
&+ \sum\nolimits_{k=1}^r \left|\int_0^t I_{(\widehat{x}_{s-}\le 0)}\widehat{\sigma}_k(s,\omega) dM_k(s)\right| \\
&+ \left|\int_0^t \int_Z I_{(\widehat{x}_{s-}\le 0)}\widehat{c}(s,z,\omega)\widetilde{N}_k(ds,dz)\right| < \infty,\ P-a.s.,
\end{aligned}$$

where we write

$$\begin{aligned}
&\widehat{x}_t = x_t^1 - x_t^2,\ \widehat{b}_p(s,\omega) = b_p^1(s,x_s^1,\omega) - b_p^2(s,x_s^2,\omega), \\
&\widehat{\sigma}_k(s,\omega) = \sigma_k(s,x_s^1,\omega) - \sigma_k(s,x_s^2,\omega), \\
&\widehat{c}(s,z,\omega) = c^1(s,x_{s-}^1,z,\omega) - c^2(s,x_{s-}^2,z,\omega);
\end{aligned}$$

$$\begin{aligned}
&2)\ \int_0^t \int_Z [|\widehat{x}_{s-} + \widehat{c}(s,z,\omega)| - |\widehat{x}_{s-}| - sgn(\widehat{x}_{s-})\widehat{c}(s,z,\omega)] N_k(ds,dz) \\
&= 2\sum\nolimits_{0<s\le t} [I_{(\widehat{x}_{s-}>0)}\widehat{x}_s^- + I_{(\widehat{x}_{s-}\le 0)}\widehat{x}_s^+] \\
&\le 4\left|\widehat{x}_t^+ - \widehat{x}_0^+\right| + 4\sum\nolimits_{p=1}^m \int_0^t \left|\widehat{b}_p(s,\omega)\right| dA_p(s) \\
&+2\sum\nolimits_{k=1}^r \left|\int_0^t I_{(\widehat{x}_{s-}>0)}\widehat{\sigma}_k(s,\omega) dM_k(s)\right| + 2\left|\int_0^t \int_Z I_{(\widehat{x}_{s-}>0)}\widehat{c}(s,z,\omega)\widetilde{N}_k(ds,dz)\right| \\
&+2\sum\nolimits_{k=1}^r \left|\int_0^t I_{(\widehat{x}_{s-}\le 0)}\widehat{\sigma}_k(s,\omega) dM_k(s)\right| + 2\left|\int_0^t \int_Z I_{(\widehat{x}_{s-}\le 0)}\widehat{c}(s,z,\omega)\widetilde{N}_k(ds,dz)\right| \\
&< \infty,\ P-a.s.
\end{aligned}$$

Proof. 2) is deduced from 1). In fact, $|x| = x^+ + x^-$. So

$$\begin{aligned}
&\left||\widehat{x}_{s-} + \widehat{c}(s,z,\omega)| - |\widehat{x}_{s-}| - sgn(\widehat{x}_{s-})\widehat{c}(s,z,\omega)\right| \\
&\le \left|(\widehat{x}_{s-} + \widehat{c}(s,z,\omega))^+ - (\widehat{x}_{s-})^+ - I_{(\widehat{x}_{s-}>0)}\widehat{c}(s,z,\omega)\right| \\
&+ \left|(\widehat{x}_{s-} + \widehat{c}(s,z,\omega))^- - (\widehat{x}_{s-})^- + I_{(\widehat{x}_{s-}\le 0)}\widehat{c}(s,z,\omega)\right|.
\end{aligned}$$

Thus 1) implies 2). Let us show 1). Notice that $x^- = (-x)^+$. So we only need to show the first conclusion in 1). Take $0 < a_n \downarrow 0$ and continuous functions $0 \le g_n(x)$ such that

$\int_0^{a_{n-1}} g_n(x)dx = 1$, $g_n(x) = 0$, as $x \notin (0, a_{n-1})$; $n = 1, 2, \cdots$
Let
$h_n(x) = \int_0^{x^+} dy \int_0^y g_n(u)du, n = 1, 2, \cdots$
Then
$h_n'(x) = I_{(x>0)} \int_0^{x^+} g_n(u)du,\ h_n''(x) = g_n(x^+)$.
Hence
$h_n(x) \in C^2(R),\ |h_n'(x)| \leq 1;\ h_n(x) \to x^+, h_n'(x) \to I_{(x>0)}$, as $n \to \infty$.
Applying Ito's formula, we find that

$$h_n(\widehat{x}_t) - h_n(\widehat{x}_0) = \int_0^t h_n'(\widehat{x}_{s-})d\widehat{x}_s + A_t^n, \tag{4.12}$$

where

$$A_t^n = \sum_{0<s\leq t} (h_n(\widehat{x}_s) - h_n(\widehat{x}_{s-}) - h_n'(\widehat{x}_{s-})\triangle\widehat{x}_s) + \frac{1}{2}\int_0^t h_n''(\widehat{x}_s)d\langle M\rangle_s. \tag{4.13}$$

Since $h_n''(x) \geq 0$, so by Taylor's formula
$h_n(\widehat{x}_s) - h_n(\widehat{x}_{s-}) - h_n'(\widehat{x}_{s-}) \triangle \widehat{x}_s = \frac{1}{2}h_n''(\widehat{x}_{s-} + \theta \triangle \widehat{x}_s)(\triangle\widehat{x}_s)^2 \geq 0$.
Hence A_t^n is an increasing process for each n. Now let

$$\widetilde{A}_t = \widehat{x}_t^+ - \widehat{x}_0^+ - \int_0^t I_{(\widehat{x}_{s-}>0)}d\widehat{x}_s. \tag{4.14}$$

By the localization technique we may assume that M_t and N_t are martingales, moreover, $M_t, \langle M\rangle_t$, and $E\langle N\rangle_t$ are bounded. Then it is easily seen that as $n \to \infty$ by (4.12)
$A_t^n \to \widetilde{A}_t$, in probability for each $t \geq 0$.
In fact, as $n \to \infty$
$h_n(\widehat{x}_t) \to \widehat{x}_t^+$, $P - a.s.$ for each $t \geq 0$,
and by Lebesgue's dominated convergence theorem as $n \to \infty$
$E\left|\int_0^t h_n'(\widehat{x}_{s-})d\widehat{x}_s - \int_0^t I_{(\widehat{x}_{s-}>0)}d\widehat{x}_s\right|^2 \to 0$.
From this it follows easily that $\widetilde{A}_t$ is also an increasing process. Notice that
$0 \leq \triangle\widetilde{A}_t = \widehat{x}_t^+ - \widehat{x}_{t-}^+ - I_{(\widehat{x}_{t-}>0)} \triangle \widehat{x}_t, \forall t \geq 0$,
where we write $\widehat{x}_{0-}^+ = \widehat{x}_0^+$. Hence $\forall t \geq 0$
$0 \leq \triangle\widetilde{A}_t = I_{(\widehat{x}_{t-}>0)}\widehat{x}_t^- + I_{(\widehat{x}_{t-}\leq 0)}\widehat{x}_t^+$,
where we have used the fact that
$I_{(\widehat{x}_{t-}>0)} \triangle \widetilde{A}_t = I_{(\widehat{x}_{t-}>0)}[\widehat{x}_t^+ - (\widehat{x}_{t-} + \triangle\widehat{x}_t)]$
$= I_{(\widehat{x}_{t-}>0)}[\widehat{x}_t^+ - \widehat{x}_t] = I_{(\widehat{x}_{t-}>0)}\widehat{x}_t^-$,
and
$I_{(\widehat{x}_{t-}\leq 0)} \triangle \widetilde{A}_t = I_{(\widehat{x}_{t-}\leq 0)}\widehat{x}_t^+$.
Therefore
$\infty > \widetilde{A}_t \geq \sum_{0<s\leq t}\left|\triangle\widetilde{A}_s\right| = \sum_{0<s\leq t}\triangle\widetilde{A}_s$
$= \sum_{0<s\leq t}(I_{(\widehat{x}_{s-}>0)}\widehat{x}_s^- + I_{(\widehat{x}_{s-}\leq 0)}\widehat{x}_s^+)$

$= 2\int_0^t\int_Z[I_{(\widehat{x}_{s-}>0)}(\widehat{x}_{s-}+\widehat{c}(s,z,\omega))^- + I_{(\widehat{x}_{s-}\leq 0)}(\widehat{x}_{s-}+\widehat{c}(s,z,\omega))^+]N_k(ds,dz)$
$= \int_0^t\int_Z[(\widehat{x}_{s-}+\widehat{c}(s,z,\omega))^+ - (\widehat{x}_{s-})^+ - I_{(\widehat{x}_{s-}>0)}\widehat{c}(s,z,\omega)]N_k(ds,dz).$

On the other hand by the definition (4.14) of $\widetilde{A}_t$ one sees that

$\widetilde{A}_t \leq \left|\widehat{x}_t^+ - \widehat{x}_0^+\right| + \sum_{p=1}^m \int_0^t \left|\widehat{b}_p(s,\omega)\right| dA_p(s)$
$+\sum_{k=1}^r \left|\int_0^t I_{(\widehat{x}_{s-}>0)}\widehat{\sigma}_k(s,\omega)dM_k(s)\right|$
$+\left|\int_0^t\int_Z I_{(\widehat{x}_{s-}>0)}\widehat{c}(s,z,\omega)\widetilde{N}_k(ds,dz)\right| < \infty,\ P-a.s.$

Therefore the first conclusion in 1) is proved. ∎

Remark 151 *1) By the proof of Theorem 150 one sees that*

$\sum_{0<s\leq t}[I_{(\widehat{x}_{s-}>0)}(\widehat{x}_s)^- + I_{(\widehat{x}_{s-}\leq 0)}(\widehat{x}_s)^+]$
$\leq \left|\widehat{x}_t^+ - \widehat{x}_0^+\right| + \overline{A}_t + \left|\int_0^t I_{(\widehat{x}_{s-}>0)}dM_s\right| < \infty,\ P-a.s.$

holds for a general real semi-martingale

$\widehat{x}_t = \widehat{x}_0 + \overline{A}_t + \widetilde{M}_t,$

where $\overline{A}_t \geq 0$ with $\overline{A}_0 = 0$ is a real continuous increasing process which is locally integrable, i.e. for each $T<\infty$

$E\overline{A}_T < \infty;$

and $\widetilde{M}_t$ with $\widetilde{M}_0 = 0$ is a real RCLL local square integrable martingale.

2) By the proof of Theorem 150 one also sees that

$\sum_{0<s\leq t}[I_{(\widehat{x}_{s-}>a)}(\widehat{x}_s - a)^- + I_{(\widehat{x}_{s-}\leq a)}(\widehat{x}_s - a)^+]$
$\leq |(\widehat{x}_t - a)^+ - (\widehat{x}_0 - a)^+| + \overline{A}_t + \left|\int_0^t I_{(\widehat{x}_{s-}>a)}dM_s\right| < \infty,\ P-a.s.$

holds for a general real semi-martingale

$\widehat{x}_t = \widehat{x}_0 + \overline{A}_t + \widetilde{M}_t$

as that in 1) and for any constant $a\in R^1$.

To see that 2) in the above corollary is true one only needs to write

$h_n(x) = \int_0^{(x-a)^+} dy \int_0^y g_n(u)du, n = 1,2,\cdots,$

and then to proceed with the proof in almost the same way as in the proof of Theorem 150, as a result of which we obtain 2).

Now from Theorem 148, Theorem 150 and their proofs we immediately get the following theorem.

Theorem 152 *Assume that the condition 2° in Theorem 148 holds, and assume that*

1°′ $\left|b^i(t,x,\omega)\right|^2 + |\sigma(t,x,\omega)|^2 + \int_Z \left|c^i(t,x,z,\omega)\right|^2 \pi(dz) \leq g(|x|),$

$i=1,2$, where $g(u)$ is a continuous non-random function on $u\geq 0$..

If $x_t^i, i=1,2$ are two solutions of (4.11) with the initial value $x_0^i, i=1,2$, respectively, then $\forall t\geq 0$

$\left(x_t^1 - x_t^2\right)^+ = \left(x_0^1 - x_0^2\right)^+ + \int_0^t I_{(x_s^1 - x_s^2)>0}d(x_s^1 - x_s^2)$
$+\int_0^t\int_Z[(x_{s-}^1 - x_{s-}^2 + c^1(s,x_{s-}^1,z,\omega) - c^2(s,x_{s-}^2,z,\omega))^+ - (x_{s-}^1 - x_{s-}^2)^+$
$-I_{(x_s^1 - x_s^2)>0}(c^1(s,x_{s-}^1,z,\omega) - c^2(s,x_{s-}^2,z,\omega))]N_k(ds,dz),$

$\left(x_t^1 - x_t^2\right)^- = \left(x_0^1 - x_0^2\right)^- + \int_0^t I_{(x_s^1 - x_s^2)\leq 0}d(x_s^1 - x_s^2)$
$+\int_0^t\int_Z[(x_{s-}^1 - x_{s-}^2 + c^1(s,x_{s-}^1,z,\omega) - c^2(s,x_{s-}^2,z,\omega))^- - (x_{s-}^1 - x_{s-}^2)^-$

$+I_{(x_s^1-x_s^2)\leq 0}(c^1(s,x_{s-}^1,z,\omega)-c^2(s,x_{s-}^2,z,\omega))]N_k(ds,dz),$
and
$\left|x_t^1-x_t^2\right|=\left|x_0^1-x_0^2\right|$
$+\int_0^t sgn(x_s^1-x_s^2)d(x_s^1-x_s^2)$
$+\int_0^t\int_Z[\left|x_{s-}^1-x_{s-}^2+c^1(s,x_{s-}^1,z,\omega)-c^2(s,x_{s-}^2,z,\omega)\right|-\left|x_{s-}^1-x_{s-}^2\right|$
$-sgn(x_{s-}^1-x_{s-}^2)\cdot(c^1(s,x_{s-}^1,z,\omega)-c^2(s,x_{s-}^2,z,\omega))]N_k(ds,dz).$
Moreover, the last terms of the above three formulas are absolutely convergent, $P-a.s.$

4.2.5 Tanaka Type Formula in The Component Form

The 1− dimensional result and Remark 151 motivates us to consider a Tanaka formula for the component $(x_{it}^1-x_{it}^2)$ of $(x_t^1-x_t^2)$, when x_t^1 and x_t^2 are solutions of $n-$dimensional SDE with jumps (4.7) under some other conditions. For this we write (4.7) in its component forms: for $i=1,2,\cdots,d$

$$x_{it}^j=x_{i0}^j+\sum_{p=1}^m\int_0^t b_i^j(s,x_s,\omega)dA_{ps}+\sum_{p=1}^r\int_0^t\sigma_{ip}(s,x_s,\omega)dM_{ps}$$
$$+\int_0^t\int_Z c_i^j(s,x_{s-},z,\omega)\widetilde{N}_k(ds,dz),\forall t\geq 0,j=1,2. \tag{4.15}$$

Thus we can apply Theorem 152 to obtain the following theorem.

Theorem 153 *Assume that*
1°′ $\left|b^j(t,x,\omega)\right|^2+\left|\sigma(t,x,\omega)\right|^2+\int_Z\left|c^j(t,x,z,\omega)\right|^2\pi(dz)\leq g(|x|),$
where $g(u)\geq 0$ is a continuous non-random function on $u\geq 0$;
2° $\left|\sigma_{ip}(t,x,\omega)-\sigma_{ip}(t,y,\omega)\right|^2\leq k_N^T(t)\rho_N^T(|x_i-y_i|)$, *as* $|x|,|y|\leq N$,
$\forall x=(x_1,\cdots,x_i,\cdots,x_d),y=(y_1,\cdots,y_i,\cdots,y_d)\in R^d,t\in[0,T]$,
$p=1,\cdots,r;$
where $0\leq k_N^T(t)$, $\sum_{k=1}^r E\int_0^T k_N^T(t)d\left\langle M_k\right\rangle_t<\infty$, *for each* $T<\infty;0\leq\rho_N^T(u)$ *is non-random, strictly increasing in* $u\geq 0$ *with* $\rho_N^T(0)=0$, *and*
$\int_{0+}du/\rho_N^T(u)=\infty.$
If $x_t^j,j=1,2$ are two solutions of (4.7) with the initial value $x_0^j,j=1,2$, respectively, then $\forall t\geq 0,i=1,2,\cdots,d$
$\left(x_{it}^1-x_{it}^2\right)^+=\left(x_{i0}^1-x_{i0}^2\right)^++\int_0^t I_{(x_{is}^1-x_{is}^2)>0}d\left(x_{is}^1-x_{is}^2\right)$
$+\int_0^t\int_Z[\left(x_{is-}^1-x_{is-}^2+c_i^1(s,x_{s-}^1,z,\omega)-c_i^2(s,x_{s-}^2,z,\omega)\right)^+-\left(x_{is-}^1-x_{is-}^2\right)^+$
$-I_{(x_{is}^1-x_{is}^2)>0}(c_i^1(s,x_{s-}^1,z,\omega)-c_i^2(s,x_{s-}^2,z,\omega))]N_k(ds,dz),$
and similarly,
$\left(x_{it}^1-x_{it}^2\right)^-=\left(x_{i0}^1-x_{i0}^2\right)^-+\int_0^t I_{(x_{is}^1-x_{is}^2)\leq 0}d\left(x_{is}^1-x_{is}^2\right)$
$+\int_0^t\int_Z[\left(x_{is-}^1-x_{is-}^2+c_i^1(s,x_{s-}^1,z,\omega)-c_i^2(s,x_{s-}^2,z,\omega)\right)^--\left(x_{is-}^1-x_{is-}^2\right)^-$
$+I_{(x_{is}^1-x_{is}^2)\leq 0}(c_i^1(s,x_{s-}^1,z,\omega)-c_i^2(s,x_{s-}^2,z,\omega))]N_k(ds,dz).$
Moreover, the last terms of the above two formulas are absolutely convergent, $P-a.s.$

This theorem is easily seen by combinning the results of Theorem 148 and Remark 151. Here the condition on $c^j(t,x,z,\omega)$ is weaker than that in Theorem 142, because here we do not need to assume that $\int_Z \left|c^j(t,x,z,\omega)\right| \pi(dz) < \infty$. However, the condition on $\sigma(s,x,\omega)$ is on one hand stronger and on another hand weaker than that in Theorem 142, because here we almost assume that $\sigma_{ip}(t,x,\omega) = \sigma_{ip}(t,x_i,\omega)$ only depends on x_i, but it is already enough if $\sigma_{ip}(t,x_i,\omega)$ is Holder continuous in x_i with index $\frac{1}{2}$, that is, $\sigma_{ip}(t,x_i,\omega)$ is not necessary Lipschitz continuous.

4.2.6 Pathwise Uniqueness of solutions

For 1−dimensional SDEs with jumps Theorem 152 can help us to get a better thoerem in some sense on the pathwise uniqueness of solutions. For convenience we write out all conditions here, and for simplicity we discuss the following SDE with jumps in 1−dimensional space.

$$\begin{aligned} x_t = x_0 &+ \int_0^t b(s,x_s,\omega)ds + \int_0^t \sigma(s,x_s,\omega)dw_s \\ &+ \int_0^t \int_Z c(s,x_{s-},z,\omega)\widetilde{N}_k(ds,dz), \forall t \geq 0, \end{aligned} \tag{4.16}$$

where w_t and $\widetilde{N}_k(ds,dz)$ are the BM and Poisson martingale measure, respectively, which have been explained before many times.

Theorem 154 *Assume that*
1° $E\left|x_0\right|^2 < \infty$, $2\langle x \cdot b(t,x,\omega)\rangle \leqslant c(t)g(|x|)$, and
$|\sigma(t,x,\omega)|^2 + \int_Z |c(t,x,z,\omega)|^2 \pi(dz) \leqslant c(t)g(|x|)$,
where $g(u) \geq 0$ is a continuous non-random function on $u \geq 0$, and $0 \leqslant c(t)$ is non-random, such that for any $0 < T < \infty$
$C_T = \int_0^T c(t)dt < \infty$,
2° $|\sigma(t,x,\omega) - \sigma(t,y,\omega)|^2 \leq k_N^T(t)\rho_N^T(|x-y|)$, as $|x|, |y| \leq N, t \in [0,T]$;
where $0 \leq k_N^T(t)$ is non-random such that $\int_0^T k_N^T(t)dt < \infty$, for each given $T < \infty; 0 \leq \rho_N^T(u)$ is non-random, strictly increasing in $u \geq 0$ with $\rho_N^T(0) = 0$, and $\int_{0+} du/\rho_N^T(u) = \infty$;
3° $sgn\,(x-y) \cdot (b(t,x,\omega) - b(t,y,\omega)) \leq k_N^T(t)\rho_N^T(|x-y|)$,
as $|x|, |y| \leq N, \forall x,y \in R^1, \forall t \in [0,T]$;
where $k_N^T(t)$ and $\rho_N^T(u)$ have the same property as that in 2°, besides, $\rho_N^T(u)$ is concave;
4° one of the following conditions is satisfied:
(i) $\int_Z |c(t,x,z,\omega)| \pi(dz) \leq c(t)g(|x|)$,
$\int_Z |c(t,x,z,\omega) - c(t,y,z,\omega)| \pi(dz) \leq k_N^T(t)\rho_N^T(|x-y|)$,
as $|x|, |y| \leq N, \forall x,y \in R^1, \forall t \in [0,T]$;
where $c(t)$ and $g(u)$ have the same property as that in 1°; moreover, $k_N^T(t)$ and $\rho_N^T(u)$ have the same property as that in 2°, besides, $\rho_N^T(u)$ is concave;

(ii) $x \geq y \implies x + c(t,x,z,\omega) \geq y + c(t,y,z,\omega)$.
Then the pathwise uniqueness holds for SDE (4.16).

Proof. Assume that $x_t^i, i = 1,2$ are two solutions of (4.16). Write
$\widehat{X}_t = x_t^1 - x_t^2$, $\tau_N = \inf\left\{t > 0 : \left|x_t^1\right| + \left|x_t^2\right| > N\right\}$.
Then by the Tanaka type formula (Theorem 145)
$E\left|\widehat{X}_{t\wedge\tau_N}\right| \leq E\int_0^{t\wedge\tau_N} k_N^T(s)\rho_N^T(\left|\widehat{X}_s\right|)ds + EJ_{t\wedge\tau_N}$,
where
$J_t = \int_0^t \int_Z [|x_{s-}^1 - x_{s-}^2 + c(s,x_{s-}^1,z,\omega) - c(s,x_{s-}^2,z,\omega)| - \left|x_{s-}^1 - x_{s-}^2\right|$
$-sgn(x_{s-}^1 - x_{s-}^2) \cdot (c(s,x_{s-}^1,z,\omega) - c(s,x_{s-}^2,z,\omega))]N_k(ds,dz)$.
In the case that the condition (ii) of 4° holds then $J_t = 0$. So we have
$E\left|\widehat{X}_{t\wedge\tau_N}\right| \leq \int_0^t k_N^T(s)\rho_N^T(E\left|\widehat{X}_{s\wedge\tau_N}\right|)ds$, as $t \in [0,T]$.
Thus $E\left|\widehat{X}_{t\wedge\tau_N}\right| = 0, t \in [0,T]$. From this one easily derives that
$\widehat{X}_t = 0, t \in [0,T]$.
Since $T < \infty$ is arbitrary given, so $\widehat{X}_t = 0, \forall t \geq 0$. In the case that the condition (i) of 4° holds then by the same Tanaka type formula
$E\left|\widehat{X}_{t\wedge\tau_N}\right| \leq E\int_0^{t\wedge\tau_N} k_N^T(s)\rho_N^T(\left|\widehat{X}_s\right|)ds$
$+2E\int_0^{t\wedge\tau_N}\int_Z \left|c(t,x_s^1,z,\omega) - c(t,x_s^2,z,\omega)\right| \pi(dz)ds$
$\leq 3E\int_0^{t\wedge\tau_N} k_N^T(s)\rho_N^T(\left|\widehat{X}_s\right|)ds$, as $t \in [0,T]$.
The conclusion now follows. ∎
Comparing the uniqueness theorem here and Theorem 115, one finds that the condition on σ here is weaker than there. Here σ can be only Hölder continuous with index $\frac{1}{2}$. However, the condition on c is different. Anyway, if the SDE is without jumps, that is $c = 0$, and then we have got a really weaker conditions on the pathwise uniqueness of solutions.

For an SDE with jumps in $d-$dimensional space we can also have a uniquenes theorem discribed by the conditions on the components of the coefficients which are different from Theorem 143 and Theorem 115. For simplicity consider the following SDE in $d-$dimensional space

$$x_t = x_0 + \int_0^t b(s,x_s,\omega)ds + \int_0^t \sigma(s,x_s,\omega)dw_s$$
$$+\int_0^t \int_Z c(s,x_{s-},z,\omega)\widetilde{N}_k(ds,dz), \forall t \geq 0, \tag{4.17}$$

Theorem 155 *Assume that*
1° $E\left|x_0\right|^2 < \infty$, *and*
$2\left\langle x \cdot b(t,x,\omega)\right\rangle \leqslant c(t)(1 + |x|^2)$,
$|\sigma(t,x,\omega)|^2 + \int_Z |c(t,x,z,\omega)|^2 \pi(dz) \leqslant c(t)(1 + |x|^2)$,
where $0 \leqslant c(t)$ *is non-random, such that for any* $0 < T < \infty$
$C_T = \int_0^T c(t)dt < \infty$,

2° $|\sigma_{ip}(t,x,\omega)-\sigma_{ip}(t,y,\omega)|^2 \leq k_N^T(t)\rho_N^T(|x_i-y_i|)$,
as $|x|,|y| \leq N, t\in[0,T]$,
$\forall x=(x_1,\cdots,x_i,\cdots,x_d), y=(y_1,\cdots,y_i,\cdots,y_d)\in R^d, t\in[0,T]$,
$p=1,\cdots,r$;
where $0\leq k_N^T(t)$ *is non-random such that* $\int_0^T k_N^T(t)dt<\infty$, *for each* $T<\infty$; $0\leq \rho_N^T(u)$ *is non-random, strictly increasing in* $u\geq 0$ *with* $\rho_N^T(0)=0$, *and*
$\int_{0+} du/\rho_N^T(u)=\infty$.
3° $sgn_i\,(x-y)\cdot(b_i(t,x,z,\omega)-b_i(t,y,z,\omega))\leq k_N^T(t)\rho_N^T(|x-y|)$,
as $|x|,|y|\leq N, t\in[0,T], \forall x,y\in R^d$;
where $sgn_i x=\frac{x_i}{|x|}$, $k_N^T(t)$ *and* $\rho_N^T(u)$ *have the same property as that in 2°, besides,* $\rho_N^T(u)$ *is concave;*
4° $x_i\geq y_i \Longrightarrow x_i+c_i(t,x,z,\omega)\geq y_i+c_i(t,y,z,\omega)$,
where $x=(x_1,\cdots,x_i,\cdots,x_d), y=(y_1,\cdots,y_i,\cdots,y_d)$.
Then the pathwise uniqueness holds for SDE (4.16).

Proof. Notice that for any real number x
$|x|=x^+ + x^-$.
Now using the notation in the proof of Theorem 154 by the Tanaka formula on components (Theorem 153) we have that as $t\in[0,T]$
$E\left|\widehat{X}_{i,t\wedge\tau_N}\right|\leq E\int_0^{t\wedge\tau_N} k_N^T(s)\rho_N^T(\left|\widehat{X}_s\right|)ds,\ \forall i=1,2,\cdots,d.$
Hence as $t\in[0,T]$
$E\left|\widehat{X}_{t\wedge\tau_N}\right|\leq \sum_{i=1}^d E\left|\widehat{X}_{i,t\wedge\tau_N}\right|\leq d\cdot E\int_0^{t\wedge\tau_N} k_N^T(s)\rho_N^T(\left|\widehat{X}_s\right|)ds.$
The conclusion now follows easily as in the proof of the previous theorem.
■

4.3 Local Time and Occupation Density Formula

Consider a real semi-martingale x_t which is given by (4.4):

$$x_t=x_0+A_t+M_t+\int_0^t\int_Z c(s,z,\omega)\widetilde{N}_k(ds,dz), t\geq 0. \tag{4.18}$$

We can introduce the following definition.

Definition 156 *The function* $L_t^a(x)$ *of two variables* (t,a) *is called the local time of semi-martingale* x_t, *if it satisfies: for* $\forall t\geq 0$ *and* $a\in R$

$$(x_t-a)^+=(x_0-a)^+ +\int_0^t I_{(x_{s-}>a)}dx_s+\frac{1}{2}L_t^a(x)$$

$$+\int_0^t\int_Z((x_{s-}+c(s,z,\omega)-a)^+-(x_{s-}-a)^+-I_{(x_{s-}>a)}c(s,z,\omega))N_k(ds,dz). \tag{4.19}$$

It is easily seen that (4.19) can be rewritten as

$$(x_t - a)^+ = (x_0 - a)^+ + \int_0^t I_{(x_{s-}>a)} dx_s + \frac{1}{2} L_t^a(x)$$
$$+ \sum_{0<s\leq t} I_{(x_{s-}>a)}(x_s - a)^- + \sum_{0<s\leq t} I_{(x_{s-}\leq a)}(x_s - a)^+.$$

Moreover, by this form one easily sees that the following lemma is true.

Lemma 157 *$L_t^a(x)$ is continuous in $t \geq 0$.*

Proof. In fact, for each $t > 0$

$$\frac{1}{2}(L_t^a(x) - L_{t-}^a(x)) = (x_t - a)^+ - (x_{t-} - a)^+ - I_{(x_{t-}>a)} \triangle x_t$$
$$-I_{(x_{t-}>a)}(x_t - a)^- - I_{(x_{t-}\leq a)}(x_t - a)^+.$$

Hence,

$$\frac{1}{2}(L_t^a(x) - L_{t-}^a(x)) I_{(x_{t-}>a)} = I_{(x_{t-}>a)}(x_t - a)^+$$
$$-I_{(x_{t-}>a)}(x_{t-} - a + \triangle x_t) - I_{(x_{t-}>a)}(x_t - a)^- = 0,$$

and

$$\frac{1}{2}(L_t^a(x) - L_{t-}^a(x)) I_{(x_{t-}\leq a)} = I_{(x_{t-}\leq a)}(x_t - a)^+ - I_{(x_{t-}\leq a)}(x_t - a)^+ = 0.$$

Therefore, $L_t^a(x)$ is left continuous in $t > 0$. However, as we said before we only consider right continuous martingales (or their right continuous versions), so it is easily seen that $L_t^a(x)$ is also right continuous in $t \geq 0$. Therefore, $L_t^a(x)$ is continuous in $t \geq 0$. ∎

By Remark 151 one immediately obtains the following lemma.

Lemma 158 *$L_t^a(x)$ defined by (4.19) exists, and it is a continuous process in t. Moreover, $\forall t \geq 0$*

$$\int_0^t \int_Z ((x_{s-} + c(s,z,\omega) - a)^+ - (x_{s-} - a)^+ - I_{(x_{s-}>a)} c(s,z,\omega)) N_k(ds,dz)$$
$$(= \int_0^t \int_Z (I_{(x_{s-}>a)}(x_{s-} + c(s,z,\omega) - a)^-$$
$$+ I_{(x_{s-}\leq a)}(x_{s-} + c(s,z,\omega) - a)^+) N_k(ds,dz))$$
$$\leq |(x_t - a)^+ - (x_0 - a)^+| + A_t + \left| \int_0^t I_{(x_s>a)} dM_s \right|$$
$$+ \left| \int_0^t \int_Z I_{(x_{s-}>a)} c(s,z,\omega) \widetilde{N}_k(ds,dz) \right| < \infty, \ P-a.s.$$

Therefore,

$$\frac{1}{2} |L_t^a(x)|$$
$$\leq | \int_0^t \int_Z ((x_{s-} + c(s,z,\omega) - a)^+ - (x_{s-} - a)^+$$
$$- I_{(x_{s-}>a)} c(s,z,\omega)) N_k(ds,dz)|$$
$$+ \left| (x_t - a)^+ - (x_0 - a)^+ - \int_0^t I_{(x_{s-}>a)} dx_s \right|$$
$$\leq 2[|(x_t - a)^+ - (x_0 - a)^+| + A_t + \left| \int_0^t I_{(x_s>a)} dM_s \right|$$
$$+ \left| \int_0^t \int_Z I_{(x_{s-}>a)} c(s,z,\omega) \widetilde{N}_k(ds,dz) \right|] < \infty, \ P-a.s.$$

and

$$E| \int_0^t \int_Z ((x_{s-} + c(s,z,\omega) - a)^+ - (x_{s-} - a)^+$$
$$- I_{(x_{s-}>a)} c(s,z,\omega)) N_k(ds,dz)| \leq \widetilde{k}_0 E[|x_t - x_0| + A_t + \langle M \rangle_t^{1/2}$$
$$+ \left(\int_0^t \int_Z |c(s,z,\omega)|^2 \pi(dz) ds \right)^{1/2}],$$
$$E|L_t^a(x)| \leq k_0 E[|x_t - x_0| + A_t + \langle M \rangle_t^{1/2}$$
$$+ \left(\int_0^t \int_Z |c(s,z,\omega)|^2 \pi(dz) ds \right)^{1/2}].$$

(Later, we will see that $L_t^a(x)$ is increasing in $t \geq 0$, thus
$L_t^a(x) = |L_t^a(x)| \geq 0)$.
Also,
$(x_t - a)^- = (x_0 - a)^- - \int_0^t I_{(x_{s-} \leq a)} dx_s + \frac{1}{2} L_t^a(x)$
$+ \int_0^t \int_Z ((x_{s-} + c(s,z,\omega) - a)^- - (x_{s-} - a)^-$
$- I_{(x_{s-} \leq a)} c(s,z,\omega)) N_k(ds,dz)$
$= (x_0 - a)^- - \int_0^t I_{(x_{s-} \leq a)} dx_s + \frac{1}{2} L_t^a(x)$
$+ \sum_{0<s\leq t} I_{(x_{s-} > a)} (x_s - a)^- + \sum_{0<s\leq t} I_{(x_{s-} \leq a)} (x_s - a)^+$,
and
$|x_t - a| = |x_0 - a| + \int_0^t sgn(x_{s-} - a) dx_s + L_t^a(x)$
$+ \int_0^t \int_Z (|x_{s-} + c(s,z,\omega) - a| - |x_{s-} - a|$
$- sgn(x_{s-} - a) c(s,z,\omega)) N_k(ds,dz)$,
where we define $sgn(0) = -1$.

Proof. The first conclusion is derived by the proof of Theorem 150 and Remark 151. Notice that
$x_t - a = x_0 - a + \int_0^t dx_s$, $x^- = x^+ - x$,
$|x| = x^+ + x^-$,
So the second and the third formulas are established. ■
The following occupation density formula on local time is very useful.

Lemma 159 *For any bounded Borel measurable function f*

$$\int_0^t f(x_s) d\langle M \rangle_s = \int_R f(a) L_t^a(x) da. \tag{4.20}$$

Proof. By approximation we may assume that $f \in C_0^\infty(R)$, the totality of functions with compact support. Let
$F(x) = \int_R f(a)(x-a)^+ da$, $\forall x \in R$.
Then it is easily seen that
$F'(x) = \int_{-\infty}^x f(a) da$, $F''(x) = f(x)$.
By Ito's formula

$$F(x_t) - F(x_0) - \int_0^t F'(x_s) dA_s - \int_0^t F'(x_s) dM_s$$
$$- \int_0^t \int_Z F'(x_{s-}) c(s,z,\omega) \widetilde{N}_k(ds,dz) - \int_0^t \int_Z [F(x_{s-} + c(s,z,\omega))$$
$$- F(x_{s-}) - F'(x_{s-}) c(s,z,\omega)] N_k(ds,dz) = \frac{1}{2} \int_0^t f(x_s) d\langle M \rangle_s. \tag{4.21}$$

By localization we may suppose that $|A|_t$, M_t, $\langle M \rangle_t$ and $\langle N \rangle_t$ are bounded, where $N_t = \int_0^t \int_Z c(s,z,\omega) \widetilde{N}_k(ds,dz)$. Denote the left side of (4.21) by $\sum_{i=1}^6 I_i$. Obviously,
$I_1 = \int_R f(a)(x_t - a)^+ da$,
$-I_3 = \int_R f(a) (\int_0^t I_{(x_s > a)} dA_S) da$,

$-I_6 = \int_R f(a) \int_0^t \int_Z ((x_{s-} + c(s,z,\omega) - a)^+ - (x_{s-} - a)^+$
$-I_{(x_{s-}>a)} c(s,z,\omega)) N_k(ds,dz)da,$
where we have applied Fubini's theorem for Lebesgue-Stieltjes integrals. By the stochastic Fubini theorem below we find that $P-a.s.$ $\forall t \geq 0$

$$I_5 = \int_0^t \int_Z F'(x_{s-}) c(s,z,\omega) \widetilde{N}_k(ds,dz)$$
$$= \int_0^t \int_Z \int_R f(a) I_{(x_{s-}>a)} c(s,z,\omega) da \widetilde{N}_k(ds,dz)$$
$$= \int_R f(a) \int_0^t \int_Z I_{(x_{s-}>a)} c(s,z,\omega)) \widetilde{N}_k(ds,dz)da,$$

and

$$I_4 = \int_0^t F'(x_s) dM_s = \int_0^t \int_R f(a) I_{(x_s>a)} da dM_s$$
$$= \int_R f(a) \int_0^t I_{(x_s>a)} dM_s da.$$

Hence (4.20) is only a rewritten form of (4.21). ■

Theorem 160 *(Stochastic Fubini Theorem). 1) Assume that* $\{f(t,a,\omega)\}_{t\geq 0}$ *is a real random process such that*

(i) $|f(t,a,\omega)| \leq g(a)$,
where $0 \leq g(a)$ *is a Borel measurable function on* $a \in R^1$;
(ii) $\{f(t,a,\omega)\}_{(t,\omega)\in[0,\infty)\times\Omega, a\in R^1}$ *is a* $\mathcal{P}\times\mathfrak{B}(R^1)-$*measurable function, where* $\mathcal{P}$ *is the predictable* $\sigma-$*field (for the definition of predictable* $\sigma-$*field and predictable function see Definition 12).*
Now suppose that $\{M_t\}_{t\geq 0}$ *is a continuous square integrable martingale, and for each* $t > 0$ $\left\{\int_0^t f(s,a,\omega)dM_s\right\}_{(a,\omega)\in R^1\times\Omega}$ *is a* $\mathfrak{B}(R^1)\times\mathfrak{F}-$*measurable function. Furthermore, suppose that* $\mu(da)$ *is a non-negative Borel measure on* R^1 *and* $\int_{R^1} g(a)\mu(da) < \infty$.
Then $P-a.s.$ $\forall t \geq 0$

$$\int_0^t [\int_{R^1} f(s,a,\omega)\mu(da)] dM_s = \int_{R^1} [\int_0^t f(s,a,\omega)dM_s] \mu(da).$$

2) Suppose that $\widetilde{N}_k(dt,dz)$ *is a Poisson martingale measure generated by a* $\mathfrak{F}_t-$*Poisson point process* $k(.)$ *with the compensator* $\pi(dz)dt$ *such that* $\widetilde{N}_k(dt,dz) = N_k(dt,dz) - \pi(dz)dt$, *where* $\pi(dz)$ *is a* $\sigma-$*finite measure on the measurable space* $(Z,\mathfrak{B}_Z)$, *and suppose that* $\mu(da)$ *is a non-negative Borel measure on* R^1. *Now assume that* $\{f(t,a,z,\omega)\}_{t\geq 0}$ *is a real random process such that*

(i) $|f(t,a,z,\omega)| \leq g(a,z)$,
where $0 \leq g(a,z)$ *is a* $\mathfrak{B}(R^1)\times\mathfrak{B}_Z-$*measurable function on* $(a,z) \in R^1 \times Z$ *such that* $\int_{R^1}\int_Z g(a,z)^2 \pi(dz)\mu(da) < \infty$;
(ii) $\{f(t,a,z,\omega)\}_{(t,z,\omega)\in[0,\infty)\times Z\times\Omega, a\in R^1}$ *is a* $\mathcal{P}\times\mathfrak{B}(R^1)-$*measurable function, where* $\mathcal{P}$ *is the predictable* $\sigma-$*field (for the definition of predictable* $\sigma-$*field and predictable function see Definition 1261). Moreover, for each* $t > 0$ $\left\{\int_0^t \int_Z f(s,a,z,\omega)\widetilde{N}_k(ds,dz)\right\}_{(a,\omega)\in R^1\times\Omega}$ *is a* $\mathfrak{B}([0,\infty))\times\mathfrak{F}-$*measurable function.*
Then $P-a.s.$ $\forall t \geq 0$

$$\int_0^t \int_Z [\int_{R^1} f(s,a,z,\omega)\mu(da)] \widetilde{N}_k(ds,dz)$$
$$= \int_{R^1} [\int_0^t \int_Z f(s,a,z,\omega)\widetilde{N}_k(ds,dz)] \mu(da).$$

Proof. 1): First, by assumption (i) and (ii) $\int_0^t f(s,a,\omega)dM_s$ exists, since
$E\int_0^t |f(s,a,\omega)|^2 d\langle M\rangle_s \leq g(a)^2 E\langle M\rangle_t < \infty$.
Now notice that
$E|\int_0^t[\int_{R^1} f(s,a,\omega)\mu(da)]dM_s - \int_{R^1}[\int_0^t f(s,a,\omega)dM_s]\mu(da)|^2$
$= E[\int_0^t |\int_R f(s,a,\omega)\mu(da)|^2 d\langle M\rangle_s$
$+E\int_R\int_R\int_0^t f(s,a,\omega)f(s,a_1,\omega)d\langle M\rangle_s \mu(da_1)\mu(da)$
$-2E\int_R(\int_0^t\int_R f(s,a,\omega)\mu(da)dM_s\int_0^t f(s,a_1,\omega)dM_S)\mu(da_1)$
$= 2E\int_0^t\int_R\int_R f(s,a,\omega)f(s,a_1,\omega)\mu(da_1)\mu(da)d\langle M\rangle_s$
$-2E\int_R\int_0^t\int_R f(s,a,\omega)\mu(da)f(s,a_1,\omega)d\langle M\rangle_s \mu(da_1) = 0,$
where we have applied Fubini's theorem for Lebesgue-Stieltjes integrals. Hence for each $t \geq 0$, $P-a.s.$

$$\int_0^t[\int_{R^1} f(s,a,\omega)\mu(da)]dM_s = \int_{R^1}[\int_0^t f(s,a,\omega)dM_s]\mu(da). \tag{4.22}$$

We still need to show that there exists a set $\Lambda \in \mathfrak{F}$ with $P(\Lambda) = 0$ such that as $\overline{\omega} \notin \Lambda, \forall t \geq 0$,
$(\int_0^t[\int_{R^1} f(s,a,\omega)\mu(da)]dM_s)(\overline{\omega}) = (\int_{R^1}[\int_0^t f(s,a,\omega)dM_s]\mu(da)(\overline{\omega})$.
(We call Λ a common $P-$null set for all $t \geq 0$). That is, $P-a.s.$ $\forall t \geq 0$
$\int_0^t[\int_{R^1} f(s,a,\omega)\mu(da)]dM_s = \int_{R^1}[\int_0^t f(s,a,\omega)dM_s]\mu(da)$.
For this let us show that $I(t) = \int_{R^1}[\int_0^t f(s,a,\omega)dM_s]\mu(da)$ is a "right continuous with a left limit" (RCLL) process. In fact,
$E\int_{R^1}\sup_{t\leq T}\left|\int_0^t f(s,a,\omega)dM_s\right|\mu(da) \leq \int_R |g(a)|\, E(\langle M\rangle_T^{1/2})\mu(da) < \infty.$
So $P-a.s.$
$\int_{R^1}\sup_{t\leq T}\left|\int_0^t f(s,a,\omega)dM_s\right|\mu(da) < \infty.$
Hence if $t_n \downarrow t$, $t < T$, then one can apply Lebesgue's dominated convergence theorem to find that as $n \to \infty$,
$I(t_n) \to I(t)$,
since $\int_0^t f(s,a,\omega)dM_s$ is right continuous in t. The proof that $I(t)$ has a left limit for any $t > 0$ is the same. Now we have that the both sides of (4.22) are RCLL, so the common $P-$null set for all $t \geq 0$ is easily constructed.

2): First, by assumption (i) and (ii) $\int_0^t\int_Z f(s,a,z,\omega)\widetilde{N}_k(ds,dz)$ exists, since
$E\int_0^t\int_Z |f(s,a,z,\omega)|^2 \pi(dz)ds \leq t\int_Z g(a,z)^2\pi(dz) < \infty.$
Now the proof is similar to that in 1). In fact,
$E|\int_0^t\int_Z[\int_{R^1} f(s,a,z,\omega)\mu(da)]\widetilde{N}_k(ds,dz)$
$-\int_{R^1}[\int_0^t\int_Z f(s,a,z,\omega)\widetilde{N}_k(ds,dz)]\mu(da)|^2$
$= E[\int_0^t\int_Z(\int_{R^2} f(s,a_1,z,\omega)f(s,a_2,z,\omega)\mu(da_1)\mu(da_2))\pi(dz)ds$
$+\int_{R^2}\int_0^t\int_Z f(s,a_1,z,\omega)f(s,a_2,z,\omega)\pi(dz)ds\mu(da_1)\mu(da_2)$
$-2\int_{R^2}\int_0^t\int_Z f(s,a_1,z,\omega)f(s,a_2,z,\omega)\pi(dz)ds\mu(da_1)\mu(da_2)] = 0,$
again by Fubini's theorem for Lebesgue-Stieltjes integrals. Hence, it is similar to 1), and we obtains 2). ■

Let us explain the pysical meaning of the local time. By (4.20) one has that for a.e. $a \in R$

$$L_t^a(x) = \lim_{\varepsilon \downarrow 0} \varepsilon^{-1} \int_0^t I_{[a,a+\varepsilon)}(x_s) d\langle x^c \rangle_s, \quad (4.23)$$

where $x^c = M$. Hence $L_t^a(x)$ can be viewed as the "measure" of occupation at the point a by the process $x(\cdot)$ during the time interval $[0,t]$. In particular, as $M_t = w_t$ is the BM, one sees that

$L_t^a(x) = \lim_{\varepsilon \downarrow 0} \varepsilon^{-1} \int_0^t I_{[a,a+\varepsilon)}(x_s) ds.$

This is exactly the Lebesgue measure of the time set $\{s \in [0,t] : x_s = a\}$. Furthermore, by (4.23) we immediately obtain the following corollary.

Corollary 161 *$L_t^a(x)$ is an increasing function on $t \geq 0$.*

4.4 Krylov Estimation

4.4.1 The case for 1−dimensional space

By means of the occupation density formula one can derive the so-called Krylov estimation with respect to the 1−dimensional SDE with jumps as follows:

Theorem 162 *Assume that x_t satisfies (4.18) with c satisfying that for each $T < \infty$*

$\int_0^T \int_Z |c(t,z,\omega)|^2 \pi(dz) < \infty,\ P-a.s.$

Let

$\tau_N = \inf\{t \geq 0 : |x_t|^2 + |A|_t^2 + |M_t|^2 + |N_t|^2 + \langle M \rangle_t + \langle N \rangle_t > n\}.$

If $f \in L^1(R)$, then for any $0 \leq T < \infty$, one has that

$E \int_0^{T \wedge \tau_N} |f(x_s)| \, d\langle M \rangle_s \leq k_{N,T} \int_R |f(y)| \, dy \widehat{=} k_{N,T} \|f\|_{L^1(R)},$

where $k_{N,T} \geq 0$ is a constant independent of f such that

$k_{N,T} = k_0 E(|A|_{T \wedge \tau_N} + \langle M \rangle_{T \wedge \tau_N}^{1/2} + \langle N \rangle_{T \wedge \tau_N}^{1/2}).$

Hence, if M_t is a square integrable martingale, c satisfies

$E \int_0^T \int_Z |c(s,z,\omega)|^2 \pi(dz) ds < \infty,$

and

$E|A|_T < \infty,$

then

$E \int_0^T |f(x_s)| \, d\langle M \rangle_s \leq k_T \|f\|_{L^1(R)},$

where $0 \leq k_T = k_{\infty,T}$.

Proof. Let $f(x) \geq 0$ and

$f_n(x) = f(x) \wedge n.$

The by the formula (4.20)

$\int_0^t |f_n(x_s)| \, d\langle M \rangle_s = \int_R |f_n(a)| \, L_t^a(x) da, \forall t \geq 0.$

By Lemma 158 and Corollary 161

$0 \leq EL^a_{T\wedge\tau_N}(x) \leq k_{N,T}$.
Thus we have
$E\int_0^{T\wedge\tau_N} |f_n(x_s)|\, d\langle M\rangle_s \leq k_{N,T}\int_R |f_n(y)|\, dy \leq k_{N,T}\int_R |f(y)|\, dy$.
Letting $n \to \infty$ and applying Fatou's lemma, one obtains
$E\int_0^{T\wedge\tau_N} |f(x_s)|\, d\langle M\rangle_s \leq k_{N,T}\, \|f\|_{L^1(R)}$ ■

Corollary 163 *If*

$M_t = \int_0^t \sigma(s,\omega)dw_s$, $A_t = \int_0^t b(s,\omega)ds$,
and $|\sigma(s,\omega)| \geq \delta_0 > 0$,
$|b(t,\omega)|^2 + |\sigma(t,\omega)|^2 + \int_Z |c(t,z,\omega)|^2\, \pi(dz) \leq k_0$,
where $0 \leq k_0$, and δ_0 are constants, and $w_t, t \geq 0$, is a standard Brownian Motion process, then for any f, which is a Borel measurable function valued in R,
$E\int_0^T |f(x_s)|\, ds \leq k_T\, \|f\|_{L^1(R)}$,
where $0 \leq k_T$ is a constant depending on k_0, δ_0 and T only.

Proof. If
$\|f\|_{L^1(R)} = \infty$,
then the result is trivial. If
$\|f\|_{L^1(R)} < \infty$,
then the result follows by Theorem 162. ■

The Krylov estimate is a useful tool in Stochastic Analysis, because it transforms the estimate on functionals of solutions for $1-$dimensional SDEs into the calculation on the usual deterministic L^p-integrals on R^1. Here we use the occupation formula of local time for the solution x_t to establish the Krylov estimate, which is different from the usual or original proof for such kinds of formula that make use of the PDE results. We can see that the technique used here has some advantages. First, in Corollary 163 the coefficients $b(t,x,\omega), \sigma(t,x,\omega)$, and $c(t,x,z,\omega)$ of SDE with jumps can depend on ω, that is, the coefficients can be random, where when the PDE results are used this cannot be so. Second, in Theorem 162 the Krylov estimate is obtained for a very general SDE, which is hard to use with the PDE technique in general.

4.4.2 The Case for d−dimensional space

However, to derive the Krylov estimate on functionals of solutions for $d-$dimensional SDEs we have to get some help from the PDE results, so the coefficients σ and c cannot be random. Assume that $(x(t))$ satisfies the following $d-$dimensional stochastic system:

$$x_t = x_0 + \widetilde{A}_t + \int_0^t \sigma(s,x_s)dw_s + \int_0^t\int_Z c(s,x_{s-},z)\widetilde{N}_k(ds,dz), \quad (4.24)$$

where $\widetilde{A}_t^i$ is a continuous locally integrable finite variational $\mathfrak{F}_t-$adapted process, $\forall i = 1, 2, \cdots, d$, and $\widetilde{A}_t = (\widetilde{A}_t^1, \cdots, \widetilde{A}_t^d)$; that is, each component $\widetilde{A}_t^i$ (of $\widetilde{A}_t$) can be expressed as the difference of two $\mathfrak{F}_t-$adapted increasing processes, and $E\left|\widetilde{A}^i\right|_T < \infty$, for each $T < \infty$, where $\left|\widetilde{A}^i\right|_T$ is the total variation of $\widetilde{A}_t^i$ on $t \in [0, T]$; moreover, w_t is a $d-$dimensional BM, and $\widetilde{N}_k(dt, dz)$ is a $1-$dimensional Poisson martingale measure with the compensator $\pi(dz)dt$ such that

$\widetilde{N}_k(dt, dz) = N_k(dt, dz) - \pi(dz)dt.$

We quote a result from Krylov (1974)[63] (or, see [3]) as the following lemma, which indicates the existence of a solution for a partial differential inequality.

Lemma 164 *Suppose we are given a bounded Borel measurable function* $0 \leq f = f(t, x) : \overline{R}_+^1 \times R^d \to \overline{R}_+^1 (= [0, \infty))$ *and* $\lambda > 0$, *then for any* $\varepsilon > 0$ *there exists a smooth function* $u^\varepsilon(t, x) : R \times R^d \to [0, \infty)$ *such that*

1) $\sum_{i,j=1}^d h_i h_j (\partial^2 / \partial x_i \partial x_j) u^\varepsilon \leq \lambda u^\varepsilon$, $\forall h \in R^d, |h| \leq 1$,

2) $\forall p \geq d + 1, (t, x) \in \overline{R}_+^1 \times R^d$

$u^\varepsilon(t, x) \leq k(p, \lambda, d, \varepsilon) \left\| e^{-\lambda s(d+1)/p} f(t + s, y) \right\|_{p, (s,y) \in \overline{R}_+^1 \times R^d}$,

where

$k(p, \lambda, d, \varepsilon) = e^{\varepsilon \lambda (d+1)/p} p^{d/p} (V_d d!)^{1/p} \lambda^{d/2p} (\lambda(d + 1))^{(1/p)-1}$,

V_d *is the volume of* $d-$*dimensional unit sphere, and*

$\|g(s, y)\|_{p, (s,y) \in \overline{R}_+^1 \times R^d} = (\int_{\overline{R}_+^1 \times R^d} |g(s, y)|^p \, ds dx)^{1/p}$

$(= \|g(s, y)\|_{p, \overline{R}_+^1 \times R^d})$,

3) $|grad_x u^\varepsilon| \leq (\lambda)^{1/2} u^\varepsilon$,

4) for all non-negative definite symmetric matrices $A = (a_{i,j})_{d \times d}$

$\sum_{i,j=1}^d a_{i,j} (\partial^2 u^\varepsilon / \partial x_i \partial x_j) - \lambda (tr.A) u^\varepsilon \leq 0$,

$\sum_{i,j=1}^d a_{i,j} (\partial^2 u^\varepsilon / \partial x_i \partial x_j) - \lambda (tr.A + 1) u^\varepsilon + (\partial / \partial t) u^\varepsilon \leq -(\det A)^{1/(d+1)} f_\varepsilon$,

where f_ε *is the smoothness function of* f, *i.e.*

$f_\varepsilon = f * \omega^\varepsilon = \int_{-\infty}^\infty ds \int_{R^d} \widetilde{f}(t - \varepsilon s, x - \varepsilon y) \omega(s, y) dy$,

$$\widetilde{f}(t, x) = \begin{cases} f(t, x), & as\ t \geq 0, \\ 0, & otherwise, \end{cases}$$

$\omega(t, x)$ *is a smooth function such that* $\omega(t, x) : R^1 \times R^d \to [0, \infty)$,

$\omega(t, x) = 0$, *as* $(t, x) \notin [-1, 1] \times [-1, 1]^d$,

$\int_{-\infty}^\infty dt \int_{R^d} \omega(t, x) dx dt = 1$.

By using the above lemma we can prove the following

Theorem 165 *(Krylov type estimate). Assume that*

$\sigma = \sigma(t, x) : [0, T] \times R^d \to R^{d \otimes d}$,

$c = c(t, x, z) : [0, T] \times R^d \times Z \to R^d$

are jointly measurable and satisfy the conditions

$$\|\sigma(t,x)\|^2 + \int_Z |c(t,x,z)|^2 \pi(dz) \leq c_1(t), \tag{4.25}$$

where $k_T = \int_0^T c_1(t)dt < \infty$, *for each* $T < \infty$. *Moreover, assume that* $E\left|\widetilde{A}^i\right|_T \leq k_T$, *for each* $T < \infty$. *Write*

$\eta_t = \int_0^t tr.A(s,x_s)ds,\ A(s,x) = \frac{1}{2}\sigma\sigma^*(s,x).$

Then for all $0 \leq f$, *which is a bounded Borel measurable function defined on* $[0,T] \times R^d$, *and* $p \geq d+1$

$$E\int_0^T (\det A(s,x_s))^{1/(d+1)} f(s,x_s)ds \leq k(p,k_T,d,T)\,\|f\|_{p,[0,T]\times R^d}\,, \tag{4.26}$$

where we write $\|f\|^p_{p,[0,T]\times R^d} = \int_0^T \int_{R^d} |f(t,x)|^p\, dxdt$, *and* $k \geq 0$ *is a constant depending only on* p, k_T, d *and* T.

Furthermore, in the case that σ *is uniformly non-degenerate or locally uniformly non-degenerate we have the following corollaries:*

1) If, in addition, for each $N = 1, 2, \cdots$ *there exists a* $\delta_N > 0$ *such that for all* $\mu \in R^d$,

$\langle A(t,x)\mu,\mu\rangle \geq |\mu|^2\,\delta_N$, *as* $|x| \leq N$,

where $\langle\cdot,\cdot\rangle$ *is the inner product in* R^d, *and* $A(t,x) = \sigma\sigma^*(t,x)$, *(that is,* σ *is locally uniformly non-degenerate), and letting*

$\tau_N = \inf\{t \in [0,T] : |x_t| > .N\}.$

then

$E\int_0^{\tau_N} f(s,x_s)ds \leq k(p,k_T,d,\delta_N,T)\,\|f\|_{p,[0,T]\times[-N,N]^{\otimes d}}\,,$

where $k \geq 0$ *is a constant only depending on* p, k_0, d, δ_N *and* T.

2) If, in addition, there exists a $\delta_0 > 0$ *such that for all* $\mu \in R^d$,

$\langle A(t,x)\mu,\mu\rangle \geq |\mu|^2\,\delta_0,$

(that is, σ *is uniformly non-degenerate), then*

$E\int_0^T f(s,x_s)ds \leq k(p,k_T,d,\delta,T)\,\|f\|_{p,[0,T]\times R^d}\,,$

where $k \geq 0$ *is a constant only depending on* p, k_T, d, δ *and* T.

Proof. Let

$f(t,x) = 0$, as $(t,x) \in (T,\infty) \times R^d$.

According to Lemma 164, for $f \geq 0$ there exists a u^ε satisfying 1)-4) in Lemma 164. Applying Ito's formula to $u^\varepsilon(t,x_t)e^{-\lambda(\eta_t+t)}$ on $[0,T]$ we find that

$E\ u^\varepsilon(T,x_T)e^{-\lambda(\eta_T+T)} - Eu^\varepsilon(0,x_0)$

$= E\int_0^T e^{-\lambda(\eta_s+s)} grad_x u^\varepsilon(s,x_s)d\widetilde{A}_s + E\int_0^T e^{-\lambda(\eta_s+s)}((\partial/\partial s)u^\varepsilon$

$+\sum_{i,j=1}^d a_{ij}(\partial^2 u^\varepsilon/\partial x_i \partial x_j) - \lambda(tr.A_s+1)u^\varepsilon)ds$

$+E\sum_{0<s\leq T} e^{-\lambda(\eta_s+s)}(u^\varepsilon(s,x_s) - u^\varepsilon(s,x_{s-})$

$-grad_x u^\varepsilon(s,x_{s-})\cdot\triangle x_s) = \sum_{i=1}^3 I_T^i.$

Now we set

$\beta = k(p, \lambda, d, \varepsilon) \left\| e^{-\lambda s(d+1)/p} f(s, y) \right\|_{p,[0,T]\times R^d}.$

By 2) in Lemma 164 $E\ u^\varepsilon(0, x_0) \leq \beta$. Notice that by 3) and 2) in Lemma 164

$I_T^1 \leq k_0 \beta \lambda^{\frac{1}{2}} E \int_0^T e^{-\lambda \eta_s - \lambda(1-(d+1)/p)s} d\left|\widetilde{A}\right|_s \leq k_0 \beta \lambda^{\frac{1}{2}} E\left|\widetilde{A}\right|_T \leq k_T k_0 \lambda^{\frac{1}{2}} \beta.$

On the other hand, notice that

$(\int_{R^d} \int_0^\infty \left| e^{-\lambda r(d+1)/p} f(r+s, y) \right|^p drdy)^{1/p}$

$\leq e^{\lambda s(d+1)/p} \left\| e^{-\lambda r(d+1)/p} f(r, y) \right\|_{p,(r,y)\in \overline{R}_+^1 \times R^d}$

$= e^{\lambda s(d+1)/p} \left\| e^{-\lambda r(d+1)/p} f(r, y) \right\|_{p,(r,y)\in [0,T]\times R^d}.$

Hence by 1) in Lemma 164 and the condition (4.25) we have

$I_T^3 \leq \frac{1}{2} E \sum_{0<s\leq T} \sum_{i,j=1}^d e^{-\lambda(\eta_s + s)}$

$\cdot(\partial^2 u^\varepsilon(s, x_{s-} + \theta \triangle x_s)/\partial x_i \partial x_j) \triangle x_i(s) \triangle x_j(s)$

$\leq \frac{1}{2} \lambda \beta e^{\lambda T(d+1)/p} E \sum_{0<s\leq T} (\triangle x(s))^2 \leq k'_{d,\lambda,T} \beta \int_0^T c_1(s) ds.$

Furthermor, by 4) in Lemma 164

$I_T^2 \leq -E \int_0^T e^{-\lambda(\eta_s + s)} (\det A(s, x_s))^{1/(d+1)} f_\varepsilon(s, x_s) ds$

Thus (4.26) is valid for $f_\varepsilon(s, x_s)$. Let $\varepsilon \to 0$. By Fatou's lemma (4.26) is derived for f.

Now let us show the corollaries. 1): By (4.26)

$E \int_0^{\tau_N} (\det A(s, x_s))^{1/(d+1)} f(s, x_s) ds$

$\leq E \int_0^T (\det A(s, x_s))^{1/(d+1)} f(s, x_s) I_{(|x_s| \leq N)} ds$

$\leq k(p, k_T, d, T) \left\| f \right\|_{p,[0,T]\times[-N,N]}.$

Therefore,

$0 \leq E \int_0^{\tau_N} f(s, x_s) ds \leq \frac{1}{\delta_N} k(p, k_T, d, T) \left\| f \right\|_{p,[0,T]\times[-N,N]}.$

1) is proved. Corollary 2) obviously follows by (4.26). ∎

4.4.3 Applications to Convergence of Solutions to SDE with Jumps

Now we are going to use the Krylov estimate to discuss the convergence of solutions to stochastic differential equations with jumps, which have some weaker conditions on the coefficients in some respect.

Suppose that $x_t^n, n = 0, 1, 2, \cdots$ satisfy the following $d-$dimensional SDE's, respectively,

$$x_t^n = x_0 + \int_0^t b^n(s, x_s^n) ds + \int_0^t \sigma^n(s, x_s^n) dw_s$$
$$+ \int_0^t \int_Z c^n(s, x_{s-}^n, z) \widetilde{N}_k(ds, dz), \forall t \geq 0, n = 0, 1, 2, \cdots; \tag{4.27}$$

where w_t is a $r-$dimensional BM, and $\widetilde{N}_k(ds, dz)$ is a $1-$dimensional Poisson martingale with the compensator $\pi(dz)dt$ as before, so $b^n, c^n \in R^d$, and $\sigma^n \in R^{d\otimes r}$. Here we write $x_t^0 = x_t, b^0 = b$.

Theorem 166 *Assume that*

1° $|b^n(s,x)|^2 + |\sigma^n(s,x)|^2 + \int_Z |c^n(s,x,z)|^2 \pi(dz) \le k_0$,
where $k_0 \ge 0$ *is a constant,*
2° $(x-y)\cdot(b(s,x)-b(s,y)) \le F(s)\rho(|x-y|^2)$,
$\|\sigma(t,x)-\sigma(t,y)\|^2 + \int_Z |c(t,x,z)-c(t,y,z)|^2 \pi(dz)$
$\le F(s)\rho(|x-y|^2)$,
where $0 \le F(s)$ *satisfies that* $\forall t \ge 0$
$\int_0^t F(s)ds < \infty$,
and $\rho(u) > 0$, *on* $u > 0$, *is strictly increasing, continuous, and concave such that* $\int_{0+} du/\ \rho(u) = \infty$;
3° $\left\| |b^n(t,x)-b(t,x)|^2 \right\|_{L^q([0,T]\times R^d)} + \left\| |\sigma^n(t,x)-\sigma(t,x)|^2 \right\|_{L^q([0,T]\times R^d)}$
$+ \left\| \int_Z |c^n(t,x,z)-c(t,x,z)|^2 \pi(dz) \right\|_{L^q([0,T]\times R^d)} \to 0$,
as $n \to \infty$, *for each* $T < \infty$,
where $q \ge d+1$, *and*
$\|b^n(t,x)-b(t,x)\|_{L^q([0,T]\times R^d)} = \int_0^T \int_{R^d} |b^n(t,x)-b(t,x)|^q\, dxdt$, *etc.;*
4° *there exists a* $\delta_0 > 0$ *such that for all* $\mu \in R^d, \forall n = 1,2,\cdots$,
$\langle A^n(t,x)\mu,\mu\rangle \ge |\mu|^2 \delta_0$,
where $A^n = \sigma^n\sigma^{n*}$;
5° $lim_{n\to\infty} E|x_0^n - x_0|^2 = 0$.
Then we have that $\forall t \ge 0$

$$\lim_{n\to\infty} E[|x_t^n - x_t|^2] = 0. \tag{4.28}$$

Proof. By Ito's formula
$E[|x_t^n - x_t|^2] - E[|x_0^n - x_0|^2] = 2E\int_0^t (x_s^n - x_s)\cdot(b^n(s,x_s^n) - b(s,x_s))ds$
$+E\int_0^t \|\sigma^n(s,x_s^n) - \sigma(s,x_s)\|^2 ds$
$+E\int_0^t \int_Z |c^n(s,x_s^n,z) - c(s,x_s,z)|^2 \pi(dz)ds = \sum_{i=1}^3 I^n(i,t)$.
Obviously, by the Krylov type estimate (Theorem 165)
$I^n(1,t) \le 2E\int_0^t (x_s^n - x_s)\cdot(b^n(s,x_s^n) - b(s,x_s^n))ds$
$+2E\int_0^t (x_s^n - x_s)\cdot(b(s,x_s^n) - b(s,x_s))ds$
$\le E\int_0^t |x_s^n - x_s|^2 ds + k_T \left\| |b^n(t,x)-b(t,x)|^2 \right\|_{L^q([0,T]\times R^d)}$
$+2\int_0^t F(s)\rho(E|x_s^n - x_s|^2)ds$.
Similarly,
$I^n(2,t) \le 2k_T \left\| |\sigma^n(t,x)-\sigma(t,x)|^2 \right\|_{L^q([0,T]\times R^d)}$
$+2\int_0^t F(s)\rho(E|x_s^n - x_s|^2)ds$,
and
$I^n(3,t) \le 2k_T \left\| \int_Z |c^n(t,x,z)-c(t,x,z)|^2 \pi(dz) \right\|_{L^q([0,T]\times R^d)}$
$+2\int_0^t F(s)\rho(E|x_s^n - x_s|^2)ds$.
Hence
$E|x_t^n - x_t|^2 \le E\int_0^t |x_s^n - x_s|^2 ds + 6\int_0^t F(s)\rho(E|x_s^n - x_s|^2)ds$

$+k_T \left\| |b^n(t,x) - b(t,x)|^2 \right\|_{L^q([0,T]\times R^d)}$
$+2k_T \left\| |\sigma^n(t,x) - \sigma(t,x)|^2 \right\|_{L^q([0,T]\times R^d)}$
$+2k_T \left\| \int_Z |c^n(t,x,z) - c(t,x,z)|^2 \pi(dz) \right\|_{L^q([0,T]\times R^d)}.$

Notice that by Lemma 114
$E\sup_{t\in[0,T]} |x_t^n|^2 \leq \widetilde{k}_0 < \infty, \forall n = 0,1,2,\cdots.$
Thus applying Fatou's lemma, one finds that
$\overline{\lim}_{n\to\infty} E|x_t^n - x_t|^2 \leq \int_0^t \overline{\lim}_{n\to\infty} E|x_s^n - x_s|^2 ds$
$+6\int_0^t F(s)\rho(\overline{\lim}_{n\to\infty} E|x_s^n - x_s|^2)ds.$
Therefore,
$\overline{\lim}_{n\to\infty} E|x_t^n - x_t|^2 = 0.$ ∎

The condition on 3°, and the condition that σ^n can change with n, are weaker than those in Theorem 145. However, the condition on 1° for b^n and c^n, and the condition that all coefficients do not depend on ω, are stronger than those in Theorem 145. By using the Tanaka type formula and the Krylov type estimate we can also obtain another theorem on the convergence of solutions.

Theorem 167 *Assume that*
1° $|b^n(s,x)| + |\sigma(s,x)|^2 + \int_Z |c^n(s,x,z)| \pi(dz)$
$+ \int_Z |c^n(s,x,z)|^2 \pi(dz) \leq k_0,$
where k_0 *is a constant;*
2° $\sum_{i=1}^d sgn_i(x-y)\cdot(b_i(s,x) - b_i(s,y)) \leq F(s)\rho(|x-y|),$
$\|\sigma(t,x) - \sigma(t,y)\|^2 \leq F(t)\rho(|x-y|)\,|x-y|,$
$|c(t,x,z) - c(t,y,z)| \leq G(t,z)\rho(|x-y|),$
where $sgn_i x = \frac{x_i}{|x|}, 0 \leq F(s), G(s,z)$ *satisfy that* $\forall t \geq 0$
$\int_0^t F(s)ds < \infty, \int_0^t \int_Z G(s,z)\pi(dz)ds < \infty,$
and $\rho(u) > 0$, *on* $u > 0$, *is strictly increasing, continuous, and concave such that* $\int_{0+} du/\,\rho(u) = \infty;$
3° $\|b^n(t,x) - b(t,x)\|_{L^q([0,T]\times R^d)}$
$+ \left\|\int_Z |c^n(t,x,z) - c(t,x,z)| \pi(dz)\right\|_{L^q([0,T]\times R^d)} \to 0,$
as $n \to \infty$, *for each* $T < \infty$,
where $q \geq d+1;$
4° there exists a $\delta_0 > 0$ *such that for all* $\mu \in R^d, \forall n = 1,2,\cdots,$
$\langle A^n(t,x)\mu,\mu\rangle \geq |\mu|^2 \delta_0,$
where $A^n = \sigma^n\sigma^{n*};$
5° $lim_{n\to\infty} E|x_0^n - x_0| = 0.$
Then we have that $\forall t \geq 0$

$$\lim_{n\to\infty} E[|x_t^n - x_t|] = 0. \tag{4.29}$$

Proof. By the Tanaka type formula (Theorem 142)
$E|x_t^n - x_t| - E|x_0^n - x_0| =$

$+E\int_0^t sgn(x_s^n - x_s)\cdot(b^n(s,x_s^n) - b(s,x_s))ds$
$+\frac{1}{2}\sum_{i,j,k=1}^{d,d,r} E\int_0^t I_{(x_s^n \neq x_s)} \frac{|x_s^n - x_s|^2\delta_{ij} - (x_{is}^n - x_{is})(x_{js}^n - x_{js})}{|x_s^n - x_s|^3}$
$\cdot(\sigma_{ik}(s,x_s^n,\omega) - \sigma_{ik}(s,x_s,\omega))(\sigma_{jk}(s,x_s^n,\omega) - \sigma_{jk}(s,x_s,\omega))ds$
$+E\int_0^t\int_Z[|x_{s-}^n - x_{s-} + c^n(s,x_{s-}^n,z,\omega) - c(s,x_{s-},z,\omega)| - |x_{s-}^n - x_{s-}|$
$-sgn(x_{s-}^n - x_{s-})\cdot(c^n(s,x_{s-}^n,z,\omega) - c(s,x_{s-},z,\omega))]N_k(ds,dz)$
$= \sum_{i=1}^3 I^n(i,t).$

Obviously,

$I^n(1,t) \leq \int_0^t F(s)\rho(E|x_s^n - x_s|)ds + k_T\|b^n(t,x) - b(t,x)\|_{L^q([0,T]\times R^d)},$
$I^n(2,t) \leq \widetilde{k}_T\int_0^t F(s)\rho(E|x_s^n - x_s|)ds,$

and

$I^n(3,t) \leq 2\int_0^t\int_Z G(s,z)\rho(E|x_s^n - x_s|)\pi(dz)ds$
$+2k_T\left\|\int_Z|c^n(t,x,z) - c(t,x,z)|\pi(dz)\right\|_{L^q([0,T]\times R^d)}.$

Now the proof can be completed just as in the last part of the proof of the previous theorem. ∎

For the convergence theorem on solutions to 1−dimensional SDE with jumps by the Krylov estimate with other conditions we can easily obtain some other theorems. Suppose that the $x_t^n, n = 0,1,2,\cdots$ satisfy the following 1−dimensional SDE's, respectively,

$$x_t^n = x_0 + \int_0^t b^n(x_s^n)ds + \int_0^t \sigma^n(x_s^n)dw_s$$
$$+\int_0^t\int_Z c^n(x_{s-}^n,z)\widetilde{N}_k(ds,dz), \forall t\geq 0, n = 0,1,2,\cdots; \tag{4.30}$$

where w_t is a r−dimensional BM, and $\widetilde{N}_k(ds,dz)$ is a 1−dimensional Poisson martingale with the compensator $\pi(dz)dt$ as before, so that $b^n, c^n \in R^1$, and $\sigma^n \in R^{1\otimes r}$. Here we write $x_t^0 = x_t, b^0 = b$.

Theorem 168 *Assume that*
1° $|b^n(x)|^2 + |\sigma^n(x)|^2 + \int_Z|c^n(x,z)|^2\pi(dz) \leq k_0(1+|x|^2)$,
where k_0 *is a constant,*
2° conditions 2°−5° in Theorem 166 are automatically reduced to the case $d = 1$, *and the* $L^q([0,T]\times R^d)$*−norm is replaced by the* $L^1(R)$*−norm.*
Then we still have that $\forall t\geq 0$

$$\lim_{n\to\infty} E[|x_t^n - x_t|^2] = 0.$$

Theorem 169 *Assume that*
1° $|b^n(x)|^2 + |\sigma(x)|^2 + \int_Z|c^n(x,z)|^2\pi(dz) \leq k_0(1+|x|^2)$,
where k_0 *is a constant,*
2° conditions 2°−5° in Theorem 167 are automatically reduced to the case $d = 1$, *and the* $L^q([0,T]\times R^d)$*−norm is replaced by the* $L^1(R)$*−norm.*
Then we still have that $\forall t\geq 0$

$$\lim_{n\to\infty} E[|x_t^n - x_t|] = 0.$$

The above two theorems can almost be proved in exactly the same way as in Theorem 166 and Theorem 167. The only difference in the proofs is that the Krylov estimate used in the proof here should refer to Corollary 163. However, we also notice that the condition in 1° here is weaker than the boundedness condition.

5

Stochastic Differential Equations with Non-Lipschitzian Coefficients

In many cases we need to minimize some target functional subject to a controlled dynamical system; for example, to minimize the energy expended by the controlled system during a period of time, like, minimizing $E\int_0^T |x_t^u|^2 \, dt$, where $u(.)$ is a control, x_t^u is the solution of the system corresponding to the applied control $u(.)$. We will find that the minimal value of the target functional will be obtained when we can apply some extreme solution of the dynamic system. For this example the idea is that at each time when the trajectory of the state process leaves the point 0, we should immediately use a feedback control to fully pull back the trajectory directed towards 0, because if the state x_t^u is closer to 0, then the energy $|x_t^u|^2$ expended is also closer to zero and so it is smaller, even though it cannot be 0. Such an extreme feedback control is called a Bang-Bang control. Obviously, such a feedback control is not Lipschitz continuous, and so it also makes the coefficients of the system non-Lipschitzian, for example, when the system is linear with respect to the control $u(.)$: the system coefficient is $A(t)x_t + B(t)u_t$. However, we need the state of the system, that is, the solution, to exist for such a control, so the system can be controlled .

Therefore, discussing solutions for stochastic differential equations (SDEs) with jumps and with non-Lipschitzian coefficients, is necessary and useful from the practical point of view. The interesting thing is also that in the ordinary differential equation (ODE) case, if its coefficients are only continuous then a solution, even when it exists, is not necessary unique. However, in the SDE case we can have a unique solution even when the coefficients are not continuous. This means that a stochastic perturbation can sometimes improve the nice properties of the solution. Furthermore, in a later chapter, we will see that the stochastic integral term is very important in the financial market. Actually, its coefficient corresponds to a part of a portfolio of investment of the stocks by an investor in the financial market. In the optimal consumption problem the SDEs with non-Lipschitzian coefficients also need to be considered.

In this chapter we will use the smoothness method and the Skorohod weak convergence technique to discuss the existence and uniqueness of strong solutions and weak solutions for SDE with jumps and with non-Lipschitzian coefficients.

5.1 Strong Solutions. Continuous Coefficients with $\rho-$ Conditions

In this section we will use the smothness method to obtain the existence and uniqueness of a strong solution for a SDE with continuous coefficients, which satisfy some so-called $\rho-$condition. Consider the following SDE with jumps:

$$x_t = x_0 + \int_0^t b(s, x_s, \omega)ds + \int_0^t \sigma(s, x_s, \omega)dw_s + \int_0^t \int_Z c(s, x_{s-}, z, \omega)\widetilde{N}_k(ds, dz), \forall t \geq 0 \tag{5.1}$$

where $\{w_t\}_{t\geq 0}$ is a $d-$dimensional BM, $\widetilde{N}_k(ds, dz)$ is the Poisson martingale measure generated by a Poisson point process $k(.)$ such that $\widetilde{N}_k(ds, dz) = N_k(ds, dz) - \pi(dz)dt$, where $N_k(ds, dz)$ is the counting measure with the compensator $\pi(dz)dt$ generated by $k(.)$, $\pi(\cdot)$ is a $\sigma-$finite measure on some measurable space $(Z, \mathfrak{B}_Z)$, and $b \in R^d, \sigma \in R^{d\otimes d}, c \in R^d$. In (5.1) if $c = 0$, we get a continuous SDE. Furthermore, if $\sigma = 0$, then (5.15) will be reduced to a continuous ODE for each fixed ω. So we will call the case $\sigma = 0$ a degenerate case, no matter if $c = 0$ or not. We will use the smoothness technique to show the results.

Theorem 170 *Assume that*
1° $b = b(t, x, \omega) : [0, \infty) \times R^d \times \Omega \to R^d$,
$\sigma = \sigma(t, x, \omega) : [0, \infty) \times R^d \times \Omega \to R^{d\otimes d}$,
$c = c(t, x, z, \omega) : [0, \infty) \times R^d \times Z \times \Omega \to R^d$,
are $\mathfrak{F}_t^{w,\widetilde{N}_k}-$ *adapted and measurable processes such that* $P-a.s.$
$|b(t, x, \omega)| \leqslant c_1(t)(1 + |x|)$,
$|\sigma(t, x, \omega)|^2 + \int_Z |c(t, x, z, \omega)|^2 \pi(dz) \leqslant c_1(t)(1 + |x|^2)$,

where $\mathfrak{F}_t^{w,\widetilde{N}_k}$ *is the* $\sigma-$*field generated by* w *and* $\widetilde{N}_k$ *up to time* t, *that is,* $\mathfrak{F}_t^{w,\widetilde{N}_k} = \sigma(w_s, \widetilde{N}_k((0,s],U), \forall U \in \mathfrak{B}_Z, s \leq t)$, *and* $c_1(t)$ *is non-negative and non-random such that for each* $T < \infty$

$\int_0^T c_1(t)dt < \infty$;

2° $b(t,x,\omega)$ *and* $\sigma(t,x,\omega)$ *are continuous in* x; *and*

$\lim_{h\to 0} \int_Z |c(t,x+h,z,\omega) - c(t,x,z,\omega)|^2 \pi(dz) = 0$;

3° for each $N = 1,2,\cdots$, *and each* $T < \infty$,

$$2\langle (x_1 - x_2), (b(t,x_1,\omega) - b(t,x_2,\omega))\rangle$$
$$+ |\sigma(t,x_1,\omega) - \sigma(t,x_2,\omega)|^2 + \int_Z |c(t,x_1,z,\omega) - c(t,x_2,z,\omega)|^2 \pi(dz)$$
$$\leqslant c_T^N(t)\rho_T^N(|x_1 - x_2|^2),$$

as $|x_i| \leqslant N, i = 1,2,\ t \in [0,T]$; *where* $\int_0^T c_T^N(t)dt < \infty$; *and* $\rho_T^N(u) \geq 0$, *as* $u \geq 0$, *is non-random, strictly increasing, continuous and concave such that*

$\int_{0+} du/\rho_T^N(u) = \infty$.

Then for any given constant $x_0 \in R^d$ *(5.1) has a pathwise unique strong solution.*

First, let us give an example of the existence of a solution to an SDE in the case that $\forall T < \infty$, $\int_0^T c_1(t)dt < \infty$, $c_1(t)$ is unbounded and, moreover, b_1 is also unbounded and non-Lipschitzian continuous in x.

Example 171 *Let* $b(t,x) = -I_{t\neq 0}I_{x\neq 0}t^{-\alpha_1}x|x|^{-\beta}$, *where* $\alpha_1 < 1, 0 < \beta < 1$, *and suppose that* σ *and* c *satisfy* 1° *and* 2° *in Theorem 170, and satisfy the condition* 3° *in Theorem 170 with* $b = 0$. *Then (5.1) has a pathwise unique strong solution.*

Obviously, $c(t) = I_{s\neq 0}s^{-\alpha_1}$, *is unbounded in* t, *and* b *is also unbounded in* t *and* x, *and is non-Lipschitz continuous in* x.

Proof. Notice that $\forall x, x' \in R^d$

$$\left\langle x - x', -x|x|^{-\beta} + x'|x'|^{-\beta}\right\rangle = -|x|^{2-\beta} - |x'|^{2-\beta}$$
$$+|x'|^{-\beta}\langle x,x'\rangle + |x|^{-\beta}\langle x,x'\rangle \leq -|x|^{2-\beta} - |x'|^{2-\beta}$$
$$+|x'|^{-\beta+1}|x| + |x|^{-\beta+1}|x'| = (|x| - |x'|)(|x'|^{1-\beta} - |x|^{1-\beta}) \leq 0.$$

Hence Theorem 170 applies. ■

Before we prove Theorem 170, let us first establish a lemma.

Lemma 172 *Under assumptions* 1° *and* 2° *in Theorem 170 there exist* b^n, σ^n *and* $c^n, n = 1,2,\cdots$, *satisfying the following conditions:*

1) $|b^n(t,x,\omega)| \leq 2c_1(t)(1+|x|)$, *as* $n \geq N_0$;

$|\sigma^n(t,x,\omega)|^2 + \int_Z |c^n(t,x,z,\omega)|^2 \pi(dz) \leqslant 8c_1(t)(1+|x|^2)$,

where $N_0 > 0$ *is a constant;*

2) as $x, x' \in R^d$

$|b^n(t,x,\omega) - b^n(t,x',\omega)| \leqslant k_n c_1(t)|x - x'|$,

$|\sigma^n(t,x,\omega) - \sigma^n(t,x',\omega)|^2 + \int_Z |c^n(t,x,z,\omega) - c^n(t,x',z,\omega)|^2 \pi(dz)$

$\leqslant k_n c_1(t)|x - x'|^2$,

where $k_n \geq 0$ *is a constant only depending on* n,

3) for any $N > 0$ *and for each* $t \geq 0, \omega \in \Omega$, *as* $n \to \infty$
$\sup_{|x| \leq N} |b^n(t,x,\omega) - b(t,x,\omega)| \to 0,$
$\sup_{|x| \leq N} |\sigma^n(t,x,\omega) - \sigma(t,x,\omega)|^2$
$+ \sup_{|x| \leq N} \int_Z |c^n(t,x,z,\omega) - c(t,x,z,\omega)|^2 \pi(dz) \to 0.$

Proof. Let us smooth out b only with respect to x to get b^n, i.e. define
$b^n(t,x,\omega) = \int_{R^d} b(t, x - n^{-1}\overline{x}, \omega) J(\overline{x}) d\overline{x},$
where for all $u \in R^d$

$$J_d(u) = \begin{cases} c_d \exp(-(1 - |u|^2)^{-1}), & \text{for } |u| < 1, \\ 0, \text{ otherwise}, \end{cases}$$

and the constant c_d satisfies $\int_{R^d} J_d(u) du = 1$. Then
$|b^n(t,x,\omega)| \leq \int_{R^d} |b(t, x - n^{-1}\overline{x}, \omega)| J(\overline{x}) d\overline{x}$
$\leq c_1(t) \int_{R^d} (1 + |x - n^{-1}\overline{x}|) J(\overline{x}) d\overline{x}$
$\leq c_1(t)(1 + |x| + n^{-1} \int_{R^d} |\overline{x}| J(\overline{x}) d\overline{x}) = c_1(t)(1 + |x| + n^{-1} k_0)$
$\leq (1 + n^{-1}k_0) c_1(t)(1 + |x|) \leq 2c_1(t)(1 + |x|)$, as $n > k_0$.
So b^n satisfies 1). On the other hand,
$|b^n(t,x,\omega) - b^n(t,x',\omega)| = |n^d \int_{R^d} b(t, \overline{x}, \omega) J(n(x - \overline{x})) d\overline{x}$
$-n^d \int_{R^d} b(t,\overline{x},\omega) J(n(x' - \overline{x})) d\overline{x}| \leq n^d \int_{R^d} |b(t,\overline{x},\omega)| \, |J(n(x - \overline{x})$
$-J(n(x' - \overline{x})| d\overline{x} \leq n^d c_1(t) |x - x'|$
$\cdot \int_{R^d} \int_0^1 (1 + |\overline{x}|) grad[J(n(x - \overline{x} + \theta(x' - x))] d\theta d\overline{x} \leq k_n c_1(t) |x - x'|.$
So 2) is established for b^n. Now by Heine-Borel's finite covering theorem for any $N > 0$ and any given $\widetilde{\varepsilon} > 0$ one can find a $\delta > 0$, δ may depend on t and ω such that as $\frac{1}{n} < \delta$, $|b(t, x - n^{-1}, \omega) - b(t,x,\omega)| < \widetilde{\varepsilon}$, $\forall |x| \leq N$; because b is continuous in x. Hence, as $n \geq \frac{1}{\delta}$,
$\sup_{|x| \leq N} |b^n(t,x,\omega) - b(t,x,\omega)|$
$\leq | \int_{R^d} \sup_{|x| \leq N} |b(t, x - n^{-1}\overline{x}, \omega) - b(t,x,\omega)| J(\overline{x}) d\overline{x}$
$= | \int_{|\overline{x}| \leq 1} \sup_{|x| \leq N} |b(t, x - n^{-1}\overline{x}, \omega) - b(t,x,\omega)| J(\overline{x}) d\overline{x} < \widetilde{\varepsilon}.$
Thus 3) is also true for b^n. Now, defining σ^n and c^n similarly, it is easily seen that $\sigma_n, n = 1, 2, \cdots$, also satisfy 1), 2) and 3). For c^n the proof is also similar. In fact,
$\int_Z |c^n(t,x,z,\omega)|^2 \pi(dz) = \int_Z |\int_{R^d} c(t, x - n^{-1}\overline{x}, z, \omega) J(\overline{x}) d\overline{x}|^2 \pi(dz)$
$\leq \int_Z \int_{R^d} |c(t, x - n^{-1}\overline{x}, z, \omega)|^2 J(\overline{x}) d\overline{x} \pi(dz)$
$\leq c_1(t) \int_{R^d} (1 + |x - n^{-1}\overline{x}|^2) J(\overline{x}) d\overline{x} \leq 2c_1(t)(1 + 2|x|^2$
$+ 2n^{-1} \int_{R^d} |\overline{x}|^2 J(\overline{x}) d\overline{x}) \leq 2c_1(t)(1 + 2|x|^2 + 2n^{-1}\widetilde{k}_0)$
$\leq 4c_1(t)(1 + |x|^2)$, as $n > 2\widetilde{k}_0$.
So 1) is proved for c^n. On the other hand,
$\int_Z |c^n(t,x,z,\omega) - c^n(t,x',z,\omega)|^2 \pi(dz)$
$\leq n^d \int_{R^d} |c(t,\overline{x},z,\omega)|^2 \pi(dz) |J(n(x - \overline{x}) - J(n(x' - \overline{x})|^2 d\overline{x}$
$\leq n^d c_1(t) |x - x'|^2 \int_{R^d} \int_0^1 (1 + |\overline{x}|^2) |grad[J(n(x - \overline{x} + \theta(x' - x))]|^2 d\theta d\overline{x}$
$\leq k_n c_1(t) |x - x'|.$
So for c^n 2) is also established. Finally, by a similar proof as in b^n one easily derives that 3) is also true for c^n. ∎

Now let us prove Theorem 170.

Proof. For b^n, σ^n and c^n obtained from the above lemma by Theorem 117, there exists a pathwise unique strong solution (x_t^n) satisfying the following SDE

$$x_t^n = x_0 + \int_0^t b^n(s, x_s^n, \omega)ds + \int_0^t \sigma^n(s, x_s^n, \omega)dw_s + \int_0^t \int_Z c^n(s, x_{s-}^n, z, \omega)\tilde{N}_k(ds, dz). \quad (5.2)$$

By Ito's formula

$E[|x_t^m - x_t^n|^2 = 2E[\int_0^t (x_s^m - x_s^n) \cdot (b^m(s, x_s^m, \omega) - b^n(s, x_s^n, \omega))\, ds$
$+ \int_0^t |\sigma^m(s, x_s^m, \omega) - \sigma^n(s, x_s^n, \omega)|^2\, ds$
$+ \int_0^t \int_Z |c^m(s, x_s^m, z, \omega) - c^n(s, x_s^n, z, \omega)|^2\, \pi(dz)ds]$
$= E[\int_0^t \int_{R^d} [2\,(x_s^m - x_s^n) \cdot (b(s, x_s^m - m^{-1}\overline{x}, \omega)$
$-b(s, x_s^n - n^{-1}\overline{x}, \omega)) + \left|\sigma(s, x_s^m - m^{-1}\overline{x}, \omega) - \sigma(s, x_s^n - n^{-1}\overline{x}, \omega)\right|^2$
$+ \int_Z \left|c(s, x_s^m - m^{-1}\overline{x}, z, \omega) - c(s, x_s^n - n^{-1}\overline{x}, z, \omega)\right|^2 \pi(dz)]J(\overline{x})d\overline{x}$
$\leqslant E \int_0^t \int_{R^d} \{c_1(s)\rho\left(\left|x_s^m - x_s^n - \left(m^{-1} - n^{-1}\right)\overline{x}\right|^2\right)$
$+20c_1(s)\left|\left(m^{-1} - n^{-1}\right)\overline{x}\right|\}J(\overline{x})d\overline{x}ds.$

Hence as $t \in [0, T]$,

$$E[|x_t^m - x_t^n|^2 \leq k_T'\left(m^{-1} + n^{-1}\right)$$
$$+k_0' \int_0^t \{c_1(s) \int_{R^d} \rho\left(E\left|x_s^m - x_s^n - \left(m^{-1} - n^{-1}\right)\overline{x}\right|^2\right) J(\overline{x})d\overline{x}\}ds \quad (5.3)$$

Since by Lemma 114 for all n and $\forall T < \infty$

$$E(\sup_{t \leqslant T} |x_t^n|^2)ds \leqslant k_T < \infty, \quad (5.4)$$

Hence, by Fatou's lemma, it is easily seen that

$\overline{\lim_{m,n\to\infty}} E[|x_t^m - x_t^n|^2] \leqslant \widetilde{k}_T \int_0^t c_1(s)\rho_1\left(\overline{\lim_{m,n\to\infty}} E\left|x_s^m - x_s^n\right|^2\right) ds,$

where $\rho_1(u) = \rho(u) + u$. Therefore, $\overline{\lim_{m,n\to\infty}} E\left|x_t^m - x_t^n\right|^2 = 0$. By (5.3) one also finds that for each $T < \infty$ $\overline{\lim_{m,n\to\infty}} E \int_0^T \left|x_t^m - x_t^n\right|^2 dt = 0$. So there exists an $(x_t) \in L_{\mathfrak{F}}^2(R^d)$ such that for each $T < \infty$

$\lim_{n\to\infty} E \int_0^T \left|x_t^n - x_t\right|^2 dt = 0.$

On the other hand, by the above result one also has that for each $t \geq 0$

$\lim_{n\to\infty} E\left|x_t^n - x_t\right|^2 = 0.$

So $x_t^n \to x_t$, in probability for each t, and one can choose a subsequence $\{n_k\}$ of $\{n\}$, denoted by $\{n\}$ again, such that $P-a.s.$ as $n \to \infty$,

$x_t^n \to x_t^0,\ \forall t = r_k, k = 1, 2, \cdots;$

where $\{r_k\}_{k=1}^{\infty} \subset [0,T]$ is the totality of rational numbers in $[0,T]$. Hence by Fatou's lemma

$$E\sup_{t\leq T}\left|x_t^0\right| \leq E[\sup_k \lim_{n\to\infty}\left|x_{r_k}^n\right|^2] \leq \underline{\lim}_{n\to\infty}E[\sup_{t\leq T}|x_t^n|^2] \leq k_T. \tag{5.5}$$

Now applying Remark 397 in Appendix B to show that as $t \in [0,T]$, when $n \to \infty$

$$\int_0^t\int_Z c^n(s,x_{s-}^n,z,\omega)\tilde{N}_k(ds,dz) \to \int_0^t\int_Z c(s,x_{s-},z,\omega)\tilde{N}_k(ds,dz), \text{ in } P \tag{5.6}$$

one may assume that $\sup_{t\leq T}|x_t^n| \leq k_0, \forall n$ and $\sup_{t\leq T}|x_t| \leq k_0$. However, in this case, as $t\in[0,T]$, for any $\varepsilon > 0$

$P(\left|\int_0^t\int_Z c^n(s,x_{s-}^n,z,\omega)\tilde{N}_k(ds,dz) - \int_0^t\int_Z c(s,x_{s-},z,\omega)\tilde{N}_k(ds,dz)\right| > \varepsilon)$
$\leq \frac{1}{4\varepsilon^2}E\int_0^T\int_Z |c^n(s,x_s^n,z,\omega) - c(s,x_s^n,z,\omega)|^2 I_{|x_t^n|\leq k_0}I_{|x_t|\leq k_0}\pi(dz)ds$
$+\frac{1}{4\varepsilon^2}E\int_0^T\int_Z |c(s,x_s^n,z,\omega) - c(s,x_s,z,\omega)|^2 I_{|x_t^n|\leq k_0}I_{|x_t|\leq k_0}\pi(dz)ds$
$\leq \frac{1}{\varepsilon^2}\int_0^T \sup_{|x|\leq k_0}\int_Z |c^n(s,x,z,\omega) - c(s,x,z,\omega)|^2\pi(dz)ds$
$+\frac{1}{4\varepsilon^2}E\int_0^T\int_Z |c(s,x_s^n,z,\omega) - c(s,x_s,z,\omega)|^2 I_{|x_t^n|\leq k_0}I_{|x_t|\leq k_0}\pi(dz)ds$
$= I^{1,n} + I^{2,n}.$

Notice from Lemma 172 one finds that
$\sup_{|x|\leq k_0}\int_Z |c^n(t,x,z)|^2\pi(dz) \leq 8c_1(t)(1+k_0^2)$, and
$\lim_{n\to\infty}\sup_{|x|\leq k_0}\int_Z |c^n(s,x,z) - c(s,x,z)|^2\pi(dz) = 0.$

Thus one can apply Lebesgue's dominated convergence theorem to get

$$\lim_{n\to\infty} I^{1,n} = 0. \tag{5.7}$$

Moreover, one also finds that as $n\to\infty$,

$$E\int_0^T\int_Z |c(s,x_s^n,z,\omega) - c(s,x_s,z,\omega)|^2 I_{|x_t^n|\leq k_0}I_{|x_t|\leq k_0}\pi(dz)ds \to 0. \tag{5.8}$$

In fact,
$P(\int_Z |c(s,x_s^n,z,\omega) - c(s,x_s,z,\omega)|^2 I_{|x_t^n|\leq k_0}I_{|x_t|\leq k_0}\pi(dz) > \varepsilon)$
$\leq P(\left|x_s^n - x_s^0\right| > \delta)$
$+P(I_{|x_s^n - x_s^0|\leq\delta,|x_s^n|\leq k_0,|x_s^0|\leq k_0}\int_Z \left|c(s,x_s^n,z) - c(s,x_s^0,z)\right|^2\pi(dz) > \varepsilon)$
$= J_1^{n,\delta} + J_2^{n,\delta}.$

Now since
$\lim_{h\to 0}\sup_{|x|\leq k_0}\int_Z |c(s,x+h,z) - c(s,x,z)|^2\pi(dz) = 0,$
one can take a small enough $\delta > 0$ such that
$\sup_{|x|\leq k_0,|h|\leq\delta}\int_Z |c(s,x+h,z) - c(s,x,z)|^2\pi(dz) < \varepsilon.$

Hence for this $\delta > 0$, $J_2^{n,\delta} = 0$. Furthermore, for arbitrary given $\widetilde{\varepsilon} > 0$ there exists a $\widetilde{N}$ such that as $n \geq \widetilde{N}$, $J_1^{n,\delta} < \widetilde{\varepsilon}$. Thus, for each s, as $n \to \infty$,

$\int_Z |c(s, x_s^n, z, \omega) - c(s, x_s, z, \omega)|^2 I_{|x_t^n| \leq k_0} I_{|x_t| \leq k_0} \pi(dz) \to 0$, in P.

Hence, Lebesgue's dominated convergence theorem applies, and (5.8) holds. Thus (5.6) follows. By the same token one easily shows that as $n \to \infty$,

$\int_0^t b^n(s, x_s^n, \omega) ds \to \int_0^t b(s, x_s, \omega) ds$, in P;

and

$\int_0^t \sigma^n(s, x_s^n, \omega) dw_s \to \int_0^t \sigma(s, x_s, \omega) dw_s$, in P.

Therefore, (x_t) is a solution of (5.1). The pathwise uniqueness follows from Lemma 115. ∎

5.2 The Skorohod Weak Convergence Technique

To discuss the existence of a weak solution for a SDE under some weak conditions the following Skorohod weak convergence technique is very useful and we will use it frequently in this chapter. Let us establish a lemma, which is very useful in the discussion of the existence of a weak solution to an SDE. In the rest of this Chapter let us assume that

$Z = R^d - \{0\}$, and $\int_Z \frac{|z|^2}{1+|z|^2} \pi(dz) < \infty$.

Lemma 173 *Suppose that*

$|b^n(t, x, \omega)| \leq c_1(t)(1 + |x|)$, *as* $n \geq N_0$;

$|\sigma^n(t, x, \omega)|^2 + \int_Z |c^n(t, x, z, \omega)|^2 \pi(dz) \leqslant c_1(t)(1 + |x|^2)$,

where $N_0 > 0$ *is a constant.*

Assume that for each $n = 1, 2, \cdots$, x_t^n *is the solution of the following SDE:*

$x_t^n = x_0 + \int_0^t b^n(s, x_s^n) ds + \int_0^t \sigma^n(s, x_s^n) dw_s$

$+ \int_0^t \int_Z c^n(s, x_{s-}^n, z) q(ds, dz)$,

where we denote $q(dt, dz) = \widetilde{N}_k(dt, dz)$ *the Poisson martingale measure with the compensator* $\pi(dz)dt$ *such that* $q(dt, dz) = p(dt, dz) - \pi(dz)dt$ *and* $p(dt, dz) = N_k(dt, dz)$.

Then the following fact holds, this fact we may call "the result of SDE from the Skorohod weak convergence technique":

There exists a probability space $(\widetilde{\Omega}, \widetilde{\mathfrak{F}}, \widetilde{P})$ *(actually,* $\widetilde{\Omega} = [0,1], \widetilde{\mathfrak{F}} = \mathfrak{B}([0,1])$*) and a sequence of RCLL processes* $(\widetilde{x}_t^n, \widetilde{w}_t^n, \widetilde{\zeta}_t^n), n = 0, 1, 2, \cdots$, *defined on it such that* $(\widetilde{x}_t^n, \widetilde{w}_t^n, \widetilde{\zeta}_t^n), n = 1, 2, \cdots$ *have the same finite probability distributions as those of* $(x_t^n, w_t, \zeta_t), n = 1, 2, \cdots$, *where*

$\zeta_t = \int_0^t \int_{|z| \leq 1} z \widetilde{N}_k(ds, dz) + \int_0^t \int_{|z|>1} z N_k(ds, dz)$,

and as $n \to \infty, \forall t \geq 0$,

$\tilde{\eta}_t^n \to \tilde{\eta}_t^0$, *in probability, as* $\tilde{\eta}_t^n = \widetilde{x}_t^n, \widetilde{w}_t^n, \widetilde{\zeta}_t^n, n = 0, 1, 2, \ldots$.

Write

$\widetilde{p}^n(dt, dz) = \sum_{s \in dt} I_{(0 \neq \Delta \tilde{\zeta}_s^n \in dz)}(s)$, $\widetilde{q}^n(dt, dz) = \widetilde{p}^n(dt, dz) - \pi(dz)dt$,

$\forall n = 0, 1, 2, ...$
Then $\widetilde{p}^n(dt, dz)$ *is a Poisson random counting measure with the compensator* $\pi(dz)dt$ *for each* $n = 0, 1, 2, ...,$ *and it satisfies the condition*
$\widetilde{\zeta}_t^n = \int_0^t \int_{|z|\leq 1} z\widetilde{q}^n(ds, dz) + \int_0^t \int_{|z|>1} z\widetilde{p}^n(ds, dz), n = 0, 1, 2, ...$
Moreover, $\widetilde{w}_s^n$ *and* $\widetilde{w}_t^0$ *are BMs on the probability space* $(\widetilde{\Omega}, \widetilde{\mathfrak{F}}, \widetilde{P})$ *and,* $\widetilde{p}^n(dt, dz)$ *and* $\widetilde{p}^0(dt, dz)$ *are Poisson martingale measures with the same compensator* $\pi(dz)dt$*. Furthermore,* $(\widetilde{x}_t^n)$ *satisfies the following SDE with* $\widetilde{w}_t^n$ *and* $\widetilde{q}^n(dt, dz)$ *on* $(\widetilde{\Omega}, \widetilde{\mathfrak{F}}, \widetilde{P})$*.*
$\widetilde{x}_t^n = x_0 + \int_0^t b^n(s, \widetilde{x}_s^n)ds + \int_0^t \sigma^n(s, \widetilde{x}_s^n)d\widetilde{w}_s^n$
$+ \int_0^t \int_Z c^n(s, \widetilde{x}_{s-}^n, z)\widetilde{q}^n(ds, dz).$

Proof. By the properties of b^n, σ^n, and c^n, applying Lemma 114 one immediately finds that
$\sup_n E(\sup_{t\leq T} |x_t^n|^2) \leq k_T$. Moreover, as $r \leq t \leq T$
$E|x_t^n - x_r^n|^2 \leq 3E\left|\int_r^t b^n(s, x_s^n)ds\right|^2 + 3E\int_r^t |\sigma^n(s, x_s^n)|^2 ds$
$+3E\int_r^t \int_Z |c^n(s, x_s^n, z)|^2 \pi(dz)ds \leq 6(t-r)\int_r^t E(1 + |x_s^n|)^2 c_1(s)ds$
$+24\int_r^t E(1 + |x_s^n|)^2 c_1(s)ds \leq k'_T(t-r).$
So
$\sup_n \sup_{t_1, t_2 \leq T; |t_1 - t_2| \leq h} E(|x_{t_1}^n - x_{t_2}^n|^2) \leq k'_T h.$
Thus for each $T \geq 0, \varepsilon > 0$
$\lim_{N\to\infty} \sup_n \sup_{t\leq T} P\{|x_t^n| > N\} \leq \lim_{N\to\infty} \frac{k_T}{N^2} = 0,$
$\lim_{h\downarrow 0} \sup_n \sup_{t_1, t_2 \leq T, |t_1 - t_2| \leq h} P\{|x_{t_1}^n - x_{t_2}^n| > \varepsilon\} \leq \lim_{h\downarrow 0} k'_T h = 0.$
Therefore, $\forall n \geq 1$

$$\begin{cases} \lim_{N\to\infty} \sup_n \sup_{t\leq T} P(|x_t^n| > N) = 0, \\ \lim_{h\downarrow 0} \sup_n \sup_{t_1, t_2 \leq T, |t_1 - t_2| \leq h} P(|x_{t_1}^n - x_{t_2}^n| > \varepsilon) = 0. \end{cases} \tag{5.9}$$

Now write
$\zeta_t = \int_0^t \int_{|z|\leq 1} z\widetilde{N}_k(ds, dz) + \int_0^t \int_{|z|>1} zN_k(ds, dz) = \zeta_t^1 + \zeta_t^2.$
Let us show that ζ_t also satisfies (5.9). In fact, by the martingale inequality
$E\sup_{t\leq T} |\zeta_t^1|^2 \leq \int_0^T \int_{|z|\leq 1} |z|^2 \pi(dz)ds \leq 2T\int_{|z|\leq 1} \frac{|z|^2}{1+|z|^2}\pi(dz) < \infty.$
Hence $\lim_{N\to\infty} \sup_{t\leq T} P(|\zeta_t^1| > N) \to 0$. Write $\bar{I}_t^2 = \int_0^t \int_{|z|>1} |z| N_k(ds, dz)$.
Since $\bar{I}_t^2$ is RCLL, $\{0 < s \leq T : \Delta\bar{I}_s^2 > 1\}$ is a finite set, so
$\sum_{0<s\leq T} \Delta\bar{I}_s^2 I_{(\Delta\bar{I}_s^2 > 1)} = \sum_{k=1}^{n(\omega)} |z_k(\omega)| I_{|z_k(\omega)|>1} < \infty.$
Hence $P(\sup_{t\leq T} |I_t^2| < \infty) = 1$. In particular, $\lim_{N\to\infty} \sup_{t\leq T} P(|\zeta_t^2| > N) = 0$. Now for arbitrary $\varepsilon > 0$
$P(|\zeta_t - \zeta_s| > \varepsilon) \leq P(\left|\int_s^t \int_{|z|\leq 1} z\widetilde{N}_k(ds, dz)\right| > \varepsilon/2)$
$+ P(\left|\int_s^t \int_{|z|>1} z\widetilde{N}_k(ds, dz)\right| > \varepsilon/2) = J_1 + J_2.$
It is evident that as $|t - s| \leq h \to 0$
$J_1 \leq (2/\varepsilon)^2 E\left|\int_s^t \int_{|z|\leq 1} z\widetilde{N}_k(ds, dz)\right|^2 \leq (2/\varepsilon)^2 \int_{|z|\leq 1} |z|^2 \pi(dz)\, |t - s|$

$\leq 2(2/\varepsilon)^2 \int_{|z|\leq 1} \frac{|z|^2}{1+|z|^2}\pi(dz)\,|t-s| \to 0$.
Notice that $N_k(dt,dz)$ is a Poisson random measure with the compensator $\pi(dz)dt$, as $|t-s| \leq h \to 0$

$J_2 \leq P(N_k((s,t],|z|>1)>0) = 1-\exp(-\int_s^t \int_{|z|>1} \pi(dz)dr)$
$\leq 1-\exp(-\pi(|z|>1)h) \to 0$,

where $\pi(|z|>1) = \int_{|z|>1}\pi(dz) = 2\int_{|z|>1}\frac{|z|^2}{1+|z|^2}\pi(dz) < \infty$. Hence ζ_t satisfies (5.9), that is,

$\lim_{N\to\infty}\sup_{t\leq T} P(|\zeta_t|>N)=0$, and
$\lim_{h\downarrow 0}\sup_{t_1,t_2\leq T,|t_1-t_2|\leq h} P(|\zeta_{t_1}-\zeta_{t_2}|>\varepsilon)=0.$.

Since $E|w_t-w_s|^2 = |t-s|$. One also easily shows that (5.9) holds for w_t.
Hence Skorohod's theorem (Theorem 398) applies to $\{x_t^n,\zeta_t,w_t\}$ and the conclusion follows by Lemma 399. ∎

Remark 174 *By this lemma one sees that if "the result of SDE from the Skorohod weak convergence technique" holds, and we can prove that*

$$\left|\int_0^t (b^n(s,\widetilde{x}_s^n)-b(s,\widetilde{x}_s^0))ds\right| \to 0, \ \textit{in probability } \widetilde{P},$$
$$\int_0^t \sigma^n(s,\widetilde{x}_s^n)d\widetilde{w}_s^n \to \int_0^t \sigma(s,\widetilde{x}_s^0)d\widetilde{w}_s^0, \ \textit{in } \widetilde{P},$$
$$\int_0^t\int_Z c^n(s,\widetilde{x}_{s-}^n,z)\widetilde{q}^n(ds,dz) \to \int_0^t\int_Z c(s,\widetilde{x}_{s-}^0,z)\widetilde{q}^0(ds,dz), \ \textit{in } \widetilde{P}, \quad (5.10)$$

then $(\widetilde{\Omega},\widetilde{\mathfrak{F}},(\widetilde{\mathfrak{F}}_t)_{t\geqslant 0},\widetilde{P};\{\widetilde{w}_t^0\}_{t\geq 0},\widetilde{q}^0(dt,dz),\{\widetilde{x}_t^0\}_{t\geq 0})$, *or say* $\widetilde{x}_t^0$, *is a weak solution of (5.11) in the next section.*

5.3 Weak Solutions. Continuous Coefficients

The technique used in proving Theorem 170 motivates us to obtain an existence theorem for weak solutions of SDE with jumps and with σ, which can be degenerate. Consider the following SDE with non-random coefficients: $\forall t\geq 0$,

$$x_t = x_0 + \int_0^t b(s,x_s)ds + \int_0^t \sigma(s,x_s)dw_s + \int_0^t\int_Z c(s,x_{s-},z)\widetilde{N}_k(ds,dz). \tag{5.11}$$

Theorem 175 *Assume that*
1° $b=b(t,x):[0,\infty)\times R^d \to R^d$,
$\sigma=\sigma(t,x):[0,\infty)\times R^d \to R^{d\otimes d}$,

$c = c(t, x, z) : [0, \infty) \times R^d \times Z \to R^d$,
are jointly Borel measurable such that $P - a.s.$
$\int_Z |c(t, x, z)|^2 \pi(dz) \leqslant c_1(t)(1 + |x|^2)$,
where $c_1(t)$ is non-negative such that for each $T < \infty$
$\int_0^T c_1(t)dt < \infty$;
2° $|b(t, x)|^2 + |\sigma(t, x)| \leqslant c_1(t)(1 + |x|^2)$,
where $c_1(t)$ has the same property as in 1°;
3° $b(t, x)$ is continuous in x and $\sigma(t, x)$ is jointly continuous in (t, x); and
$\lim_{h,h' \to 0} \int_Z |c(t + h', x + h, z) - c(t, x, z)|^2 \pi(dz) = 0$;
4° $Z = R^d - \{0\}$, $\int_Z \frac{|z|^2}{1+|z|^2}\pi(dz) < \infty$.
Then for any given constant $x_0 \in R^d$ (5.11) has a weak solution.

Proof. By Lemma 172 we can smooth out b, σ and c only with respect to x to get b^n, σ^n, and c^n, respectively. Then we have a pathwise unique strong solution x_t^n satisfying a SDE similar to (5.2), but here all coefficients b^n, σ^n, and c^n do not directly depend on ω. Now applying Lemma 173 "the resul of SDE from the Skorohod weak convergence technique" holds. So we only need to show (5.10) in Remark 174 holds. However, since $\forall t \geq 0$,
$\widetilde{x}_t^0 \to \widetilde{x}_t^0$, in probability $\widetilde{P}$, as $n \to \infty$,
as in the proof of Theorem 170 one finds that (5.4) and (5.5) hold. So by Remark 397 in the Appendix we may assume that all $\{\widetilde{x}_t^n, t \in [0, T]\}_{n=0}^{\infty}$ are uniformly bounded, that is, $|\widetilde{x}_t^n| \leq k_0, \forall t \in [0, T], \forall n = 0, 1, 2, \cdots$ in all following discussion on the convergence in probability. Now for an arbitrary given $\varepsilon > 0$

$$\widetilde{P}(\left|\int_0^t \int_Z c^n(s, \widetilde{x}_{s-}^n, z)\widetilde{q}^n(ds, dz) - \int_0^t \int_Z c(s, \widetilde{x}_{s-}^0, z)\widetilde{q}^0(ds, dz)\right| > \varepsilon)$$
$$\leq \widetilde{P}(\left|\int_0^t \int_Z (c^n(s, \widetilde{x}_{s-}^n, z) - c(s, \widetilde{x}_{s-}^n, z))\widetilde{q}^n(ds, dz)\right| > \varepsilon/3)$$
$$+\widetilde{P}(\left|\int_0^t \int_Z (c(s, \widetilde{x}_{s-}^n, z) - c(s, \widetilde{x}_{s-}^0, z))\widetilde{q}^n(ds, dz)\right| > \varepsilon/3)$$
$$+\widetilde{P}\left|\int_0^t \int_Z c(s, \widetilde{x}_{s-}^0, z)\widetilde{q}^n(ds, dz) - \int_0^t \int_Z c(s, \widetilde{x}_{s-}^0, z)\widetilde{q}^0(ds, dz)\right| > \varepsilon/3)$$
$$= \sum_{i=1}^3 I_i^n.$$

Obviously,

$$I_1^n \leq \frac{9}{\varepsilon^2} E^{\widetilde{P}} \int_0^t \sup_{|x| \leq k_0} \int_Z |c^n(s, x, z) - c(s, x, z)|^2 \pi(dz)ds = I_{11}^n, \text{ and}$$
$$I_2^n \leq \frac{9}{\varepsilon^2} E^{\widetilde{P}} \int_0^t |c(s, \widetilde{x}_s^n, z, \omega) - c(s, \widetilde{x}_s, z, \omega)|^2 I_{|\widetilde{x}_t^n| \leq k_0} I_{|\widetilde{x}_t| \leq k_0} \pi(dz)ds = I_{21}^n$$

Now as the proof of (5.7) and (5.8) one finds that

$$\lim_{n \to \infty} I_1^n \leq \lim_{n \to \infty} I_{11}^n = 0, \text{ and}$$
$$\lim_{n \to \infty} I_2^n \leq \lim_{n \to \infty} I_{21}^n = 0.$$

Let us show that $\lim_{n \to \infty} I_3^n = 0$. In fact, for any $0 < T < \infty$,

$$I_3^n \leq 2\left(\tfrac{12}{\varepsilon}\right)^2 E \int_0^T \int_{0<|z|<\delta} |c(s, \widetilde{x}_s^0, z)|^2 \pi(dz)ds$$
$$+\widetilde{P}(|\int_0^t \int_{|z| \geq \delta} I_{|\widetilde{x}_s^0| \leq k_0} c(s, \widetilde{x}_{s-}^0, z)\widetilde{q}^n(ds, dz)$$
$$- \int_0^t \int_{|z| \geq \delta} I_{|\widetilde{x}_s^0| \leq k_0} c(s, \widetilde{x}_{s-}^0, z)\widetilde{q}^0(ds, dz)| > \tfrac{\varepsilon}{6}) = I_2^\delta + I_3^{n,\delta}.$$

Notice that as $\delta \downarrow 0$,

$E\int_0^T \int_{\{0<|z|<\delta\}} \left|c(s,\widetilde{x}_s^0,z)\right|^2 \pi(dz)ds < \infty$, and $\{0<|z|<\delta\} \downarrow \phi$.
So one can take a small enough $\delta > 0$ such that $I_2^\delta < \widetilde{\varepsilon}/3$. Observe that

$$I_3^{n,\delta} \leq 2\left(\tfrac{18}{\varepsilon}\right)^2 E\int_0^T \int_{|z|\geq\delta} I_{\sup_{s\leq T}|\widetilde{x}_s^0|\leq k_0}|c(s,\widetilde{x}_{s-}^0,z)$$
$$-\sum_{i=0}^{2^m-1} c(\tfrac{iT}{2^m},\widetilde{x}^0_{\frac{iT}{2^m}},z)I_{(\frac{iT}{2^m},\frac{(i+1)T}{2^m}]}(s)|^2\pi(dz)ds$$
$$+\widetilde{P}(\sum_{i=0}^{2^m-1} I_{\sup_{s\leq T}|\widetilde{x}_s^0|\leq k_0}|\int_{\frac{iT}{2^m}}^{\frac{(i+1)T}{2^m}} \int_{|z|\geq\delta} c(\tfrac{iT}{2^m},\widetilde{x}^0_{\frac{iT}{2^m}},z)\widetilde{p}^n(ds,dz)$$
$$-\int_{\frac{iT}{2^m}}^{\frac{(i+1)T}{2^m}} \int_{|z|\geq\delta} c(\tfrac{iT}{2^m},\widetilde{x}^0_{\frac{iT}{2^m}},z)\widetilde{p}^0(ds,dz)| > \tfrac{\varepsilon}{6}) = I_{31}^m + I_{32}^{m,\delta},$$

where $0 < \frac{T}{2^m} < \frac{2T}{2^m} < \cdots < \frac{iT}{2^m} < \cdots < T$ is a division on $[0,T]$. Since by conditions 1° - 3° $\lim_{m\to\infty} I_{31}^m = 0$, one can choose a large enough m such that $I_{31}^m < \widetilde{\varepsilon}/6$. On the other hand, by the conclusion 1) of Lemma 400 in Appendix C for these given m,δ there exists a $\widetilde{N}$ such that as $n \geq \widetilde{N}$, $I_{32}^{m,\delta} < \widetilde{\varepsilon}/6$. So, we have proved that $\lim_{n\to\infty} I_3^n = 0$, and eventually we obtain that

$$\lim_{n\to\infty}\sum_{i=1}^3 I_i^n = 0.$$

That is, the third limit in (5.10) holds. The proofs of the remaining results are similar and even simpler. Thus $\widetilde{x}_t^0$ is a weak solution. ∎

For that the coefficient b can be greater than linear growth we can establish the following thoerem.

Theorem 176 *Assume that conditions* 1°, 3° *and* 4° *in Theorem 183 hold and assume that*

5° $|b(t,x)| \leq c_1(t)(1+|x|\prod_{k=1}^m g_k(x))$,
$|\sigma(t,x)|^2 \leq k_0(1+|x|^2\prod_{k=1}^m g_k(x))$,

where $c_1(t) \geq 0$ *has the same property as that in the condition 1° of Theorem 183, and* $g_k(x)$ *is such that*

$$g_k(x) = 1 + \underbrace{\ln(1+\ln(1+\cdots\ln}_{k-times}(1+|x|^{2n_0}))),$$

(n_0 *is some natural number).*
Then for any given constant $x_0 \in R^d$ *(5.11) has a weak solution on* $t \geq 0$.

Proof. For each $n = 1,2,\cdots$ introduce a real smooth function $W^n(x), x \in R^d$, such that $0 \leq W^n(x) \leq 1$ and $W^n(x) = 1$, as $|x| \leq n$; $W^n(x) = 0$, as $|x| \geq n+1$. Write

$$b^n(t,x) = b(t,x)W^n(x),\ \sigma^n(t,x) = \sigma(t,x)W^n(x).$$

Then by Theorem 175 for eact n there exists a weak solution x_t^n with a BM w_t^n and a Poisson martingale measure $\widetilde{N}_{k^n}(dt,dz)$, which has the same compensator $\pi(dz)dt$, defined on some probability space $(\Omega^n,\mathfrak{F}^n,\{\mathfrak{F}_t^n\},P^n)$ such that $P^n - a.s.$ $\forall t \geq 0$,

$$x_t^n = x_0 + \int_0^t b^n(s,x_s^n)ds + \int_0^t \sigma^n(s,x_s^n)dw_s^n + \int_0^t\int_Z c(s,x_{s-}^n,z)\widetilde{N}_{k^n}(ds,dz).$$

Construct a space $\Omega^n = D\times W_0\times D$, where D and W_0 are the totality of all RCLL real functions and all real continuous functions $f(t)$ with $f(0) = 0$, defined on $[0,\infty)$, respectively. Map $(x^n(.,\omega),w^n(.,\omega),\varsigma^n(.,\omega))$ into the Ω, where

$\zeta_t^n = \int_0^t \int_{|z|<1} z\tilde{N}_{k^n}(ds,dz) + \int_0^t \int_{|z|\geq 1} zN_{k^n}(ds,dz),$

and

$N_{k^n}((0,t],U) = \sum_{0<s\leq t} I_{0\neq \Delta\zeta_s^n \in U}$, for $t \geq 0, U \in \mathfrak{B}(Z)$,

$\tilde{N}_{k^n}(dt,dz) = N_{k^n}(dt,dz) - \pi(dz)dt.$

From this map we get a probability law $P_{x_0}^n$ on Ω^n. Now let

$\Omega = \times_{n=1}^\infty \Omega^n$, $\mathfrak{F} = \times_{n=1}^\infty \mathfrak{F}^n$, $P = \times_{n=1}^\infty P_{x_0}^n$,

where $\mathfrak{F}^n = \mathfrak{B}_D \times \mathfrak{B}_{W_0} \times \mathfrak{B}_D$, and define $\forall \omega = (\omega^1, \cdots, \omega^n, \cdots) \in \Omega$,

$\widetilde{x}_t^1(\omega) = x_t^1(\omega^1)$, $\widetilde{w}_t^1(\omega) = w_t^1(\omega^1), \widetilde{\zeta}_t^1(\omega) = \zeta_t^1(\omega^1)$,

$\cdots\cdots\cdots\cdots$,

$\widetilde{x}_t^n(\omega) = x_t^n(\omega^n)$, $\widetilde{w}_t^n(\omega) = w_t^n(\omega^n), \widetilde{\zeta}_t^n(\omega) = \zeta_t^n(\omega^n)$,

$\cdots\cdots\cdots\cdots$.

Then one finds that for each n, $\widetilde{x}_t^n$ satisfies the following SDE: $P-a.s.$

$\widetilde{x}_t^n = x_0 + \int_0^t b^n(s,\widetilde{x}_s^n)ds + \int_0^t \sigma^n(s,\widetilde{x}_s^n)d\widetilde{w}_s^n$

$+ \int_0^t \int_Z c(s,\widetilde{x}_{s-}^n,z)\tilde{N}_{k'^n}(ds,dz), \forall t \geq 0,$

where

$N_{k'^n}((0,t],U) = \sum_{0<s\leq t} I_{0\neq \Delta\widetilde{\zeta}_s^n \in U}$, for $t \geq 0, U \in \mathfrak{B}(Z)$,

$\tilde{N}_{k'^n}(dt,dz) = N_{k'^n}(dt,dz) - \pi(dz)dt,$

and

$\widetilde{\zeta}_t^n = \int_0^t \int_{|z|<1} z\tilde{N}_{k'^n}(ds,dz) + \int_0^t \int_{|z|\geq 1} zN_{k'^n}(ds,dz).$

Let us show that the following facts hold for $\eta_t^n = \widetilde{\zeta}_t^n, \widetilde{w}_t^n$,and $\widetilde{x}_t^n$:

$$\lim_{N\to\infty} \sup_n \sup_{t\leq T} P(|\eta_t^n| > N) = 0,$$

$$\lim_{h\downarrow 0} \sup_n \sup_{t_1,t_2\leq T, |t_1-t_2|\leq h} P(|\eta_{t_1}^n - \eta_{t_2}^n| > \varepsilon) = 0, \tag{5.12}$$

In fact, as the proof of Lemma 173 one easily sees that ζ_t^n satisfies the condition:

$\lim_{N\to\infty} \sup_n \sup_{t\leq T} P(\left|\widetilde{\zeta}_t^n\right| > N)$

$= \lim_{N\to\infty} \sup_n \sup_{t\leq T} P_{x_0}^n(|\zeta_t^n| > N)$

$= \lim_{N\to\infty} \sup_{t\leq T} P_{x_0}^1(|\zeta_t^1| > N) = 0,$

and

$\lim_{h\downarrow 0} \sup_n \sup_{t_1,t_2\leq T,|t_1-t_2|\leq h} P(\left|\widetilde{\zeta}_{t_1}^n - \widetilde{\zeta}_{t_2}^n\right| > \varepsilon)$

$= \lim_{h\downarrow 0} \sup_n \sup_{t_1,t_2\leq T,|t_1-t_2|\leq h} P_{x_0}^n(|\zeta_{t_1}^n - \zeta_{t_2}^n| > \varepsilon)$

$= \lim_{h\downarrow 0} \sup_{t_1,t_2\leq T,|t_1-t_2|\leq h} P_{x_0}^1(|\zeta_{t_1}^1 - \zeta_{t_2}^1| > \varepsilon) = 0,$

because all $\{\zeta_t^n\}_{t\geq 0}, n = 1, 2, \cdots$ have the same probability laws. So (5.12) holds for ζ_t^n. Simialrly, $\widetilde{w}_t^n$ also satisfies the (5.12). Now applying Ito's formula to $g_{m+1}(\widetilde{x}_t^n)$, one finds that $P-a.s.$

$g_{m+1}(\widetilde{x}_t^n) = g_{m+1}(x_0) + \int_0^t g'_{m+1}(\widetilde{x}_s^n)b^n(s,\widetilde{x}_s^n)ds$

$+ \int_0^t g'_{m+1}(\widetilde{x}_s^n)\sigma(s,\widetilde{x}_s^n)dw_s + \frac{1}{2}\int_0^t \left\|g''_{m+1}(\widetilde{x}_s^n)\sigma(s,\widetilde{x}_s^n)\right\|^2 ds$

$+ \int_0^t \int_Z g'_{m+1}(\widetilde{x}_{s-}^n)c(s,\widetilde{x}_{s-}^n,z)\tilde{N}_{k'^n}(ds,dz)$

$+ \int_0^t \int_Z [g_{m+1}(\widetilde{x}_{s-}^n + c(s,\widetilde{x}_{s-}^n,z)) - g_{m+1}(\widetilde{x}_{s-}^n)$

$-g'_{m+1}(\widetilde{x}_{s-}^n)c(s,\widetilde{x}_{s-}^n,z)]N_{k'^n}(ds,dz),$

where we write $g'_{m+1}(x)$ =grad $g_{m+1}(x)$, and $g''_{m+1}(x) = \left[\frac{\partial^2}{\partial x_i \partial x_j} g_{m+1}(x)\right]_{i,j=1}^d$.
By evaluation and from the assumption one sees that

$\left|g'_{m+1}(x)b^n(s,x)\right| \leq \prod_{k=1}^m g_k^{-1}(x)\frac{2n_0|x|^{2n_0-2}}{1+|x|^{2n_0}}\left|x \cdot b^n(s,x)\right| \leq k_0 c_1(t),$

$\left\|g''_{m+1}(x)\sigma(s,x)\right\|^2$

$\leq \sum_{i,j,l=1}^d \{\prod_{k=1}^m g_k^{-1}(x)[\frac{2n_0\delta_{ij}|x|^{2n_0-2}+4n_0(n_0-1)x_i x_j|x|^{2n_0-4}}{1+|x|^{2n}} - \frac{4n_0^2 x_i x_j |x|^{4n_0-4}}{(1+|x|^{2n_0})^2}]$

$-\prod_{k=1}^m g_k^{-1}(x)\frac{4n_0^2 x_i x_j|x|^{4n_0-4}}{(1+|x|^{2n_0})^2}\sum_{k=0}^m \prod_{l=1}^m g_l^{-1}(x)\}(\sigma_{il}\sigma_{jl})(t,x) \leq k_0,$

$\left\|g''_{m+1}(x)\right\|^2 \leq k_0,$

where $k_0 > 0$ is a constant, and we write $g_0(x) = 1$. Hence using the fact that

$\sup_{t\leq T}\ln(1+|x_t|) = \ln(1+\sup_{t\leq T}|x_t|)$

one finds that as $T < \infty, \forall n$, when $N \to \infty$,

$P(\sup_{t\leq T}|\widetilde{x}_t^n| > N) \leq \frac{1}{g_{m+1}(N)}Eg_{m+1}(\sup_{t\leq T}|\widetilde{x}_t^n|)$

$= \frac{1}{g_{m+1}(N)}E\sup_{t\leq T}g_{m+1}(|\widetilde{x}_t^n|) \leq k'_0(1+\int_0^T c_1(t)dt)/g_{m+1}(N) \to 0.$

This means that

$$\lim_{N\to\infty}\sup_{n=1,2\cdots}P(\sup_{t\leq T}|\widetilde{x}_t^n| > N) = 0. \tag{5.13}$$

Furthermore, by Ito's formula to $g_{m+1}(\widetilde{x}_s^n - \widetilde{x}_r^n)$, $s \in (r,t]$, one finds that $P-a.s.$

$g_{m+1}(\widetilde{x}_t^n - \widetilde{x}_r^n) = 1 + \int_r^t g'_{m+1}(\widetilde{x}_s^n - \widetilde{x}_r^n)b^n(s,\widetilde{x}_s^n)ds$

$+\int_t^t g'_{m+1}(\widetilde{x}_s^n - \widetilde{x}_r^n)\sigma(s,\widetilde{x}_s^n)dw_s + \frac{1}{2}\int_r^t \left\|g''_{m+1}(\widetilde{x}_s^n - \widetilde{x}_r^n)\sigma(s,\widetilde{x}_s^n)\right\|^2 ds$

$+\int_r^t\int_Z g'_{m+1}(\widetilde{x}_{s-}^n - \widetilde{x}_r^n)c(s,\widetilde{x}_{s-}^n,z)\widetilde{N}_{k'^n}(ds,dz)$

$+\int_r^t\int_Z[g_{m+1}(\widetilde{x}_{s-}^n - \widetilde{x}_r^n + c(s,\widetilde{x}_{s-}^n,z)) - g_{m+1}(\widetilde{x}_{s-}^n - \widetilde{x}_r^n)$

$-g'_{m+1}(\widetilde{x}_{s-}^n - \widetilde{x}_r^n)c(s,\widetilde{x}_{s-}^n,z)]N_{k'^n}(ds,dz).$

Thus one similarly has that $\forall\varepsilon > 0$, $0 \leq t-r \leq h, t \leq T, \forall n$,

$P(|\widetilde{x}_t^n - \widetilde{x}_r^n| > \varepsilon) \leq \sup_n P(\sup_{t\leq T}|\widetilde{x}_t^n| > N)$

$+\frac{1}{g_{m+1}(\varepsilon)-1}E(g_{m+1}(\widetilde{x}_t^n - \widetilde{x}_r^n) - 1)I_{\sup_{t\leq T}|\widetilde{x}_t^n|\leq N}$

$\leq k'_0[\int_r^t c_1(t)dt + (t-r)]/(1-g_{m+1}(\varepsilon)) \to 0$, when $h \to 0$.

Therefore, (5.12) holds for $\eta_t^n = \widetilde{x}_t^n$. Hence Skorohod's theorem (Theorem 398) applies. By this and by Lemma 399 "the result of SDE from the Skorohod weak convergence technique" holds. (See Lemma 173 and Remark 174). For simplicity we still use the same notation as in Lemma 399 to denote the result. So we only need to show that (5.10) in Remark 174 holds. For this let us first show that $\lim_{N\to\infty}P(\sup_{t\leq T}\left|\widetilde{x}_t^0\right| > N) = 0$. In fact,

$P(\sup_{t\leq T}\left|\widetilde{x}_t^0\right| > N) = P(\sup_{k=1,2,\cdots}\left|\widetilde{x}_{r_k}^0\right| > N)$

$\leq P(\sup_{k=1,2,\cdots}\left|\widetilde{x}_{r_k}^0 - \widetilde{x}_{r_k}^{n_k}\right| > \frac{N}{2}) + P(\sup_{k=1,2,\cdots}\left|\widetilde{x}_{r_k}^{n_k}\right| > \frac{N}{2})$

$\leq P(\sup_{k=1,2,\cdots}\left|\widetilde{x}_{r_k}^0 - \widetilde{x}_{r_k}^{n_k}\right| > \frac{1}{2}) + P(\sup_{k=1,2,\cdots}\left|\widetilde{x}_{r_k}^{n_k}\right| > \frac{N}{2}) = I_1 + I_2^N,$

where $\{r_k\}_{k=1}^\infty$ is the set of all rational numbers in $[0,T]$. However, for arbitrary given $\widetilde{\varepsilon} > 0$ and for each r_k we may take an n_k large enough such that $P(|\widetilde{x}^0_{r_k} - \widetilde{x}^{n_k}_{r_k}| > \frac{1}{2}) < \frac{\widetilde{\varepsilon}}{2^{k+1}}$, $k = 1, 2, \cdots$. Hence $I_1 \leq \sum_{k=1}^{\infty} \frac{\widetilde{\varepsilon}}{2^{k+1}} = \frac{\widetilde{\varepsilon}}{2}$. On the other hand, by (5.13) there exists a $\widetilde{N}$ such that as $N \geq \widetilde{N}$, $I_2^N < \frac{\widetilde{\varepsilon}}{2}$. Therefore, $\lim_{N\to\infty} P(\sup_{t\leq T} |\widetilde{x}^0_t| > N) = 0$ holds true. Now let us prove the second limit in (5.10). Notice that by Remark 397 in the Appendix and from (5.13) and the result just proved we may assume that $|\widetilde{x}^n_t| \leq k_0, \forall t \in [0,T], \forall n = 0, 1, 2, \cdots$. Now for any given $\varepsilon > 0$

$$\widetilde{P}(\left|\int_0^t \sigma^n(s,\widetilde{x}^n_s)d\widetilde{w}^n_s - \int_0^t \sigma(s,\widetilde{x}^0_s)d\widetilde{w}^0_s\right| > \varepsilon)$$
$$\leq \left(\tfrac{2}{\varepsilon}\right)^2 E\int_0^t \left|\sigma^n(s,\widetilde{x}^n_s) - \sigma(s,\widetilde{x}^0_s)\right|^2 I_{|\widetilde{x}^n_s|\leq k_0} I_{|\widetilde{x}^0_s|\leq k_0} ds$$
$$+P(\left|\int_0^t I_{|\widetilde{x}^0_s|\leq k_0}\sigma(s,\widetilde{x}^0_s)d\widetilde{w}^n_s - \int_0^t I_{|\widetilde{x}^0_s|\leq k_0}\sigma(s,\widetilde{x}^0_s)d\widetilde{w}^0_s\right| > \tfrac{\varepsilon}{2})$$
$$= I_3^n + I_4^n.$$

Notice that for any $\varepsilon > 0$ as $n \geq k_0$,

$$\widetilde{P}(\left|\sigma^n(s,\widetilde{x}^n_s) - \sigma(s,\widetilde{x}^0_s)\right|^2 I_{|\widetilde{x}^n_s|\leq k_0} I_{|\widetilde{x}^0_s|\leq k_0} > \varepsilon)$$
$$= \widetilde{P}(\left|\sigma(s,\widetilde{x}^n_s) - \sigma(s,\widetilde{x}^0_s)\right|^2 I_{|\widetilde{x}^n_s|\leq k_0} I_{|\widetilde{x}^0_s|\leq k_0} > \varepsilon)$$
$$\leq \widetilde{P}(\left|\widetilde{x}^n_s - \widetilde{x}^0_s\right| > \eta)$$
$$+\widetilde{P}(\left|\sigma(s,\widetilde{x}^n_s) - \sigma(s,\widetilde{x}^0_s)\right|^2 I_{|\widetilde{x}^n_s|\leq k_0} I_{|\widetilde{x}^0_s|\leq k_0} I_{|\widetilde{x}^n_s-\widetilde{x}^0_s|\leq\eta} > \varepsilon).$$

Since $\sigma(s,x)$ is continuous in x, so it is uniformly continuous in $|x| \leq k_0$. Hence one can choose a small enough $\eta > 0$ (which can depend on s) such that as $|x' - x''| \leq \eta$ and $|x'|, |x''| \leq k_0$, $|\sigma(s,x') - \sigma(s,x'')| < \varepsilon$. This means that we can have the result that as $n \to \infty$,

$$\widetilde{P}(\left|\sigma^n(s,\widetilde{x}^n_s) - \sigma(s,\widetilde{x}^0_s)\right|^2 I_{|\widetilde{x}^n_s|\leq k_0} I_{|\widetilde{x}^0_s|\leq k_0} > \varepsilon) \leq \widetilde{P}(\left|\widetilde{x}^n_s - \widetilde{x}^0_s\right| > \eta) \to 0.$$

So, by Lebesgue's dominated convergence theorem as $n \to \infty$, $I_3^n \to 0$. Now notice that $\sigma(t,x)$ is jointly continuous, so if we write $\sigma_m(t,x)$ as its smooth functions, then

$$\lim_{m\to\infty} |\sigma_m(t,x) - \sigma(t,x)|^2 = 0, \forall t, x;$$

and

$$|\sigma_m(t,x) - \sigma_m(s,y)| \leq k_m[|t-s| + |x-y|],$$

where $k_m \geq 0$ is a constant depending only on m. Observe that

$$I_4^n \leq 2\left(\tfrac{6}{\varepsilon}\right)^2 E\int_0^T \left|\sigma(s,\widetilde{x}^0_s) - \sigma_m(s,\widetilde{x}^0_s)\right|^2 I_{|\widetilde{x}^0_s|\leq k_0} ds$$
$$+P(\left|\int_0^t I_{|\widetilde{x}^0_s|\leq k_0}\sigma_m(s,\widetilde{x}^0_s)d\widetilde{w}^n_s - \int_0^t I_{|\widetilde{x}^0_s|\leq k_0}\sigma_m(s,\widetilde{x}^0_s)d\widetilde{w}^0_s\right| > \tfrac{\varepsilon}{3})$$
$$= I_{41}^m + I_{42}^{m,n}.$$

So for any given $\widetilde{\varepsilon} > 0$ by Lebesgue's dominated convergence theorem we can choose a large enough m such that $I_{41}^m < \widetilde{\varepsilon}/2$. Then applying Lemma 401 in Appendix C we can have $\lim_{n\to\infty} I_{42}^{m,n} = 0$. Thus we obtain that $\lim_{n\to\infty} I_4^n = 0$, and the second limit in (5.10) is established. The proof for the remaining results are similar. ■

5.4 Existence of Strong Solutions and Applications to ODE

Applying the above results and using the Yamada-Watanabe type theorem (Theorem 137) we immediately obtain the following theorems on the existence of a pathwise unique strong solution to SDE (5.15).

Theorem 177 *Under the assumption of Theorem 176 if, in addition, the following condition for the pathwise uniqueness holds:*
(PWU1) for each $N = 1, 2, \cdots$, *and each* $T < \infty$,
$2\langle (x_1 - x_2), (b(t, x_1) - b(t, x_2)\rangle$
$+ |\sigma(t, x_1) - \sigma(t, x_2)|^2 + \int_Z |c(t, x_1, z) - c(t, x_2, z)|^2 \pi(dz)$
$\leqslant c_T^N(t)\rho_T^N(|x_1 - x_2|^2)$,
as $|x_i| \leqslant N, i = 1, 2,\ t \in [0, T]$; *where* $c_T^N(t) \geq 0$ *such that* $\int_0^T c_T^N(t)dt < \infty$; *and* $\rho_T^N(u) \geq 0$, *as* $u \geq 0$, *is strictly increasing, continuous and concave such that*
$\int_{0+} du/\rho_T^N(u) = \infty$;
then (5.11) has a pathwise unique strong solution.

Furthermore, by using Theorem 176 and Theorem 177 we immediately obtain a result on the ODE.

Theorem 178 *1) If* $b(t, x)$ *is jointly Borel measurable and continuous in* x *such that*
$|b(t, x)| \leq c_1(t)(1 + |x| \prod_{k=1}^m g_k(x))$,
where $c_1(t)$ *and* $g_k(x)$ *have the same properties as in Theorem 176, then the ODE*

$$x_t = x_0 + \int_0^t b(s, x_s)ds, t \geq 0, \tag{5.14}$$

has a solution. (It is not necessary unique).

2) In addition, if $b(t, x)$ *is such the (PWU1) condition only for* b *in Theorem 177, then ODE (5.14) has a unique solution.*

Proof. 1) is obtained by Theorem 176 by setting $\sigma = 0, c = 0$; and 2) follows from Theorem 177 by letting $\sigma = 0, c = 0$. ■

Example 179 *Let* $b(t, x) = I_{t\neq 0}I_{x\neq 0}t^{-\alpha_1}x|x|^{-\beta}$, *where* $\alpha_1 < 1, 0 < \beta < 1$. *Then ODE (5.14) has a solution. Let* $b(t, x) = -I_{t\neq 0}I_{x\neq 0}t^{-\alpha_1}x|x|^{-\beta}$, *where* $\alpha_1 < 1, 0 < \beta < 1$. *Then ODE (5.14) has a unique solution.*

5.5 Weak Solutions. Measurable Coefficient Case

In this section we will discusss the existence of weak solutions of SDEs with measurable coefficients. In this case we have to assume that the SDEs are non-degenerate. In this case one sees that the Krylov type estimate is a very powerful tool for establishing the existence theorem for weak solutions of SDE with jumps under very weak conditions.

Consider a $d-$dimensional SDE with jumps as follows: $\forall t \geq 0$,

$$x_t = x_0 + \int_0^t b(s, x_s)ds + \int_0^t \sigma(s, x_s)dw_s + \int_0^t \int_Z c(s, x_{s-}, z)\widetilde{N}_k(ds, dz), \tag{5.15}$$

where w_t and $\widetilde{N}_k(dt, dz)$ have the same meaning as in (5.1) and all coefficients b, σ and c are non-random.

Theorem 180 *Assume that*
1° $Z = R^d - \{0\}$, and $\pi(dz) = dz/|z|^{d+1}$;
2° $|b(t,x)| + |\sigma(t,x)| + \int_Z |c(t,x,z)|^2 \pi(dz) \leq k_0$,
where $k_0 > 0$ is a constant, b, σ and c are Borel measurable functions;
3° there exists a constant $\delta_0 > 0$ such that for all $\mu \in R^d$,
$\langle \sigma(t,x)\mu, \mu\rangle \geq |\mu|^2 \delta_0$.
Then there exists a weak solution for (5.15).

Before we prove this theorem let us establish the following lemma.

Lemma 181 *Under assumption of Theorem 180 there exist smooth functions: $\forall n = 1, 2, \cdots$*
$b^n(t,x), \sigma^n(t,x),\ (t,x) \in [0,\infty) \times R^d$,
which are the smoothness functions of $b(t,x), \sigma(t,x)$, on $[0,\infty) \times R^d$, respectively; and there exist smooth functions: $\forall n = 1, 2, \cdots$
$\widetilde{c}^n(t,x,z),\ (t,x,z) \in [0,\infty) \times R^d \times \{\varepsilon_n \leq |z| \leq \varepsilon_n^{-1}\} = A_n$,
which are the smoothness functions of $c(t,x,z)$, on A_n, where $\varepsilon_n \downarrow 0$, such that set $c^n(t,x,z) = \widetilde{c}^n(t,x,z) I_{\{\varepsilon_n \leq |z| \leq \varepsilon_n^{-1}\}}$, then
1) $|b^n(t,x)| \leq k_0, |\sigma^n(t,x)| \leq k_0$,
$\int_Z |c^n(t,x,z)|^2 \pi(dz) \leq 2k_0$, $\forall n = 1, 2, \cdots$;
2) $\forall\ \mu \in R^d, \forall (t,x) \in [0,\infty) \times R^d$,
$\langle \sigma^n(t,x)\mu, \mu\rangle \geq |\mu|^2 \delta_0$;
3) $\forall n = 1, 2, \cdots, \forall t \geq 0, \forall x, x' \in R^d$,
$|b^n(t,x) - b^n(t',x')| + |\sigma^n(t,x) - \sigma^n(t',x')| \leqslant k_n k_0[|x - x'| + |t - t'|]$,
$\int_{\{\varepsilon_n \leq |z| \leq \varepsilon_n^{-1}\}} |c^n(s,x,z) - c^n(s,x\prime,z)|^2 \pi(dz) \leq k_n k_0[|x - x'|^2 + |t - t'|^2]$,
4) for each $T < \infty$, $\forall N = 1, 2, \cdots, \forall q \geq 1$, as $n \to \infty$,
$\|b^n - b\|_{q,[0,T]\times S_N} + \|\sigma^n - \sigma\|_{q,[0,T]\times S_N} \to 0$, *and*
$\left\| \int_{\{\varepsilon_n \leq |z| \leq \varepsilon_n^{-1}\}} |c^n - c|^2 \pi(dz) \right\|_{q,[0,n]\times S_n} < \frac{1}{2^n}$,
where $S_n = \{x \in R^d : |x| \leq n\}$.
5) $\left\| \int_Z |c^n - c|^2 \pi(dz) \right\|_{q,[0,T]\times S_N} \to 0$, *as $n \to \infty; \forall T, N < \infty$.*

Proof. Let us smooth out σ to get σ^n, i.e. let for all $u \in R^{d+1}$

$$J_{d+1}(u) = \begin{cases} c_{d+1}\exp(-(1-|u|^2)^{-1}), & \text{for } |u|<1, \\ 0, & \text{otherwise,} \end{cases}$$

such that the constant c_{d+1} satisfies the condition

$\int_{R^{d+1}} J_{d+1}(u)du = 1.$

and write for $(t,x) \in [0,\infty) \times R^d, n = 1,2,\cdots$

$\sigma^n(t,x) = \int_{R^{d+1}} \sigma(t-n^{-1}\overline{t}, x-n^{-1}\overline{x})J(\overline{t},\overline{x})d\overline{t}d\overline{x}$

$= \int_{R^d}\int_{R^1} \sigma(t-n^{-1}\overline{t}, x-n^{-1}\overline{x})J(\overline{t},\overline{x})d\overline{t}d\overline{x},$

where we define $\sigma(t,x) = 0$, for $t<0$. σ^n are usually called the smoothness functions of σ on $[0,\infty) \times R^d$. For these $\sigma^n, n=1,2,\cdots$ let us show that they satisfy 1) - 4). In fact, $\forall \mu \in R^d$

$|\sigma^n(t,x)| \leq \int_{R^{d+1}} |\sigma(t-n^{-1}\overline{t}, x-n^{-1}\overline{x})| J(\overline{t},\overline{x})d\overline{t}d\overline{x} \leq k_0,$

$\langle \sigma^n\mu,\mu\rangle = \int_{R^{d+1}} \langle \sigma(t-n^{-1}\overline{t}, x-n^{-1}\overline{x})\mu,\mu\rangle J(\overline{t},\overline{x})d\overline{t}d\overline{x} \geq |\mu|^2\delta_0,$

Moreover, because the $\sigma^n(t,x)$ are the smoothness functions of $\sigma(t,x)$, so $\sigma^n(t,x) \to \sigma(t,x), a.e.(t,x)$.[7],[28] Hence for any $q>0$, for each $T<\infty$ and $N = 1,2,\cdots$, as $n\to\infty$,

$$\|\sigma^n - \sigma\|_{q,[0,T]\times[-N,N]^{\otimes d}} \to 0. \tag{5.16}$$

Furthermore, one easily sees that for each $n = 1,2,\cdots$ as $x,x' \in R^d$

$|\sigma^n(t,x) - \sigma^n(t,x')| \leqslant k_n k_0 |x-x'|.$

Thus $\sigma^n, n=1,2,\cdots$, satisfy 1) - 4). In the same way one can construct $b^n(t,x), n=1,2,\cdots$, such that they satisfy 1), 3) and 4).

However, for the smothness of c, to meet our purpose we need more discussion. First we take a sequence $\varepsilon_n \downarrow 0$. Set $c_{\varepsilon_n} = cI_{\{\varepsilon_n\leq|z|\leq\varepsilon_n^{-1}\}}(z)$, and

$c^m(t,x,z) = I_{\{\varepsilon_n\leq|z|\leq\varepsilon_n^{-1}\}}(z)\cdot\widetilde{c}^m(t,x,z),$

where $\widetilde{c}^m(t,x,z) =$

$\cdot\int_{R^1\times R^d\times Z} c_{\varepsilon_n}(t-m^{-1}\overline{t}, x-m^{-1}\overline{x}, z-m^{-1}\overline{z})J(\overline{t},\overline{x},\overline{z})d\overline{t}d\overline{x}d\overline{z},$

where we define $c(t,x,z) = 0$, as $t<0$. That is, $\widetilde{c}^m(t,x,z)$ is the smoothness function of $c(t,x,z)$ on $A_n = [0,\infty)\times R^d \times \{\varepsilon_n \leq |z| \leq \varepsilon_n^{-1}\}$. Then

$\int_Z |c^m|^2\pi(dz) \leq \int_{\{\varepsilon_n\leq|\overline{z}|\leq\varepsilon_n^{-1}\}}\int_{R^d}\int_{R^1}\int_{\{\varepsilon_n\leq|z|\leq\varepsilon_n^{-1}\}}\cdot$

$\cdot\frac{|c(t-m^{-1}\overline{t},x-m^{-1}\overline{x},z-m^{-1}\overline{z})|^2}{|z-m_n^{-1}\overline{z}|^{d+1}}dz\frac{|z-m^{-1}\overline{z}|^{d+1}}{|z|^{d+1}}J(\overline{t},\overline{x},\overline{z})d\overline{t}d\overline{x}d\overline{z}$

$\leq k_0\int_{\{\varepsilon_n\leq|\overline{z}|\leq\varepsilon_n^{-1}\}}\int_{R^d}\int_{R^1} 2J(\overline{t},\overline{x},\overline{z})d\overline{t}d\overline{x}d\overline{z} \leq 2k_0,$

where we have used the fact that for $\varepsilon_n \leq |\overline{z}| \leq \varepsilon_n^{-1}$, and $\varepsilon_n \leq |z| \leq \varepsilon_n^{-1}$

$\frac{|z-m^{-1}\overline{z}|^{d+1}}{|z|^{d+1}} \leq (1+\left|\frac{\overline{z}/m}{z}\right|)^{d+1} \leq 2,$

if we take $m > \frac{1}{\varepsilon_0}\varepsilon_n^{-2}$, and $\varepsilon_0 > 0$ is a constant such that $(1+\varepsilon_0)^{d+1} \leq 2$. Thus we have proved that $\int_Z |c^m|^2\pi(dz) \leq 2k_0$, as $m > \frac{1}{\varepsilon_0}\varepsilon_n^{-2}$. Now for each ε_n by assumption

$k_0 \geq \int_{\{\varepsilon_n\leq|z|\leq\varepsilon_n^{-1}\}} |c|^2\frac{dz}{|z|^{d+1}} \geq \int_{\{\varepsilon_n\leq|z|\leq\varepsilon_n^{-1}\}} |c|^2\varepsilon_n^{d+1}dz.$

So for each ε_n, S_N and $T<\infty$ $\int_{[0,T]\times S_N\times\{\varepsilon_n\leq|z|\leq\varepsilon_n^{-1}\}} |c|^2 dtdxdz < \infty$. Thus by the property of the smoothness functions[7],[28],[64] as $m\to\infty$,

$\int_{[0,T]\times S_N\times\{\varepsilon_n\le|z|\le\varepsilon_n^{-1}\}}|c-c^m|^2\,dtdxdz$
$=\int_{[0,T]\times S_N\times\{\varepsilon_n\le|z|\le\varepsilon_n^{-1}\}}|c-\widetilde{c}^m|^2\,dtdxdz\to 0.$
Hence as $m\to\infty$,
$\int_{[0,T]\times S_N\times\{\varepsilon_n\le|z|\le\varepsilon_n^{-1}\}}|c-c^m|^2\frac{dtdxdz}{|z|^{d+1}}$
$\le\int_{[0,T]\times S_N\times\{\varepsilon_n\le|z|\le\varepsilon_n^{-1}\}}|c-c^m|^2\frac{dtdxdz}{\varepsilon_n^{d+1}}\to 0,$
for each fixed n,N and $T<\infty$. This deduces that as $m\to\infty$,
$\int_{\{\varepsilon_n\le|z|\le\varepsilon_n^{-1}\}}|c-c^m|^2\frac{dz}{|z|^{d+1}}\to 0,\ a.e.(t,x)\in[0,T]\times S_N.$
(Otherwise, a contradiction is easily derived). Now applying Lebesgue's dominated convergence theorem one finds that for any $q\ge 1$ as $m\to\infty$,
$\left\|\int_{\{\varepsilon_n\le|z|\le\varepsilon_n^{-1}\}}|c-c^m|^2\frac{dz}{|z|^{d+1}}\right\|_{L^q([0,T]\times S_N)}\to 0$, for each n,N and T.
From this for each n one easily choose a m_n such that $m_n>\frac{1}{\varepsilon_0}\varepsilon_n^{-2}$ and
$\left\|\int_{\{\varepsilon_n\le|z|\le\varepsilon_n^{-1}\}}|c-c^{m_n}|^2\frac{dz}{|z|^{d+1}}\right\|_{L^q([0,n]\times S_n)}<\frac{1}{2^n}.$
For simplicity write c^n for c^{m_n}. Since c^n is smooth in t and x, so
$\int_{\{\varepsilon_n\le|z|\le\varepsilon_n^{-1}\}}|c^n(t,x_1,z)-c^n(t,x_2,z)|^2\frac{dz}{|z|^{d+1}}$
$\le\int_{\{\varepsilon_n\le|z|\le\varepsilon_n^{-1}\}}k_n[|x_1-x_2|^2+|t_1-t_2|^2]\frac{dz}{|z|^{d+1}}$
$\le k_n'[|x_1-x_2|^2+|t_1-t_2|^2].$
Finally, as $n\to\infty$,
$\left\|\int_Z|c^n-c|^2\pi(dz)\right\|_{q,[0,T]\times S_N}\le\left\|\int_{\{0<|z|<\varepsilon_n\}\cup\{\varepsilon_n^{-1}<|z|\}}|c|^2\pi(dz)\right\|_{q,[0,T]\times S_N}$
$+\left\|\int_{\{\varepsilon_n\le|z|\le\varepsilon_n^{-1}\}}|c^n-c|^2\pi(dz)\right\|_{q,[0,T]\times S_N}\to 0.$
The Proof is complete. ■

Now let us prove Theorem 180

Proof. For b^n,σ^n and c^n, which are constructed in the previous lemma, by Theorem 117 for each $n=1,2,\cdots$ there exists a unique strong solution (x_t^n) of the following SDE: $t\ge 0$

$$\begin{aligned}x_t^n &= x_0+\int_0^t b^n(s,x_s^n)ds+\int_0^t\sigma^n(s,x_s^n)dw_s\\ &\quad+\int_0^t\int_Z c^n(s,x_{s-}^n,z)\widetilde{N}_k(ds,dz).\end{aligned}\tag{5.17}$$

Now applying Lemma 173 "the result of SDE from the Skorohod weak convergence technique" holds. So we only needs to show (5.10) in Remark 174 holds. As in the proof of Theorem 175 we may assume that $|\widetilde{x}_t^n|\le k_0,\forall t\in[0,T],\forall n=0,1,2,\cdots$. Notice that for any given $\varepsilon>0$
$\widetilde{P}(\left|\int_0^t(b^n(s,\widetilde{x}_s^n)-b(s,\widetilde{x}_s^0))ds\right|>\varepsilon)$
$\le\frac{3}{\varepsilon}E^{\widetilde{P}}[\int_0^t|(b^n-b^{n_0})(s,\widetilde{x}_s^n)|\,I_{|\widetilde{x}_s^n|\le k_0}ds$
$+\widetilde{P}(\left|\int_0^t(b^{n_0}(s,\widetilde{x}_s^n)-b^{n_0}(s,\widetilde{x}_s^0)ds\right|>\varepsilon/3)$
$+\frac{3}{\varepsilon}E^{\widetilde{P}}\int_0^t|(b^{n_0}-b)(s,\widetilde{x}_s^0)|\,I_{|\widetilde{x}_s^0|\le k_0}ds]$

$= I_1^{n,n_0} + I_2^{n_0,n} + I_3^{n_0}.$

Obviously, by 4) in Lemma 181 and by the Krylov type estimate (Theorem 165) there exists a $\widetilde{N}$ such that as $n \geq \widetilde{N}, n_0 \geq \widetilde{N}$,

$I_1^{n,n_0} + I_3^{n_0} \leq 2 \cdot \frac{3}{\varepsilon} \|b^n - b^{n_0}\|_{d+1,[0,T]\times S_{k_0}} < 2\overline{\varepsilon}/4.$

Now for each $n_0 \geq \widetilde{N}$, by (5.9) and by 3) in Lemma 181 as $n \to \infty, \forall t \in [0,T]$,

$I_2^{n_0,n} \to 0.$

Thus the first limit in (5.10) is proved. Now notice that for each $n^0 = 1, 2, ...$

$\widetilde{P}(\left|\int_0^t \int_Z c^n(s, \widetilde{x}_{s-}^n, z)\widetilde{q}^n(ds, dz) - \int_0^t \int_Z c(s, \widetilde{x}_{s-}^0, z)\widetilde{q}^0(ds, dz)\right| > \varepsilon)$

$\leq \left(\frac{3}{\varepsilon}\right)^2 E \int_0^t \int_Z \left|(c^n - c^{n^0})(s, \widetilde{x}_s^n, z)\right|^2 I_{|\widetilde{x}_s^n| \leq k_0} \pi(dz)ds$

$+\widetilde{P}(|\int_0^t \int_Z c^{n^0}(s, \widetilde{x}_{s-}^n, z)\widetilde{q}^n(ds, dz)$

$- \int_0^t \int_Z c^{n^0}(s, \widetilde{x}_{s-}^0, z)\widetilde{q}^0(ds, dz)| I_{\sup_{t\in[0,T]}|\widetilde{x}_t^n| + \sup_{t\in[0,T]}|\widetilde{x}_t^0| \leq 2k_0} > \frac{\varepsilon}{3})$

$+ \left(\frac{3}{\varepsilon}\right)^2 E \int_0^t \int_Z \left|(c^{n^0} - c)(s, \widetilde{x}_s^0, z)\right|^2 I_{|\widetilde{x}_s^0| \leq k_0} \pi(dz)ds$

$= I_2^{n,n^0} + I_3^{n^0,n} + I_4^{n^0}.$

For an arbitrary given $\overline{\varepsilon} > 0$, as above (by using the Krylov estimate) one can show that there exist a large enough $\widetilde{N}$ such that for any fixed $n^0 \geq \widetilde{N}$, as $n \geq \widetilde{N}$

$I_2^{n,n^0} + I_4^{n^0} < \frac{3}{4}\overline{\varepsilon}.$

On the other hand, by using Lemma 400 in the Appendix one also finds that as $n \to \infty$, $\forall n_0$

$$I_3^{n^0,n} \to 0. \tag{5.18}$$

In fact,

$I_3^{n^0,n}$

$\leq \widetilde{P}(|\int_0^t \int_Z (c^{n^0}(s, \widetilde{x}_{s-}^n, z) - c^{n^0}(s, \widetilde{x}_{s-}^0, z)) I_{|\widetilde{x}_s^n| \leq k_0, |\widetilde{x}_s^0| \leq k_0} \widetilde{q}^n(ds, dz)| > \frac{\varepsilon}{6})$

$+\widetilde{P}(\left|\int_0^t \int_Z c^{n^0}(s, \widetilde{x}_{s-}^0, z)\widetilde{q}^n(ds, dz) - \int_0^t \int_Z c^{n^0}(s, \widetilde{x}_{s-}^0, z)\widetilde{q}^0(ds, dz)\right| > \frac{\varepsilon}{6})$

$\leq \left(\frac{6}{\varepsilon}\right)^2 E \int_0^t \int_Z \left|c^{n^0}(s, \widetilde{x}_s^n, z) - c^{n^0}(s, \widetilde{x}_s^0, z)\right|^2 I_{|\widetilde{x}_s^n| \leq k_0, |\widetilde{x}_s^0| \leq k_0} \pi(dz)ds$

$+ \left(\frac{12}{\varepsilon}\right)^2 E \int_0^t \int_{\{|z| \leq \delta\} \cup \{|z| \geq \delta^{-1}\}} \left|c^{n^0}(s, \widetilde{x}_s^0, z)\right|^2 \pi(dz)ds$

$+\widetilde{P}(|\int_0^t \int_{\delta < |z| < \delta^{-1}} (I_{|\widetilde{x}_s^0| \leq k_0} c^{n^0}(s, \widetilde{x}_{s-}^0, z)\widetilde{p}^n(ds, dz)$

$- \int_0^t \int_{\delta < |z| < \delta^{-1}} (I_{|\widetilde{x}_s^0| \leq k_0} c^{n^0}(s, \widetilde{x}_{s-}^0, z)\widetilde{p}^0(ds, dz)| > \frac{\varepsilon}{12})$

$= \sum_{i=1}^3 I_{3i}^{n^0,n}.$

Notice that $\forall s \geq 0, \widetilde{x}_s^n \to \widetilde{x}_s^0$, in probability, as $n \to \infty$, so applying Lebesgue's dominated convergence theorem,

$I_{31}^{n^0,n} \leq k_{n^0} E \int_0^t \left|\widetilde{x}_s^n - \widetilde{x}_s^0\right|^2 I_{|\widetilde{x}_s^n| \leq k_0, |\widetilde{x}_s^0| \leq k_0} ds \to 0$, as $n \to \infty$.

Now by $E \int_0^t \int_Z \left|c^{n^0}(s, \widetilde{x}_s^0, z)\right|^2 \pi(dz)ds \leq k_0 t < \infty$ for any $\overline{\varepsilon} > 0$ one can choose a small enough $\delta > 0$ such that

$I_{32}^{n^0,n} < \overline{\varepsilon}/3.$

Let us show that for any $\delta > 0$ and n^0, as $n \to \infty$,

$$I_{33}^{n^0,n} \to 0. \tag{5.19}$$

To show this we make a division: $0 = s_0 < s_1 < \cdots < s_{m+1} = t$. Then

$I_{33}^{n^0,n,N} \leq \widetilde{P}(|\int_0^t \int_{\delta<|z|<\delta^{-1}} I_{|\widetilde{x}_s^0|\leq k_0} c^{n^0}(s, \widetilde{x}_{s-}^0, z)\widetilde{p}^n(ds, dz)$
$-\sum_{i=0}^k \int_{s_i}^{s_{i+1}} \int_{\delta<|z|<\delta^{-1}} I_{|\widetilde{x}_s^0|\leq k_0} c^{n^0}(s_i, \widetilde{x}_{s_i}^0, z)\widetilde{p}^n(ds, dz)| > \frac{\varepsilon}{12})$
$+\widetilde{P}(|\sum_{i=0}^k \int_{s_i}^{s_{i+1}} \int_{\delta<|z|<\delta^{-1}} I_{|\widetilde{x}_s^0|\leq k_0} c^{n^0}(s_i, \widetilde{x}_{s_i}^0, z)\widetilde{p}^n(ds, dz)$
$-\sum_{i=0}^k \int_{s_i}^{s_{i+1}} \int_{\delta<|z|<\delta^{-1}} I_{|\widetilde{x}_s^0|\leq k_0} c^{n^0}(s_i, \widetilde{x}_{s_i}^0, z)\widetilde{p}^0(ds, dz)| > \frac{\varepsilon}{12})$
$+\widetilde{P}(|\int_0^t \int_{\delta<|z|<\delta^{-1}} I_{|\widetilde{x}_s^0|\leq k_0} c^{n^0}(s, \widetilde{x}_{s-}^0, z)\widetilde{p}^0(ds, dz)$
$-\sum_{i=0}^k \int_{s_i}^{s_{i+1}} \int_{\delta<|z|<\delta^{-1}} I_{|\widetilde{x}_s^0|\leq k_0} c^{n^0}(s_i, \widetilde{x}_{s_i}^0, z)\widetilde{p}^0(ds, dz)| > \frac{\varepsilon}{12})$
$= I_{331}^{n^0,n} + I_{332}^{n^0,n} + I_{333}^{n^0,n}$.

Because c^{n^0} is a smooth function satisfying the condition that as $\varepsilon_{n_0} \leq \delta$
$\int_{\delta<|z|<\delta^{-1}} |c^{n_0}(s, x, z) - c^{n_0}(s', x\prime, z)|^2 \pi(dz)$
$\leq k_{n_0} \widetilde{k}_0[|x - x'|^2 + |s - s'|]$, and
$|c^{n_0}(s, x, z)| \leq \widetilde{k}_{n_0,\delta,k_0}$,
as $(s, x, z) \in [0, T] \times \{|x| \leq k_0\} \times \{\delta < |z| < \delta^{-1}\}$,

where $k_{n_0,\delta} > 0$ is a constant only depending on n_0 and δ; $\widetilde{k}_{n_0,\delta,k_0} > 0$ is a constant only depending on n_0, δ and k_0; and $\widetilde{x}_s^0$ is right continuous such that $s \downarrow s_i \Longrightarrow x_s \to x_{s_i}$. So by using Lebesgue's dominated convergence theorem, one finds that as $\lambda = \max_{i=0,\cdots,m}(s_{i+1} - s_i) \to 0$

$I_{331}^{n^0,n}, I_{33}^{n^0,n} \leq \left(\frac{12}{\varepsilon}\right)^2 E\int_0^t \int_{\delta<|z|<\delta^{-1}} |I_{|\widetilde{x}_s^0|\leq k_0} c^{n^0}(s, \widetilde{x}_{s-}^0, z)$
$-\sum_{i=0}^k I_{|\widetilde{x}_s^0|\leq k_0} c^{n^0}(s_i, \widetilde{x}_{s_i}^0, z) I_{(s_i,s_{i+1}]}(s)|^2 \pi(dz)ds \to 0$

Furthermore by Lemma 400 in the Appendix, for any given division,
$\lim_{n\to\infty} I_{332}^{n^0,n} = 0$.

Therefore, (5.19) holds. Thus (5.18) is proved, and the third limit in (5.10) is also established. Finally, the second limit in (5.10) can be similarly proved. In fact, for arbitrary $\varepsilon > 0$

$P(\left|\int_0^t \sigma^n(s, \widetilde{x}_s^n)d\widetilde{w}_s^n - \int_0^t \sigma(s, \widetilde{x}_s^0)d\widetilde{w}_s^0\right| > \varepsilon)$
$\leq \left(\frac{3}{\varepsilon}\right)^2 E\int_0^t \left|(\sigma^n - \sigma^{n^0})(s, \widetilde{x}_s^n)\right|^2 I_{|\widetilde{x}_s^n|\leq k_0} ds$
$+\widetilde{P}(|\int_0^t \sigma^{n^0}(s, \widetilde{x}_s^n)d\widetilde{w}_s^n - \int_0^t \sigma^{n^0}(s, \widetilde{x}_s^0)d\widetilde{w}_s^0|$
$\cdot I_{\sup_{t\in[0,T]}|\widetilde{x}_t^n| + \sup_{t\in[0,T]}|\widetilde{x}_t^0| \leq 2k_0} > \frac{\varepsilon}{3})$
$+\left(\frac{3}{\varepsilon}\right)^2 E\int_0^t \int_Z \left|(\sigma^{n^0} - \sigma)(s, \widetilde{x}_s^0)\right|^2 I_{|\widetilde{x}_s^0|\leq k_0} ds$
$= I_2^{n,n^0} + I_3^{n^0,n} + I_4^{n^0}$.

Now the proof can be completed in the same way as that of the third limit in (5.10) by using Lemma 401. So the second limit in (5.10) is established. Thus we have proved that $\{\widetilde{x}_t^0\}_{t\geq 0}$ satisfies the following SDE on probability space $(\widetilde{\Omega}, \widetilde{\mathfrak{F}}, \widetilde{P})$ for any $T < \infty$ as $t \in [0, T]$

$\widetilde{x}_t^0 = x_0 + \int_0^t b(s,\widetilde{x}_s^0)ds + \int_0^t \sigma(s,\widetilde{x}_s^0)d\widetilde{w}_s^0 + \int_0^t \int_Z c(s,\widetilde{x}_{s-}^0,z)\widetilde{q}^0(ds,dz),$
where $\widetilde{w}_t^0$ and $\widetilde{q}^0(dt,dz)$ are a BM and a Poisson martingale with the compensator $\pi(dz)dt$, respectively. ∎

By using Theorem 180 and the Girsanov type thoerem we can obtain the existence of a weak solution to a BSDE with jumps under much weaker conditions.

Theorem 182 *Assume that b, σ and c are Borel measurable functions such that*
1° $Z = R^d - \{0\}$, and $\pi(dz) = dz/|z|^{d+1}$;
2° $|\sigma(t,x)| + \int_Z |c(t,x,z)|^2 \pi(dz) \leq k_0$,
where $k_0 > 0$ is a constant;
3° there exists a $\delta_0 > 0$ such that for all $\mu \in R^d$,
$\langle\sigma(t,x)\mu,\mu\rangle \geq |\mu|^2 \delta_0$.
4° $\langle x, b(t,x)\rangle \leq c_1(t)(1 + |x|^2 \prod_{k=1}^m g_k(x))$,
where $c_1(t)$ and $g_k(x)$ have the same properties as in Theorem 176; furthermore, $b(t,x)$ is locally bounded for x, that is, for each $r > 0$, as $|x| \leq r$,
$|b(t,x)| \leq k_r$,
where $k_r > 0$ is a constant only depending on r.
Then there exists a weak solution for (5.15).

Proof. The proof involves a combination of the results of Theorem 180 and Theorem 133. In fact, by Theorem 180 there exists a weak solution for the following SDE with jumps: $\forall t \geq 0$,
$x_t = x_0 + \int_0^t \sigma(s,x_s)dw_s + \int_0^t \int_Z c(s,x_{s-},z)\widetilde{N}_k(ds,dz).$

Notice that by Skorohod theorem (Theorem 398 in the Appendix) we know that the above SDE holds in a probability space $(\Omega,\mathfrak{F},P)$, where $\Omega = [0,1], \mathfrak{F} = \mathfrak{B}([0,1]), P =$Lebesgue measure on $[0,1]$. Since such $(\Omega,\mathfrak{F})$ is a standard measurable space. So applying Theorem 133 the conclusion follows on $\forall t \geq 0$. ∎
In the above theorem we assume that σ is bounded. Now we relax the coefficient σ to be less than linear growth, (so, it can be unbounded). In this case we have to assume that σ and c are jointly continuous.

Theorem 183 *Assume that conditions 1°, 3° and 4° in the previous theorem hold, and assume that*
5° $\int_Z |c(t,x,z)|^2 \pi(dz) \leqslant c_1(t)$,
6° $\sigma(t,x)$ is jointly continuous in (t,x); and
$\lim_{h,h'\to 0} \int_Z |c(t+h',x+h,z) - c(t,x,z)|^2 \pi(dz) = 0$;
7° there exists a $\delta_0 > 0$ such that $|\sigma(t,x)| \geq \delta_0$, and
$|\sigma(t,x)|^2 \leqslant c_1(t)(1+|x|^2)$
Then for any given constant $x_0 \in R^d$ (5.15) has a weak solution on $t \geq 0$.

Example 184 *If*
$b(t,x) = -x|x|^{2n_1} + x \prod_{k=1}^m g_k(x),$

where n_1 *is any natural number, and* $g_k(x)$ *is defined in 2° of Theorem 176, then* $b(t,x)$ *satisfies all conditions in Theorem 183. However,* $b(t,x)$ *is very much greater than linear growth in* x.

Now let us prove Theorem 183.

Proof. The proof can be completed by applying Theorem 175 and the Girsanov type theorem (Theorem 133). Since it is completely similar to the proof of the previous theorem. We do not repeate it. ■

Finally, applying the above results and applying the Yamada-Watanabe type theorem (Theorem 137) we immediately obtain the following theorems on the existence of a pathwise unique strong solution to SDE (5.15).

Theorem 185 *Under the assumption of Theorem 183 if, in addition, the (PWU1) condition in Theorem 177 holds, then (5.15) has a pathwise unique strong solution.*

Part II

Applications

6

How to Use the Stochastic Calculus to Solve SDE

To help the reader who wants to quickly know how to use the stochastic calculus to solve stochastic differential equations (SDE) we offer a short introductory chapter. Some instructive examples are also presented. Actually, some of the material only represent special cases of the general results in the first part of the book. However, these simpler cases can help explain the ideas more directly and clearly, and so may help the reader master the main ideas. A reader who is already familiar with Ito's formula, Girsanov's theorem and their applications may skip this chapter.

6.1 The Foundation of Applications: Ito's Formula and Girsanov's Theorem

In solving ordinary differential equations (ODEs) the following technique is frequently used: If we can guess the form of a solution, we will use the differentiation to check whether the guess is true solution, or to make some changes to make it true. Such idea can also be applied to the solution of SDE. However, the rule for differentiation in the Stochastic Calculus is different, and for this we need the Ito differential rule, that is, the Ito formula.

Another frequently used technique is that if a tranformation can be made to simplify the ODE, then this will always be done first to make finding the solution easier. Such a technique is also applied to SDE. However, to use a trasformation in an SDE is much harder than in the case of an ODE. The problem is that, after the transformation, is the new differential equation still an SDE that we can understand; that is, does the new stochastic integral still makes sense? To answer this question we need the Girsanov's transformation, or say, a Girsanov type theorem.

As for the existence and uniqueness of solutions of an SDE we need a related theorem in SDE, and that is the theory in the first part of the book. However, in this case the discussion of the theory also needs the help of Ito's formula.

Let us look at the following examples.

Example 186 *Find the solution of the following* $1-$*dimensional SDE*

$dx_t = ax_t dt + bx_t dw_t,\ x_0 = c,\ t \geq 0,$

where a, b, c *are constants,* $w_t, t \geq 0,$ *is a* $1-$*dimensional BM.*

If $b = 0$, then by the usual differential rule, one easily checks that $x_t = ce^{at}$ satisfies the ODE

$dx_t = ax_t dt,\ x_0 = c,\ t \geq 0.$

However, setting $x_t = ce^{at+bw_t}$, by Ito's differential rule (Ito's formula) this does not satisfy the above SDE, because it only satisfies the following SDE:

$dx_t = cde^{at+bw_t} = cdf(at + bw_t) = cf'(at + bw_t)d(at + bw_t)$

$+\frac{1}{2}cf''(at + bw_t)d\langle bw\rangle_t = ce^{at+bw_t}(adt + bdw_t) + \frac{1}{2}ce^{at+bw_t}d(b^2 t)$

$= x_t(adt + bdw_t) + \frac{1}{2}b^2 x_t dt = ax_t dt + bx_t dw_t + \frac{1}{2}b^2 x_t dt.$

That is $x_t = ce^{at+bw_t}$ only satisfies the SDE

$dx_t = ax_t dt + bx_t dw_t + \frac{1}{2}b^2 x_t dt,$

$x_0 = c,\ t \geq 0.$

Here we have applied the following Ito's formula, which is a special case of Thoerem 93 in the first part of the book, and we set $y_t = at+bw_t$, $f(y) = e^y$ to get the above result. (Recall that by the notation before Lemma 62 $\langle M\rangle_t = \langle M, M\rangle_t$ is the characteristic process of the locally square integrable martingale M_t such that $\langle M\rangle_t$ comes from the Doob-Meyer decomposition: M_t^2 =a local martingale $+\langle M\rangle_t$. Now for a BM we have that $\langle bw\rangle_t = \langle bw, bw\rangle_t = b^2\langle w, w\rangle_t = b^2\langle w\rangle_t = b^2 t$)

Theorem 187 *(Ito's formula). Suppose that*

$y_t = y_0 + A_t + M_t,$

where $y_0 \in \mathfrak{F}_0$, $\{A_t\}_{t\geq 0}$ *is a continuous finite variational* ($\mathfrak{F}_t-$*adapted) process with* $A_0 = 0$, $\{M_t\}_{t\geq 0} \in \mathcal{M}^{2,loc,c}$.

If $f(x) \in C^2(R)$, *then*

$$f(y_t) - f(y_0) = \int_0^t f'(y_s)dA_s + \int_0^t f'(y_s)dM_s + \frac{1}{2}\int_0^t f''(y_s)d\langle M\rangle_s. \quad (6.1)$$

Or, we can write it in the differntial form:

$df(y_t) = f'(y_t)dA_t + f'(y_t)dM_t + \frac{1}{2}f''(y_t)d\langle M\rangle_t.$

Recall that the differntial form is only a form, and its exact meaning is that the integral equality holds.

Remark 188 *To use the Ito formula one should know how to calculate the characteristic process* $\langle M \rangle_t$ *of the given locally square integrable martingale* M_t. *For the convenience of the reader we recall the following fact here: If* $M_t = \int_0^t \sigma(s,\omega)dw_s, t \in [0,T]$, *where* $w_t, t \in [0,T]$ *is a BM, and* $E\int_0^T |\sigma(s,\omega)|^2 ds < \infty$, *then* $\langle M \rangle_t = \int_0^t |\sigma(s,\omega)|^2 ds, \forall t \in [0,T]$.

Now we can solve Example 186 by making a small change to the guess form as follows:

Solution 189 *Set* $x_t = ce^{at+bw_t-\frac{1}{2}b^2t}$. *Write* $y_t = at + bw_t - \frac{1}{2}b^2t$, *and* $f(y) = e^y$. *Then by (6.1)*
$dx_t = cdf(y_t) = cf'(y_t)dy_t + \frac{1}{2}cf''(y_t)d\langle bw \rangle_t$
$= ce^{y_t}(adt + bdw_t - \frac{1}{2}b^2dt) + \frac{1}{2}b^2ce^{y_t}dt = ce^{y_t}(adt + bdw_t)$
$= ax_tdt + bx_tdw_t$.
So $x_t = ce^{at+bw_t-\frac{1}{2}b^2t}$ *is a solution of Example 186. Moreover, it is the unique solution, because the coefficient of the SDE is Lipschitz continuous and less than linear growth (Theorem 117).*

Example 190 *Find the solution of the following* $1-$*dimensional SDE*

$$dx_t = ax_tdt + f(t,x_t)dt + bdw_t, x_0 = c, t \in [0,T], \tag{6.2}$$

where $b > 0, a, c$ *are constants,* $w_t, t \geq 0$, *is a* $1-$*dimensional BM, in the case that*

1) $f(t,x)$ *is bounded and jointly measurable;*

2) $f(t,x)$ *is bounded and satisfies the Lipschitzian condition, that is, there exists a constant* $k_0 > 0$ *such that*
$|f(t,x)| \leq k_0, \forall t \in [0,T], \forall x \in R^1$,
$|f(t,x) - f(t,y)| \leq k_0 |x-y|, \forall t \in [0,T], \forall x,y \in R^1$.

In this example in case 1), without the help of Girsanov's theorem we cannot even know that (6.2) will have a weak solution. Here, a "weak solution" means that the solution $x_t, t \in [0,T]$, with a BM $w_t, t \in [0,T]$, on some probability space $(\Omega, \mathfrak{F}, \{\mathfrak{F}_t\}_{t\in[0,T]}, P)$ satisfies (6.2), but $\mathfrak{F}_t \neq \mathfrak{F}_t^w$. So x_t is $\mathfrak{F}_t-$adapted, but is not necessary $\mathfrak{F}_t^w-$adapted. (See Definition 127). Intuitively we see that (6.2) equals
$dx_t = ax_tdt + f(t,x_t)dt + bdw_t = ax_tdt + b(b^{-1}f(t,x_t)dt + dw_t)$
$= ax_tdt + bd\widetilde{w}_t$,
where we write $d\widetilde{w}_t = dw_t + b^{-1}f(t,x_t)dt$. However, the existence of a solution x_t of (6.2) is not yet known. So we should solve (6.2) in an different way. First we solve the simpler SDE $dx_t = ax_tdt + bdw_t$ for a given BM w_t to find a solution x_t. Secondly, we let $d\widetilde{w}_t = dw_t - b^{-1}f(t,x_t)dt$. If there is a theorem (we call it a transformation theorem) to guarantee that such a $\widetilde{w}_t$ is still a BM, but under some new probability measure $\widetilde{P}_T$, then we can arrive at $\widetilde{P}_T - a.s.$
$dx_t = ax_tdt + f(t,x_t)dt + bd\widetilde{w}_t, x_0 = c, t \in [0,T]$.

That is, we have that $x_t, t \in [0,T]$, with a BM $\widetilde{w}_t, t \in [0,T]$, on the probability space $(\Omega, \mathfrak{F}, \{\mathfrak{F}_t\}_{t\in[0,T]}, \widetilde{P}_T)$ satisfies (6.2). So (6.2) has a weak solution. Fortunately, such a useful transformation theorem exists, and it is the so-called Girsanov type theorem that can be stated as follows, which is a special case of Theorem 124 in the First Part of this book.

Theorem 191 *If on a given probability space $(\Omega, \mathfrak{F}, \{\mathfrak{F}_t\}_{t\in[0,T]}, P)$ a 1−dimensional process θ_t^w is $\mathfrak{F}_t$−adapted such that*

$$|\theta_t^w|^2 \leqslant c(t),$$

where $c(t) \geqslant 0$ is non-random such that $\int_0^T c(t)dt < \infty$, then defining as
$d\hat{P}_T = \exp[\int_0^T \theta_s^w dw_s - \frac{1}{2}\int_0^T |\theta_s^w|^2 ds]dP$,
where $w_t, t \in [0,T]$ is a BM on this probability space, $\hat{P}_T$ is a new probability measure, and

$$w_t' = w_t - \int_0^t \theta_s^w ds, t \in [0,T],$$

is a new BM under the probability $\hat{P}_T$.

Now let us use this theorem to solve SDE (6.2) by using this approach.

Solution 192 *For a given probability space $(\Omega, \mathfrak{F}, \{\mathfrak{F}_t\}_{t\in[0,T]}, P)$ and a given BM $w_t, t \in [0,T]$ defined on it, we can solve the simpler SDE*
$dx_t = ax_t dt + bdw_t, x_0 = c, t \in [0,T]$,
to get a unique $\mathfrak{F}_t^w$−adapted solution $P - a.s.$
$x_t = e^{at}c + b\int_0^t e^{a(t-s)}dw_s, \forall t \in [0,T]$.
(In fact, by Ito's formula one can easily checks that it satisfies the simpler SDE, and by Theorem 117 it is the unique $\mathfrak{F}_t^w$−adapted solution). Now applying the above Girsanov type thoerem (Theorem 191)
$\widetilde{w}_t = w_t - \int_0^t b^{-1}f(s,x_s)ds,\ t \in [0,T]$,
is a new BM under the new probability measure $\hat{P}_T$, where
$d\hat{P}_T = \exp[\int_0^T b^{-1}f(s,x_s)dw_s - \frac{1}{2}\int_0^T \left|b^{-1}f(s,x_s)\right|^2 ds]dP$.
So we have that $\hat{P}_T - a.s.$

$$dx_t = ax_t dt + f(t,x_t)dt + bd\widetilde{w}_t, x_0 = c, t \in [0,T]. \tag{6.3}$$

where
$x_t = e^{at}c + b\int_0^t e^{a(t-s)}dw_s$,
$x_t \in \mathfrak{F}_t^w \subset \mathfrak{F}_t$, but $x_t \notin \mathfrak{F}_t^{\widetilde{w}}$. So $(x_t, \widetilde{w}_t)_{t\in[0,T]}$ is only a weak solution of (6.2) in case 1).

Next we discuss case 2). In this case the pathwise uniqueness of solutions of (6.3) holds. Hence one can apply the Yamada-Watanabe theorem (Theorem 137) to get the result that $(x_t, \widetilde{w}_t)_{t\in[0,T]}$ is actually a strong solution of SDE (6.3); that is, $x_t \in \mathfrak{F}_t^{\widetilde{w}}$ (x_t is $\mathfrak{F}_t^{\widetilde{w}}$−adapted). So there *exists a Baire function* F *such that* $x_t = F(\widetilde{w}_s, s \leq t)$. *Therefore, let* $\widetilde{x}_t = F(w_s, s \leq t)$, *then* $(\widetilde{x}_t, w_t)_{t\in[0,T]}$ *satisfies (6.2) on the original probability space* $(\Omega, \mathfrak{F}, \{\mathfrak{F}_t\}_{t\in[0,T]}, P)$ *with the original BM* $w_t, t \in [0,T]$, *and* $(\widetilde{x}_t)_{t\in[0,T]}$ *is the pathwise unique strong solution, that is it is unique and* $\widetilde{x}_t \in \mathfrak{F}_t^w$.

6.2 More Useful Examples

In the later Chapter we will meet a Stock price SDE as follows:

$dP_t^1 = P_t^1[r_t dt + \sigma_t d\widetilde{w}_t)],\ P_0^1 = P_0^1;\ \forall t \in [0,T], \widetilde{P} - a.s.,$

where for simplicity we assume that all of the processes that occur here are real-valued, and $\widetilde{w}_t, t \in [0,T]$, is a BM under the probability measure $\widetilde{P}$.

Example 193 *Under assumption that* r_t, σ_t *are non-random and*

$\int_0^T [|r_t| + |\sigma_t|^2]dt < \infty$

the solution of the stock price SDE is: $\widetilde{P} - a.s. \forall t \in [0,T]$

$P_t^1 = P_0^1 \exp[\int_0^t r_s ds + \int_0^t \sigma_s d\widetilde{w}_s - \frac{1}{2}\int_0^t |\sigma_s|^2 ds].$

Proof. Write $y_t = \int_0^t r_s ds + \int_0^t \sigma_s d\widetilde{w}_s - \frac{1}{2}\int_0^t |\sigma_s|^2 ds$, and $f(y) = e^y$. Applying Ito's formula (Theorem 187) to $P_t^1 = P_0^1 e^{y_t}$ we find that $\widetilde{P} - a.s.$

$dP_t^1 = P_0^1 e^{y_t} dy_t + \frac{1}{2} P_0^1 e^{y_t} |\sigma_t|^2 dt$

$= P_0^1 e^{y_t}[r_t dt + \sigma_t d\widetilde{w}_t - \frac{1}{2}|\sigma_t|^2 dt] + \frac{1}{2} P_0^1 e^{y_t} |\sigma_t|^2 dt = P_t^1[r_t dt + \sigma_t d\widetilde{w}_t].$

So $P_t^1 = P_0^1 e^{y_t}$ solves the stock price SDE. Moreover, by Theorem 117 it is the unique strong solution of the stock price SDE. ∎

In the later Chapter we will also meet a wealth process SDE as follows:

$dx_t = r_t x_t dt + \pi_t \sigma_t d\widetilde{w}_t, x_0 = x_0,\ \widetilde{P} - a.s.,$

where for simplicity we assume that all of the processes that occur here are real-valued, and $\widetilde{w}_t, t \in [0,T]$, is a BM under the probability measure $\widetilde{P}$.

Example 194 *Under assumption that* r_t *is non-random and* $|\sigma_t| \leq k_0$

$\int_0^T |r_t| dt < \infty$, *and* $E\int_0^T |\pi_t|^2 dt < \infty$,

the solution of the wealth process SDE is: $\widetilde{P} - a.s. \forall t \in [0,T]$

$x_t = \exp[\int_0^t r_s ds] x_0 + \int_0^t \exp[\int_s^t r_u du] \pi_s \sigma_s d\widetilde{w}_s.$

Proof. Applying Ito's formula (Theorem 187) to the above x_t we find that $\widetilde{P} - a.s.$

$dx_t = x_0 d(\exp[\int_0^t r_s ds]) + d\int_0^t \exp[\int_s^t r_u du]\pi_s \sigma_s d\widetilde{w}_s]$

$= r_t x_0 \exp[\int_0^t r_s ds]dt + \pi_t \sigma_t d\widetilde{w}_t + r_t dt \int_0^t \exp[\int_s^t r_u du]\pi_s \sigma_s d\widetilde{w}_s$

$= r_t[\exp[\int_0^t r_s ds] x_0 + \int_0^t \exp[\int_s^t r_u du]\pi_s \sigma_s d\widetilde{w}_s]dt + \pi_t \sigma_t d\widetilde{w}_t$

$= r_t x_t dt + \pi_t \sigma_t d\widetilde{w}_t,$

where we have used the following result:

$d\int_0^t \exp[\int_s^t r_u du]\pi_s \sigma_s d\widetilde{w}_s] = \pi_t \sigma_t d\widetilde{w}_t + r_t dt \int_0^t \exp[\int_s^t r_u du]\pi_s \sigma_s d\widetilde{w}_s.$

Indeed, let $y_t = \int_0^t \exp[-\int_0^s r_u du]\pi_s \sigma_s d\widetilde{w}_s]$. Then by Ito's formula

$d\int_0^t \exp[\int_s^t r_u du]\pi_s\sigma_s d\widetilde{w}_s] = de^{\int_0^t r_u du} y_t$
$= r_t e^{\int_0^t r_u du} y_t dt + e^{\int_0^t r_u du} dy_t$
$= r_t dt e^{\int_0^t r_u du} \int_0^t e^{-\int_0^s r_u du} \pi_s\sigma_s d\widetilde{w}_s + e^{\int_0^t r_u du} e^{-\int_0^t r_u du} \pi_t\sigma_t d\widetilde{w}_t$
$= r_t dt \int_0^t e^{\int_s^t r_u du} \pi_s\sigma_s d\widetilde{w}_s + \pi_t\sigma_t d\widetilde{w}_t.$

So the result is true. Thus

$x_t = \exp[\int_0^t r_s ds]x_0 + \int_0^t \exp[\int_s^t r_u du]\pi_s\sigma_s d\widetilde{w}_s$

solves the wealth process SDE. Moreover, by Theorem 117 it is the unique strong solution of the wealth process SDE. ∎

7

Linear and Non-linear Filtering

In many cases the signal process which we want to examine cannot be observed directly, and we can only examine a different observable process which is related to the signal process. This poses the question of how we estimate the true signal process by using the information obtained from the observable process? Such an estimate of the present signal obtained by using the information from an observable process up to the present time is called a filter. In this chapter we will discuss the fitering equation for both the non-linear case and the linear case. When the linear case is without jumps, we will derive and solve the continuous linear filtering equation - the famous Kalman-Bucy equation, and we will also consider the non-linear case when we will derive the Zakai equation and discuss its solutions.

7.1 Solutions of SDE with Functional Coefficients and Girsanov Theorems

In discussing filtering problems we need to consider a SDE with functional coefficients. Let us begin by introducing some notation. Let

$D = D([0,\infty); R^d)$ = The totality of RCLL maps from $[0,\infty)$ to R^d, with the Skorohod metric (see Lemma 388 in the Appendix),

$D_T = D([0,T]; R^d)$,

$C = C([0,\infty); R^d)$ = the totality of continuous maps from $[0,\infty)$ to R^d,

$C_T = C([0,T]); R^d)$,

$\mathfrak{B}(D)$ = The topological $\sigma-$field on D,

(that is, the σ−field is generated by the totality of open sets),
$\mathfrak{B}_t(D) = \mathfrak{B}(D_t)$.

Consider the following SDE with jumps and functional coefficients in d−dimensional space : $\forall t \geq 0$,

$$\begin{aligned} x_t &= x_0 + \int_0^t b(s, \{x_r\}_{r \leq s}, \omega)ds + \int_0^t \sigma(s, \{x_r\}_{r \leq s}, \omega)dw_s \\ &\quad + \int_0^t \int_Z c(s, \{x_r\}_{r<s}, z, \omega)\tilde{N}_k(ds, dz), \end{aligned} \tag{7.1}$$

where $w_t^* = (w_t^1, \cdots, w_t^{d_1}), 0 \leqslant t$, is a d_1-dimensional $\mathfrak{F}_t$−adapted standard Brownian motion (BM), w_t^* is the transpose of w_t; $k^* = (k_1, \cdots, k_{d_2})$ is a d_2−dimensional $\mathfrak{F}_t$−adapted stationary Poisson point process with independent components, and $\tilde{N}_{k_i}(ds, dz)$ is the Poisson martingale measure generated by k_i satisfying

$$\tilde{N}_{k_i}(ds, dz) = N_{k_i}(ds, dz) - \pi(dz)ds, \quad i = 1, \cdots, d_2,$$

where $\pi(.)$ is a σ-finite measure on a measurable space $(Z, \mathfrak{B}(Z))$, and $N_{k_i}(ds, dz)$ is the Poisson counting measure generated by k_i, where for simplicity we always assume that

$N_{k_i}(\{t\}, U)N_{k_j}(\{t\}, U) = 0$, as $i \neq j$, for all $U \in \mathfrak{B}(Z)$, $\pi(U) < \infty$.

For simplicity we also write

$b(s, x(.), \omega) = b(s, x_r, r \leq s, \omega) = b(s, \{x_r\}_{r \leq s}, \omega)$,

$\sigma(s, x(.), \omega) = \sigma(s, x_r, r \leq s, \omega) = \sigma(s, \{x_r\}_{r \leq s}, \omega)$,

and

$c(s-, x(.), z, \omega) = c(s, x_r, r < s, z, \omega) = c(s, \{x_r\}_{r<s}, z, \omega)$.

The definitions of a solution and the pathwise uniqueness of solutions to (7.1) are completely the same as those of (3.1). Furthermore, by using almost completely the same proof as that in Lemma 195 we have an a priori estimate of the solution to such an SDE.

Lemma 195 *Assume that x_t is a solution of (7.1), and assume that*
1° $E|x_0|^2 < \infty$, and

$$2\langle x_t \cdot b(t, x(.), \omega)\rangle \leqslant c(t)(1 + \|x\|_t^2),$$

$$|\sigma(t, x(.), \omega)|^2 + \int_Z |c(t, x(.), z, \omega)|^2 \pi(dz) \leqslant c(t)(1 + \|x\|_t^2),$$

where $\|x\|_t = \sup_{s \leq t} |x_s|$, and $0 \leqslant c(t)$ is non-random, such that

$$C_T = \int_0^T c(t)dt < \infty,$$

for any $0 < T < \infty$.

Then

$$E(\|x\|_T^2) \leqslant k_T < \infty,$$

where $k_T \geqslant 0$ *is a constant depending only on* C_T *and* $E|x_0|^2$.

For the uniqueness of the solution we have the following similar lemma.

Lemma 196 *Under the assumption in Lemma 195 if for each* $N = 1, 2, \cdots$ *there exist non-random functions* $c^N(t)$ *and* $\rho^N(u)$ *such that as* $\|x_1\|_t$ *and* $\|x_2\|_t \leq N$, $\forall x_1(.), x_2(.) \in D$

$$\begin{aligned}
&2(x_1(t) - x_2(t)) \cdot (b(t, x_1(.), \omega) - b(t, x_2(.), \omega)) \leqslant c(t)\rho(\|x_1 - x_2\|_t^2), \\
&|\sigma(t, x_1(.), \omega) - \sigma(t, x_2(.), \omega)|^2 + \int_Z |c(t, x_1(.), z, \omega) - c(t, x_2(.), z, \omega)|^2 \pi(dz) \\
&\leqslant c^N(t)\rho^N(\|x_1 - x_2\|_t^2),
\end{aligned}$$

where $c^N(t) \geq 0$ *satisfies the condition* $\int_0^T c^N(t)dt < \infty$ *for each* $T < \infty$, *and* $\rho(u) \geqslant 0$, *as* $u \geqslant 0$, *is increasing, continuous and concave such that* $\int_{0+} du/\rho(u) = \infty$, *then the solution of (7.1) is pathwise unique.*

Since the conditions above are a little different from Lemma 115, we give a proof here.

Proof. Asume that $\{x_t^i\}_{t\geq 0}, i = 1, 2$ are two solutions of (7.1) with the same BM $\{w_t\}_{t\geq 0}$ and Poisson martingale measure $\widetilde{N}_k(dt, dz)$. Let

$$X_t = x_t^1 - x_t^2,$$

and

$\tau_N = \inf\{t \geq 0 : \|x_1\|_t + \|x_2\|_t > N\}$.

Then by Ito's formula and by the martingale inequality one sees that

$$\begin{aligned}
Z_{t\wedge\tau_N} &= E\|X\|_{t\wedge\tau_N}^2 \leqslant k_0' E \int_0^{t\wedge\tau_N} c(s)\rho\left(\|X\|_s^2\right) ds + \frac{1}{2} E\|X\|_{t\wedge\tau_N}^2 \\
&\leq \int_0^t c(s)\rho\left(Z_{s\wedge\tau_N}\right) ds + \frac{1}{2} Z_{t\wedge\tau_N}.
\end{aligned}$$

Hence by Lemma 116 for any $T < \infty$, $P - a.s.$ $Z_{t\wedge\tau_N} = 0, \forall t \in [0, T]$. Letting $N \to \infty$ one finds that $P - a.s.$

$$Z_t = 0, \forall t \in [0, T].$$

By the RCLL property of $\{x_t^i\}_{t\geq 0}, i = 1, 2$ the conclusion now follows. ■

The proof of the following theorem is also completely the same as that of Theorem 117. So we will not repeat it.

Theorem 197 *Assume that*
1°

$$b(t, x(.), \omega) : [0, \infty) \times D \times \Omega \longmapsto R^d$$

is a $\mathfrak{F}_t-$ adapted and measurable process such that $P - a.s.$

$$|b(t, x(.), \omega)| \leqslant c(t)(1 + \|x\|_t),$$

$$|\sigma(t, x(.), \omega)|^2 + \int_Z |c(t, x(.), z, \omega)|^2 \pi(dz) \leqslant c(t)(1 + \|x\|_t^2),$$

where $c(t)$ is non-negative and non-random such that for each $T < \infty$

$$\int_0^T c(t)dt < \infty;$$

2° $\forall x_1(.), x_2(.) \in D$

$$|b(t, x_1(.), \omega) - b(t, x_2(.), \omega)| \leqslant c(t) \|x_1 - x_2\|_t^2,$$

$$|\sigma(t, x_1(.), \omega) - \sigma(t, x_2(.), \omega)|^2 + \int_Z |c(t, x_1(.), z, \omega) - c(t, x_2(.), z, \omega)|^2 \pi(dz)$$
$$\leqslant c(t) \|x_1 - x_2\|_t^2,$$

where $c(t)$ satisfies the same conditions as in 1°;
3°

$$x_0 \in \mathfrak{F}_0, E|x_0|^2 < \infty.$$

Then (7.1) has a pathwise unique $\mathfrak{F}_t-$ adapted solution. In the case that

$b(t, x, \omega)$ and $\sigma(t, x, \omega)$ are $\mathfrak{F}_t^{w,\tilde{N}_k}-$ adapted, and $c(t, x, z, \omega)$ is $\mathfrak{F}_t^{w,\tilde{N}_k}-$ predictable, then the solution is also $\mathfrak{F}_t^{w,\tilde{N}_k}-$ adapted, i.e. it is a strong solution.

Similar to Theorem 120 we also have the following theorem.

Theorem 198 *If the condition 3° in Theorem 197 is weakened to be*
3°′ there is a given process $\xi_t \in \mathfrak{F}_t^{w,\tilde{N}_k}$ such that $E|\xi_t|^2 < \infty$;
and assume that $b(t, x(.), \omega)$ and $\sigma(t, x(.), \omega)$ are $\mathfrak{F}_t^{w,\tilde{N}_k}-$ adapted, and $c(t, x(.), z, \omega)$ is $\mathfrak{F}_t^{w,\tilde{N}_k}-$ predictable, then there exists a pathwise unique strong solution $x_t, t \geq 0$, satisfying the following SDE:

$$\begin{aligned} x_t &= \xi_t + \int_0^t b(s, x(.), \omega)ds + \int_0^t \sigma(s, x(.), \omega)dw_s \\ &+ \int_0^t \int_Z c(s-, x(.), z, \omega)\tilde{N}_k(ds, dz), \ t \geq 0 \end{aligned} \quad (7.2)$$

Later, when deriving a non-linear filtering equation, we will need a martingale representation theorem for a SDE with functional coefficients. However, such a representation theorem depends on the Girsanov type theorems discussed below. So we discuss them here. Suppose that x_t solves the following SDE with jumps and with functional coefficients on $(\Omega, \mathfrak{F}, (\mathfrak{F}_t), P)$:

$$\begin{cases} x_t = x_0 + \int_0^t b(s, x(.))ds + \int_0^t \sigma(s, x(.))dw_s \\ \qquad + \int_0^t \int_Z c(s-, x(.), z)\widetilde{N}_k(ds, dz), \end{cases} \tag{7.3}$$

where we make the following assumption that
(H)$_2$ $b(s, x) : [0, \infty) \times D \to R^d$
is jointly measurable and $\mathfrak{F}_t$−adapted.
Thus we have

Theorem 199 *Assume that* $\forall x = x(.) \in D$,
1° $|b(t,x)|^2 + |\sigma(t,x)|^2 + \int_Z |c(t,x,z)|^2 \pi(dz) \leq c_0(t)(1 + \|x\|_t^2)$,
where $c_0(t) \geq 0$ *is non-random such that* $\int_0^T c_0(t)dt < \infty$, *for each* $T < \infty$, *and write* $\|x\|_t = \sup_{s \leq t} |x(s)|$,
2° $\sigma^{-1}(t, x)$ *exists and is bounded; that is,* $|\sigma^{-1}(t,x)| \leq k_0$,
3° $E|x_0|^2 < \infty$.
Let
$d\widetilde{P}_t = z_t(\sigma^{-1}b)dP$,
where

$$z_t(-\sigma^{-1}b) = \exp[-\int_0^t \sigma^{-1}b(s,x) \cdot dw_s - \frac{1}{2}\int_0^t |\sigma^{-1}b(s,x)|^2 ds], \tag{7.4}$$

and x_t *is the solution of (7.3).*
Then
1) $Ez_t(\sigma^{-1}b) = 1$, $\forall t \geq 0$,
2) there exists a probability measure $\widetilde{P}$ *defined on* $(\Omega, \mathfrak{F})$ *such that*
$(d\widetilde{P}/dP)|_{\mathfrak{F}_t} = z_t(-\sigma^{-1}b)$
3) let
$\widetilde{w}_t = w_t + \int_0^t \sigma^{-1}b(s,x)ds$,
then $\widetilde{w}_t$ *is a* $\widetilde{P}$ *-BM,*
4) $\widetilde{N}_k(dt, dz)$ *is still a* $\widetilde{P}$ *-Poisson martingale measure with the same compensator* $\pi(dz)dt$,
5) x_t *also solves the following SDE on* $(\Omega, \mathfrak{F}, (\mathfrak{F}_t), \widetilde{P})$*:* $\forall t \geq 0$

$$x_t = x_0 + \int_0^t \sigma(s, x(.))d\widetilde{w}_s + \int_0^t \int_Z c(s-, x(.), z)\widetilde{N}_k(ds, dz). \tag{7.5}$$

The proof of this theorem is almost completely the same as that of Theorem 132. So we still omit it.

Now we discuss some other conditions on $b(t, x)$ to get the Girsanov type theorem.

Theorem 200 *If we replace the condition 1° in Theorem 199 by*
$1^{\circ\prime}$ $|\sigma(t,x)|^2 + \int_Z |c(t,x,z)|^2 \pi(dz) \leq c(t)(1 + \|x\|_t^2), \forall x \in D,$
where $c(t) \geq 0$ *is non-random such that* $\int_0^T c(t)dt < \infty$, *for each* $T < \infty$; *and*
$\int_0^T |b(t,x(.))|^2 dt \leq k_T < \infty, \forall x \in D,$
where $k_T > 0$ *is a constant only depending on* T;
then the conclusion of Theorem 199 still holds.

Remark 201 *If we introduce Novikov's theorem as follows:*[99]
$Ee^{\frac{1}{2}\int_0^T |\widetilde{b}(s,\omega)|^2 ds} < \infty \Longrightarrow Ez_T(\widetilde{b}) = 1,$
where
$z_T(\widetilde{b}) = \exp[\int_0^t \widetilde{b}(s,\omega) \cdot dw_s - \frac{1}{2}\int_0^t \left|\widetilde{b}(s,\omega)\right|^2 ds],$
then the conclusion of Theorem 200 becomes trivial. However, to prove Novikov's theorem needs more pages of detailed analysis and also some properties of special functions, so we will give a direct proof of Theorem 200.

Proof. We only need to show that
$Ez_t(-\sigma^{-1}b) = 1, \forall t \geq 0.$
For any given $T < \infty$ let $b_N = bI_{|b|\leq N}$. Then by the previous theorem $Ez_t(-\sigma^{-1}b_N)(x) = 1, \forall t \geq 0$, where
$z_t(-\sigma^{-1}b_N)(x) = \exp[-\int_0^t \sigma^{-1}b_N(s,x) \cdot dw_s - \frac{1}{2}\int_0^t |\sigma^{-1}b_N(s,x)|^2 ds],$
and x_t is the solution of (7.3). However, by Lebesgue's dominated convergence theorem as $N \to \infty$,
$E\int_0^T |\sigma^{-1}b_N(s,x) - \sigma^{-1}b(s,x)|^2 ds \to 0.$
Hence as $N \to \infty$,
$E\sup_{t\leq T}\left|\int_0^t \sigma^{-1}b_N(s,x)\cdot dw_s - \int_0^t \sigma^{-1}b(s,x)\cdot dw_s\right|^2 \to 0.$
Take a subsequence denoted by $\{N\}$ again, such that as $N \to \infty$,
$\left|\int_0^t \sigma^{-1}b_N(s,x)\cdot dw_s - \int_0^t \sigma^{-1}b(s,x)\cdot dw_s\right| \to 0,$
and
$\left|\int_0^t |\sigma^{-1}b_N(s,x)|^2 ds - \int_0^t |\sigma^{-1}b(s,x)|^2 ds\right| \to 0,$
uniformly w.r.t. $t \in [0,T]$, $P-a.s.$ Notice that
$\sup_n E\left|z_T(-\sigma^{-1}b_N)\right|^2$
$= Ee^{-\int_0^t 2\sigma^{-1}b_N(s,x)\cdot dw_s - \frac{1}{2}\int_0^t 4|\sigma^{-1}b_N(s,x)|^2 ds} \cdot e^{\int_0^t |\sigma^{-1}b_N(s,x)|^2 ds}$
$\leq e^{k_0k_T}Ee^{-\int_0^t 2\sigma^{-1}b_N(s,x)\cdot dw_s - \frac{1}{2}\int_0^t 4|\sigma^{-1}b_N(s,x)|^2 ds} \leq e^{k_0k_T}.$
Hence $\left\{z_T(-\sigma^{-1}b_N)\right\}_{N=1}^{\infty}$ is uniformly integrable. Therefore as $N \to \infty$,
$Ez_T(-\sigma^{-1}b_N) \to Ez_T(-\sigma^{-1}b).$
So $Ez_T(-\sigma^{-1}b) = 1.$ ∎

Later, when deriving the non-linear filtering equation we also need some martingale representation theorems for the SDE with functional coefficients as follows:

$$\begin{cases} x_t = x_0 + \int_0^t b(s, x(.))ds + \int_0^t \sigma(s, x(.))dw_s \\ \qquad + \int_0^t c(s-, x(.))d\widetilde{N}_s, \end{cases} \tag{7.6}$$

where for simplicity assume that $\widetilde{N}_t$ is a 1−dimensional centralized Poisson process with constant density λ such that $EN_t = E\widetilde{N}_t + \lambda t$. (In this case we also say that $\widetilde{N}_t$ has a compensator λt). This SDE can be seen as a special case of (7.1). In fact, in (7.1) if we assume that $\pi(Z) = \lambda$, $c(s-, x(.)) = c(s-, x(.), z)$ does not depend on z, then from (7.1) we get (7.6).

For such a SDE we can relax condition $1^{\circ'}$ in Theorem 200 to get a better Girsanov type theorem as follows:

Theorem 202 *If we relax the condition $1^{\circ'}$ in Theorem 200 by*
$1^{\circ''}$ $|\sigma(t,x)|^2 + |c(t,x,z)|^2 \leq c(t)(1 + \|x\|_t^2), \forall x \in D,$
where $c(t) \geq 0$ is non-random such that $\int_0^T c(t)dt < \infty$, for each $T < \infty$; and
$P(\int_0^T |b(t,x(.))|^2 dt < \infty) = 1, \forall x \in D;$
and assume that the condition 2° in Theorem 197 holds; then the conclusion of Theorem 199 still holds.

Remark 203 *This theorem cannot be derived from Novikov's theorem. (See remark 201).*

Proof. By assumptions there exists a pathwise unique strong solution y_t satisfying
$y_t = x_0 + \int_0^t \sigma(s, y(.))dw_s + \int_0^t c(s-, y(.))d\widetilde{N}_s, t \geq 0.$
For each $T < \infty$ and $z \in C$ let
$\tau_n(z) = T \wedge \inf\left\{t > 0 : \int_0^t |(\sigma^{-1}b)(s, z(.))|^2 dt > n\right\},$
Then by Theorem 198 there exists a pathwise unique strong solution x_t^n of the following SDE: $\forall t \in [0,T]$
$x_t^n = x_{\int_0^t I_{s\leq\tau_n(x)}ds} + \int_0^t (1 - I_{s\leq\tau_n(x)})\sigma(s, x^n(.))dw_s$
$+ \int_0^t (1 - I_{s\leq\tau_n(x)})c(s-, x^n(.))d\widetilde{N}_s,$
where x_t is the solution of (7.3). Obviously, $x_t^n = x_t$, as $t \leq \tau_n(x)$. On the other hand, by Ito's formula one easily sees that as $T \geq t > \tau_n(x)$
$dx_t^n = \sigma(t, x^n(.))dw_t + c(t-, x^n(.))d\widetilde{N}_t.$
However, if $\tau_n(x^n) > \tau_n(x)$, then by the definition of $\tau_n(x^n)$ we must have $b(s, x_s^n) = 0$, as $s \in [\tau_n(x), \tau_n(x^n)]$. Hence in any case one has that
$dx_t^n = b(t, x^n(.))I_{t\leq\tau_n(x^n)}dt + \sigma(t, x^n(.))dw_t + c(t-, x^n(.))d\widetilde{N}_t.$
Now since $\forall T < \infty$
$\int_0^T |(\sigma^{-1}b)(t, x^n(.))I_{t\leq\tau_n(x^n)}|^2 dt \leq n.$
Hence by the previous theorem
$E(Z_{T\wedge\tau_n(x^n)}(-\sigma^{-1}b)(x^n)) = E(Z_T(-\sigma^{-1}bI_{t\leq\tau_n(x^n)})(x^n)) = 1,$
where
$Z_T(-\sigma^{-1}b)(x^n)$

$= e^{-\int_0^{T\wedge\tau_n(x^n)}\sigma^{-1}b(s,x^n)\cdot dw_s - \frac{1}{2}\int_0^{T\wedge\tau_n(x^n)}\left|\sigma^{-1}b(s,x^n)\right|^2 ds}$
$= \exp[-\int_0^{T\wedge\tau_n(x^n)}\sigma^{-1}b(s,x^n)\cdot\sigma^{-1}(s,x^n)dx_s^n$
$+\frac{1}{2}\int_0^{T\wedge\tau_n(x^n)}\left|\sigma^{-1}b(s,x^n)\right|^2 ds + \sum_{0<s\leq T\wedge\tau_n(x^n)}\sigma^{-1}(s,x^n)\triangle x_s^n$
$-\lambda\int_0^{T\wedge\tau_n(x^n)}(\sigma^{-1}c)(s,x^n)ds]$.
Moreover, x_t^n satisfies the following SDE: $\forall t\in[0,T]$
$dx_t^n = \sigma(t,x^n(.))d\widetilde{w}_t^n + c(t-,x^n(.))d\widetilde{N}_t,\ x_0^n = x_0;\ \widetilde{P}^n - a.s.$
where $\widetilde{w}_t^n = w_t + \int_0^t(\sigma^{-1}b)(s,x^n(.))I_{s\leq\tau_n(x)}ds$ is a $\widetilde{P}^n$–BM, and $\widetilde{N}_t$ is a $\widetilde{P}^n$–centralized Poisson process still with the same compensator λt, and $d\widetilde{P}^n = Z_{T\wedge\tau_n(x^n)}(-\sigma^{-1}b)(x^n)dP$. This means that
$\frac{d\widetilde{P}^n}{dP} = Z_{T\wedge\tau_n(x^n)}(-\sigma^{-1}b)(x^n)$.
Notice that the pathwise uniqueness implies the uniqueness in probability laws (the weak uniqueness). One finds that $Py^{-1} = \widetilde{P}^n(x^n)^{-1}$, where Py^{-1} is the probability measure on the measurable space $(C_T,\mathfrak{B}(C_T))$ induced by the process $\{y_t\}$ with respect to the probability P, etc. Let us write $\mu_y = Py^{-1}, \mu^n = \widetilde{P}^n(x^n)^{-1}$, $\mu_{x^n} = P(x^n)^{-1}$ and $\mu_x = Px^{-1}$, where x_t is the solution of (7.3) with $c(t,x,z)=0$. Then, $\forall\Gamma\in\mathfrak{B}(C_T)$,
$\mu_y(\Gamma\cap\{z\in C_T:\tau_n(z)=T\}) = \mu^n(\Gamma\cap\{z\in C_T:\tau_n(z)=T\})$
$= \widetilde{P}^n\{\omega: x^n\in\Gamma,\tau_n(x^n)=T\} = \int_{\{\omega:x^n\in\Gamma,\tau_n(x^n)=T\}}\frac{d\widetilde{P}^n}{dP}(x^n)dP(\omega)$
$= \int_{\Gamma\cap\{z\in C_T:\tau_n(z)=T\}} Z_{T\wedge\tau_n(z)}(-\sigma^{-1}b)(z)d\mu_{x^n}(z)$.
So $\frac{d\mu_y}{d\mu_{x^n}}(z) = Z_{T\wedge\tau_n(z)}(-\sigma^{-1}b)(z)$, on $z\in\{z\in C_T:\tau_n(z)=T\}$. Furthermore, $\forall\Gamma\in\mathfrak{B}(C_T)$,
$\mu_x(\Gamma) = \lim_{n\to\infty}\mu_x(\Gamma\cap\{z\in C_T:\tau_n(z)=T\})$
$= \lim_{n\to\infty}\mu_{x^n}(\Gamma\cap\{z\in C_T:\tau_n(z)=T\})$
$= \lim_{n\to\infty}\int_{\Gamma\cap\{z\in C_T:\tau_n(z)=T\})}(Z_{T\wedge\tau_n(z)}(-\sigma^{-1}b)(z))^{-1}\mu_y(dz)$
$= \lim_{n\to\infty}\int_{\Gamma\cap\{z\in C_T:\tau_n(z)=T\})}(Z_T(-\sigma^{-1}b)(z))^{-1}\mu_y(dz)$
$= \int_\Gamma(Z_T(-\sigma^{-1}b)(z))^{-1}\mu_y(dz)$.
This shows that $\frac{d\mu_x}{d\mu_y}(y) = (Z_T(-\sigma^{-1}b)(y))^{-1}$, where
$Z_T(-\sigma^{-1}b)(y) = \exp[-\int_0^T\sigma^{-1}b(s,y)\cdot\sigma^{-1}(s,y)dy_s + \frac{1}{2}\int_0^T\left|\sigma^{-1}b(s,y)\right|^2 ds$
$+\sum_{0<s\leq T\wedge\tau_n(x^n)}\sigma^{-1}(s,y)\triangle y_s - \lambda\int_0^{T\wedge\tau_n(x^n)}(\sigma^{-1}c)(s,y)ds]$.
Notice that
$EZ_T(-\sigma^{-1}b)(x) = \int_\Omega Z_T(-\sigma^{-1}b)(x)dP = \int_{C_T}Z_T(-\sigma^{-1}b)(z)\mu_x(dz)$
$= \int_{C_T}\frac{d\mu_y}{d\mu_x}(z)\mu_x(dz) = \int_{C_T}\mu_y(dz) = 1.$
Hence, let for $\forall t\geq 0$
$(d\widetilde{P}/dP)|_{\mathfrak{F}t} = z_t(-\sigma^{-1}b)(x)$, and
$\widetilde{w}_t = w_t + \int_0^t\sigma^{-1}b(s,x)ds$,
then $\widetilde{w}_t, t\geq 0$ is a $\widetilde{P}$ -BM, and $\widetilde{N}_t$ is a $\widetilde{P}$–centralized Poisson process still with the same compensator λt. Therefore, SDE
$x_t = x_0 + \int_0^t b(s,x(.))ds + \int_0^t\sigma(s,x(.))dw_s$
$+\int_0^t c(s-,x(.))d\widetilde{N}_s, t\geq 0;\ P - a.s.$

can also be rewritten as
$x_t = x_0 + \int_0^t \sigma(s, x(.))d\widetilde{w}_s \int_0^t c(s-, x(.))d\widetilde{N}_s, t \geq 0;\ \widetilde{P} - a.s.$ ∎

7.2 Martingale Representation Theorems (Functional Coefficient Case)

Consider SDE (7.6). We have the

Theorem 204 *Assume that*
1° $\|\sigma(t,x)\|^2 + |c(t,x)|^2 \leq c_0(t)(1 + \|x\|_t^2),\ \forall x = x(.) \in D,$
where $c_0(t) \geq 0$ *is non-random such that* $\int_0^T c_0(t)dt < \infty$*, for each* $T < \infty$*; and* $c^{-1}(t,x)$ *exists, and denote the norm of matrix* $\sigma(t,x)$ *by* $\|\sigma(t,x)\|$*; and for any* $T < \infty, \forall x = x(.) \in D,$

$$P(\int_0^T |b(t,x)|^2\, dt < \infty) = 1 \tag{7.7}$$

where $k_T \geq 0$ *is a constant depending on* T *only;*
2° σ^{-1} *exists such that*
$\|\sigma^{-1}\| \leq k_0,$
3° $\|\sigma(t,x) - \sigma(t,y)\|^2 + |c(t,x) - c(t,y)|^2 \leq k_0(\|x - y\|_t^2),$
for all $x, y \in D,\ t \geq 0,$
Now assume that x_t *satisfies (7.6). If* ξ_t *is a RCLL* $R^{d_1}-$*valued square integrable* $\mathfrak{F}_t^x-$*martingale, where* $\mathfrak{F}_t^x = \sigma(x_s, s \leq t)$*, then there exists an*
$f(t,\omega) : [0,\infty) \times \Omega \to R^{d_1 \otimes d};\ g(t,\omega) : [0,\infty) \times \Omega \to R^{d_1},$
which is $\mathfrak{F}_t^x-$*adapted,* $\mathfrak{F}_t^x-$*predictable, respectively, satisfying for any* $0 \leq T < \infty$
$E[\int_0^T \|f(t,\omega)\|^2\, dt + \int_0^T |g(t,\omega)|^2\, dt] < \infty,$
such that for all $t \geq 0$
$\xi_t = \xi_0 + \int_0^t f(s,\omega)dw_s + \int_0^t g(s,\omega)d\widetilde{N}_s.$
Moreover, the representation is unique.

Proof. Let
$z_{0,t}^w = z_{0,t}^w(-\sigma^{-1}b) = \exp(-\int_0^t \sigma^{-1}b(s,x) \cdot dw_s - \frac{1}{2}\int_0^t \|\sigma^{-1}b(s,x)\|^2\, ds),$
$d\widetilde{P}_t = z_{0,t}^w dP,$
where x_t is the solution of (7.6) given in the assumption. Then by Theorem 202 above there exists a probability measure $\widetilde{P}$ on $(\Omega, \mathfrak{F})$ such that
$\widetilde{P}\big|_{\mathfrak{F}_t^\xi} = \widetilde{P}_t,$
and
$\widetilde{w}_t = w_t + \int_0^t \sigma^{-1}b(s,x)ds,\ \ 0 \leq t$
is a BM under the probability measure $\widetilde{P}$, moreover, $\widetilde{N}_t$ is still a centralized Poisson process under $\widetilde{P}$ with the same compensator λt. Hence x_t also

satisfies that $\widetilde{P}-a.s.\forall t\geq 0$

$$x_t=x_0+\int_0^t\sigma(s,x(.))d\widetilde{w}_s+\int_0^t c(s-,x(.))d\widetilde{N}_s. \qquad (7.8)$$

Since by Theorem 197 the solution of (7.8) is a (pathwise unique) strong solution. Hence

$\mathfrak{F}_t^x\subset\mathfrak{F}_t^{\widetilde{w},N}$, for all $t\geq 0$.

On the other hand, by (7.8)

$N_t=\sum_{0<s\leq t}c_{s-}(x)^{-1}\triangle x_s$

is $\mathfrak{F}_t^x-$adapted, and so is $\widetilde{N}_t=N_t-\lambda t$. However, $\widetilde{w}_t$ is obviously $\mathfrak{F}_t^{x,\widetilde{N}}-$adapted. Hence

$\mathfrak{F}_t^x=\mathfrak{F}_t^{\widetilde{w},\widetilde{N}}$.

Let us show that $\xi_t\cdot(z_{0,t}^w)^{-1}$ is a $\mathfrak{F}_t^{\widetilde{w},\widetilde{N}}-$ adapted locally square integrable martingale under probability $\widetilde{P}$. Indeed, for all $A\in\mathfrak{F}_s^{\widetilde{w},\widetilde{N}}, s\leq t$

$\int_A\xi_t\cdot(z_{0,t}^w)^{-1}d\widetilde{P}=\int_\Omega I_A\xi_t\cdot E(z_{t,T}^w\left|\mathfrak{F}_t^{\widetilde{w},\widetilde{N}}\right.)\,dP=\int_A\xi_s\cdot dP$

$=\int_A\xi_s\cdot(z_{0,s}^w)^{-1}d\widetilde{P}.$

This shows that $\left\{\xi_t\cdot(z_{0,t}^w)^{-1},\mathfrak{F}_t^{\widetilde{w},\widetilde{N}}\right\}_{t\geq 0}$ is a $\widetilde{P}-$ martingale. On the other hand,

$\sup_{t\leq T}E_{\widetilde{P}}\left|\xi_t\cdot(z_{0,t}^w)^{-1}\right|=\sup_{t\leq T}E\left|\xi_t\right|=\sup_{t\leq T}E\left|E(\xi_T\left|\mathfrak{F}_t^{\widetilde{w},\widetilde{N}}\right.)\right|$

$\leq E\left|\xi_T\right|<\infty,$

since ξ_t is a $(\mathfrak{F}_t^{\widetilde{w},\widetilde{N}},P)-$martingale, and we have applied the Jensen inequality, and

$E(z_{t,T}^w\left|\mathfrak{F}_t^{\widetilde{w},\widetilde{N}}\right.)=E[E(z_{t,T}^w\left|\mathfrak{F}_t^\xi\right.)\left|\mathfrak{F}_t^{\widetilde{w},\widetilde{N}}\right.]=1$, for all $0\leq t\leq T$.

Moreover, let

$\sigma_N=\inf\left\{t\in[0,T]:\left|(z_{0,t}^w)^{-1}\right|+\left|\xi_t\right|>N\right\}.$

Then $\sigma_N\uparrow\infty$, as $N\uparrow\infty$, since $(z_{0,t}^w)^{-1}$ is a continuous process, and ξ_t is a RCLL (right continuous with left limit) process. Now since ξ_t is a square integrable $(\mathfrak{F}_t^{\widetilde{w},\widetilde{N}},P)-$martingale, we also have

$\sup_{t\leq T}E_{\widetilde{P}}\left|\xi_{t\wedge\sigma_N}\cdot(z_{0,t\wedge\sigma_N}^w)^{-1}\right|^2\leq N\sup_{t\leq T}E_{\widetilde{P}}[\left|\xi_{t\wedge\sigma_N}\right|^2\cdot(z_{0,t\wedge\sigma_N}^w)^{-1}]$

$=N\sup_{t\leq T}E\left|\xi_{t\wedge\sigma_N}\right|^2\leq N\sup_{t\leq T}E\left|\xi_t\right|^2<\infty.$

This shows that $\xi_t\cdot(z_{0,t}^w)^{-1}$ is locally $\widetilde{P}-$ square integrable. Therefore $\xi_t\cdot(z_{0,t}^w)^{-1}$ is a locally square integrable $(\mathfrak{F}_t^{\widetilde{w},\widetilde{N}},\widetilde{P})-$martingale. Now by the martingale representation theorem for each given $0\leq T<\infty$ there exists an

$f_T(t,\omega):[0,T]\times\Omega\to R^{d_1\otimes d},$

which is $\mathfrak{F}_t^{\widetilde{w},\widetilde{N}}-$adapted, satisfying $\widetilde{P}(\int_0^T\left\|f_T(t,\omega)\right\|^2dt<\infty)=1$, and there exists a

$g_T(t,\omega):[0,T]\times\Omega\to R^{d_1},$

which is $\mathfrak{F}_t^{\widetilde{w},\widetilde{N}}$-predictable, and satisfies $\widetilde{P}(\int_0^T |g_T(t,\omega)|^2 dt < \infty) = 1$, such that $\widetilde{P} - a.s.$

$\overline{\xi}_t = \xi_t \cdot (z_{0,t}^w)^{-1} = \overline{\xi}_0 + \int_0^t f_T(s,\omega)d\widetilde{w}_s + \int_0^t g_T(s,\omega)d\widetilde{N}_s$, for all $t \in [0,T]$.

Since $\widetilde{P} \sim P$. Hence $P - a.s.$

$\overline{\xi}_t = \overline{\xi}_0 + \int_0^t f_T(s,\omega) \cdot (\sigma^{-1}b(s,x))ds + \int_0^t f_T(s,\omega)dw_s$
$+ \int_0^t g_T(s,\omega)d\widetilde{N}_s$, for all $t \in [0,T]$.

Notice that $\xi_t = \overline{\xi}_t \cdot (z_{0,t}^w)$. Applying Ito's formula, one verifies that $P - a.s$ for all $t \in [0,T]$

$\xi_t = \xi_0 + \int_0^t z_{0,s}^w f_T(s,\omega)dw_s$
$+ \int_0^t z_{0,s}^w f_T(s,\omega)(\sigma^{-1}b(s,x))ds - \int_0^t z_{0,s}^w f_T(s,\omega)(\sigma^{-1}b(s,x))ds$
$+ \int_0^t z_{0,s}^w g_T(s,\omega)d\widetilde{N}_s - \int_0^t \xi_s(\sigma^{-1}b(s,x) \cdot dw_s)$.

Hence

$$\xi_t = \xi_0 + \int_0^t \overline{f}_T(s,\omega) \cdot dw_s + \int_0^t \overline{g}_T(s,\omega)d\widetilde{N}_s, \tag{7.9}$$

where (denote $A^* =$ the transpose of A)

$\overline{f}_T(s,\omega) = z_{0,s}^w f_T(s,\omega) - \xi_s \cdot (\sigma^{-1}b(s,x))^*$,
$\overline{g}_T(s,\omega) = z_{0,s}^w g_T(s,\omega)$.

Moreover, it is evident that $\overline{f}_T(t,\omega)$ is $\mathfrak{F}_t^x$-adapted,

$P(\int_0^T \|\overline{f}_T(t,\omega)\|^2 dt < \infty) = 1$,

and $\overline{g}_T(t,\omega)$ is $\mathfrak{F}_t^x$-predictable,

$P(\int_0^T |\overline{g}_T(t,\omega)|^2 dt < \infty) = 1$.

However, by assumption, ξ_t is R^{d_1}-valued and square integrable. Hence applying the representation (7.9) yields

$E\int_0^{t\wedge\tau_N} [\|\overline{f}_T(t,\omega)\|^2 + |\overline{g}_T(t,\omega)|^2]dt$
$= E|\xi_{t\wedge\tau_N} - \xi_0|^2 \leq 2[E|\xi_T|^2 + E|\xi_0|^2] < \infty$,

where

$\tau_N = \inf\left\{t \in [0,T] : \int_0^t \|\overline{f}_T(s,\omega)\|^2 + |\overline{g}_T(s,\omega)|^2]ds > N\right\}$,
$\tau_N = T$, for $\inf\{\Phi\}$.

Now as $N \to \infty$, $\tau_N \to T$. Hence by Fatou's lemma

$$E\int_0^T [\|\overline{f}_T(t,\omega)\|^2 + |\overline{g}_T(t,\omega)|^2]dt \leq 2[E|\xi_T|^2 + E|\xi_0|^2] < \infty. \tag{7.10}$$

Recall that, by the definition of a stochastic integral, if $f^i(t,\omega), g^i(t,z,\omega)$, $i = 1,2$, are such that

$$E\int_0^T [\|f^1(t,\omega) - f^2(t,\omega)\|^2 + |g^1(t,\omega) - g^2(t,\omega)|^2]dt = 0, \tag{7.11}$$

then $P - a.s.$ for all $t \in [0,T]$

$$\begin{cases} \int_0^t f^1(s,\omega)dw_s = \int_0^t f^1(s,\omega)dw_s, \\ \int_0^t g^1(s,\omega)d\widetilde{N}_s = \int_0^t g^2(s,\omega)d\widetilde{N}_s, \end{cases}$$

and f^1, g^1 are not distinguished from f^2, g^2 in the above integrals, and so we denoted by

$f^1 \doteq f^2,\ g^1 \doteq g^2.$

Now if there exist $f_T^i, g_T^i, i = 1,2$, for each i, with both satisfying (7.9) and (7.10), then (7.11) holds. Hence

$f_T^1 \doteq f_T^2,\ g_T^1 \doteq g_T^2.$

Now define

$$f(t,\omega) = \begin{cases} \overline{f}_1(t,\omega), & \text{as } t \in [0,1], \\ \overline{f}_n(t,\omega), & \text{as } t \in (n, n+1], n = 1,2,...., \end{cases}$$

$$g(t,\omega) = \begin{cases} \overline{g}_1(t,\omega), & \text{as } t \in [0,1], \\ \overline{g}_n(t,\omega), & \text{as } t \in (n, n+1], n = 1,2,.... \end{cases}$$

Clearly, $f(t,\omega)$ is $\mathfrak{F}_t^x$–adapted, and $g(t,\omega)$ is $\mathfrak{F}_t^x$–predictable such that $P-a.s$ for all $t \geq 0$

$\xi_t = \xi_0 + \int_0^t f(s,\omega) \cdot dw_s + \int_0^t g(s,\omega) d\widetilde{N}_s.$

Moreover, for any $0 \leq T < \infty$

$E[\int_0^T \|f(t,\omega)\|^2 dt + \int_0^T |g(t,\omega)|^2 dt] < \infty.$ ∎

7.3 Non-linear Filtering Equation

After the above preparation now we can discuss the filtering problems. Let us consider the following partially observed Stochastic Differential System:

$$\begin{cases} h_t = h_0 + \int_0^t H_s ds + y_t, \\ \xi_t = \xi_0 + \int_0^t A_s(\omega) ds + \int_0^t B_s(\xi) dw_s + \int_0^t C_{s-}(\xi) d\widetilde{N}_s, \end{cases} \tag{7.12}$$

where y_t is a d–dimensional RCLL (right continuous with left limit) square integrable martingale with $y_0 = 0$, and we call h_t a signal process, ξ_t an observable process, and they are all $d-$ dimensional processes.

In this section we are going to derive the SDE satisfied by the non-linear filtering processes $E(h_t | \mathfrak{F}_t^\xi)$, where $\mathfrak{F}_t^\xi = \sigma(\xi_s; s \leq t)$. Naturally, in (7.12) we assume that h, H, y, ξ, A and C are d–dimensional vectors, B is a $d \times d$ matrix, and on a given probability space $(\Omega, \mathfrak{F}, (\mathfrak{F}_t), P)$ w_t is a d dimensional standard $\mathfrak{F}_t$–Brownian Motion process, and for simplicity we assume that $\widetilde{N}_t$ is a 1 dimensional $\mathfrak{F}_t$–centralized Poisson process with the intensity λdt such that

$\widetilde{N}_t = N_t - \lambda dt,$

where $\lambda > 0$ is a constant, N_t is a Poisson process with $EN_t = \lambda dt$. We always make the following assumption

(H1) $A_t(\omega) : [0,\infty) \times \Omega \to R^d$

is jointly measurable and $\mathfrak{F}_t$–adapted,

$B_t(x) = B(t, x(r), r \leq t) : [0,\infty) \times D \to R^{d \otimes d}$

is $\mathfrak{B}([0,\infty)) \times \mathfrak{B}(D)/\mathfrak{B}(R^{d\otimes d})$–measurable, and for each $t \geq 0$ $B_t(x)$ is $\mathfrak{B}_t(D)/\mathfrak{B}(R^d)$–measurable;

$C_t(x) = C(t, x(r), r \leq t) : [0, \infty) \times D \to R^d$,
has the same properties as those of $B_t(x)$;

(Hence. if we set $C_{t-}(\xi) = C(t, \xi(r), r < t)$, where $\xi(r)$ is a $\mathfrak{F}_t-$adapted RCLL process, then $C_{t-}(\xi)$ is $\mathfrak{F}_t-$predictable. Similar notation is also made for $B_{t-}(\xi)$).

We have the following

Theorem 205 *Assume that (h_t, ξ_t) is a $\mathfrak{F}_t-$adapted solution of (7.12) and for all $T > 0$*

$P(\int_0^T |A_s(\omega)| \, ds < \infty) = 1$, $P(\int_0^T |B_s(\xi)|^2 \, ds < \infty) = 1$,

$P(\int_0^T |C_s(\xi)|^2 \, \lambda ds < \infty) = 1$,

and B^{-1} exists such that

$|B^{-1}| \leq k_0$.

Then

1) $\overline{M}_t = \int_0^t B_{s-}^{-1}(\xi)[d\xi_s - E(A_s \,|\mathfrak{F}_s^\xi)ds]$,

is a locally square integrable martingale on the probability space $(\Omega, \mathfrak{F}, (\mathfrak{F}_t^\xi), P)$. (Recall that $\mathfrak{F}_t^\xi = \sigma(\xi_s, s \leq t)$). The 2nd equation of (7.12) can be rewritten on the probability space $(\Omega, \mathfrak{F}, (\mathfrak{F}_t^\xi), P)$ as

$$\xi_t = \xi_0 + \int_0^t E(A_s \,|\mathfrak{F}_s^\xi)ds + \int_0^t B_{s-}(\xi)d\overline{M}_s. \tag{7.13}$$

2) Furthermore, if $C_{s-}^{-1}(\xi)$ exists, then

$\overline{M}_t = \overline{w}_t + \int_0^t B_{s-}^{-1}(\xi)C_{s-}(\xi)d\widetilde{N}_s$,

where

$\overline{w}_t = \int_0^t B_s^{-1}(\xi)(A_s - E(A_s \,|\mathfrak{F}_s^\xi))ds + w_t = \overline{M}_t^c$,

the continuous martingale part of $\overline{M}_t$, is a BM on $(\Omega, \mathfrak{F}, (\mathfrak{F}_t^\xi), P)$; and

$\widetilde{N}_t = \int_0^t C_{s-}^{-1}(\xi)B_{s-}(\xi)d\overline{M}_s^d$,

where $\overline{M}_t^d$ is the pure discontinuous martingale part of $\overline{M}_t$, and $\widetilde{N}_t$ is still a centralized Poisson random process with intensity function λt on $(\Omega, \mathfrak{F}, (\mathfrak{F}_t^\xi), P)$ such that

$\widetilde{N}_t = N_t - \lambda dt$,

that is, N_t is a Poisson random process such that $EN_t = \lambda dt$. Usually, we call $\overline{w}_t$ and $\widetilde{N}_t$ the innovation processes for w_t and $\widetilde{N}_t$. The 2nd equation of (7.12) can be rewritten on the probability space $(\Omega, \mathfrak{F}, (\mathfrak{F}_t^\xi), P)$ as

$$\xi_t = \xi_0 + \int_0^t E(A_s \,|\mathfrak{F}_s^\xi)ds + \int_0^t B_s(\xi)d\overline{w}_s + \int_0^t C_{s-}(\xi)d\widetilde{N}_s, \tag{7.14}$$

First we notice that

$E \int_0^T |B_s(\xi) - B_{s-}(\xi)|^2 \, ds$

$= E \int_0^T |B(s, \xi(r), r \leq s) - B(s, \xi(r), r < s)|^2 \, ds = 0$,

since ξ_t is a RCLL function only with countable discontinuous points in t for a.s. ω. Secondly, we also notice that the coefficient $E(A_s \,|\mathfrak{F}_s^\xi)$ is always a functional of $\xi_s, s \leq t$. That is why we need to consider the SDE with functional coefficients in the previous sections. Now let us prove Theorem 205. From now on let us write $\pi_t(A) = E(A_t|\mathfrak{F}_t^\xi)$, etc.

Proof. 1): First, we make a stronger assumption that

$$E\int_0^T |A_s(\omega)|\, ds < \infty, E\int_0^T |B_s(\xi)|^2\, ds < \infty, E\int_0^T |C_s(\xi)|^2 \lambda ds < \infty. \tag{7.15}$$

Let us show that

$$d\xi_t = \pi_t(A)dt + dm_t, \tag{7.16}$$

where $(m_t, \mathfrak{F}_t^\xi)_{t\geq 0} \in \mathcal{M}^{2,loc}$. In fact, comparing this equation with the 2nd equation of (7.12), we have

$$\begin{aligned} m_t &= \int_0^t [E(A_s(\omega)|\mathfrak{F}_t^\xi) - \pi_s(A)]ds + E((\int_0^t B_s(\xi)dw_s \\ &\quad + \int_0^t C_{s-}(\xi)d\widetilde{N}_s)|\mathfrak{F}_t^\xi) = m_t^1 + m_t^2. \end{aligned} \tag{7.17}$$

Notice that for any $t \leq u \leq T < \infty$

$E[m_u^1 - m_t^1|\mathfrak{F}_t^\xi] = E[\int_0^u [E(A_s(\omega)|\mathfrak{F}_u^\xi) - \pi_s(A)]ds$
$- \int_0^t [E(A_s(\omega)|\mathfrak{F}_t^\xi) - \pi_s(A)]ds|\mathfrak{F}_t^\xi] = E[\int_0^u E(A_s(\omega)|\mathfrak{F}_u^\xi)ds$
$- \int_0^t E(A_s(\omega)|\mathfrak{F}_t^\xi)ds|\mathfrak{F}_t^\xi] - E[\int_0^u \pi_s(A)ds - \int_0^t \pi_s(A)ds|\mathfrak{F}_t^\xi] = I_1 - I_2.$

However,

$I_1 = E[\int_0^u E(A_s(\omega)|\mathfrak{F}_u^\xi)ds - \int_0^t E(A_s(\omega)|\mathfrak{F}_t^\xi)ds|\mathfrak{F}_t^\xi]$
$= \int_0^u E(A_s(\omega)|\mathfrak{F}_t^\xi)ds - \int_0^t E(A_s(\omega)|\mathfrak{F}_t^\xi)ds = \int_t^u E(A_s(\omega)|\mathfrak{F}_t^\xi)ds$
$= \int_t^u E[E(A_s(\omega)|\mathfrak{F}_s^\xi)|\mathfrak{F}_t^\xi]ds = \int_t^u E[\pi_s(A)|\mathfrak{F}_t^\xi]ds,$

and

$I_2 = E[\int_t^u \pi_s(A)ds|\mathfrak{F}_t^\xi] = \int_t^u E[\pi_s(A)|\mathfrak{F}_t^\xi]ds.$

Hence

$E[m_u^1 - m_t^1|\mathfrak{F}_t^\xi] = 0.$

On the other hand,

$E[m_u^2 - m_t^2|\mathfrak{F}_t^\xi] = E[E(\int_0^u B_s(\xi)dw_s + \int_0^u C_{s-}(\xi)d\widetilde{N}_s|\mathfrak{F}_u^\xi)$
$-(E(\int_0^t B_s(\xi)dw_s + \int_0^t C_{s-}(\xi)d\widetilde{N}_s)|\mathfrak{F}_t^\xi)|\mathfrak{F}_t^\xi]$
$= E[E(\int_0^u B_s(\xi)dw_s|\mathfrak{F}_u^\xi) - E(\int_0^t B_s(\xi)dw_s|\mathfrak{F}_t^\xi)|\mathfrak{F}_t^\xi]$
$+E[E(\int_0^u C_{s-}(\xi)d\widetilde{N}_s|\mathfrak{F}_u^\xi) - E(\int_0^t C_{s-}(\xi)d\widetilde{N}_s)|\mathfrak{F}_t^\xi)|\mathfrak{F}_t^\xi] = I^1 + I^2.$

However, by $\mathfrak{F}_t^\xi \subset \mathfrak{F}_t$

$I^1 = E(\int_0^u B_s(\xi)dw_s|\mathfrak{F}_t^\xi) - E(\int_0^t B_s(\xi)dw_s|\mathfrak{F}_t^\xi)$
$= E(E(\int_t^u B_s(\xi)dw_s|\mathfrak{F}_t)|\mathfrak{F}_t^\xi) = 0,$

and

$I^2 = E(\int_0^u C_{s-}(\xi)d\widetilde{N}_s|\mathfrak{F}_t^\xi) - E(\int_0^t C_{s-}(\xi)d\widetilde{N}_s)|\mathfrak{F}_t^\xi)$
$= E(E(\int_t^u C_{s-}(\xi)d\widetilde{N}_s|\mathfrak{F}_t)|\mathfrak{F}_t^\xi) = 0.$

Hence

$E[m_u^2 - m_t^2|\mathfrak{F}_t^\xi] = 0.$
So, for any $t \leq u \leq T < \infty$
$E[m_u - m_t|\mathfrak{F}_t^\xi] = 0.$
Hence m_t, $t \in [0,T]$ is a $\mathfrak{F}_t^\xi-$ martingale. Since $T < \infty$ can be chosen arbitrarily, we have that $(m_t, \mathfrak{F}_t^\xi)_{t\geq 0}$ is a martingale. The result $E|m_t|^2 < \infty$ can be derived by working the stronger assumption (7.15). Now let

$$\overline{M}_t = \int_0^t B_{s-}^{-1}(\xi)dm_s. \tag{7.18}$$

Since $|B_t^{-1}(\xi)| \leq k_0$, we have $(\overline{M}_t, \mathfrak{F}_t^\xi)_{t\geq 0} \in \mathcal{M}^2$. Therefore $d\xi_t = \pi_t(A)dt + B_{t-}(\xi)d\overline{M}_t$, and 1) is proved under the stronger assumption (7.15). However, in the general case for all $T > 0$
$P(\int_0^T |A_s(\omega)|\, ds < \infty) = 1$, $P(\int_0^T |B_s(\xi)|^2\, ds < \infty) = 1$,
$P(\int_0^T |C_s(\xi)|^2 \lambda ds < \infty) = 1$,
1) is also true by taking the limit through a series of stopping times $\sigma_n \uparrow \infty$.
2): Without losing any generality we may assume that the stronger assumption (7.15) holds. Notice that
$\triangle\xi_t = C_{t-}(\xi) \triangle N_t.$
Hence
$N_t = \sum_{0<s\leq t} C_{s-}(\xi)^{-1} \triangle \xi_s$
is $\mathfrak{F}_t^\xi-$adapted, and so is $\widetilde{N}_t$. However, we already know that $(\widetilde{N}_t, \mathfrak{F}_t)_{t\geq 0}$ is a centralized Poisson process with intensity λt, and so is $(\widetilde{N}_t, \mathfrak{F}_t^\xi)_{t\geq 0}$ with the same intensity. In fact, $\forall s < t$,
$E[(N_t - \lambda t|\mathfrak{F}_s^\xi] = E[E[(N_t - \lambda t|\mathfrak{F}_s]|\mathfrak{F}_s^\xi] = E[N_s - \lambda s)|\mathfrak{F}_s^\xi]$
$= N_s - \lambda s,$
where we have used the fact that $\widetilde{N}_t \in \mathfrak{F}_t^\xi$, $\mathfrak{F}_t^\xi \subset \mathfrak{F}_t$, and $\widetilde{N}_t = N_t - \lambda t$. So $N_t = \widetilde{N}_t + \lambda t$ is still a $\mathfrak{F}_t^\xi-$Poisson process with the same density λ, (or say the same intensity λt). Now write
$\overline{w}_t = \int_0^t B_{s-}^{-1}(\xi)[d\xi_s - E(A_s|\mathfrak{F}_s^\xi)\, ds - C_{s-}(\xi)d\widetilde{N}_s]$
$= \int_0^t B_s^{-1}(\xi)(A_s - E(A_s|\mathfrak{F}_s^\xi))ds + w_t.$
Obviously, $\overline{w}_t$ is $\mathfrak{F}_t^\xi-$ adapted by the first equality. By Ito's formula $\forall\widetilde{\lambda} \in R^d$,
$e^{i\langle\widetilde{\lambda},\overline{w}_t-\overline{w}_s\rangle} - 1 = i\left\langle\widetilde{\lambda}, \int_s^t e^{i\langle\widetilde{\lambda},\overline{w}_r-\overline{w}_s\rangle}[B_r^{-1}(\xi)(A_r - E(A_r|\mathfrak{F}_r^\xi))dr + dw_r],\right\rangle$
$-\frac{|\widetilde{\lambda}|^2}{2}\int_s^t e^{i\langle\widetilde{\lambda},\overline{w}_r-\overline{w}_s\rangle}dr = i\left\langle\widetilde{\lambda}, \int_s^t e^{i\langle\widetilde{\lambda},\overline{w}_r-\overline{w}_s\rangle}[d\overline{M}_r - B_{s-}^{-1}(\xi)C_{r-}(\xi)d\widetilde{N}_r]\right\rangle$
$-\frac{|\widetilde{\lambda}|^2}{2}\int_s^t e^{i\langle\widetilde{\lambda},\overline{w}_r-\overline{w}_s\rangle}dr,$
where $\overline{M}_t$ and $\widetilde{N}_t$ are both square integrable $\mathfrak{F}_t^\xi-$martingales. Hence for any $G \in \mathfrak{F}_s^\xi, s \leq t$, $EI_G e^{i\langle\widetilde{\lambda},\overline{w}_t-\overline{w}_s\rangle}$ satisfies the following ordinary differential equation:
$dy_t = -\frac{|\widetilde{\lambda}|^2}{2}y_t dt,\ t \geq s;\ y_s = P(G).$
Solving it gives

$EI_G e^{i\langle \lambda, \overline{w}_t - \overline{w}_s \rangle} = P(G) e^{-|\lambda|^2 (t-s)/2}$,
that is, $\overline{w}_t$ is a $\mathfrak{F}_t^\xi$–BM. By the definitions of $\overline{w}_t$ and $\overline{M}_t$, and also from (7.16) one immediately sees that
$\overline{M}_t = \overline{w}_t + \int_0^t B_s^{-1}(\xi) C_s(\xi) d\widetilde{N}_s$.
Now (7.14) is easily derived. ■

In the following we are going to derive the non-linear filtering equation for the partially observed SDE system (7.12). More precisely, we are going to show the following theorem. (Recall that $\|x\|_t^2 = \sup_{s \leq t} |x_s|$).

Theorem 206 . *Assume that*
1° $|B_t(x)|^2 + |C(t,x)|^2 \leq c_0(t)(1 + \|x\|_t^2)$, $\forall x = x(.) \in D$,
where $c_0(t) \geq 0$ *is non-random such that* $\int_0^T c_0(t) dt < \infty$, *for each* $T < \infty$; *and* $C^{-1}(t,x)$ *exist and we still denote the norm of matrix* $\sigma(t,x)$ *by* $|\sigma(t,x)|$; *and for any* $T < \infty$,
$E \int_0^T |A_t(\omega)|^2 dt < \infty$;
2° B^{-1} *exists such that*
$|B^{-1}| \leq k_0$;
3° $|B(t,x) - B(t,y)|^2 + |C(t,x) - C(t,y)|^2 \leq k_0 \|x - y\|_t^2$,
for all $x, y \in D$, $t \geq 0$;
4° $E|h_0|^2 < \infty$, $E|\xi_0|^2 < \infty$,
$E \int_0^T |H_t|^2 dt < \infty$, $sup_{t \in [0,T]} E |h_t|^2 < \infty$, $\forall T < \infty$.
Then the optimal filter $\pi_t(h) = E[h_t | \mathfrak{F}_t^\xi]$, *or say, the best estimate in* $L^2(\Omega, \mathfrak{F}, (\mathfrak{F}_t^\xi)_{t \geq 0}, P)$, *for estimating* h_t *based on the information* $\mathfrak{F}_t^\xi$ *(that is, it is* $\mathfrak{F}_t^\xi$*– measurable), will satisfy the following non-linear filtering equation (*$P - a.s.$*)*

$$\Pi_t(h) = \Pi_0(h) + \int_0^t \Pi_s(H) ds + \int_0^t \{\Pi_s(D^c)$$

$$+ [\Pi_s(hA^*) - \Pi_s(h)\Pi_s(A^*)] B_s^{-1}(\xi)^*\} d\overline{w}_s + \frac{1}{\lambda} \int_0^t \pi_{s-}(D^d) d\widetilde{N}_s, \qquad (7.19)$$

where A^* *means the transpose of* A, $\overline{w}_t$ *is the innovation process defined in the previous Theorem 205, and*

$$D_t^c dt = d\langle y^c, w \rangle_t = d \begin{bmatrix} \langle y_1^c, w_1 \rangle_t & \cdots & \langle y_1^c, w_d \rangle_t \\ \cdots & \cdots & \cdots \\ \langle y_d^c, w_1 \rangle_t & \cdots & \langle y_d^c, w_d \rangle_t \end{bmatrix},$$

$$D_t^d dt = d \left[\left[y_1^d, \widetilde{N} \right]_t \quad \cdots \quad \left[y_d^d, \widetilde{N} \right]_t \right]^*,$$

$\langle y^c, w \rangle_t$ *is the (predictable) quadratic variational* $\mathfrak{F}_t$*–adapted process corresponding to the martingale* y^c *and* w, *and* $\left[y_i^d, \widetilde{N} \right]_t = \sum_{s \leq t} \triangle y_s^d \triangle \widetilde{N}_s$. *Moreover, the adapted Radon-Nikodym derivatives* D_t^c *and* D_t^d *exist.*

Let us first show that $E[h_t | \mathfrak{F}_t^\xi]$ is the best estimate of h_t in $L^2(\Omega, \mathfrak{F}, (\mathfrak{F}_t^\xi)_{t \geq 0}, P)$. Suppose that $\widetilde{h}_{[0,t]} \in \mathfrak{F}_t^\xi$ is any given estimate.

$$E\left|\widetilde{h}_{[0,t]}-h_t\right|^2 = E\left|\widetilde{h}_{[0,t]}\right|^2 + E\left|h_t\right|^2 - 2E[\widetilde{h}_{[0,t]}h_t]$$
$$= E\left|\widetilde{h}_{[0,t]}\right|^2 + E\left|h_t\right|^2 - 2E[\widetilde{h}_{[0,t]}E(h_t|\mathfrak{F}_t^\xi)]$$
$$= E\left|E(h_t|\mathfrak{F}_t^\xi) - \widetilde{h}_{[0,t]}\right|^2 - E\left|E(h_t|\mathfrak{F}_t^\xi)\right|^2 + E\left|h_t\right|^2$$
$$\geq E\left|h_t\right|^2 - E\left|E(h_t|\mathfrak{F}_t^\xi)\right|^2.$$

So, if $\widetilde{h}_{[0,t]} = E(h_t|\mathfrak{F}_t^\xi)$, then the estimate error $E\left|\widetilde{h}_{[0,t]}-h_t\right|^2$ attains its minimum $E\left|h_t\right|^2 - E\left|E(h_t|\mathfrak{F}_t^\xi)\right|^2$. Therefore $E[h_t|\mathfrak{F}_t^\xi]$ is the best estimate of h_t in $L^2(\Omega, \mathfrak{F}, (\mathfrak{F}_t^\xi)_{t\geq 0}, P)$.

However, to derive the non-linear filtering equation (7.19) we need to establish a series of lemmas.

First we see that by (7.12)

$$\pi_t(h) = E(h_0|\mathfrak{F}_t^\xi) + E(\int_0^t H_s ds|\mathfrak{F}_t^\xi) + E(y_t|\mathfrak{F}_t^\xi). \tag{7.20}$$

Let us discuss (7.20) term by term.

Lemma 207 $\left\{E(y_t|\mathfrak{F}_t^\xi), \mathfrak{F}_t^\xi\right\}_{t\geq 0} \in \mathcal{M}^2$.

Proof. By Jensen's inequality it is obviously square integrable. Moreover, $\forall A \in \mathfrak{F}_s^\xi, s < t$,

$E[I_A E(y_t|\mathfrak{F}_t^\xi)] = E[E(I_A y_t|\mathfrak{F}_t^\xi)] = E(I_A y_t)$

$= E(I_A y_s) = E(I_A E(y_s|\mathfrak{F}_s^\xi))$,

since y_t is a $\mathfrak{F}_t$−martingale. This shows that as $s < t$,

$E(E(y_t|\mathfrak{F}_t^\xi)|\mathfrak{F}_s^\xi) = E(y_s|\mathfrak{F}_s^\xi)$.

Hence $\left\{E(y_t|\mathfrak{F}_t^\xi)\right\}_{t\geq 0}$ is a $\mathfrak{F}_t^\xi$− martingale. ∎

Lemma 208 $\left\{E(\int_0^t H_s ds|\mathfrak{F}_t^\xi) - \int_0^t \pi_s(H)ds, \mathfrak{F}_t^\xi\right\}_{t\geq 0} \in \mathcal{M}^2$.

Proof. In fact, for any $0 \leq t \leq u$

$E[E(\int_0^u H_s ds|\mathfrak{F}_u^\xi) - \int_0^u \pi_s(H)ds|\mathfrak{F}_t^\xi]$

$= E(\int_0^u H_s ds|\mathfrak{F}_t^\xi) - E(\int_0^u \pi_s(H)ds|\mathfrak{F}_t^\xi) = E(\int_0^t H_s ds|\mathfrak{F}_t^\xi) - \int_0^t \pi_s(H)ds$

$+E(\int_t^u H_s ds|\mathfrak{F}_t^\xi) - E(\int_t^u \pi_s(H)ds|\mathfrak{F}_t^\xi) = E(\int_0^t H_s ds|\mathfrak{F}_t^\xi) - \int_0^t \pi_s(H)ds + 0$,

where we have applied the fact that

$E(\int_t^u H_s ds|\mathfrak{F}_t^\xi) - E(\int_t^u \pi_s(H)ds|\mathfrak{F}_t^\xi)$

$= E(\int_t^u \pi_s(H)ds|\mathfrak{F}_t^\xi) - E(\int_t^u \pi_s(H)ds|\mathfrak{F}_t^\xi) = 0.$

Hence $E(\int_0^t H_s ds|\mathfrak{F}_t^\xi) - \int_0^t \pi_s(H)ds, t \geq 0$ is a $\mathfrak{F}_t^\xi$− martingale. ∎

Applying Lemma 207, 208, Theorem 205 and Theorem 204 we immediately obtain the following Lemma.

Lemma 209 *There exist a* $f(t,\omega) \in L^2_{\mathfrak{F}}(R^d)$ *and a* $g(t,\omega) \in L^2_{\mathfrak{F}}(R^d)$ *(*$f(t,\omega)$ *and* $g(t,\omega)$ *are both* $\mathfrak{F}_t$*−predictable) such that* $(P - a.s.)$
$E(h_0|\mathfrak{F}_t^\xi) + E(\int_0^t H_s ds|\mathfrak{F}_t^\xi) + E(y_t|\mathfrak{F}_t^\xi) - \int_0^t \pi_s(H)ds$
$= \pi_0(h) + \int_0^t f(s,\omega)d\overline{w}_s + \int_0^t g(s,\omega)d\widetilde{N}_s.$
Hence
$\pi_t(h) = E(h_0|\mathfrak{F}_t^\xi) + E(\int_0^t H_s ds|\mathfrak{F}_t^\xi) + E(y_t|\mathfrak{F}_t^\xi)$
$= \pi_0(h) + \int_0^t \pi_s(H)ds + \widetilde{y}_t,$
where $\widetilde{y}_t = \int_0^t f(s,\omega)d\overline{w}_s + \int_0^t g(s,\omega)d\widetilde{N}_s.$

Proof. Since by Theorem 205 (7.14) holds. Hence, applying Lemma 207, 208, and Theorem 204, the conclusion is derived. ■

Now we are in a position to prove Theorem 206. We see that to derive the non-linear filtering equation (7.19) we only need to show that
$f(s,\omega) = \Pi_s(D^c) + [\Pi_s(hA^*) - \Pi_s(h)\Pi_s(A^*)]B_s^{-1}(\xi)^*,$
$g(s,\omega) = \frac{1}{\lambda}\Pi_{s-}(D^d).$

Proof. Let $z_t = \int_0^t \lambda_s^1(\xi)d\overline{w}_s + \int_0^t \lambda_s^2(\xi)d\widetilde{N}_s$, where $\lambda_t^1(\xi)$ is any given $d \times d$ matrix - valued bounded $\mathfrak{F}_t^\xi$− adapted random process, and $\lambda_t^2(\xi)$ is any given d− dimensional bounded $\mathfrak{F}_t^\xi$− predictable random process. Then $\left\{z_t, \mathfrak{F}_t^\xi\right\}_{t\geq 0} \in \mathcal{M}^2$. Moreover,

$$E\widetilde{y}_t z_t = \int_0^t tr.(\lambda_s^1(\xi)f(s,\omega)^*)ds + \lambda\int_0^t \lambda_s^2(\xi)g(s,\omega)ds. \tag{7.21}$$

On the other hand, since $\widetilde{y}_t = \pi_t(h) - \pi_0(h) - \int_0^t \pi_s(H)ds \in \mathfrak{F}_t^\xi$. Hence

$$E\widetilde{y}_t z_t = E[z_t h_t - \int_0^t z_s H_s ds], \tag{7.22}$$

where we have used the facts that
$E\pi_0(h)z_t = E[E(z_t|\mathfrak{F}_0^\xi)\pi_0(h)] = E[z_0\pi_0(h)] = 0,$
$E\pi_t(h)z_t = E[E(h_t|\mathfrak{F}_t^\xi)z_t] = Ez_t h_t,$
$E(z_t\int_0^t \pi_s(H)ds) = E[\int_0^t E(z_t|\mathfrak{F}_s^\xi)\pi_s(H)ds] = E[\int_0^t z_s\pi_s(H)ds]$
$= E\int_0^t z_s H_s ds.$
Notice that now by Theorem 205
$\overline{w}_t = \int_0^t B_s^{-1}(\xi)(A_s - E(A_s|\mathfrak{F}_s^\xi))ds + w_t.$
Hence
$z_t = \widetilde{z}_t + \int_0^t \lambda_s^1(\xi)B_s^{-1}(\xi)(A_s - E(A_s|\mathfrak{F}_s^\xi))ds$,
where
$\widetilde{z}_t = \int_0^t \lambda_s^1(\xi)dw_s + \int_0^t \lambda_s^2(\xi)d\widetilde{N}_s.$

Now one finds that by (7.22)

$$E\widetilde{y}_t \cdot z_t = E[\widetilde{z}_t \cdot h_t - \int_0^t \widetilde{z}_s \cdot H_s ds]$$
$$+Eh_t \cdot \int_0^t \lambda_s^1(\xi) B_s^{-1}(\xi)(A_s - E(A_s | \mathfrak{F}_s^\xi))ds$$
$$-E\int_0^t H_u \cdot \int_0^u \lambda_s^1(\xi) B_s^{-1}(\xi)(A_s - E(A_s | \mathfrak{F}_s^\xi))ds\, du. \tag{7.23}$$

However, $\widetilde{z}_t$ is a $\mathfrak{F}_t$−adapted square integrable martingale with respect to the probability measure P. Hence
$E\widetilde{z}_t \cdot h_0 = E[h_0 \cdot E(\widetilde{z}_t | \mathfrak{F}_0)] = E[h_0 \cdot \widetilde{z}_0] = 0.,$
and
$E\int_0^t \widetilde{z}_s \cdot H_s ds = E\int_0^t [E(\widetilde{z}_t | \mathfrak{F}_s) \cdot H_s]ds = E(\widetilde{z}_t \cdot \int_0^t H_s ds).$
Therefore
$E[\widetilde{z}_t \cdot h_t - \int_0^t \widetilde{z}_s \cdot H_s ds] = E[\widetilde{z}_t \cdot (h_t - h_0 - \int_0^t H_s ds)]$
$= E[\widetilde{z}_t \cdot y_t] = E\langle \widetilde{z}^c, y^c \rangle_t + E[\widetilde{z}^d, y^d]_t.$
Hence

$$E[\widetilde{z}_t \cdot h_t - \int_0^t \widetilde{z}_s \cdot H_s ds] = E\int_0^t tr.(\lambda_s^1(\xi) D^c(s)^*)ds$$
$$+E\int_0^t \lambda_s^2(\xi) D^d(s)ds, \tag{7.24}$$

where

$D_t^c = \frac{d\langle y^c, w\rangle_t}{dt} = d\begin{bmatrix} \langle y_1^c, w_1\rangle_t & \cdots & \langle y_1^c, w_d\rangle_t \\ \cdots & \cdots & \cdots \\ \langle y_d^c, w_1\rangle_t & \cdots & \langle y_d^c, w_d\rangle_t \end{bmatrix} /dt,$

$D_t^d = d\begin{bmatrix} [y_1^d, \widetilde{N}]_t & \cdots & [y_d^d, \widetilde{N}]_t \end{bmatrix}^* /dt,$

are some versions of the Radon- Nikodym derivatives, which exist and are $\mathfrak{F}_t$−adapted by Lemma 210 and 211 below. Now computing the second item on the right side of (7.23), one obtains
$Eh_t \cdot \int_0^t \lambda_s^1(\xi) B_s^{-1}(\xi)(A_s - \pi_s(A))ds$
$= E\int_0^t \lambda_s^1(\xi) B_s^{-1}(\xi)(A_s - \pi_s(A)) \cdot h_s ds$
$+E\int_0^t \lambda_s^1(\xi) B_s^{-1}(\xi)(A_s - \pi_s(A))) \cdot (h_t - h_s)ds.$
Hence

$$Eh_t \cdot \int_0^t \lambda_s^1(\xi) B_s^{-1}(\xi)(A_s - \pi_s(A))ds$$
$$= E\int_0^t tr.(\lambda_s^1(\xi)(\pi_s[B_s^{-1}(\xi)(A_s h_s^*)] - [B_s^{-1}(\xi)(\pi_s(A)\pi_s[h_s]^*))ds$$
$$+E\int_0^t \lambda_s^1(\xi) B_s^{-1}(\xi)(A_s - \pi_s(A)) \cdot (h_t - h_s)ds.$$

Notice that
$h_t - h_s = \int_s^t H_u du + (y_t - y_s)$,
and $E(y_t - y_s |\mathfrak{F}_s) = 0$. Hence
$E\int_0^t \lambda_s^1(\xi)B_s^{-1}(\xi)(A_s - \pi_s(A)) \cdot (h_t - h_s)ds$
$= E\int_0^t \lambda_s^1(\xi)B_s^{-1}(\xi)(A_s - \pi_s(A)) \cdot (y_t - y_s)ds$
$+E\int_0^t \lambda_s^1(\xi)B_s^{-1}(\xi)(A_s - \pi_s(A)) \cdot \int_s^t H_u duds$
$= E\int_0^t \int_0^u \lambda_s^1(\xi)B_s^{-1}(\xi)(A_s - \pi_s(A))ds \cdot H_u du.$
From this it follows that

$$Eh_t \cdot \int_0^t \lambda_s^1(\xi)B_s^{-1}(\xi)(A_s - \pi_s(A))ds$$
$$-E\int_0^t H_u \cdot \int_0^u \lambda_s^1(\xi)B_s^{-1}(\xi)(A_s - E(A_s |\mathfrak{F}_s^\xi))ds\, du$$
$$= E\int_0^t tr.\lambda_s^1(\xi)[B_s^{-1}(\xi)(\pi_s(Ah^*) - (\pi_s(A)\pi_s(h)^*))]ds. \tag{7.25}$$

Hence by (7.23), (7.24) and (7.25) we get
$E\widetilde{y}_t \cdot z_t = E\int_0^t \lambda_s^2(\xi) \cdot D^d(s)ds$
$+E\int_0^t tr.\lambda_s^1(\xi)[D^c(s)^* + B_s^{-1}(\xi)(\pi_s(Ah^*) - (\pi_s(A)\pi_s(h)^*))]ds$
$= E\int_0^t \lambda_s^2(\xi) \cdot \pi_s(D^d)ds$
$+E\int_0^t tr.\lambda_s^1(\xi)[\pi_s(D^{c*}) + B_s^{-1}(\xi)(\pi_s(Ah^*) - (\pi_s(A)\pi_s(h)^*))]ds$
Comparing (7.21) and the above equation, one finds that
$f_s(\xi)^* = \pi_s(D^{c*}) + B_s^{-1}(\xi)(\pi_s(Ah^*) - (\pi_s(A)\pi_s(h)^*))$,
$g_s(\xi) = \frac{1}{\lambda}\pi_{s-}(D^d)$.
The conclusion now follows. ∎

The following two lemmas show the existence of Radon-Nikodym derivatives.

Lemma 210 *Consider the* $1-$ *dimensional random processes. If* $\{x_t, \mathfrak{F}_t\}_{t\geq 0} \in \mathcal{M}^2$, *and* $\{w_t, \mathfrak{F}_t\}_{t\geq 0}$ *is a BM, then there exists a* $\{a_t(\omega), \mathfrak{F}_t\}_{t\geq 0} \in L_{\mathfrak{F}}^{2,loc}(R)$ *such that* ($P-a.s.$)
$\langle x, w\rangle_t = \int_0^t a_s(\omega)ds, \quad \forall t \geq 0.$
Naturally, we write $\frac{d\langle x,w\rangle_t}{dt} = a_t(\omega)$.

Proof. Fix an arbitrary $T < \infty$. For $\forall S \times A \in \mathfrak{B}([0,T]) \times \mathfrak{F}_T$ let
$Q(S \times A) = E[I_A \int_S d\langle x, w\rangle_s]$
Then $dQ << dt \times dP$. (That is, dQ is absolutely continuous with respect to $dt \times dP$, which is equivalent to requiring that $\forall S \times A \in \mathfrak{B}([0,T]) \times \mathfrak{F}_T$
$\int_0^t I_{S\times A}(dt \times dP) = 0 \Longrightarrow \int_0^t I_{S\times A}dQ = Q(S \times A) = 0$).
In fact, if $S \in \mathfrak{B}([0,T])$ is such that $\int_0^t I_S ds = 0$, let $y_t = \int_0^t I_S dw_s$, then $E|y_t|^2 = E\left|\int_0^t I_S dw_s\right|^2 = E\int_0^t I_S ds = 0$. $y_t = 0$, $P-a.s.$ Thus $0 = E\langle x, y\rangle_t = E\int_0^t I_S d\langle x, w\rangle_s$. So $dQ << dt \times dP$ easily follows. Hence by

the Radon-Nikodym theorem[28] there exists a $\mathfrak{B}([0,T]) \times \mathfrak{F}_T-$measurable function $f(t,\omega)$ such that

$Q(S \times A) = E[I_A \int_S f(t,\omega)dt]$.

So $E[I_A \langle x, w\rangle_t] = E[I_A \int_0^t f(s,\omega)ds]$, and furthermore, $P - a.s.$

$\langle x, w\rangle_t = \int_0^t f(s,\omega)ds, \ \forall t \in [0,T]$.

Now let the smoothness of f be

$f_n(t,\omega) = n \int_0^t e^{-n(t-s)} f(s,\omega)ds$.

Then it is easily checked that there exists a subsequence $\{n_k\}$ of $\{n\}$, denoted by $\{n\}$ again, such that as $n \to \infty$

$\int_0^T |f_n(t,\omega) - f(t,\omega)|\, dt \to 0, \ \ P - a.s.$

$|f_n(t,\omega) - f(t,\omega)| \to 0, \ \ dt \times dP - a.e.$

Now let us show that the solution x_t^n of the following O.D.E. for each fixed $\omega \in \Omega$

$\frac{dx_t^n}{dt} = -nx_t^n + n\langle x, w\rangle_t, x_0^n = 0, \, t \in [0,T]$.

satisfies

$\frac{dx_t^n}{dt} = f_n(t,\omega)$.

In fact, $x_t^n = n \int_0^t e^{-n(t-s)} \langle x, w\rangle_s ds$. Hence we have that

$\frac{dx_t^n}{dt} = n[\langle x, w\rangle_t - x_t^n] = n[\int_0^t f(s,\omega)ds - \int_0^t f(s,\omega)(n\int_s^t e^{-n(t-u)}du)ds]$

$= n\int_0^t e^{-n(t-s)} f(s,\omega)ds = f_n(t,\omega)$.

Since x_t^n is $\mathfrak{F}_t-$adapted, it follows that $f_n(t,\omega)$ is also $\mathfrak{F}_t-$adapted, and so is its limit $\widetilde{f}(t,\omega) = \lim_{n\to\infty} f_n(t,\omega)$. Hence $\widetilde{f}(t,\omega) = f(t,\omega)$, $dt \times dP - a.e.$ on $[0,T] \times \Omega$. By the continuity of $\langle x, w\rangle_t$ one has that $P - a.s.$

$\langle x, w\rangle_t = \int_0^t \widetilde{f}(s,\omega)ds, \ \forall t \in [0,T]$.

Finally, let $y_n(t) = \int_0^t \widetilde{f}(s,\omega) I_{|\widetilde{f}(s,\omega)|<n} dw_s$. Notice that

$0 \leq E(x_T - y_n(T))^2$

$= Ex_T^2 + E\int_0^T \left|\widetilde{f}(s,\omega)\right|^2 I_{|\widetilde{f}(s,\omega)|<n} ds - 2E\int_0^T \widetilde{f}(s,\omega) I_{|\widetilde{f}(s,\omega)|<n} d\langle x, w\rangle_s$

$= Ex_T^2 - E\int_0^T \left|\widetilde{f}(s,\omega)\right|^2 I_{|\widetilde{f}(s,\omega)|<n} ds$.

Therefore,

$E\int_0^T \left|\widetilde{f}(s,\omega)\right|^2 I_{|\widetilde{f}(s,\omega)|<n} ds \leq Ex_T^2 < \infty$.

This gives that

$E\int_0^T \left|\widetilde{f}(s,\omega)\right|^2 ds < \infty$.

Since $T < \infty$ is arbitrary, one easily obtains the result. ■

Let us establish a similar result for a pure discontinuous martingale and a centralized Poisson process.

Lemma 211 *Consider a* $1-$ *dimensional random process. If* $\{x_t^d, \mathfrak{F}_t\}_{t\geq 0} \in \mathcal{M}^2$ *is a pure dicontinuous square integrable martingale, and* $\left\{\widetilde{N}_t, \mathfrak{F}_t\right\}_{t\geq 0}$ *is a centralized Poisson process with the compensator intensity* λt*, where* $\lambda > 0$ *is a constant, that is,* $N_t = \widetilde{N}_t + \lambda t$ *is a Poisson process such that*

$EN_t = \lambda t$, then there exists a $\{b_t(\omega), \mathfrak{F}_t\}_{t\geq 0}$ with $E\int_0^T |b_s(\omega)|^2 ds < \infty$, $\forall T < \infty$, such that ($P-a.s.$)

$$\left[x^d, \widetilde{N}\right]_t = \int_0^t b_s(\omega)ds, \quad \forall t \geq 0.$$

Naturally, we write $\frac{d[x^d,\widetilde{N}]_t}{dt} = b_t(\omega)$.

Proof. Fix an arbitrary $T < \infty$. For $\forall S \times A \in \mathfrak{B}([0,T]) \times \mathfrak{F}_T$ let

$$Q(S \times A) = E[I_A \int_S d\left[x^d, \widetilde{N}\right]_s]$$

Then $dQ << dt \times dP$. In fact, if $S \in \mathfrak{B}([0,T])$ is such that $\int_0^t I_S ds = 0$, let $y_t = \int_0^t I_S d\widetilde{N}_s$, then $E|y_t|^2 = E\left|\int_0^t I_S d\widetilde{N}_s\right|^2 = E\lambda \int_0^t I_S ds = 0$. $y_t = 0$, $P-a.s.$ Thus $0 = E\left[x^d, y\right]_t = E\int_0^t I_S d\left[x^d, \widetilde{N}\right]_s$. So $dQ << dt \times dP$ easily follows. Hence by the Radon-Nikodym theorem there exists a $\mathfrak{B}([0,T]) \times \mathfrak{F}_T$-measurable function $f(t,\omega)$ such that

$$Q(S \times A) = E[I_A \int_S f(t,\omega)dt].$$

So $E[I_A \left[x^d, \widetilde{N}\right]_t] = E[I_A \int_0^t f(s,\omega)ds]$, and furthermore, $P-a.s.$

$$\left[x^d, \widetilde{N}\right]_t = \int_0^t f(s,\omega)ds, \ \forall t \in [0,T].$$

Now the proof can be completed in exactly the same way as the proof in the previous lemma. The only thing we need to do is to replace $\langle x, w\rangle_t$ there by $\left[x^d, \widetilde{N}\right]_t$ here. ■

Here we should mention that even in the filtering equation (7.19) the jump term coefficient $C(t, x_s(.), s \leq t)$ of the observable equation does not appear directly. However, equation (7.19) does depend on $C(t, x_s(.), s \leq t)$, because the solution ξ_t of the observable equation does.

Corollary 212 *If*

$$y_t = \int_0^t b_s^1(\omega)dw_s^1 + \int_0^t b_s(\omega)dw_s + \int_0^t c_s^1(\omega)d\widetilde{N}_s^1 + \int_0^t c_s(\omega)d\widetilde{N}_s,$$

where $\{w_t^1, \mathfrak{F}_t\}_{t\geq 0}$ is another $d-$ dimensional BM independent of $\{w_t, \mathfrak{F}_t\}_{t\geq 0}$, $\left\{\widetilde{N}_t^1, \mathfrak{F}_t\right\}_{t\geq 0}$ is another $1-$ dimensional centralized Poisson process with the intensity $EN_t^1 = \lambda^1 t$, and $\widetilde{N}_t^1 = N_t^1 - \lambda^1 t$ such that it has no common jump times with $\widetilde{N}_t$; that is, $\Delta\widetilde{N}_t^1 \,\Delta\, \widetilde{N}_t = 0, \forall t \geq 0$, where $\lambda^1 > 0$ is a constant; and for simplicity we assume that for all $T < \infty$,

$$E[\int_0^T |b_s^1(\omega)|^2 ds + \int_0^T |b_s(\omega)|^2 ds + \int_0^T |c_s^1(\omega)|^2 ds + \int_0^T |c_s(\omega)|^2 ds] < \infty,$$

then the non-linear filtering equation (7.19) becomes

$$\pi_t(h) = \pi_0(h) + \int_0^t \pi_s(H)ds + \int_0^t (\pi_s(b)+$$

$$\cdot(\pi_s(hA^*) - \pi_s(h)\pi_s(A^*))B_s^{-1}(\xi)^* d\overline{w}_s + \int_0^t \pi_{s-}(c)d\widetilde{N}_s. \qquad (7.26)$$

Corollary 213 *1) In the case that* $\triangle y_t \equiv 0$*, that is, for* y_t *without jumps, then the non-linear filtering equation (7.19) becomes*

$$\pi_t(h) = \pi_0(h) + \int_0^t \pi_s(H)ds + \int_0^t (\pi_s(D^c) + \cdot(\pi_s(hA^*) - \pi_s(h)\pi_s(A^*))B_s^{-1}(\xi)^* d\overline{w}_s. \tag{7.27}$$

2) In the case that $C_t(\xi, z) \equiv 0$*, that is, the observation process* ξ_t *has no jumps, one also finds that (7.27) is the filtering equation.*

7.4 Optimal Linear Filtering

Let us apply the previous results to a linear system. For simplicity suppose that we have an 1−dimensional signal process $\{\theta_t, \mathfrak{F}_t\}_{t\geq 0}$ satisfying the following SDE:

$$\begin{aligned} d\theta_t &= a(t)\theta_t dt + b^1(t)dw_t^1 + b(t)dw_t \\ &\quad + c^1(t)d\widetilde{N}_t^1 + c(t)d\widetilde{N}_t, \end{aligned} \tag{7.28}$$

which cannot be observed, and we have a 1−dimensional observable process $\{\xi_t, \mathfrak{F}_t\}_{t\geq 0}$ related to the signal process satisfying the SDE as follows:

$$d\xi_t = A(t)\theta_t dt + B(t)dw_t + C(t)d\widetilde{N}_t, \tag{7.29}$$

where $\{w_t^1, \mathfrak{F}_t\}_{t\geq 0}$ and $\{w_t, \mathfrak{F}_t\}_{t\geq 0}$ are two independent BMs, and $\left\{\widetilde{N}_t^1, \mathfrak{F}_t\right\}_{t\geq 0}$ and $\left\{\widetilde{N}_t, \mathfrak{F}_t\right\}_{t\geq 0}$ are two centralized Poisson processes such that
$\widetilde{N}_t^1 = N_t^1 - \lambda^1 t,\ \widetilde{N}_t = N_t - \lambda t$
and λ^1, $\lambda > 0$ are constants, moreover, $\Delta\widetilde{N}_t^1 \triangle \widetilde{N}_t = 0, \forall t \geq 0$; that is, $\widetilde{N}_t^1$ and $\widetilde{N}_t$ have no common jump times. Let us assume that all coefficients are non-random such that $\left|B^{-1}(t)\right| \leq k_0$, $\left|C^{-1}(t)\right| \leq k_0$, and
$\int_0^T [|a(t)| + |A(t)| + |b(t)|^2 + \left|b^1(t)\right|^2 + |B(t)|^2$
$+ |c(t)|^2 + \left|c^1(t)\right|^2 + |C(t)|^2]dt < \infty.$
By theorem 117 (7.28) has a pathwise unique strong solution
$\theta_t = e^{\int_0^t a(u)du}[\theta_0 + \int_0^t e^{-\int_0^s a(u)du}(b^1(s)dw_s^1 + b(s)dw_s + c^1(s)d\widetilde{N}_s^1 + c(s)d\widetilde{N}_s]$.
Applying Corollary 212 one immediately finds that

$$\pi_t(\theta) = \pi_0(\theta) + \int_0^t a(s)\pi_s(\theta)ds + \int_0^t [b(s) + B^{-1}(s)(A(s)\pi_s(\theta^2) - A(s)\pi_s(\theta)^2)]d\overline{w}_s + \int_0^t c(s)d\widetilde{N}_s. \tag{7.30}$$

However, (7.30) is not very convenient, because there is another term $\pi_s(\theta^2)$ in the equation. Notice that

$$\pi_s(\theta^2) = E[\theta_s^2|\mathfrak{F}_s^\xi] = E[(\theta_s - \pi_s(\theta) + \pi_s(\theta))^2|\mathfrak{F}_s^\xi]$$
$$= E[(\theta_s - \pi_s(\theta))^2|\mathfrak{F}_s^\xi] + \pi_s(\theta)^2 - 2\pi_s(\theta)E[\theta_s - \pi_s(\theta)|\mathfrak{F}_s^\xi]$$
$$= E[(\theta_s - \pi_s(\theta))^2|\mathfrak{F}_s^\xi] + \pi_s(\theta)^2.$$

So if we write $\widetilde{\gamma}_s = E[(\theta_s - \pi_s(\theta))^2|\mathfrak{F}_s^\xi]$, then (7.30) can be rewritten as

$$\pi_t(\theta) = \pi_0(\theta) + \int_0^t a(s)\pi_s(\theta)ds$$
$$+ \int_0^t [b(s) + B^{-1}(s)A(s)\widetilde{\gamma}_s]d\overline{w}_s + \int_0^t c(s)d\widetilde{N}_s. \tag{7.31}$$

Notice that $\widetilde{\gamma}_t = \pi_t(\theta^2) - \pi_t(\theta)^2$. Applying Ito's formula to $(\theta_t)^2$ and $\pi_t(\theta)^2$, respectively, one has that

$$(\theta_t)^2 = (\theta_0)^2 + \int_0^t [2(b^1(s)\theta_s dw_s^1 + b(s)\theta_s dw_s)$$
$$+(2c^1(s)\theta_{s-} + c^1(s)^2)d\widetilde{N}_s^1 + (2c(s)\theta_{s-} + c(s)^2)d\widetilde{N}_s]$$
$$+ \int_0^t [2a(s)\theta_s^2 + b^1(s)^2 + b(s)^2 + \lambda^1 c^1(s)^2 + \lambda c(s)^2)]ds, \tag{7.32}$$

and

$$\pi_t(\theta)^2 = \pi_0(\theta)^2 + \int_0^t [2a(s)\pi_s(\theta)^2 + \left|b(s) + B^{-1}(s)A(s)\widetilde{\gamma}_s\right|^2 + \lambda \left|c(s)\right|^2]ds$$
$$+2\int_0^t (b(s) + B^{-1}(s)A(s)\widetilde{\gamma}_s)\pi_s(\theta)d\overline{w}_s + \int_0^t [2\pi_{s-}(\theta)c(s) + |c(s)|^2]d\widetilde{N}_s, \tag{7.33}$$

respectively. Now applying the non-linear filtering equation (7.26) to $(\theta_t)^2$ one has that

$$\pi_t(\theta^2) = \pi_0(\theta^2) + \int_0^t [2a(s)\pi_s(\theta^2) + b^1(s)^2 + b(s)^2 + \lambda^1 c^1(s)^2 + \lambda c(s)^2]ds$$
$$+ \int_0^t [2b(s)\pi_s(\theta) + B^{-1}(s)[A(s)\pi_s(\theta^3) - A(s)\pi_s(\theta^2)\pi_s(\theta)]d\overline{w}_s$$
$$+ \int_0^t (2c(s)\pi_{s-}(\theta) + c(s)^2)d\widetilde{N}_s. \tag{7.34}$$

Therefore, substracting (7.34) by (7.33) one obtains

$$\widetilde{\gamma}_t = \widetilde{\gamma}_0 + \int_0^t [2a(s)\widetilde{\gamma}_s + b^1(s)^2 + \lambda^1 c^1(s)^2 - 2b(s)B^{-1}(s)A(s)\widetilde{\gamma}_s$$
$$- \left|B^{-1}(s)A(s)\widetilde{\gamma}_s\right|^2]ds + \int_0^t B^{-1}(s)A(s)[\pi_s(\theta^3) - \pi_s(\theta^2)\pi_s(\theta) - 2\widetilde{\gamma}_s\pi_s(\theta)]d\overline{w}_s. \tag{7.35}$$

Equations (7.31) and (7.35) are the equations for the (optimal) filtering and the conditional mean square error of filtering, respectively. The interesting thing is that there is no jumps in (7.35). Thus we arrive at the following theorem.

Theorem 214 *Suppose that a 1−dimensional signal process $\{\theta_t, \mathfrak{F}_t\}_{t\geq 0}$ and an 1−dimensional observable process $\{\xi_t, \mathfrak{F}_t\}_{t\geq 0}$ are given by (7.28) and (7.13). Assume that $\left|B^{-1}(t)\right| \leq k_0$, $\left|C^{-1}(t)\right| \leq k_0$, and*
$\int_0^T [|a(t)| + |A(t)| + |b(t)|^2 + \left|b^1(t)\right|^2 + |B(t)|^2$
$+ |c(t)|^2 + \left|c^1(t)\right|^2 + |C(t)|^2]dt < \infty.$
Write $\pi_t(\theta) = E[\theta_t|\mathfrak{F}_t^\xi], \widetilde{\gamma}_t = E[(\theta_t - \pi_t(\theta))^2|\mathfrak{F}_t^\xi]$. Then they satisfy the following filtering equation and equation for conditional mean square error:

$$\pi_t(\theta) = \pi_0(\theta) + \int_0^t a(s)\pi_s(\theta)ds$$
$$+ \int_0^t [b(s) + B^{-1}(s)A(s)\widetilde{\gamma}_s]d\overline{w}_s + \int_0^t c(s)d\widetilde{N}_s, \tag{7.36}$$

$$\widetilde{\gamma}_t = \widetilde{\gamma}_0 + \int_0^t [2a(s)\widetilde{\gamma}_s + b^1(s)^2 + \lambda^1 c^1(s)^2 - 2b(s)B^{-1}(s)A(s)\widetilde{\gamma}_s$$
$$- \left|B^{-1}(s)A(s)\widetilde{\gamma}_s\right|^2]ds + \int_0^t B^{-1}(s)A(s)[\pi_s(\theta^3) - \pi_s(\theta^2)\pi_s(\theta) - 2\widetilde{\gamma}_s\pi_s(\theta)]d\overline{w}_s. \tag{7.37}$$

where $\left\{\overline{w}_t, \mathfrak{F}_t^\xi\right\}_{t\geq 0}$ is a BM, and $\left\{\widetilde{N}_t, \mathfrak{F}_t^\xi\right\}_{t\geq 0}$ is still a centralized Poisson process such that $\widetilde{N}_t = N_t - \lambda t$, (where N_t is a Poisson process with the intensity $EN_t = \lambda t$), both under the original probability P such that $P-a.s.$
$d\xi_t = A(t)\pi_t(\theta)dt + B(t)d\overline{w}_t + C(t)d\widetilde{N}_t.$

The interesting thing here is even the filter $\pi_t(\theta)$ is RCLL (right continuous with left limit), that is, it can be discontinuous; however, the conditional mean square error $\widetilde{\gamma}_t$ of the filter is continuous, and its SDE is without jump terms.

7.5 Continuous Linear Filtering. Kalman-Bucy Equation

In the case that $b(t) = c(t) = c^1(t) = C(t) = 0$, the signal process and the observable process (7.28) and (7.29) become

$$\begin{cases} d\theta_t = a(t)\theta_t dt + b^1(t)dw_t^1, \\ d\xi_t = A(t)\theta_t dt + B(t)dw_t, \end{cases} \tag{7.38}$$

where $\{w_t^1, \mathfrak{F}_t\}_{t\geq 0}$ and $\{w_t, \mathfrak{F}_t\}_{t\geq 0}$ are two independent BMs and (θ_0, ξ_0) is $\mathfrak{F}_0-$measurable. Moreover, the filtering equation and the conditional mean square error equation in Theorem 214 become

$$\pi_t(\theta) = \pi_0(\theta) + \int_0^t a(s)\pi_s(\theta)ds + \int_0^t B^{-1}(s)A(s)\widetilde{\gamma}_s d\overline{w}_s, \tag{7.39}$$

$$\begin{aligned} \widetilde{\gamma}_t = \widetilde{\gamma}_0 &+ \int_0^t [2a(s)\widetilde{\gamma}_s + b^1(s)^2]ds - \int_0^t B^{-2}(s)A(s)^2\widetilde{\gamma}_s^2 ds \\ &+ \int_0^t B^{-1}(s)A(s)[(\pi_s(\theta^3) - \pi_s(\theta^2)\pi_s(\theta)) - 2\widetilde{\gamma}_s\pi_s(\theta)]d\overline{w}_s, \end{aligned} \tag{7.40}$$

where $\overline{w}_t = B^{-1}(t)(d\xi_t - A(t)\pi_t(\theta)dt)$, and $\left\{\overline{w}_t, \mathfrak{F}_t^\xi\right\}_{t\geq 0}$ is a BM. Recall that $\widetilde{\gamma}_s = E[(\theta_s - \pi_s(\theta))^2|\mathfrak{F}_s^\xi]$. We have the following theorem.

Theorem 215 *Under the assumption in Theorem 214 and the assumption that $(\theta_t, \xi_t)_{t\geq 0}$ is a jointly Gaussian process, (θ_0, ξ_0) is $\mathfrak{F}_0-$measurable, $\{w_t^1, \mathfrak{F}_t\}_{t\geq 0}$ and $\{w_t, \mathfrak{F}_t\}_{t\geq 0}$ are two independent BMs, and $b(t) = c(t) = c^1(t) = C(t) = 0$ one has that*

1) $\widetilde{\gamma}_s = E[(\theta_s - \pi_s(\theta))^2]$.

2) $\widetilde{\gamma}_t$ satisfies the following ODE (ordinary differential equation)

$$\widetilde{\gamma}_t = \widetilde{\gamma}_0 + \int_0^t [2a(s)\widetilde{\gamma}_s + b^1(s)^2]ds - \int_0^t B^{-2}(s)A(s)^2\widetilde{\gamma}_s^2 ds. \tag{7.41}$$

Equations (7.39) and (7.41) are called the Kalman-Bucy filtering equations. Equation (7.41) is also a Riccati equation.

To establish this theorem we need the following lemma.

Lemma 216 *If (ξ, θ) are jointly Guassian (both can be multi-dimensional), then*

$$E(\theta|\xi) = E\theta + D_{\theta\xi}D_{\xi\xi}^+(\xi - E\xi), \tag{7.42}$$

$$cov(\theta, \theta|\xi) \widehat{=} E[(\theta - E(\theta|\xi))(\theta - E(\theta|\xi))^*|\xi] = D_{\theta\theta} - D_{\theta\xi}D_{\xi\xi}^+ D_{\theta\xi}^*, \tag{7.43}$$

where $D_{\theta\xi} = E[(\theta - E\theta)(\xi - E\xi)^*]$, and $D_{\theta\theta}, D_{\xi\xi}$ are similarly defined and, moreover, we write $D_{\xi\xi}^+ = D_{\xi\xi}^{-1}$, if $D_{\xi\xi} > 0$; and $D_{\xi\xi}^+ = 0$, otherwise.

Proof. Notice that if $\eta = (\theta - E\theta) + C(\xi - E\xi)$, then
$C = -D_{\theta\xi}D^+_{\xi\xi} \implies E\eta(\xi - E\xi)^* = 0.$
In fact, if $D_{\xi\xi} > 0$, then $D^+_{\xi\xi} = D^{-1}_{\xi\xi}$. Thus the above statements are obvious. If $D_{\xi\xi} = 0$, then $\xi = E\xi$. In this case any constant C will make $E\eta(\xi - E\xi)^* = 0$. For $C = -D_{\theta\xi}D^+_{\xi\xi}$ one has that $\{\eta, \xi\}$ is an independent system, since η and ξ are jointly Gaussian and not linearly correlated. So,

$$\eta = (\theta - E\theta) - D_{\theta\xi}D^+_{\xi\xi}(\xi - E\xi), \tag{7.44}$$

$0 = E(\eta) = E(\eta|\xi) = E(\theta|\xi) - E\theta - D_{\theta\xi}D^+_{\xi\xi}(\xi - E\xi).$
Thus (7.42) is obtained. On the other hand, substracting (7.42) from θ, one finds that $\eta = \theta - E(\theta|\xi)$. By the independence of $\{\eta, \xi\}$
$E[(\theta - E(\theta|\xi))(\theta - E(\theta|\xi))^*|\xi] = E\eta\eta^*.$
Using (7.44) one has that as $D_{\xi\xi} > 0$
$E\eta\eta^* = D_{\theta\theta} + D_{\theta\xi}D^{-1}_{\xi\xi}D_{\xi\xi}D^{*-1}_{\xi\xi}D^*_{\theta\xi} - D_{\theta\xi}D^{*-1}_{\xi\xi}D^*_{\theta\xi} - D_{\theta\xi}D^{-1}_{\xi\xi}D^*_{\theta\xi}$
$= D_{\theta\theta} - D_{\theta\xi}D^{*-1}_{\xi\xi}D^*_{\theta\xi}.$
In the case $D_{\xi\xi} = 0$, so then $\xi = E\xi$, $D^+_{\xi\xi} = 0$, and
$E\eta\eta^* = D_{\theta\theta} = D_{\theta\theta} - D_{\theta\xi}D^+_{\xi\xi}D^*_{\theta\xi}$. The proof is complete. ■

Now we are in a position to prove Theorem 215.

Proof. Let us show that $\widetilde{\gamma}_t = E[(\theta_t - \pi_t(\theta))^2]$. For this let us make a subdivision on $[0, t]$ by
$0 = t^n_0 < t^n_1 = \frac{t1}{2^n} < \cdots < t^n_k = \frac{kt}{2^n} < \cdots < t^n_{2^n} = t.$
Write
$\mathfrak{F}^\xi_{t,n} = \sigma(\xi_{t^n_0}, \cdots, \xi_{t^n_{2^n}}) = \sigma(\xi_{t^n_0}, \xi_{t^n_1} - \xi_{t^n_0}, \cdots, \xi_{t^n_{2^n}} - \xi_{t^n_{2^n-1}}),$
$m^n_t = E[\theta_t|\mathfrak{F}^\xi_{t,n}], \gamma^n_t = E[|\theta_t - m^n_t|^2 |\mathfrak{F}^\xi_{t,n}],$
and

$m_t = E[\theta_t|\mathfrak{F}^\xi_t] = \pi_t(\theta), \gamma_t = E[|\theta_t - m_t|^2 |\mathfrak{F}^\xi_t] = E[|\theta_t - \pi_t(\theta)|^2 |\mathfrak{F}^\xi_t].$
Since $(\theta_t, (\xi_{t^n_0}, \cdots, \xi_{t^n_{2^n}}))$ is Gaussian by Lemma 216
$\gamma^n_t = E[|\theta_t - m^n_t|^2 |\mathfrak{F}^\xi_{t,n}] = D_{\theta\theta} - D_{\theta\xi^n}D^+_{\xi^n\xi^n}D^*_{\theta\xi^n}$
$= E[D_{\theta\theta} - D_{\theta\xi^n}D^+_{\xi^n\xi^n}D^*_{\theta\xi^n}] = E|\theta_t - m^n_t|^2 = \widetilde{\gamma}^n_t,$
where $\xi^n = (\xi_{t^n_0}, \cdots, \xi_{t^n_{2^n}})$. If we can show that as $n \to \infty$,
$\gamma^n_t \to \gamma_t = E[|\theta_t - m_t|^2 |\mathfrak{F}^\xi_t],$
$\widetilde{\gamma}^n_t \to \widetilde{\gamma}_t = E|\theta_t - m_t|^2,$
then $\gamma_t = \widetilde{\gamma}_t$, and taking the expectation in both sides of (7.40) we obtain (7.41). However, by Levi's theorem as $n \to \infty$,
$\gamma^n_t = E[|\theta_t - m^n_t|^2 |\mathfrak{F}^\xi_{t,n}]$
$= E[|\theta_t|^2 |\mathfrak{F}^\xi_{t,n}] - 2E[\theta_t|\mathfrak{F}^\xi_{t,n}]m^{n*}_t + |m^n_t|^2 = E[|\theta_t|^2 |\mathfrak{F}^\xi_{t,n}] - |m^n_t|^2$
$\to E[|\theta_t|^2 |\mathfrak{F}^\xi_t] - |m_t|^2 = E[|\theta_t - m_t|^2 |\mathfrak{F}^\xi_t] = \gamma_t.$
On the other hand,
$\widetilde{\gamma}^n_t = E|\theta_t - m^n_t|^2 = E|(\theta_t - m_t) + (m_t - m^n_t)|^2$
$= \widetilde{\gamma}_t + E|m_t - m^n_t|^2 + 2E(\theta_t - m_t)(m_t - m^n_t)^*.$
By this we easily see that as $n \to \infty$, so $|\widetilde{\gamma}^n_t - \widetilde{\gamma}_t| \to 0$. Hence the proof is complete. ■

Corollary 217 *Under the assumptions in Theorem 215, the Riccati equation (7.41) has a unique solution. Hence (7.39) also has a unique solution. Furthermore, if* $\widetilde{\gamma}_0 > 0$, *then we also have* $\widetilde{\gamma}_t > 0, \forall t > 0$.

Proof. By Theorem 215 it is already known that (7.41) has a solution $\widetilde{\gamma}_t = E[(\theta_t - \pi_t(\theta))^2]$. Let us show the uniqueness. First, by Ito's formula one easily sees that the solution (θ_t, ξ_t) of SDE (7.38) has the following property: for any $T < \infty$

$E[\sup_{t\in[0,T]} |\theta_t|^2 + \sup_{t\in[0,T]} |\xi_t|^2] \leq k_T < \infty,$

where k_T is a constant depending on T only. Hence as $t \leq T$

$\widetilde{\gamma}_t = E[(\theta_t - \pi_t(\theta))^2] \leq 4k_T < \infty.$

Now suppose that $\widetilde{\gamma}_t^i, i = 1, 2$ are two solution of (7.41). Set $y_t = \left|\widetilde{\gamma}_t^1 - \widetilde{\gamma}_t^2\right|$.

By (7.41) one has that

$0 \leq y_t \leq \int_0^t [2a(s)y_s + 4k_T k_0^2 A(s)^2] y_s ds \leq 8k_T \int_0^t [a(s) + k_0^2 A(s)^2] y_s ds.$

So applying Gronwall's inequality one finds that $y_t = 0$, i.e. the solution of the Riccati equation (7.41) is unique. By the Lipschitzian condition, (7.39) also has a unique solution. Finally, let us show that $\widetilde{\gamma}_t > 0, \forall t > 0$. In fact, let $\delta_t = \theta_t - \pi_t(\theta)$. Then by (7.38) and (7.39) it satisfies the following SDE

$d\delta_t = a(t)\delta_t dt + b(t)dw_t^1 - A(t)\widetilde{\gamma}_t B^{-2}(t)(d\xi_s - A(t)\pi_t(\theta)dt)$

$= a(t)\delta_t dt + b(t)dw_t^1 - \widetilde{\gamma}_t B^{-2}(t)A^2(t)\delta_t dt - \widetilde{\gamma}_t B^{-1}(t)A(t)dw_t^2.$

Applying Ito's formula one finds that

$|\delta_t|^2 = |\delta_0|^2 + 2\int_0^t \widetilde{A}_s |\delta_s|^2 ds + \int_0^t [b^2(s) + |\widetilde{\gamma}_s B^{-1}(s)A(s)|^2] ds$

$+2\int_0^t \delta_s b(s) dw_s^1 - 2\int_0^t \delta_s B^{-1}(s)A(s)\widetilde{\gamma}_s dw_s^2,$

where $\widetilde{A}_s = a(s) - \widetilde{\gamma}_s B^{-2}(s)A^2(s)$. Hence as $\widetilde{\gamma}_0 > 0$,

$\widetilde{\gamma}_t = E|\delta_t|^2 = \widetilde{\gamma}_0 e^{2\int_0^t \widetilde{A}_s ds} + \int_0^t e^{\int_s^t 2\widetilde{A}_r dr}[b(s)^2 + B^{-2}(s)A(s)^2\widetilde{\gamma}_s^2] ds > 0.$

■

7.6 Kalman-Bucy Equation in Multi-Dimensional Case

For the continuous linear filtering problem in multi-dimensional case we have the following corresponding theorem.

Theorem 218 *Suppose that a* $k-$*dimensional signal process* $\{\theta_t, \mathfrak{F}_t\}_{t\geq 0}$ *and an* $l-$*dimensional observable process* $\{\xi_t, \mathfrak{F}_t\}_{t\geq 0}$ *are given by the following SDE:*

$$\begin{cases} d\theta_t = a(t)\theta_t dt + b^1(t)dw_t^1, \\ d\xi_t = A(t)\theta_t dt + B(t)dw_t, \end{cases}$$

where $\{w_t^1, \mathfrak{F}_t\}_{t\geq 0}$ *and* $\{w_t, \mathfrak{F}_t\}_{t\geq 0}$ *are two independent BMs, the first one is* $k-$*dimensional, and the second one is* $l-$*dimensional; moreover,* $(\theta_t, \xi_t)_{t\geq 0}$ *is a jointly Gaussian process,* (θ_0, ξ_0) *is* $\mathfrak{F}_0-$*measurable, and* $a(t), b^1(t) \in R^{k\otimes k}$; $A(t), B(t) \in R^{l\otimes l}$.

Assume that $\left|B^{-1}(t)\right| \leq k_0$, *and*
$\int_0^T [|a(t)| + |A(t)| + \left|b^1(t)\right|^2 + |B(t)|^2]dt < \infty$.
Denote $\pi_t(\theta) = E[\theta_t|\mathfrak{F}_t^\xi], \widetilde{\gamma}_t = E[(\theta_t - \pi_t(\theta))^2]$. *Then they are the unique solutions of the following filtering equation and equation for mean square error:*
$\pi_t(\theta) = \pi_0(\theta) + \int_0^t a(s)\pi_s(\theta)ds + \int_0^t \widetilde{\gamma}_s A^*(s)B^{*-1}(s)d\overline{w}_s$,
$\widetilde{\gamma}_t = \widetilde{\gamma}_0 + \int_0^t [a(s)\widetilde{\gamma}_s + \widetilde{\gamma}_s a^*(s) + b^1(s)b^{1*}(s)]ds$
$- \int_0^t \widetilde{\gamma}_s A^*(s)(B(s)B^*(s))^{-1}A(s)\widetilde{\gamma}_s ds$,
where $\overline{w}_t = B^{-1}(t)(d\xi_t - A(t)\pi_t(\theta)dt)$, *and* $\left\{\overline{w}_t, \mathfrak{F}_t^\xi\right\}_{t\geq 0}$ *is a* $P-BM$. *Furthermore, if* $\widetilde{\gamma}_0 > 0$, *that is, it is positive definite, then* $\widetilde{\gamma}_t > 0, \forall t \geq 0$.

Since the proof is almost completely the same as in the 1-dimensional case, we will not repeat it. This theorem actually gives us a practically closed form method for solving the filtering equation. In fact, we can first solve the second ODE (the so-called Riccati equation) to get the mean square error $\widetilde{\gamma}_t$ for the filter, and then put it into the first equation and solve the first linear SDE to obtain the filter $\pi_t(\theta)$.

7.7 More General Continuous Linear Filtering

In this section we will consider the filtering problem on a more general continuous partially observed system:

$$\begin{cases} d\theta_t = (a^0(t) + a^1(t)\theta_t + a^2(t)\xi_t)dt + b^1(t)dw_t^1 + b^2(t)dw_t^2, \\ d\xi_t = (A^0(t) + A^1(t)\theta_t + A^2(t)\xi_t)dt + B^1(t)dw_t^1 + B^2(t)dw_t^2, \end{cases} \tag{7.45}$$

where $\left\{w_t^1, \mathfrak{F}_t\right\}_{t\geq 0}$ and $\left\{w_t^2, \mathfrak{F}_t\right\}_{t\geq 0}$ are $k-$dimensional and $l-$dimensional BMs, respectively, and they are independent; moreover, $\{\theta_t, \mathfrak{F}_t\}_{t\geq 0}$ and $\{\xi_t, \mathfrak{F}_t\}_{t\geq 0}$ are $k-$dimensional and $l-$dimensional random processes, respectively. Naturally, we assume that
$a^0(t) \in R^{k\otimes 1}; a^1(t), b^1(t) \in R^{k\otimes k}; a^2(t), b^2(t) \in R^{k\otimes l}$;
$A^0(t) \in R^{l\otimes 1}; A^1(t), B^1(t) \in R^{l\otimes k}; A^2(t), B^2(t) \in R^{l\otimes l}$,
$\int_0^T |h(t)|\,dt < \infty,\ h = a^0, a^1, a^2, A^0, A^1, A^2$;
$\int_0^T |g(t)|^2\,dt < \infty,\ g = b^1, b^2, B^1, B^2$;
and assume that (θ_0, ξ_0) is Guassian independent of $\left\{w_t^1\right\}_{t\geq 0}$ and $\left\{w_t^2\right\}_{t\geq 0}$.
The idea for getting a closed form for the filtering of this system is as follows: 1) Find a new $l+k-$dimensional BM vector $\left\{\overline{w}_t^1, \mathfrak{F}_t\right\}_{t\geq 0}$ and $\left\{\overline{w}_t^2, \mathfrak{F}_t\right\}_{t\geq 0}$ and a new $(l+k) \times (l+k)$ matrix D_t with a non-singular $D_2(t)$, that is, $D_2^{-1}(t)$ exists, such that
$D_t d\begin{pmatrix} \overline{w}_t^1 \\ \overline{w}_t^2 \end{pmatrix} = \begin{pmatrix} d_1(t) & d_2(t) \\ 0 & D_2(t) \end{pmatrix}\begin{pmatrix} d\overline{w}_t^1 \\ d\overline{w}_t^2 \end{pmatrix}$

$$= \begin{pmatrix} b^1(t) & b^2(t) \\ B^1(t) & B^2(t) \end{pmatrix} \begin{pmatrix} dw_t^1 \\ dw_t^2 \end{pmatrix}.$$

So that we can rewrite the partially observed system as the simpler one:

$$\begin{cases} d\theta_t = (a^0(t) + a^1(t)\theta_t + a^2(t)\xi_t)dt + d_1(t)d\overline{w}_t^1 + d_2(t)d\overline{w}_t^2, \\ d\xi_t = (A^0(t) + A^1(t)\theta_t + A^2(t)\xi_t)dt + D_2(t)d\overline{w}_t^2. \end{cases}$$

2) Suppose 1) is done. If we write the second SDE as

$$d\overline{\xi}_t = A^1(t)\overline{\theta}_t dt + D_2(t)d\overline{w}_t^2, \tag{7.46}$$

and let $v_t = \theta_t - \overline{\theta}_t$, then we should have

$d\overline{\xi}_t = d\xi_t - (A^0(t) + A^1(t)v_t + A^2(t)\xi_t)dt,$

and

$d\overline{w}_t^2 = D_2^{-1}(t)[d\xi_t - (A^0(t) + A^1(t)\theta_t + A^2(t)\xi_t)dt].$

Moreover, substituting this expression for $d\overline{w}_t^2$, and $\theta_t = \overline{\theta}_t + v_t$, into the SDE for θ_t, we find that

$$\begin{aligned} &d\overline{\theta}_t + dv_t = (a^0(t) + a^1(t)(\overline{\theta}_t + v_t) + a^2(t)\xi_t)dt + d_1(t)d\overline{w}_t^1 \\ &+d_2(t)D_2^{-1}(t)[d\xi_t - (A^0(t) + A^1(t)(\overline{\theta}_t + v_t) + A^2(t)\xi_t)dt] \\ &= [a^1(t) - d_2(t)D_2^{-1}(t)A^1(t)]\overline{\theta}_t dt + d_1(t)d\overline{w}_t^1 \\ &+(a^0(t) + a^1(t)v_t + a^2(t)\xi_t)dt + d_2(t)D_2^{-1}(t)[d\xi_t - (A^0(t) + A^1(t)v_t) \\ &+A^2(t)\xi_t]dt. \end{aligned}$$

So that if we set

$$\begin{aligned} &dv_t = [(a^0(t) - d_2(t)D_2^{-1}(t)A^0(t)) + (a^1(t) - d_2(t)D_2^{-1}(t)A^1(t))v_t \\ &+(a^2(t) - d_2(t)D_2^{-1}(t)A^2(t))\xi_t]dt + d_2(t)D_2^{-1}(t)d\xi_t, \end{aligned}$$

(this linear SDE, when ξ_t is given, always has a solution v_t, which is $\mathfrak{F}_t^\xi$–adapted), then

$$d\overline{\theta}_t = [a^1(t) - d_2(t)D_2^{-1}(t)A^1(t)]\overline{\theta}_t dt + d_1(t)d\overline{w}_t^1. \tag{7.47}$$

Now the partially observed SDE system (7.46) and (7.47) is similar to the type considered in Theorem 218. If we can show that $\mathfrak{F}_t^\xi = \mathfrak{F}_t^{\overline{\xi}}, \forall t \geq 0$, then $E[\cdot|\mathfrak{F}_t^\xi] = E[\cdot|\mathfrak{F}_t^{\overline{\xi}}]$. Thus by means of $\theta_t = v_t + \overline{\theta}_t$ we have $m_t = v_t + \overline{m}_t$, where $m_t = E[\theta_t|\mathfrak{F}_t^\xi]$, and $\overline{m}_t = E[\overline{\theta}_t|\mathfrak{F}_t^{\overline{\xi}}]$. Furthermore, we also have $\widetilde{\gamma}_t = \widetilde{\overline{\gamma}}_t$, since $E[(\theta_t - m_t)^2|\mathfrak{F}_t^\xi] = E[(\overline{\theta}_t - \overline{m}_t)^2|\mathfrak{F}_t^{\overline{\xi}}]$. So the filtering equation can be derived by using Theorem 218. Let us show the following lemmas.

Lemma 219 $\mathfrak{F}_t^\xi = \mathfrak{F}_t^{\overline{\xi}}, \forall t \geq 0.$

Proof. Obviously, $\mathfrak{F}_t^{\overline{\xi}} \subset \mathfrak{F}_t^\xi$, since v_t is $\mathfrak{F}_t^\xi$–adapted, and $\overline{\xi}_t$ is expressed by a function of $(\xi_s, v_s), s \leq t$. However, by the above discussion we can also write

$$d\xi_t = (A^0(t) + A^1(t)v_t + A^2(t)\xi_t)dt + d\overline{\xi}_t, \xi_0 = \overline{\xi}_0; \tag{7.48}$$

$$dv_t = (a^0(t) + a^1(t)v_t + a^2(t)\xi_t)dt + d_2(t)D_2^{-1}(t)d\overline{\xi}_t.v_0 = 0. \tag{7.49}$$

Hence, by this linear matrix equation when $\overline{\xi}_s, s \leq t$ is given, the solution (ξ_t, v_t) is $\mathfrak{F}_t^{\overline{\xi}}$−adapted. (Actually, it can be expressed as some exponential function of $\overline{\xi}_s, s \leq t$). So $\mathfrak{F}_t^{\overline{\xi}} \supset \mathfrak{F}_t^{\xi}$. ∎

Applying Theorem 218 to the partially observed SDE system (7.46) and (7.47) we obtain the following lemma.

Lemma 220 *The filtering equation for (7.46) and (7.47) is*
$d\overline{m}_t = [a^1(t) - d_2(t)D_2^{-1}(t)A^1(t)]\overline{m}_t dt$
$+ \widetilde{\overline{\gamma}}_t A^{1*}(t)(D_2(t)D_2^*(t))^{-1}(d\overline{\xi}_t - A^1(t)\overline{m}_t dt),$
$d\widetilde{\overline{\gamma}}_t = [a^1(t) - d_2(t)D_2^{-1}(t)A^1(t)]\widetilde{\overline{\gamma}}_t + \widetilde{\overline{\gamma}}_t[a^1(t) - d_2(t)D_2^{-1}(t)A^1(t)]^*$
$-\widetilde{\overline{\gamma}}_t A^{1*}(t)(D_2(t)D_2^*(t))^{-1}A^1(t)\widetilde{\overline{\gamma}}_t + d_1(t)d_1^*(t).$

However, to obtain the filtering equation for the original partially observed SDE system (7.45) we need to find out the concrete form of D_t; that is, we should complete step 1). For this we see that if step 1) is true, then
$E\int_0^t D_s D_s^* ds = E\int_0^t B_s B_s^* ds, \forall t \geq 0,$
where $B_t = \begin{pmatrix} b^1(t) & b^2(t) \\ B^1(t) & B^2(t) \end{pmatrix}$. So we may use the following approach to find D_t. Let

$$D_t D_t^* = B_t B_t^*. \tag{7.50}$$

Then we have the following matrix equations
$d_1(t)d_1^*(t) + d_2(t)d_2^*(t) = b_1(t)b_1^*(t) + b_2(t)b_2^*(t) = (b \circ b)(t),$
$d_2(t)D_2^*(t) = b_1(t)B_1^*(t) + b_2(t)B_2^*(t) = (b \circ B)(t),$
$D_2(t)D_2^*(t) = B_1(t)B_1^*(t) + B_2(t)B_2^*(t) = (B \circ B)(t).$
Thus, by using such definitions for $(b \circ b)(t), (b \circ B)(t)$ and $(B \circ B)(t)$, we obtain the following lemma.

Lemma 221 *Assume that $(B \circ B)^{-1}(t)$ exists and is bounded $|(B \circ B)^{-1}(t)| \leq k_0, \forall t \geq 0$. Then*
1) the sysmetric matrix
$(b \circ b)(t) - (b \circ B)(t)(B \circ B)^{-1}(t)(b \circ B)^*(t) \geq 0,$
that is, it is non-negative;
2) if we set
$D_2(t) = D_2^*(t) = (B \circ B)^{1/2}(t),$
$d_2(t) = (b \circ B)(B \circ B)^{-1/2}(t),$
$d_1(t) = d_1^*(t) = [(b \circ b)(t) - (b \circ B)(t)(B \circ B)^{-1}(t)(b \circ B)^*]^{1/2}(t),$
then $D_t = \begin{pmatrix} d_1(t) & d_2(t) \\ 0 & D_2(t) \end{pmatrix}$ *solves (7.50);*
3) let $\overline{w}_t = \int_0^t D_s^{-1} B_s dw_s$, (since $D_t D_t^* = B_t B_t^* > 0$, this definition makes sense), where $w_t = (w_t^1, w_t^2)$ is the original $k + l-$dimensional BM given in (7.45), then $\{\overline{w}_t, \mathfrak{F}_t\}_{t \geq 0}$ is a BM, and $\int_0^t D_s d\overline{w}_s = \int_0^t B_s dw_s$.

Proof. Recall that a conditional mean square error is always non-negative. So we will use Lemma 216 to prove result 1). For this we introduce $k+l$-dimensional Gaussian vectors $(\varepsilon^1, \varepsilon^2)^* = ((\varepsilon_1^1, \cdots, \varepsilon_k^1), (\varepsilon_1^2, \cdots, \varepsilon_l^2))$ with independent components and $E\varepsilon_j^i = 0$, $D\varepsilon_j^i = E(\varepsilon_j^i \varepsilon_j^i) = 1$. Set

$\theta = b_1\varepsilon^1 + b_2\varepsilon^2,$
$\xi = B_1\varepsilon^1 + B_2\varepsilon^2.$

Applying Lemma 216 to these two random vectors, we have

$(b \circ b)(t) - (b \circ B)(t)(B \circ B)^{-1}(t)(b \circ B)^*(t) = cov(\theta, \theta|\xi)$
$= E[(\theta - E(\theta|\xi))(\theta - E(\theta|\xi))^*] \geq 0,$

where

$(b \circ b)(t) = b_1(t)b_1^*(t) + b_2(t)b_2^*(t),$
$(b \circ B)(t) = b_1(t)B_1^*(t) + b_2(t)B_2^*(t),$
$(B \circ B)(t) = B_1(t)B_1^*(t) + B_2(t)B_2^*(t).$

Thus 1) is proved. 2) is obviously true. Let us prove 3). Notice hat $\forall t > u$

$E[(\overline{w}_t - \overline{w}_u)(\overline{w}_t - \overline{w}_u)^*|\mathfrak{F}_u] = E[\int_u^t D_s^{-1}B_sB_s^*(D_s^{-1})^*ds|\mathfrak{F}_s]$
$= E[\int_u^t D_s^{-1}D_sD_s^*(D_s^{-1})^*ds|\mathfrak{F}_s] = t - u.$

Applying Theorem 97 we find that $\{\overline{w}_t, \mathfrak{F}_t\}_{t\geq 0}$ is a BM. By definition $\int_0^t D_s d\overline{w}_s = \int_0^t B_s dw_s$. The proof is complete. ■

Finally, we can deduce the filtering equation for the original partially observed SDE system (7.45). Let us write out the final result as follows.

Theorem 222 *Suppose that a $k+l$-dimensional partially observed processes $\{\theta_t, \mathfrak{F}_t\}_{t\geq 0}$ and $\{\xi_t, \mathfrak{F}_t\}_{t\geq 0}$ are given by the SDE (7.45), and suppose that the assumption made in the beginning of this section holds. Moreover, assume that $(B \circ B)^{-1}(t)$ exists and is bounded $|(B \circ B)^{-1}(t)| \leq k_0, \forall t \geq 0$, and $(\theta_t, \xi_t)_{t\geq 0}$ is a jointly Gaussian process, (θ_0, ξ_0) is $\mathfrak{F}_0$-measurable. Write $\pi_t(\theta) = E[\theta_t|\mathfrak{F}_t^\xi], \widetilde{\gamma}_t = E[(\theta_t - \pi_t(\theta))^2]$, then they are the unique solutions of the following filtering equation and equation for mean square error:*

$\pi_t(\theta) = \pi_0(\theta) + \int_0^t [a^0(s) + a^1(s)\pi_s(\theta) + a^2(s)\xi_s)ds$
$+ \int_0^t [(b \circ B)(s) + \widetilde{\gamma}_s A^{1*}(s)](B \circ B)^{-1}(s)$
$\cdot [d\xi_s - (A^0(s) + A^1(s)\pi_s(\theta) + A^2(s)\xi_s)ds],$
$\widetilde{\gamma}_t = \widetilde{\gamma}_0 + \int_0^t [a^1(s)\widetilde{\gamma}_s + \widetilde{\gamma}_s a^{1*}(s) + b \circ b(s)]ds$
$- \int_0^t [(b \circ B)(s) + \widetilde{\gamma}_s A^{1*}(s)](B \circ B)^{-1}(s)[(b \circ B)(s) + \widetilde{\gamma}_s A^{1*}(s)]^* ds.$

Furthermore, if $\widetilde{\gamma}_0 > 0$, that is, if it is positive definite, then $\widetilde{\gamma}_t > 0, \forall t \geq 0$.

Proof. Notice that $\pi_t(\theta) = m_t = v_t + \overline{m}_t$, and $\widetilde{\gamma}_t = \widetilde{\widetilde{\gamma}}_t$. Hence by using Lemma 220 and by using 2) of Lemma 221, and (7.48), (7.49) we easily derive the final result. ■

7.8 Zakai Equation

In this section we are going to derive the Zakai equation for some concrete partially observed system. Suppose that we are given a signal process x_t satisfying the following SDE :

$$\begin{aligned} x_t &= x_0 + \int_0^t b(s, x(.))ds + \int_0^t \sigma(s, x(.))dw_s' \\ &\quad + \int_0^t \int_Z c_{s-}(x(.), z)\widetilde{N}_{k'}(ds, dz) \\ &= x_0 + \int_0^t b(s, x(.))ds + \int_0^t \sigma(s, x(.))dw_s' + y_t, \end{aligned} \tag{7.51}$$

where $w_t', t \geq 0$, is a $d-$dimensional BM, $\widetilde{N}_{k'}(ds, dz)$ is a Poisson martingale measure with the compensator $\pi'(dz)ds$ such that
$\widetilde{N}_{k'}(ds, dz) = N_{k'}(ds, dz) - \pi'(dz)ds$,
where $\pi'(dz)$ is a $\sigma-$finite measure on the measurable space $(Z, \mathfrak{B}(Z))$, $N_{k'}(ds, dz)$ is the counting measure generated by the Poisson point process k_t' and x_0 is supposed to be independent of $\mathfrak{F}_0^{w',k'}$, and suppose that the observation ξ_t is given as follows:

$$\xi_t = \xi_0 + \int_0^t A_s(x(.))ds + \int_0^t B_s(\xi(.))dw_s + \int_0^t C_{s-}(\xi(.))d\widetilde{N}_s, \tag{7.52}$$

where A_t is assumed to be a bounded function, B, C satisfy the same conditions stated in Theorem 206, and $\widetilde{N}_t$ is a centralized Poisson process like that given in the same Theorem. Write
$\langle w'^i, w^j \rangle = \int_0^t \rho_s^{ij} ds,\ 1 \leq i, j \leq d.$

For the random process $A_t(x(.))$ write $\pi_t(A) = E(A_t(x(.)) \big| \mathfrak{F}_t^\xi)$, etc. Then we have

Theorem 223 . *Suppose $f \in C_b^2([0, \infty); R^1)$. Then by Ito's formula*
$f(x_t) = f(x_0) + \int_0^t Lf(x, s)ds + \int_0^t \nabla f(x_s) \cdot \sigma(s, x(.))dw_s'$
$+ \int_0^t \int_Z L^{(1)} f(x, s-)\widetilde{N}_{k'}(ds, dz),$
where for each $x(.) \in D([0, \infty), R^d)$,
$Lf(x, s) = b(s, x(.)) \cdot \nabla f(x_s) + \frac{1}{2} tr.(\sigma(s, x(.))^*(\partial^2 f(x_s)/\partial x^2)\sigma(s, x(.)))$
$+ \int_Z (f(x_{s-} + c(s-, x(.), z)) - f(x_{s-}) - \nabla f(x_s) \cdot c(s-, x(.), z))\pi'(dz),$
$L^{(1)} f(x, s-) = f(x_{s-} + c(s-, x(.), z)) - f(x_{s-}).$
Furthermore, if
$\langle w_i', w_j \rangle_t = \int_0^t \rho_s^{ij} ds.$

then

$$\pi_t(f) = \pi_0(f) + \int_0^t \pi_s(Lf)ds + \int_0^t [\pi_s(\bigtriangledown f \cdot \sigma \cdot \rho) +$$

$$\cdot(\pi_s(fA^*) - \pi_s(f)\pi_s(A^*))B_s^{-1}(\xi)^*]d\overline{w}_s + \frac{1}{\lambda}\int_0^t \pi_{s-}(\widetilde{\overline{D}}^d)d\widetilde{N}_s, \quad (7.53)$$

where $\overline{D}_t^d dt = d[y^d, \widetilde{N}]_t$, *and* $y_t^d = \int_0^t \int_Z L^{(1)} f(x, s-)\widetilde{N}_{k'}(ds, dz)$.

Theorem 223 is a direct corollary of Theorem 206. One only needs to see $f(x_t)$ as the signal process h_t in (7.12), then applying Theorem 206, the conclusion is derived immediately.

Corollary 224 *Assume that in (7.51) and (7.52) the signal and observation noise are independent, and the jump noise in the signal and observation have no common jump time, i.e.* $\rho \equiv 0, \triangle y_t \cdot \triangle \widetilde{N}_t \equiv 0$; *and* $P[x_t \leq x \big| \mathfrak{F}_t^\xi]$, *(which is the conditional distribution of* x_t *under given* $\mathfrak{F}_t^\xi$, *and* $\{x_t \leq x\}$ *is the set* $\{x_t^1 \leq x^1, \cdots, x_t^d \leq x^d\}$*) has a density* $\widehat{p}(t,x) = dP[x_t \leq x \big| \mathfrak{F}_t^\xi]/dx$, *which satisfies suitable differential hypothesis, and* $\widetilde{A}(x_t) = A_t(x(.))$ *only depends on* x_t, *and it is a bounded function. Then one has the following Zakai equation satisfied by the conditional density* $\widehat{p}(t,x)$

$$d\widehat{p} = L^*\widehat{p}dt + \widehat{p}(\widetilde{A}^* - \Pi_t(\widetilde{A}^*))B_t^{-1*}d\overline{w}_t, \quad (7.54)$$

where L^* *is the adjoint operator of* L.

Proof. To show Corollary 224 one only needs to notice that
$\pi_t(\widetilde{A}^*) = \int_{R^d} \widetilde{A}^*(x)\widehat{p}(t,x)dx = (\widetilde{A}^*\widehat{p}(t,\cdot)) = (\widetilde{A}^*, \widehat{p})$,
and to apply integration by parts. Indeed, by assumption $\rho \equiv 0, \triangle y_t \cdot \triangle \widetilde{N}_t \equiv 0$, hence (7.53) becomes $\forall f \in C_0^2([0,\infty); R^1)$ (where $C_0^2([0,\infty); R^1)$ is the totality of functions $f : [0,\infty) \to R^1$, with continuous derivatives up to the second order and with compact support),
$\pi_t(f) = \pi_0(f) + \int_0^t \pi_s(Lf)ds + \int_0^t (\pi_s(fA^*) - \pi_s(f)\pi_s(A^*))B_s^{-1}(\xi)^* d\overline{w}_s$.
Or, $\forall f \in C_0^2([0,\infty); R^1)$
$(f, d_t\widehat{p}(t,\cdot)) = d_t(f, \widehat{p}(t,\cdot))$
$= (f, L^*\widehat{p}(t,\cdot))dt + (f, \widehat{p}(t,\cdot)(A^* - \Pi_t(A^*))B_t^{-1*}d\overline{w}_t$.
Hence (7.54) now follows. ■

The advantage of Zakai's equation is that it is a linear partial stochastic differential equation (PSDE) for $\widehat{p}(t,x)$, and usually a linear PSDE is much easier to handle. As soon as the solution $\widehat{p}$ is obtained, then the non-linear filter $\pi_t(x) = E(x_t|\mathfrak{F}_t^\xi) = \int_{R^d} x\widehat{p}(t,x)dx$ is also obtained.

7.9 Examples on Linear Filtering

In many cases, or in an ideal case, we will consider the original signal system to be non-random, that is, the coefficients of the signal dynamics are non-random. However, the initial value of the signal process may be random. For example, the initial value of the population of fish in a large pond actually is random. Moreover, since the signal process itself usually cannot be observed directly, we can only estimate it and understand it through an observable process that is related to it. Obviously, the observed results will usually be disturbed by many stochastic perturbations. So the appropriate assumption is that we have a pair of a signal process θ_t and an observable process ξ_t as follows, where for simplicity we consider them both in 1−dimensional space:

$$\begin{cases} d\theta_t(\omega) = a(t)\theta_t(\omega)dt, \theta_0(\omega) = \theta_0(\omega); \\ d\xi_t = A(t)\theta_t dt + B(t)dw_t, \xi_0 = \xi_0, \end{cases} \tag{7.55}$$

where all coefficients $a(t), A(t)$ and $B(t)$ are non-random. By Theorem 215 we have that

$$\begin{cases} \pi_t(\theta) = \pi_0(\theta) + \int_0^t a(s)\pi_s(\theta)ds + \int_0^t B^{-1}(s)A(s)\widetilde{\gamma}_s d\overline{w}_s, \\ \widetilde{\gamma}_t = \widetilde{\gamma}_0 + \int_0^t 2a(s)\widetilde{\gamma}_s ds - \int_0^t B^{-2}(s)A(s)^2\widetilde{\gamma}_s^2 ds, \end{cases} \tag{7.56}$$

where $\pi_t(\theta) = E[\theta_t|\mathfrak{F}_t^\xi]$ is the estimate of θ_t based on the information given by the observation $\xi_s, s \leq t$, and $\widetilde{\gamma}_s = E[(\theta_s - \pi_s(\theta))^2]$ is the mean square error of the estimate with $\widetilde{\gamma}_0 = E[(\theta_0 - \pi_0(\theta))^2]$. Again by Theorem 215 it is already known that (7.56) has a unique solution $(\pi_t(\theta), \widetilde{\gamma}_t), \forall t \geq 0$. Here, we are interested in how to get the explicit formulas for the solution. First, from the practical point of view, let us replace $\{\overline{w}_t\}$ by $\{\xi_t\}$ from the formula:
$d\xi_t = A(t)\pi_t(\theta)dt + B(t)d\overline{w}_t$,
because our observation is $\{\xi_t\}$. Thus we get the following filtering SDE system:

$$\begin{cases} \pi_t(\theta) = \pi_0(\theta) + \int_0^t (a(s) - B^{-2}(s)A^2(s)\widetilde{\gamma}_s)\pi_s(\theta)ds + \int_0^t B^{-2}(s)A(s)\widetilde{\gamma}_s d\xi_s, \\ \widetilde{\gamma}_t = \widetilde{\gamma}_0 + \int_0^t 2a(s)\widetilde{\gamma}_s ds - \int_0^t B^{-2}(s)A(s)^2\widetilde{\gamma}_s^2 ds. \end{cases} \tag{7.57}$$

Obviously, if we can find a formula for $\widetilde{\gamma}_t$ solving the second ordinary differential equation - the so-called Riccati equation in (7.57) then the estimate, or say, the filter $\pi_t(\theta)$, can also be obtained from the following formula:

$$\pi_t(\theta) = e^{\int_0^t (a(s) - B^{-2}(s)A^2(s)\widetilde{\gamma}_s)ds} \pi_0(\theta) \tag{7.58}$$
$$+ \int_0^t e^{\int_s^t (a(r) - B^{-2}(r)A^2(r)\widetilde{\gamma}_r)dr} B^{-2}(s)A(s)\widetilde{\gamma}_s d\xi_s$$

(In fact, one can use the Ito formula to check that $\pi_t(\theta)$ defined above satisfies the first SDE in (7.57)). Fortunately, the solution of the Riccati equation does have an explicit formula, if we make some further assumptions that

$a(s) = a_0$, $A(s) = A_0$, $B(s) = B_0 > 0$, $\widetilde{\gamma}_0 = E[(\theta_0 - \pi_0(\theta))^2] > 0$

are all constants. In this case one easily checks that the following $\widetilde{\gamma}_t$ satisfies the second Riccati - ODE in (7.57):

$$\widetilde{\gamma}_t = 2a_0 B_0^{-2} A_0^2 / [1 + (\frac{2a_0 B_0^{-2} A_0^2}{\widetilde{\gamma}_0^2} - 1)e^{-2a_0 t}]. \tag{7.59}$$

These tell us that if we can know "the mean square error of the initial estimate" $\widetilde{\gamma}_0$, which is larger than zero, then the estimate $\pi_t(\theta) = E[\theta_t | \mathfrak{F}_t^\xi]$ by observation, and $\widetilde{\gamma}_t$—the mean square error of the estimate, can be calculated by formulas (7.58) and (7.59), respectively. One naturally asks how about $\widetilde{\gamma}_0 = 0$. In this case $\pi_0(\theta) = \theta_0$ and one finds that $\widetilde{\gamma}_t = 0, \forall t \geq 0$ is the unique solution of the Riccati equation, that is the second equation of (7.57). So $\pi_t(\theta)$ satisfies the following equation

$\pi_t(\theta) = \theta_0 + \int_0^t a(s)\pi_s(\theta)ds,$

that is, the same equation as the signal equation. So one immediately gets the solution formulated by

$\pi_t(\theta) = \theta_t = \theta_0 e^{\int_0^t a(s)ds}.$

This means that the estimate is exactly equal to the signal process. This is quite reasonable. Because the initial value can be explicitly observed, so one can directly use the known signal dynamics to get the explicit signal process.

However, one should notice that if the signal process satisfies an SDE (Not an ODE!)

$d\theta_t = a(t)\theta_t dt + b^1(t)dw_t^1, \theta_0 = \theta_0,$

and the obsevable process ξ_t still satisfies the second SDE in (7.55), then by Theorem 215 $\widetilde{\gamma}_t$ will satisfy a more complicated Riccati equation:

$\widetilde{\gamma}_t = \widetilde{\gamma}_0 + \int_0^t [2a(s)\widetilde{\gamma}_s + b^1(s)^2]ds - \int_0^t B^{-2}(s)A(s)^2 \widetilde{\gamma}_s^2 ds.$

In this case even if $\widetilde{\gamma}_0 = 0$, we still cannot get $\widetilde{\gamma}_t = 0$.

So for continuous linear filtering problems one asks in what cases we can get explicit formulas for the filterings, then by the above discussion one sees that this completely depends on how many explicit formulas we have for solutions of the Riccati equations.

8

Option Pricing in a Financial Market and BSDE

In this chapter we will discuss option pricing in the financial market and how this problem will draw us to study the backward stochastic differential equations (BSDEs) and how the problem can be solved by a BSDE. Furthermore, we will also use the partial differentail equation (PDE) technique to solve the option pricing problem and to establish the famous Black-Scholes formula.

8.1 Introduction

1. Hedging contingent claim, option pricing and BSDE

In a financial market there are two kinds of securities: one kind is without risk. We call it a bond and if, for example, you deposit your money in a bank, you will get a bond that will pay some interest at an agreed rate. It is natural to assume that the bond price equation is:

$$dP^0(t) = P^0(t)r(t)dt, P^0(0) = 1, \tag{8.1}$$

where $r(t)$ is the rate function. Another kind of security in the financial market is with risk. We call it a stock. Since in the market there can be many stocks, say, for example, N different kinds of stocks, and they will usually be disturbed by some stochastic perturbations, for simplicity we assume that the stochastic perturbations are continuous, so it is also natural to assume that the stochastic differential equations for the prices of stocks are:

$$
\begin{aligned}
dP^i(t,\omega) &= P^i(t,\omega)[(b^i(t,\omega)dt + \sum_{k=1}^{d} \sigma^{ik}(t,\omega)dw_t^k], \qquad (8.2)\\
P^i(0,\omega) &= P^i(0,\omega), \quad i = 1,2,\cdots,N;
\end{aligned}
$$

where $w_t = (w_t^1, w_t^2, \cdots, w_t^d)^*$ is a standard $d-$dimensional Brownian Motion process, and A^* means the transpose of A. Now suppose that a small investor who wants his money (or say his wealth) from the market at the future time T when it reaches X. (Notice that X is not necessary a constant, for example, $X = c_0 + c_1 P_T^1$, where c_0 and c_1 are non-negative constants, and P_T^1 is the price for the first stock at the furture time T, because the investor has confidence that the first stock can help him to earn money). How much money x_t should he invest in the market, and how could he choose the right investment portfolio at time t? Suppose the right portfolio $(\pi_t^0, \pi_t^1 \cdots, \pi_t^N)$ exists, where π_t^0 is the money invested in the bond, and π_t^i is the money invested in the ith stock. Then he should have

$$
\begin{cases} x_t = \sum_{i=0}^N \pi_t^i = \sum_{i=0}^N \eta_t^i P_t^i, \\ x_T = X, \end{cases} \qquad (8.3)
$$

where η_t^0 is the number of bond units bought by the investor, and η_t^i is the amount of units for the ith stock. We call x_t the wealth process for this investor in the market. Now let us derive intuitively the stochastic differential equation (SDE) for the wealth process as follows: Suppose the portfolio is self-financed, i.e. in a short time dt the investor doesn't put in or withdraw any money from the market. He only lets the money x_t change in the market due to the market own performance, i.e. self-finance produces $dx_t = \eta_t^0 dP_t^0 + \sum_{i=1}^N \eta_t^i dP_t^i$.

Now substituting (8.1) and (8.2) into the above equation, after a simple calculation we arrive at the following backward SDE (BSDE), where the wealth process x_t and the portfolio $\pi_t = (\pi_t^1 \cdots, \pi_t^N)$ (actually it is the risk part of the portfolio) should satisfy:

$$
\begin{cases} dx_t = r_t x_t dt + \pi_t(b_t - r_t \underline{1})dt + \pi_t \sigma_t dw_t, \\ x_T = X, \ t \in [0,T], \end{cases} \qquad (8.4)
$$

where $\underline{1} = (1, \cdots, 1)^*$ is an $N-$dimensional constant vector.

In a financial market if we let X be a contingent claim, then the solution (x_t, π_t) of the BSDE (8.4) actually tells us the following fact:

At any time t, let us invest a total amount of money x_t, and dividing it into two parts: One part of the money is for the non-risky investment; that is we invest the money $\pi_t^0 = x_t - \sum_{i=1}^N \pi_t^i$ into the bond. The other part of the money is for the risky investment $\pi_t = (\pi_t^1 \cdots, \pi_t^N)$; that is we invest the money π_t^i into the the $i-$th stock, $i = 1, 2, \cdots, N$. Then, eventually, at the terminal time T our total money $x_t, t = T$, will arrive at the contingent claim X; that is, such an x_t with $\pi_t, 0 \leq t \leq T$, can produce the contingent claim X, and we will say that $(x_t, \pi_t), 0 \leq t \leq T$, hedges the contingent claim X.

In a financial market we can see that to price an option will also lead to the solution of a BSDE like (8.4). The so-called "option" actually is a contract, which provides the right for the contract owner to do the following thing.

"At the given future time T, by the contract, one can use a fixed price K to buy a unit of some fixed goods or some fixed stock in the market."

Suppose that the price of the fixed goods or the fixed stock for one unit at the future time T becomes P_T. Then at the future time T the option will help its owner to earn the money $x_T = (P_T - K)^+$. Therefore, if somebody wants to own this option now he has to pay for it. That is to say, at the time t, the option should have a price x_t. This raises an important and interesting problem: how can we price the option at the time t as x_t, in a way that is reasonable and fair to both the seller and the buyer? To solve this problem, and to show the idea involved, for simplicity we assume that in the market there is only one stock which we mentioned and one bond. We can imagine that if we put the money x_t into the market as follows: one part π_t^0 of it is deposited in the bank (with no risk), and another part π_t^1 of it is used to buy the fixed stock, or say, to buy the stock of the fixed goods (whose price changes by some stochastic perturbation, so it is with risk). Then as time t evolves to T, if the option price x_t is a fair price, it should arrive at $x_T = (P_T - K)^+$ at the time T. So, actually, we should require that

$$\begin{cases} x_t = \pi_t^0 + \pi_t^1, t \in [0, T), \\ x_T = (P_T - K)^+. \end{cases}$$

Thus we have the same condition as (8.3) for $X = (P_T - K)^+$ and $N = d = 1$. ($N = 1$ means there is only one stock in the market, and $d = 1$ means there is only one continuous real-valued stochastic perturbation or, say, the dimension of the BM is 1). So a samilar discussion will lead to $(x_t, \pi_t), 0 \leq t \leq T$, satisfying BSDE (8.4) for $N = d = 1$.

2. Difference between forward SDE and BSDE.

We see that (8.4) is actually a backward SDE, that is, the terminal condition $x_T = X$ is given, and we want to find a solution to satisfy (8.4). So actually we need to solve the SDE backwards. However, there is a big difference between solving the forward SDE (given the initial condition $x_0 = x$ and then solving the SDE to find the solution $x_t, 0 \leq t \leq T$) and solving the backward SDE. In fact, even in the continuous case, if we want to solve the following forward SDE (FSDE):

$$dx_t = b(t, x_t)dt + \sigma(t, x_t)dw_t, \; x_0 = x_0, \; t \in [0, T],$$

then by the standard SDE theory in Chapter 3 under a less than linear growth condition and a Lipschitzian condition on $b(t,x)$ and $\sigma(t,x)$ with respect to x, this SDE will have a pathwise unique $\mathfrak{F}_t^w$−adapted solution $x_t, t \in [0, T]$. So there is no difficulty at all in solving an FSDE under the usual standard condition. However, if we want to solve an SDE similar to the one above, but bakwards:

$$dx_t = b(t, x_t)dt + \sigma(t, x_t)dw_t, x_T = X, t \in [0, T], \tag{8.5a}$$

a big difficulties arise. In fact, since the terminal condition $x_T = X$ is given first, and usually X is $\mathfrak{F}_T$−measurable, one can imagine that the solution $x_t, t \in [0, T]$, if it exists, will depend on the given X, so x_t will also be $\mathfrak{F}_T$−measurable for all $t \in [0, T]$. However, this will make the stochastic integral $\int_0^t \sigma(s, x_s)dw_s$ have no meaning in general, because the integrand of a stochastic integral should be at least $\mathfrak{F}_s$−adapted. (But now $\sigma(s, x_s)$ is $\mathfrak{F}_T$−measurable!) So, usually we cannot solve the BSDE in (8.5a) even when b and σ are bounded and Lipschitz continuous! Fortunately, for the BSDE (8.4) in the financial market, we know that actually the solution is a one-pair radom processes (x_t, π_t), and from practical financial experience it seems that it can exist and satisfy the BSDE (8.4). This motivates us to discuss the general BSDE theory and its applications to Finanacial markets.

3. The developement of BSDE

The adapted solution for a linear backward stochastic differential equation (BSDE), which appeared as the adjoint process for a stochastic control problem, was first investigated by Bismut (1973), and then by Bensoussan (1982), and others, while the first result for the existence of an adapted solution to a continuous non-linear BSDE with Lipschitzian-coefficient was obtained by Pardoux and Peng (1990). Today the BSDE has become a powerful tool in the study of financial markets and stochastic control. (See, for example, El Karoui et al. 1997, and References in [166]). However, one also finds that almost all interesting problems studied in BSDE can be initiated by mathematical finance.

8.2 A More Detailed Derivation of the BSDE for Option Pricing

To help the reader understand clearly what is involved we give the detail of the derivation of the BSDE satisfied by the option pricing problem. As in the introduction, we have an equation for the bond price $P^0(t)$ and an SDE for the stock price $P(t, \omega)$ as follows:

$dP^0(t) = P^0(t)r(t)dt, P^0(0) = 1,$

$dP(t, \omega) = P(t, \omega)[(b(t, \omega)dt + \sigma(t, \omega)dw_t], P(0, \omega) = P(0, \omega).$

The fair price x_t for the option at any time t should satisfy

$$\begin{cases} x_t = \pi_t^0 + \pi_t^1 = \eta_t^0 P_t^0 + \eta_t^1 P_t, \\ x_T = (P_T - K)^+, \end{cases}$$

where π_t^0 is the investment for the bond and π_t^1 for the stock, and η_t^0 is the amount of bond units bought by the investor, and η_t^1 is the amount of units of the stock. In a self-financed market we have

$dx_t = \eta_t^0 dP_t^0 + \eta_t^1 dP_t^1.$

Now replacing dP_t^0 and dP_t^1 by their equations we find that

$dx_t = \eta_t^0 P^0(t) r(t)dt + \eta_t^1 P(t,\omega)[(b(t,\omega)dt + \sigma(t,\omega)dw_t]$

$= \pi_t^0 r(t)dt + \pi_t^1[(b(t,\omega)dt + \sigma(t,\omega)dw_t]$

$= (\pi_t^0 + \pi_t^1)r(t)dt + \pi_t^1(b(t,\omega) - r(t))dt + \pi_t^1\sigma(t,\omega)dw_t$

$= r_t x_t dt + \pi_t(b_t - r_t)dt + \pi_t\sigma_t dw_t.$

So x_t–the price for the option, with π_t^1–the portfolio, should satisfy the following BSDE

$$\begin{cases} dx_t = r_t x_t dt + \pi_t(b_t - r_t)dt + \pi_t\sigma_t dw_t, \\ x_T = (P_T - K)^+, \ 0 \le t \le T. \end{cases} \tag{8.6}$$

More generally, we may consider the BSDE

$$\begin{cases} dx_t = \widetilde{b}(t, x_t, \pi_t)dt + \widetilde{\sigma}(t, \pi_t)dw_t, \\ x_T = X, \ 0 \le t \le T, \end{cases}$$

and find its $\mathfrak{F}_t$–adapted solution (x_t, π_t). Or, to simplify the notation, we set $q_t = \widetilde{\sigma}(t, \pi_t)$, and if this equation can be solved to give $\pi_t = f(t, q_t)$, then we may consider the BSDE

$$\begin{cases} dx_t = b(t, x_t, q_t)dt + q_t dw_t, \\ x_T = X, \ 0 \le t \le T, \end{cases} \tag{8.7}$$

where $b(t, x_t, q_t) = \widetilde{b}(t, x_t, f(t, q_t))$, and then find the $\mathfrak{F}_t$–adapted solution (x_t, q_t). After that, setting $\pi_t = f(t, q_t)$, we know that (x_t, π_t) will satisfy the original BSDE. In the next section we will consider the BSDE in (8.7) and also a more general case.

8.3 Existence of Solutions with Bounded Stopping Times

8.3.1 The General Model and its Explanation

For more general usage let us consider the following backward stochastic differential equation (BSDE) with jumps in R^d :

$$\begin{aligned} x_t = X + \int_{t\wedge\tau}^{\tau} b(s, x_s, q_s, p_s, \omega)ds - \int_{t\wedge\tau}^{\tau} q_s dw_s \\ - \int_{t\wedge\tau}^{\tau}\int_Z p_s(z)\tilde{N}_k(ds, dz), \ 0 \leqslant t, \end{aligned} \tag{8.8}$$

where $w_t^T = (w_t^1, \cdots, w_t^{d_1}), 0 \leqslant t$, is a d_1-dimensional standard Brownian motion (BM), w_t^T is the transpose of w_t; $k^T = (k_1, \cdots, k_{d_2})$ is a d_2–dimensional stationary Poisson point process with independent components, $\tilde{N}_{k_i}(ds, dz)$ is the Poisson martingale measure generated by k_i satisfying

$\tilde{N}_{k_i}(ds, dz) = N_{k_i}(ds, dz) - \pi(dz)ds, \ i = 1, \cdots, d_2,$
where $\pi(.)$ is a σ-finite measure on a measurable space $(Z, \mathfrak{B}(Z))$, $N_{k_i}(ds, dz)$ is the Poisson counting measure generated by k_i, and τ is a bounded $\mathfrak{F}_t$ -stopping time, where $\mathfrak{F}_t$ is the $\sigma-$algebra generated (and completed) by $w_s, k_s, s \leqslant t$.

Remark 225 *1) If $X \in \mathfrak{F}^w_\tau$, and τ is a $\mathfrak{F}^w_t-$stopping time, then BSDE (8.8) reduces to a continuous BSDE*

$$x_t = X + \int_{t\wedge\tau}^{\tau} b(s, x_s, q_s, 0, \omega)ds - \int_{t\wedge\tau}^{\tau} q_s dw_s. \tag{8.9}$$

In fact, if (x_t, q_t) solves (8.9), then $(x_t, q_t, 0)$ solves (8.8).
2) Set $\tau = T$. If X =a constant k_0, and the coefficient $b = b(t, x, q, p)$ does not depend on ω, then BSDE (8.8) reduces to a usual continuous ordinary differential equation (ODE)

$$x_t = k_0 + \int_t^T b(s, x_s, 0, 0)ds, \tag{8.10}$$

In fact, if (x_t) solves (8.10), then $(x_t, 0, 0)$ solves (8.8).
So (8.8) is a general form which includes the ODE, the continuous BSDE and the terminal time which is non-random.
The reason we consider a jump term, as in (8.8), is that the perturbations usually occur at finite discontinuities. If the perturbation is a random process, then the number of the points where jumps occur is finite but random. Moreover, the jump values taken at the jump points are also random. For such a kind of random point process the simplest and most frequently encountered one is the Poisson point process. (See the explanation below). So it is natural to consider that the jump term is an integral over a martingale measure (it will have a nice property) generated by a Poisson point process $k(\cdot)$.
Now let us explain why a Poisson point process is the one most frequently met among all point processes. In fact, if $p(\cdot)$ is a finite Poisson point process, let
$N(t) = N^U(t) = N_p((0, t], U),$
where $N_p((0, t], U)$ is the counting measure generated by $p(\cdot)$, that is, it is the numbers of $p(s)$ dropped in U during $s \in (0, t]$, then $N(t), t \geq 0$, is a Poisson process, which has the following nice properties:
1) It has an independent increments, that is, if $0 < t_1 < t_2 < \cdots < t_n$, then all increnents $N(t_i) - N(t_{i-1}), i = 1, \cdots, n$, are independent. In other words, the increments of $N(t)$ in disjoint intervals occur independently.
2) The probability of $\{N(t + h) - N(t) \geq 2\}$ is a higher degree infinitesmal of h. Or, in other words, in a sufficiently small time interval the likelihood that more than two jumps happen is almost impossible (the probability is very small compared to the event that only one jump happens).

3) $N(t)$ only takes values $0, 1, 2, \cdots$. (Because $N(t)$ is the numbers of jumps counted). So in any finite interval there are only finite jumps. (Otherwise, there exists an interval $(t_1, t_2]$ such that $N(t_2) - N(t_1) = \infty$. This is impossible. In fact, in this case $N(t_2) = N(t_1) + (N(t_2) - N(t_1)) = \infty$. This is a contradiction).
4) The probability of $\{N(a+t) - N(a) = k\}$ is the same as that of $\{N(t) = k\}$, for all $k = 0, 1, 2, \cdots$, and all $t \geq 0$, and, moreover, it cannot be that $P\{N(t) = 0\} \equiv 1, \forall t \geq 0$. In other words, in any time interval $(a, a+t]$ the probability of that k-jumps happen only depends on the length of the interval t, and not on the starting point a. Moreover, it cannot always be that no jumps occur, $\forall t \geq 0$. (Otherwise, $N(t) = 0$, $\forall t \geq 0$, and we cannot call it a Poisson (jump) process. At most it can only be called a degenerate Poisson process or a constant process).

One can prove that the above four properties characterize a (stationary) Poisson process. (See [26]). That is, it is a necessary and sufficient conditions for a (stationary) Poisson process.

Furthermore, if we are also concerned with the jump value falling in a set U, then we will meet the counting measure $N_p((0,t], U)$ generated by a random point process $p(\cdot)$. Suppose that $N^U(t) = N_p((0,t], U)$ is a (stationary) Poisson process for any $U \in \mathfrak{B}_Z$ (that is, U is a measurable set). The by the above property 4) there exists a function $\lambda(U) > 0$ such that $EN_p((0,t], U) = \lambda(U)t$. Suppose that, $N_p((0,t], U)$ also has the following property:

5) $N_p((0,t], U)$ is a non-negative finite measure on $U \in \mathfrak{B}_Z$, and as $U_1, \cdots, U_n \in \mathfrak{B}_Z$, the random variables $N_p((0,t], U_1), \cdots, N_p((0,t], U_n)$ are independent. In other words, the events that the jump values fall in disjoint sets occur independently.
Then $N_p((0,t], U)$ is a finite Poisson random measure, so $p(\cdot)$ is a finite Poisson random point process. (See Definition 51 in Chapter 1).

However, since in some cases we will also meet the case that the function $\lambda(U) > 0$ is only σ-finite, (that is, there exists a sequence of measurable sets $Z_1 \subset \cdots \subset Z_n \subset \cdots$ such that $Z_n \in \mathfrak{B}_Z, \cup_{n=1}^\infty Z_n = Z$, and $\lambda(Z_n) < \infty$); and for $\forall U \in \mathfrak{B}_Z$ such that $\lambda(U) < \infty$, then properties 1) - 5) hold for $N_p((0,t], U)$. For such a general case we will call the random point process a σ-finite Poisson point process, or simply just a Poisson Point process. (Actually, if $\lambda(U) = \infty$, then one understands that $N_p((0,t], U) \equiv \infty, \forall t \geq 0$. So U is only a very extreme special set. We do not need to check its other property).

Now let us assume that $0 \leqslant \tau \leqslant T_0$,where T_0 is a fixed number, and b in (8.8) is a R^d-valued function, and $X \in \mathfrak{F}_\tau$.

It is known that the study of (8.8) is useful for the option pricing in the financial market (See the Introduction above and [8]).

For the precise definition of solution (8.8) we need the following notation.

$$S^2_{\mathfrak{F}}(R^d) = \left\{ \begin{array}{c} f(t,\omega) : f(t,\omega) \text{ is } \mathfrak{F}_t - \text{adapted}, R^d - \text{valued such that} \\ E\sup_{t\geqslant 0} |f(t\wedge\tau,\omega)|^2 < \infty \end{array} \right\},$$

$$L^2_{\mathfrak{F}}(R^{d\otimes d_1}) = \left\{ \begin{array}{c} f(t,\omega) : f(t,\omega) \text{ is } \mathfrak{F}_t - \text{adapted}, R^{d\otimes d_1} - \text{valued} \\ \text{such that } E\int_0^\tau |f(t,\omega)|^2 dt < \infty \end{array} \right\},$$

and

$$F^2_{\mathfrak{F}}(R^{d\otimes d_2}) = \left\{ \begin{array}{c} f(t,z,\omega) : f(t,z,\omega) \text{ is } R^{d\otimes d_2} - \text{valued}, \mathfrak{F}_t\text{-predictable} \\ \text{such that } E\int_0^\tau \int_Z |f(t,z,\omega)|^2 \pi(dz)dt < \infty \end{array} \right\}.$$

Definition 226 . *(x_t, q_t, p_t) is said to be a solution of* (8.8), *iff* $(x_t, q_t, p_t) \in S^2_{\mathfrak{F}}(R^d) \times L^2_{\mathfrak{F}}(R^{d\otimes d_1}) \times F^2_{\mathfrak{F}}(R^{d\otimes d_2})$, *and it satisfies* (8.8).

Remark 227 *As in Remark 225, if $X \in \mathfrak{F}^w_\tau$, and τ is a $\mathfrak{F}^w_t$−stopping time, then (8.8) will reduce to a continuous BSDE* (8.9) *with the coefficient $b(t,x,q,0,\omega)$. So the solution for a continuous BSDE also reduces to $(x_t, q_t) \in S^2_{\mathfrak{F}}(R^d) \times L^2_{\mathfrak{F}}(R^{d\otimes d_1})$, and it satisfies* (8.9) *with the coefficient $b(t,x,q,0,\omega)$..*

From Definition 226 it is seen that when discussing the solution of (8.8) we always need to assume that b satisfies the following assumption
$(\mathrm{A})_1$ $b : [0,T_0] \times R^d \times R^{d\otimes d_1} \times L^2_{\pi(\cdot)}(R^{d\otimes d_2}) \times \Omega \to R^d$
is jointly measurable, $\mathfrak{F}_t$−adapted, where

$$L^2_{\pi(\cdot)}(R^{d\otimes d_2}) = \left\{ \begin{array}{c} f(z) : f(z) \text{ is } R^{d\otimes d_2} - \text{valued, and} \\ \|f\|^2 = \int_Z |f(z)|^2 \pi(dz) < \infty \end{array} \right\}$$

Moreover, to simplify the discussion of (8.8) we also suppose that
all $N_{k_i}(ds,Z), 1 \leqslant i \leqslant d_2$, have no common jump time;
i.e. we always make the following assumption
$(\mathrm{A})_2$ $N_{k_i}(\{t\},U)N_{k_j}(\{t\},U) = 0$, as $i \neq j$, for all $U \in \mathfrak{B}(Z)$, such that $\pi(U) < \infty$.

By Definition 226 it is also seen that the following remark holds:

Remark 228 . *If (x_t, q_t, p_t) satisfies* (8.8) , *and*
$E\left[\left(\int_0^\tau |b(s,x_s,q_s,p_s,\omega)|\, ds\right)^2\right] < \infty, E|X|^2 < \infty,$
then
$(x_t, q_t, p_t) \in S^2_{\mathfrak{F}}(R^d) \times L^2_{\mathfrak{F}}(R^{d\otimes d_1}) \times F^2_{\mathfrak{F}}(R^{d\otimes d_2})$
$\Longleftrightarrow (x_t, q_t, p_t) \in L^2_{\mathfrak{F}}(R^d) \times L^2_{\mathfrak{F}}(R^{d\otimes d_1}) \times F^2_{\mathfrak{F}}(R^{d\otimes d_2})$

In fact, since $\tau \leqslant T_0$, $S^2_{\mathfrak{F}}(R^d) \subset L^2_{\mathfrak{F}}(R^d)$. Hence the implication is true. Conversely, by (8.8) and by the martingale inequality, one also easily sees that under the assumption of this remark
$E\sup_{t\geqslant 0} |x_{t\wedge\tau}|^2 \leqslant 4(E|X|^2 + E\left(\int_0^\tau |b(s,x_s,q_s,p_s,\omega)|\, ds\right)^2$
$+4E\int_0^\tau |q_s|^2 ds + 4E\int_0^\tau \int_Z |p_s(z)|^2 \pi(dz)ds) < \infty.$
Hence

$L^2_{\mathfrak{F}}(R^d) \times L^2_{\mathfrak{F}}(R^{d\otimes d_1}) \times F^2_{\mathfrak{F}}(R^{d\otimes d_2})$
$\subset S^2_{\mathfrak{F}}(R^d) \times L^2_{\mathfrak{F}}(R^{d\otimes d_1}) \times F^2_{\mathfrak{F}}(R^{d\otimes d_2})$.

8.3.2 A Priori Estimate and Uniqueness of a Solution

As in the SDE case we have the a priori estimate and the uniqueness results on solutions.

Lemma 229 *Assume that (x_t, q_t, p_t) solves (8.8), $\tau \leqslant T_0, E|X|^2 < \infty$ and assume that b satisfies that*
1° $x \cdot b(t,x,q,p,\omega) \leqslant c_1(t)(1+|x|+|x|^2) + c_2(t)|x|(|q|+\|p\|)$,
where $a \cdot b$ means the inner product of $a, b \in R^d, 0 \leqslant c_i(t), i = 1,2,$ are non-random, such that
$c_0 = \int_0^{T_0} c_1(t)dt + \int_0^{T_0} (c_2(t))^2 dt < \infty.$
Then
$E(\sup_{t\geqslant 0} |x_{t\wedge\tau}|^2 + \int_0^\tau (|q_t|^2 + \|p_t\|^2)dt) \leqslant k_0 < \infty,$
where $k_0 \geqslant 0$ is a constant depending on c_0 and $E|X|^2$ only.

Proof. By Ito's formula

$$\begin{aligned}
&|x_t|^2 + \int_{t\wedge\tau}^\tau |q_s|^2 ds + \int_{t\wedge\tau}^\tau \int_Z |p_s(z)|^2 N(ds,dz) \\
&= |X|^2 + 2\int_{t\wedge\tau}^\tau x_s \cdot b(s,x_s,q_s,p_s,\omega)ds \\
&- 2\int_{t\wedge\tau}^\tau x_s \cdot q_s dw_s - 2\int_{t\wedge\tau}^\tau \int_Z x_s \cdot p_s(z)\tilde{N}_k(ds,dz). \qquad (8.11)
\end{aligned}$$

Since
$E[\int_0^\tau |x_s \cdot q_s|^2 ds]^{1/2} \leqslant 2E\sup_{0\leqslant t} |x_{t\wedge\tau}|^2 + 2E\int_0^\tau |q_s|^2 ds < \infty.$
$\int_0^{t\wedge\tau} x_s \cdot q_s dw_s$ is a martingale. In fact, let
$\sigma_N = \inf\left\{t \geqslant 0 : \int_0^\tau |x_s \cdot q_s|^2 ds > N\right\}.$
Then for each $N = 1,2,\cdots$, we have that $\left\{\int_0^{t\wedge\tau\wedge\sigma_N} x_s \cdot q_s dw_s\right\}_{t\geqslant 0}$ is a square integrable martingale, i.e. $\forall A \in \mathfrak{F}_r$, $0 \leqslant r \leqslant t$,
$E\int_0^{t\wedge\tau\wedge\sigma_N} x_s \cdot q_s dw_s I_A = E\int_0^{r\wedge\tau\wedge\sigma_N} x_s \cdot q_s dw_s I_A.$ $(*)$
So by the Fatou lemma
$E\left|\int_0^{t\wedge\tau} x_s \cdot q_s dw_s\right| = E\underline{\lim}_{N\to\infty}\left|\int_0^{t\wedge\tau\wedge\sigma_N} x_s \cdot q_s dw_s\right|$
$\leqslant c_0\underline{\lim}_{N\to\infty} E[\int_0^{t\wedge\tau\wedge\sigma_N} |x_s \cdot q_s|^2 ds]^{1/2}$
$\leqslant 2E\sup_{0\leqslant t} |x_{t\wedge\tau}|^2 + 2E\int_0^\tau |q_s|^2 ds < \infty.$
Moreover, by the Lebesgue dominated convergence theorem as $N \to \infty$
$E\left|\int_0^{t\wedge\tau} x_s \cdot q_s I_A(I_{s\leqslant\sigma_N} - 1)dw_s\right|$
$\leqslant c_0' E\left|\int_0^{t\wedge\tau} |x_s \cdot q_s I_A(I_{s\leqslant\sigma_N} - 1)|^2 ds\right|^{1/2} \to 0.$

Hence letting $N \to \infty$ in $(*)$ one finds that $\int_0^{t\wedge\tau} x_s \cdot q_s dw_s$ is a martingale. A similar conclusion holds for $\int_0^{t\wedge\tau}\int_Z x_s \cdot p_s(z)\tilde{N}_k(ds,dz)$. Hence

$E[|x_t|^2 + \frac{1}{2}\int_{t\wedge\tau}^{\tau}|q_s|^2 ds + \frac{1}{2}\int_{t\wedge\tau}^{\tau}\int_Z |p_s(z)|^2\pi(dz)ds]$
$\leqslant E|X|^2 + E\int_{t\wedge\tau}^{\tau}(2c_1(s)+4c_2(s)^2)|x_s|^2 ds$
$+E\int_{t\wedge\tau}^{\tau}c_1(s)|x_s|^2 ds + 2E\int_{t\wedge\tau}^{\tau}c_1(s)ds,$

where we have used the result that

$2\int_{t\wedge\tau}^{\tau}c_1(s)|x_s|ds \leqslant \int_{t\wedge\tau}^{\tau}c_1(s)|x_s|^2 ds + \int_{t\wedge\tau}^{\tau}c_1(s)ds.$

Write

$y_t = \begin{cases} x_t, & \text{as } t\in[0,\tau], \\ x_\tau, & \text{as } t>\tau. \end{cases}$

$E[|y_t|^2 + \frac{1}{2}\int_{t\wedge\tau}^{\tau}|q_s|^2 ds + \frac{1}{2}\int_{t\wedge\tau}^{\tau}\int_Z |p_s(z)|^2\pi(dz)ds]$
$\leqslant k_0' + \int_t^{T_0}(3c_1(s)+4c_2(s)^2)E|y_s|^2 ds,$

where

$k_0' = E|X|^2 + 2\int_0^{T_0}c_1(s)ds.$

By Gronwall's inequality one easily finds that

$$E[|y_t|^2 + \frac{1}{2}\int_{t\wedge\tau}^{\tau}|q_s|^2 ds + \frac{1}{2}\int_{t\wedge\tau}^{\tau}\int_Z |p_s(z)|^2\pi(dz)ds]$$
$$\leqslant k_0' e^{\int_0^{T_0}(3c_1(s)+4c_2(s)^2)ds} = k_0'' < \infty. \tag{8.12}$$

Moreover,

$\sup_{s\geqslant 0}E|x_{s\wedge\tau}|^2 \leqslant k_0''.$

Now by (8.11), using the martingale inequality, one has

$\frac{1}{2}E\sup_{s\geqslant 0}|x_{s\wedge\tau}|^2 \leqslant E|X|^2 + k_0''\int_0^{T_0}[3c_1(s)+4c_2(s)^2]ds$
$+k_0'E\int_0^{\tau}(|q_t|^2 + \|p_t\|^2)dt + 2\int_0^{T_0}c_1(s)ds$
$\leqslant E|X|^2 + (k_0''+1)\int_0^{T_0}[5c_1(s)+4c_2(s)^2]ds + 2k_0'k_0'',$

where we have used the result that $\int_{t\wedge\tau}^{\tau}x_s\cdot q_s dw_s = (\int_0^{\tau} - \int_0^{r\wedge\tau})x_s\cdot q_s dw_s,$ and

$E\sup_{r\in[0,t]}\left|\int_0^{r\wedge\tau}x_s\cdot q_s dw_s\right| \leqslant \tilde{k}_0 E[\int_0^{t\wedge\tau}|x_s\cdot q_s|^2 ds]^{1/2}$
$\leqslant \frac{1}{4}E\sup_{s\geqslant 0}|x_{s\wedge\tau}|^2 + 4\tilde{k}_0 E[\int_0^{t\wedge\tau}|q_s|^2 ds],$

etc. Hence one finds that

$E\sup_{s\in[0,t]}|x_{s\wedge\tau}|^2 \leqslant k_0''' < \infty.$ ■

Lemma 230 *Under the assumption in Lemma 229, if*

$(x_1 - x_2)\cdot(b(t,x_1,q_1,p_1\omega) - b(t,x_2,q_2,p_2,\omega))$
$\leqslant c_1(t)\rho(|x_1-x_2|^2) + c_2(t)|x_1-x_2|(|q_1-q_2| + \|p_1-p_2\|)],$

where $c_i(t), i=1,2,$ satisfy the same condition as that in Lemma 229, and $\rho(u)\geqslant 0$, as $u\geqslant 0$ is non-random, increasing, continuous and concave such that

$\int_{0+}du/\rho(u) = \infty,$

then the solution of (8.8) is unique.

Proof. Let

$X_t = x_t^1 - x_t^2, \quad Q_t = q_t^1 - q_t^2, \quad P_t = p_t^1 - p_t^1.$

Then as in (3.2) by Ito's formula one finds that

$Z_t = E[|X_t|^2 + \frac{1}{2}\int_{t\wedge\tau}^{\tau}|Q_s|^2 ds + \frac{1}{2}\int_{t\wedge\tau}^{\tau}\int_Z |P_s(z)|^2 \pi(dz)ds]$

$\leqslant E\int_{t\wedge\tau}^{\tau}\left[2c_1(s)\rho\left(|X_s|^2\right) + 4c_2(s)^2|X_s|^2\right]ds$

$\leqslant 4\int_t^{T_0}(c_1(s)+c_2(s)^2)\rho_1\left(|Z_s|^2\right)ds$

where we define $Z_s = 0$, for $s > \tau$, and $\rho_1(u) = \rho(u) + u$. Hence by Lemma 235 below $P - a.s.$

$Z_t = 0, \forall t \in [0,\tau].$ ∎

Unlike the SDE theory, an interesting thing is that for BSDE we have the following result: If the terminal variable X is bounded, then the solution is always bounded.

Lemma 231 *Assume that 1° in Lemma 229 holds, $\tau \leqslant T_0$, and there exists a constant $k_0 \geqslant 0$ such that*

$|X| \leqslant k_0, P - a.s.$

If (x_t, q_t, p_t) is a solution of (8.8), then $P - a.s.$

$|x_t| \leqslant N_0, \forall t \in [0,\tau], \omega \in \Omega,$

where $N_0 \geqslant 0$ is a constant depending on $\int_0^{T_0} c_1(t)dt$, $\int_0^{T_0}(c_1(t))^2 dt$ and k_0 only.

Proof. Similar to (8.12) one obtains

$|x_t|^2 \leqslant E^{\mathfrak{F}_t}[|x_t|^2 + \frac{1}{2}\int_{t\wedge\tau}^{\tau}|q_s|^2 ds + \frac{1}{2}\int_{t\wedge\tau}^{\tau}\int_Z |p_s(z)|^2 \pi(dz)ds]$

$\leqslant [E^{\mathfrak{F}_t}|X|^2 + \int_0^{T_0} c_1(s)ds]e^{\int_0^{T_0}(3c_1(s)+4c_2(s)^2)ds},$

where we write $E^{\mathfrak{F}_t}[\cdot] = E[\cdot|\mathfrak{F}_t]$. Hence the conclusion now follows. ∎

Remark 232 *By the proof one sees that the condition in Lemma 231 can be weakened to $E^{\mathfrak{F}_t}|X|^2 \leq k_0, \forall t \in [0,\tau]$.*

8.3.3 Existence of Solutions for the Lipschitzian Case

In this section we are going to discuss the existence of a solution to BSDE (8.8). As usual, to show the existence of a solution for some equation, if the coefficient in this equation is Lipschitzian, one can use the contraction principle. However, to let the contraction principle work, we first need to discuss the following simple BSDE:

$$\begin{aligned} x_t = X &+ \int_{t\wedge\tau}^{\tau} f(s,\omega)ds - \int_{t\wedge\tau}^{\tau} q_s dw_s \\ &- \int_{t\wedge\tau}^{\tau}\int_Z p_s(z)\tilde{N}_k(ds,dz), 0 \leqslant t. \end{aligned} \tag{8.13}$$

We have

Lemma 233 . *If τ is a bounded $\mathfrak{F}_t-$ stopping time, X is $\mathfrak{F}_\tau-$ measurable and R^d- valued, $f(t,\omega)$ is $\mathfrak{F}_t-$ adapted and R^d- valued such that*
$E|X|^2 < \infty, E\left|\int_0^\tau |f(s,\omega)|\,ds\right|^2 < \infty,$
then (8.13) *has a unique solution.*

Proof. Let
$x_t = E^{\mathfrak{F}_t}(X + \int_{t\wedge\tau}^\tau f(s)ds).$
This makes sense, since by assumption $E^{\mathfrak{F}_t}(\left|X + \int_{t\wedge\tau}^\tau f(s)ds\right|^2) < \infty, \forall t \geqslant 0.$
Hence $x_0 = E^{\mathfrak{F}_0}(X + \int_0^\tau f(s)ds)$. Notice that

$M_t = E^{\mathfrak{F}_t}(X + \int_0^\tau f(s)ds)$ is a square integrable martingale. Hence by the martingale representation theorem (see Theorem 103) there exists a unique $(q_t, p_t) \in L^2_{\mathfrak{F}}(R^{d\otimes d_1}) \times F^2_{\mathfrak{F}}(R^{d\otimes d_2})$ such that

$$M_t = M_0 + \int_0^t q_s dw_s + \int_0^t \int_Z p_s(z)\tilde{N}_k(ds,dz), 0 \leqslant t. \tag{8.14}$$

However
$M_{t\wedge\tau} = E^{\mathfrak{F}_{t\wedge\tau}}(X + \int_0^\tau f(s)ds) = x_t + \int_0^{t\wedge\tau} f(s)ds,$
and by (8.14)
$X + \int_0^\tau f(s)ds = x_0 + \int_0^\tau q_s dw_s + \int_0^\tau \int_Z p_s(z)\tilde{N}_k(ds,dz).$
Hence
$X + \int_{t\wedge\tau}^\tau f(s,\omega)ds - \int_{t\wedge\tau}^\tau q_s dw_s - \int_{t\wedge\tau}^\tau \int_Z p_s(z)\tilde{N}_k(ds,dz)$
$= x_0 + \int_0^{t\wedge\tau} q_s dw_s + \int_0^{t\wedge\tau} \int_Z p_s(z)\tilde{N}_k(ds,dz) - \int_0^{t\wedge\tau} f(s)ds$
$= M_{t\wedge\tau} - \int_0^{t\wedge\tau} f(s)ds = x_t.$
We have now that (x_t, q_t, p_t) satisfies (8.13) . Now by (8.13)
$E\sup_{t\in[0,\tau]} |x_t|^2 \leqslant E[3\left||X| + \left|\int_0^\tau |f(s)|\,ds\right|\right|^2 + 6\int_0^\tau (|q_s|^2 + \|p_s\|^2)ds]$
$< \infty.$
Hence $x_t \in S^2_{\mathfrak{F}}(R^d)$. Therefore (x_t, q_t, p_t) is a solution of (8.13) .
The uniqueness of the solution to (8.13) follows from Ito's formula (Theorem 93).LabI ■

Now we can show an existence and uniqueness result for the solution to (8.8) by using the contraction principle.

Theorem 234 *Assume that* $\tau \leqslant T_0$
1° $b(t,x,q,p,\omega) : [0,T_0] \times R^d \times R^{d\otimes d_1} \times L^2(R^{d\otimes d_2}) \times \Omega \longmapsto R^d$
is a $\mathfrak{F}_t-$ adapted and measurable process such that $P-a.s.$
$|b(t,x,q,p,\omega)| \leqslant c_1(t)(1+|x|) + c_2(t)(1+|q|+\|p\|),$
where $c_1(t)$ and $c_2(t)$ are non-negative and non-random such that
$\int_0^{T_0} c_1(t)dt + \int_0^{T_0} c_2(t)^2 dt < \infty;$
2° $|b(t,x_1,q_1,p_1,\omega) - b(t,x_2,q_2,p_2,\omega)| \leqslant c_1(t)|x_1 - x_2|$
$+c_2(t)[|q_1 - q_2| + \|p_1 - p_2\|],$
where $c_1(t)$ and $c_2(t)$ satisfy the same conditions in 1°;
3° $X \in \mathfrak{F}_\tau, E|X|^2 < \infty.$
Then (8.8) has a unique solution.

Proof. Let us use the contractive mapping principle to show this result. Introduce a new norm as follows:

For $(x_\cdot, q_\cdot, p_\cdot) \in \widetilde{B} = L^2_{\mathfrak{F}}(R^d) \times L^2_{\mathfrak{F}}(R^{d\otimes d_1}) \times F^2_{\mathfrak{F}}(R^{d\otimes d_2})$ let

$\|(x_\cdot, q_\cdot, p_\cdot)\|^2_M = \sup_{t\leqslant T_0} e^{-bA(t)} E\,|x_{t\wedge\tau}|^2$

$+ \sup_{t\leqslant T_0} e^{-bA(t)} E \int_{t\wedge\tau}^{\tau} |q_t|^2\, dt + \sup_{t\leqslant T_0} e^{-bA(t)} E \int_{t\wedge\tau}^{\tau} \|p_t\|^2\, dt,$

where $b \geqslant 0$ is a constant, which will be determined later, and $A(t) = \int_t^{T_0}(c_1(s) + c_2(s)^2)ds$. Write

$H = \left\{(x_\cdot, q_\cdot, p_\cdot) \in \widetilde{B} : \|(x_\cdot, q_\cdot, p_\cdot)\|_M < \infty\right\}.$

Then H is a Banach space. Now by Lemma 233 for any $\left(\overline{x}^i_\cdot, \overline{q}^i_\cdot, \overline{p}^i_\cdot\right) \in H, i = 1, 2,$ there exist unique solutions $\left(x^i_\cdot, q^i_\cdot, p^i_\cdot\right), i = 1, 2,$ of the following BSDE

$x^i_t = X + \int_{t\wedge\tau}^{\tau} b(s, \overline{x}^i_s, \overline{q}^i_s, \overline{p}^i_s, \omega)ds - \int_{t\wedge\tau}^{\tau} q^i_s dw_s$

$- \int_{t\wedge\tau}^{\tau} \int_Z p^i_s(z)\tilde{N}_k(ds, dz),\ 0 \leqslant t,\ i = 1, 2;$

moreover, by Lemma 229 $\left(x^i_\cdot, q^i_\cdot, p^i_\cdot\right) \in H, i = 1, 2$. Let

$X_t = x^1_t - x^2_t,\quad Q_t = q^1_t - q^2_t,\quad P_t = p^1_t - p^1_t.$

Similarly, define $\overline{X}_t, \overline{Q}_t$, and $\overline{P}_t$. Then by Ito's formula as in (3.2) one gets that

$E[|X_t|^2 + \int_{t\wedge\tau}^{\tau} |Q_s|^2\, ds + \int_{t\wedge\tau}^{\tau} \|P_s\|^2\, ds]$

$\leqslant \gamma^{-1} E[\int_{t\wedge\tau}^{\tau} c_1(s) \left|\overline{X}_s\right|^2 ds + \int_{t\wedge\tau}^{\tau} \left|\overline{Q}_s\right|^2 ds + \int_{t\wedge\tau}^{\tau} \left\|\overline{P}_s\right\|^2 ds]$

$+\gamma E \int_{t\wedge\tau}^{\tau}(c_1(s) + c_2(s)^2)\, |X_s|^2\, ds$

$= \gamma^{-1} I^1_t + \gamma E \int_{t\wedge\tau}^{\tau}(c_1(s) + c_2(s)^2)\, |X_s|^2\, ds$

$\leqslant \gamma^{-1} I^1_t + \gamma \int_t^{T_0}(c_1(s) + c_2(s)^2) E\, |X_{s\wedge\tau}|^2\, ds$

$\leqslant \gamma^{-1} I^1_t + \int_t^{T_0} \exp(\gamma(A(t) - A(s)))(c_1(s) + c_2(s)^2) I^1_s ds,$

where we have applied Lemma 235 below. Notice that $0 \leqslant A(t)$ is decreasing, so

$e^{-bA(t)} E \int_{t\wedge\tau}^{\tau} c_1(s) \left|\overline{X}_s\right|^2 ds \leqslant e^{-bA(t)} \int_t^{T_0} c_1(s) E \left|\overline{X}_{s\wedge\tau}\right|^2 ds$

$\leqslant \sup_{s\leqslant T_0} e^{-bA(s)} E \left|\overline{X}_{s\wedge\tau}\right|^2 \int_0^{T_0} c_1(s)ds = \sup_{s\leqslant T_0} e^{-bA(s)} E \left|\overline{X}_{s\wedge\tau}\right|^2 \widetilde{k}_0,$

$e^{-bA(t)} \int_t^{T_0} \exp(\gamma(A(t) - A(s)))(c_1(s) + c_2(s)^2) E \int_{s\wedge\tau}^{\tau} c_1(r) \left|\overline{X}_r\right|^2 drds$

$\leqslant \int_t^{T_0} e^{-(b-\gamma)(A(t)-A(s))}(c_1(s) + c_2(s)^2) ds \sup_{s\leqslant T_0} e^{-bA(s)} E \left|\overline{X}_{s\wedge\tau}\right|^2 \widetilde{k}_0$

$\leqslant \sup_{s\leqslant T_0} e^{-bA(s)} E \left|\overline{X}_{s\wedge\tau}\right|^2 \widetilde{k}_0 (b-\gamma)^{-1}, for 0 < \gamma < b.$

Now write $u_t = E \int_{t\wedge\tau}^{\tau} |Q_s|^2\, ds$ or $E \int_{t\wedge\tau}^{\tau} \left\|\overline{P}_s\right\|^2 ds$. Then

$e^{-bA(t)} \int_t^{T_0} \exp(\gamma(A(t) - A(s)))(c_1(s) + c_2(s)^2) u(s) ds$

$\leqslant \sup_{s\leqslant T_0} e^{-bA(s)} u(s) \int_t^{T_0} e^{-(b-\gamma)(A(t)-A(s))}(c_1(s) + c_2(s)^2) ds$

$\leqslant \sup_{s\leqslant T_0} e^{-bA(s)} u(s) (b-\gamma)^{-1}.$

Hence

$\|(X_\cdot, Q_\cdot, P_\cdot)\|^2_M \leqslant \max(\gamma^{-1}(\widetilde{k}_0 + 2), (b-\gamma)^{-1}(\widetilde{k}_0 + 2) \left\|\left(\overline{X}_\cdot, \overline{Q}_\cdot, \overline{P}_\cdot\right)\right\|^2_M.$

After an appropriate choise of γ and b, one easily completes the proof. ∎

Lemma 235 . *(Gronwall's inequality). If* $0 \leqslant y_t \leqslant \gamma v_t + \int_t^{T_0} c(s) y_s ds, \forall t \geqslant 0$, *where* $\gamma > 0$ *is a constant, and* $c(s) \geqslant 0$, *then*

$y_t \leqslant \gamma v_t + \gamma \int_t^{T_0} \exp\left(\int_t^s c(r)dr\right) c(s) v_s ds.$

Proof. Let $0 \leqslant z_t = \int_t^{T_0} c(s) y_s ds$, $g_t = y_t - \gamma v_t - z_t \leqslant 0$. Then
$-\dot{z}_t = c(t) y_t = c(t)\left(g_t + \gamma v_t + z_t\right).$
From this one gets
$z_t = \int_t^{T_0} \exp\left(\int_t^s c(r)dr\right) c(s)\left(g_s + \gamma v_s\right) ds$
$\leqslant \int_t^{T_0} \exp\left(\int_t^s c(r)dr\right) c(s) \gamma v_s ds.$
Hence
$y_t \leqslant \gamma v_t + z_t \leqslant \gamma v_t + \gamma \int_t^{T_0} \exp\left(\int_t^s c(r)dr\right) c(s) v_s ds.$ ∎

Now let us give some examples to show that the conditions on $c_1(t)$ and $c_2(t)$ in Theorem 234 cannot be weakened.

Example 236 . *(Condition $\int_0^{T_0} c_1(t)dt < \infty$ cannot be weakened). Consider the following BSDE $0 \leqslant t \leqslant T$,*
$x_t = 1 + \int_t^T I_{s\neq 0} s^{-\alpha} x_s ds - \int_t^T q_s dw_s - \int_t^T \int_Z p_s(z) \widetilde{N}_k(ds, dz).$
Obviously, if $\alpha < 1$, then by Theorem 234 it has a unique solution $(x_t, 0, 0)$, where x_t is non-random. However, if $\alpha \geqslant 1$, it has no solution. Otherwise, for the solution (x_t, q_t, p_t) one has
$Ex_t^i = 1 + \int_t^T I_{s\neq 0} s^{-\alpha} Ex_s^i ds = e^{\int_t^T I_{s\neq 0} s^{-\alpha} ds}, \forall i = 1, 2, \cdots, d.$
Hence
$Ex_0^i = \infty, \forall i = 1, 2, \cdots, d,$ *as* $\alpha \geqslant 1$,
which is a contradiction.

Example 237 *(Condition $\int_0^{T_0} c_2(t)^2 dt < \infty$ cannot be weakened). Now suppose that all processes n BSDE (8.8) are real-valued. Let*
$X = \int_0^T I_{s\neq 0}(1+s)^{-1/2}(\log(1+s))^{-\alpha_2} dw_s$
$+ \int_0^T \int_Z I_{s\neq 0}(1+s)^{-1/2}(\log(1+s))^{-\alpha_3} I_U(z) \widetilde{N}_k(ds, dz),$
$b = I_{s\neq 0}(1+s)^{-1/2}(\log(1+s))^{-\alpha_1}(\widetilde{k}_1 |q| + \widetilde{k}_2 \int_Z |p(z)| I_U(z)\pi(dz)),$
where $\widetilde{k}_1, \widetilde{k}_2 \geqslant 0$ are constants, and we assume that $0 < \alpha_2, \alpha_3 < \frac{1}{2}$, and $0 < \pi(U) < \infty$. Obviously, if $0 < \alpha_1 < \frac{1}{2}$, then Theorem 234 applies, and (8.8) has a unique solution. However, if $\alpha_1 > \frac{1}{2}$, and $\widetilde{k}_1 > 0, \alpha_1 + \alpha_2 \geqslant 1$, or $\widetilde{k}_2 > 0, \alpha_1 + \alpha_3 \geqslant 1$, then (8.8) has no solution. Otherwise, the solution (x_t, q_t, p_t) should satisfy
$x_t = \int_0^t I_{s\neq 0}(1+s)^{-1/2}(\log(1+s))^{-\alpha_2} dw_s$
$+ \int_0^t \int_Z I_{s\neq 0}(1+s)^{-1/2}(\log(1+s))^{-\alpha_3} I_U(z) \widetilde{N}_k(ds, dz)$
$+ \int_t^T I_{s\neq 0}(1+s)^{-1}[\widetilde{k}_1(\log(1+s))^{-(\alpha_1+\alpha_2)} + \widetilde{k}_2(\log(1+s))^{-(\alpha_1+\alpha_3)}\pi(U)]ds,$
$q_s = I_{s\neq 0}(1+s)^{-1/2}(\log(1+s))^{-\alpha_2},$
$p_s(z) = I_{s\neq 0}(1+s)^{-1/2}(\log(1+s))^{-\alpha_3} I_U(z).$
Hence one has that
$Ex_0 = \int_0^T I_{s\neq 0}(1+s)^{-1}[\widetilde{k}_1(\log(1+s))^{-(\alpha_1+\alpha_2)}$
$+ \widetilde{k}_2(\log(1+s))^{-(\alpha_1+\alpha_3)}\pi(U)]ds = \infty,$

in the case $\widetilde{k}_1 > 0, \alpha_1 + \alpha_2 \geqslant 1$, *or* $\widetilde{k}_2 > 0, \alpha_1 + \alpha_3 \geqslant 1$, *which is a contradiction.*

8.4 Explanation of the Solution of BSDE to Option Pricing

8.4.1 Continuous Case

1. Option pricing in the continuous case.

Now suppose that in a financial market there is a bond and only one stock, and this stock is only disturbed by a continuou stochastic perturbation; that is, it is a 1−dimensional BM. Now we are going to price an option related to this stock. As in the previous sections we know that x_t (the price for the option) with π_t^1 (the portfolio) should satisfy the following BSDE in (8.6):

$$\begin{cases} dx_t = r_t x_t dt + \pi_t(b_t - r_t)dt + \pi_t \sigma_t dw_t, \\ x_T = (P_T - K)^+, \ 0 \leq t \leq T. \end{cases} \tag{8.15}$$

Now let $q_t = \pi_t \sigma_t$. Then (8.15) becomes

$$\begin{cases} dx_t = r_t x_t dt + q_t \sigma_t^{-1}(b_t - r_t)dt + q_t dw_t, \\ x_T = (P_T - K)^+, \ 0 \leq t \leq T. \end{cases}$$

By Theorem 234 this BSDE has a unique solution $(x_t, q_t), 0 \leq t \leq T$, provided the following conditions hold:

(A) r_t, b_t, and σ_t are all non-random such that σ_t^{-1} exists and
$\int_0^T [|r_t| + |\sigma_t^{-1}(b_t - r_t)|^2]dt < \infty$;
moreover, $E[(P_T - K)^+]^2 < \infty$.

Hence (8.15) also has a unique solution $(x_t, \pi_t^1), 0 \leq t \leq T$, where $\pi_t = q_t \sigma_t^{-1}$, provided assumption (A) holds. A particular case for assumption (A) to be true is that r_t, b_t, and σ_t^{-1} are all non-random and bounded. This means precisely that under assumption (A) the option can be priced as x_t.

The same idea can be applied to use the BSDE solution to hedge the contingent claim.

2. Hedging contingent claims in the continuous case.

For simplicity, as in the Introduction of this chapter, suppose that in a financial market there is a bond, its price equation is (8.1), and there are two kinds of stocks disturbed by two independent contiuous stochastic perturbations, with their price SDEs described by (8.2) (that is, $N = 2$, and $d = 2$ in (8.2)):

$$dP^i(t,\omega) = P^i(t,\omega)[(b^i(t,\omega)dt + \textstyle\sum_{k=1}^2 \sigma^{ik}(t,\omega)dw_t^k],$$
$$P^i(0,\omega) = P^i(0,\omega), \quad i = 1,2;$$

where $w_t = (w_t^1, w_t^2)^T$ is a 2−dimensional BM. Then by means of a derivation similar to that in (8.6), hedging a contingent claim $X \in \mathfrak{F}_T^w$ will lead to finding a $\mathfrak{F}_t^w$−adapted solution $(x_t, \pi_t) \in R^1 \times R^2$ (that is, $\pi_t = (\pi_t^1, \pi_t^2)$) solving the following BSDE

$$\begin{cases} dx_t = r_t x_t dt + \pi_t(b_t - r_t \underline{1})dt + \pi_t \sigma_t dw_t, \\ x_T = X, \ t \in [0, T], \end{cases} \tag{8.16}$$

where $\underline{1} = (1,1)^*$ is a 2−dimensional constant vector, and
$b_t = \begin{pmatrix} b_t^1 \\ b_t^2 \end{pmatrix} \in R^{2\otimes 1}, \sigma_t = \begin{pmatrix} \sigma_t^{11} & \sigma_t^{12} \\ \sigma_t^{21} & \sigma_t^{22} \end{pmatrix} \in R^{2\otimes 2}.$

In other words, if the solution $(x_t, \pi_t), 0 \leq t \leq T$ of (8.16) exists, then it can reproduce X. Now let $q_t = \pi_t \sigma_t$. Then (8.16) becomes
$\begin{cases} dx_t = r_t x_t dt + q_t \sigma_t^{-1}(b_t - r_t \underline{1})dt + q_t dw_t, \\ x_T = X, \ 0 \leq t \leq T. \end{cases}$

By Theorem 234 this BSDE has a unique $\mathfrak{F}_t^w$−adapted solution $(x_t, q_t) \in R^1 \times R^2, 0 \leq t \leq T$, provided that the following conditions hold:
(A) r_t, b_t, and σ_t are all non-random such that σ_t^{-1} exists and
$\int_0^T [|r_t| + \left|\sigma_t^{-1}(b_t - r_t \underline{1})\right|^2]dt < \infty;$
moreover, $E|X|^2 < \infty$.

Hence (8.16) also has a unique solution $(x_t, \pi_t) \in R^1 \times R^2, 0 \leq t \leq T$, where $\pi_t = q_t \sigma_t^{-1}$, provided that assumption (A) holds. Notice that here $b_t \in R^{2\otimes 1}, \sigma_t \in R^{2\otimes 2}$ are not real numbers. A particular case, when assumption (A) is true, is when r_t, b_t, and σ_t^{-1} are all non-random and bounded. This means precisely that under assumption (A) the claim can be priced as x_t, and can be hedged by $(x_t, \pi_t), 0 \leq t \leq T$. The conclusion is easily generalized to the case $N = d > 2$.

In a financial market, if every contingent claim can be hedged, (that is, can be reproduced), then this market is called a complete (financial) market. By the discussion above one sees that a necessary condition for a continuous market to be complete is that $N \geq d$, that is, the number of kinds of stocks should at least be equal to the number of continuous independent stochastic perturbations. (That is, the dimension of the BM). Otherwise, σ_t^{-1} does not exist, and we cannot transform (8.16) to a solvable BSDE as here.

8.4.2 Discontinuous Case

1. Option pricing in markets with jumps.

It is easily imagined that in a financial market on many occasions the stochastic perturbation will also have discontinuities. So we need to discuss a market with jumps. Suppose that in a market there is a bond, and that its price equation is still (8.1), and there are two kinds of stocks, each disturbed by a contiuous stochastic perturbation (i.e. a BM) and a jump stochastic perturbation (i.e. a centralized Poisson process), where their price SDEs are described as follows:

$dP^i(t,\omega) = P^i(t,\omega)[(b^i(t,\omega)dt + \sigma^i(t,\omega)dw_t + \rho^i(t,\omega)d\widetilde{N}_t],$

$P^i(0,\omega) = P^i(0,\omega), \quad i = 1,2;$
where w_t is a 1−dimensional BM, and $\widetilde{N}_t$ is a centralized Poisson process, which can be seen as
$\widetilde{N}_t = \widetilde{N}_k((0,t],Z) = N_k((0,t],Z) - t\pi(Z) = N_k((0,t],Z) - t,$
where $k(\cdot)$ is a 1−dimensional Poisson point process, and for simplicity we assume that $\pi(Z) = 1$. Recall that if $k(\cdot)$ is a 1−dimensional Poisson point process, then $N_k((0,t],Z)$ is a Poisson process with the intensity function $EN_k((0,t],Z) = t\pi(Z) = t$. Here the term "centralized" will mean to make it become a martingale process. Based on these ideas we see that the above stock price SDE can actually be rewritten as
$dP^i(t,\omega) = P^i(t,\omega)[(b^i(t,\omega)dt + \sigma^i(t,\omega)dw_t + \rho^i(t,\omega)\int_Z \widetilde{N}_k(dt,dz)],$
$P^i(0,\omega) = P^i(0,\omega), \quad i = 1,2.$
So it is a special case of our SDE with jumps. (See (3.1)). Again by means of a similar derivation, in a self-financed market, pricing an option of the first stock will lead to the solution of the following BSDE

$$\begin{cases} dx_t = r_t x_t dt + \pi_t(b_t - r_t \underline{1})dt + \pi_t\sigma_t dw_t + \pi_t\rho_t d\widetilde{N}_t, \\ x_T = (P_T^1 - K)^+, \ t \in [0,T], \end{cases} \tag{8.17}$$

where $\underline{1}= (1,1)^*$ is a 2−dimensional constant vector,
$b_t = \begin{pmatrix} b_t^1 \\ b_t^2 \end{pmatrix}, \sigma_t = \begin{pmatrix} \sigma_t^1 \\ \sigma_t^2 \end{pmatrix}, \rho_t = \begin{pmatrix} \rho_t^1 \\ \rho_t^2 \end{pmatrix} \in R^{2\otimes 1},$
and $\pi_t = (\pi_t^1, \pi_t^2)$. Now let $q_t = \pi_t\sigma_t, p_t = \pi_t\rho_t$. Then (8.17) becomes
$\begin{cases} dx_t = r_t x_t dt + (q_t \ p_t)\widetilde{\sigma}_t^{-1}(b_t - r_t\underline{1})dt + q_t dw_t + p_t d\widetilde{N}_t, \\ x_T = (P_T^1 - K)^+, \ 0 \le t \le T. \end{cases}$
where
$\widetilde{\sigma}_t = (\sigma_t, \rho_t) = \begin{pmatrix} \sigma_t^1 & \rho_t^1 \\ \sigma_t^2 & \rho_t^2 \end{pmatrix}.$
Since by Theorem 234 this BSDE has a unique $\mathfrak{F}_t^{w,\widetilde{N}}$−adapted solution $(x_t, q_t, p_t) \in R^1 \times R^1 \times R^1, 0 \le t \le T$, provided the following conditions hold:
(A) r_t, b_t, σ_t and ρ_t are all non-random and such that $\widetilde{\sigma}_t^{-1}$ exists, and
$\int_0^T [|r_t| + \left|\widetilde{\sigma}_t^{-1}(b_t - r_t\underline{1})\right|^2]dt < \infty;$
and, moreover, $E[(P_T^1 - K)^+]^2 < \infty$.

Hence (8.17) also has a unique solution $(x_t, \pi_t) \in R^1 \times R^2, 0 \le t \le T$, where $\pi_t = (q_t, p_t)\widetilde{\sigma}_t^{-1}$, and $(q_t, p_t) \in R^{1\otimes 2}$, provided assumption (A) holds. Notice that here $b_t, \sigma_t, \rho_t \in R^{2\otimes 1}$, so they are not real numbers. A particular case in which assumption (A) is true is that r_t, b_t, σ_t^{-1} and ρ_t are all non-random and bounded. This means that under assumption (A) the option of the first stock can be priced as x_t. However, in a financial market with jumps, unlike the continuous case, the option price x_t actually depends on the prices of two stocks, because in the BSDE (8.17) for the option price b_t^2, σ_t^2, and ρ_t^2 are involved. Moreover, if there is only one stock and one bond in the financial market with jumps, then in general the option for this stock cannot be priced by this way because, in general, $q_t = \pi_t^1 \sigma_t, p_t = \pi_t^1 \rho_t$ has no solution for π_t^1.

2. Hedging contingent claims in a market with jumps.

Now suppose that in a market there is a bond, its price equation is still (8.1), and there are three kinds of stocks disturbed by a contiuous and two independent jump stochastic perturbations, and that their price SDEs are described as follows:

$dP^i(t,\omega) = P^i(t,\omega)[(b^i(t,\omega)dt + \sigma^i(t,\omega)dw_t + \rho_1^i(t,\omega)d\widetilde{N}_t^1 + \rho_2^i(t,\omega)d\widetilde{N}_t^2],$
$P^i(0,\omega) = P^i(0,\omega), \quad i = 1,2,3;$

where w_t is a 1−dimensional BM, and $\widetilde{N}_t = (\widetilde{N}_t^1, \widetilde{N}_t^2)$ is a 2−dimensional centralized Poisson process with the same density function $E\widetilde{N}_t^i = t$; that is, for each i, $(i = 1,2)$, $\widetilde{N}_t^i$ is a Poisson process with the same density function t, and $\widetilde{N}_t^1, t \geq 0$ is independent of $\widetilde{N}_t^2, t \geq 0$; moreover, they have no common jump times, that is, $\triangle\widetilde{N}_t^1 \cdot \triangle\widetilde{N}_t^2 = 0, \forall t \geq 0$. For a given contingent claim $X \in \mathfrak{F}_T^{w,\widetilde{N}}$ in this market, if the market is self-financed and we want to hedge it, then this will lead to the solution of the following BSDE

$$\begin{cases} dx_t = r_t x_t dt + \pi_t(b_t - r_t \underline{1})dt + \pi_t \sigma_t dw_t + \sum_{j=1}^2 \pi_t \rho_{jt} d\widetilde{N}_t^j, \\ x_T = X, \ t \in [0,T], \end{cases} \tag{8.18}$$

where $\underline{1} = (1,1,1)^*$ is a 3−dimensional constant vector,

$$b_t = \begin{pmatrix} b_t^1 \\ b_t^2 \\ b_t^3 \end{pmatrix}, \sigma_t = \begin{pmatrix} \sigma_t^1 \\ \sigma_t^2 \\ \sigma_t^3 \end{pmatrix} \in R^{2\otimes 1}, \rho_t = \begin{pmatrix} \rho_{1t}^1 & \rho_{2t}^1 \\ \rho_{1t}^2 & \rho_{2t}^2 \\ \rho_{1t}^3 & \rho_{2t}^3 \end{pmatrix} \in R^{2\otimes 2},$$

and $\pi_t = (\pi_t^1, \pi_t^2, \pi_t^3)$, and $(x_t, \pi_t) \in \mathfrak{F}_t^{w,\widetilde{N}}$. Now let $q_t = \pi_t \sigma_t, p_t = \pi_t \rho_t$. Then (8.17) becomes

$$\begin{cases} dx_t = r_t x_t dt + (q_t \ p_t)\widetilde{\sigma}_t^{-1}(b_t - r_t \underline{1})dt + q_t dw_t + p_t d\widetilde{N}_t, \\ x_T = X, \ 0 \leq t \leq T. \end{cases}$$

where $(q_t \ p_t) = (q_t, p_{1t}, p_{2t}) \in R^3$

$$\widetilde{\sigma}_t = (\sigma_t \ \rho_t) = \begin{pmatrix} \sigma_t^1 & \rho_{1t}^1 & \rho_{2t}^1 \\ \sigma_t^2 & \rho_{1t}^2 & \rho_{2t}^2 \\ \sigma_t^3 & \rho_{1t}^3 & \rho_{2t}^3 \end{pmatrix} \in R^{3\otimes 3}.$$

Since by Theorem 234 this BSDE has a unique $\mathfrak{F}_t^{w,\widetilde{N}}$−adapted solution $(x_t, q_t, p_t) \in R^1 \times R^1 \times R^2, 0 \leq t \leq T$, provided the following conditions hold:

(A) r_t, b_t, σ_t and ρ_t are all non-random such that $\widetilde{\sigma}_t^{-1}$ exists, and

$\int_0^T [|r_t| + \left|\widetilde{\sigma}_t^{-1}(b_t - r_t \underline{1})\right|^2]dt < \infty;$

mooreover, $E|X|^2 < \infty$.

Hence (8.18) also has a unique $\mathfrak{F}_t^{w,\widetilde{N}}$−adapted solution $(x_t, \pi_t) \in R^1 \times R^3, 0 \leq t \leq T$, where $\pi_t = (q_t \ p_t)\widetilde{\sigma}_t^{-1}$, and

$(q_t \ p_t) = (q_t, p_{1t}, p_{2t}) \in R^{1\otimes 3},$

provided assumption (A) holds. Notice here that $b_t, \sigma_t \in R^{3\otimes 1}, \rho_t \in R^{3\otimes 2}$ are not real numbers. A particular case in which assumption (A) is true is that r_t, b_t, σ_t^{-1} and ρ_t are all non-random and bounded. This means that under assumption (A) the contingent claim can be hedged by (x_t, q_t, p_t). By the same token one easily obtains the following theorem.

Theorem 238 *In a self-financed market, suppose that there is one bond, whose price satisfies (8.1), and that there exist $N-$stocks, whose prices satisfy the following SDEs, respectively:*

$$\begin{aligned} dP^i(t,\omega) &= P^i(t,\omega)[(b^i(t,\omega)dt + \sum_{k=1}^{d} \sigma^{ik}(t,\omega)dw_t^k + \sum_{k=1}^{m} \rho^{ik}(t,\omega)d\widetilde{N}_t^k], \\ P^i(0,\omega) &= P^i(0,\omega), \quad i = 1, \cdots, N; \end{aligned} \tag{8.19}$$

where $w_t = (w_t^1, \cdots, w_t^d)$ is a $d-$dimensional BM, and $\widetilde{N}_t = (\widetilde{N}_t^1, \cdots, \widetilde{N}_t^m)$ is a $m-$dimensional centralized Poisson process with the same density function $E\widetilde{N}_t^i = t$, $i = 1, \cdots, m$.

Assume that

(A_1) $N = d + m$, and r_t, b_t, σ_t and ρ_t are all non-random such that $\widetilde{\sigma}_t^{-1}$ exists, and

$\int_0^T [|r_t| + \left|\widetilde{\sigma}_t^{-1}(b_t - r_t\underline{1})\right|^2]dt < \infty,$

where

$\sigma_t \in R^{N\otimes d}, \rho_t \in R^{N\otimes m}, \widetilde{\sigma}_t = (\sigma_t \ \rho_t) \in R^{N\otimes(d+m)};$

moreover, $X \in \mathfrak{F}_t^{w,\widetilde{N}}$ and $E|X|^2 < \infty$.

Then the following BSDE has a unique $\mathfrak{F}_t^{w,\widetilde{N}}-$adapted solution $(x_t, \pi_t) \in R^1 \times R^N, 0 \le t \le T$,

$$\begin{cases} dx_t = r_t x_t dt + \pi_t(b_t - r_t\underline{1})dt + \pi_t\sigma_t dw_t + \pi_t\rho d\widetilde{N}_t, \\ x_T = X, \ 0 \le t \le T. \end{cases}$$

Moreover, $(x_t, \pi_t), 0 \le t \le T$, hedges the contingent claim $X \in \mathfrak{F}_T$.

Obviously the above theorem is very general, and it includes all previous results in the financial market obtained previously in this book. For example, set

$d = 1, m = 0, X = (P_T - K)^+,$

then we get the option pricing result for the continuous case at the beginning of this section.

8.5 Black-Scholes Formula for Option Pricing. Two Approaches

In this section we are going to price an option using a probaility approach, and also by the partial differential equation (PDE) technique to establish the famous Black-Scholes formula. First, we see that the stock price SDE (8.2) for $i = 1, N = d = 1$ equals

$dP_t^1 = P_{t-}^1[r_t dt + \sigma_t(\sigma_t^{-1}(b_t - r_t)dt + dw_t)],\ P_0^1 = P_0^1;\ P - a.s.$

and the SDE (8.16) for hedging the contingent claim X equals the

$dx_t = r_t x_t dt + \pi_t \sigma_t(\sigma_t^{-1}(b_t - r_t)dt + dw_t), x_T = X,\ P - a.s.$

If we take $X = (P_T - K)^+$, then it becomes the option price SDE. Hence, applying Girsanov's theorem (Theorem 124) we immediately obtain the following lemma.

Lemma 239 *Assume that*
(B) σ_s^{-1} exists, and that there exists a constant $k_0 > 0$ such that
$\left|\sigma_s^{-1}(b_s - r_s)\right| \leq k_0$
Then 1) the stock price SDE can be rewritten as: $\widetilde{P}_T - a.s.$

$$\begin{aligned} dP_t^1 &= P_t^1[r_t dt + \sigma_t d\widetilde{w}_t], \\ P_0^1 &= P_0^1,\ 0 \leq t \leq T; \end{aligned} \tag{8.20}$$

and the SDE for hedging the contingent claim X also can be rewritten to be a $\widetilde{P}_T - a.s.$

$$\begin{cases} dx_t = r_t x_t dt + \pi_t \sigma_t d\widetilde{w}_t, \\ x_T = X,\ 0 \leq t \leq T. \end{cases} \tag{8.21}$$

where
$\widetilde{w}_t = w_t - \int_0^t (-\sigma_s^{-1}(b_s - r_s))ds,\ t \in [0, T],$
is a new BM under the new probability measure
$d\widetilde{P}_T = \exp[\int_0^T (-\sigma_s^{-1}(b_s - r_s))dw_s - \frac{1}{2}\int_0^T \left|\sigma_s^{-1}(b_s - r_s)\right|^2 ds]dP.$

2) $\widetilde{P}_T - a.s. \forall t \in [0, T]$,

$$\begin{aligned} P_t^1 &= P_0^1 \exp[\int_0^t r_s ds + \int_0^t \sigma_s d\widetilde{w}_s - \frac{1}{2}\int_0^t |\sigma_s|^2 ds], \\ x_t &= \exp[-\int_t^T r_s ds]X - \int_t^T \exp[-\int_t^s r_u du]\pi_s \sigma_s d\widetilde{w}_s. \end{aligned} \tag{8.22}$$

3) Since the bond price equation is non-random
$dP_t^0 = P_t^0 r_t dt,\ P_0^0 = 1,$
where for simplicity we take the initial value equal to one unit, so that he solution of the bond price equation is
$P_t^0 = e^{\int_0^t r_s ds}.$
4) If the contingent claim is $X \in \mathfrak{F}_T$, then by (8.22) the price of this contingent claim x_t is
$x_t = E^{\widetilde{P}_T}[X \exp[-\int_t^T r_s ds]|\mathfrak{F}_t].$
This pricing formula, even though is theoretical, has a nice physical meaning: it simply says that if we want the contingent claim to be X at the future time T, at time t we can expect (that is, we take the conditional rexpectation) that deposite money $X \exp[-\int_t^T r_s ds]$ will produce the claim X at time T.

Proof. 3) is obvious. 4) is true by taking the conditional expectation under the probability measure $\widetilde{P}_T$ on both sides of (8.22). So we only need to prove 1) and 2). However, 1) follows immediately from Girsanov's theorem (Theorem 124 or Theorem 191).

2): Applying Ito's formula to (8.22) one sees that the $\mathfrak{F}_t^{\widetilde{w}}$-adapted processes P_t^1 and x_t satisfy (8.20) and (8.21), respectively. ■

Now we are going to establish the famous Black-Scholes formula for the price of option related to the one stock price $P_t^1, t \in [0,T]$. First, we will use the probability approach. One will see that this approach has more probability meaning (hence also more physical meaning). From this approach we can guess the result; that is, the formula, can actually be constructed.

Theorem 240 *(Black-Scholes formula). Assume that r, b, σ are all constants, then the option price related to the first stock is*

$x_t = P_t^1 N(d_0(P_t^1) + \sqrt{T-t}\sigma) - Ke^{-r(T-t)} N(d_0(P_t^1)),$

where

$d_0(P_t^1) = \frac{1}{\sigma\sqrt{T-t}} \ln \frac{P_t^1}{Ke^{-r(T-t)}} - \frac{1}{2}\sigma\sqrt{T-t},$

$N(x) = \int_{-\infty}^{x} \frac{1}{\sqrt{2\pi(T-t)}} e^{-\frac{|y|^2}{2(T-t)}} dy.$

This can also be written in the more symetric form:

$x_t = P_t^1 N(\rho_+(T-t, P_t^1)) - Ke^{-r(T-t)} N(\rho_-(T-t, P_t^1)),$

where

$\rho_\pm(T-t, P_t^1) = \frac{1}{\sigma\sqrt{T-t}}[\ln \frac{P_t^1}{K} + (r \pm \frac{\sigma^2}{2})\sqrt{T-t})].$

Proof. By (8.22)

$P_t^1 = P_0^1 \exp[\int_0^t r_s ds + \int_0^t \sigma_s d\widetilde{w}_s - \frac{1}{2}\int_0^t |\sigma_s|^2 ds],$

$P_T^1 = P_0^1 \exp[\int_0^T r_s ds + \int_0^T \sigma_s d\widetilde{w}_s - \frac{1}{2}\int_0^T |\sigma_s|^2 ds].$

Evaluating $\frac{P_T^1}{P_t^1}$, and noticing that r, b, σ are all constants, one finds that $\widetilde{P}_T - a.s. \forall t \in [0,T]$,

$$P_T^1 = P_t^1 \exp[r(T-t) + \sigma(\widetilde{w}_T - \widetilde{w}_t) - \frac{1}{2}|\sigma|^2 (T-t)]. \tag{8.23}$$

Hence one sees that the price of the option is (Write $h(P_T^1) = (P_T^1 - K)^+$)

$x_t = E^{\widetilde{P}}(e^{-r(T-t)} h(P_T^1)|\mathfrak{F}_t)$

$= E^{\widetilde{P}}(e^{-r(T-t)} h(P_T^1)|\widetilde{w}_t)$

$= e^{-r(T-t)} \int_{R^1} h(P_t^1 e^{[\sigma y - \frac{1}{2}(|\sigma|^2 - r)(T-t)]}) \frac{1}{\sqrt{2\pi(T-t)}} e^{-\frac{|y|^2}{2(T-t)}} dy$

$= e^{-r(T-t)} \int_{R^1} h(P_t^1 e^{[\sigma y\sqrt{T-t} - \frac{1}{2}(|\sigma|^2 - r)(T-t)]}) \frac{1}{\sqrt{2\pi}} e^{-\frac{|y|^2}{2}} dy,$

where we have applied the facts that $\mathfrak{F}_t = \mathfrak{F}_t^{\widetilde{w}}$, and $\widetilde{w}_t, t \in [0,T]$, is a stationary Markov process such that $\widetilde{w}_T - \widetilde{w}_t \sim N(0,(T-t))$. That is, $\widetilde{w}_T - \widetilde{w}_t$ is a Normally distributed random variable with the mean 0 and variance $T-t$, because $\widetilde{w}_t, t \in [0,T]$, is also a BM under the probability $\widetilde{P}_T$ (See Corollary 102 in the First Part). Hence we have found a formula for the evaluation of the conditional expectation:

$$
\begin{aligned}
& E^{\widetilde{P}}(e^{-r(T-t)}h(P_T^1)|\mathfrak{F}_t) \\
= \; & e^{-r(T-t)}\int_{R^1} h(P_t^1 e^{[\sigma y\sqrt{T-t}-\frac{1}{2}(|\sigma|^2-r)(T-t)]})\frac{1}{\sqrt{2\pi}}e^{-\frac{|y|^2}{2}}dy. \quad (8.24)
\end{aligned}
$$

Actually this formula is also true for any Borel measurable function $h(.)$. Notice that $P_T^1 > K$ is equivalent to

$-(\widetilde{w}_T - \widetilde{w}_t) < -\ln\frac{K}{P_t^1} - \frac{1}{2}|\sigma|^2(T-t) + r(T-t)$

$= (\frac{1}{\sigma\sqrt{T-t}}\ln\frac{P_t^1}{Ke^{-r(T-t)}} - \frac{1}{2}\sigma\sqrt{T-t})\sigma\sqrt{T-t} = d_0(P_t^1)\sigma\sqrt{T-t}.$

So

$\mathcal{E} = \{P_T^1 > K\} = \left\{-\frac{(\widetilde{w}_T-\widetilde{w}_t)}{\sigma\sqrt{T-t}} < d_0(P_t^1)\right\},$

where

$d_0(P_t^1) = \frac{1}{\sigma\sqrt{T-t}}\ln\frac{P_t^1}{Ke^{-r(T-t)}} - \frac{1}{2}\sigma\sqrt{T-t}.$

Thus the option price is

$x_t = E^{\widetilde{P}}(e^{-r(T-t)}(P_T^1 - K)^+|\mathfrak{F}_t)$

$= e^{-r(T-t)}E^{\widetilde{P}}[P_T^1 I_{\mathcal{E}}|\mathfrak{F}_t] - Ke^{-r(T-t)}E^{\widetilde{P}}[I_{\mathcal{E}}|\mathfrak{F}_t].$

Notice that by (8.24)

$E^{\widetilde{P}}[I_{\mathcal{E}}|\mathfrak{F}_t] = \int_{R^1} I_{y>-d_0(P_t^1)}\frac{1}{\sqrt{2\pi}}e^{-\frac{|y|^2}{2}}dy = \int_{-d_0(P_t^1)}^{\infty}\frac{1}{\sqrt{2\pi}}e^{-\frac{|y|^2}{2}}dy$

$= \int_{-\infty}^{d_0(P_t^1)}\frac{1}{\sqrt{2\pi}}e^{-\frac{|y|^2}{2}}dy = N(d_0(P_t^1)),$

where as usual we write the standard Normal distribution function as

$N(x) = \int_{-\infty}^{x}\frac{1}{\sqrt{2\pi(T-t)}}e^{-\frac{|y|^2}{2(T-t)}}dy.$

On the other hand, again by (8.24)

$e^{-r(T-t)}E^{\widetilde{P}}[P_T^1 I_{\mathcal{E}}|\mathfrak{F}_t]$

$= e^{-r(T-t)}\int_{R^1} P_t^1 e^{[\sigma\sqrt{T-t}y-\frac{1}{2}(|\sigma|^2-r)(T-t)]}I_{y>-d_0(P_t^1)}\frac{1}{\sqrt{2\pi}}e^{-\frac{|y|^2}{2}}dy.$

By the technique of completing the square one finds that

$\sigma\sqrt{T-t}y - \frac{1}{2}(|\sigma|^2 - r)(T-t) - \frac{|y|^2}{2} = -\frac{1}{2}(y-\sqrt{T-t}\sigma)^2$

$+\frac{1}{2}(T-t)|\sigma|^2 - \frac{1}{2}(|\sigma|^2 - r)(T-t) = -\frac{1}{2}(y-\sqrt{T-t}\sigma)^2$

$+\frac{1}{2}r(T-t).$

Hence let $\widetilde{y} = y - \sqrt{T-t}\sigma$,

$e^{-r(T-t)}E^{\widetilde{P}}[P_T^1 I_{\mathcal{E}}|\mathfrak{F}_t] = \int_{R^1} P_t^1\frac{1}{\sqrt{2\pi}}e^{-\frac{|\widetilde{y}|^2}{2}}I_{-\widetilde{y}<d_0(P_t^1)+\sqrt{T-t}\sigma}d\widetilde{y}$

$= P_t^1 N(d_0(P_t^1) + \sqrt{T-t}\sigma).$

Therefore,

$x_t = P_t^1 N(d_0(P_t^1) + \sqrt{T-t}\sigma) - Ke^{-r(T-t)}N(d_0(P_t^1)),$

and the Black-Scholes formula is proved. This can also be written in the more symetric form.

$d_0(P_t^1) = \frac{1}{\sigma\sqrt{T-t}}\ln\frac{P_t^1}{Ke^{-r(T-t)}} - \frac{1}{2}\sigma\sqrt{T-t}$

$= \frac{1}{\sigma\sqrt{T-t}}[\ln\frac{P_t^1}{K} + (r - \frac{\sigma^2}{2})\sqrt{T-t})] = \rho_-(T-t, P_t^1),$

$d_0(P_t^1) + \sqrt{T-t}\sigma = \frac{1}{\sigma\sqrt{T-t}}[\ln\frac{P_t^1}{K} + (r - \frac{\sigma^2}{2})\sqrt{T-t})]$

$+\sqrt{T-t}\sigma = \frac{1}{\sigma\sqrt{T-t}}[\ln \frac{P_t^1}{K} + (r + \frac{\sigma^2}{2})\sqrt{T-t})] = \rho_+(T-t, P_t^1).$
Thus we also have that the option price equals
$x_t = P_t^1 N(\rho_+(T-t, P_t^1)) - Ke^{-r(T-t)} N(\rho_-(T-t, P_t^1)).$ ■

Now we are going to establish the Black-Scholes formula using the partial differential equation (PDE) approach. For this we will first use Ito's formula and the theory of BSDE to establish the relation between the solution of some PDE and the price of a more general contingent claim in a self-financed market. Suppose that σ_t is non-random, and $X = g(P_T^1)$. If we introduce a function $u(t,y) \in C^{1,2}(R_+ \times R^1)$, and we require that $u(t, P_t^1) = x_t$, then by applying Ito's formula to $u(t, P_t^1)$ on $t \in [0,T]$ with (8.20) one easily derives the following lemma.

Lemma 241 *Under assumption (B) in Lemma 239 and the condition that* $\int_0^t |r_t|\, dt < \infty$, *if* $u(t,y) \in C^{1,2}([0,T) \times R^1)$ *satisfies the following partial differential equation*

$$\begin{cases} -\frac{\partial u}{\partial t} + r_t u = \frac{1}{2}|\sigma_t|^2 |y|^2 \frac{\partial^2 u}{\partial y^2} + r_t y \frac{\partial u}{\partial y}, \text{ on } [0,T) \times R^1 \\ u(T,y) = g(y), \forall y \in R^1, \end{cases} \tag{8.25}$$

and $g(y)$ *is continuous such that* $E\left|g(P_T^1)\right|^2 < \infty$, *then* $u(t, P_t^1) = x_t$ *and* $x_T = u(T, P_T^1) = g(P_T^1)$ *holds true, where* $g(P_T^1) \in \mathfrak{F}_T^w$ *is a contingent claim and* x_t *is its price at time* t. *Moreover, if we let* $\pi_t = \frac{\partial u}{\partial y}(t, P_t^1)P_t^1$, *then* $(x_t, \pi_t) = (u(t, P_t^1), \frac{\partial u}{\partial y}(t, P_t^1)P_t^1), t \in [0,t]$, *hedges the contingent claim* $g(P_T^1)$.

Proof. Applying Ito's formula to $u(t, P_t^1)$ on $t \in [0,T]$, where P_t^1 satisfies (8.20), one finds that $\widetilde{P}_T - a.s. \forall t \in [0,T]$,
$u(t, P_t^1) = u(T, P_T^1) - \int_t^T [\frac{\partial u}{\partial s}(s, P_s^1) + \frac{\partial u}{\partial y}(s, P_s^1)P_s^1 r_s$
$+\frac{1}{2}|\sigma_s|^2 \left|P_s^1\right|^2 \frac{\partial^2 u}{\partial y^2}]ds - \int_t^T \frac{\partial u}{\partial y}(s, P_s^1)P_s^1 \sigma_s d\widetilde{w}_s$
$= g(P_T^1) - \int_t^T r_s u(s, P_s^1)ds - \int_t^T \frac{\partial u}{\partial y}(s, P_s^1)P_s^1 \sigma_s d\widetilde{w}_s.$
On the other hand, the wealth process x_t (or say, the price for the contingent claim $X = g(P_T^1)$) satisfies the BSDE: $\widetilde{P}_T - a.s.$
$x_t = g(P_T^1) - \int_t^T r_s x_s ds - \int_t^T \pi_s \sigma_s d\widetilde{w}_s, \forall t \in [0,T].$
However, the above BSDE has a unique $\mathfrak{F}_t^{\widetilde{w}}$-adapted solution $(x_t, q_t) = (x_t, \pi_t \sigma_t)$. Hence by the uniqueness of the solution of a BSDE, comparing the two BSDEs one finds that
$x_t = u(t, P_t^1), \pi_t = \frac{\partial u}{\partial y}(t, P_t^1)P_t^1.$
In particular, $x_T = u(T, P_T^1) = g(P_T^1)$ holds. ■

This lemma tells us that the option pricing problem can be solved by the PDE technique. Actually we can obtain the Black-Scholes formula by this approach. However, from such a PDE approach it is not easy to guess the formula, though we can always verify whether a given formula is true.

Theorem 242 *Assume that* $r > 0, b, \sigma$ *are all constants, and* $g(y) = (y - K)^+, y \geq 0$. *Let* $u(t,y) = BS(t,T,y,K,\sigma,r)$, *as* $0 \leqslant t < T, y \geq 0$; *and* $u(t,y) = g(y)$, *as* $t = T, y \geq 0$; *where*
$BS(t,T,y,K,\sigma,r) = yN(\rho_+(T-t,y)) - Ke^{-r(T-t)}N(\rho_-(T-t,y))$,
and
$\rho_\pm(t,y) = \frac{1}{\sigma\sqrt{t}}[\log\frac{y}{K} + t(r \pm \frac{\sigma^2}{2})]$, $N(y) = \frac{1}{\sqrt{2\pi}}\int_{-\infty}^{y} e^{-z^2/2}dz$.
Then $u(t,y) \in C^{1,2}([0,T) \times (0,\infty))$ *and satisfies (8.21). Furthermore,* $u(t,P_t^1)$ *is the price of the option, and* $\frac{\partial u}{\partial y}(t,P_t^1)P_t^1$ *is the portfolio such that*
$(u(t,P_t^1), \frac{\partial u}{\partial y}(t,P_t^1)P_t^1)$ *can duplicate the money* $(P_T^1 - K)^+$ *which the option owner can earn at the future time* T. *Such an explicit formula* $u(t,P_t^1) = BS(t,T,P_t^1,K,\sigma,r)$ *for the option price is called the Black-Scholes formula.*

Proof. It is a rutine matter to show that $u(t,y) \in C^{1,2}([0,T) \times (0,\infty))$ and that it satisfies (8.21). The reader can verify those results for $u(t,y)$ by himself. Now let us check the terminal condition $u(T,y) = (y-K)^+, \forall y \geq 0$ also holds true. In fact,
$u(t,y) = [yN(\rho_+(T-t,y)) - Ke^{-r(T-t)}N(\rho_-(T-t,y))]$
$= (y-K)\frac{1}{\sqrt{2\pi}}\int_{-\infty}^{\rho_+(T-t,y)} e^{-z^2/2}dz + K(1-e^{-r(T-t)})\frac{1}{\sqrt{2\pi}}\int_{-\infty}^{\rho_-(T-t,y)} e^{-z^2/2}dz$
$= I_t^1 + I_t^2$.
Obviously, as $t \uparrow T, 0 \leq I_t^2 \leq K(1 - e^{-r(T-t)}) \to 0$. Observe that in the expression for $\rho_+(T-t,y)$, if $y > K$ then as $t \uparrow T, \rho_+(T-t,y) \to \infty$; on the other hand, if $y \leq K$, then as $t \uparrow T, \rho_+(T-t,y) \to -\infty$. Therefore, as $t \uparrow T$, $I_t^1 \to (y-K)^+$. The terminal condition holds for $u(t,y)$. Finally, let us show that the proof of Lemma 241 goes through. In fact, notice that by definition $u(t,0) = 0$. So $\frac{\partial}{\partial t}u(t,0) = 0$. Now applying Ito's formula to $u(t,P_t^1)$ on $t \in [0,T]$, where P_t^1 satisfies (8.20), one finds that $\widetilde{P}_T - a.s.\forall t \in [0,T]$,
$u(t,P_t^1) = u(T,P_T^1) - \int_t^T [\frac{\partial u}{\partial s}(s,P_s^1) + \frac{\partial u}{\partial y}(s,P_s^1)P_s^1 r_s$
$+\frac{1}{2}|\sigma_s|^2 |P_s^1|^2 \frac{\partial^2 u}{\partial y^2}]I_{P_s^1 \neq 0}ds - \int_t^T \frac{\partial u}{\partial y}(s,P_s^1)P_s^1\sigma_s d\widetilde{w}_s$
$= g(P_T^1) - \int_t^T r_s u(s,P_s^1)I_{P_s^1 \neq 0}ds - \int_t^T \frac{\partial u}{\partial y}(s,P_s^1)P_s^1\sigma_s d\widetilde{w}_s$
$= g(P_T^1) - \int_t^T r_s u(s,P_s^1)ds - \int_t^T \frac{\partial u}{\partial y}(s,P_s^1)P_s^1\sigma_s d\widetilde{w}_s$.
So the proof of Lemma 241 still goes through. ■

The Black-Scholes formula is easily evaluated pratically, because in Statistics books there are tables for the values of $N(x)$. Furthermore, the Black-Scholes formula can also be converted into a computer program, so the option price can be found immediately from the computer, if it is provided with the necessary data.

8.6 Black-Scholes Formula for Markets with Jumps

Since financial markets with jumps are more realistic, in this section we will briefly discuss how to obtain a Black-Scholes formula for option pricing in such markets.

Recall that from Theorem 238 we have considered a self-financed market. There is a bond, whose price satisfies the bond price equation:

$$dP^0(t) = P^0(t)r(t)dt, P^0(0) = 1, \tag{8.26}$$

where $r(t)$ is the rate function, and there are N different kinds of stocks, with their prices satisfying the following general stochastic differential equations:

$$\begin{aligned} dP^i(t,\omega) &= P^i(t-,\omega)[b^i(t,\omega)dt + \sum_{k=1}^{d} \sigma^{ik}(t,\omega)dw_t^k + \sum_{k=1}^{m} \rho^{ik}(t,\omega)d\widetilde{N}_t^k], \\ P^i(0,\omega) &= P^i(0,\omega), \quad i = 1, 2, \cdots, N; \end{aligned} \tag{8.27}$$

where $w_t = (w_t^1, w_t^2, \cdots, w_t^d)^T$ is a standard $d-$dimensional Brownian Motion process, $\widetilde{N}_t = (\widetilde{N}_t^1, \cdots, \widetilde{N}_t^m)^T$ is a $m-$dimensional centralized Poisson process, i.e. all components are independent and any two components have non-common jumps, such that $d\widetilde{N}_t^i = dN_t^i - dt$, and $\sigma = (\sigma^{ik})_{i,k=1}^{N,d} \in R^{N\otimes d}, \rho = (\rho^{i,k})_{i,k=1}^{N,m} \in R^{N\otimes m}$. In such a market if $X \in \mathfrak{F}_T = \mathfrak{F}_T^{w,\widetilde{N}}$ is a contingent claim, and $(x_t, \pi_t), t \in [0,T]$ can hedge this contingent claim, then they should satisfy:

$$\begin{cases} dx_t = r_t x_t dt + \pi_t(b_t - r_t \underline{1})dt + \pi_t \sigma_t dw_t + \pi_t \rho_t d\widetilde{N}_t \\ \qquad x_T = X, \ t \in [0,T], \end{cases} \tag{8.28}$$

where $\underline{1} = (1, \cdots, 1)^T$ is a $N-$dimensional constant vector, and x_t is called the price of the contingent claim X, (it is also called a wealth process in many cases), and $\pi_t = (\pi_t^1 \cdots, \pi_t^N)$ is called a portfolio, (actually it is the risk part of the portfolio). According to Theorem 238, under the condition (A_1), (8.28) has a unique $\mathfrak{F}_t^{w,\widetilde{N}}-$adapted solution (x_t, π_t). Now let
$[q_t, p_t] = \pi_t[\sigma_t, \rho_t] = \pi_t \widetilde{\sigma}_t,$
where $q_t = (q_t^1, \cdots, q_t^d), p_t = (p_t^1, \cdots, p_t^m)$.

If $\widetilde{\sigma}_t^{-1} = [\sigma_t, \rho_t]^{-1}$exists, (for simplicity, to ensure this, let us assume that $N = d + m$), then (8.28) can be rewritten as

$$\begin{aligned} x_t = X &- \int_t^T r_s x_s ds - \int_t^T (q_s \theta_s^w + p_s \theta_s^{\widetilde{N}})ds \\ &- \int_t^T q_s dw_s - \int_t^T p_s d\widetilde{N}(s), 0 \leqslant t \leqslant T, \end{aligned} \tag{8.29}$$

where we set

$$\theta_t = \begin{bmatrix} \theta_t^w \\ \theta_t^{\widetilde{N}} \end{bmatrix} = \widetilde{\sigma}_t^{-1}[b_t - r_t \underline{1}], \tag{8.30}$$

with $\theta_t^w \in R^{d\otimes 1}, \theta_t^{\widetilde{N}} \in R^{m\otimes 1}$. Then the existence of the solution (x_t, q_t, p_t) for BSDE (8.29) , or, more precisely, the existence of a solution (x_t, π_t) for BSDE (8.28) will help the investor to make his wealth x_T arrive at X at the future time T (or say x_T can duplicate the contingent claim X), if at time t by using the portfolio π_t he invests the money x_t. Moreover, if one calls X the value of some contingent claim, then the existence of x_t also shows that one can give the contingent claim a price x_t at time t. We can give another existence theorem (the sufficient conditions are a little different from Theorem 238) for BSDE (8.29) and BSDE (8.28) as follows (its proof is similar to that of Theorem 238):

Theorem 243 .[166] *Suppose that b, r, σ, ρ are all bounded $\mathfrak{F}_t^{w,\widetilde{N}}$ −predictable processes and $\widetilde{\sigma}_t^{-1}$ exists and is also bounded, and assume that $N = d + m$, $X \in \mathfrak{F}_T^{w,\widetilde{N}}$ and $E|X|^2 < \infty$. Then (8.29) has a unique solution $(x_t, q_t, p_t) \in L^2_{\mathfrak{F}}(R^1) \times L^2_{\mathfrak{F}}(R^{1\otimes d}) \times F^2_{\mathfrak{F}}(R^{1\otimes m})$, and there also exist a unique wealth process $x_t \in L^2_{\mathfrak{F}}(R^1)$ and a unique portfolio process $\pi_t \in L^2_{\mathfrak{F}}(R^N)$ satisfying (8.28), where*
$x_t = x_t,\ \pi_t^i = \sum_{j=1}^d q_s^j(\widetilde{\sigma}_s^{-1})_{ji} + \sum_{j=1}^m p_s^j(\widetilde{\sigma}_s^{-1})_{(j+d)i},\ i = 1, \cdots, N.$

Recall that if the contingent claim X can be duplicated by (x_t, π_t) (i.e. putting such (x_t, π_t) into the market, when t evolves to time T, then $x_T = X$), then we will say that (x_t, π_t) hedges the contingent claim X. So we see that under the conditions of Theorem 243 any contingent claim can be hedged. Such a market can be called a complete market.

Now we will use the Girsanov type thoerm to simplify these equations: (8.27) and (8.28).

First we know that (8.27) also can be rewritten simply as

$$\begin{aligned} dP_t &= P_{t-}[r_t dt + (b_t - r_t\underline{1})dt + \sigma_t dw_t + \rho_t d\widetilde{N}_t], \\ P_0 &= P_0, \quad t \in [0, T], \end{aligned} \tag{8.31}$$

where $P_t \in R^N, r_t \in R^1, \sigma_t \in R^{N\otimes d}, \rho \in R^{N\otimes m}$. So we can establish the following theorem by using the Girsanov type transformation.

Theorem 244 *Under the assumptions of the above theorem if, in addition,*
$\mu_t^i = -(\theta_t^{\widetilde{N}})^i > -1, \forall i = 1, \cdots, m;$
then 1) the stock price SDE can be rewritten as: $\widetilde{P}_T - a.s.$

$$\begin{aligned} dP_t &= P_t[r_t dt + \sigma_t d\widetilde{w}_t + \rho_t d\widetilde{N}_t], \\ P_0 &= P_0, \ 0 \leq t \leq T; \end{aligned} \tag{8.32}$$

and the SDE for hedging the contingent claim X also can be rewritten as $\widetilde{P}_T - a.s.$

$$\begin{cases} dx_t = r_t x_t dt + \pi_t \sigma_t d\widetilde{w}_t + \pi_t \rho_t d\overline{N}_t, \\ x_T = X, \ 0 \leq t \leq T. \end{cases} \tag{8.33}$$

where we write

$\widetilde{P}_{t,T} = \exp[-\int_t^T \langle \theta_s^w, dw_s \rangle - \frac{1}{2}\int_t^T |\theta_s^w|^2 ds - \int_t^T \langle \mu_s, \underline{1} \rangle ds] \cdot$
$\cdot \prod_{t<s\leqslant T}(1 + \langle \mu_s, \triangle N_s \rangle) dP,$

and $\widetilde{P}_T = \widetilde{P}_{0,T}$ *is a new probability measure. Moreover, where* $\widetilde{w}_t$ *and* $\overline{N}_t$ *are a new BM and a new centralized pure jump process, respectively, such that*

$\widetilde{w}_t^i = w_t^i + \int_0^t (\theta_s^w)^i ds, \ i = 1, \cdots, d;$

$\overline{N}_t^i = \widetilde{N}_t^i + \int_0^t (\theta_s^{\widetilde{N}})^i ds = \widetilde{N}_t^i - \int_0^t \mu_s^i ds, i = 1, \cdots, m;$

and $\overline{N}_t^i$ *is with the new compensator* $\int_0^t (1 + \mu_s^i) ds.$

2) $\widetilde{P}_T - a.s. \forall t \in [0, T],$

$$\begin{aligned} P_t^i &= P_0^i \exp[\int_0^t r_s ds + \sum_{k=1}^d \int_0^t \sigma_s^{ik} d\widetilde{w}_s^k - \frac{1}{2}\sum_{k=1}^d \int_0^t \left|\sigma_s^{ik}\right|^2 ds] \cdot \\ &\cdot \prod_{k=1}^m \prod_{0<s\leqslant t} (1 + \rho_s^{ik} \triangle N_s^k) e^{-\int_0^t \widetilde{\lambda}_s^k \rho_s^{ik} ds}, \ i = 1, 2, \cdots, N. \end{aligned} \tag{8.34}$$

$$\begin{aligned} x_t = X \exp[-\int_t^T r_s ds] - \int_t^T \pi_s \exp[-\int_t^s r_u du] \sigma_s d\widetilde{w}_s \\ - \int_t^T \pi_s \exp[-\int_t^s r_u du] \rho_s d\overline{N}_s, \ t \in [0, T]; \end{aligned} \tag{8.35}$$

where the compensator of $\triangle N_t^i$ *is* $\int_0^t \widetilde{\lambda}_s^i ds = \int_0^t (1 + \mu_s^i) ds, \widetilde{P} - a.s.$ *and* $\mu_s^i = -(\theta_s^{\widetilde{N}})^i, \widetilde{\lambda}_s^i = 1 + \mu_s^i.$

3) Since the bond price equation is still

$dP_t^0 = P_t^0 r_t dt, \ P_0^0 = 1,$

where for simplicity we take the initial value equal to one unit, the solution of the bond price equation is

$P_t^0 = e^{\int_0^t r_s ds}.$

4) If the contingent claim is $X \in \mathfrak{F}_T^{\widetilde{w}, \widetilde{N}}$, *then by (8.34) the price of this contingent claim* x_t *is*

$$x_t = E^{\widetilde{P}_T}[X \exp[-\int_t^T r_s ds] | \mathfrak{F}_t] = E[(X \exp[-\int_t^T r_s ds] \widetilde{P}_{t,T} | \mathfrak{F}_t] \,. \tag{8.36}$$

Eeven though this pricing formula is theoretical, it has a nice physical meaning: It says that if we want the contingent claim to be X at the future time

T, then now at time t we can expect (that is, take the conditional expectation) that depositing the money $X\exp[-\int_t^T r_s ds]$ will produce the claim X at time T.

Proof. Theorem 244 can be proved by the Girsanov theorem for SDE with jumps. In fact, formally, (8.29) can be rewritten as

$$x_t = X - \int_t^T r_s x_s ds - \int_t^T q_s(\theta_s^w ds + dw_s) - \int_t^T p_s(\theta_s^{\tilde{N}} ds + d\tilde{N}(s))$$
$$= X - \int_t^T r_s x_s ds - \int_t^T q_s d\widetilde{w}_s - \int_t^T p_s d\overline{N}(s), 0 \leqslant t \leqslant T,$$

where we write

$$\widetilde{w}_t^i = w_t^i + \int_0^t (\theta_s^w)^i ds,\ i = 1, \cdots, d;$$
$$\overline{N}_t^i = \tilde{N}_t^i + \int_0^t (\theta_s^{\tilde{N}})^i ds = \tilde{N}_t^i - \int_0^t \mu_s^i ds, i = 1, \cdots, m.$$

However, by the Girsanov type theorem (Theorem 124) $\widetilde{w}_t, t \in [0,T]$, and $\overline{N}_t, t \in [0,T]$, are a BM and a centralized pure jump process under the new probability measure $\widetilde{P}_T$ defined in the conclusion 1) of Theorem 244, respectively. Notice that $[q_t, p_t] = \pi_t[\sigma_t, \rho_t]$. Hence (8.33) now follows. (8.32) is similarly established. So 1) is proved. Notice that by Theorem 122 the solution of the linear SDE (8.32) is $\widetilde{P} - a.s.$

$$P_t^i = \exp[\int_0^t r_s ds + \sum_{k=1}^d \int_0^t \sigma_s^{ik} d\widetilde{w}_s^k - \frac{1}{2}\sum_{k=1}^d \int_0^t \left|\sigma_s^{ik}\right|^2 ds] \cdot$$
$$\cdot \prod_{0<s\leqslant t}[(1 + \sum_{k=1}^m \rho_s^{ik} \triangle N_s^k) e^{-\sum_{k=1}^m \int_0^t \widetilde{\lambda}_s^k \rho_s^{ik} ds}].$$

Hence (8.34) holds. On the other hand, applying Ito's formula (Theorem 94) one sees that (8.35) satisfies (8.33). So 2) now follows. 3) is obvious. Finally, the first equality sign is obtained by taking the conditional expectation $E^{\widetilde{P}_T}[\cdot|\mathfrak{F}_t]$ on both sides of (8.33). The second equality sign can be seen as follows: $\forall A \in \mathfrak{F}_t$,

$$\int_A E^{\widetilde{P}_T}[X\exp[-\int_t^T r_s ds]|\mathfrak{F}_t] d\widetilde{P}_T = \int_A X\exp[-\int_t^T r_s ds] d\widetilde{P}_T$$
$$= \int_A X\exp[-\int_t^T r_s ds] \frac{d\widetilde{P}_T}{dP} dP = \int_A E[X\exp[-\int_t^T r_s ds]\widetilde{P}_{0,T}|\mathfrak{F}_t] dP$$
$$= \int_A E[X\exp[-\int_t^T r_s ds] \left(\widetilde{P}_{0,t}\right)\left(\widetilde{P}_{t,T}\right)|\mathfrak{F}_t] dP$$
$$= \int_A \left(\widetilde{P}_{0,t}\right) E[X\exp[-\int_t^T r_s ds]\left(\widetilde{P}_{t,T}\right)|\mathfrak{F}_t] dP.$$

On the other hand,

$$\int_A E^{\widetilde{P}_T}[X\exp[-\int_t^T r_s ds]|\mathfrak{F}_t] d\widetilde{P}_T = \int_A E^{\widetilde{P}_T}[X\exp[-\int_t^T r_s ds]|\mathfrak{F}_t]\widetilde{P}_{0,T} dP$$
$$= \int_A \left(\widetilde{P}_{0,t}\right) E^{\widetilde{P}_T}[X\exp[-\int_t^T r_s ds]|\mathfrak{F}_t] E[\widetilde{P}_{t,T}|\mathfrak{F}_t] dP.$$

Notice that $E[\widetilde{P}_{t,T}|\mathfrak{F}_t] = 1$. Hence

$$E^{\widetilde{P}_T}[X\exp[-\int_t^T r_s ds]|\mathfrak{F}_t] = E[X\exp[-\int_t^T r_s ds]\widetilde{P}_{t,T}|\mathfrak{F}_t]. \blacksquare$$

Now suppose that σ, ρ, λ are all non-random, and $X = g(P_T)$. If we introduce a function $u(t,y) \in C^{1,2}(R_+ \times R^N)$, and we require that $u(t, P_t) = x_t$, then by using Ito's formula to $e^{-\int_0^t r_s ds} u(t, P_t)$ on $t \in [0,T]$ with (8.32) one easily derives the following fact: if $u(t,y)$ satisfies the following partial differential equation

$$\begin{cases} -\frac{\partial u}{\partial t} + r_t u = \frac{1}{2}\sum_{i.j=1}^N \sum_{k=1}^d \sigma_t^{ik}\sigma_t^{jk} y^i y^j \frac{\partial^2 u}{\partial y^i \partial y^j} + r_t \sum_{i=1}^N y^i \frac{\partial u}{\partial y^i} \\ + \sum_{k=1}^m \sum_{i=1}^N [u(t, y + (0, \cdots, 0, \rho_t^{ik}, 0, \cdots, 0)^T) - u(t, y) \\ -y^i \rho_t^{ik} \frac{\partial u(t,y)}{\partial y^i}]\widetilde{\lambda}_t^k, \ u(T, y) = g(y), t \in [0, T], \end{cases} \tag{8.37}$$

then $u(t, P_t) = x_t$ and $x_T = u(T, P_T) = g(P_T)$ holds true. This tells us that the option pricing problem in the market with jumps still can be solved by the PDE technique. To get a more explicit formula for the price of the option we should assume more. Suppose, further, that $g(P_T) = (P_T^1 - K)^+$, that is, the contingent claim only depends on the 1st stock, or say we want to price a call option on stock 1, and suppose that $N = d + m = 1 + 1 = 2$, and $r > 0, \sigma, \rho$ are all non-random constants. Then by (8.30) θ^N also is a constant. Under these conditions $\overline{N}_t = N_t - (1 - \theta^N)t$ becomes a centralized Poisson process with the intensity function $(1 - \theta^N)t$ with respect to the probability measure $\widetilde{P}_T$. Since by (8.34) $\widetilde{P}_T - a.s.$

$$\begin{aligned} P_T^1 &= P_t^1 \exp[r(T - t) + \sigma(\widetilde{w}_T - \widetilde{w}_t) - \frac{1}{2}|\sigma|^2 (T - t)] \cdot \\ &\cdot \prod_{t < s \leqslant T} (1 + \rho \triangle N_s) e^{-\widetilde{\lambda}\rho(T-t)}. \end{aligned} \tag{8.38}$$

Notice that now $\{\widetilde{w}_t\}$ and $\{N_s\}$ are independent under the probability measure $\widetilde{P}_T$, and that they have independent increments. Hence by using Ito's formula it is not difficult to verify the fact that if we write

$$u(t, y) = \sum_{n=0}^\infty e^{-\widetilde{\lambda}(T-t)} \frac{[\widetilde{\lambda}(T - t)]^n}{n!} [BS(t, T; y(1 + \rho)^n e^{-\widetilde{\lambda}\rho(T-t)}, K, \sigma, r)], \tag{8.39}$$

where

$$BS(t, T, y, K, \sigma, r) = y\Phi(\rho_+(T - t, y)) - Ke^{-r(T-t)}\Phi(\rho_-(T - t, y)), \tag{8.40}$$

as $0 \leqslant t < T, y \geq 0$, and

$\rho_\pm(t, y) = \frac{1}{\sigma\sqrt{t}}[\log \frac{y}{K} + t(r \pm \frac{\sigma^2}{2})]$, $\Phi(y) = \frac{1}{\sqrt{2\pi}} \int_{-\infty}^y e^{-z^2/2} dz$,

then $u(t, y) \in C^{1,2}([0, T) \times (0, \infty))$ and $u(t, y)$ satisfies (8.37) with $g(y) = (y - K)^+$ and $k = 1; i, j = 1$. So after a discussion similar to the one in the previous section one sees that

$$\begin{aligned} x_t &= E^{\widetilde{P}}(e^{-r(T-t)}(P_T^1 - K)^+ | \mathfrak{F}_t) = u(t, P_t^1) \\ &= \sum_{n=0}^\infty e^{-\widetilde{\lambda}(T-t)} \frac{[\widetilde{\lambda}(T - t)]^n}{n!} [BS(t, T; P_t^1(1 + \rho)^n e^{-\widetilde{\lambda}\rho(T-t)}, K, \sigma, r)], \end{aligned} \tag{8.41}$$

is the price of the option for the first stock.

Noticing that in (8.41) the price of the option for stock 1 actually depends on both stocks, because $\widetilde{\lambda}$ does so. (Actually, $\widetilde{\lambda}$ depends on $\theta^{\widetilde{N}}$. By (8.30) $\theta^{\widetilde{N}}$ is solved by both coefficients of the two stocks). Hence, if there exists only one stock with a continuous and a jump perturbations, then usually we cannot have a formula, because in this case $\widetilde{\lambda}$ usually does not exist. Formula (8.41) is a generalization of the Black-Scholes formula for option pricing. In the case that $\rho = 0$ and $N = 1$, i.e. the stock is without any jump perturbation, and we only consider one stock in the market, then formula (8.41) is reduced to

$$x_t = E^{\widetilde{P}}((P_T^1 - K)^+|\mathfrak{F}_t) = BS(t, T; P_t^1, K, \sigma, r). \tag{8.42}$$

That is the standard Black-Scholes formula which we found in the previous section.

So the option pricing formula also can be evaluated by solving a PDE, even in a self-financed market with jumps, but under the assumption that $2 = N = d + m = 1 + 1$ and all coefficients are constants. That is, we arrive at the following theorem.

Theorem 245 *Assume that $r > 0, b, \sigma$ are all constants, $d = m = 1, N = 2$ and $g(y) = (y - K)^+, y \geq 0$. Let $u(t, y)$ define by (8.39)as $0 \leqslant t < T, y \geq 0$. Then $u(t, y) \in C^{1,2}([0, T) \times (0, \infty))$ and satisfies (8.37), with $u(T, y) = (y - K)^+, \forall y \geq 0$. Furthermore, $u(t, P_t^1)$ is the price of the option, $\frac{\partial u}{\partial y}(t, P_t^1)P_t^1$ is the portfolio, such that $(u(t, P_t^1), \frac{\partial u}{\partial y}(t, P_t^1)P_t^1)$ can duplicate the money $(P_T^1 - K)^+$, which the option owner can earn at the future time T. Such an explicit formula $u(t, P_t^1)$ (defined by (8.41)) for the option price is called a generalized Black-Scholes formula.*

8.7 More General Wealth Processes and BSDEs

In a financial market sometimes we will meet more general wealth processes. Suppose that during the time interval $[0, T]$ the investor also consumes the money $dC_t \geqslant 0$ in a small time dt with $C_0 = 0$, and $C_T < \infty$, and borrows the money from the bank with the interest rate R_t at time t. Then his wealth process x_t developed in the financial market should satisfy the following SDE:

$$dx_t = r_t x_t dt - dC_t + \pi_t[b_t + \delta_t - r_t \underline{1}]dt$$
$$-\widetilde{k}_0(R_t - r_t)(x_t - \sum_{i=1}^{N} \pi_t^i)^- dt + \pi_t \sigma_t dw_t + \pi_t \rho_t d\widetilde{N}_t, 0 \leqslant t \leqslant T, \tag{8.43}$$

where $\widetilde{\sigma}_t = [\sigma_t, \rho_t]$ is the $N \times (d+m)$ volatility matrix process, $\delta_t = (\delta_{1t}, ..., \delta_N)^T$ is a dividend rate process, and $\widetilde{k}_0 \geqslant 0$ is a constant, and we write $a^- = \max(-a, 0)$.

(8.43) is a general model in the financial market, which includes many interesting cases. For example, let $\widetilde{k}_0 = 0$, then we get the model in [8]; and if we let $\widetilde{k}_0 = 1$, and $\rho_t^i \equiv 0, 1 \leqslant i \leqslant m$, (i.e. no jumps), we get a model discussed in [22],[33].

Suppose now the investor expects that his wealth will arrive at X at the future time T, then his present wealth (or say, his present investment) x_t should satisfy the following BSDE from (8.43):

$$x_t = X - \int_t^T r_s x_s ds + C_T - C_t - \int_t^T \pi_s[b_s + \delta_s - r_s \underline{1}]ds$$
$$+\widetilde{k}_0 \int_t^T (R_s - r_s)(x_s - \sum_{i=1}^N \pi_s^i)^- ds - \int_t^T \pi_s \sigma_s dw_s - \int_t^T \pi_s^T \rho_s d\tilde{N}(s). \tag{8.44}$$

As in the previous section we write

$[q_t, p_t] = \pi_t[\sigma_t, \rho_t] = \pi_t \widetilde{\sigma}_t,$

where $q_t = (q_t^1, \cdots, q_t^d), p_t = (p_t^1, \cdots, p_t^m)$, and assume that $N = d + m$. If $\widetilde{\sigma}_t^{-1} = [\sigma_t, \rho_t]^{-1}$exists, then the above BSDE can be rewritten as

$$x_t = X - \int_t^T r_s x_s ds + C_T - C_t - \int_t^T (q_s\theta_s^w + p_s\theta_s^{\tilde{N}})ds$$
$$+\widetilde{k}_0 \int_t^T (R_s - r_s)(x_s - \sum_{i=1}^N (\sum_{j=1}^d q_s^j(\widetilde{\sigma}_s^{-1})_{ji} + \sum_{j=1}^m p_s^j(\widetilde{\sigma}_s^{-1})_{(j+d)i}))^- ds$$
$$- \int_t^T q_s dw_s - \int_t^T p_s d\tilde{N}(s), 0 \leqslant t \leqslant T, \tag{8.45}$$

where as before we set

$\theta_t = \begin{bmatrix} \theta_t^w \\ \theta_t^{\tilde{N}} \end{bmatrix} = \widetilde{\sigma}_t^{-1}[b_t + \delta_t - r_t \underline{1}].$

Now suppose that the investor wants his consumption to satisfy the rule:

$$C_t = \int_0^t c(s, x_s, q_s, p_s)ds. \tag{8.46}$$

Then the existence of the solution (x_t, q_t, p_t) for BSDE (8.45) or, more precisely, the existence of the solution (x_t, π_t) for BSDE (8.44), will help the investor to make his wealth x_T reach X at the future time T, if he at time t by using the portfolio π_t invests the money x_t. Moreover, if one takes X as the value of some contingent claim, then the existence of x_t also shows that one can give the contingent claim a price x_t at time t. However, later one will see that the consumption density function $c(s, x, q, p)$ may, in some cases, be non-Lipschitzian: in particular if we consider the optimal consumption problem. So we need to consider the BSDEs with more general coefficients.

8.8 Existence of Solutions for Non-Lipschitzian Case

Consider a general BSDE in R^d as (8.8). We have the following theorem for the existence of solution of BSDE with a non-Lipschitzian coefficient. The method of deriving such a kind of theorem involves using the smoothness and limit technique.

Theorem 246 *Assume that* $\tau \leqslant T_0,\ b = b_1 + b_2,$
1° $b_i = b_i(t, x, q, p, \omega) : [0, T_0] \times R^d \times R^{d\otimes d_1} \times L^2_{\pi(\cdot)}(R^{d\otimes d_2}) \times \Omega \to R^d$, $i = 1, 2$, *are* $\mathfrak{F}_t-$ *adapted and measurable processes such that* $P-a.s.$
$|b_1(t, x, q, p, \omega)| \leqslant c_1(t)(1 + |x|)$,
$|b_2(t, x, q, p, \omega)| \leqslant c_1(t)(1 + |x|) + c_2(t)(1 + |q| + \|p\|)$,
where $c_1(t)$ *and* $c_2(t)$ *are non-negative and non-random such that*
$\int_0^{T_0} c_1(t)dt + \int_0^{T_0} c_2(t)^2 dt < \infty$;
2° $b_1(t, x, q, p, \omega)$ *is continuous in* (x, q, p);
3° one of the following conditions is satisfied:
(i) $X \in \mathfrak{F}_\tau, |X| \leqslant k_0$,
and
$(x_1 - x_2) \cdot (b_1(t, x_1, q_1, p_1, \omega) - b_1(t, x_2, q_2, p_2, \omega))$
$\leqslant c_1^N(t)\rho^N(|x_1 - x_2|^2) + c_2^N(t)\, |x_1 - x_2| \cdot (|q_1 - q_2| + \|p_1 - p_2\|)$,
$|b_1(t, x, q, p_1, \omega) - b_1(t, x, q, p_2, \omega)| \leqslant c_2^N(t)\, \|p_1 - p_2\|$,
$|b_2(t, x_1, q_1, p_1, \omega) - b_2(t, x_2, q_2, p_2, \omega)| \leqslant c_1^N(t)\, |x_1 - x_2|$
$+ c_2^N(t)[|q_1 - q_2| + \|p_1 - p_2\|]$,
as $|x| \leqslant N, |x_i| \leqslant N,\ i = 1, 2;\ N = 1, 2, \cdots$; *where for each* N, $c_1^N(t)$ *and* $c_2^N(t)$ *satisfy the same conditions for* $c_1(t)$ *and* $c_2(t)$ *as in 1°; and for each* N, $\rho^N(u) \geqslant 0$, *as* $u \geqslant 0$, *is non-random, increasing, continuous and concave such that*
$\int_{0+} du/\rho^N(u) = \infty$;
(ii) $X \in \mathfrak{F}_\tau, E|X|^2 < \infty$,
and coefficiewnts b_1 *and* b_2 *satisfy conditions in (i) above such that* $c_1^N(t)$, $c_2^N(t)$, $\rho^N(u)$ *do not depend on* N.
Then (8.8) has a unique solution.

First, let us give an example for the existence of a solution to BSDE in the case that $\int_0^{T_0} c_1(t)dt + \int_0^{T_0} c_2(t)^2 dt < \infty$, $c_1(t)$ and $c_2(t)$ are unbounded, and moreover, b_1 is also unbounded and non-Lipschitzian continuous in x.

Example 247 *Let*
$b = -I_{s\neq 0}s^{-\alpha_1}x\,|x|^{-\beta} + I_{s\neq 0}s^{-\alpha_2}q + I_{s\neq 0}s^{-\alpha_2}\int_Z p(z)I_U(z)\pi(dz),$
where $\alpha_1 < 1; \alpha_2 < 1/2; 0 < \beta < 1; \pi(U) < \infty$; *and assume that* $X \in \mathfrak{F}_T, E\,|X|^2 < \infty$.

Obviously, by Theorem 246 (8.8) has a unique solution. However, $c_1(s) = I_{s\neq 0}s^{-\alpha_1}, c_2(s) = I_{s\neq 0}s^{-\alpha_2}$ *are unbounded in* s, *and* $b_1 = -I_{s\neq 0}s^{-\alpha_1}x\,|x|^{-\beta}$ *is also unbounded in* s *and* x, *and is non-Lipschitzian continuous in* x. *Notice that here we have not assumed that* X *is bounded. (cf. [153]).*

Now let us show Theorem 246.

Proof. We can show this Theorem by using the smoothness technique. However, for simplicity we can assume that $b_2 = 0$. (Otherwise we can smooth out b_1 and then proceed the same way, as follows.)

Case 1. Under condition (i) in 3°

$$|X| \leqslant k_0. \tag{8.47}$$

In this case we can make a stronger assumption on b_1 :

$$|b_1| \leqslant c_1(t), \tag{8.48}$$

where $c_1(t)$ is non-negative and non-random such that
$\int_0^{T_0} c_1(t)dt < \infty$;
and make a stronger global condition than (i) of 3° for b_1 and b_2, i.e. assume that in (i) of 3°

$$c_1^N(t) = c_1(t), c_2^N(t) = c_2(t) \text{ and } \rho^N(u) = \rho(u) \text{ do not depend on N.} \tag{8.49}$$

Indeed, if the conclusion is proved under these stronger conditions, then one can derive the same conclusion under the weaker condition
$|b_1(t,x,q,p,\omega)| \leqslant c_1(t)(1+|x|),$
where $c_1(t)$ is non-negative and non-random such that $\int_0^{T_0} c_1(t)dt < \infty$, and under the weaker local condition as (i) of 3°.

For this for each N let $W^N(x) \in C_0^\infty(R^d)$ be such that
$W^N(x) = \begin{cases} 1, & \text{for } |x| \leqslant N+2, \\ 0, & \text{for } |x| \geqslant N+3, \end{cases}$
$\left|W^N(x) - W^N(y)\right| \leqslant \widetilde{k}_0\,|x-y|\,, \forall x,y \in R^d,$
and $0 \leqslant W^N(x) \leqslant 1$, where $\widetilde{k}_0$ is a constant. Define
$b_i^N(t,x,q,p,\omega) = b_i(t,x,q,p,\omega)W^N(x), i = 1,2.$
Then it is easily seen that if one lets
$\widetilde{c}_N^i(t) = c_i(t)(1+N+3)\widetilde{k}_0 + c_i^{N+3}(t), i = 1,2,$
$\widetilde{\rho}_N(u) = \rho^{N+3}(u) + u,$
then for each N b_1^N satisfies the stronger assumption
$\left|b_1^N\right| \leqslant c(t)(4+N).$

Moreover, for each N the condition 3° holds globally for $b_i^N, i=1,2$, with these new $\widetilde{c}_N^i(t), i=1,2$, and $\widetilde{\rho}_N(u)$. Hence there exists a unique solution (x_t^N, q_t^N, p_t^N) of (8.8)with such b^N for each N. Applying Lemma 231, one sees that there exists a natural number N_0, which depends on k_0 in (8.47), $\int_0^{T_0} c_1(t)dt$ and $\int_0^{T_0} c_2(t)^2 dt$ in condition 1° only, such that $|x_t^N| \leqslant N_0 + 2, \forall N = 1, 2, \cdots$.

Therefore $\left(x_t^{N_0}, q_t^{N_0}, p_t^{N_0}\right)$ is a solution of (8.8). The uniqueness is easily derived by Lemma 231 and 115.

Now let us show the result under additional conditions (8.47), (8.48), and (8.49).

Recall that we have already assumed that $b_2 = 0$. In this case for simplicity denote b_1 by b. Let us smooth out b to get b^n, i.e. let for all $u \in R^d$

$$J_d(u) = \begin{cases} c_d \exp(-(1-|u|^2)^{-1}), & \text{for } |u| < 1, \\ 0, & \text{otherwise,} \end{cases}$$

such that the constant c_d satisfies

$$\int_{R^d} J_d(u)du = 1;$$

and for $u \in R^{d\otimes d_1}$ (it is viewed as a $d \cdot d_1-$dimensional vector), $J_{d\cdot r}(u)$ is similarly defined. Set for $(x,q) \in R^d \times R^{d\otimes d_1}$

$$J(x,q) = J_d(x) \cdot J_{d\cdot d_1}(q).$$

Define

$$b^n(t,x,q,p,\omega) = \int_{R^d \times R^{d\otimes d_1}} b(t, x - n^{-1}\overline{x}, q - n^{-1}\overline{q}, p, \omega) J(\overline{x}, \overline{q}) d\overline{x} d\overline{q}.$$

Since for each $n = 1, 2, \cdots$ it is easily seen that as $(x,q,p), (x',q',p') \in R^d \times R^{d\otimes d_1} \times L^2_{\pi(.)}(R^d)$

$$|b^n(t,x,q,p,\omega) - b^n(t,x',q',p',\omega)| \leqslant k_n c_1(t) |x-x'| \\ + c_2(t)(|q-q'| + \|p-p'\|).$$

Hence by Theorem 234 there exists a unique solution (x_t^n, q_t^n, p_t^n) of the following BSDE (denote $t' = t \wedge \tau$ below)

$$\begin{aligned} x_t^n &= X + \int_{t\wedge\tau}^{\tau} b^n(s, x_s^n, q_s^n, p_s^n, \omega) ds - \int_{t\wedge\tau}^{\tau} q_s^n dw_s \\ &\quad - \int_{t\wedge\tau}^{\tau} \int_Z p_s^n(z) \tilde{N}_k(ds, dz). \end{aligned} \tag{8.50}$$

By Ito's formula

$$E[|x_t^m - x_t^n|^2 + \int_{t\wedge\tau}^{\tau} (|q_s^m - q_s^n|^2 + \|p_s^m - p_s^n\|^2) ds$$
$$= 2E\int_{t\wedge\tau}^{\tau} (x_s^m - x_s^n) \cdot (b^m(s, x_s^m, q_s^m, p_s^m, \omega) - b^n(s, x_s^n, q_s^m, p_s^n, \omega)) ds$$
$$\leqslant 2E\int_{t\wedge\tau}^{\tau} \int_0^t \{c_1(s)\rho\left(\left|x_s^m - x_s^n - \left(m^{-1} - n^{-1}\right)\overline{x}\right|^2\right)$$
$$+ c_2(s)\left|x_s^m - x_s^n - \left(m^{-1} - n^{-1}\right)\overline{x}\right| \cdot \left[\left|q_s^m - q_s^n - \left(m^{-1} - n^{-1}\right)\overline{q}\right|\right.$$
$$\left. + \|p_s^m - p_s^n\|\right] + 2c_1(s)\left|\left(m^{-1} - n^{-1}\right)\overline{x}\right|\} J(\overline{x}, \overline{q}) d\overline{x} d\overline{q} ds.$$

Hence

$$E[|x_t^m - x_t^n|^2 + \tfrac{1}{2}\int_{t\wedge\tau}^{\tau} (|q_s^m - q_s^n|^2 + \|p_s^m - p_s^n\|^2) ds$$

$\leqslant k_0' \int_t^{T_0} \{c_1(s) \int_0^t \rho \left(E \left| x_s^m - x_s^n - \left(m^{-1} - n^{-1}\right) \overline{x} \right|^2 \right) J(\overline{x}, \overline{q}) d\overline{x} d\overline{q}$
$+ c_2(s)^2 E \left| x_s^m - x_s^n \right|^2 \} ds + k_0' \left(m^{-1} + n^{-1}\right).$

Since by Lemma 114 for all n

$E(\sup_{t \leqslant \tau} |x_t^n|^2 + \int_0^\tau (|q_s^n|^2 + \|p_s^n\|^2) ds \leqslant k_T < \infty.$

Hence by Fatou's lemma it is easily seen that

$\overline{\lim_{m,n \to \infty}} E[|x_t^m - x_t^n|^2 + \overline{\lim_{m,n \to \infty}} \int_{t\wedge\tau}^\tau (|q_s^m - q_s^n|^2 + \|p_s^m - p_s^n\|^2) ds]$

$\leqslant \widetilde{k}_T \int_t^{T_0} (c_1(s) + c_2(s)^2) \rho_1 \left(\overline{\lim_{m,n \to \infty}} E \left| x_s^m - x_s^n \right|^2 \right) ds,$

where $\rho_1(u) = \rho(u) + u$. Therefore,

$\overline{\lim_{m,n \to \infty}} E[|x_t^m - x_t^n|^2 + \overline{\lim_{m,n \to \infty}} \int_{t\wedge\tau}^\tau (|q_s^m - q_s^n|^2 + \|p_s^m - p_s^n\|^2) ds = 0.$

From this it is not difficult to show that there exists a

$(x_t, q_t, p_t) \in S^2_{\mathfrak{F}_t}(R^d) \times L^2_{\mathfrak{F}_t}(R^{d\otimes d_1}) \times F^2_{\mathfrak{F}_t}(R^{d\otimes d_2})$

such that as $n \to \infty$,

$(x_t^n, q_t^n, p_t^n) \to (x_t, q_t, p_t)$, in $S^2_{\mathfrak{F}_t}(R^d) \times L^2_{\mathfrak{F}_t}(R^{d\otimes d_1}) \times F^2_{\mathfrak{F}_t}(R^{d\otimes d_2})$.

Notice that

$|b^n(s, x_s^n, q_s^n, p_s^n, \omega) - b(s, x_s, q_s, p_s, \omega)|$

$\leqslant \int_0^t |b(s, x_s^n - n^{-1}\overline{x}, q_s^n - n^{-1}\overline{x}, p_s^n, \omega) - b(s, x_s, q_s, p_s, \omega) \mid J(\overline{x}, \overline{q}) d\overline{x} d\overline{q}.$

Hence, letting $n \to \infty$ (take a subsequence, if necessary) in (8.50) one finds that (x_t, q_t, p_t) is a solution of (8.8). The uniqueness can be derived by Lemma 231 and 115.

Case 2. Suppose that we have $X \in \mathfrak{F}_\tau, E|X|^2 < \infty$. Let

$$X^n = X I_{|X| \leqslant n}.$$

Then by step 1 there exists a unique solution (x_t^n, q_t^n, p_t^n) satisfying as $0 \leqslant t$,

$$\begin{aligned} x_t^n &= X^n + \int_{t\wedge\tau}^\tau b(s, x_s^n, q_s^n, p_s^n, \omega) ds - \int_{t\wedge\tau}^\tau q_s^n dw_s \\ &\quad - \int_{t\wedge\tau}^\tau \int_Z p_s^n(z) \tilde{N}_k(ds, dz), \end{aligned} \tag{8.51}$$

with coefficient b and terminal random variable X^n. By Ito's formula it is easily seen (cf. the proof of step 1 above) that (x_t^n, q_t^n, p_t^n) is a Cauchy sequence in $S^2_{\mathfrak{F}_t}(R^d) \times L^2_{\mathfrak{F}_t}(R^{d\otimes d_1}) \times F^2_{\mathfrak{F}_t}(R^{d\otimes d_2})$. Hence there exists a limit $(x_t, q_t, p_t) \in S^2_{\mathfrak{F}_t}(R^d) \times L^2_{\mathfrak{F}_t}(R^{d\otimes d_1}) \times F^2_{\mathfrak{F}_t}(R^{d\otimes d_2})$. Taking limit in (8.51) by letting $n \to \infty$ (if necessary taking a subsequence), one finds that (x_t, q_t, p_t) is a solution of (8.8). The uniqueness can be derived by Lemma 230. ■

8.9 Convergence of Solutions

We have the following convergence theorems for solutions of BSDE (8.8).

Theorem 248 *Assume that τ is a $\mathfrak{F}_t-$stopping time for $n = 0, 1, 2, \cdots$*
1° $b^n = b^n(t, x, q, p, \omega) : [0, T] \times R^d \times R^{d\otimes d_1} \times L^2_{\pi(.)}\left(R^{d\otimes d_2}\right) \to R^d$
are $\mathfrak{F}_t-$adapted such that $P - a.s.$
$\langle x, b^n(t, x, q, p, \omega)\rangle \leqslant c_1(t)(1 + |x|) + c_2(t)|x|(1 + |q| + \|p\|),$
where $c_1(t)$ and $c_2(t)$ have the same properties as that in 1° of Theorem 246;
2° $\left\langle (x_1 - x_2), (b^0(t, x_1, q_1, p_1, \omega) - b^0(t, x_2, q_2, p_2, \omega))\right\rangle$
$\leqslant c_1(t)\rho(|x_1 - x_2|^2) + c_2(t)|x_1 - x_2|(|q_1 - q_2| + \|p_1 - p_2\|),$
where $\rho(u)$ has the same property as that in 2° of Theorem 246;
3° X^n is $\mathfrak{F}_{T\wedge\tau}-$measurable such that $E|X^n|^2 < \infty$, for all $n = 0, 1, 2, \cdots$, and
$\lim_{n\to\infty} E\left|X^n - X^0\right|^2 = 0;$
4° $\lim_{n\to\infty} \int_0^T \sup_{\substack{(x,q)\in R^d\times R^{d\otimes d_1} \\ p\in L^2_{\pi(.)}\left(R^{d\otimes d_2}\right) \\ \omega\in\Omega}} |b^n(t, x, q, p, \omega) - b^0(t, x, q, p, \omega)|^2 dt = 0.$
If (x_t^n, q_t^n, p_t^n) is a solution of the following BSDE: $0 \leqslant t \leqslant T$
$x_t^n = X^n + \int_{t\wedge\tau}^{\tau\wedge T} b^n(s, x_s^n, q_s^n, p_s^n, \omega)ds - \int_{t\wedge\tau}^{\tau\wedge T} q_s^n dw_s$
$- \int_{t\wedge\tau}^{\tau\wedge T}\int_Z p_s^n(z)\tilde{N}_k(ds, dz),$
$n = 0, 1, 2, \cdots$, respectively, then
$\lim_{n\to\infty} E[\sup_{t\in[0,\tau\wedge T]}\left|x_t^n - x_t^0\right|^2 + \int_0^{\tau\wedge T}(\left|q_s^n - q_s^0\right|^2 + \left\|p_s^n - p_s^0\right\|^2)ds] = 0.$

Proof. By Ito's formula

$$E[\left|x_{t\wedge\tau}^n - x_{t\wedge\tau}^0\right|^2 + \tfrac{1}{2}\int_{t\wedge\tau}^{\tau\wedge T}(\left|q_s^n - q_s^0\right|^2 + \left\|p_s^n - p_s^0\right\|^2)ds$$
$$\leqslant k_0\int_t^T\{c_1(s)\rho\left(E\left|x_{s\wedge\tau}^n - x_{s\wedge\tau}^0\right|^2\right) + (c_2(s)^2 + 1)E\left|x_{s\wedge\tau}^n - x_{s\wedge\tau}^0\right|^2\}ds$$
$$+k_0 E\int_t^T\left|b^n(s, x_{s\wedge\tau}^n, q_{s\wedge\tau}^n, p_{s\wedge\tau}^n, \omega) - b^0(s, x_{s\wedge\tau}^n, q_{s\wedge\tau}^n, p_{s\wedge\tau}^n, \omega)\right|^2 ds$$
$$+E\left|X^n - X^0\right|^2.$$

Hence one easily shows that for all $t \in [0, T]$

$$\lim_{n\to\infty} E[\left|x_{t\wedge\tau}^n - x_{t\wedge\tau}^0\right|^2 + \tfrac{1}{2}\int_{t\wedge\tau}^{\tau\wedge T}(\left|q_s^n - q_s^0\right|^2 + \left\|p_s^n - p_s^0\right\|^2)ds] = 0.$$

Applying Ito's formula to $\left|x_{t\wedge\tau}^n - x_{t\wedge\tau}^0\right|^2$ for the forward SDE and using the martingale inequality one also easily shows that

$$\lim_{n\to\infty} E\sup_{t\in[0,\tau\wedge T]}\left|x_t^n - x_t^0\right|^2 = 0.$$

■

We also have some other useful convergence theorem as follows:

Theorem 249 *Assume that*
1°′ $|b^n(t, x, q, p, \omega)| \leqslant c_1(t)(1 + |x|) + c_2(t)(1 + |q| + \|p\|),$
$\forall n = 1, 2, \cdots;$
$b^0(t, x, q, p, \omega) = b^{01}(t, x, q, p, \omega) + b^{02}(t, x, q, p, \omega)$
$\left|b^{01}(t, x, q, p, \omega)\right| \leqslant c_1(t)(1 + |x|),$
$\left|b^{02}(t, x, q, p, \omega)\right| \leqslant c_1(t)(1 + |x|) + c_2(t)(1 + |q| + \|p\|),$
where $c_1(t)$ and $c_2(t)$ have the same properties as that in 1° of Theorem 246;

2°′ $\langle (x_1 - x_2), (b^n(t, x_1, q_1, p_1, \omega) - b^n(t, x_2, q_2, p_2, \omega))\rangle$
$\leqslant c_1(t)\rho(|x_1 - x_2|^2) + c_2(t)\,|x_1 - x_2|\,(|q_1 - q_2| + \|p_1 - p_2\|),$
$\forall n = 1, 2, \cdots,$
$|b^{01}(t, x, q, p_1, \omega) - b^{01}(t, x, q, p_2, \omega)| \leqslant c_2(t)\,\|p_1 - p_2\|,$
$|b^{02}(t, x_1, q_1, p_1, \omega) - b^{02}(t, x_2, q_2, p_2, \omega)|$
$\leqslant c_1(t)\,|x_1 - x_2| + c_2(t)(|q_1 - q_2| + \|p_1 - p_2\|),$
where $\rho(u)$ *has the same property as that in 2° of Theorem 248;*
3°′ The same as 3° in Theorem 248;
4°′ $\lim_{n\to\infty} b^n(t, x, q, p, \omega) = b^0(t, x, q, p, \omega), P - a.s.$
5°′ $b^{01}(t, x, q, p, \omega)$ *is continuous in* (x, q, p).
If (x_t^n, q_t^n, p_t^n) *is a solution of the following BSDE:* $0 \leqslant t \leqslant T$
$x_t^n = X^n + \int_{t\wedge\tau}^{\tau\wedge T} b^n(s, x_s^n, q_s^n, p_s^n, \omega)ds - \int_{t\wedge\tau}^{\tau\wedge T} q_s^n dw_s$
$- \int_{t\wedge\tau}^{\tau\wedge T} \int_Z p_s^n(z)\tilde{N}_k(ds, dz),$
$n = 1, 2, \cdots,$ *respectively, then*
1) the above SDE for $n = 0$ *also has a solution* (x_t^0, q_t^0, q_t^0), *which is unique;*
2) $\lim_{n\to\infty} E[\sup_{t\in[0,\tau\wedge T]} |x_t^n - x_t^0|^2 + \int_0^{\tau\wedge T}(|q_s^n - q_s^0|^2 + \|p_s^n - p_s^0\|^2)ds] = 0.$

Proof. 1) In the condition $2^{\circ\prime}$ letting $n \to \infty$ one sees that $2^{\circ\prime}$ is also true for $n = 0$. Furthermore, by $b^{01} = b^0 - b^{02}$ one sees $2^{\circ\prime}$ is also true for b^{01} at most with some other $c_1(t)$ and $c_2(t)$. Thus by Theorem 246 1) is obtained.

2): By Ito's formula
$E[|x_{t\wedge\tau}^n - x_{t\wedge\tau}^0|^2 + \frac{1}{2}\int_{t\wedge\tau}^{\tau\wedge T}(|q_s^n - q_s^0|^2 + \|p_s^n - p_s^0\|^2)ds$
$\leqslant E\,|X^n - X^0|^2 + k_0 \int_t^T \int_0^t \{c_1(s)\rho\left(E\,|x_{s\wedge\tau}^n - x_{s\wedge\tau}^0|^2\right)$
$+(c_2(s)^2 + 1)E\,|x_{s\wedge\tau}^n - x_{s\wedge\tau}^0|^2\}ds$
$+k_0 E \int_t^T |b^n(s, x_{s\wedge\tau}^0, q_{s\wedge\tau}^0, p_{s\wedge\tau}^0, \omega) - b^0(s, x_{s\wedge\tau}^0, q_{s\wedge\tau}^0, p_{s\wedge\tau}^0, \omega)|^2 ds.$
Hence one can still complete the proof in a same way as above. ■

8.10 Explanation of Solutions of BSDEs to Financial Markets

In the previous sections we have already introduced the following BSDE for a general wealth process x_t :

$$dx_t = r_t x_t dt - dC_t + \pi_t[b_t + \delta_t - r_t\underline{1}]dt$$
$$-\widetilde{k}_0(R_t - r_t)(x_t - \sum_{i=1}^N \pi_t^i)^- dt + \pi_t\sigma_t dw_t + \pi_t\rho_t d\widetilde{N}_t,$$
$$x_T = X,\ 0 \leqslant t \leqslant T, \tag{8.52}$$

where x_t is the wealth of a small investor at time t, $\widetilde{\sigma}_t = [\sigma_t, \rho_t]$ is the $N \times (d+m)$ volatility matrix process, $\delta_t = (\delta_{1t}, ..., \delta_N)^T$ is a dividend rate process, and $\widetilde{k}_0 \geqslant 0$ is a constant; and during the time interval $[0,T]$ the investor spends the money $dC_t \geqslant 0$ in a small time dt with $C_0 = 0$, and $C_T < \infty$, and borrows the money from the bank with the interest rate R_t at time t. Now by using the existence theorem of BSDE (Theorem 246) we can show the following fact, that under appropriate conditions (8.52) can be solved. That is, there exists a solution (x_t, π_t), where x_t is the total invest money of the investor at time t and π_t is the portfolio of the investment (actually it is the part of investment for the stocks) at time t, such that as the time evolves they can produce the money X in the market at the future time T. Or say, if the small investor invests his money x_t with the portfolio π_t at each time t in this market, even if he spends the money dC_t at a small time dt, and he borrows the money from the bank with the rate R_t, his wealth can still eventually arrive at the amount X in the market at the future time T.

Actually we have the following theorem.

Theorem 250 *In a financial market suppose that there is one bond, which price satisfies (8.26), and there exist $N-$stocks, whose prices satisfying the SDEs (8.27).*

Assume that

(A_1) $N = d+m$, and $r_t, b_t, \delta_t, \sigma_t, R_t$ and ρ_t are all non-random such that $\widetilde{\sigma}_t^{-1}$ exists and is bounded, that is, $\left|\widetilde{\sigma}_t^{-1}\right| \leq k_0$, and

$\int_0^T [|r_t| + \left|\widetilde{\sigma}_t^{-1}(b_t + \delta_t - r_t\underline{1})\right|^2 + |R_t - r_t|^2] dt < \infty,$

where

$\sigma_t \in R^{N\otimes d}, \rho_t \in R^{N\otimes m}, \widetilde{\sigma}_t = \begin{pmatrix} \sigma_t & \rho_t \end{pmatrix} \in R^{N\otimes(d+m)};$

$X \in \mathfrak{F}_t^{w,\widetilde{N}}$ and $E|X|^2 < \infty$;

and, moreover, one of the following conditions is satisfied:

(i) $0 \leq C_t \uparrow$ is $\mathfrak{F}_t-$adapted such that $EC_T^2 < \infty$;

(ii) $0 \leq C_t = \int_0^t \widetilde{c}(s, x_s) ds$,

where $0 \leq \widetilde{c}(t,x)$ satisfies the properties: $\widetilde{c}(t,x)$ is continuous in x, and

$|\widetilde{c}(t,x)| \leq c_0(t)(1+|x|),$

$(\widetilde{c}(t,x) - \widetilde{c}(t,y))(x-y) \leq c_0(t)\,|x-y|\,, \forall t \in [0,T], \forall x, y \in R^1,$

and $0 \leq c_0(t)$, $\int_0^T c_0(t)dt < \infty$.

Then the following BSDE has a unique $\mathfrak{F}_t = \mathfrak{F}_t^{w,\widetilde{N}}-$adapted solution $(x_t, \pi_t) \in R^1 \times R^N, 0 \leq t \leq T$,

$$\begin{cases} dx_t = r_t x_t dt + \pi_t(b_t - r_t\underline{1})dt - dC_t \\ -\widetilde{k}_0(R_t - r_t)(x_t - \sum_{i=1}^N \pi_t^i)^- dt + \pi_t^T \sigma_t dw_t + \pi_t^T \rho_t d\widetilde{N}_t, \\ x_T = X, \ 0 \leq t \leq T. \end{cases}$$

Moreover, $(x_t, \pi_t), 0 \leq t \leq T$, hedges the contingent claim $X \in \mathfrak{F}_T$, if we explain X as a contingent claim.

Proof. Write $[q_t, p_t] = \pi_t[\sigma_t, \rho_t] = \pi_t\widetilde{\sigma}_t$, where $q_t = (q_t^1, \cdots, q_t^d), p_t = (p_t^1, \cdots, p_t^m)$.

Then the above BSDE can be rewritten as

$$x_t = X - \int_t^T r_s x_s ds + C_T - C_t - \int_t^T (q_s\theta_s^w + p_s\theta_s^{\tilde{N}})ds$$
$$+\widetilde{k}_0 \int_t^T (R_s - r_s)(x_s - \sum_{i=1}^N (\sum_{j=1}^d q_s^j(\widetilde{\sigma}_s^{-1})_{ji} + \sum_{j=1}^m p_s^j(\widetilde{\sigma}_s^{-1})_{(j+d)i}))^- ds$$
$$- \int_t^T q_s dw_s - \int_t^T p_s d\tilde{N}(s), 0 \leqslant t \leqslant T, \qquad (8.53)$$

where we write

$$\theta_t = \begin{bmatrix} \theta_t^w \\ \theta_t^{\tilde{N}} \end{bmatrix} = \widetilde{\sigma}_t^{-1}[b_t + \delta_t - r_t\underline{1}].$$

Moreover, in case (ii) one sees that

$C_T - C_t = \int_t^T \widetilde{c}(s, x_s)ds.$

Hence, for either of cases (i) and (ii), applying Theorem 246, one finds that (8.53) has a unique $\mathfrak{F}_t$−adapted solution (x_t, q_t, p_t). Let $\pi_t = [q_t, p_t]\widetilde{\sigma}_t^{-1}$. Then (x_t, π_t) is the unique $\mathfrak{F}_t$−adapted solution of (8.52). ∎

8.11 Comparison Theorem for BSDE with Jumps.

Comparison theorems for solutions of BSDEs are very important in both theory and applications. For example, in a financial market we naturally require the following facts: If an investor wants his money to increase to a higher level at the future time T, or during the time interval $s \in [0, T]$ he wants to consume more, then now, at the present time t, should he invest more in the market. That is, $X^1 \geq X^2, C_s^1 \geq C_s^2, s \in [0, T] \Longrightarrow x_t^1 \geq x_t^2, t \in [0, T]$. This is exactly the content of the comparison theorem of solutions of BSDEs. In this section we will discuss such kind of problem. We will see that the answers to thess problems can not only be applied to the financial market, but can also be used to discuss some deeper theorectical problems in BSDEs.

Suppose that $(x_t^i, q_t^i, p_t^i), i = 1, 2$, are solutions of the following BSDEs in 1−dimensional space for $t \geqslant 0; i = 1, 2$

$$x_t^i = X^i + \int_{t\wedge\tau}^{\tau} b^i(s, x_s^i, q_s^i, p_s^i, \omega)ds - \int_{t\wedge\tau}^{\tau} q_s^i dw_s - \int_{t\wedge\tau}^{\tau}\int_Z p_s^i(z)\tilde{N}_k(ds, dz), \qquad (8.54)$$

where b^i in (8.54) is a real function, $i = 1, 2$, $q_s^i \in R^{1\otimes d_1}, p_s^i(z) \in R^{1\otimes d_2}$, and for simplicity we assume that τ is a bounded $\mathfrak{F}_t = \mathfrak{F}_t^{w,\tilde{N}_k}$−stopping time. Then we have the following

Theorem 251 *Assume that*
1° $b^1(t,x,q,p,\omega) \geqslant b^2(t,x,q,p,\omega)$;
2° $|b^2(t,x_1,q_1,p_1,\omega) - b^2(t,x_2,q_2,p_2,\omega)| \leqslant k_0(|x_1 - x_2| + |q_1 - q_2|)$
$+ \int_Z |C_t(z,\omega)|\,|(p_1(z) - p_2(z))|\,\pi(dz)$,
where $k_0 \geqslant 0$ *is a constant, and* $C_t(z,\omega)$ *satisfies the condition*
(B) $|C_{jt}(z,\omega)| \leqslant 1, j = 1,\cdots,d_2$;
$C_t(z) = (C_{1t}(z),\cdots,C_{d_2t}(z)) \in F^2_{\mathfrak{F}}(R^{1\otimes d_2})$;
3° $X^i \in \mathfrak{F}_\tau, E|X^i|^2 < \infty, i = 1,2$,
$X^1 \geqslant X^2$.
Then $P-a.s.$
$x^1_t \geqslant x^2_t, \forall \in [0,\tau]$.

Theorem 251 can be applied to explain the option pricing problem in a financial market. If we comsider X^i to be the contingent claim with respect to a given option at the future time T, and let x^i_t be the present option price, then Theorem 251 tells us that for a bigger contingent claim it is reasonable to make the option price bigger at the present time.

One notices that in assumption 2° of Theorem 251 the condition for b with respect to p is stronger than the usual Lipschitz condition, and later we will give an example to show that such a stronger condition cannot be weakened to the usual Lipschitzian condition for p. Now let us give a simpler comparison theorem. However, from this simpler comparison theorem we can derive the above Theorem 251.

Theorem 252 *Under assumption 1° and 3° in Theorem 251 and the assumption that*
2°′ one of the following conditions is satisfied:
(i) $b^1(t,x,q,p,\omega) = \beta(t,x,q,\omega) + \int_Z C^1_t(z,\omega)p(z)^T\pi(dz)$,
$|\beta(t,x,q,\omega) - \beta(t,x',q',\omega)| \leqslant k_0[|x - x'| + |q - q'|]$,
where $k_0 \geqslant 0$ *is a constant, and* $C^1_t(z,\omega)$ *satisfies the condition*
$(B)_1$ $|C^1_t(z,\omega)| \leqslant k_0, C^1_{it}(z) \geqslant -1, i = 1,\cdots,d_2$;
$C^1_t(z) = (C^1_{1t}(z),\cdots,C^1_{d_2t}(z)) \in F^2_{\mathfrak{F}}(R^{1\otimes d_2})$;
(ii) $b^2(t,x,q,p,\omega) = \beta(t,x,q,\omega) + \int_Z C^2_t(z,\omega)p(z)^T\pi(dz)$,
$|\beta(t,x,q,\omega) - \beta(t,x',q',\omega)| \leqslant k_0[|x - x'| + |q - q'|]$,
where $k_0 \geqslant 0$ *is a constant, and* $C^2_t(z,\omega)$ *satisfies the condition*
$(B)_2$ $|C^2_t(z,\omega)| \leqslant k_0, C^2_{it}(z) \leqslant 1, i = 1,\cdots,d_2$;
$C^2_t(z) = (C^2_{1t}(z),\cdots,C^2_{d_2t}(z)) \in F^2_{\mathfrak{F}}(R^{1\otimes d_2})$;
the conclusion of Theorem 251 still holds.

Before we establish Theorem 251 by using Theorem 252, let us explain the following fact: The condition 2°′ in Theorem 252 actually cannot be weakened. Intuitively, the condition $C^2_{it}(z,\omega) \leqslant 1$ or $C^1_{it}(z,\omega) \geqslant -1$ means that the coefficient cannot be influenced too much by the jumps. Otherwise, the conclusion of the above theorems would fail. The following is a counter example when the condition is violated.

Example 253 *Assume that* $0 < \pi(U) < \infty, \varepsilon_0 > 0$. *Notice that*
$P(N_k([0,t],U) = 0) = e^{-t\pi(U)} > 0, \forall t \geqslant 0.$
Hence we can construct an example as follows: Consider BSDE (with all random processes in R^1*)*

$$x_t = X + \int_t^{T_0} b(p_s)ds - \int_t^{T_0} q_s dw_s - \int_t^{T_0} \int_Z p_s(z)\tilde{N}_k(ds,dz), \quad (8.55)$$

with
$b(p) = (1+\varepsilon_0)\int_Z I_U(z)p(z)\pi(dz).$
Obviously, $(0,0,0)$ *solves the above BSDE with* $X = 0$, *and*
$(x_t, q_t, p_t) = (-N_k([0,t],U) + \varepsilon_0(T_0 - t)\pi(U), 0, -I_{z\in U})$
solves the above BSDE with $X = -N_k([0,T_0],U) \leqslant 0$. *In fact, by*
$\widetilde{N}_k([t,T_0],U) = N_k([t,T_0],U) - (T_0 - t)\pi(U)$
one easily checks this result. However,
$C_t^2(z,\omega) = (1+\varepsilon_0)I_U(z) > 1,$ *as* $z \in U$;
and
$P(x_t > 0) \geq P(N_k([0,t],U) = 0) > 0, \forall t \in [0,T_0).$

A counter example for
$C_t^1(z,\omega) = -(1+\varepsilon_0) < -1$
can also be similarly constructed.

Now we will give an example to show that the condition in 2° of Theorem 251 for b with respect to p cannot be weakened to the usual Lipschitz condition.

Example 254 . *Assume that* $\pi(U) = 1/2, \pi(V) = 1, U \subset V$. *Consider the BSDE* (8.55) *with*
$b(p) = \int_Z I_V(z)p(z)\pi(dz).$
Obviously, $(0,0,0)$ *solves the above BSDE with* $X = 0$, *and*
$(x_t, q_t, p_t) = (-N_k([0,t],U), 0, -I_{z\in U})$
solves the above BSDE with $X = -N_k([0,T_0],U) \leqslant 0$. *Moreover, here* $C_t^2(z,\omega) = I_V(z)$ *and*
$P(x_t \leqslant 0) = 1, \forall t \in [0,T_0].$
Hence the comparison theorem holds. Now if we consider the BSDE (8.55) *with*
$b(p) = (\int_Z |p(z)|^2 \pi(dz))^{1/2}.$
then the coefficient b satisfies that

$$|b(p_1) - b(p_2)| \leqslant \|C_t(.)\| \, \|p_1 - p_2\|, \quad (8.56)$$

with $C_t(z) = I_V(z), \|p\| = \sqrt{\int_Z |p(z)|^2 \pi(dz)}, \pi(U) = \frac{1}{2}, \pi(V) = 1, U \subset V$. *(Notice that the condition (8.56) is weaker than that* $|b(p_1) - b(p_2)| \leqslant$

$\int_Z |C_t(z)| \, |(p_1(z) - p_2(z))| \, \pi(dz))$. *Obviously,* $(0,0,0)$ *still solves the above BSDE with* $X = 0$, *and*

$(x_t, q_t, p_t) = \left(-N_k([0,t],U) + [\sqrt{\pi(U)} - \pi(U)](T_0 - t), 0, -I_{z\in U}\right)$

solves the above BSDE with $X = -N_k([0,T_0],U) \leqslant 0$. *However,*

$P(x_t > 0) \geq P(N_k([0,t],U) = 0) > 0, \forall t \in [0,T_0)$,

and so the comparison theorem fails.

Now, let us establish Theorem 251 by using Theorem 252.

Proof. Notice that $(\widehat{x}_t, \widehat{q}_t, \widehat{p}_t) = (x_t^2 - x_t^1, q_t^2 - q_t^1, p_t^2 - p_t^1)$ satisfies the BSDE as follows:

$$\begin{aligned} \widehat{x}_t &= X + \int_{t\wedge\tau}^{\tau} (a_s(\omega)\widehat{x}_s + \widetilde{b}_s(\omega)\widehat{q}_s^T + \int_Z \widetilde{C}_s(z,\omega)\widehat{p}_s^T(z)\pi(dz) \\ &\quad + f_0(s,\omega))ds - \int_{t\wedge\tau}^{\tau} \widehat{q}_s dw_s - \int_{t\wedge\tau}^{\tau}\int_Z \widehat{p}_s(z)\widetilde{N}(ds,dz), \end{aligned} \quad (8.57)$$

where $\widehat{q}_s^T$ is the transpose of $\widehat{q}_s$, etc., and

$X = X^2 - X^1$,

$a_t(\omega) = I_{(\widehat{x}_t \neq 0)}(b^2(t, x_t^2, q_t^2, p_t^2, \omega) - b^2(t, x_t^1, q_t^2, p_t^2, \omega))(\widehat{x}_t)^{-1}$,

$\widetilde{b}_{it}(\omega) = I_{(\widehat{q}_{it} \neq 0)}(b^2(t, x_t^1, \widetilde{q}_{i-1,t}^1, p_t^2, \omega) - b^2(t, x_t^1, \widetilde{q}_{i,t}^1, p_t^2, \omega))(\widehat{q}_{it})^{-1}$,

$i = 1, ..., d_1$;

$\widehat{C}_{it}(\omega) = I_{(\int_Z C_{it}(z,\omega)\widehat{p}_{it}(z)\pi(dz) \neq 0)} (b^2(t, y_t^1, q_t^1, \widetilde{p}_{i-1,t}^1, \omega)$

$-b^2(t, y_t^1, q_t^1, \widetilde{p}_{i,t}^1, \omega)) \cdot (\int_Z |C_{it}(z,\omega)| \, |\widehat{p}_{i,t}(z)| \, \pi(dz))^{-1}$,

$\widetilde{C}_{it}(z,\omega) = \widehat{C}_{it}(\omega) \, |C_{it}(z,\omega)| \, sgn(\widehat{p}_{i,t}(z)), \left|\widehat{C}_{it}(\omega)\right| \leqslant 1, i = 1, ..., d_2$;

$f_0(t,\omega) = b^2(t, y_t^1, q_t^1, p_t^1, \omega) - b^1(t, y_t^1, q_t^1, p_t^1, \omega) \leqslant 0$,

and

$\widetilde{q}_{i,t}^1 = (q_{1t}^1, ..., q_{it}^1, q_{(i+1)t}^2, ..., q_{d_1 t}^2)$,

$q_t^1 = (q_{1t}^1, ..., q_{d_1 t}^1)$, $\widehat{q}_t = (\widehat{q}_{1t}, \cdots, \widehat{q}_{d_1 t})$,

$C_t(z,\omega) = (C_{1t}(z,\omega), \cdots, C_{d_2 t}(z,\omega))$, etc.

Obviously, condition (ii) of $2^{\circ\prime}$ in Theorem 252 is satisfied. Hence Theorem 252 applies. One has $P - a.s.$

$\widehat{x}_t = x_t^2 - x_t^1 \leqslant 0, \forall t \in [0,\tau]$.

The proof using Theorem 252 to derive Theorem 251 is complete. ∎

Before we establish Theorem 252, some preparation is necessary.

Introduce another condition for $C_t(z)$:

(C)$_1$ $C_{it}(z) \geqslant -1, i = 1, \cdots, d_2$;

$C_t(z) = (C_{1t}(z), \cdots, C_{d_2 t}(z)) \in F_{\mathfrak{F}}^2(R^{1\otimes d_2})$,

$\int_Z \left|1 - \sqrt{1 + C_{it}(z)}\right|^2 \pi(dz) \leqslant k_0 < \infty$.

Suppose that (x_t, q_t, p_t) solves the following BSDE for $0 \leqslant t$

$$\begin{aligned} x_t &= X + \int_{t\wedge\tau}^{\tau} (a_s(\omega)x_s + b_s(\omega)q_s^T + \int_Z C_s(z,\omega)p_s(z)^T\pi(dz) \\ &\quad + f_0(s,\omega))ds - \int_{t\wedge\tau}^{\tau} q_s dw_s - \int_{t\wedge\tau}^{\tau}\int_Z p_s(z)\tilde{N}_k(ds,dz), \quad (8.58) \end{aligned}$$

where $a_t \in L^2_{\mathfrak{F}}(R^1), b_t \in L^2_{\mathfrak{F}}(R^{1\otimes d_1}), f_0 \in L^1_{\mathfrak{F}}(R^1)$, moreover, a_t and b_t are bounded, i.e.$|a_t| + |b_t| \leqslant k_0$.

We have the following simple comparison lemma. However, this is a basic comparison result for the solutions of BSDEs, and actually, all other comparison theorems can be derived from it.

Lemma 255 *If $f_0 \geqslant 0, X \geqslant 0$, and $C_t(z,\omega)$ satisfies the condition $(B)_1$, then $P-a.s.$*
$x_t \geqslant 0$, for all $t \in [0,\tau]$.

The comparison result for solutions of BSDEs cannot be derived by Tanaka's formula as in the forward SDE case. Since now we cannot know whether the coefficients q_t satisfies the Holder-continuity with the index $\frac{1}{2}$ required by the Tanaka formula. However, there are two methods that can be used to solve such kinds of problem. One is by using Girsanov's theorem to simplify the BSDE to make it become the summation of a non-negative process and a new martingale under a new probability measure, and then to take the conditional expectation under this new probability measure to obtain the result. This method will also be seen later in the dicussion of arbitrage-free markets. Here we will first use "a duality method" to derive the comparison result. The idea is that we introduce a relative forward SDE. Then, by using Ito' formula to the multiplication of the solutions of BSDE and the forward SDE, we can show that this multiplication process is actually the summation of a non-negative process and a martingale under the original probability measure. Thus we can then take the conditional expectation under the original probability measure to obtain the result. We will also use the Girsanov's theorem to again establish the result. Let us now discuss the details of such approaches. We now give the proof of Lemma 255.

Proof. Method 1. (A Duality Method).

First of all, suppose that the condition $(C)_1$ holds. For any given $(t,x) \in [0,\infty) \times R^1$ assume that $y_s, t \leqslant s \leqslant T$, solves the following forward SDE
$y_s = x + \frac{1}{2}\int_{t\wedge\tau}^{s\wedge\tau} a_r y_r dr + \frac{1}{2}\int_{t\wedge\tau}^{s\wedge\tau} b_r y_r dw_r + \frac{1}{2}\int_{t\wedge\tau}^{s\wedge\tau}\int_Z D_r y_{r-}\tilde{N}_k(dr,dz)$
$-\frac{1}{8}\int_{t\wedge\tau}^{s\wedge\tau} |b_r|^2 y_r dr - \frac{1}{8}\int_{t\wedge\tau}^{s\wedge\tau}\int_Z |D_r|^2 y_r \pi(dz)dr$,
where $D_r \in R^{1\otimes d_2}$ will be determined later, $b_r \in R^{1\otimes d_1}$ and
$\int_{t\wedge\tau}^{s\wedge\tau} b_r y_r dw_r = \int_t^s b_r y_r I_{r\leqslant\tau} dw_r$,
etc. Then $|y_s|^2$ satisfies the following forward SDE:
$|y_s|^2 = |x|^2 + \int_{t\wedge\tau}^{s\wedge\tau} a_r |y_r|^2 dr + \int_{t\wedge\tau}^{s\wedge\tau} b_r |y_r|^2 dw_r + \int_{t\wedge\tau}^{s\wedge\tau}\int_Z D_r |y_{r-}|^2 \tilde{N}_k(dr,dz)$

$+\frac{1}{4}\sum_{i=1}^{d_2}\int_{t\wedge\tau}^{s\wedge\tau}\int_Z D_{ir}^2 |y_{r-}|^2 N_{k_i}(dr,dz) - \frac{1}{4}\int_{t\wedge\tau}^{s\wedge\tau}\int_Z |D_r|^2 |y_r|^2 \pi(dz)dr$
$= |x|^2 + \int_{t\wedge\tau}^{s\wedge\tau} a_r |y_r|^2 dr + \int_{t\wedge\tau}^{s\wedge\tau} b_r |y_r|^2 dw_r$
$+\sum_{i=1}^{d_2}\int_{t\wedge\tau}^{s\wedge\tau}\int_Z [D_{ir} + \frac{1}{4}D_{ir}^2] |y_r|^2 \tilde{N}_{k_i}(dr,dz).$

Applying Ito's formula to $x_s |y_s|^2$, $s \in [t\wedge\tau, \tau]$, one finds that

$X|y_\tau|^2 = x_t |x|^2 - \int_{t\wedge\tau}^{\tau} |y_r|^2 f_0(r,\omega)dr + M_\tau - M_{t\wedge\tau}$
$+\sum_{i=1}^{d_2}\int_{t\wedge\tau}^{\tau}\int_Z (\frac{1}{4}|D_{ir}|^2 + D_{ir} - C_{ir})p_{ir} |y_r|^2 \pi(dz)dr,$

where M_t is a martingale under assumption $(C)_1$ determined by the stochastic integrals with respect to w_s, and $\tilde{N}_k(ds,dz), s \leqslant t$. Hence, if we want D_r to satisfy

$\frac{1}{4}|D_{ir}|^2 + D_{ir} - C_{ir} = 0, i = 1,\cdots,d_2;$

we can let

$D_{ir}(z) = -2 + 2\sqrt{1 + C_{ir}(z)}, i = 1,\cdots,d_2.$

Thus one has that

$x_t |x|^2 = E[X|y_\tau|^2 + \int_{t\wedge\tau}^{\tau} |y_r|^2 f_0(r,\omega)dr \,|\mathfrak{F}_{t\wedge\tau}] \geqslant 0.$

The conclusion is now derived under condition $(C)_1$.

Now, for the general condition $(B)_1$, one can take $Z_n \uparrow Z$ such that $\pi(Z_n) < \infty$. By assumption there exist a unique solution (x_t^n, q_t^n, p_t^n) that solves BSDE (8.58) with Z_n replacing Z. Obviously, the result applies to x_t^n. Letting $n \to \infty$, one arrives at the required conclusion under condition $(B)_1$.

Method 2. (Girsanov's Transformation).

By the Girsanov type theorem (8.58) can be rewritten as $\widetilde{P} - a.s.$

$x_t = X + \int_{t\wedge\tau}^{\tau}(a_s(\omega)x_s + f_0(s,\omega))ds - \int_{t\wedge\tau}^{\tau} q_s d\widetilde{w}_s - \int_{t\wedge\tau}^{\tau}\int_Z p_s(z)\bar{N}_k(ds,dz),$

where

$\widetilde{w}_t = w_t - \int_0^t b_s(\omega)ds$, and

$\bar{N}_{k_i}(dt,dz) = \widetilde{N}_{k_i}(dt,dz) - C_{it}(z)\pi(dz),\ i = 1,2,\cdots,d_2,$

are a new BM and new martingale measures under the new probability measure $\widetilde{P}$:

$$\frac{d\widetilde{P}}{dP}|_{\mathfrak{F}_t} = (\exp[\int_0^t \langle b_s, dw_s\rangle - \frac{1}{2}\int_0^t |b_s|^2 ds]\cdot$$
$$\prod_{i=1}^{d_2}\prod_{0<s\leqslant t}(1 + \int_Z C_{it}(z)N_{k_i}(\{s\},dz))e^{-\int_0^t\int_Z C_{it}(z)\pi(dz)ds})dP\ ;$$

moreover, $\bar{N}_{k_i}(dt,dz) = N_{k_i}(dt,dz) - (1 + C_{it}(z))\pi(dz),\ i = 1,2,\cdots,d_2;$ provided furthermore that $C_{it}(z) > -1$ and $\pi(Z) < \infty$. (In this case

$\int_Z |C_{it}(z)|^2 \pi(dz) \le k_0\pi(Z)$, and for any $0 \le T < \infty$

$\int_0^T |k_0\pi(Z)|^2 dt < \infty.$

So the Girsanov type theorem (Theorem 124) can be applied). (Recall that by assumption all k_i have no common points in their domains, so all counting measures $N_{k_i}(dt,dz)$ have no common jump points). Now applying Ito's formula to $e^{\int_0^t a_s ds}x_t$, one easily sees that

$x_t e^{\int_0^{t\wedge\tau} a_s(\omega)ds} = e^{\int_0^{\tau} a_s(\omega)ds}X + \int_{t\wedge\tau}^{\tau} e^{\int_0^s a_s(\omega)ds} f_0(s,\omega)ds$
$-\int_{t\wedge\tau}^{\tau} e^{\int_0^s a_s(\omega)ds} q_s d\widetilde{w}_s - \int_{t\wedge\tau}^{\tau}\int_Z e^{\int_0^s a_s(\omega)ds} p_s(z)\overline{N}(ds,dz),\ \widetilde{P} - a.s.$

Hence, taking the conditional expectation under the new probability measure $\widetilde{P}$, one obtains that $\widetilde{P} - a.s., t \in [0, \tau]$,

$$x_t = x_{t\wedge\tau} = E^{\widetilde{P}}\left[x_{t\wedge\tau}|\,\mathfrak{F}_{t\wedge\tau}\right]$$
$$= E^{\overline{P}}\left[e^{\int_{t\wedge\tau}^{\tau} a_s ds}X + \int_{t\wedge\tau}^{\tau} e^{\int_t^s a_r dr} f_0(s,\omega)ds\middle|\,\mathfrak{F}_{t\wedge\tau}\right] \geq 0. \quad (8.59)$$

Notice that $\widetilde{P} \sim P$. One also has that $P - a.s., t \in [0, \tau], x_t \geq 0$.

Now for the case that $C_{it}(z) \geq -1$ and $\pi(Z) < \infty$ we can let
$C_{it}^n(z) = C_{it}(z) + \frac{1}{n} > -1$.
Then by Theorem 234 there exists a unique solution (x_t^n, q_t^n, p_t^n) solving the following BSDE
$x_t^n = X + \int_{t\wedge\tau}^{\tau}(a_s(\omega)x_s^n + b_s(\omega)\cdot(q_s^n)^T + \int_Z C_s^n(z,\omega)p_s^n(z)^T\pi(dz)$
$+f_0(s,\omega))ds - \int_{t\wedge\tau}^{\tau} q_s^n dw_s - \int_{t\wedge\tau}^{\tau}\int_Z p_s^n \tilde{N}_k(ds,dz)$.

Then by the result just proved $P - a.s., t \in [0, \tau]$,
$x_t^n \geq 0$.
Letting $n \to \infty$, by Theorem 249 one obtains that $E\sup_{t\in[0,\tau\wedge T]}|x_t^n - x_t|^2 \to 0$, for any $T < \infty$. Hence there exists a subsequence $\{n_k\}$ of $\{n\}$, that for simplicity we again denote by $\{n\}$ again, such that $P - a.s.$
$\lim_{n\to\infty}\sup_{t\in[0,\tau\wedge T]}|x_t^n - x_t|^2 = 0$.
This shows that $P - a.s., t \in [0, \tau], x_t \geq 0$.

Finally, for the case that $C_{it}(z) \geq -1$, letting $Z_n \uparrow Z$ such that $\pi(Z_n) < \infty$, then by Theorem 234 there exists a unique solution (x_t^n, q_t^n, p_t^n) solving the following BSDE
$x_t^n = X + \int_{t\wedge\tau}^{\tau}(a_s(\omega)x_s^n + b_s(\omega)\cdot(q_s^n)^T + \int_{Z_n} C_s(z,\omega)p_s^n(z)^T\pi(dz)$
$+f_0(s,\omega))ds - \int_{t\wedge\tau}^{\tau} q_s^n dw_s - \int_{t\wedge\tau}^{\tau}\int_{Z_n} p_s^n \tilde{N}_k(ds,dz)$,

and by the result just proved $P - a.s., t \in [0, \tau]$,
$x_t^n \geq 0$.
Notice that $(x_t^n, q_t^n, p_t^n(\cdot)I_{Z_n}(\cdot))$ solves the following BSDE
$x_t^n = X + \int_{t\wedge\tau}^{\tau}(a_s(\omega)x_s^n + b_s(\omega)\cdot(q_s^n)^T + \int_Z C_s(z,\omega)p_s^n(z)^T I_{Z_n}(z)\pi(dz)$
$+f_0(s,\omega))ds - \int_{t\wedge\tau}^{\tau} q_s^n dw_s - \int_{t\wedge\tau}^{\tau}\int_{Z_n} p_s^n I_{Z_n}(z)\tilde{N}_k(ds,dz)$,
and
$|a_t| + |b_t| \leq k_0$, $E\int_0^{T\wedge\tau}\int_Z |C_s(z,\omega)|^2\pi(dz)ds \leq k_T < \infty$,
$E\int_0^{T\wedge\tau}\int_Z |p_s^n(z,\omega)|^2\pi(dz)ds \leq k_T < \infty, \forall n$.
Hence applying Ito's formula to $|x_t^n - x_t|^2$, and proceeding as in the proof of Theorem 234, one easily sees that as $n \to \infty$, $E\sup_{t\in[0,T\wedge\tau]}|x_t^n - x_t|^2 \to 0$. From this one easily finds that $P - a.s., t \in [0, \tau]$,
$x_t \geq 0$. ∎

Now let us use Lemma 255 to establish Theorem 252.

Proof. One can establish Theorem 252 by constructing a BSDE similar to the one in (8.57), and then using Lemma 255 under the condition $(B)_1$.

The proof for condition $(\mathrm{B})_2$ is similar. Thus the proof of Theorem 252 is complete. ■

Corollary 256 *Under the assumption of Theorem 251*

$$x_0^1 = x_0^2 \Longleftrightarrow X^1 = X^2, b^1(s, x_s^1, q_s^1, p_s^1, \omega) = b^2(s, x_s^1, q_s^1, p_s^1, \omega).$$

Proof. Recall that $(\widehat{x}_t, \widehat{q}_t, \widehat{p}_t) = (x_t^2 - x_t^1, q_t^2 - q_t^1, p_t^2 - p_t^1)$ satisfies

$$\begin{aligned} \widehat{x}_t &= \widehat{X} + \int_{t\wedge\tau}^{\tau} (a_s(\omega)\widehat{x}_s + \widetilde{b}_s(\omega)\widehat{q}_s^T + \int_Z \widetilde{C}_s(z,\omega)\widehat{p}_s(z)^T \pi(dz) \\ &\quad + f_0(s,\omega))ds - \int_{t\wedge\tau}^{\tau} \widehat{q}_s dw_s - \int_{t\wedge\tau}^{\tau}\int_Z \widehat{p}_s(z)\widetilde{N}(ds,dz), \end{aligned} \quad (8.60)$$

where
$\widetilde{C}_t(z,\omega) = \widehat{C}_t(\omega)\left|C_t(z,\omega)\right| sgn(\widehat{p}_t(z)),$
$f_0(s,\omega) = b^2(s, x_s^1, q_s^1, p_s^1, \omega) - b^1(s, x_s^1, q_s^1, p_s^1, \omega).$
See the notation in (8.57).

Comparing (8.60) with 8.58, from (8.59) one finds that

$$\widehat{x}_t = E^{\overline{P}}\left[e^{\int_{t\wedge\tau}^{\tau} a_s ds}\widehat{X} + \int_{t\wedge\tau}^{\tau} e^{\int_t^s a_r dr} f_0(s,\omega)ds \middle| \mathfrak{F}_{t\wedge\tau} \right], \quad (8.61)$$

where
$\frac{d\overline{P}}{dP}|_{\mathfrak{F}_T} = [\exp(\int_0^T \widetilde{b}_s dw_s - \frac{1}{2}\int_0^T \left|\widetilde{b}_s\right|^2 ds)\cdot$
$\cdot \prod_{0<s\leqslant T}(1 + \int_Z \widetilde{C}_s(z,\omega)N(\{s\},dz))e^{-\int_0^T\int_Z \widetilde{C}_s(z,\omega)\pi(dz)ds}dP$
provided that $\widetilde{C}_s(z,\omega) < 1$, and $\pi(Z) < \infty$. Notice that $\widehat{X} \leq 0$ and $f_0(s,\omega) \leq 0$. Hence $\overline{P} - a,s,$
$\widehat{x}_0 = 0 \Longleftrightarrow \widehat{X} = 0, f_0(s,\omega) = 0,$
provided $\widetilde{C}_s(z,\omega) < 1$, and $\pi(Z) < \infty$. Since $P \sim \overline{P}$, the conclusion is true provided that $\widetilde{C}_s(z,\omega) < 1$, and $\pi(Z) < \infty$. Now for the general case the conclusion can be drived just as in the proof of Lemma 255. ■

A similar corollary can also be obtained for Theorem 252.

8.12 Explanation of Comparison Theorem. Arbitrage-Free Market

Now let us apply Theorem 251 to the financial market. Suppose we have two wealth processes $x_t^i, i = 1,2$, satisfying the BSDEs as (8.52) in a financial market, as in the previous sections, respectively: $i = 1,2,$

$$dx_t^i = r_t x_t^i dt - dC_t^i + \pi_t^i [b_t + \delta_t - r_t \underline{1}] dt$$
$$-\widetilde{k}_0 (R_t - r_t)(x_t^i - \sum_{j=1}^N \pi_t^{ij})^- dt + \pi_t^i \sigma_t dw_t + \pi_t^i \rho_t d\tilde{N}_t,$$
$$x_T^i = X^i, \; 0 \leqslant t \leqslant T; \tag{8.62}$$

where (x_t^i, π_t^i) is the solution of (8.62) corresponding to the given terminal wealth X^i at the time T and the given consumption process $C_t^i, t \in [0, T]$. Write

$[q_t, p_t] = \pi_t [\sigma_t, \rho_t] = \pi_t \widetilde{\sigma}_t,$

where $q_t = (q_t^1, \cdots, q_t^d), p_t = (p_t^1, \cdots, p_t^m)$, and assume that $N = d + m$. If $\widetilde{\sigma}_t^{-1} = [\sigma_t, \rho_t]^{-1}$exists, then the above BSDE can be rewritten as $i = 1, 2,$

$$x_t^i = X^i - \int_t^T r_s x_s^i ds + C_T^i - C_t^i - \int_t^T (q_s^i \theta_s^w + p_s^i \theta_s^{\tilde{N}}) ds$$
$$+\widetilde{k}_0 \int_t^T (R_s - r_s)(x_s^i - \sum_{k=1}^N (\sum_{j=1}^d q_s^{ij} (\widetilde{\sigma}_s^{-1})_{ji} + \sum_{j=1}^m p_s^{ij} (\widetilde{\sigma}_s^{-1})_{(j+d)k}))^- ds$$
$$- \int_t^T q_s^i dw_s - \int_t^T p_s^i d\tilde{N}(s), 0 \leqslant t \leqslant T, \tag{8.63}$$

where as before we write

$\theta_t = \begin{bmatrix} \theta_t^w \\ \theta_t^{\tilde{N}} \end{bmatrix} = \widetilde{\sigma}_t^{-1} [b_t + \delta_t - r_t \underline{1}].$

Thus, by using Theorem 251, we immediately obtain the following theorem.

Theorem 257 *Under assumptions of Theorem 250 if*

$X^1 \geq X^2$, $C_t^1 \geq C_t^2$, $\forall t \in [0, T]$, *and*

$\left|\theta_{js}^{\tilde{N}}\right| + \widetilde{k}_0 (R_s - r_s) \sum_{i=1}^N \left|(\widetilde{\sigma}_s^{-1})_{(j+d)i}\right| \leqslant 1, j = 1, 2, \cdots, m;$

then $P - a.s.$

$x_t^1 \geq x_t^2, \forall t \in [0, T]$.

In particular, if $X^1 \geq 0$, *and* $C_t^1 \geq 0$, $\forall t \in [0, T]$, *then* $P - a.s.$

$x_t^1 \geq 0, \forall t \in [0, T]$.

Furthermore, if $\widetilde{k}_0 = 0$, *all coefficients are bounded, and* $\widetilde{\sigma}_s^{-1}$ *is also bounded, and* $P(X^1 > 0) > 0, X^1 \geq 0$, *then* $P - a.s.$

$x_t^1 > 0, \forall t \in [0, T], \forall \omega \in \{\omega : X^1(\omega) > 0\}$.

Proof. We only need to show the last conclusion. Suppose that $P(X^1 > 0) > 0$. Now applying Girsanov's theorem we find that (8.63) can be rewritten as $\widetilde{P}_T - a.s.$ $0 \leqslant t \leqslant T,$

$x_t^1 = X^1 - \int_t^T r_s x_s^1 ds + C_T^1 - C_t^1 - \int_t^T q_s^1 d\widetilde{w}_s - \int_t^T p_s^1 d\overline{N}_s,$

where

$\widetilde{w}_t = w_t + \int_0^t \theta_s^w ds,$
$\overline{N}_t = \widetilde{N}_t + \int_0^t \theta_s^{\widetilde{N}} ds,$
are a new BM and a new centralized (not necessary stationary) Poisson process under the new probability measure $\widetilde{P}_T$, respectivvely, such that
$\widetilde{P}_T = \exp[-\int_0^T \langle \theta_s^w, dw_s \rangle - \frac{1}{2}\int_0^T |\theta_s^w|^2 ds - \int_0^T \langle \mu_s, \underline{1} \rangle ds] \cdot$
$\cdot \prod_{0<s\leqslant T}(1 + \langle \mu_s, \triangle N_s \rangle) dP,$
$\mu_s = -\theta_s^{\widetilde{N}},$
$\overline{N}_t = N_t - \int_0^t (1 - \theta_s^{\widetilde{N}}) ds, \ \widetilde{P}_T - a.s.$
Hence applying Ito's formula to $x_t^1 e^{-\int_0^t r_s ds}$, one sees that $\widetilde{P}_T - a.s.$
$x_t^1 e^{-\int_0^t r_s ds} = X^1 e^{-\int_0^T r_s ds} + \int_t^T e^{-\int_0^u r_s ds} dC_u^1$
$- \int_t^T e^{-\int_0^u r_s ds} q_u^1 d\widetilde{w}_u - \int_t^T e^{-\int_0^u r_s ds} p_u^1 d\overline{N}_u.$
Now taking the conditional expectation $E^{\widetilde{P}_T}(\cdot|\mathfrak{F}_t)$, one finds that $\widetilde{P}_T - a.s.$ $\forall t \in [0,T]$,
$x_t^1 e^{-\int_0^t r_s ds} = E^{\widetilde{P}_T}(X^1 e^{-\int_0^T r_s ds} + \int_t^T e^{-\int_0^u r_s ds} dC_u^1 | \mathfrak{F}_t)$
$\geq E^{\widetilde{P}_T}(X^1 e_u^{-\int_0^T r_s ds} | \mathfrak{F}_t) \geq E^{\widetilde{P}_T}(X^1 I_{X^1>0} e_u^{-\int_0^T r_s ds} | \mathfrak{F}_t).$
So $\widetilde{P}_T - a.s.$ $\forall t \in [0,T]$,
$x_t^1 \geq E^{\widetilde{P}_T}(X^1 e_u^{-\int_t^T r_s ds} I_{X^1>0} | \mathfrak{F}_t).$
Now let $A = \{\omega : X^1(\omega) > 0\}$, and $P^A = P(\cdot|A)$. Then P^A is still a probability measure. Replacing P by P^A, and discussing as in the above case, one finds that $\widetilde{PA}_T - a.s.$
$x_t^1 \geq E^{\widetilde{PA}_T}(X^1 e_u^{-\int_t^T r_s ds} I_{X^1>0} | \mathfrak{F}_t) > 0.$
Since $\widetilde{PA}_T \sim P^A$. So $P^A - a.s.$ $\forall t \in [0,T]$, $x_t^1 > 0$. That is, $P - a.s.$ $\forall t \in [0,T]$, $\forall \omega \in A$,
$x_t^1 > 0.$ ∎

This theorem has a nice physical meaning. Actually, it tells us that in a financial market the following facts are true.

1) If a small investor wants his wealth in the market at the future time T to reach a higher level, and his consumption rule does not change, then he must invest much money now; that is,
$X^1 \geq X^2; C_t^1 = C_t^2, \forall t \in [0,T] \Longrightarrow x_t^1 \geq x_t^2, \forall t \in [0,T].$

2) If a small investor wants to consume more during the time interval $[0,T]$, and his terminal target does not change, then he also must invest more money now; that is,
$C_t^1 \geq C_t^2, \ \forall t \in [0,T]; X^1 = X^2 \Longrightarrow x_t^1 \geq x_t^2, \forall t \in [0,T].$

3) If someone wants his money in the future in the market to be non-negative, and he also wants to consume money in the market, then he must invest now.

Naturally, if a zero investment in a market can produce a positive gain with a positive chance (that is, with a positive probability), then such a market has an arbitrage chance, and we will call it an arbitrage market. Otherwise, we will call the market an arbitrage-free market.

To explain the concept of "arbitrage" more clearly, let us simplify the market. Suppose that in the market there is one bond, and its price P_t^0 satisfies the equation as before:

$dP_t^0 = P_t^0 r_t dt,\ P_0^0 = 1.$

There are $N-$stocks, and as before they satisfy the following SDE:

$dP_t^i = P_t^i[(b_t^i dt + \sum_{k=1}^d \sigma_t^{ik} dw_t^k + \sum_{k=1}^m \rho_t^{ik} d\widetilde{N}_t^k],$

$P_0^i = P_0^i \geq 0, \quad i = 1, \cdots, N.$

Applying Ito's formula, one can verify that the solution of the bond price is

$P_t^0 = e^{\int_0^t r_s ds} > 0$, as $t \geq 0$,

and the solutions of stock prices are

$P_t^i = P_0^i \exp[\int_0^t b_s ds + \sum_{k=1}^d \int_0^t \sigma_s^{ik} dw_s^k - \frac{1}{2}\sum_{k=1}^d \int_0^t \left|\sigma_s^{ik}\right|^2 ds]\cdot$

$\cdot \prod_{k=1}^m \prod_{0<s\leqslant t}(1 + \rho_s^{ik} \vartriangle N_s^k)e^{-\int_0^t \rho_s^{ik} ds} > 0,\ i = 1, 2, \cdots, N,$

provided that $\rho_s^{ik} > -1$, and $P_0^i > 0$. Obviously, the wealth process of a small investor in the market under the condition that $\rho_s^{ik} > -1$ and $P_0^i \geq 0$, is

$x_t = \sum_{i=0}^N \eta_t^i P_t^i = \sum_{i=0}^N \pi_t^i \geq 0,$

where η_t^0 is the number of bond units bought by the investor, and $\eta_t^i-$the amount of units for the $i-$th stock. Now suppose that the market is self-financed, then the wealth process x_t should satisfy

$dx_t = \sum_{i=0}^N \eta_t^i dP_t^i.$

This deduces that

$$dx_t = r_t x_t dt + \pi_t(b_t - r_t \underline{1})dt + \pi_t \sigma_t dw_t + \pi_t \rho_t d\tilde{N}_t. \tag{8.64}$$

Now let us give the explicit definition of an arbitrage market and an arbitrage-free market.

Definition 258 *Suppose in a market the wealth process is described by (8.64), and the conditions stated above are satisfied. Thus, if for* $x_0 = 0, P-a.s.$, *there exists a portfolio* $\pi_t, t \in [0, T]$ *such that the wealth process* x_t *developed by SDE (8.64) reaches* X *at the future time* T, *such that one of the following results holds:*

(i) $X \geq 0$ *and* $P(X > 0) > 0$;

(ii) $X \leq 0$ *and* $P(X < 0) > 0$;

then $\pi_t, t \in [0, T]$, *is called an arbitrage portfolio, and the market is called an arbitrage market. Otherwise, the market is called an arbitrage-free market.*

An arbitrage-free market actually means that in such a market, without any investment; that is, $x_0 = 0$, we cannot have a gain with a positive chance; that is, $X \geq 0$ and $P(X > 0) > 0$; and we also cannot have a loss with a positive chance; that is, $X \leq 0$ and $P(X < 0) > 0$. Otherwise, somebody can use such a chance to make money. Hence applying Theorem 250 and the above theorem we immediately obtain the following result.

Theorem 259 *If for a financial market all of the assumptions in Theorem 257 hold and* $\widetilde{k}_0 = 0$, $\rho_t^{ik} > -1, \forall i, k, t$, *then this market is a complete market, (that is, any contingent claim in this market can be duplicated), and moreover, it is an arbitrage-free market.*

Notice that $x_t^1 e^{-\int_0^t r_s ds} = x_t^1 / P_t^0$. So we may call $x_t^1 e^{-\int_0^t r_s ds}$ a discount (w.r.t the bond) wealth process. By the proof of the arbitrage-free market, or say, the proof of the last conclusion of Theorem 257, one immediately sees that the following theorem is true.

Theorem 260 *For a self-financed market, if there exists an equivalent probaility measure* $\widetilde{P}_T$ *(that is,* $\widetilde{P}_T$ *is a probability measure and* $\widetilde{P}_T \sim P$*) such that under this measure* $\widetilde{P}_T$ *the discount wealth process becomes a martingale, then this self-financed market is an arbitrage-free market.*

In fact in this case $x_t^1 e^{-\int_0^t r_s ds} = E^{\widetilde{P}_T}(X^1 e^{-\int_0^T r_s ds} | \mathfrak{F}_t)$. So the proof of the last conclusion of Theorem 257 still goes through.

8.13 Solutions for Unbounded (Terminal) Stopping Times

It is quite possible that in a financial market some investor will stop his all activities at some random time according to his rule for investment. However, it is not known beforehand when he will stop. Moreover, sometimes it is also unreasonable to assume that he will stop before some fixed time; that is, the terminal stopping time is not neccessary bounded. So we need to consider the BSDE with an unbounded stopping time as a terminal time.

Now let us discuss BSDE (8.8) with an unbounded stopping time $\tau \in [0, \infty]$. In this case we still use Definition 226 as the definition of the solution to (8.8). Moreover, one easily sees that Lemma 229, Lemma 231, and Lemma 230 still hold, if we substitute T_0 by ∞ in their proofs. However, the proof of Lemma 233 needs a little more discussion. Let us rewrite it as follows:

Lemma 261 *If* τ *is a* $\mathfrak{F}_t-$ *stopping time,* X *is* $\mathfrak{F}_\tau-$ *measurable and* R^d- *valued,* $f(t, \omega)$ *is* $\mathfrak{F}_t-$ *adapted and* R^d- *valued such that*

$E|X|^2 < \infty, E\left|\int_0^\tau |f(s, \omega)| ds\right|^2 < \infty,$

then (8.8) *has a unique solution.*

Proof. Let

$x_t = E^{\mathfrak{F}_t}(X + \int_{t\wedge\tau}^\tau f(s)ds), \forall t \geqslant 0.$

Then, in the same way as before, one sees that it makes sense. Similarly,

$M_t = E^{\mathfrak{F}_t}(X + \int_0^\tau f(s)ds), t \geqslant 0,$

is a square integrable martingale. Hence by the martingale representation theorem (see Theorem 103 in Chapter 2) there exists a unique

$(q_t, p_t) \in L^{2,loc}_{\mathfrak{F}}(R^{d\otimes d_1}) \times F^{2,loc}_{\mathfrak{F}}(R^{d\otimes d_2})$
such that

$$M_t = M_0 + \int_0^t q_s dw_s + \int_0^t \int_Z p_s(z)\tilde{N}_k(ds, dz), 0 \leqslant t. \tag{8.65}$$

where

$$L^{2,loc}_{\mathfrak{F}}(R^{d\otimes d_1}) = \left\{ \begin{array}{c} f(t,\omega) : f(t,\omega) \text{ is } \mathfrak{F}_t - \text{adapted}, \quad R^{d\otimes d_1} - \text{valued} \\ \text{such that } E\int_0^T |f(t,\omega)|^2 dt < \infty, \forall T < \infty \end{array} \right\},$$

etc. Let us show that $E\int_0^\tau |q_t|^2 dt < \infty$, and $E\int_0^\tau \int_Z |p_t(z)|^2 \pi(dz)dt < \infty$.
In fact, by assumption one finds that $E|M_\tau|^2 < \infty$ and $E|M_0|^2 < \infty$. So by (8.65) one sees that

$E\int_0^{\tau\wedge T} |q(t,\omega)|^2 dt + E\int_0^{\tau\wedge T} \int_Z |p_t(z)|^2 \pi(dz)dt = E|M_{\tau\wedge T} - M_0|^2$
$\leq 2[E|M_{\tau\wedge T}|^2 + E|M_0|^2] \leq 2[E|M_\tau|^2 + E|M_0|^2] < \infty.$

Letting $T \uparrow \infty$. By Fatou's lemma one finds that

$E\int_0^\tau |q(t,\omega)|^2 dt + E\int_0^\tau \int_Z |p_t(z)|^2 \pi(dz)dt \leq 2[E|M_\tau|^2 + E|M_0|^2] < \infty.$

Now the rest of the proof is just the same as that of Lemma 233. ■

Theorem 234 is also true when τ is unbounded. For this we only need to put $T_0 = \infty$ in the proof of Theorem 234, and introduce a new norm as follows: For $(x_\cdot, q_\cdot, p_\cdot) \in \widetilde{B}$ let

$\|(x_\cdot, q_\cdot, p_\cdot)\|^2_M = \sup_{t\geqslant 0} e^{-bA(t)} E|x_{t\wedge\tau}|^2 + \sup_{t\geqslant 0} e^{-bA(t)} E\int_{t\wedge\tau}^\tau |q_t|^2 dt$
$\quad + \sup_{t\geqslant 0} e^{-bA(t)} E\int_{t\wedge\tau}^\tau \|p_t\|^2 dt,$

where

$\widetilde{B} = L^2_{\mathfrak{F}}(0,\tau; R^d) \times L^2_{\mathfrak{F}}(0,\tau; R^{d\otimes d_1}) \times F^2_{\mathfrak{F}}(0,\tau; R^{d\otimes d_2}),$
$b \geqslant 0$ is a constant, and $A(t) = \int_t^\infty (c_1(s) + c_2(s)^2)ds$. and

$$L^{2,loc}_{\mathfrak{F}}(0,\tau; R^{d\otimes d_1}) = \left\{ \begin{array}{c} f(t,\omega) : f(t,\omega) \text{ is } \mathfrak{F}_t - \text{adapted}, \quad R^{d\otimes d_1} - \text{valued} \\ \text{such that } E\int_0^\tau |f(t,\omega)|^2 dt < \infty, \forall T < \infty \end{array} \right\},$$

then the proof can be completed as that of Theorem 234.

Similarly, if one takes $T_0 = \infty$, and changes the interval $[0, T_0]$ to be $[0, \infty)$, then Theorem 246 is also true.

Furthermore, we also have examples to show that the condition $\int_0^\infty c_1(t)dt + \int_0^\infty c_2(t)^2 dt < \infty$ can not be weakened.

Example 262 . *(Condition $\int_0^\infty c_2(t)^2 dt < \infty$ can not be weakened). Consider BSDE in 1- dimensional space:*

$x_t = X + \int_t^\infty |q_s| ds - \int_t^\infty q_s dw_s - \int_t^\infty \int_Z p_s(z)\widetilde{N}_k(ds, dz),$

where

$X = \int_0^\infty f(s)dw_s + \int_0^\infty \int_Z f(s)I_U(z)\widetilde{N}_k(ds, dz) = I_2(0,\infty) + I_3(0,\infty),$
$f(t) = [(2+t)\ln(2+t)]^{-1},$

and $\pi(U) < \infty, 0 \leqslant T < \infty$. Then $c_2(t) = 1$ and the BSDE has no solution.

Indeed, if it has a solution, it should be equal to

$x_t = I_1(t,\infty) + I_2(0,t) + I_3(0,t),\ q_t = f(t), p_t(z) = f(t)I_U(z),$
where $I_1(t,T) = \int_t^T f(s)ds$. *However, this is impossible, since in this case we have*
$\int_0^\infty |q_s|\, ds = \int_0^\infty f(s)ds = \int_0^\infty [(2+t)\ln(2+t)]^{-1}dt = \infty,$
which is a contradiction.

Example 263 *(Condition* $\int_0^\infty c_1(t)dt < \infty$ *can not be weakened). Use the above notation. Then BSDE*
$x_t = 1 + \int_t^\infty f(s)x_s ds - \int_t^\infty q_s dw_s - \int_t^\infty \int_Z p_s(z)\widetilde{N}_k(ds,dz), \forall t \geqslant 0,$
where $c_1(t) = |f(t)| = [(2+t)\ln(2+t)]^{-1}$, *has no solution. Indeed, if it has a solution, it should satisfy*
$Ex_t = 1 + \int_t^\infty f(s)Ex_s ds = e^{\int_t^\infty f(s)ds} = \infty,$
which is a contradiction.

Now we have that all of the existence theorems of solutions including Lipschitzian and non-Lipschitzian conditions, the a priori estimates of solutions, and the uniqueness of solutions for BSDEs in this case still hold true. However, the proofs of the convergence theorems require more discussion.

Actually, we will have the following convergence theorems:

Theorem 264 . *Assume that conditions* $1^\circ - 3^\circ$ *in Theorem 248 hold, and assume that*
$4^{\circ'}$ $\lim_{n\to\infty} \int_0^\infty \sup_{\substack{(x,q)\in R^d\times R^{d\otimes d_1} \\ p\in L^2_{\pi(.)}(R^{d\otimes d_2}) \\ \omega\in\Omega}} |b^n(t,x,q,p,\omega) - b^0(t,x,q,p,\omega)\,|\,dt = 0.$
Then the following conclusion holds:
$\lim_{n\to\infty} E[\sup_{t\geqslant 0} |x^n_{t\wedge\tau} - x^0_{t\wedge\tau}|^2 + \int_0^\tau (|q^n_s - q^0_s|^2 + \|p^n_s - p^0_s\|^2)ds] = 0.$

Remark 265 *If*
$\lim_{n\to\infty} \sup_{\substack{(x,q)\in R^d\times R^{d\otimes d_1} \\ p\in L^2_{\pi(.)}(R^{d\otimes d_2}) \\ \omega\in\Omega}} |b^n(t,x,q,p,\omega) - b^0(t,x,q,p,\omega)\,| = 0,$
then under the condition 1°, $4^{\circ'}$ *is satisfied.*

In fact, in this case Lebesgue's dominated convergence theorem applies.

Now let us prove Theorem 264.

Proof. We use the previous notation. Note that
$2(x^n_s - x^0_s)(b^n(s,x^n_s,q^n_s,p^n_s,\omega) - b^0(s,x^0_s,q^0_s,p^0_s,\omega))$
$= 2(x^n_s - x^0_s)(b^n(s,x^n_s,q^n_s,p^n_s,\omega) - b^0(s,x^n_s,q^n_s,p^n_s,\omega))$
$+2(x^n_s - x^0_s)(b^0(s,x^n_s,q^n_s,p^n_s,\omega) - b^0(s,x^0_s,q^0_s,p^0_s,\omega))$
$= I^{1,n}_s + I^{2,n}_s \leqslant I^{1,n}_s + 4(c_1(s) + c_2(s)^2)\rho_1(|(x^n_s - x^0_s)|^2)$
$+2^{-1}(|q^n_s - q^0_s|^2 + \|p^n_s - p^0_s\|^2),$
where $\rho_1(u) = \rho(u) + u$. However, one can show that
$\lim_{n\to\infty} E\int_0^\tau I^{1,n}_s ds = 0.$

Indeed, by the Schwarz inequality and Lemma 229 for unbounded stopping times

$E\int_0^{\tau} I_s^{1,n}ds \leqslant 2(E\sup_{t\geqslant 0}\left|x_{t\wedge\tau}^n - x_{t\wedge\tau}^0\right|^2)^{\frac{1}{2}}$

$\cdot(E[(\int_0^{\tau} |b^n(t,x_s^n,q_s^n,p_s^n,\omega) - b^0(t,x_s^n,q_s^n,p_s^n,\omega)|dt)^2])^{\frac{1}{2}}$

$\leqslant 2(2k_0)^{\frac{1}{2}}\int_0^{\infty} \sup_{\substack{(x,q)\in R^d\times R^{d\otimes d_1} \\ p\in L^2_{\pi(.)}(R^{d\otimes d_2}) \\ \omega\in\Omega}} |b^n(t,x,q,p,\omega) - b^0(t,x,q,p,\omega)|dt$

$\to 0, \text{ as } n\to\infty.$

From these results one easily finds that

$\lim_{n\to\infty} E[|x_{t\wedge\tau}^n - x_{t\wedge\tau}^0|^2 + \int_0^{\tau}(|q_s^n - q_s^0|^2 + \|p_s^n - p_s^0\|^2)ds] = 0.$

Applying the martingale inequality to the FSDE one easily obtains the conclusion of the theorem. ■

Theorem 266 *All assumptions are the same as in Theorem 249 for solutions to BSDE with bounded stopping times, but now $\tau\in[0,\infty]$ and $T = T_0 = \infty$. Then the conclusion of Theorem 264 still holds.*

Proof. Notice that now we have

$$\begin{aligned}
& 2(x_s^n - x_s^0)(b^n(s,x_s^n,q_s^n,p_s^n,\omega) - b^0(s,x_s^0,q_s^0,p_s^0,\omega)) \\
= \;& 2(x_s^n - x_s^0)(b^n(s,x_s^n,q_s^n,p_s^n,\omega) - b^n(s,x_s^0,q_s^0,p_s^0,\omega)) \\
& +2(x_s^n - x_s^0)(b^n(s,x_s^0,q_s^0,p_s^0,\omega) - b^0(s,x_s^0,q_s^0,p_s^0,\omega)) \\
= \;& I_s^{1,n} + I_s^{2,n} \leqslant 4(c_1(s) + c_2(s)^2)\rho_1(|(x_s^n - x_s^0)|^2) \\
& +2^{-1}(|q_s^n - q_s^0|^2 + \|p_s^n - p_s^0\|^2) + I_s^{2,n},
\end{aligned}$$

However, now one finds that by Lebesgue's dominated convergence theorem

$$\begin{aligned}
& E\int_0^{\tau} I_s^{2,n}ds \leqslant 2(E\sup_{t\geqslant 0}\left|x_{t\wedge\tau}^n - x_{t\wedge\tau}^0\right|^2)^{\frac{1}{2}} \\
& \cdot(E[(\int_0^{\tau} |b^n(s,x_s^0,q_s^0,p_s^0,\omega) - b^0(s,x_s^0,q_s^0,p_s^0,\omega)|dt)^2])^{\frac{1}{2}} \\
& \leqslant 2(2k_0)^{\frac{1}{2}}(E[(\int_0^{\tau} |b^n(s,x_s^0,q_s^0,p_s^0,\omega) - b^0(s,x_s^0,q_s^0,p_s^0,\omega)|dt)^2])^{\frac{1}{2}} \\
& \to 0, \quad \text{as } n\to\infty.
\end{aligned}$$

The proof is now completed as above. ■

The results of the BSDEs for the terminal time being an unbounded stopping time can also be explained in the financial market. One only needs to replace the terminal time T by the stopping time τ and to replace $X\in\mathfrak{F}_T$ by $X\in\mathfrak{F}_\tau$, then a similar explanation can still be made. We leave this as an exercise for the reader.

8.14 Minimal Solution for BSDE with Discontinuous Drift

Applying Theorem 251 we can derive a new existence theorem for solutions to BSDE (8.8) with discontinuous coefficient in R^1 (i.e. we assume that $d = 1$ in (8.8)), where for simplicity we also assume that τ is a bounded stopping time: $\tau \leq T_0 < \infty$.

Theorem 267 *Consider (8.8). Assume that*
1° $b(t, x, q, p, \omega)$ is jointly continuous in $(x, q, p) \in R^1 \times R^{1\otimes d_1} \times L^2_{\pi(.)}(R^{1\otimes d_2}) \backslash G$, where $G \subset R^1 \times R^{1\otimes d_1} \times L^2_{\pi(.)}(R^{1\otimes d_2})$ is a Borel measurable set such that $(x, q, p) \in G \Longrightarrow |q| > 0$, and $m_1 G_1 = 0$, where $G_1 = \{x : (x, q, p) \in G\}$, and m_1 is the Lebesgue measure in R^1; moreover, $b(t, x, q, p, \omega)$ is a separable process with respect to (x, q), (i.e. there exists a countable set $\{(x_i, q_i)\}_{i=1}^\infty$ such that for any Borel set $A \subset R^1$ and any open rectangle $B \subset R^1 \times R^{1\otimes d_1}$ the $\omega-$sets
$\{\omega : b(t, x, q, p, \omega) \in A, (x, q) \in B\}$, *and*
$\{\omega : b(t, x, q, p, \omega) \in A, (x_i, q_i) \in B, \forall i\}$
only differ a zero-probability $\omega-$set;
2° $|b(t, x, q, p, \omega)| \leqslant c_1(t)(1 + |x|) + c_2(t)(|q| + \|p\|)),$
$|b(t, x, q, p_1, \omega) - b(t, x, q, p_2, \omega)| \leqslant \int_Z |C_t(z, \omega)| \, |p_1(z) - p_2(z)| \, \pi(dz),$
where $c_1(t), c_2(t) \geqslant 0$ are non-random such that $\int_0^{T_0} (c_1(t) + c_2(t)^2)dt < \infty$, and $C_t(z, \omega)$ satisfies the condition (B) in 2° of Theorem 251.
Then (8.8) has a minimal solution. (A minimal solution x_t is one such that for any solution y_t it holds true that $P-a.s.$ $y_t \geq x_t, \forall t \in [0, \tau]$).
Furthermore, if b satisfies
3° $(x_1 - x_2)(b(t, x_1, q_1, p_1, \omega) - b(t, x_2, q_2, p_2, \omega))$
$\leqslant c_1(t)\rho(|x_1 - x_2|^2) + c_2(t) |x_1 - x_2| (|q_1 - q_2| + \|p_1 - p_2\|)],$
where $c_1(t), c_2(t) \geqslant 0$ have the same properties as those in 2°; and $\rho(u) \geqslant 0$, as $u \geqslant 0$, is non-random, increasing, continuous and concave such that $\int_{0+} du/\rho(u) = \infty$;
then the solution of (8.8) is unique.

Remark 268 *In the case that $\pi(\cdot) = 0$ (a zero measure) and $X \in \mathfrak{F}^w_\tau$ (8.8) is reduced to the following continuous BSDE:*
$x_t = X + \int_{t\wedge\tau}^\tau b(s, x_s, q_s, \omega)ds - \int_{t\wedge\tau}^\tau q_s dw_s, 0 \leqslant t,$
where τ is a $\mathfrak{F}_t-$stopping time such that $0 \leqslant \tau \leqslant T_0$, and T_0 is a fixed number.

The reader easily obtains the existence of solutions to the continuous BSDE with discontinuous coefficients b from Theorem 267.

Before proving Theorem 267 let us give some examples.

Example 269 . *Let for $0 < \alpha \leqslant 1$*
$b'_\alpha(t, x, q, p, \omega) = k_0 |q|^\alpha (x/|x|) I_{x\neq 0} + k'_0 I_{x\neq 0} \frac{x}{|x|^\delta}$

$+k_0''x + k_1 q^T + \int_Z C_t(z)p(z)\pi(dz),$
where $k_1 = (k_1^1, ..., k_1^{d_1}); k_0, k_0', k_0'' \geqslant 0, k_1^i, 1 \leqslant i \leqslant d_1, 0 < \delta < 1$ *are all constants, and* $C_t(z)$ *satisfies the condition (B).*

Then conditions $1^\circ - 2^\circ$ *in Theorem 267 are satisfied. Hence (8.8) has a minimal solution. However,* b'_α *is discontinuous at* (x, q, p), *where* $x = 0, q \neq 0$, *and* p *is arbitrary given. In particular, for* $0 < \alpha < 1$ b'_α *is also non-Lipschitzian continuous in* q. *However, if we let* $\alpha = 1$, *and*

$b(t, x, q, p, \omega) = -b'_1(t, x, q, p, \omega),$

then 3° *in Theorem 267 is also satisfied. Hence (8.8) has a unique solution.*

Example 270 *Assume that*

$\sum_{k=1}^{\infty} a_k < \infty,$

where $a_k > 0$. *Let*

$\{r_k\}_{k=1}^{\infty} =$ *the totality of rational numbers in* R^1,

$f(x) = \sum_{n:r_n<x} a_n.$

It is not difficult to show that $f(x)$ *is strictly increasing, continuous at each irrational point and discontinuous at each rational point. For* $0 < \alpha \leqslant 1$ *let*

$b_\alpha(t, x, q, p, \omega) = k_0 |q|^\alpha f(x) + k_0'x + k_1 q^T + \int_Z C_t(z)p(z)\pi(dz),$

where k_1, k_0, k_0' *and* $C_t(z)$ *are given as the same in Example 269. Then (8.8) has a minimal solution. Moreover, for* $0 < \alpha < 1$ b_α *is non-Lipschitzian continuous in* q.

To show Theorem 2 we need the following lemma.

Lemma 271 *Assume that* $b(t, x, q, p, \omega)$ *satisfies conditions* 1° *and* 2° *for* b *in Theorem 267. Let* $n \geqslant 1$,

$b_n(t, x, q, p, \omega) = \inf_{y,\widetilde{q}}(b(t, y, \widetilde{q}, p, \omega) + (n \vee c_1(t)) |y - x| + (n \vee c_2(t)) |q - \widetilde{q}|)$

Then $\forall n$, b_n *is still* $\mathfrak{F}_t = \mathfrak{F}_t^{w,\widetilde{N}_k}$*−adapted, and*

1) $b_n(t, x, q, p, \omega) \uparrow$, *as* $n \uparrow, \forall t \geqslant 0, (x, q, p) \notin G$;

$|b_n(t, x, q, p, \omega)| \leqslant c_1(t)(1 + |x|) + c_2(t)(|q| + \|p\|)$;

2) $|b_n(t, x, q, p, \omega) - b_n(t, y, q', p, \omega)| \leqslant (n \vee c_1(t))(|y - x|$
$+ (n \vee c_2(t)) |q - q'|)$,

3) $b_n(t, x, q, p, \omega)$ *still satisfies condition* 2° *in Theorem 267, for all* $n \geqslant 1$;

4) $(x_n, q_n, p_n) \to (x, q, p)$ *in* $R^1 \times R^{1 \otimes d_1} \times L^1_{\pi(.)}(R^{1 \otimes d_2})$

$\Rightarrow b_n(t, x_n, q_n, p_n, \omega) \to b(t, x, q, p, \omega). \forall t \geqslant 0, (x, q, p) \notin G.$

Proof. By condition 1° one easily sees that $b_n(t, x, q, p, \omega)$ is jointly measurable and $\mathfrak{F}_t-$ adapted. Obviously, $b_n \leqslant b_{n+1}$and

$b_n \leqslant b \leqslant c_1(t)(1 + |x|) + c_2(t)(1 + |q| + \|p\|)$

Moreover, since $(n \vee c_1(t)) \geqslant c_1(t), (n \vee c_2(t)) \geqslant c_2(t)$,

$b_n \geqslant \inf_{y,\widetilde{q}}(-c_1(t)(1 + |y|) - c_2(t)(1 + |\widetilde{q}| + \|p\|)$
$+ (n \vee c_1(t)) |y - x| + (n \vee c_2(t)) |q - \widetilde{q}|)$
$\geqslant \inf_{y,\widetilde{q}}(-c_1(t)(1 + |y|) - c_2(t)(1 + |\widetilde{q}| + \|p\|)$
$+ c_1(t)) |y - x| + c_2(t)) |q - \widetilde{q}|)$
$\geqslant -c_1(t)(1 + |x|) - c_2(t)(1 + |q| + \|p\|).$

Notice that from the elemantary inequality

$\inf_x f(x) - \inf_x g(x) \leqslant \sup_x |f(x) - g(x)|$

one has that

$|\inf_x f(x) - \inf_x g(x)| \leqslant \sup_x |f(x) - g(x)|$.

From this 2) is easily proved. Moreover, one also has that

$|b_n(t,x,q,p,\omega) - b_n(t,x,q,p',\omega)| \leqslant \sup_{x,q} |b(t,x,q,p,\omega) - b(t,x,q,p',\omega)|$.

Hence 3) is also obtained. Finally, notice that

$|b_n(t,x_n,q_n,p_n,\omega) - b(t,x,q,p,\omega)|$
$\leqslant |b_n(t,x_n,q_n,p_n,\omega) - b_n(t,x_n,q_n,p,\omega)|$
$+ |b_n(t,x_n,q_n,p,\omega) - b(t,x,q,p,\omega)|$
$\leqslant \left|\int_Z C_t(z,\omega)(p_n(z) - p(z))\pi(dz)\right| + |b_n(t,x_n,q_n,p,\omega) - b(t,x,q,p,\omega)|$.

If $(x_n,q_n) \to (x,q)$, by that $b(t,x_n,q_n,p,\omega)$ has a less than linear growth in (x_n,q_n,p), so $b(t,x_n,q_n,p,\omega)$ is bounded as ω is fixed. By the definition of an infimum there exists a sequence $(y_n,\widetilde{q}_n)$ such that

$b(t,x_n,q_n,p,\omega) \geqslant b_n(t,x_n,q_n,p,\omega) \geqslant b(t,y_n,\widetilde{q}_n,p,\omega)$
$+(n \vee c_1(t))\,|y_n - x_n| + (n \vee c_2(t))\,|\widetilde{q}_n - q_n| - \frac{1}{n} \geqslant b(t,y_n,\widetilde{q}_n,p,\omega) - \frac{1}{n}$.

Notice that we have already shown that $b(t,x_n,q_n,p,\omega)$ is bounded, as t and ω are fixed, so by the above inequaity one finds that $|y_n - x_n| \to 0, |\widetilde{q}_n - q_n| \to 0$, as $n \to \infty$. From this one has that $\lim_{n\to\infty} b_n(t,x_n,q_n,p,\omega) = \lim_{n\to\infty} b(t,y_n,\widetilde{q}_n,p,\omega) = \lim_{n\to\infty} b(t,x_n,q_n,p,\omega) = b(t,x,q,p,\omega)$. Now applying condition 2° in Theorem 267 4) then follows. The proof is complete. ■

By using Lemma 271 and the local time technique of treatment for discontinuous points of b with respect to x in [135] one can derive Theorem 267.

Proof. Take $b_n(t,x,q,p,\omega)$ from Lemma 271. Obviously, $b_n(t,x,q,p,\omega)$ is jointly Lipschitzian continuous in $(x,q,p) \in R^1 \times R^{1\otimes d_1} \times L^1_{\pi(.)}(R^{1\otimes d_2})$. Hence by Theorem 234 there exists a unique solution (x_t^n, q_t^n, p_t^n) satisfying (8.8) with coefficients b_n for each $n = 1, 2, \cdots$. Set

$h(t,x,q,p,\omega) = k_0(1 + |x| + |q| + \|p\|)$.

Again by Theorem 234 there exists a unique solution (ξ_t, u_t, v_t) satisfying the BSDE

$\xi_t = X + \int_{t\wedge\tau}^{\tau} h(s,\xi_s,u_s,v_s,\omega)ds - \int_{t\wedge\tau}^{\tau} u_s dw_s - \int_{t\wedge\tau}^{\tau}\int_Z v_s(z)\tilde{N}_k(ds,dz)$.

Moreover, by property 1)-3) of Lemma 271, applying the comparison Theorem 251 one finds that $x_t^n \leqslant x_t^{n+1} \leqslant \xi_t, \forall n$. Hence there exists a limit $P - a.s.$

$x_t = \lim_{n\to\infty} x_t^n, \forall t \geqslant 0$.

Moreover, by Lebesgue's dominated convergence theorem one also has that

$\lim_{n\to\infty} E\int_0^{\tau} |x_t^n - x_t|^2 ds = 0$.

Applying Ito's formula, one easily shows that there exists a unique $(q_s, p_s) \in L^2_{\mathfrak{F}_t}(R^{1\otimes d_1}) \times F^2_{\mathfrak{F}_t}(R^{1\otimes d_2})$ such that

$\lim_{n\to\infty} E(\int_0^{\tau} |q_s^n - q_s|^2 ds + \int_0^{\tau}\int_Z |p_s^n - p_s|^2 \pi(dz)ds) = 0$.

From these one can take a subsequence $\{n_k\}$ of $\{n\}$, denoted again by $\{n\}$ again, such that $dt \times dP - a.e.$

$q_s^n \to q_s$, in $R^{1\otimes d_1}$; $p_s^n \to p_s$, in $L^2_{\pi(.)}(R^{1\otimes d_2})$;

and

$E(\int_0^T |q_s^n - q_s|^2 ds + \int_0^T \int_Z |p_s^n - p_s|^2 \pi(dz)ds) \leqslant 1/2^n, n = 1, 2, ...$

Hence $E(\int_0^T \sup_n(|q_t^n| + \|p_t^n\|)^2 dt < \infty$. A similar conclusion holds for x_t^n. Now notice that

$x_t^n = x_0^n + A_t^n + \int_0^t q_s^n dw_s + \int_0^t \int_Z p_s^n(z)q(ds, dz),$

where $A_t^n = \int_0^t b_n(s, x_s^n, q_s^n, p_s^n, \omega)ds$. Hence there exists a limit (if necessary, take a subsequence)

$A_t = \lim_{n\to\infty} A_t^n,$

and we have

$x_t = x_0 + A_t + \int_0^t q_s dw_s + \int_0^t \int_Z p_s(z)q(ds, dz).$

Notice that

$|b_n(s, x_s^n, q_s^n, p_s^n, \omega)| \leqslant k_0[1 + \sup_n(|x_t^n| + |q_t^n| + \|p_t^n\|)].$

Hence A_t^n is a finite variational process, and so also is A_t. Let us show that

$$A_t = \int_0^t b(s, x_s, q_s, p_s, \omega)ds. \tag{8.66}$$

Indeed, from $m_1(a : (a, q, p) \in G) = 0$, applying the occupation density formula for the local time of a semi-martingale (Lemma 159), one finds that

$0 = \int_{G_1} L_t^a(x)da = \int_0^t I_{\{x_s \in G_1\}} |q_s|^2 ds \geqslant \int_0^t I_{\{(x_s, q_s, p_s) \in G\}} |q_s|^2 ds.$

Now by assumption $|q_s|^2 > 0$, as $(x_s, q_s, p_s) \in G$, so

$m_1(s \in [0, t] : (x_s, q_s, p_s) \in G) = 0.$

Therefore, by conclusion 4) of Lemma 271,

$\int_0^t |b(s, x_s, q_s, p_s, \omega) - b_n(s, x_s^n, q_s^n, p_s^n, \omega)| ds$

$\leqslant \int_0^t I_{((x_s, q_s, p_s) \notin G)} |b(s, x_s, q_s, p_s, \omega) - b_n(s, x_s^n, q_s^n, p_s^n, \omega)| ds \to 0,$

and (8.66) is derived. Now suppose that (x_t', q_t', p_t') is another solution of (8.8). By Lemma 271

$b_n(t, x, q, p, \omega) \leqslant b(t, x, q, p, \omega).$

Hence applying Theorem 251 one has that $x_t^n \leqslant x_t', \forall n$. Letting $n \to \infty$, we obtain $x_t \leqslant x_t'$. Therefore x_t is a minimal solution of (8.8). By Lemma 230, under condition 3° the solution is also unique. ■

Applying Theorem 267 we can get a better comparison theorem for solutions to (8.8) with a discontinuous coefficient $b^2(t, x, q, \omega)$. Actually, we have

Theorem 272 *Assume that conditions 1° and 3° of Theorem 251 hold and, moreover, assume that b^2 satisfies all conditions in Theorem 267. Then the conclusion in Theorem 251 still holds.*

Proof. As the proof of Theorem 267 we construct b_n from b^2 by Lemma 271. Denote by $(x_t'^n, q_t'^n, p_t'^n)$ the unique solution for (8.8) with the coefficients b_n. Since $b_n \leqslant b^2 \leqslant b^1$, by comparison Theorem 251 $x_t^{\prime n} \leqslant x_t^1$. Hence

$\lim_{n\to\infty} x_t^{\prime n} \leqslant x_t^1$.

However, by the uniqueness of solutions to (8.8) with coefficient b^2, one has $x_t^2 = \lim_{n\to\infty} x_t^{\prime n}$. Hence $x_t^2 \leqslant x_t^1$, and the proof is complete. ∎

8.15 Existence of Non-Lipschitzian Optimal Control. BSDE Case

In some practical cases one needs to consider the optimal consumption problem. Suppose the wealth process of a small investor satisfies the standard BSDE:

$$\begin{aligned} x_t = X - \int_t^T r_s x_s ds + C_T - C_t - \int_t^T \pi_s [b_s + \delta_s - r_s \underline{1}] ds \\ - \int_t^T \pi_s \sigma_s dw_s - \int_t^T \pi_s \rho_s d\tilde{N}(s), \end{aligned} \tag{8.67}$$

which is a special case of (8.52) with $\widetilde{k}_0 = 0$. Then under appropriate conditions (for example under the conditions in Theorem 244 with $\delta_s = 0$), we can rewrite it as (8.32). Or we can rewrite it as $\widetilde{P}_T - a.s$

$$\widetilde{x}_t = \widetilde{X} + \int_t^T \widetilde{c}(s) ds - \int_t^T \widetilde{\pi}_s \sigma_s d\widetilde{w}_s - \int_t^T \widetilde{\pi}_s \rho_s d\overline{N}(s), \tag{8.68}$$

if $dC_t = c_s ds$, and we denote $\widetilde{x}_t = e^{-\int_0^t r_s ds} x_t$, etc. Therefore, for simplicity, we can consider the simpler BSDE (8.68) directly. To simplify the notation let us still write P for $\widetilde{P}_T$, (x_t, π_t) for $(\widetilde{x}_t, \widetilde{\pi}_t)$, and $(w_t, \widetilde{N}_t)$ for $(\widetilde{w}_t, \overline{N}_t)$. Then we can consider the following wealth process for a small investor: For $0 \leq t \leq T, P - a.s.$

$$x_t = X + \int_t^T c(s) ds - \int_t^T \pi_s \sigma_s dw_s - \int_t^T \pi_s \rho_s d\widetilde{N}(s), \ t \in [0, T] \tag{8.69}$$

Furthermore, if we denote $q_t = \pi_t \sigma_t, p_t = \pi_t \rho_t$, then (8.69) can be rewritten as

$$x_t = X + \int_t^T u(s) ds - \int_t^T q_s dw_s - \int_t^T p_s d\widetilde{N}(s), t \in [0, T], \tag{8.70}$$

where we write $u(t) = c(t)$.

In this section we are going to apply the results found above to solve some special kinds of optimal stochastic control problems with respect to (8.70). We obtain results on the existence of some non-Lipschitzian optimal controls for some special stochastic control problems with respect to such BSDE systems with jumps. Some of the results can also be explained in the financial market.

Let us consider the following more general $d-$dimensional BSDE system as (8.8): for $0 \leqslant t \leqslant T$,

$$\begin{aligned} x_t^u &= X + \int_t^T u(s, x_s^u, q_s^u, p_s^u)ds - \int_t^T q_s^u dw_s \\ &\quad - \int_t^T \int_Z p_s^u(z)\tilde{N}_k(ds, dz), \end{aligned} \tag{8.71}$$

where $x_s^u, u(s, x_s^u, q_s^u, p_s^u) \in R^d, q_s^u \in R^{d\otimes d_1}, p_s^u(z) \in R^{d\otimes d_2}$, and $u \in U$,

$$U = \left\{ \begin{array}{c} u = u(t,x,q,p) : u(t,x,q,p) \text{ is jointly measurable such that} \\ (8.71) \text{ has a unique solution } (x_t, q_t, p_t), \text{ and } |u(t,x,q,p)| \leqslant |x|^\beta \end{array} \right\} \tag{8.72}$$

where $0 < \beta \leqslant 1$ is a given fixed constant.

We have the following

Lemma 273 . *Let*
$V(t,x) = \frac{1}{2}|x|^2 + T - t,$
for $x \in R^d$, $t \in [0,T]$, *then* $V(t,x)$ *satisfies*

$$\left\{ \begin{array}{l} (\partial/\partial t)V - \inf_{u\in R^d, |u|\leqslant |x|^\beta}(u \cdot \frac{\partial}{\partial x}V) + 1 - |x|^{1+\beta} = 0, \ 0 \leqslant t \leqslant T, \\ \qquad V(T,x) = \frac{1}{2}|x|^2. \end{array} \right.$$

Moreover, if $u^0(x) = -I_{x\neq 0}x/|x|^{1-\beta}$, *then* $|u^0(x)| \leq |x|^\beta$, *and*
$u^0(x) \cdot \frac{\partial}{\partial x}V(t,x) = \inf_{u\in R^d, |u|\leqslant |x|^\beta}(u \cdot \frac{\partial}{\partial x}V) = -|x|^{1+\beta}.$

Proof. The conclusion can be checked directly. ∎

Now let

$$u^0(x) = u^0(t,x,q,p) = -I_{x\neq 0}x/|x|^{1-\beta}. \tag{8.73}$$

Then $u^0(x)$ is continuous and such that
$(x-y)\cdot(u^0(x) - u^0(y)) \leqslant -(|x| - |y|)(|x|^\beta - |y|^\beta) \leqslant 0.$
Hence, by Theorem 246, (8.71) has a unique solution for such coefficient u^0. However, in the case that β is given and satisfies $0 < \beta < 1$, then such u^0 is non-Lipschitzian in x. Denote this solution by (x_t^0, q_t^0, p_t^0). Then by Ito's formula
$V(T, x_T^0) - V(t.x_t^0) = \int_t^T (\frac{\partial}{\partial s}V - u^0(x_s^0)\cdot\frac{\partial}{\partial x}V + 1 - |x_s^0|^{1+\beta})ds$

$+\text{martingale}+\frac{1}{2}\int_t^T(|q_s^0|^2+\int_Z|p_s^0|^2\pi(dz))ds-\int_t^T(1-|x_s^0|^{1+\beta})ds$
$=\text{martingale}+\frac{1}{2}\int_t^T(|q_s^0|^2+\int_Z|p_s^0|^2\pi(dz))ds-\int_t^T(1-|x_s^0|^{1+\beta})ds.$
Similarly, for any $u\in U$
$V(T,x_T^u)-V(t.x_t^u)$
$\leqslant\text{martingale}+\frac{1}{2}\int_t^T(|q_s^u|^2+\int_Z|p_s^u|^2\pi(dz))ds-\int_t^T(1-|x_s^u|^{1+\beta})ds.$
Therefore, for all $u\in U$
$\text{martingale}+\frac{1}{2}|x_t^u|^2+\frac{1}{2}\int_t^T(|q_s^u|^2+\int_Z|p_s^u|^2\pi(dz))ds$
$+\int_t^T(-1+|x_s^u|^{1+\beta})ds\geqslant\frac{|X|^2}{2}-T+t=\text{martingale}$
$+\frac{1}{2}|x_t^0|^2+\frac{1}{2}\int_t^T(|q_s^0|^2+\int_Z|p_s^0|^2\pi(dz))ds+\int_t^T(-1+|x_s^0|^{1+\beta})ds.$
Hence
$E^{\mathfrak{F}_t}(\frac{1}{2}|x_t^u|^2+\frac{1}{2}\int_t^T(|q_s^u|^2+\int_Z|p_s^u|^2\pi(dz))ds+\int_t^T|x_s^u|^{1+\beta}ds)$
$\geqslant E^{\mathfrak{F}_t}(\frac{1}{2}|x_t^0|^2+\frac{1}{2}\int_t^T(|q_s^0|^2+\int_Z|p_s^0|^2\pi(dz))ds+\int_t^T|x_s^0|^{1+\beta}ds).$
Thus we have proved the following Theorem 274, which shows that a non-Lipschitzian feedback optimal stochastic control exists.

Theorem 274 . *Define $u^0(x)$ by (8.73), and for any $u\in U$, where U is defined by (8.72), let*
$J(u)=E(\frac{1}{2}|x_0^u|^2+\frac{1}{2}\int_0^T(|q_s^u|^2+\int_Z|p_s^u|^2\pi(dz))ds+\int_0^T|x_s^u|^{1+\beta}ds),$
where (x_t^u,q_t^u,p_t^u) is the unique solution of (8.71) for $u\in U$.
Then
1) $u^0\in U$,
2) $J(u)\geqslant J(u^0)$, for all $u\in U$.

The above target functional $J(u)$ can be explained as an energy functional. We can also discuss another kind of optimal control problem which can be explained in the financial market. Still consider the BSDE system (8.71), but this time with $x_t^u,u(s,x_s^u,q_s^u,p_s^u)\in R^1,q_s^u\in R^{1\otimes d_1},p_s^u(z)\in R^{1\otimes d_2}$, and consider the admissible control set

$$U=\left\{\begin{array}{c} u=u(t,x,q,p):u(t,x,q,p)\text{ is jointly measurable such that}\\ (8.71)\text{ has at least a solution }(x_t,q_t,p_t),\text{ and }|u(t,x,q,p)|\leqslant|x|^\beta\end{array}\right\}. \tag{8.74}$$

where $0<\beta\leqslant1$ is a given fixed constant. Denote the target functional by
$J(u)=\sup\{E(2\int_0^T|x_s^u|^{1+\beta}ds-|x_0^u|^2-\int_0^T(|q_s^u|^2+\int_Z|p_s^u|^2\pi(dz))ds)\}$
$=\sup\{E[\int_0^T|x_s^u|^{1+\beta}ds-|x_0^u|^2$
$+(\int_0^T|x_s^u|^{1+\beta}ds-\int_0^T(|q_s^u|^2+\int_Z|p_s^u|^2\pi(dz))ds)]\},$
for each $u\in U$, where (x_t^u,q_t^u,p_t^u) is any solution corresponding to the same u. We have the following

Theorem 275 *Let*
$u^0(x)=I_{x\neq0}x/|x|^{1-\beta}.$
Then
1) $u^0\in U$,

2) $J(u) \leqslant J(u^0)$, $\forall u \in U$.

To prove Theorem 275 one only needs to use the following

Lemma 276 $V(t,x) = \frac{1}{2}|x|^2 + T - t$ *satisfies*

$$\begin{cases} (\partial/\partial t)V - \sup_{u \in R^1, |u| \leqslant |x|^\beta} (u \cdot \frac{\partial}{\partial x} V) + 1 + |x|^{1+\beta} = 0, 0 \leqslant t \leqslant T, \\ V(T,x) = \frac{1}{2}|x|^2 . \end{cases}$$

Moreover, if $u^0(x) = I_{x \neq 0} x / |x|^{1-\beta}$, *then* $|u^0(x)| \leq |x|^\beta$, *and*
$u^0(x) \cdot \frac{\partial}{\partial x} V(t,x) = \sup_{u \in R^1, |u| \leqslant |x|^\beta} (u \cdot \frac{\partial}{\partial x} V) = |x|^{1+\beta}$.

The proof for 2) of Theorem 275 is similar to the above. However, since now the BSDE is in R^1, and $u^0(x)$ is continuous and depends on x only, by Theorem 267 it is seen that (8.71) has a solution for $u = u^0$. Hence 1) of Theorem 275 is proved.

Now let us explain Theorem 275 in terms of the financial market. If we see (8.71) as the equation for the wealth process x_t of a small investor, the control $u(t, x_t, q_t, p_t)$ can be explained as his generalized feedback consumption process, and $(q_t, p_t(\cdot))$ as some generalized portfolio process for stocks, then in the case $\beta = 1$ we can interprete the target functional $J(u)$ as a subtraction of the total summation of the squares of wealth and the squares of money of bonds in the total time interval from the square of initial wealth (or say initial investment) of the investor. So $J(u)$ can be seen as some generalized utility functional for the investor. Theorem 275 tells the investor that he can get a maximum utility if he choose the consumption law to be $u(x_t) = x_t / |x_t|^{1-\beta}$, i.e. when the wealth process $x_t \geqslant 0$, he should consume the money $x_t / |x_t|^{1-\beta}$; and when $x_t \leqslant 0$, he should borrow the money $-x_t / |x_t|^{1-\beta}$.

Let us explain this in more detail. Suppose that in the market there is a bond with a price that satisfies an equation:
$dP_t^0 = P_t^0 r_t dt$, $P_0^0 = 1$;
and there are $N = d + m$ stocks, with prices that satisfy the following SDEs
$dP_t^i = P_{t-}^i [b_t^i dt + \sum_{k=1}^d \sigma_t^{ik} dw_t^k + \sum_{k=1}^m \rho_t^{ik} d\tilde{N}_t^k]$,
$P_0^i = P_0^i, \quad i = 1, 2, \cdots, N$.
If the market is self-financed, then the wealth process of a small investor will satisfy
$dx_t = r_t x_t dt + \pi_t [b_t - r_t \underline{1}] dt + \pi_t \sigma_t dw_t + \pi_t \rho_t d\tilde{N}_t, 0 \leqslant t \leqslant T$,
where $\pi_t = (\pi_t^1, \cdots, \pi_t^N)$ is the profolio of the investment of the investor. (See (8.4)). Now suppose that the investor also consumes the money $dC_t = c_t dt$ during the time dt, then his wealth process will satisfy

$$\begin{aligned} dx_t &= r_t x_t dt + \pi_t [b_t - r_t \underline{1}] dt - c_t dt + \pi_t \sigma_t dw_t + \pi_t \rho_t d\tilde{N}_t, \\ x_T &= X, 0 \leqslant t \leqslant T. \end{aligned} \tag{8.75}$$

According to Theorem 243 we have the following fact that assumes b, r, σ, ρ are all bounded $\mathfrak{F}_t^{w,\widetilde{N}}$-predictable processes, and $\widetilde{\sigma}_t^{-1}$ exists and is also bounded and, furthermore, that c_t is predictable such that $E\int_0^T |c_t|^2 dt < \infty$, and we will assume that $N = d + m$, $X \in \mathfrak{F}_T^{w,\widetilde{N}}$ and $E|X|^2 < \infty$. Then there exists a unique wealth process $x_t \in L^2_{\mathfrak{F}}(R^1)$ and a unique portfolio process $\pi_t \in L^2_{\mathfrak{F}}(R^N)$ satisfying (8.75). Furthermore, assume $(\theta_t^{\widetilde{N}})^i < 1, \forall i = 1, \cdots, m$, where $\theta_t = \begin{bmatrix} \theta_t^w \\ \theta_t^{\widetilde{N}} \end{bmatrix} = \widetilde{\sigma}_t^{-1}[b_t - r_t \underline{1}]$, then by Theorem 244 (8.75) can be simplified to $\widetilde{P}_T - a.s.$

$$\begin{cases} dx_t = r_t x_t dt - c_t dt + \pi_t \sigma_t d\widetilde{w}_t + \pi_t \rho_t d\overline{N}_t, \\ x_T = X, \ 0 \leq t \leq T, \end{cases}$$

where $\widetilde{P}_T \sim P$ and, $\widetilde{w}_t$ and $\overline{N}_t$ are respectively, a d-dimensional BM and an m-dimensional pure jump martingale with a new compensator $\int_0^t (\underline{1} - \theta_s) ds$. (See (8.33)). In the case that θ_t is non-random, then $\overline{N}_t$ is an m-dimensional centralized Poisson process. Furthermore, if $\theta = \theta_t$ is an m-dimensional constant vector, which does not depend on t, then $\overline{N}_t$ is a m-dimensional centralized stationary Poisson process with the conpensator $(\underline{1} - \theta)t$. Now let $\widetilde{x}_t = e^{-\int_0^t r_s ds} x_t$. Then the above BSDE can be rewritten as $\widetilde{P}_T - a.s$

$$\widetilde{x}_t = \widetilde{X} + \int_t^T \widetilde{c}(s) ds - \int_t^T \widetilde{\pi}_s \sigma_s d\widetilde{w}_s - \int_t^T \widetilde{\pi}_s \rho_s d\overline{N}(s), \ t \in [0, T].$$

For simplicity of notation we still write P for $\widetilde{P}_T$, (x_t, π_t) for $(\widetilde{x}_t, \widetilde{\pi}_t)$, and $(w_t, \widetilde{N}_t)$ for $(\widetilde{w}_t, \overline{N}_t)$. Then we can write the above BSDE as $P - a.s.$

$$x_t = X + \int_t^T c(s) ds - \int_t^T \pi_s \sigma_s dw_s - \int_t^T \pi_s \rho_s d\widetilde{N}(s), t \in [0, T], \quad (8.76)$$

where we set $u(s) = c(s)$. Again by Theorem 244 let

$$[q_t, p_t] = \pi_t[\sigma_t, \rho_t] = \pi_t \widetilde{\sigma}_t.$$

Then (8.76) is equivalent to

$$x_t = X + \int_t^T c(s) ds - \int_t^T q_s dw_s - \int_t^T p_s d\widetilde{N}(s), \ t \in [0, T]; \quad (8.77)$$

and

$$x_t = x_t, \ \pi_t^i = \sum_{j=1}^d q_s^j (\widetilde{\sigma}_s^{-1})_{ji} + \sum_{j=1}^m p_s^j (\widetilde{\sigma}_s^{-1})_{(j+d)i}, \ i = 1, \cdots, N.$$

Thus the admissible control set in (8.74) is

$$U = \left\{ \begin{array}{l} u = u(t, x, \pi\sigma, \pi\rho) : u(t, x, \pi\sigma, \pi\rho) \text{ is jointly measurable such that} \\ (8.76) \text{ has at least a solution } (x_t, \pi_t), \text{ and } |u(t, x, \pi\sigma, \pi\rho)| \leqslant |x|^\beta \end{array} \right\}. \quad (8.78)$$

where $0 < \beta \leqslant 1$ is a given fixed constant. The target functional in Theorem 275 is

$J(u) = \sup\{E(2\int_0^T |x_s^u|^{1+\beta} ds - |x_0^u|^2 - \int_0^T (|\pi_s^u \sigma_s|^2 + \int_Z |\pi_s^u \rho_s|^2 \pi(dz))ds)\}$
$= \sup\{E[\int_0^T |x_s^u|^{1+\beta} ds - |x_0^u|^2 + (\int_0^T |x_s^u|^{1+\beta} ds$
$- \int_0^T (|\pi_s^u \sigma_s|^2 + \int_Z |\pi_s^u \rho_s|^2 \pi(dz))ds)]\}$
for each $u \in U$, where (x_t^u, π_t^u) is any solution of (8.76) corresponding to the same u, and Theorem 275 tells us that the following theorem holds true.

Theorem 277 *Let*
$u^0(x) = I_{x \neq 0} x / |x|^{1-\beta}$.
Then
1) $u^0 \in U$,
2) $J(u) \leqslant J(u^0)$, $\forall u \in U$.

For the special case that $\beta = 1, \rho = 0, \sigma_s = I_{d \times d}$, where $I_{d \times d}$ is the $d \times d$ unit matrix, and $N = d, X \in \mathfrak{F}_T$, the target functional becomes
$J(u) = \sup\{E[(\int_0^T |x_s^u|^2 ds - |x_0^u|^2) + (\int_0^T (|x_s^u|^2 - |\pi_s^u|^2)ds)]\}$.
We can explain this as an ultility functional for the investor. Obviously, this ultility has contributions from two terms: the first term is the subtraction of the summation of the wealth square $|x_s^u|^2$ in the total time interval from the initial wealth square $|x_0^u|^2$ of the investor, while the second term is the summation of the subtraction of the wealth square $|x_s^u|^2$ from the porfolio square $|\pi_s^u|^2$ in the total time interval. Or, roughly speaking, the ultility is positively proportional to the money square deposited in the bank and the total money square subtracted from the initial investment square. The surprising thing is that for such a ultility functional the zero consumption is not the best. To make this ultility attain its supremum the optimal consumption law should be: $u(x_t) = x_t / |x_t|^{1-\beta}$, i.e. when the wealth process $x_t \geqslant 0$, the investor should consume the money $x_t / |x_t|^{1-\beta}$ (a positive consumption $x_t / |x_t|^{1-\beta}$); and when $x_t \leqslant 0$, he should borrow the money $-x_t / |x_t|^{1-\beta}$ or, say, he should sell some stocks to get the money $-x_t / |x_t|^{1-\beta}$ (a negative consumption $x_t / |x_t|^{1-\beta}$) (Notice that here η_t^i–the amount of units for the i–th stock bought by the investor can be negative, that is, he may sell it to someone else, so the investment at time t may be negative: $x_t \leq 0$).

8.16 Existence of Discontinuous Optimal Control. BSDEs in R^1

Consider the following BSDE system: for $0 \leqslant t \leqslant T$,

$$
\begin{aligned}
x_t^u &= X + \int_t^T [-k_0 x_s^u + |q_s^u|\, u(s, x_s^u, q_s^u, p_s^u)]ds - \int_t^T q_s^u dw_s \\
&\quad - \int_t^T \int_Z p_s^u(z) \tilde{N}_k(ds, dz),
\end{aligned}
\tag{8.79}
$$

where k_0 is a constant, $x_t^u, u(s, x_s^u, q_s^u, p_s^u) \in R^1, q_s^u \in R^{1\otimes d_1}, p_s^u(z) \in R^{1\otimes d_2}$, $X \in \mathfrak{F}_T^{w,k}$, and $u \in U$,

$$
U = \left\{ \begin{array}{c} u = u(t,x,q,p) : u(t,x,q,p) \text{ is jointly measurable such that} \\ (8.79) \text{ has a unique solution } (x_t, q_t, p_t), \text{ and} \\ |u(t,x,q,p)| \leqslant 1. \end{array} \right\}
\tag{8.80}
$$

We have the following

Lemma 278 . *Let*
$V(t,x) = \frac{1}{2}|x|^2 + T - t,$
for $x \in R$, $t \in [0,T]$, *then* $V(t,x)$ *satisfies*

$$
\left\{ \begin{array}{l} (\partial/\partial t)V - \inf_{u \in R, |u| \leqslant 1} [(|q|\, u - k_0 x) \cdot \frac{\partial}{\partial x} V] + 1 \\ - |q|\,|x| - k_0 |x|^2 = 0, 0 \leqslant t < T; \ V(T,x) = \frac{1}{2}|x|^2. \end{array} \right.
$$

Moreover, if $u^0(x) = -I_{x \neq 0} x / |x|$, *then* $|u^0(x)| \leq 1$, *and*
$u^0(x) \cdot \frac{\partial}{\partial x} V(t,x) = \inf_{u \in R^d, |u| \leqslant 1} (u \cdot \frac{\partial}{\partial x} V) = -|x|.$

Proof. The conclusion can be checked directly. ■
Now let

$$
u^0(x) = u^0(t,x,q,p) = -I_{x \neq 0} x / |x|. \tag{8.81}
$$

Then $u^0(x)$ is discontinuous in x and does not depend on q, p such that
$(x - x') \cdot (|q|\, u^0(x) - |q'|\, u^0(x')) \leqslant |q|\,(x - x') \cdot (u^0(x) - u^0(x'))$
$+ |x - x'|\,|q - q'| \leqslant |x - x'|\,|q - q'|.$
Hence, by Theorem 267, (8.79) has a unique solution for such a coefficient u^0. Denote this solution by (x_t^0, q_t^0, p_t^0). Then by Ito's formula
$V(T, x_T^0) - V(t.x_t^0) = \int_t^T (\frac{\partial}{\partial s} V - (|q_s^0|\, u^0(x_s^0) - k_0 x_s^0) \frac{\partial}{\partial x} V + 1$
$- |q_s^0|\,|x_s^0| - k_0 |x_s^0|^2) ds + \text{martingale} + \frac{1}{2}\int_t^T (|q_s^0|^2 + \int_Z |p_s^0|^2 \pi(dz)) ds$
$- \int_t^T (1 - |q_s^0|\,|x_s^0| - k_0 |x_s^0|^2) ds = \text{martingale}$
$+ \frac{1}{2}\int_t^T (|q_s^0|^2 + \int_Z |p_s^0|^2 \pi(dz)) ds - \int_t^T (1 - |q_s^0|\,|x_s^0| - k_0 |x_s^0|^2) ds.$
Similarly, for any $u \in U$
$V(T, x_T^u) - V(t.x_t^u) \leqslant \text{martingale} + \frac{1}{2}\int_t^T (|q_s^u|^2 + \int_Z |p_s^u|^2 \pi(dz)) ds$
$+ \frac{1}{2}\int_t^T (|q_s^0|^2 + \int_Z |p_s^0|^2 \pi(dz)) ds - \int_t^T (1 - |q_s^0|\,|x_s^0| - k_0 |x_s^0|^2) ds.$

Therefore, for all $u \in U$
martingale$+\frac{1}{2}\left|x_t^u\right|^2+\frac{1}{2}\int_t^T(\left|q_s^u\right|^2+\int_Z\left|p_s^u\right|^2\pi(dz))ds$
$+\int_t^T(-1+\left|q_s^u\right|\left|x_s^u\right|+k_0\left|x_s^u\right|^2)ds \geqslant \frac{|X|^2}{2}-T+t$
$=$martingale$+\frac{1}{2}\left|x_t^0\right|^2+\frac{1}{2}\int_t^T(\left|q_s^0\right|^2+\int_Z\left|p_s^0\right|^2\pi(dz))ds$
$+\int_t^T(-1+\left|q_s^0\right|\left|x_s^0\right|+k_0\left|x_s^0\right|^2)ds$
Hence
$E^{\mathfrak{F}_t}(\frac{1}{2}\left|x_t^u\right|^2+\frac{1}{2}\int_t^T(\left|q_s^u\right|^2+\int_Z\left|p_s^u\right|^2\pi(dz))ds$
$+\int_t^T(\left|q_s^u\right|\left|x_s^u\right|+k_0\left|x_s^u\right|^2)ds) \geqslant E^{\mathfrak{F}_t}(\frac{1}{2}\left|x_t^0\right|^2$
$+\frac{1}{2}\int_t^T(\left|q_s^0\right|^2+\int_Z\left|p_s^0\right|^2\pi(dz))ds+\int_t^T(\left|q_s^0\right|\left|x_s^0\right|+k_0\left|x_s^0\right|^2)ds).$
Thus we have proved the following Theorem 279, which shows that an optimal feedback Bang-Bang stochastic control exists.

Theorem 279 . *Define $u^0(x)$ by (8.81), and for any $u \in U$, where U is defined by (8.80), let*
$J(u)=E(\frac{1}{2}\left|x_0^u\right|^2+\frac{1}{2}\int_0^T(\left|q_s^u\right|^2+\int_Z\left|p_s^u\right|^2\pi(dz))ds$
$+\int_0^T(\left|q_s^u\right|\left|x_s^u\right|+k_0\left|x_s^u\right|^2)ds),$
where (x_t^u, q_t^u, p_t^u) is the unique solution of (8.79) for $u \in U$.
Then
1) $u^0 \in U$,
2) $J(u) \geqslant J(u^0)$, for all $u \in U$.

Similarly, Consider the following BSDE system: for $0 \leqslant t \leqslant T$,

$$\begin{aligned} x_t^u &= X+\int_t^T[-k_0x_s^u+\left|q_s^u\right|u(s,x_s^u,q_s^u,p_s^u)]ds-\int_t^T q_s^u dw_s \\ &\quad -\int_t^T\int_Z p_s^u(z)\tilde{N}_k(ds,dz), \end{aligned} \tag{8.82}$$

where k_0 is a constant, $x_t^u, u(s,x_s^u,q_s^u,p_s^u) \in R^1, q_s^u \in R^{1\otimes d_1}, p_s^u(z) \in R^{1\otimes d_2}$, $X \in \mathfrak{F}_T^{w,k}$, and $u \in U$,

$$U=\left\{\begin{array}{c} u=u(t,x,q,p): u(t,x,q,p) \text{ is jointly measurable such that} \\ (8.79) \text{ has a unique solution } (x_t,q_t,p_t), \text{ and} \\ |u(t,x,q,p)| \leqslant 1. \end{array}\right\} \tag{8.83}$$

We have the following

Lemma 280 . *Let*
$V(t,x)=\frac{1}{2}|x|^2+T-t,$
for $x \in R$, $t \in [0,T]$, then $V(t,x)$ satisfies

$$\left\{\begin{array}{l} (\partial/\partial t)V+\inf_{u\in R,|u|\leqslant 1}[(|q|\,u-k_0x)\cdot\frac{\partial}{\partial x}V]+1 \\ +|q|\,|x|+k_0|x|^2=0, 0\leqslant t<T;\ V(T,x)=\frac{1}{2}|x|^2. \end{array}\right.$$

Moreover, if $u^0(x) = -I_{x\neq 0}x/|x|$, *then* $|u^0(x)| \leq 1$, *and*
$u^0(x) \cdot \frac{\partial}{\partial x}V(t,x) = \inf_{u\in R^d, |u|\leqslant 1}(u \cdot \frac{\partial}{\partial x}V) = -|x|.$

Proof. The conclusion can be checked directly. ■
Now let

$$u^0(x) = u^0(t,x,q,p) = -I_{x\neq 0}x/|x|. \tag{8.84}$$

Then $u^0(x)$ is discontinuous at $x=0$ and does not depend on q,p. Hence, by Theorem 267, (8.82) has a unique solution for such a coefficient u^0. Denote this solution by (x_t^0, q_t^0, p_t^0). Then by Ito's formula

$V(T,x_T^0) - V(t.x_t^0) = \int_t^T(\frac{\partial}{\partial s}V + (|q_s^0|\, u^0(x_s^0) - k_0x_s^0)\frac{\partial}{\partial x}V + 1$
$+ |q_s^0|\,|x_s^0| + k_0|x_s^0|^2)ds +$martingale$+\frac{1}{2}\int_t^T(|q_s^0|^2 + \int_Z |p_s^0|^2\pi(dz))ds$
$- \int_t^T(1 + |q_s^0|\,|x_s^0| + k_0|x_s^0|^2)ds =$martingale
$+\frac{1}{2}\int_t^T(|q_s^0|^2 + \int_Z |p_s^0|^2\pi(dz))ds - \int_t^T(1 + |q_s^0|\,|x_s^0| + k_0|x_s^0|^2)ds.$

Similarly, for any $u \in U$

$V(T,x_T^u) - V(t.x_t^u) \geq$martingale$+\frac{1}{2}\int_t^T(|q_s^u|^2 + \int_Z |p_s^u|^2\pi(dz))ds$
$- \int_t^T(1 + |q_s^u|\,|x_s^u| + k_0|x_s^u|^2)ds.$

Therefore, for all $u \in U$

martingale$+\frac{1}{2}|x_t^u|^2 + \frac{1}{2}\int_t^T(|q_s^u|^2 + \int_Z |p_s^u|^2\pi(dz))ds$
$- \int_t^T(1 + |q_s^u|\,|x_s^u| + k_0|x_s^u|^2)ds \leq \frac{|X|^2}{2} - T + t$
$=$martingale$+\frac{1}{2}|x_t^0|^2 + \frac{1}{2}\int_t^T(|q_s^0|^2 + \int_Z |p_s^0|^2\pi(dz))ds$
$- \int_t^T(1 + |q_s^0|\,|x_s^0| + k_0|x_s^0|^2)ds$

Hence

$E^{\mathfrak{F}_t}(\frac{1}{2}|x_t^u|^2 + \frac{1}{2}\int_t^T(|q_s^u|^2 + \int_Z |p_s^u|^2\pi(dz))ds - \int_t^T(|q_s^u|\,|x_s^u| + k_0|x_s^u|^2)ds)$
$\leq E^{\mathfrak{F}_t}(\frac{1}{2}|x_t^0|^2 + \frac{1}{2}\int_t^T(|q_s^0|^2 + \int_Z |p_s^0|^2\pi(dz))ds - \int_t^T(|q_s^0|\,|x_s^0| + k_0|x_s^0|^2)ds).$

Thus we have proved the following Theorem 281, which shows that an optimal feedback Bang-Bang stochastic control exists.

Theorem 281 . *Define* $u^0(x)$ *by (8.84), and for any* $u \in U$, *where* U *is defined by (8.83), let*

$J(u) = E(\frac{1}{2}|x_0^u|^2 + \frac{1}{2}\int_0^T(|q_s^u|^2 + \int_Z |p_s^u|^2\pi(dz))ds$
$- \int_0^T(|q_s^u|\,|x_s^u| + k_0|x_s^u|^2)ds),$

where (x_t^u, q_t^u, p_t^u) *is the unique solution of (8.82) for* $u \in U$.
Then
1) $u^0 \in U$,
2) $J(u) \leq J(u^0)$, *for all* $u \in U$.

The target functionals in Theorem 279 and Theorem 281 can be explained as some energy functionals.

8.17 Application to PDE. Feynman-Kac Formula

In this section we will discuss the following problem. If we have a classical solution for some partial differential equation and with integral term, how could we express it as the solution of some BSDE or FSDE under some appropriate conditions. Such an expression is usually called a Feyman-Kac formula. Suppose that $D \subset R^d$ is a bounded open region, ∂D is its boundary thatt is supposed to be sufficiently smooth, and set $D^c = R^d - D$. Consider the following integral-differential equation (IDE)

$$\begin{aligned}
\pounds_{b,\sigma,c} u(t,x) &\equiv \frac{\partial}{\partial t} u(t,x) + \sum_{i=1}^{d} b_i(t,x) \frac{\partial}{\partial x_i} u(t,x) \\
&+ \frac{1}{2} \sum_{i,j=1}^{d} a_{ij}(t,x) \frac{\partial^2}{\partial x_i \partial x_j} u(t,x) \\
+ \int_Z (u(t, x + c(t,x,z)) &- u(t,x) - \sum_{i=1}^{d} c_i(t,x,z) \frac{\partial}{\partial x_i} u(t,x)) \pi(dz) \\
= f(t,x,u(t,x),u'_x(t,x) &\cdot \sigma(t,x), u(t,x+c(t,x,.)) - u(t,x)),
\end{aligned} \tag{8.85}$$

$$u(T,x) = \phi(x), u(t,x) \mid_{D^c} = \psi(t,x), \psi(T,x) = \phi(x) \mid_{D^c}, \tag{8.86}$$

$$u \in C^{1,2}([0,T] \times R^d) (= C^{1,2}([0,T] \times R^d, R^1))$$

where for the coefficients b, σ, c we make the following assumptions (C):
C.1 $b : [0,T] \times R^d \to R^d$ is bounded and measurable;
C.2 $\sigma : [0,T] \times R^d \to R^{d \otimes r}$ is bounded, measurable and, moreover, there exist constants $\delta_0, k_0 > 0$ such that $\delta_0 |\lambda|^2 \leqslant \langle \lambda, a\lambda \rangle \leqslant k_0 |\lambda|^2$, for all $\lambda \in R^d$,where $a = \sigma\sigma^* = (a_{ij})_{i,j=1}^d$,and $\langle .,. \rangle$ is the inner product in R^d;
C.3 $c(t,x,z) : [0,T] \times R^d \times Z \to R^d$ is bounded, measurable, and
$\int_Z \left| |c(s,x,z)|^2 \right| \pi(dz) \leq k_0$
C.4 $2 \langle y_1 - y_2, b(s,y_1) - b(s,y_2) \rangle + |\sigma(s,y_1) - \sigma(s,y_2)|^2$
$+ \int_Z |c(s,y_1,z) - c(s,y_2,z)|^2 \pi(dz) \leqslant c_{T,N}(s) \rho_{T,N}(|y_1 - y_2|^2)$,
as $|y_1|, |y_2| \leqslant N$ and $t \in [0,T]$,
where $c_{T,N}(t) \geqslant 0$ is non-random; and $\rho_N(u) \geqslant 0$ defined on $u \geqslant 0$ is non-random, increasing, continuous and concave, and such that
$\int_0^T c_{T,N}(t)^2 dt < \infty, \int_{0+} du/\rho_{T,N}(u) = \infty$,
for each $N = 1, 2, ...$ and each $T < \infty$.

For functions f, ϕ, ψ we make the following assumption (D):
D.1. $f : [0,T] \times D \times R^1 \times R^{1 \otimes r} \times L^2_{\pi(.)}(R^1) \to R^1$ is measurable such that

$g^u \in L_p([0,T] \times D)$,
where

$$g^u(t,x) = f(t,x,u(t,x),u'_x(t,x)\cdot\sigma(t,x),u(t,x+c(t,x,.))-u(t,x)), \quad (8.87)$$

for all $u \in C^{1,2}([0,T] \times R^d)$ satisfying (8.86);
D.2. $|f(t,x,r_1,q_1,p_1) - f(t,x,r_2,q_2,p_2)|$
$\leqslant k_0(|r_1 - r_2| + |q_1 - q_2| + \|p_1 - p_2\|)$,
for all $r_1, r_2 \in R^1; q_1, q_2 \in R^{1\otimes r}, p_1, p_2 \in L^2_{\pi(.)}(R^1)$;
D.3. $\phi \in C(R^d); (\phi : R^d \to R^1)$;
$\psi \in C([0,T] \times D^c); (\psi : [0,T] \times D^c \to R^1)$;
D.4. the boundary ∂D is sufficiently smooth;
D.5. $|f(t,x,r,q,p)| \leqslant k_0(1 + |r| + |q| + \|p\|)$.
Consider now the following forward SDE with Poisson jumps in R^d for any given $(t,x) \in [0,T] \times D$:

$$y_s = x + \int_t^s b(r,y_r)dr + \int_t^s \sigma(r,y_r)dw_r + \int_t^s \int_Z c(r,y_{r-},z)\tilde{N}_k(dr,dz), \quad (8.88)$$

as $t \leqslant s \leqslant T$, where $w_s, 0 \leqslant s \leqslant T$, is an r-dimensional BM, and for simplicity $\tilde{N}_k(dr,dz)$ is 1-dimensional Poisson martingale measure defined in (8.8). Under condition (C) by Theorem 185, (8.88) has a pathwise unique strong solution. Here "strong" means that the solution $y_s^{t,x}$ is $\Im_{t,s}-$measurable, where $\Im_{t,s} = \sigma\{w(r) - w(t), N_k((t,r],U); \text{for all } t \leqslant r \leqslant s, U \in \Re(Z)\}$.
Now for any given $(t,x) \in [0,T] \times D$ let
$\tau = \tau_{t,x} = \inf\{s > t : y_s^{t,x} \notin D\}$,and
$\tau = \tau_{t,x} = T$, for $\inf\{\emptyset\}$.
Write

$$S^2_{\Im_{t,s}}(R^1) = \left\{ \begin{array}{c} f(s,\omega) : f(s,\omega) \text{ is } \mathfrak{F}_{t,s} - \text{adapted}, R^d - \text{valued} \\ \text{such that } E\sup_{s\geqslant t}|f(s\wedge\tau,\omega)|^2 < \infty \end{array} \right\},$$

$$L^2_{\Im_{t,s}}(R^{d\otimes d_1}) = \left\{ \begin{array}{c} f(s,\omega) : f(s,\omega) \text{ is } \mathfrak{F}_{t,s} - \text{adapted}, R^{d\otimes r} - \text{valued} \\ \text{such that } E\int_{t\wedge\tau}^T |f(t,\omega)|^2 dt < \infty \end{array} \right\},$$

etc.
Then we have the following

Theorem 282 *Under assumption (C) and (D) for (f,ϕ,ψ) there exists a unique $(x_s,q_s,p_s) \in S^2_{\Im_{t,s}}(R^1) \times L^2_{\Im_{t,s}}(R^{1\otimes r}) \times F^2_{\Im_{t,s}}(R^1)$, which solves the* *following BSDE*

$$x_s = I_{\tau<T}\psi(\tau, y_\tau) + I_{\tau=T}\phi(y_T) - \int_{s\wedge\tau}^{\tau} f(r, y_r, x_r, q_r, p_r)dr - \int_{s\wedge\tau}^{\tau} q_r dw_r$$
$$- \int_{s\wedge\tau}^{\tau} \int_Z p_r(z)\tilde{N}_k(dr, dz), \text{ as } t \leqslant s \leqslant T, \quad (8.89)$$

where $y_s = y_s^{t,x}, t \leqslant s \leqslant T$, *is the solution of (8.88).*

Theorem 282 is true by virtue of Theorem 246 in this Chapter. Now by Ito's formula one finds that

$$u(s \wedge \tau_G, y_{s\wedge\tau_G}) - u(t, x) = \int_t^{s\wedge\tau_G} \pounds_{b,\sigma,c} u(r, y_r)dr$$
$$+ \int_t^{s\wedge\tau_G} u'_y(r, y_r) \cdot \sigma(r, y_r)dw_r$$
$$+ \int_t^{s\wedge\tau_G} (u(r, y_r + c(r, y_{r-}, z)) - u(r, y_r))\widetilde{N}_k(dr, dz), \quad (8.90)$$

Hence we can arrive at a new Feynman-Kac formula as follows:

Theorem 283 *(Feynman-Kac formula). Under assumption (C) and (D)*
$x_s = u(s, y_s), q_s = \sigma(s, y_s)\partial_x u(s, y_s),$
$p_s(\cdot) = u(s, y_{s-} + c(s, y_{s-}, \cdot)) - u(s, y_{s-}),$
where y_s *is the unique strong solution of the FSDE (8.88), and*
$(x_s, q_s, p_s) \in S^2_{\Im_{t,s}}(R^1) \times L^2_{\Im_{t,s}}(R^{1\otimes r}) \times F^2_{\Im_{t,s}}(R^1)$
is the unique solution of the BSDE (8.89). Hence x_s *has a "Markov property". Furthermore, we have a Feynman-Kac formula*
$u(t, x) = x_s|_{s=t} = E_{t,x}x_t,$
where $u(t, x) \in C^{1,2}([0, T] \times R^d, R^m)$ *is the unique classical solution of (8.85) and (8.86), if it exists, we denote by*
$E_{t,x}F(y_r, \omega) = EF(y_r, \omega) = EF(y_r^{t,x}, \omega).$

Proof. *Applying Ito's formula to* $u(r, y_r)$ *on* $r \in [s \wedge \tau, \tau]$, *and using Theorem 282 the conclusion is derived.* ■

Remark 284 . *We will say that* $x_s, t \leq s$, *has a "Markov property", if it can be expressed by some function of* y_s, *where* y_s *is a Markov process, i.e. there exist Borel functions* f_1 *such that* $x_s = f_1(s, y_s), \forall t \leq s$.

Naturally, we can also have another Feynman-Kac formula for the classical solution of (8.85) and (8.86), which only needs to know the existence of the solution of the FSDE (8.88). However, in this case the function f in the free term (in some cases we also call it a "force") cannot be so general. In fact, let us suppose that in all of the above discussion f only depends on t and x. That is, $f = f(t, x)$.In this case (8.85) becomes

$$\pounds_{b,\sigma,c} u(t, x) = f(t, x), \quad (8.91)$$

Assumptions (D) becomes (D)′:
D.1'. $f : [0,T] \times D \to R^1$ is bounded and measurable;
D.2'. the same as D.3;
D.3'. the same as D.4.
Then applying Ito's formula to $u(s, y_s)$ we immediately obtain the following theorem.

Theorem 285 *(Feynman-Kac formula). Under assumptions (C) and (D)′ if $u(t,x) \in C^{1,2}([0,T] \times R^d, R^m)$ is a classical solution of (8.91) and (8.86), then*

$$u(t,x) = E_{t,x}[I_{\tau<T}\psi(\tau, y_\tau) + I_{\tau=T}\phi(y_T) - \int_{t\wedge\tau}^{\tau} f(r, y_r)dr],$$

where $y_s, s \geq t$ is the unique strong solution of the FSDE (8.88).

Notice that the conditions in Theorem 285 are weaker than those in Theorem 283, so the conclusion of Theorem 283 still holds in this case. However, obtaining a Sobolev solution for (8.85) and (8.86) is much easier than finding a classical solution. So one natually asks can we have similar Feyman-Kac formulas for the Sobolev solutions of (8.85) and (8.86)? The answer is positive. For this we first change the condition on ϕ and ψ to satisfy the following Assumption (D)$''$
D.1$''$. $f : [0,T] \times D \times R^1 \times R^{1\otimes r} \times L^2_{\pi(.)}(R^1) \to R^1$ is measurable such that $g^u \in L_p([0,T] \times D)$,
where

$$g^u(t,x) = f(t,x,u(t,x), u'_x(t,x)\cdot\sigma(t,x), u(t,x+c(t,x,.)) - u(t,x)), \quad (8.92)$$

for all $u \in W^{1,2}_p([0,T] \times R^d)$ satisfying (8.86);
D.2$''$. The same as D.2.
D.3$''$. $\phi \in W_p^{2(1-\frac{1}{p})}(R^d)$; $(\phi : R^d \to R^1)$;
$\psi \in W^{1,2}_p([0,T] \times D^c)$; $(\psi : [0,T] \times D^c \to R^1)$;
D.4. $\partial D \in O^2$; (for the definition of O^2 see [68]).
Then we have the following theorem.

Theorem 286 *(Feynman-Kac formula). Under assumption (C) and (D)$''$ if $p > 2d + 4$ then*

$$x_s = u(s, y_s), \; q_s = \sigma(s, y_s)\partial_x u(s, y_s),$$
$$p_s(\cdot) = u(s, y_{s-} + c(s, y_{s-}, \cdot)) - u(s, y_{s-}),$$

where y_s is the unique strong solution of the FSDE (8.88), and

$$(x_s, q_s, p_s) \in S^2_{\Im}([t,\tau]; R^m) \times L^2_{\Im}([t,\tau]; R^{m\otimes d_1}) \times F^2_{\Im}([t,\tau]; R^{m\otimes d_2})$$

is the unique solution of the BSDE (8.89), which exists. Hence (x_s) has a "Markov property". Furthermore, we have a Feynman-Kac formula

$$u(t,x) = x_s\,|_{s=t} = E_{t,x}x_t,$$

where $u(t,x) \in W^{1,2}_p([0,T] \times R^d, R^m)$ is the unique Sobolev solution of (8.85) and (8.86), which exists.

However, to prove this theorem we first need to generalize the Ito formula to functions with Sobolev derivatives, and then to prove it.[153],[166] Moreover, we also need to explain the meaning of a Sobolev space.[68] These matters need more space than we have here, so they will not be discussed here.

9
Optimal Consumption by H-J-B Equation and Lagrange Method

In this chapter we will show how to use the forward Hamilton-Jacobi-Bellman (H-J-B) equation technique to solve the the optimal consumption problem. We also establish the necessary conditions for the optimal consumption by a stochastic Lagrange method, which also helps us to find the optimal consumption. Some examples for such applications are also given.

9.1 Optimal Consumption

In the last sections of the previous chapter we have already seen that in a financial market the wealth process of a small investor under appropriate conditions can be written as: $P - a.s.$

$$x_t = X + \int_t^T c(s)ds - \int_t^T \pi_s \sigma_s dw_s - \int_t^T \pi_s \rho_s d\widetilde{N}(s), \ t \in [0,T] \quad (9.1)$$

Suppose that the investor has a target functional

$J(c) = E[g(x_0) + \int_0^T f(s, x_s, \pi_s, c_s)ds]$,

and he has an admissible consumption set:

$U = \left\{ c = c(t) : c(t) \text{ is } \mathfrak{F}_t - \text{adapted such that } E\int_0^T |c(t)|^2 dt < \infty. \right\}$

He wants to find out an optimal consumption $c^0(.) \in U$ such that it can minimize the target functional:

$J(c^0) = \inf_{c(.) \in U} J(c)$.

Let us consider the following forward Hamilton-Jacobi-Bellman equation:

$$\inf_c[-\frac{\partial}{\partial t}V(t,x)+ c\frac{\partial}{\partial x}V(t,x)-\frac{\partial^2}{\partial x^2}V(t,x)\,|\pi\sigma_t|^2$$
$$-\sum_{k=1}^m(V(t,x+\sum_{i=1}^N\pi^i\rho_t^{ik})-V(t,x)-\frac{\partial V(t,x)}{\partial x}\sum_{i=1}^N\pi^i\rho_t^{ik})$$
$$+f(t,x,\pi,c)]=0,\ V(0,x)=g(x);\ \ 0\leqslant t\leqslant T; \tag{9.2}$$

where we assume that $f(t,x,\pi,c)$ is jointly measurable, and
$|\pi\sigma_t|^2=\sum_{k=1}^d\left|\sum_{i=1}^N\pi^i\sigma^{ik}\right|^2.$
(9.2) will give us an approach by which to find the optimal consumption if (9.2) has a solution $V(t,x)\in C^{1,2}$. In fact, we have the following

Theorem 287 *Suppose that (9.2) has a solution $V(t,x)\in C^{1,2}$, by the measurable version theorem[189] there exists a jointly measurable function $c^0=c^0(t,x,\pi)$ such that*

$-\frac{\partial}{\partial t}V(t,x)+\ c^0(t,x,\pi)\frac{\partial}{\partial x}V(t,x)-\frac{\partial^2}{\partial x^2}V(t,x)\,|\pi\sigma_t|^2$
$-\sum_{k=1}^m(V(t,x+\sum_{i=1}^N\pi^i\rho_t^{ik})-V(t,x)-\frac{\partial V(t,x)}{\partial x}\sum_{i=1}^N\pi^i\rho_t^{ik})$
$+f(t,x,\pi,c^0(t,x,\pi))$
$=\inf_c[-\frac{\partial}{\partial t}V(t,x)+\ c\frac{\partial}{\partial x}V(t,x)-\frac{\partial^2}{\partial x^2}V(t,x)(|\pi\sigma_t|^2$
$-\sum_{k=1}^m(V(t,x+\sum_{i=1}^N\pi^i\rho_t^{ik})-V(t,x)-\frac{\partial V(t,x)}{\partial x}\sum_{i=1}^N\pi^i\rho_t^{ik})$
$+f(t,x,\pi,c)].$

Moreover, if there exists a unique solution (x_t^0,π_t^0) satisfying
$x_t=X+\int_t^T c^0(s,x_s,\pi_s)ds-\int_t^T\pi_s\sigma_s dw_s-\int_t^T\pi_s\rho_s d\widetilde{N}(s),\ t\in[0,T],$
then x_t^0 and π_t^0 are the optimal trajectory and optimal portfolio, respectively, such that
$E[g(x_0^0)+\int_0^T f(s,x_s^0,\pi_s^0,c^0(s,x_s^0,\pi_s^0))ds]$
$=\inf_{c(.)\in U}E[g(x_0)+\int_0^T f(s,x_s,\pi_s,c_s)ds].$

Proof. Applying Ito's formula to $V(t,x_t^0),t\in[0,T]$, where (x_t^0,π_t^0) satisfies
$x_t^0=X+\int_t^T c^0(s,x_s^0,\pi_s^0)ds-\int_t^T\pi_s^0\sigma_s dw_s-\int_t^T\pi_s^0\rho_s d\widetilde{N}(s),\ t\in[0,T],$
one finds that
$E[V(T,X)-V(0,x_0^0)]=E\int_0^T f(s,x_s^0,\pi_s^0,c^0(s,x_s^0,\pi_s^0))ds,$
where we have applied the forward H-J-B equation (9.2). So
$E[V(T,X)=g(x_0^0)+E\int_0^T f(s,x_s^0,\pi_s^0,c^0(s,x_s^0,\pi_s^0))ds.$
Applying Ito's formula to $V(t,x_t^c),t\in[0,T]$, where (x_t^c,π_t^c) satisfies
$x_t^c=X+\int_t^T c_s ds-\int_t^T\pi_s^c\sigma_s dw_s-\int_t^T\pi_s^c\rho_s d\widetilde{N}(s),\ t\in[0,T],$
one also has that $\forall c(.)\in U$
$E[V(T,X)-V(0,x_0^c)]\leqslant E\int_0^T f(s,x_s^c,\pi_s^c,c_s)ds.$
where we have also applied the forward H-J-B equation (9.2). So
$E[V(T,X)\leq g(x_0^c)+E\int_0^T f(s,x_s^c,\pi_s^c,c_s)ds.$
The conclusion follows. ■

In general the solution of (9.2) is not easy to obtain. However, for some special cases we can solve the problem. Let us always assume that $\widetilde{\sigma}_t^{-1} = [\sigma_t, \rho_t]^{-1}$ exists and is bounded. Write

$$U = \left\{ \begin{array}{c} c = c(t,x,\pi) : c(t,x,\pi) \text{ is jointly measurable such that} \\ (9.1) \text{ has a unique solution } (x_t, \pi_t), \text{ and } |c(t,x,\pi)| \leq |x|^\beta \end{array} \right\},$$

where $0 < \beta \leq 1$ is a given fixed constant, and U is an admissible consumption set. The following Theorem shows that a non-Lipschitzian feedback optimal stochastic consumption exists.

Theorem 288 . *Define* $c^0(x) = c^0(t,x,\pi) = -I_{x\neq 0}x/|x|^{1-\beta}$ *.and let*

$$J(c) = E(\tfrac{1}{2}|x_0^c|^2 + \tfrac{1}{2}\int_0^T e^{-\alpha s}(|\pi_s^c \sigma_s|^2 + \sum_{k=1}^m \left|\sum_{i=1}^N \pi_s^{ci}\rho_s^{ik}\right|^2)ds$$
$$+\int_0^T e^{-\alpha s}(-\tfrac{\alpha}{2}|x_s^c|^2 + |x_s^c|^{1+\beta})ds),$$

where α *is a constant,* (x_t^c, π_t^c) *is the unique solution of (9.1) for* $c \in U$. *Then*

1) $c^0 \in U$,

2) $J(c) \geq J(c^0)$, *for all* $c \in U$.

The above target functional $J(c)$ can be explained as an energy functional, provided $\alpha \leqslant 0$. To show Theorem 288 we need the following lemma.

Lemma 289 $V(t,x) = \frac{1}{2}e^{-\alpha t}|x|^2 + t$ *satisfies*

$$-(\partial/\partial t)V + \inf_{u\in R^1, |u|\leqslant |x|^\beta}(u\cdot \tfrac{\partial}{\partial x}V) - \tfrac{1}{2}ae^{-\alpha t}|x|^2 + 1 + e^{-\alpha t}|x|^{1+\beta} = 0,$$
$$V(0,x) = \tfrac{1}{2}|x|^2, 0 \leqslant t \leqslant T.$$

Moreover, if $u^0(x) = -I_{x\neq 0}x/|x|^{1-\beta}$, *then* $|u^0(x)| \leq |x|^\beta$, *and*

$$u^0(x)\cdot \tfrac{\partial}{\partial x}V(t,x) = \inf_{u\in R^1, |u|\leqslant |x|^\beta}(u\cdot \tfrac{\partial}{\partial x}V) = -e^{-\alpha t}|x|^{1+\beta}.$$

Lemma 289 can be checked directly. Now to show Theorem 288 one only needs to apply Ito's formula to $V(t, x_t^c)$ and $V(t, x_t^0)$, where x_t^c satisfies (9.1), and x_t^0 satisfies (9.1) with $c^0(x)$, respectively, and then using Lemma 289, in exactly the same way as in the proof of Theorem 274, one arrives at the conclusion of Theorem 288.

We can also consider some other target functional similar to Theorem 275, but to save space we omit it.

9.2 Optimization for a Financial Market with Jumps by the Lagrange Method

In this section we discuss the optimal consumption problems for a financial market with jumps. We provide a proof of the necessary conditions for optimal consumption by using a Lagrange method for models driven by Brownian motions as well as Poisson jump processes. Such a method was introduced by Chow (1997) for continuous models.

9.2.1 *Introduction*

As we know, the optimal consumption problems for financial models described by stochastic differential equations driven by Brownian motion can be discussed by many methods, e.g. by the martingale method (Karatzas et al., 1991), by the backward stochastic differential equation technique (El Karoui et al., 1997), by the dynamic programming method - Bellman equation (Cox et al., 1985), and others (Merton, 1971). However, for practical use more direct ways which can be easier to evaluate are preferred. In this sense the Lagrange method may be an appropriate one (Chow, 1997). Because by this method the necessary conditions are much easier to derive, and the partial differential equations in necessary conditions for finding the optimal consumption are easier to handle in many cases. The idea of deriving the necessary conditions is rather intuitive and simple: Suppose we know that a utility functional arrives at its maximum at some optimal consumption process, which with its corresponding optimal wealth process are subject to some stochastic differential equation. What are its necessary conditions? Obviously, it is a question like the maximum problem with constraints in the Calculus. Hence we try to introduce a Lagrange multiplier to recast this problem into a maximum problem without constraints. Then, after calculating the directional derivatives of the recast functional at the maximum point, we obtain the necessary conditions for optimal consumption. Here the only difficulty to be overcome by us is to choose an appropriate Lagrange multiplier, to reconstruct a new utility functional without constraints correctly, and to use the Ito formula carefully.

9.2.2 *Models*

Suppose that in a financial market there is one kind of bond and m kinds of stock, and their price processes satisfy the following stochastic differential equations (SDE in short), respectively (Bardhan and Chao, 1993): $i = 1, 2, ..., m$; $t \geqslant 0$,

$$\begin{aligned} dP_0(t) &= P_0(t)r(t)dt, \ P_0(0) = 1; \\ dP_i(t) &= P_i(t-)(b_i(t) + \sum_{j=1}^{d} \sigma_{ij}(t)dw_j(t) + \sum_{j=1}^{m-d} \rho_{ij}(t)d\widetilde{Q}_j(t)), \quad (9.3) \end{aligned}$$

where $w(t) = (w_1(t), \cdots, w_d(t))^T$ is a $d-$dimensional standard Brownian motion process (BM), (α^T means the transpose of α); $\widetilde{Q}(t) = (\widetilde{Q}_1(t), \cdots, \widetilde{Q}_{m-d}(t))^T$ is a $m - d$ - dimensional centralized random point process, i.e.
$d\widetilde{Q}_i(t) = dQ_i(t) - \nu_i(t, \omega)dt$,
where $Q_i(t)$ is a random point process with jump density $\nu_i(t, \omega)$, i.e. $E\, Q_i(t) = E \int_0^t \nu_i(t, \omega)dt$. The Model (9.3) can be explained as follows: $P_0(t)$ is the price of the bond at time $t, P_i(t)$ is the price of the ith stock at

time t, $r(t)$ is the instantaneous rate of interest, $b(t) = (b_1(t), \cdots, b_m(t))$ is the vector of instantaneous appreciation rate for the stocks, and $\widetilde{\sigma}(t) = [\sigma(t), \rho(t)]$ is the $m \times m$ volatility matrix process. In particular, the jump term can be explained as the stochastic jump perturbation caused by, for example, the announcement of some important economic policy to this financial market, the performance of an important decision made by some big company, or the occurrence of some big accident like an earthquake, a big fire and others. Obviously, if one sets $m = d$, then (9.3) reduces to the usual continuous financial market.

Suppose now a small investor with initial capital $x > 0$ uses his wealth for consumption and investment in the financial market. His investment policy is described by a portfolio process $\pi(t), 0 \leqslant t$, an R^m-valued process that represents the dollar investment that the investor maintains in the m stocks. Assume that his consumption at time $t \geqslant 0$ is $C(t) \geqslant 0$, with $C(0) = 0$. Then, after a simple calculation the investor's wealth process will satisfy the following SDE (Bardhan and Chao, 1993): $t \geqslant 0$,

$$dx_t = r_t x_t dt - dC_t + \pi_t^T \sigma_t dw_t + \pi_t^T \rho_t d\widetilde{Q}_t, x_0 = x, \tag{9.4}$$

where $\delta_t = (\delta_{1t}, \cdots, \delta_{mt})^T$ is a dividend rate process, $\underline{1} = [1, \cdots, 1]^T$ is the $m-$dimensional vector with all components equaling 1. The investor would like to get a maximum utility from his investment. Let the utility U be a function of the consumption C_t and the wealth x_t, and solve the following optimization problem: Find C_t such that

$$E\int_0^\infty e^{-\beta t}U(x_t, C_t)dt = \sup_{C(.)} E\int_0^\infty e^{-\beta t}U(x_t, C_t)dt, \tag{9.5}$$

where $\beta > 0$ is some discount factor, one can get the optimal consumption process. To solve problem (9.4) and (9.5) let us consider a more general optimal stochastic control problem: Find $u(.) \in \widetilde{U}$ such that

$$E\int_0^\infty e^{-\int_0^t \beta_s ds}U(t, x_t, u_t)dt = \sup_{u(.)\in\widetilde{U}} E\int_0^\infty e^{-\int_0^t \beta_s ds}U(t, x_t, u_t)dt, \tag{9.6}$$

where $\beta_t \geqslant 0$ is a non-random and bounded function, the admissible control set is

$$\widetilde{U}^r = \left\{ \begin{array}{l} u(.) : u(t,\omega) \in R^r \text{ is a } \mathfrak{F}_t\text{-predictable process such} \\ \text{that } E\int_0^T |u_t|^2 dt < \infty, \text{ for arbitrary } 0 \leqslant T < \infty \end{array} \right\} \tag{9.7}$$

and x_t satisfies the following SDE: for all $u(.) \in \widetilde{U}$, denote $x_t = x_t^u$,

$$\begin{aligned} dx_t &= f(t, x_t, u_t)dt + S^{(1)}(t, x_t, u_t)dw_t + S^{(2)}(t, x_{t-}, u_t)d\widetilde{Q}_t, \quad (9.8) \\ x_0 &= x, \ t \geqslant 0; \end{aligned}$$

where $w_t = (w_{1t}, \cdots, w_{d_1 t})^T$ is a d_1–dimensional BM, $\widetilde{Q}_t = (\widetilde{Q}_{1t}, \cdots, \widetilde{Q}_{d_2 t})^T$ is, without loss of generality, a d_2–dimensional centralized random point process satisfying

$d\widetilde{Q}_{it} = dQ_{it} - dt,\ 1 \leqslant i \leqslant d_2,$

i.e. Q_{it} is a random point process with jump density 1 ($E\ Q_i(t) = \int_0^t ds = t$); and $x_t = (x_{1t}, \cdots, x_{nt})^T \in R^n, S^{(1)} \in R^{n \times d_1}, S^{(2)} \in R^{n \times d_2}$,and we assume that $d_1 + d_2 = n$.

9.2.3 *Main Theorem and Proof*

To solve problem (9.6)-(9.8) let us introduce a Lagrange functional as follows: for any $u(.) \in \widetilde{U}$

$$L = E\int_0^\infty \{e^{-\int_0^t \beta_s ds} U(t, x_t, u_t)dt - e^{-\int_0^t \beta_s ds}\lambda(t,\omega)^T dy_t - e^{-\int_0^t \beta_s ds} d[\lambda^T, y]_t\}, \tag{9.9}$$

where $\lambda(t,\omega) \in R^n$ is the Lagrange multiplier process, which is assumed to be a right continuous with left limit (RCLL in short) semi-martingale, and

$y_t = x_t - x - \int_0^t f(s, x_s, u_s)ds - \int_0^t S^{(1)}(s, x_s, u_s)dw_s - \int_0^t S^{(2)}(s, x_{s-}, u_s)d\widetilde{Q}_s,$

where x_t is also a RCLL semi-martingale. Here we denote the quadratic variation process connected with the RCLL semi-martingales M_t and N_t by $[M^T, N]_t$, which is defined by

$[M^T, N]_t = \langle M^{cT}, N^c \rangle_t + \sum_{0<s\leqslant t} \triangle M_s^T \triangle N_s,$

where M^c is the continuous martingale part of M, $\triangle M_s = M_s - M_{s-}$. Then by Ito's formula (Liptser and Shiryayev, 1989)

$d(e^{-\int_0^t \beta_s ds}\lambda(t,\omega)^T x_t) = e^{-\int_0^t \beta_s ds}(-\beta_t \lambda_t^T x_t dt + \lambda_t^T dx_t$
$+ x_t^T d\lambda_t + d[\lambda^T, x]_t).$

Hence (9.9) can be rewritten as follows:

$L = E\int_0^\infty e^{-\int_0^t \beta_s ds}\{\ (U(t, x_t, u_t) + \lambda_t^T f(t, x_t, u_t) - \beta_t \lambda_t^T x_t)dt$
$+ d[\lambda(.)^T, \int_0^{\cdot} S^{(1)}(s, x_s, u_s)dw_s + \int_0^{\cdot} S^{(2)}(s, x_s, u_s)d\widetilde{Q}_s]_t$
$+ x_t^T d\lambda_t\ \} + E(\lambda_0^T x) - \lim_{t\to+\infty} Ee^{-\int_0^t \beta_s ds}\lambda_t^T x_t,$

or

$$\begin{aligned} L &= E\int_0^\infty e^{-\int_0^t \beta_s ds}\{\ (U(t, x_t, u_t) + \lambda_t^T f(t, x_t, u_t) - \beta_t \lambda_t^T x_t)dt \\ &\quad + S^{(1)}(t, x_t, u_t)d\langle \lambda^c(.)^T, w.\rangle_t + S^{(2)}(t, x_{t-}, u_t)d[\lambda^d(.)^T, \widetilde{Q}.]_t \\ &\quad + x_t^T d\lambda_t\ \} + E(\lambda_0^T x) - \lim_{t\to+\infty} Ee^{-\int_0^t \beta_s ds}\lambda_t^T x_t, \end{aligned} \tag{9.10}$$

where λ^c and λ^d is the continuous martingale part and the pure discontinuous martingale part of λ, respectively, and we make the following assumption:

Assumption A. 1. $\lim_{t\to+\infty} Ee^{-\int_0^t \beta_s ds}\lambda_t^T x_t$ exists and is finite;

2. $U, f, S^{(1)}, S^{(2)}$ have all bounded, continuous first and second partial derivatives.

Now after a simple calculation one obtains by (9.10) for arbitrary $h \in \widetilde{U}^r$

$$\frac{\partial L}{\partial u(.)}(u;h) \widehat{=} \lim_{\varepsilon \to 0} \frac{L(u+\varepsilon h) - L(u)}{\varepsilon}$$

$$= E \int_0^\infty e^{-\int_0^t \beta_s ds} \{ h_t^T (\frac{\partial U}{\partial u}(t, x_t, u_t) + \frac{\partial f}{\partial u}(t, x_t, u_t)^T \lambda_t) dt$$

$$+ d[\lambda(.)^T, \int_0^\cdot h_s^T (\frac{\partial S^{(1)}(s, x_s, u_s)}{\partial u} dw_s + \frac{\partial S^{(2)}(s, x_{s-}, u_s)}{\partial u} d\widetilde{Q}_s)]_t \}; \quad (9.11)$$

for arbitrary $\widetilde{h} \in \widetilde{U}^n$

$$\frac{\partial L}{\partial x(.)}(x; \widetilde{h}) = E \int_0^\infty e^{-\int_0^t \beta_s ds} \{ \widetilde{h}_t^T (\frac{\partial U}{\partial x}(t, x_t, u_t) + \frac{\partial f}{\partial x}(t, x_t, u_t)^T \lambda_t$$

$$- \beta_t \lambda_t) dt + \widetilde{h}_t^T d\lambda_t + d[\lambda(.)^T, \int_0^\cdot \widetilde{h}_s^T (\frac{\partial S^{(1)}(s, x_s, u_s)}{\partial x} dw_s$$

$$+ \frac{\partial S^{(2)}(s, x_{s-}, u_s)}{\partial x} d\widetilde{Q}_s)]_t \} + E\lambda_0^T \tilde{h}_0 - \lim_{t \to +\infty} E e^{-\int_0^t \beta_s ds} \lambda_t^T \tilde{h}_t; \quad (9.12)$$

and by (9.9) for arbitrary $\overline{h} \in \widetilde{U}^n$

$$\frac{\partial L}{\partial \lambda(.)}(\lambda; \overline{h}) = -E \int_0^\infty e^{-\int_0^t \beta_s ds} \{ \overline{h}_t^T (dx_t - f(t, x_t, u_t) dt$$

$$- S^{(1)}(t, x_t, u_t) dw_t + S^{(2)}(t, x_{t-}, u_t) d\widetilde{Q}_t) + d[\overline{h}_\cdot^T, x_\cdot - x - \int_0^\cdot f(s, x_s, u_s) ds$$

$$- \int_0^\cdot S^{(1)}(s, x_s, u_s) dw - \int_0^\cdot S^{(2)}(s, x_{s-}, u_s) d\widetilde{Q}_s]_t \}; \quad (9.13)$$

where we denote

$$h_s^T \frac{\partial S^{(1)}(s,x_s,u_s)}{\partial u} = \begin{bmatrix} h_s^T \frac{\partial S_{11}^{(1)}(s,x_s,u_s)}{\partial u} & \cdots & h_s^T \frac{\partial S_{1d_1}^{(1)}(s,x_s,u_s)}{\partial u} \\ \cdots & \cdots & \cdots \\ h_s^T \frac{\partial S_{n1}^{(1)}(s,x_s,u_s)}{\partial u} & \cdots & h_s^T \frac{\partial S_{nd_1}^{(1)}(s,x_s,u_s)}{\partial u} \end{bmatrix}, \text{ etc.}$$

Suppose that $(u^0(\cdot), x^0(\cdot), \lambda^0(\cdot))$ is a maximum point of L defined by (9.9), then from (9.11)-(9.13) one finds that

$\frac{\partial L}{\partial u(.)}(u^0; h) \leqslant 0$, $\frac{\partial L}{\partial x(.)}(x^0; \widetilde{h}) \leqslant 0$, $\frac{\partial L}{\partial \lambda(.)}(\lambda^0; \overline{h}) \leqslant 0$,
for all $(h, \widetilde{h}, \overline{h}) \in \widetilde{U}^r \times \widetilde{U}^n \times \widetilde{U}^n$. Hence one has that for all $(h, \widetilde{h}, \overline{h}) \in \widetilde{U}^r \times \widetilde{U}^n \times \widetilde{U}^n$

$$\text{The right side of (9.11), (9.12) and (9.13)} \,\Big|_{(u(\cdot),x(\cdot),\lambda(\cdot))=(u^0(\cdot),x^0(\cdot),\lambda^0(\cdot))} = 0. \quad (9.14)$$

Therefore, by appropriately choosing $\bar{h}$, one has that by (9.13)

$$x_t^0 = x + \int_0^t f(s, x_s^0, u_s^0)ds + \int_0^t S^{(1)}(s, x_s^0, u_s^0)dw_s + \int_0^t S^{(2)}(s, x_{s-}^0, u_s^0)d\widetilde{Q}_s. \tag{9.15}$$

Now assume that $\lambda_t^0 = \lambda^0(t, x_t^0), \lambda^0(t, x) \in C^{1,2}$. Moreover, for the more precise expression of the necessary condition on the maximum point of L let us make the following assumption:

$$S_{ij}^{(2)}(t, x, u)S_{ij'}^{(2)}(t, x, u) = 0, \text{ as } j \neq j', \tag{9.16}$$

where $S^{(2)}(t, x, u) = \left[S_{ij}^{(2)}\right]_{i,j=1}^{n,d_2}$.

Then by Ito's formula for $\lambda^0(t, x_t^0)$ and the arbitrariness of $(h, \widetilde{h}) \in \widetilde{U}^r \times \widetilde{U}^n$ one obtains that *a.s. a.e.*

$$\begin{cases} \frac{\partial U}{\partial u_i}(t, x_t^0, u_t^0) + \frac{\partial f}{\partial u_i}(t, x_t^0, u_t^0)^T \lambda_t^0 + tr(\frac{\partial S^{(1)}(t,x_t^0,u_t^0)^T}{\partial u_i} \frac{\partial \lambda^0}{\partial x^T} S^{(1)}(t, x_t^0, u_t^0) \\ + \frac{\partial S^{(2)}(t,x_t^0,u_t^0)^T}{\partial u_i} \frac{\partial \lambda^0}{\partial x^T} S^{(2)}(t, x_t^0, u_t^0)) = 0, \ 1 \leqslant i \leqslant r, \\ \frac{\partial U}{\partial x_i}(t, x_t^0, u_t^0) + \frac{\partial f}{\partial x_i}(t, x_t^0, u_t^0)^T \lambda_t^0 + \frac{\partial \lambda_i^0}{\partial x^T}^T f(t, x_t^0, u_t^0) - \beta_t \lambda_{it}^0 + \frac{\partial \lambda_i^0}{\partial t} \\ + \frac{1}{2} tr(\frac{\partial^2 \lambda_i^0}{\partial x \partial x^T} S^{(1)}(t, x_t^0, u_t^0) S^{(1)}(t, x_t^0, u_t^0)^T) + tr(\frac{\partial S^{(1)}(t,x_t^0,u_t^0)^T}{\partial x_i} \frac{\partial \lambda^0}{\partial x^T} \\ \cdot S^{(1)}(t, x_t^0, u_t^0) + \frac{\partial S^{(2)}(t,x_t^0,u_t^0)^T}{\partial x_i} \frac{\partial \lambda^0}{\partial x^T} S^{(2)}(t, x_t^0, u_t^0)) \\ + \lambda_i^0(t, x_t^0 + S^{(2)}(t, x_t^0, u_t^0)) - \lambda_i^0(t, x_t^0) \\ - \sum_{k,j=1}^{n,d_2} \frac{\partial \lambda_i^0(t,x_t^0)}{\partial x_k} S_{kj}^{(2)}(t, x_t^0, u_t^0) = 0, \ 1 \leqslant i \leqslant n. \end{cases} \tag{9.17}$$

From above we have proved the following

Theorem 290 . *Consider the optimization problem (9.6) - (9.8). Assume that Assumption A.2 holds, (x_t^0, λ_t^0) satisfies Assumption A.1. Then we have the following conclusions:*

1) If $(u^0(\cdot), x^0(\cdot), \lambda^0(\cdot))$ is a maximum point of L defined by (9.9), then (9.14) and (9.15) hold. Hence $(u^0(\cdot), x^0(\cdot))$ is also a maximum point of (9.6) - (9.8).

2) Under assumption in 1) if (9.16) holds and $\lambda_t^0 = \lambda^0(t, x_t^0)$, $\lambda^0(t, x) \in C^{1,2}$, then (9.17) holds.

Furthermore, we can also show that under appropriate conditions for a maximum point $(u^0(\cdot), x^0(\cdot))$ of (9.6) - (9.8) there will exist a $\lambda^0(\cdot)$ such that they still satisfy (9.17). However, $(u^0(\cdot), x^0(\cdot), \lambda^0(\cdot))$ may not be a maximum point of L defined by (9.9). We now need to use the conception of a regular point (Curtain and Pritchard, 1977). Assume that X, Y are two Banach spaces, $\widehat{\Omega} \subset X$ is an open subset, $A : \widehat{\Omega} \to Y$ is Fréchet differentiable with continuous Fréchet differential dA. $x_0 \in \widehat{\Omega}$ is called a regular point of A, iff $dA(x_0)$ maps $\widehat{\Omega}$ onto Y. Set

$$X = \left\{ \begin{array}{c} (x(.), u(.)) : x_t \text{ is } \mathfrak{F}_t - \text{adapted}, \ u_t \text{ is } \mathfrak{F}_t - \text{predictable} \\ \text{such that } E\int_0^\infty e^{-\int_0^t \beta_s ds}(|x_t|^2 + |u_t|^2)dt < \infty, \end{array} \right\}$$

where $0 < \delta_0 \leqslant \beta_s \leqslant k_0$ is non-random, k_0 and δ_0 are constants, and write

$y_t = y(x(.), u(.))(t) = x_t - x_0 - \int_0^t f(s, x_s, u_s)ds$
$- \int_0^t S^{(1)}(s, x_s, u_s)dw_s - \int_0^t S^{(2)}(s, x_{s-}, u_s)d\widetilde{Q}_s,$

$Y = \left\{ y(.) : y_t \text{ is } \mathfrak{F}_t - \text{adapted such that } E\int_0^\infty e^{-\int_0^t \beta_s ds} |y_t|^2 dt < \infty \right\}.$

We have the following

Theorem 291 . *Assume that Assumption A.2 holds, $(x^0(\cdot), u^0(\cdot))$ is a maximum point of (9.6) - (9.8), and assume that $(x^0(\cdot), u^0(\cdot))$ is a regular point of $y(x(.), u(.))$. Then,*

1) there exists a $z^0(.) \in Y$ such that $dL(x^0(\cdot), u^0(\cdot)) = 0$, where $L(x(\cdot), u(\cdot)) = E\int_0^\infty e^{-\int_0^t \beta_s ds}(U(t, x_t, u_t) - z^0(t)^T y_t)dt$;

2) in addition, if there exists a neighborhood $\widetilde{V}_{(x^0(\cdot),u^0(\cdot))}$ of $(x^0(\cdot), u^0(\cdot))$ such that as $(x(\cdot), u(\cdot)) \in \widetilde{V}_{(x^0(\cdot),u^0(\cdot))}$, $\lim_{t\to\infty} E(e^{-\int_0^t \beta_s ds} y(x(.), u(.))(t)^T \lambda_t^0)$ exists, where $\lambda_t^0 = e^{\int_0^t \beta_s ds} E[\int_t^\infty e^{-\int_0^s \beta_r dr} z^0(s)ds \mid \mathfrak{F}_t]$, then

$L(x(\cdot), u(\cdot)) = E\int_0^\infty e^{-\int_0^t \beta_s ds}\{U(t, x_t, u_t)dt - \lambda^0(t, \omega)^T dy_t - d[(\lambda^0)^T, y]_t\}$
$- \lim_{t\to\infty} E(e^{-\int_0^t \beta_s ds} y(x(.), u(.))(t)^T \lambda_t^0).$

Therefore, in case 2) if (x_t^0, λ_t^0) satisfies Assumption A.1,(9.16) holds, and $\lambda_t^0 = \lambda^0(t, x_t^0)$, $\lambda^0(t, x) \in C^{1,2}$, then (9.17) is still derived.

Proof. 1) is derived from Theorem 12.3 of (Curtain and Pritchard, 1977)[21] and the Riesz's representation theorem for a linear functional, since now Y is a Hilbert space. Let us show 2): By the martingale representation theorem (Theorem 103) there exists a $\mathfrak{F}_t = \mathfrak{F}_t^{w,\widetilde{Q}}$−predictable processes b_t and a c_t satisfying $E\int_0^T (|b_t|^2 + |c_t|^2)dt < \infty$, for all $T < \infty$, such that

$E[\int_0^\infty e^{-\int_0^t \beta_s ds} z^0(t)dt \mid \mathfrak{F}_t] = E\int_0^\infty e^{-\int_0^t \beta_s ds} z^0(t)dt + \int_0^t b_s dw_s + \int_0^t c_s d\widetilde{Q}_s,$

for all $t \geqslant 0$. Write $\widetilde{\lambda_t^0} = E[\int_t^\infty e^{-\int_0^s \beta_r dr} z^0(s)ds \mid \mathfrak{F}_t]$. Then

$\widetilde{\lambda_t^0} = -\int_0^t e^{-\int_0^s \beta_r dr} z^0(s)ds + E\int_0^\infty e^{-\int_0^t \beta_s ds} z^0(t)dt + \int_0^t b_s dw_s + \int_0^t c_s d\widetilde{Q}_s.$

Hence

$L(x(\cdot), u(\cdot)) = E\int_0^\infty (e^{-\int_0^t \beta_s ds} U(t, x_t, u_t)dt + y_t^T d\widetilde{\lambda_t^0}).$

By an application of Ito's formula to $y_t^T \lambda_t^0$ the conclusion of 2) is derived.
■

For the practical evaluation one can solve the following equation

$$\begin{cases} \frac{\partial U}{\partial u_i}(t,x,u) + \frac{\partial f}{\partial u_i}(t,x,u)^T\lambda + tr(\frac{\partial S^{(1)}(t,x,u)^T}{\partial u_i}\frac{\partial \lambda}{\partial x^T}S^{(1)}(t,x,u) \\ \quad +\frac{\partial S^{(2)}(t,x,u)^T}{\partial u_i}\frac{\partial \lambda}{\partial x^T}S^{(2)}(t,x,u)) = 0, \ 1 \leqslant i \leqslant r, \\ \frac{\partial U}{\partial x_i}(t,x,u) + \frac{\partial f}{\partial x_i}(t,x,u)^T\lambda - \beta_t\lambda_i + \frac{\partial \lambda_i^T}{\partial x^T}f(t,x,u) \\ +\frac{1}{2}tr(\frac{\partial^2 \lambda_i}{\partial x \partial x^T}S^{(1)}(t,x,u)S^{(1)}(t,x,u)^T) + tr(\frac{\partial S^{(1)}(t,x,u)^T}{\partial x_i}\frac{\partial \lambda}{\partial x^T}S^{(1)}(t,x,u) \\ +\frac{\partial S^{(2)}(t,x,u)^T}{\partial x_i}\frac{\partial \lambda}{\partial x^T}S^{(2)}(t,x,u)) + \frac{\partial \lambda_i}{\partial t} + (\lambda_i(t,x+S^{(2)}(t,x,u)) - \lambda_i(t,x) \\ \quad - \sum_{k,j=1}^{n,d_2}\frac{\partial \lambda_i(t,x)}{\partial x_k}S_{kj}^{(2)}(t,x,u) = 0, \ 1 \leqslant i \leqslant n. \end{cases} \tag{9.18}$$

From the solution $u = u(t,x)$ of (9.18) one solves the following feedback SDE

$$\begin{aligned} dx_t &= f(t,x_t,u(t,x_t))dt + S^{(1)}(t,x_t,u(t,x_t))dw_t \\ &+S^{(2)}(t,x_{t-},u(t,x_{t-}))d\widetilde{Q}_t, \ x_0 = x, \ t \geqslant 0; \end{aligned} \tag{9.19}$$

to get the solution x_t. Then the feedback control $u(t,x_{t-})$ may be an optimal control of (9.6) - (9.8).

Note that if the coefficients $f(x,u), S^{(1)}(x,u)$ and $S^{(2)}(x,u)$ of (9.8) and the utility function $U(x,u)$ do not depend on t, then in 2) of theorem 1 we only need to assume that the optimal Lagrange multiplier λ^0 depends on x, i.e. $\lambda^0 = \lambda^0(x_t^0)$, $\lambda^0 \in C^2$. In this case in (9.17) and (9.18) the term $\frac{\partial \lambda}{\partial t}$ disappears and the equation becomes simpler.

9.2.4 Applications

In what follows we are going to give some simple examples based on the financial market with only jump stochastic perturbation, and the optimal consumption problem for special utility functions $U(c) = c^\delta/\delta$, $\delta < 1, \delta \neq 0$;and $U(c) = \log c$, which are considered in Karatzas, et al. 1991, for a continuous financial market.

Example 292 . *Consider a simple financial market with jumps. In (9.3) assume that* $b_i(t) = r_t, i = 1,2,\cdots,m$; $d = 0$, *and in (9.4) assume that* $\delta_t \equiv 0$, $dC_t = c_t dt$, *i.e. assume the wealth process (9.4) satisfies the following 1-dimensional SDE:*

$$dx_t = r_t x_t dt - c_t dt + \pi_t^T \rho_t d\widetilde{Q}_t, x_0 = x, t \geqslant 0. \tag{9.20}$$

where $\widetilde{Q}_t^T = (\widetilde{Q}_{1t},...,\widetilde{Q}_{d_2t})$ *is a* $m-$*dimensional centralized Poisson process such that*
$d\widetilde{Q}_{it} = dQ_{it} - dt, 1 \leqslant i \leqslant m,$

that is, Q_{it} is a Poisson process with jump density 1 ($EQ_{it} = \int_0^t ds = t$), and we also assume that any two $Q_{it}, Q_{jt}, i \neq j$, have no common jump time. Naturally, we assume that $\rho_t \in R^{m\otimes m}$, which is bounded, is deterministic and its inverse matrix ρ_t^{-1} exists and is also bounded. Our object is to find

$\sup_{(c(.),\pi(.))\in\widetilde{U}} E\int_0^\infty e^{-\int_0^t \beta_s ds} U(t, x_t, c_t)dt,$
where
$U(t,x,c) = c^\delta/\delta, \delta < 1, \delta \neq 0,$
$\overline{k}_0 \geqslant \beta_s \geqslant r_s \geqslant \delta_0 > 0, \beta_s - \delta r_s \geqslant \delta_0 > 0;$
δ_0 is a constant, and the admissible control set is

$$\widetilde{U} = \left\{ \begin{array}{c} (c(t),\pi(t)) : \text{they are } \mathfrak{F}_t\text{-predictable, } c(t) \geqslant 0 \text{ and} \\ \pi(t)^T = (\ \pi_1(t),\cdots,\pi_m(t)) \text{ satisfy that for any } 0 \leqslant T < \infty \\ E\int_0^T (|\pi_s|^2 + c(s)^2)ds < \infty, \text{ such that there exists a} \\ \text{solution } x_t \text{ of (9.20) satisfying that } x_t\ \geqslant 0, \text{ for all } t \geqslant 0. \end{array} \right\}$$

To simplify (9.20) let us set
$\widetilde{x}_t = e^{-\int_0^t r_s ds} x_t.$
Then (9.20) becomes

$$d\widetilde{x}_t = -e^{-\int_0^t r_s ds} c_t dt + e^{-\int_0^t r_s ds} \pi_t^T \rho_t d\widetilde{Q}_t, \widetilde{x}_0 = x,\ t \geqslant 0. \tag{9.21}$$

Since $U(s,\widetilde{x}_s,c_s) = U(s,x_s,c_s)$, the problem becomes finding
$\sup_{(c(.),\pi(.))\in\widetilde{U}'} E\int_0^\infty e^{-\int_0^s \beta_r dr} U(s,\widetilde{x}_s,c_s)ds,$
where

$$\widetilde{U}' = \left\{ \begin{array}{c} (c(t),\pi(t)) : \text{they are } \mathfrak{F}_t\text{-predictable, } c(t) \geqslant 0 \text{ and} \\ \pi(t)\ = (\ \pi_1(t),\cdots,\pi_m(t)) \text{ satisfy that for any } 0 \leqslant T < \infty \\ E\int_0^T (|\pi_s|^2 + c(s)^2)ds < \infty, \text{ such that there exists a} \\ \text{solution } \widetilde{x}_t \text{ of (9.21) satisfying that } \widetilde{x}_t\ \geqslant 0, \text{ for all } t \geqslant 0. \end{array} \right\}$$

Now by (9.18) one finds that if
$\rho_{ij}(t)\rho_{ij'}(t) = 0, for j \neq j',$
then

$$\left\{ \begin{array}{l} c^{\delta-1} - e^{-\int_0^t r_s ds}\lambda(t,x) = 0, \\ \lambda_x'(t,x)e^{-2\int_0^t r_s ds}\sum_{i,j=1}^m \rho_{i_0 j}(t)\rho_{ij}(t)\pi_i = 0,\ i_0 = 1,2,...,m; \\ -\beta_t\lambda(t,x) - \lambda_x'(t,x)e^{-\int_0^t r_s ds}c + \lambda(t, x + e^{-\int_0^t r_s ds}\sum_{i,j=1}^m \pi_i\rho_{ij}(t)) \\ -\lambda(t,x)\ - \lambda_x'(t,x)e^{-\int_0^t r_s ds}\sum_{i,j=1}^m \pi_i\rho_{ij}(t) + \lambda_t'(t,x) = 0. \end{array} \right. \tag{9.22}$$

In the special case

$$m = 1 \tag{9.23}$$

we can get a more precise answer. Indeed, solving (9.22), one has that

$$\lambda(t,x) = e^{\int_0^t \beta_s ds} k_0, t \geqslant 0,$$

$$c = c(t,x) = (e^{-\int_0^t r_s ds}\lambda)^{-1/(1-\delta)} = (e^{\int_0^t (\beta_s - r_s)ds}k_0)^{-1/(1-\delta)}, \tag{9.24}$$

where k_0 is any constant. To show that the optimal consumption process c_t^0 and optimal portfolio process π_t^0 do exist, we need the following proposition 1 even for $m \geqslant 1$.

Proposition 293 . *Suppose that $x \geqslant 0$ and a consumption process $c_t \geqslant 0$ satisfying $E\int_0^\infty e^{-\int_0^t r_s ds}c(t)^2 dt < \infty$ is given such that*
$ED = E\int_0^\infty e^{-\int_0^t r_s ds}c_t dt \leqslant x$,
where $D = \int_0^\infty e^{-\int_0^t r_s ds}c_t dt$. Then there exists a portfolio process π_t such that the pair $(c(.), \pi(.)) \in \widetilde{U}'$ for the initial endowment x. Conversely, if $x \geqslant 0$ and a consumption process $c_t \geqslant 0$ with a portfolio process π_t are given such that $(c(.), \pi(.)) \in \widetilde{U}'$, then
$ED = E\int_0^\infty e^{-\int_0^t r_s ds}c_t dt \leqslant x$.

Proof. Let us show the first part. Set
$\xi_t = E[\int_t^\infty c_s e^{-\int_0^s r_v dv} ds \mid \mathfrak{F}_t] + (x - ED)$,
where $\mathfrak{F}_t = \mathfrak{F}_t^{\widetilde{Q}}$. Then $\xi_t \geqslant 0$, and $\xi_t = \left\{x + m_t - \int_0^t c_s e^{-\int_0^s r_v dv} ds\right\}$, where $m_t = E[D \mid \mathfrak{F}_t] - ED$. By the martingale representation theorem (Theorem 103) there exists a $\mathfrak{F}_t-$predictable process ϕ_t such that
$m_t = \int_0^t \phi_s^T d\widetilde{Q}_s$, $E\int_0^t |\phi_s|^2 ds = E|m_t|^2 \leqslant E(D^2) \leqslant \widetilde{k}_0/\delta_0$,
where we assume that $E\int_0^\infty e^{-\int_0^t r_s ds}c(t)^2 ds = \widetilde{k}_0$. Hence $E\int_0^\infty |\phi_s|^2 ds < \infty$.

Now let

$$\pi_t = e^{\int_0^t r_s ds}(\rho_t^{-1})^T \phi_t. \tag{9.25}$$

Then one easily verifies that $(c(.), \pi(.)) \in \widetilde{U}'$. The inverse conclusion is also easily verified by $\widetilde{x}_t \geqslant 0$, for all $t \geqslant 0$. ∎

Now we return to our problem just discussed. Under assumption (9.23) applying Proposition 1, we get that the optimal consumption is given by (9.24), we denote it by c^0, where the constant k_0 satisfies
$E\int_0^\infty e^{-\int_0^t r_s ds}(e^{\int_0^t (\beta_s - r_s)ds}k_0)^{-1/(1-\delta)}dt = x$;
or

$$k_0 = x^{-(1-\delta)}[E\int_0^\infty e^{-\frac{1}{1-\delta}\int_0^t (\beta_s - \delta r_s)ds}dt]^{(1-\delta)}, \tag{9.26}$$

and the optimal portfolio is given by (9.25) determined by ϕ_t, hence by c^0. Moreover, the optimal wealth process is

$$\begin{aligned} x_t^0 &= e^{\int_0^t r_s ds} x - e^{\int_0^t r_s ds}[\int_0^t e^{-\int_0^s r_v dv} c_s^0 ds - \int_0^t e^{-\int_0^s r_v dv} \pi_s^0 \rho_s d\widetilde{Q}_s] \\ &= e^{\int_0^t r_s ds} x - k_0{}^{-1/(1-\delta)} e^{\int_0^t r_s ds}[\int_0^t e^{-[1/(1-\delta)]\int_0^s (\beta_v - \delta r_v)dv} ds \\ &\quad - \int_0^t e^{-\int_0^s r_v dv} \pi_s^0 \rho_s d\widetilde{Q}_s]. \end{aligned} \tag{9.27}$$

Finally, the optimal value functional is

$$\begin{aligned} &\sup_{(c(.),\pi(.))\in\widetilde{U}} E\int_0^\infty e^{-\int_0^t \beta_s ds} U(t, x_t, c_t)dt \\ &= k_0^{-\delta/(1-\delta)} \delta^{-1} E\int_0^\infty e^{-[1/(1-\delta)]\int_0^t (\beta_s - \delta r_s)ds} dt \\ &= x^\delta \delta^{-1}[E\int_0^\infty e^{-[1/(1-\delta)]\int_0^t (\beta_s - \delta r_s)ds} dt]^{1-\delta}. \end{aligned} \tag{9.28}$$

From the discussion above we see that for the case $m = 1$ the optimal consumption, the optimal portfolio, the optimal wealth, and the optimal value functional are (9.24), (9.25), (9.27) and (9.28), respectively. In particular, under condition (9.23) as r_t is non-random, so are the optimal consumption c_t and D. Hence by the proof of Proposition 1 the optimal portfolio $\pi_t \equiv 0$. This is reasonable, because in SDE (9.20) the appreciation rate r_t for the stock now is the same as the interest rate of the bond. In this case it means that it is not necessary to take a risk to invest money in the stock. However, as r_t is random and $\mathfrak{F}_t-$ predictable, the above analysis tells us that it is necessary and possible to take an optimal portfolio $\pi_t \neq 0$ like (9.25) with an optimal consumption c_t like (9.24) in order to get an optimal value functional like (9.28).

Example 294 . *Consider the same financial market and the same wealth process as (9.20). However, here our object is to find*

$\sup_{(c(\cdot),\pi(\cdot))\in\widetilde{U}} E\int_0^\infty e^{-\int_0^t \beta_s ds} U(t, x_t, c_t)dt,$

where

$U(t, x, c) = \log c,$

$k_0 \geqslant \beta_s \geqslant r_s \geqslant \delta_0 > 0;$ δ_0 *is a constant.*

Then one easily sees that in (9.22) only the first equation is changed to the following

$c^{-1} - e^{-\int_0^t r_s ds}\lambda = 0.$

Hence under assumption (9.23), a conclusion similar to that in example 1 is still valid, only now

$\lambda^0(t, x) = e^{\int_0^t \beta_s ds} k_0,\ c^0 = c^0(t, x) = (e^{\int_0^t (\beta_s - r_s)ds} k_0)^{-1},$

where k_0 satisfies $E\int_0^\infty (e^{\int_0^t \beta_s ds} k_0)^{-1} dt = x$, or $k_0 = x^{-1} E\int_0^\infty e^{-\int_0^t \beta_s ds} dt$. Thus the optimal wealth process is

$x_t^0 = e^{\int_0^t r_s ds} x - e^{\int_0^t r_v dv}[\int_0^t e^{-\int_0^s r_v dv} c_s^0 ds + \int_0^t e^{-\int_0^s r_v dv} \pi_s^0 \rho_s d\widetilde{Q}_s]$

$= e^{\int_0^t r_s ds} x - e^{\int_0^t r_v dv}[k_0^{-1} \int_0^t e^{-\int_0^s \beta_v dv} ds + \int_0^t e^{-\int_0^s r_v dv} \pi_s^0 \rho_s d\widetilde{Q}_s],$

where the optimal portfolio π_t^0 is determined by formula (9.25) with c_t^0 given here, and the optimal value functional is

$\sup_{(c(.),\pi(.))\in\widetilde{U}} E\int_0^\infty e^{-\int_0^t \beta_s ds} U(t, x_t, c_t)dt$

$$= E \int_0^\infty e^{-\int_0^t \beta_s ds}[-\log(e^{\int_0^t (\beta_s - r_s)ds} k_0)]dt$$
$$= -E \int_0^\infty e^{-\int_0^t \beta_s ds} \int_0^t (\beta_s - r_s)dsdt$$
$$-[\log(x^{-1}E \int_0^\infty e^{-\int_0^t \beta_s ds}dt)]E \int_0^\infty e^{-\int_0^t \beta_s ds}dt.$$

9.2.5 Concluding Remarks

Finally, let us point out that as in §7.5 of Chow (1997), after a simple modification to the system (9.8) with jumps we can similarly derive the optimal portfolio percentages given in Chow (equation (7.50)) when all securities are risky, or in equation (7.56) when there is one riskless security, and a similar mutual fund theorem that then follows. To economize space, the derivations are omitted.

Let us also note that when $S^{(2)} = 0$ (i.e. the system has no jumps), if the discount factor β in (9.6) and the coefficients $f(x, u), S^{(1)}(x, u)$ of (9.8) all do not depend on t, then (9.18) implies (7.21) and (7.22) in Chow (1997).

10

Comparison Theorem and Stochastic Pathwise Control

In a practical case involving the control of a real engineering dynamical system we need to control the system pathwise. So, for a stochastic dynamical system from the practical point of view, pathwise control appears to be more useful. However, one can easily imagine that to minimize (or maximize) a target functional subject to the solutions of some stochastic system through the controls can actually be achieved if we can compare the values taken by the target functional as a result of different controls. So the comparison results are very important and useful for us. In this chapter we will discuss the comparison theorem for solutions of SDE first. Then we will see that they can be used to get the optimal stochastic pathwise Bang-Bang control under some appropriate conditions. Since the optimal trajectory (that is, the solution of the stochastic system when applying the optimal control) should be pathwise (that is, a strong solution). Otherwise, it is hard to apply it from an engineering point of view. So we will also discuss the existence of a strong solution when the coefficient is non-Lipschitzian. (That is, the case that the optimal control makes the system coefficients non-Lipschitzian).

10.1 Comparison for Solutions of Stochastic Differential Equations

10.1.1 1−Dimensional Space Case

The Tanaka formula is a very useful tool for discussing the comparison of solutions to SDE. Assume that $x_t^i, i = 1, 2$, satisfy the following 1−dimensional SDEs

$$x_t^i = x_0^i + \int_0^t \beta^i(s,\omega)dA_s + \int_0^t \sigma(s, x_s^i, \omega)dM_s$$
$$+ \int_0^t \int_Z c(s, x_{s-}^i, z, \omega)\widetilde{N}_k(ds, dz), i = 1, 2, \forall t \geq 0, \qquad (10.1)$$

where A_t, M_t and $\widetilde{N}_k(dt, dz)$ have the same meaning as that in (4.11). We have the following comparison theorem for solutions of these SDEs.

Theorem 295 *Assume that*
1° $\int_0^t |\beta^i(s,\omega)| ds < \infty, i = 1, 2, \forall t \geq 0, P-a.s.$
2° there exist $b^1(t, x, \omega)$ *and* $b^2(t, x, \omega)$ *such that*
$b^1(t, x, \omega) \geq b^2(t, x, \omega)$,
$\beta^1(t, \omega) \geq b^1(t, x_t^1, \omega), b^2(t, x_t^2, \omega) \geq \beta^2(t, \omega)$,
$sgn(x - y) \cdot (b^2(t, x, \omega) - b^2(t, y, \omega))$
$\leq k_N^T(t)\rho_N^T(|x - y|)$, *as* $|x|, |y| \leq N, \forall t \in [0, T]$,
where $N, T > 0$ *are arbitrarily given, and* $0 \leq k_N^T(t)$ *is non-random such that* $\int_0^T k_N^T(t)dt < \infty; 0 \leq \rho_N^T(u)$ *is non-random, strictly increasing in* $u \geq 0$ *with* $\rho_N(0) = 0$, *and* $\int_{0+} du/\rho_N(u) = \infty$;
3° $|\sigma(t, x, \omega) - \sigma(t, y, \omega)|^2 \leq k_N^T(t)\rho_N^T(|x - y|)$, *as* $|x|, |y| \leq N, t \in [0, T]$;
where $k_N^T(t)$ *and* $\rho_N^T(u)$ *have the same property as that in 2°;*
4° $x \geq y \implies x + c(t, x, z, \omega) \geq y + c(t, y, z, \omega)$.
If $x_0^1 \geq x_0^2$, *then* $P-a.s.$
$x_t^1 \geq x_t^2, \forall t \geq 0$,
where $x_t^i, i = 1, 2$, *satisfy (10.1), respectively.*

Proof. Define
$\tau_N = \inf\{t > 0 : |x_t^1| + |x_t^1| > N\}$.
By the Tanaka type formula (Theorem 152) for any given $N, T > 0$ we have
$E(x_{t\wedge\tau_N}^2 - x_{t\wedge\tau_N}^1)^+ \leq E\int_0^{t\wedge\tau_N} I_{x_s^2 > x_s^1}(\beta^2(s,\omega) - \beta^1(s,\omega))dA_s$
$\leq E\int_0^{t\wedge\tau_N} I_{x_s^2 > x_s^1}(b^2(s, x_s^2, \omega) - b^1(s, x_s^1, \omega))dA_s$
$\leq E\int_0^{t\wedge\tau_N} k_{N,T}(s)\rho_{N,T}\left((x_s^2 - x_s^1)^+\right) dA_s, \forall t \in [0, T]$.
Set
$\overline{A}_t = \int_0^t k_{N,T}(s)d(A_s + s), \forall t \geq 0$,
$T_t = \overline{A}_t^{-1}$,
$$Y_t = \begin{cases} (x_t^2 - x_t^1)^+, & \text{as } t \in [0, T), \\ 0, & \text{otherwise.} \end{cases}$$

Then we have that $\forall t \geq 0$

$E\left|Y(T_{t\wedge\tau_N})\right| \leq \int_0^t \rho_{N,T}(E\left|Y(T_{s\wedge\tau_N})\right|)ds.$

Hence, $P-a.s.$

$Y(t) = 0, \forall t \geq 0,$

i.e. $P-a.s.$

$x_t^1 \geq x_t^2, \forall t \in [0,T).$

Since T is arbitrary, the desired result is obtained. ∎

10.1.2 Component Comparison in d−Dimensional Space

Motivated by the 1−dimensional case we can discuss the comparison of the components of two $d-$dimensional solutions for two $d-$dimensional SDEs. Assume that $x_t^i, i = 1,2,$ satisfy the following $d-$dimensional SDE: $j = 1, 2, \cdots, d$

$$x_{jt}^i = x_0^i + \sum_{p=1}^m \int_0^t \beta_{jp}^i(s,\omega)dA_{ps} + \sum_{p=1}^r \int_0^t \sigma_{jp}(s, x_s^i, \omega)dM_{ps}$$
$$+ \int_0^t \int_Z c_j(s, x_{s-}^i, z, \omega)\widetilde{N}_k(ds,dz), i = 1,2, \forall t \geq 0, \tag{10.2}$$

where A_t, M_t and $\widetilde{N}_k(dt,dz)$ have the same meaning as in (4.4). We have the following comparison theorem for the components of solutions fo these SDEs.

Theorem 296 *Assume that*

1° $\int_0^t \left|\beta^i(s,\omega)\right| ds < \infty, i = 1,2, \forall t \geq 0, P-a.s.$

2° there exist $\left\{b_{jp}^1(t,x,\omega)\right\}_{j,p=1}^{d,m}$ *and* $\left\{b_{jp}^2(t,x,\omega)\right\}_{j,p=1}^{d,m}$ *such that*

$b_{jp}^1(t,x,\omega) \geq b_{jp}^2(t,x,\omega),$

$\beta_{jp}^1(t,\omega) \geq b_{jp}^1(t,x_t^1,\omega), b_{jp}^2(t,x_t^2,\omega) \geq \beta_{jp}^2(t,\omega),$

$sgn(x_j - y_j) \cdot (\ b_{jp}^2(t,x,\omega) - b_{jp}^2(t,y,\omega))$

$\leq k_N^T(t)\rho_N^T(|x_j - y_j|)$, *as* $|x|, |y| \leq N, \forall t \in [0,T]$,

where $N, T > 0$ *are arbitrarily given, and* $0 \leq k_N^T(t)$ *is non-random such that* $\int_0^T k_N^T(t)dt < \infty; 0 \leq \rho_N^T(u)$ *is non-random, strictly increasing in* $u \geq 0$ *with* $\rho_N(0) = 0$, *and* $\int_{0+} du/\rho_N(u) = \infty$;

3° $\left|\sigma_{jp}(t,x,\omega) - \sigma_{jp}(t,y,\omega)\right|^2 \leq k_N^T(t)\rho_N^T(|x_j - y_j|)$,
as $|x|, |y| \leq N, t \in [0,T]$; *where* $k_N^T(t)$ *and* $\rho_N^T(u)$ *have the same property as that in 2°*;

4° $x_j \geq y_j \implies x_j + c_{jp}(t,x,z,\omega) \geq y_j + c_{jp}(t,y,z,\omega)$,
where x_j *and* y_j *are the* $j-th$ *components of* $x = (x_1, x_2, \cdots, x_d)$ *and* $y = (y_1, y_2, \cdots, y_d)$, *respectively.*

If $x_{j0}^1 \geq x_{j0}^2, \forall j = 1, 2, \cdots, d$, *then* $P-a.s.$

$x_{jt}^1 \geq x_{jt}^2, \forall t \geq 0, \forall j = 1, 2, \cdots, d.$

Proof. By using the notation in the proof of the previous theorem and applying Theorem 153 one finds that for any given $N, T > 0$

$$E(x^2_{j,t\wedge\tau_N} - x^1_{j,t\wedge\tau_N})^+ \leq \sum_{p=1}^m E\int_0^{t\wedge\tau_N} I_{x^2_{js}>x^1_{js}}(\beta^2_{jp}(s,\omega) - \beta^1_{jp}(s,\omega))dA_{ps}$$
$$\leq \sum_{p=1}^m E\int_0^{t\wedge\tau_N} I_{x^2_{js}>x^1_{js}}(b^2_{jp}(s,x^2_s,\omega) - b^1_{jp}(s,x^1_s,\omega))dA_{ps}$$
$$\leq \sum_{p=1}^m E\int_0^{t\wedge\tau_N} k_{N,T}(s)\rho_{N,T}\left((x^2_{js} - x^1_{js})^+\right)dA_{ps}, \forall t \in [0,T].$$

Set

$$\overline{A}_t = \int_0^t k_{N,T}(s)d\sum_{p=1}^m(A_{ps} + s), \forall t \geq 0,$$
$$T_t = \overline{A}_t^{-1},$$
$$Y_{jt} = \begin{cases} (x^2_{jt} - x^1_{jt})^+, & \text{as } t \in [0,T), \\ 0, & \text{otherwise.} \end{cases}$$

Then we have that $\forall t \geq 0, \forall j = 1,2,\cdots,d$

$$E\left|Y(T_{j,t\wedge\tau_N})\right| \leq \int_0^t \rho_{N,T}(E\left|Y(T_{j,s\wedge\tau_N})\right|)ds.$$

Hence, as in the proof of the previous theorem, we have that $P-a.s.$ $\forall j = 1,2,\cdots,d$

$$x^1_{jt} \geq x^2_{jt}, \forall t \in [0,T).$$

Since T is arbitrary the desired result is obtained. ∎

As an application we give the following example.

Example 297 *Consider the following SDE with jumps*

$$x_{jt} = x_{j0} + \int_0^t b_j(s,x_{js},\omega)ds + \sum_{k=1}^r \int_0^t \sigma_{jk}(s,\omega)dw_k(s)$$
$$+ \int_0^t \int_Z c_j(s,x_{js},\omega)\widetilde{N}_k(ds,dz),\ j = 1,2,\cdots,d.$$

Suppose that

$$sgn(x_j - y_j)\cdot(\ b^2_j(t,x_j,\omega) - b^2_j(t,y_j,\omega))$$
$$\leq k^T_N(t)\rho^T_N(|x_j - y_j|),\ as\ |x_j|,|y_j| \leq N, \forall t \in [0,T],$$
$$x_j \geq y_j \implies x_j + c_j(t,x_j,z,\omega) \geq y_j + c_j(t,y_j,z,\omega),$$

where $k^T_N(t)$ and $\rho^T_N(u)$ satisfy the same conditions as in Theorem 296. If $x^i_t, i = 1,2,$ are two solutions corresponding to the given initial conditions $x^i_0, i = 1,2,$ respectively, then

$$x^1_{j0} \geq x^2_{j0}, j = 1,2,\cdots,d$$
$$\implies x^1_{jt} \geq x^2_{jt}, \forall t \geq 0,\ j = 1,2,\cdots,d.$$

Notice that in the above example the $\sigma_{jk}(s,\omega)$ do not depend on x.

10.1.3 Applications to Existence of Strong Solutions. Weaker Conditions.

For a 1−dimensional SDE one can use the comparison theorem (Theorem 295) to derive the existence of a strong solution under weaker condition in some sense. Let us consider the following SDE with jumps in 1−dimensional space.

$$x_t = x_0 + \int_0^t b(s, x_s)ds + \int_0^t \sigma(s, x_s)dw_s$$
$$+ \int_0^t \int_Z c(s, x_{s-}, z)\widetilde{N}_k(ds, dz), \forall t \geq 0, \quad (10.3)$$

First we establish a theorem on the existence of a pathwise unique strong solution by combinning three results: the existence of a weak solution (through Theorem 175), the validity of the pathwise uniqueness of solutions (through Theorem 170), and then the existence of a pathwise unique strong solution (through Theorem 137).

Theorem 298 *Assume that*
1° $b = b(t, x) : [0, \infty) \times R^d \to R^d$,
$\sigma = \sigma(t, x) : [0, \infty) \times R^d \to R^{d \otimes d}$,
$c = c(t, x, z) : [0, \infty) \times R^d \times Z \to R^d$,
are Borel measurable processes such that $P-a.s.$
$|b(t, x)| \leqslant c_1(t)(1 + |x|)$,
$|\sigma(t, x)|^2 + \int_Z |c(t, x, z)|^2 \pi(dz) \leqslant c_1(t)(1 + |x|^2)$,
where $c_1(t)$ *is non-negative and non-random such that for each* $T < \infty$
$\int_0^T c_1(t)dt < \infty$;
2° $b(t, x)$ *and* $\sigma(t, x)$ *are continuous in* x; *and*
$\lim_{h \to 0} \int_Z |c(t, x + h, z) - c(t, x, z)|^2 \pi(dz) = 0$;
3° for each $N = 1, 2, \cdots$, *and each* $T < \infty$,
$|\sigma(t, x_1) - \sigma(t, x_2)|^2 \leqslant c_T^N(t)\rho_T^N(|x_1 - x_2|)$,
$x \geq y \Longrightarrow x + c(t, x, z) \geq y + c(t, y, z)$,
as $|x_i| \leqslant N, i = 1, 2,\ t \in [0, T]$; *where* $\int_0^T c_T^N(t)dt < \infty$; *and* $\rho_T^N(u) \geq 0$, *as* $u \geq 0$, *is non-random, strictly increasing, continuous and concave such that*
$\int_{0+} du/\rho_T^N(u) = \infty$;
4° $Z = R^d - \{0\}$, *and* $\int_{R^d - \{0\}} \frac{|z|^2}{1+|z|^2} \pi(dz) < \infty$,
5° $\langle x_1 - x_2, b(t, x_1) - b(t, x_2) \rangle \leqslant c_T^N(t)\rho_T^N(|x_1 - x_2|^2)$,
as $|x_i| \leqslant N, i = 1, 2,\ t \in [0, T]$; *where* $c_T^N(t)$ *and* $\rho_T^N(u)$ *have the same properties as that in 3°.*
Then (10.3) has a pathwise unique strong solution.

Proof. By conditions 1° and 2°, by applying Theorem 175 one finds that SDE (10.3) has a weak solution. From conditions 3° and 5°, by applying Theorem 170, one finds that pathwise uniqueness holds for SDE (10.3). Therefore applying the Yamada-Watanabe theorem (Theorem 137) there exists a pathwise unique strong solution for SDE (10.3). ■

Now let us use the comparison theorem to establish an existence result on strong solutions under conditions that in some sense are weaker.

Theorem 299 *In the above theorem (Theorem 298) if we remove condition 5° and, in addition, assume that $b(t,x)$ is jointly continuous, then SDE (10.3) still has a strong solution. (But it is not necessary unique).*

Proof. Since $b(t,x)$ is jointly continuous, for each $n = 1, 2, \cdots$, one can find a jointly Lipschitz continuous function $b^n(t,x)$ such that
$b(t,x) + \frac{1}{n+1} < b^n(t,x) < b(t,x) + \frac{1}{n}$.
Applying Theorem 298, for coefficients $(b^n(t,x), \sigma(t,x), c(t,x,z))$ there exists a pathwise unique strong solution x_t^n satisfying the following SDE

$$x_t^n = x_0 + \int_0^t b^n(s, x_s^n)ds + \int_0^t \sigma(s, x_s^n)dw_s$$
$$+ \int_0^t \int_Z c(s, x_{s-}^n, z)\widetilde{N}_k(ds, dz), \forall t \geq 0. \tag{10.4}$$

By the comparison theorem (Theorem 295) one finds that
$\widetilde{x}_t \leq x_t^{n+1} \leq x_t^n, \forall n,$
where $\widetilde{x}_t$ is the pathwise unique strong solution of the SDE with coefficients $(\widetilde{b}(t,x), \sigma(t,x), c(t,x,z))$, where $\widetilde{b}(t,x) = -c_1(t)(1+|x|)$. Hence the limit
$x_t = \lim_{n\to\infty} x_t^n, P - a.s.$ exists. However, it is not difficult to show that as $n \to \infty$,
(1) $\int_0^t b^n(s, x_s^n)ds \to \int_0^t b(s, x_s)ds$,
(2) $\int_0^t \sigma(s, x_s^n)dw_s \to \int_0^t \sigma(s, x_s)dw_s$,
(3) $\int_0^t \int_Z c(s, x_{s-}^n, z)\widetilde{N}_k(ds, dz) \to \int_0^t \int_Z c(s, x_{s-}, z)\widetilde{N}_k(ds, dz)$,
in probability. Let us show the first result. In fact, for arbitrary given $\varepsilon > 0$,
$P(\left|\int_0^t b^n(s, x_s^n)ds - \int_0^t b(s, x_s)ds\right| > \varepsilon) \leq P(|x_s| > N)$
$+ \sup_n P(|x_s^n| > N) + \frac{1}{\varepsilon} E \int_0^t [|b(s, x_s^n) - b(s, x_s)| + \frac{1}{n}] I_{|x_s^n| \leq N} ds$
$= I_1^N + I_2^N + I_3^{N,n}$.
Now, for any $\overline{\varepsilon} > 0$ take a large enough $N > 0$ such that
$I_1^N + I_2^N < \frac{\overline{\varepsilon}}{2}$.
Notice that $I_{|x_s^n| \leq N} |b(t, x_t^n)|^2 \leq c_1(t)(1+N)$. Then, by Lebesgue's domoinated convergence theorem, for each N, as $n \to \infty, I_3^{N,n} \to 0$. Hence, there exists an $\widetilde{N}$ such that for $n \geq \widetilde{N}$, $I_3^{N,n} < \frac{\overline{\varepsilon}}{2}$. Therefore (1) is established. Let us prove (3). For arbitrary given $\varepsilon > 0$,
$P(|\int_0^t \int_Z c(s, x_{s-}^n, z)\widetilde{N}_k(ds, dz) - \int_0^t \int_Z c(s, x_{s-}, z)\widetilde{N}_k(ds, dz)| > \varepsilon|)$
$\leq P(|x_s| > N) + \sup_n P(|x_s^n| > N)$
$+ \frac{1}{\varepsilon^2} E \int_0^t \int_Z |c(s, x_s^n, z) - c(s, x_s, z)|^2 I_{|x_s^n| \leq N} \pi(dz)ds$
$= I_1^N + I_2^N + I_3^{N,n}$.
By condition 2° as $n \to \infty$,
$\int_Z |c(s, x_s^n, z) - c(s, x_s, z)|^2 \pi(dz) \to 0$,
since $x_s^n \to x_s$. On the other hand,
$\int_Z |c(s, x_s^n, z)|^2 I_{|x_s^n| \leq N} \pi(dz) \leq c_1(t)(1+N^2)$.
So, by Lebesgue's dominated convergence theorem, for each N as $n \to \infty$,

$I_3^{N,n} \to 0.$

Therefore, (3) is similarly established. The proof of (2) is even simpler. ■

By using the method by which Theorem 299 was proved, one can easily obtain a comparison theorem under different conditions for the following SDEs. Suppose that $x_t^i, i = 1, 2$, satisfy the following 1−dimensional SDEs

$$x_t^i = x_0^i + \int_0^t b^i(s, x_s^i)ds + \int_0^t \sigma(s, x_s^i, \omega)dw_s$$
$$+ \int_0^t \int_Z c(s, x_{s-}^i, z, \omega)\widetilde{N}_k(ds, dz), i = 1, 2, \forall t \geq 0, \qquad (10.5)$$

respectively.

Theorem 300 *Assume that*

1° $b^i = b^i(t, x) : [0, \infty) \times R^d \to R^d,$
$\sigma = \sigma(t, x, \omega) : [0, \infty) \times R^d \times \Omega \to R^{d\otimes d},$
$c = c(t, x, z, \omega) : [0, \infty) \times R^d \times Z \times \Omega \to R^d,$

are $\mathfrak{F}_t^{w,\widetilde{N}_k}-$ *adapted and measurable processes, (where* $b^i, i = 1, 2$, *only need to be jointly Borel measurable) such that* $P-a.s.$
$|b^i(t, x)| \leqslant c_1(t)(1 + |x|),\ i = 1, 2,$
$|\sigma(t, x, \omega)|^2 + \int_Z |c(t, x, z, \omega)|^2 \pi(dz) \leqslant c_1(t)(1 + |x|^2),$
where $c_1(t)$ *is non-negative and non-random such that for each* $T < \infty$
$\int_0^T c_1(t)dt < \infty;$

2° σ *and* c *satisfy that*
$|\sigma(t, x, \omega) - \sigma(t, y, \omega)|^2 \leq k_N^T(t)\rho_N^T(|x - y|),$ *as* $|x|, |y| \leq N, t \in [0, T];$
$\lim_{h\to 0} \int_Z |c(t, x + h, z, \omega) - c(t, x, z, \omega)|^2 \pi(dz) = 0,$
where $0 \leq k_N^T(t)$ *is non-random such that* $\int_0^T k_N^T(t)dt < \infty$, *for each given* $T < \infty; 0 \leq \rho_N^T(u)$ *is non-random, strictly increasing in* $u \geq 0$ *with* $\rho_N^T(0) = 0$, *and* $\int_{0+} du/\rho_N^T(u) = \infty;$

3° $b^1(t, x) \geq b^2(t, x),$

4° $b^2(t, x)$ *(or* $b^1(t, x)$*) is jointly continuous such that the pathwise uniqueness for solutions of (10.5) with coefficients* (b^2, σ, c) *(or* (b^1, σ, c)*) holds,*

5° $x \geq y \Longrightarrow x + c(t, x, z, \omega) \geq y + c(t, y, z, \omega).$

If $x_0^1 \geq x_0^2$, *then* $P-a.s.$
$x_t^1 \geq x_t^2,\ \forall t \geq 0,$
where $x_t^i, i = 1, 2$, *satisfy (10.5), respectively.*

Proof. Take Lipschitz continuous functions $b^{2,n}$ such that
$b^1(t, x) \geq b^2(t, x) > b^2(t, x) - \frac{1}{n+1} > b^{2,n}(t, x) > b^2(t, x) - \frac{1}{n}, n = 1, 2, \cdots.$
Then by Theorem 170 there exists a pathwise unique strong solution $x_t^{2,n}$ satisfying the following SDE
$x_t^{2,n} = x_0^2 + \int_0^t b^{2,n}(s, x_s^{2,N})ds + \int_0^t \sigma(s, x_s^{2,N}, \omega)dw_s$
$+ \int_0^t \int_Z c(s, x_{s-}^{2,N}, z, \omega)\widetilde{N}_k(ds, dz), \forall t \geq 0.$
Applying the comparison theorem (Theorem 295) one also finds that

$x_t^1 \geq x_t^{2,n} \geq x_t^{2,n-1}$.
So the limit $\widetilde{x}_t^2 = \lim_{n\to\infty} x_t^{2,n} \leq x_t^1$ exists. Moreover, as before one easily shows that $\widetilde{x}_t^2$ is a solution of the BSDE with coefficients (b^2, σ, c). Since by assumption the pathwise uniqueness hold for this SDE, $\widetilde{x}_t^2$ coincides with x_t^2. Hence $x_t^2 \leq x_t^1, \forall t \geq 0$. ■

10.2 Weak and Pathwise Uniqueness for 1-Dimensional SDE with Jumps

By Corollary 140 we know that the pathwise uniqueness of solutions implies the weak uniqueness of weak solutions for $d-$dimensional SDEs with jumps. Here for $1-$dimensional SDEs with jumps. we will use the local time technique to conclude that under some general conditions the inverse statement is also true. Consider the following $1-$dimensional SDEs

$$x_t = x_0 + \int_0^t b(s, x_s)ds + \int_0^t \sigma(s, x_s)dw_s + \int_0^t \int_Z c(s, x_{s-}, z)\widetilde{N}_k(ds, dz), \forall t \geq 0. \tag{10.6}$$

Theorem 301 *Assume that the coefficients $b(t,x), \sigma(t,x)$, and $c(t,x,z)$ in (10.6) do not depend on ω, $d = 1$ and*
$1°$ $\int_0^t \int_Z |c(s,x,z)|^2 \pi(dz)ds < \infty$,
$2°$ $x \geq y \implies x + c(t,x,z) \geq y + c(t,y,z)$;
$3°$ $L_t^0(x^1(.) - x^2(.)) = 0$,
where $x^i(t), i = 1,2$, satisfy (10.6) on the same probability space with the same BM w_t and the same Poisson martingale measure $\widetilde{N}_k(ds,dz)$.

Then the weak uniqueness of weak solutions to (10.6) implies the pathwise uniqueness of solutions to (10.6).

Proof. Let
$Y_t = x^1(t) \vee x^2(t) = \max(x^1(t), x^2(t))$
$= x^2(t) + \max(x^1(t) - x^2(t), 0) = x_t^2 + (x_t^1 - x_t^2)^+$.
Then by (4.19)
$Y_t = x_t^2 + \int_0^t I_{x_s^1 - x_s^2 > 0} d(x_s^1 - x_s^2)$
$= x_0 + \int_0^t b(s, Y_s)ds + \int_0^t \sigma(s, Y_s)dw_s + \int_Z c(s, Y_{s-}, z)q(ds, dz)$,
where we have applied the fact that
$b(s, x_s^2) + I_{x_s^1 - x_s^2 > 0}(b(s, x_s^1) - b(s, x_s^2))$
$= I_{x_s^1 - x_s^2 > 0} b(s, x_s^1) + I_{x_s^1 - x_s^2 \leq 0}(b(s, x_s^2) = b(s, Y_s)$,
etc. Hence Y_t satisfies (10.6). By the assumption of weak uniqueness and $Y_t \geq x_t^i, i = 1,2$, we must have that
$P(\sup_{t \geq 0}(Y_t - x_t^i) = 0) = 1$, $i = 1,2$.
In fact, if

$P(\sup_{t\geq 0}(Y_t - x_t^2) > 0) > 0,$
then it is easy to see that there will exist a rational number r, a positive number ε and a $t > 0$ such that
$P(Y_t > r \geq x_t^2) \geq \varepsilon > 0.$
Since $Y_t \geq x_t^2$, and
$\{Y_t > r\} = \{x_t^2 > r\} \cup \{Y_t > r \geq x_t^2\},$
we have
$P\{Y_t > r\} \geq P\{x_t^2 > r\} + \varepsilon > P\{x_t^2 > r\}.$
However, by the weak uniqueness of weak solutions to (10.6) the probability laws of $\{Y_t, t \geq 0\}$ and $\{x_t^2, t \geq 0\}$ should coincide. This is a contradiction. So
$P(\sup_{t\geq 0}(Y_t - x_t^2) = 0) = 1.$
Similarly, we also have
$P(\sup_{t\geq 0}(Y_t - x_t^1) = 0) = 1.$
Hence
$P(\sup_{t\geq 0}|x_t^1 - x_t^2| = 0) = 1.$
That is, the pathwise uniqueness of solutions to (10.6) holds. ∎

From the proof of Theorem 301 one immediately sees that the following corollary holds.

Corollary 302 *Under the same assumption as those in Theorem 301 we have that*
$\overline{x}(t) = \max(x^1(t), x^2(t))$ *and* $\underline{x}(t) = \min(x^1(t), x^2(t))$ *both satisfy (10.6).*

Furthermore, we have the following sufficient condition for $L_t^0(x^1(.) - x^2(.)) = 0$.

Proposition 303 *Assume that the conditions 1° in Theorem 301 holds. and assume that*
4° for each $N = 1, 2, \cdots$ *there exists a* $k_N^T(t)$ *and* $\rho_N^T(u) \geq 0$, *as* $t, u \geq 0$, *such that*
$|\sigma(t,x) - \sigma(t,y)|^2 \leq k_N^T(t)\rho_N^T(|x-y|)$, *as* $|x|, |y| \leq N, t \in [0,T]$,
where $N, T > 0$ *are arbitrarily given, and* $0 \leq k_N^T(t)$ *is non-random such that* $\int_0^T k_N^T(t)dt < \infty; 0 \leq \rho_N^T(u)$ *is non-random, strictly increasing in* $u \geq 0$ *with* $\rho_N(0) = 0$, *and* $\int_{0+} du/\rho_N(u) = \infty$.
Then
$L_t^0(x^1(.) - x^2(.)) = 0.$

Proof. Applying Theorem 148 one finds that
$|x_t^1 - x_t^2| = \int_0^t sgn(x_s^1 - x_s^2)d(x_s^1 - x_s^2)$
$+ \int_0^t \int_Z [|x_{s-}^1 - x_{s-}^2 + c^1(s, x_{s-}^1, z, \omega) - c^2(s, x_{s-}^2, z, \omega)| - |x_{s-}^1 - x_{s-}^2|$
$-sgn(x_{s-}^1 - x_{s-}^2) \cdot (c^1(s, x_{s-}^1, z, \omega) - c^2(s, x_{s-}^2, z, \omega))] N_k(ds, dz).$
On the other hand, by Lemma 158
$|x_t^1 - x_t^2| = \int_0^t sgn(x_{s-}^1 - x_{s-}^2)d(x_s^1 - x_s^2) + L_t^0(x^1(.) - x^2(.))$
$+ \int_0^t \int_Z (|x_{s-}^1 - x_{s-}^2 + c(s, x_{s-}^1, z, \omega) - c(s, x_{s-}^2, z, \omega)| - |x_{s-}^1 - x_{s-}^2|$
$-sgn(x_{s-}^1 - x_{s-}^2)c(s, z, \omega)) N_k(ds, dz).$

Therefore,
$L_t^0(x^1(.) - x^2(.)) = 0.$ ∎
Combinning the results of Theorem 301 and Proposition 303 one immediately obtains the following theorem.

Theorem 304 *Under conditions 1° and 2° in Theorem 301 and the condition 4° in Proposition 303 the conclusion of Theorem 301 holds, that is, the weak uniqueness of weak solutions to (10.6) implies the pathwise uniqueness of solutions to (10.6).*

10.3 Strong Solutions for 1-Dimensional SDE with Jumps

10.3.1 Non-Degenerate Case

Recall that the Yamada-Watababe type theorem says that
the existence of a weak solution + the condition that pathwise uniqueness holds
$\Longrightarrow$ the existence of a pathwise unique strong solution.

So first we can use the Girsanov type theorem to find the existence of a weak solution under very weak conditions. Then as applying the result on the weak uniqueness implying the pathwise uniqueness under some general conditions, we can obtain a pathwise unique strong solution for a 1-dimensional SDE under very weak conditions. Consider a 1−dimensional SDE with jumps as follows: $\forall t \geq 0$,

$$x_t = x_0 + \int_0^t b(s, x_s)ds + \int_0^t \sigma(s, x_s)dw_s + \int_0^t \int_Z c(s, x_{s-}, z)\widetilde{N}_k(ds, dz), \tag{10.7}$$

where $x_t, b(t,x), \sigma(t,x), c(t,x,z) \in R^1$, w_t and $\widetilde{N}_k(dt,dz)$ are $d-$dimensional BM and Poisson martingale measure, respectively. Moreover, $\widetilde{N}_k(dt,dz) = N_k(dt,dz) - \pi(dz)dt$, where $N_k(dt,dz)$ is the Poisson counting measure generated by a $\mathfrak{F}_t-$Poisson point process $k(\cdot)$ and $\pi(dz)dt$ is its compensator, where $\pi(dz)$ is a $\sigma-$finite measure on some measurable space $(Z, \mathfrak{B}_Z)$.

Theorem 305 *Assume that b, σ and c are Borel measurable functions such that*
1° $Z = R^1 - \{0\}$;
2° one of the following conditions holds:
(i) $|\sigma(t,x)| + \int_Z |c(t,x,z)|^2 \pi(dz) \leq k_0$,
where $k_0 \geq 0$ *is a constant, and* $\pi(dz) = dz/|z|^2$;
(ii) $|\sigma(t,x)| \leq c_1(t)(1 + |x|^2)$,
$\int_Z |c(t,x,z)|^2 \pi(dz) \leq c_1(t)$

where $c_1(t) \geq 0$ satisfies that for any $T < \infty$, $\int_0^T c_1(t)dt < \infty$, moreover, $\sigma(t,x)$ is jointly continuous in (t,x) and
$\lim_{h,h' \to 0} \int_Z |c(t+h', x+h, z) - c(t,x,z)|^2 \pi(dz) = 0.$
and $\int_Z \frac{|z|^2}{1+|z|^2} \pi(dz) < \infty$;
3° there exists a $\delta_0 > 0$ such that
$|\sigma(t,x)| \geq \delta_0,$
and for each $N = 1, 2, \cdots, T < \infty$, there exists a $k_N^T(t)$ and $\rho_N^T(u) \geq 0$, such that
$|\sigma(t,x) - \sigma(t,y)|^2 \leq k_N^T(t)\rho_N^T(|x-y|)$, *as* $|x|, |y| \leq N, t \in [0,T]$,
where $N, T > 0$ are arbitrarily given, and $0 \leq k_N^T(t)$ is non-random such that $\int_0^T k_N^T(t)dt < \infty$; $0 \leq \rho_N^T(u)$ is non-random, strictly increasing in $u \geq 0$ with $\rho_N(0) = 0$, and $\int_{0+} du/\rho_N(u) = \infty$;
4° $x \geq y \Longrightarrow x + c(t,x,z) \geq y + c(t,y,z)$;
5° $xb(t,x) \leq c_1(t)(1 + |x|^2 \prod_{k=1}^m g_k(x))$,
where
$$g_k(x) = 1 + \underbrace{\ln(1 + \ln(1 + \cdots \ln}_{k-times}(1 + |x|^{2n_0}))),$$
(n_0 is some natural number), and $c_1(t) \geq 0$ is non-random such that for each $T < \infty$, $\int_0^T c_1(t)dt < \infty$; furthermore, $b(t,x)$ is locally bounded in x, that is, for each $r > 0$, as $|x| \leq r$,
$|b(t,x)| \leq k_r,$
where $k_r > 0$ is a constant depending only on r.
Then there exists a pathwise unique strong solution for (10.7).

Proof. Under the condition (i) of 2°, applying Theorem 182 one finds that (10.7) has a weak solution. In the case that the condition (ii) of 2° holds, by Theorem 183 one still finds that (10.7) has a weak solution. Now by Theorem 304 one sees that if the weak uniqueness holds for SDE (10.7) then the pathwiae uniqueness also holds for this SDE. Thus we can use the Y-W type theorem (Theorem 137) to obtain a pathwise unique strong solution and the conclusion of our theorem is proved. So for the rest of the result we only need to prove that the weak uniqueness holds for SDE (10.7) on $t \geq 0$. However, this can be reduced to showing that for each $T < \infty$ the weak uniqueness of weak solutions for (10.7) holds on $t \in [0,T]$. In fact, if this is true, then for each $T < \infty$ we will have a pathwise unique strong solution on $t \in [0,T]$. This immediately yields a pathwise unique strong solution on $t \geq 0$.

Now let us show that the weak uniqueness of weak solutions for (10.7) holds on $t \in [0,T]$. Suppose that $x_t^i, i = 1,2$, are two weak solutions of SDE (10.7), defined on probability spaces $(\Omega^i, \mathfrak{F}^i, \{\mathfrak{F}_t^i\}_{t \geq 0}, P^i), i = 1,2$, respectively. That is, $x_t^i, i = 1,2$, satisfy that $P^i - a.s. \forall t \geq 0$,
$x_t^i = x_0 + \int_0^t b(s, x_s^i)ds + \int_0^t \sigma(s, xi_s^i)dw_s^i + \int_0^t \int_Z c(s, x_{s-}^i, z)\widetilde{N}_{k^i}(ds, dz),$

where w_t^i and $\widetilde{N}_{k^i}(ds,dz)$ are the BM and Poisson martingale measure, defined on probability spaces $(\Omega^i, \mathfrak{F}^i, \{\mathfrak{F}_t^i\}_{t\geq 0}, P^i)$, respectively, such that $\widetilde{N}_{k^i}(ds,dz) = N_{k^i}(ds,dz) - \pi(dz)dt$, that is, the Poisson martingale measures still have the same compensator $\pi(dz)dt$. Now for any given $T<\infty$ as $i=1,2$, set

$z_t^i = \exp[-\int_0^t(\sigma^{-1}b)(s,x_s^i)dw_s^i - \frac{1}{2}\int_0^t \left|(\sigma^{-1}b)(s,x_s^i)\right|^2 ds]$,

$d\widetilde{P}_T^i = z_T^i dP^i$,

$\widetilde{w}_t^i = \int_0^t b(s,x_s^i)ds + w_t^i$,

Then one finds that $\widetilde{P}_T^i - a.s. \forall t\in[0,T]$,

$$dx_t^i = \sigma(t,x_t^i)d\widetilde{w}_t + \int_Z c(t,x_{t-}^i,z)\tilde{N}_{k^i}(dt,dz), x_t^i = x_0. \tag{10.8}$$

where $\widetilde{w}_t^i, t\in[0,T]$, is a BM under the new probability measure $\widetilde{P}_T^i$, and $\tilde{N}_{k^i}((0,t],dz)$, $t\in[0,T]$, is a Poisson martingale measure with the same compensator $\pi(dz)t$, also under the new probability measure $\widetilde{P}_T^i$. However, by Theorem 154 the pathwise uniqueness holds for SDE (10.8). So, applying Corollary 140, one finds that the weak uniqueness holds for SDE (10.8). Furthermore, take any non-negative $\mathfrak{B}_D\times\mathfrak{B}_{W_0}\times\mathfrak{B}_D-$measurable function

$f:(D\times W_0\times D, \mathfrak{B}_D\times\mathfrak{B}_{W_0}\times\mathfrak{B}_D)\to(R^1,\mathfrak{B}_{R^1})$,

where $\mathfrak{B}_D, \mathfrak{B}_{W_0}$ and $\mathfrak{B}_{R^1}$ are the Borel field in D, W_0 and R^1, respectively. Moreover, D and W_0 are the totality of all RCLL real functions and all real continuous functions $f(t)$ with $f(0)=0$, defined on $[0,T]$, respectively. By Corollary 140 one finds that

$\int_\Omega f(x^1(\cdot),\widetilde{w}^1(\cdot),\varsigma^1(\cdot))d\widetilde{P}_T^1 = \int_\Omega f(x^2(\cdot),\widetilde{w}^2(\cdot),\varsigma^2(\cdot))d\widetilde{P}_T^2$,

where

$\zeta_t^i = \int_0^t\int_{|z|<1} z\tilde{N}_{k^i}(ds,dz) + \int_0^t\int_{|z|\geq 1} zN_{k^i}(ds,dz)$,

and

$N_{k^i}((0,t],U) = \sum_{0<s\leq t} I_{0\neq\triangle\zeta_s^i\in U}$, for $t\geq 0, U\in\mathfrak{B}(Z)$,

$\tilde{N}_{k^i}(dt,dz) = N_{k^i}(dt,dz) - \pi(dz)dt$.

In particular, take

$\widetilde{f}(x^i(\cdot),\varsigma^i(\cdot)) = f(x^i(\cdot))\exp[\int_0^T(\sigma^{-1}b)(s,x_s^i)dx_s^i$

$-\frac{1}{2}\int_0^T\left|(\sigma^{-1}b)(s,x_s^i)\right|^2 ds - \int_0^T\int_Z \sigma^{-1}(s,x_{s-}^i)c(s,x_{s-}^i,z)\tilde{N}_{k^i}(ds,dz)]$.

One sees that

$\int_\Omega \widetilde{f}(x^1(\cdot),\varsigma^1(\cdot))d\widetilde{P}_T^1 = \int_\Omega \widetilde{f}(x^2(\cdot),\varsigma^2(\cdot))d\widetilde{P}_T^2$.

This is equivalent to saying that

$\int_\Omega f(x^1(\cdot))(z_T^1)^{-1}d\widetilde{P}_T^1 = \int_\Omega f(x^2(\cdot))(z_T^2)^{-1}d\widetilde{P}_T^2$.

Thus,

$\int_\Omega f(x^1(\cdot))dP_T^1 = \int_\Omega f(x^2(\cdot))dP_T^2$.

That is, the weak uniqueness of solutions to (10.8) holds. ■

Let us give an example for Theorem 305 on the existence of a pathwise unique strong solution to a 1−dimensional SDE with jumps and with the coefficients such that $b(t,x)$ is discontinuous and very much greater than the linear growth in x, $\sigma(t,x)$ is only Holder continuous with index $\frac{1}{2}$ in x, and $c(t,x,z)$ is also not Lipschitz continuous in x.

Example 306 *Set*
$\sigma(t,x) = t^{-\alpha/2}[k_0\sqrt{|x|} + k_1|x|] + k_2,\ \alpha < 1,$
where $k_0, k_1 \geq 0$ *and* $k_2 > 0$ *are constants. Then* $\sigma(t,x)$ *satisfies all conditions in Theorem 305. However,* $\sigma(t,x)$ *is only Holder continuous in* x *with index* $\frac{1}{2}$*, so it is not Lipschitz continuous in* x*. Wrirte*
$b(t,x) = t^{-\alpha}[k_0' + k_1' I_{x\neq 0}\frac{x}{|x|} - k_2' x^{2n_0+1} + \widetilde{k}_0 x$
$+\widetilde{k}_1 x g_1(x) + \widetilde{k}_2 x g_2(x) + \cdots + \widetilde{k}_m x \prod_{k=1}^m g_k(x)],\ \alpha < 1,$
where all $k_i', i = 0,1,2,$ *and* $\widetilde{k}_i, i = 0,1,2,\cdots,m,$ *are non-negative constants,* n_0 *is any natural number, and*
$g_k(x) = 1 + \underbrace{\ln(1+\ln(1+\cdots\ln}_{k-times}(1+|x|^2))).$
Obviously, $b(t,x)$ *is not continuous in* x*, its absolute value is very much greater than linear growth in* x*. However,* $b(t,x)$ *satisfies all conditions in Theorem 305. Let*
$c(t,x,z) = t^{-\alpha/2}[(\widetilde{k}_0' z + \widetilde{k}_1' |z|^{\beta_1}) I_{|z|\leq N_0} + \widetilde{k}_2' |z|^{\beta_2} I_{|z|>N_1}]$
$\cdot(\widetilde{k}_0'' + \widetilde{k}_1'' x I_{|x|\leq N_2} + \widetilde{k}_2'' \frac{x}{|x|} N_1 I_{|x|>N_2} + \widetilde{k}_3''(x)^{\frac{1}{2n_1+1}} I_{|x|\leq 1} + \widetilde{k}_4'' \frac{x}{|x|} I_{|x|>1}),$
where $\alpha < 1, \beta_1 > \frac{1}{2}, \beta_2 < \frac{1}{2}, 0 < \beta_3 < \frac{1}{2}$*, all of them are constants, and all* $\widetilde{k}_i', i = 0,1,2,$ *and* $\widetilde{k}_i'', i = 0,1,2,3,4,$ *are non-negative constants,* $N_0, N_1, N_2 > 0$ *also are constants. Apparently,* $c(t,x,z)$ *may not satisfy the Lipschitz condition:*
$\int_Z |c(t,x,z) - c(t,y,z)|^2 \pi(dz) \leq k_0 |x-y|^2,$ *as* $|x|, |y| < 1.$
However, it satisfies all conditions in Theorem 305. For the coefficients b, σ, c *Theorem 305 applies, so (10.7) has a pathwise unique strong solution.*

10.3.2 Degenerate and Partially-Degenerate Case

First for the 1−dimensional SDE with jumps we can weaken the condition on the pathwsie uniqueness. So we can obtain an existence theorem on a pathwise unique strong solution for 1−dimensional SDE with jumps under weaker conditions.

Theorem 307 *Assume that*
1° $Z = R^d - \{0\}$, *and* $\int_Z \frac{|z|^2}{1+|z|^2}\pi(dz) < \infty$
2° $b = b(t,x) : [0,\infty) \times R \to R,$
$\sigma = \sigma(t,x) : [0,\infty) \times R \to R^{1\otimes d},$
$c = c(t,x,z) : [0,\infty) \times R^d \times Z \to R,$
are jointly Borel measurable such that
$|b(t,x)| \leq c_1(t)(1 + |x| \prod_{k=1}^m g_k(x)),$
$|\sigma(t,x)|^2 \leq k_0(1 + |x|^2 \prod_{k=1}^m g_k(x)),$
$\int_Z |c(t,x,z)|^2 \pi(dz) \leqslant c_1(t),$

where

$g_k(x) = 1 + \underbrace{\ln(1+\ln(1+\cdots\ln}_{k-times}(1+|x|^{2n_0}))),$

(n_0 is some natural number), $k_0 \geq 0$ is a constant, and $c_1(t) \geq 0$ is non-random such that for each $T < \infty$, $\int_0^T c_1(t)dt < \infty$; furthermore, $b(t,x)$ is locally bounded for x, that is, for each $r > 0$, as $|x| \leq r$,

$|b(t,x)| \leq k_r,$

where $k_r > 0$ is a constant depending only on r;

3° $b(t,x)$ is continuous in x, $\sigma(t,x)$ is jointly continuous in (t,x); and

$\lim_{h,h'\to 0} \int_Z |c(t+h', x+h, z) - c(t,x,z)|^2 \pi(dz) = 0;$

4° $|\sigma(t,x,\omega) - \sigma(t,y,\omega)|^2 \leq k_N^T(t)\rho_N^T(|x-y|)$, as $|x|, |y| \leq N, t \in [0,T]$;

where $0 \leq k_N^T(t)$ is non-random such that $\int_0^T k_N^T(t)dt < \infty$, for each given $T < \infty$; $0 \leq \rho_N^T(u)$ is non-random, strictly increasing in $u \geq 0$ with $\rho_N^T(0) = 0$, and $\int_{0+} du/\rho_N^T(u) = \infty$;

5° $sgn\,(x-y)\cdot(b(t,x) - b(t,y)) \leq k_N^T(t)\rho_N^T(|x-y|)$,

as $|x|, |y| \leq N, \forall x, y \in R^1, \forall t \in [0,T]$;

where $k_N^T(t)$ and $\rho_N^T(u)$ have the same property as that in 4°, besides, $\rho_N^T(u)$ is concave;

6° one of the following conditions is satisfied:

(i) $\int_Z |c(t,x,z)|\, \pi(dz) \leq c(t)g(|x|)$,

$\int_Z |c(t,x,z) - c(t,y,z)|\, \pi(dz) \leq k_N^T(t)\rho_N^T(|x-y|)$,

as $|x|, |y| \leq N, \forall x, y \in R^1, \forall t \in [0,T]$;

where $c(t)$ and $g(u)$ satisfy the following properties:

$\int_0^T c(t)dt < \infty$, for each $T < \infty$; $g(|x|)$ is locally bounded, that is $|g(|x|)| \leq k_r$, as $|x| \leq r$, for each $r < \infty$; moreover, $k_N^T(t)$ and $\rho_N^T(u)$ have the same property as that in 4°, besides, $\rho_N^T(u)$ is concave;

(ii) $x \geq y \implies x + c(t,x,z) \geq y + c(t,y,z)$.

Then, for any given constant $x_0 \in R$, (10.7) has a pathwise unique strong solution on $t \geq 0$.

Proof. First, by Theorem 176 (10.7) has a weak solution. Second, from Theorem 154 the pathwise uniqueness for solutions of (10.7) holds. So we can apply the Y-W theorem (Theorem 137) to show that (10.7) has a pathwise unique strong solution on $t \geq 0$. ∎

From the conditions of Theorem 307 one sees that to have a pathwise unique strong solution for the σ to be possibly degenerate, in the 1−dimensional case the coefficient σ only needs to be Holder-continuous in index $\frac{1}{2}$, which is weaker than the condition on σ in Theorem ?? for $d-$dimensional SDEs.

Furthermore, we can also have the existence of a strong solution for 1−dimensional SDEs with jumps and with discontinuous coefficients b and with σ that may possibly be degenerate. For this let us first introduce a lemma, which is a special case of Lemma 271.

Lemma 308 *Assume that*
1° $b(t,x,\omega)$ *is continuous in* $x \in R^1 \backslash G$, *where* $G \subset R^1$ *is a Borel measurable set such that* $x \in G \Longrightarrow \sigma(t,x) > 0, \forall t \geq 0$, *and* $m_1 G = 0$, *where* m_1 *is the Lebesgue measure in* R^1; *moreover,* $b(t,x,\omega)$ *is a separable process with respect to* x, *(i.e. there exists a countable set* $\{x_i\}_{i=1}^{\infty}$ *such that for any Borel set* $A \subset R^1$ *and any open set* $B \subset R^1$ *the* $\omega-$*sets*
$\{\omega : b(t,x,\omega) \in A, x \in B\}$,
$\{\omega : b(t,x,\omega) \in A, x_i \in B, \forall i\}$
only differs a zero-probability $\omega-$*set;*
2° $|b(t,x,\omega)| \leqslant c_1(t)(1+|x|)$,
where $c_1(t) \geqslant 0$ *is non-random such that* $\int_0^{T_0} c_1(t)dt < \infty$.
Let
$b_n(t,x,\omega) = \inf_y (b(t,y,\omega) + (n \vee c_1(t))\,|y-x|), n \geqslant 1.$
Then $\forall n$, b_n *is still* $\mathfrak{F}_t-$ *adapted, and*
1) $b_n(t,x,\omega) \uparrow$, *as* $n \uparrow, \forall t \geqslant 0, x \notin G$;
$|b_n(t,x,\omega)| \leqslant c_1(t)(1+|x|)$;
2) $|b_n(t,x,\omega) - b_n(t,y,\omega)| \leqslant (n \vee c_1(t))\,|y-x|$,
3) $x_n \to x$ *in* R^1
$\Rightarrow b_n(t,x_n,\omega) \to b(t,x,\omega). \forall t \geqslant 0, x \notin G.$

Now we can get a result on the existence of a strong solution for SDEs with discontinuous b and with partially degenerate σ.

Theorem 309 *Assume that*
1° $b = b(t,x,\omega) : [0,\infty) \times R \times \Omega \to R$,
$\sigma = \sigma(t,x,\omega) : [0,\infty) \times R \times \Omega \to R^{1 \otimes d}$,
$c = c(t,x,z,\omega) : [0,\infty) \times R \times Z \times \Omega \to R$,
are $\mathfrak{F}_t^{w,\widetilde{N}_k}-$ *adapted and measurable processes such that* $P-a.s.$
$|b(t,x,\omega)| \leqslant c_1(t)(1+|x|)$,
$|\sigma(t,x,\omega)|^2 + \int_Z |c(t,x,z,\omega)|^2 \pi(dz) \leqslant c_1(t)(1+|x|^2)$,
where $\mathfrak{F}_t^{w,\widetilde{N}_k}$ *is the* $\sigma-$*field generated by* w *and* $\widetilde{N}_k$ *up to time* t, *that is,* $\mathfrak{F}_t^{w,\widetilde{N}_k} = \sigma(w_s, \widetilde{N}_k((0,s],U), \forall U \in \mathfrak{B}_Z, s \leq t)$, *and* $c_1(t)$ *is non-negative and non-random such that for each* $T < \infty$
$\int_0^T c_1(t)dt < \infty$;
2° $b(t,x,\omega)$ *is continuous in* $x \in R^1 \backslash G$, *where* $G \subset R^1$ *is a Borel measurable set such that* $x \in G \Longrightarrow \sigma(t,x,\omega) > 0, \forall t \geq 0, \forall \omega$ *and* $m_1 G = 0$, *where* m_1 *is the Lebesgue measure in* R^1; *moreover,* $b(t,x,\omega)$ *is a separable process with respect to* x, *(i.e. there exists a countable set* $\{x_i\}_{i=1}^{\infty}$ *such that for any Borel set* $A \subset R^1$ *and any open set* $B \subset R^1$ *the* $\omega-$*sets*
$\{\omega : b(t,x,\omega) \in A, x \in B\}$,
$\{\omega : b(t,x,\omega) \in A, x_i \in B, \forall i\}$
only differs a zero-probability $\omega-$*set;*
3° $\sigma(t,x,\omega)$ *is continuous in* x; *and*
$\lim_{h \to 0} \int_Z |c(t,x+h,z,\omega) - c(t,x,z,\omega)|^2 \pi(dz) = 0;$

4° for each $N = 1, 2, \cdots$, and each $T < \infty$,
$|\sigma(t, x_1, \omega) - \sigma(t, x_2, \omega)|^2$
$+ \int_Z |c(t, x_1, z, \omega) - c(t, x_2, z, \omega)|^2 \pi(dz) \leqslant c_T^N(t)\rho_T^N(|x_1 - x_2|^2)$,
as $|x_i| \leqslant N, i = 1, 2,\ t \in [0, T]$; *where* $\int_0^T c_T^N(t)dt < \infty$; *and* $\rho_T^N(u) \geq 0$, *as* $u \geq 0$, *is non-random, strictly increasing, continuous and concave such that* $\int_{0+} du/\rho_T^N(u) = \infty$;
5° $x \geq y \implies x + c(t, x, z, \omega) \geq y + c(t, y, z, \omega)$.
Then (10.7) has a minimal strong solution. (A solution x_t is called minimal, if all other solutions y_t are greater than it, that is $y_t \geq x_t, \forall t \geq 0$). Furthermore, if, in addition, the following condition also holds:
6° $(x - y) \cdot (b(t, x, \omega) - b(t, y, \omega)) \leq k_N^T(t)\rho_N^T(|x - y|^2)$,
as $|x|, |y| \leq N, \forall x, y \in R^1, \forall t \in [0, T]$;
where $k_N^T(t)$ and $\rho_N^T(u)$ have the same property as that in 4°;
then (10.7) has a pathwise unique strong solution.

Proof. Let
$b_n(t, x, \omega) = \inf_y(b(t, y, \omega) + (n \vee c_1(t))\,|y - x|), n \geqslant 1.$
Then by Theorem 170 (10.7) has a pathwise unique strong solution x_t^n for coefficients (b_n, σ, c).

Write
$h(t, x, \omega) = c_1(t)(1 + |x|)$.
Again, by Theorem 170, there exists a pathwise unique strong solution ξ_t satisfying (10.7) with the coefficients (h, σ, c) : .
$\xi_t = x_0 + \int_0^t h(s, \xi_s)ds + \int_0^t \sigma(s, \xi_s, \omega)dw_s + \int_0^t \int_Z c(s, \xi_{s-}, z, \omega)\widetilde{N}_k(ds, dz)$.
Moreover, by the conclusions 1)-3) of Lemma 308, applying the comparison theorem (Theorem 295) one finds that $x_t^n \leqslant x_t^{n+1} \leqslant \xi_t, \forall n$. Hence there exists a limit $P - a.s.$
$x_t = \lim_{n \to \infty} x_t^n, \forall t \geqslant 0$.

On the other hand, $|x_t^n|^2 \leq |x_t^1|^2 + |\xi_t|^2$, and by Lemma 114
$E \sup_{t \leq T}[|x_t^1|^2 + |\xi_t|^2] \leq k_T < \infty$.

Hence by Lebesgue's dominated convergence theorem one finds that
$\lim_{n \to \infty} E \int_0^T |x_t^n - x_t|^2 ds = 0$.
From these one can take a subsequence $\{n_k\}$ of $\{n\}$, denote it by $\{n\}$ again, such that $dt \times dP - a.e.$
$x_t^n \to x_t$, in R^d
and
$E(\int_0^T |x_s^n - x_s|^2 ds \leqslant 1/2^n, n = 1, 2, ...$
Hence $E \int_0^T \sup_n |x_t^n|^2 dt < \infty$. Now notice that

$x_t^n = x_0^n + A_t^n + \int_0^t \sigma_n(s, x_s^n, \omega)dw_s + \int_0^t \int_Z c_n(s, x_{s-}^n, z, \omega)\widetilde{N}_k(ds, dz)$,
where $A_t^n = \int_0^t b_n(s, x_s^n, \omega)ds$. Hence there exists a limit (if necessary, take a subsequence)
$A_t = \lim_{n \to \infty} A_t^n$,

and we have

$x_t = x_0 + A_t + \int_0^t \sigma(s, x_s, \omega)dw_s + \int_0^t \int_Z c(s, x_{s-}, z, \omega)\widetilde{N}_k(ds, dz),$

since σ is continuious in x, c satisfies condition 3°, and they are less than linear growth. Notice that

$|b_n(s, x_s^n, \omega)| \leqslant c_1(t)[1 + \sup_n(|x_t^n|)].$

Hence, one finds that A_t^n is a finite variational process, and so also is A_t. Let us show that

$$A_t = \int_0^t b(s, x_s, q_s, p_s, \omega)ds. \tag{10.9}$$

Indeed, from $m_1 G = 0$ applying the occupation density formula for the local time of semi-martingale (Lemma 159) one finds that

$0 = \int_G L_t^a(x)da = \int_0^t I_{\{x_s \in G\}} |\sigma(s, x_s, \omega)|^2 ds.$

Since by assumption $|\sigma(s, x_s, \omega)|^2 > 0$, as $x_s \in G$. Hence

$m_1(s \in [0, t] : x_s \in G) = 0.$

Therefore, by conclusion 4) of Lemma 308

$\int_0^t |b(s, x_s, \omega) - b_n(s, x_s^n, \omega)| ds$

$\leqslant \int_0^t I_{(x_s \notin G)} |b(s, x_s, \omega) - b_n(s, x_s^n, \omega)| ds \to 0.$

Thus (10.9) is derived. Now suppose that x_t' is another strong solution of (10.7). By Lemma 308

$b_n(t, x, \omega) \leqslant b(t, x, \omega).$

Hence applying again Theorem 295 one has that $x_t^n \leqslant x_t', \forall n$. Letting $n \to \infty$. It is obtained that $x_t \leqslant x_t'$. Therefore x_t is a minimal solution of (10.7). In addition, if 6° holds, then by Lemma 115 the solution is also pathwise unique. ∎

Furthermore, in the case that b, σ and c are non-random, then the condition on σ can also be weakened. Actually, we will have the following theorem.

Theorem 310 *Assume that*

1° $b = b(t, x) : [0, \infty) \times R \to R,$

$\sigma = \sigma(t, x) : [0, \infty) \times R \to R^{1 \otimes d},$

$c = c(t, x, z) : [0, \infty) \times R \times Z \to R,$

are jointly Borel measurable processes such that

$|b(t, x)| \leqslant c_1(t)(1 + |x|),$

$|\sigma(t, x)|^2 + \int_Z |c(t, x, z)|^2 \pi(dz) \leqslant c_1(t)(1 + |x|^2),$

where $c_1(t)$ *is non-negative and non-random such that for each* $T < \infty$

$\int_0^T c_1(t)dt < \infty;$

2° $b(t, x)$ *is continuous in* $x \in R^1 \backslash G$, *where* $G \subset R^1$ *is a Borel measurable set such that* $x \in G \Longrightarrow \sigma(t, x) > 0, \forall t \geq 0$, *and* $m_1 G = 0$, *where* m_1 *is the Lebesgue measure in* R^1;

3° $\sigma(t, x)$ *is jointly continuous in* (t, x); *and*

$\lim_{hh' \to 0} \int_Z |c(t + h', x + h, z) - c(t, x, z)|^2 \pi(dz) = 0;$

4° for each $N = 1, 2, \cdots$, *and each* $T < \infty$,

$|\sigma(t, x_1, \omega) - \sigma(t, x_2, \omega)|^2 \leqslant c_T^N(t)\rho_T^N(|x_1 - x_2|),$

as $|x_i| \leqslant N, i = 1, 2,\ t \in [0, T]$; *where* $\int_0^T c_T^N(t)dt < \infty$; *and* $\rho_T^N(u) \geq 0$, *as* $u \geq 0$, *is non-random, strictly increasing, continuous and concave such that* $\int_{0+} du/\rho_T^N(u) = \infty$;
5° $x \geq y \implies x + c(t, x, z, \omega) \geq y + c(t, y, z, \omega)$,
6° $Z = R^1 - \{0\}$, $\int_Z \frac{|z|^2}{1+|z|^2}\pi(dz) < \infty$.
Then (10.7) has a minimal stromg solution. Furthermore, if, in addition, the following condition also holds:
7° $sgn\,(x - y) \cdot (b(t, x) - b(t, y)) \leq k_N^T(t)\rho_N^T(|x - y|)$,
as $|x|, |y| \leq N, \forall x, y \in R^1, \forall t \in [0, T]$;
where $k_N^T(t)$ *and* $\rho_N^T(u)$ *have the same property as that in 4°;*
then (10.7) has a pathwise unique strong solution.

Proof. Let
$b_n(t, x) = \inf_y(b(t, y) + (n \vee c_1(t))\, |y - x|), n \geqslant 1.$
Then b_n is Lipschitz continuous in x. By Theorem 170 SDE

$$x_t = x_0 + \int_0^t b_n(s, x_s)ds + \int_0^t \sigma(s, x_s)dw_s + \int_0^t \int_Z c(s, x_{s-}, z)\widetilde{N}_k(ds, dz) \tag{10.10}$$

has a weak solution x_t^n. Notice that by conditions 4° and 5° the following Tanaka formual holds:
$\left|x_t^{n1} - x_t^{n2}\right| = \int_0^t sgn(x_s^{n1} - x_s^{n2})d(x_s^{n1} - x_s^{n2})$,
if $\left\{x_t^{ni}\right\}_{t\geq 0}, i = 1, 2$, are two solutions of (10.10). This means that the pathwise uniqueness of solutions to (10.10) holds, because b_n is Lipschitz continuous in x. Hence by the Yamada-Watanabe type theorem (Theorem 137) (10.10) has a pathwise unique strong solution x_t^n. Now the remaining proof follows in exactly the same way as in Theorem 309. ∎

Example 311 *Set*
$\sigma(t, x) = (I_{t\neq 0}t^{-\alpha_0} + \delta_0)[k_0 + k_1 x]$,
$c(t, x, z) = I_{t\neq 0}t^{-\alpha_1}(\widetilde{k}_0 + \widetilde{k}_1 x)c_0(z)$,
where $\alpha_i < 1, \delta_0 > 0, k_i, \widetilde{k}_i, i = 0, 1, \widetilde{k}_1 \geq 0$ *all are constants,* $k_0 > 0$ *and* $\int_Z |c_0(z)|^2\,\pi(dz) < \infty$. *Write*
$b(t, x) = I_{t\neq 0}t^{-\alpha_3}[k_0' + k_1' x + k_2' \frac{x}{|x|} I_{x\neq 0}]$,
where $\alpha_3 < 1$, $k_i', i = 0, 1, 2$, *all are constants. Then (10.7) has a strong solution for the coefficients* (b, σ, c). *In the case that* $k_2' < 0$ *then (10.7) has a pathwise unique strong solution.*

The results cannot be applied to ODEs, because that $|\sigma(t, x)| > 0$ at the discontinuous points x for $b(t, x)$. (However from SDE, returning to an ODE, one needs to put $\sigma(t, x) = 0$).

Furthermore, we can also generalize the result to the case that b has not necessary a less than linear growth.

Theorem 312 *Assume that*

1° $b = b(t,x) : [0,\infty) \times R \to R,$
$\sigma = \sigma(t,x) : [0,\infty) \times R \to R^{1\otimes d},$
$c = c(t,x,z) : [0,\infty) \times R \times Z \to R,$
are jointly Borel measurable processes such that
$|b(t,x)| \leq c_1(t)(1 + |x| \prod_{k=1}^m g_k(x)),$
$|\sigma(t,x)|^2 \leq c_1(t)(1 + |x|^2),$
$\int_Z |c(t,x,z)|^2 \pi(dz) \leqslant c_1(t),$
where
$$g_k(x) = 1 + \underbrace{\ln(1 + \ln(1 + \cdots \ln}_{k-times}(1 + |x|^{2n_0}))),$$
(n_0 is some natural number), $k_0 \geq 0$ is a constant, and $c_1(t) \geq 0$ is such that for each $T < \infty$, $\int_0^T c_1(t)dt < \infty$;
2° $b(t,x)$ is continuous in $x \in R^1 \backslash F$, where $F \subset R^1$ is a compact set such that $m_1 F = 0$, where m_1 is the Lebesgue measure in R^1; moreover, there exist a $\delta_0 > 0$ and a open set $G_1 \supset G$ such that $x \in G_1 \Longrightarrow \sigma(t,x) \geq \delta_0 > 0, \forall t \geq 0$;
3° $\sigma(t,x)$ is jointly continuous in (t,x); and
$\lim_{hh' \to 0} \int_Z |c(t+h', x+h, z) - c(t,x,z)|^2 \pi(dz) = 0;$
4° for each $N = 1,2,\cdots$, and each $T < \infty$,
$|\sigma(t,x_1,\omega) - \sigma(t,x_2,\omega)|^2 \leqslant c_T^N(t)\rho_T^N(|x_1 - x_2|),$
as $|x_i| \leqslant N, i = 1,2,\ t \in [0,T]$; where $\int_0^T c_T^N(t)dt < \infty$; and $\rho_T^N(u) \geq 0$, as $u \geq 0$, is non-random, strictly increasing, continuous and concave such that
$\int_{0+} du/\rho_T^N(u) = \infty;$
5° $x \geq y \Longrightarrow x + c(t,x,z,\omega) \geq y + c(t,y,z,\omega)$,
6° $Z = R^1 - \{0\}$, $\int_Z \frac{|z|^2}{1+|z|^2} \pi(dz) < \infty$.
Then (10.7) has a weak solution. Furthermore, if, in addition, the following condition also holds:
7° $sgn(x-y) \cdot (b(t,x) - b(t,y)) \leq k_N^T(t)\rho_N^T(|x-y|)$,
as $|x|, |y| \leq N, \forall x,y \in R^1, \forall t \in [0,T]$;
where $k_N^T(t)$ and $\rho_N^T(u)$ have the same property as that in 4°;
then (10.7) has a pathwise unique strong solution.

Proof. Let
$b^n(t,x) = b(t,x)W^n(x),$
where $W^n(x)$ is a smooth function such that $0 \leq W^n(x) \leq 1$, $W^n(x) = 1$, as $|x| \leq n$; and $W^n(x) = 0$, as $|x| \geq n+1$. Then by the previous theorem (Theorem 310) for each n there exists a pathwise unique strong solution x_t^n satisfying the following SDE: $\forall t \geq 0$,
$$x_t = x_0 + \int_0^t b^n(s,x_s)ds + \int_0^t \sigma(s,x_s)dw_s + \int_0^t \int_Z c(s,x_{s-},z)\widetilde{N}_k(ds,dz).$$
Now in the same way as in Theorem 176 one finds that "the result of SDE from the Skorohod weak convergence technique" holds. In particular, we have that $(\widetilde{x}_t'^n)$ satisfies the following SDE with $\widetilde{w}_t'^n$ and $\widetilde{q}^n(dt,dz)$ on $(\widetilde{\Omega}, \widetilde{\mathfrak{F}}, \widetilde{P})$, where $(\widetilde{x}_t'^n, \widetilde{w}_t'^n, \widetilde{q}^n(dt,dz))$ come from "the result of SDE from

the Skorohod weak convergence technique", etc:

$$\begin{aligned}\widetilde{x}_t'^n &= x_0 + \int_0^t b^n(s,\widetilde{x}_s'^n)ds + \int_0^t \sigma(s,\widetilde{x}_s'^n)d\widetilde{w}_s'^n \\ &+ \int_0^t \int_Z c(s,\widetilde{x}_{s-}'^n,z)\widetilde{q}^n(ds,dz). \end{aligned} \tag{10.11}$$

Also in exactly the same way one can prove that as $n \to \infty$
$\int_0^t \sigma(s,\widetilde{x}_s'^n)d\widetilde{w}_s^n \to \int_0^t \sigma(s,\widetilde{x}_s'^0)d\widetilde{w}_s'^0$, in probability $\widetilde{P}$,
and
$\int_0^t \int_Z c(s,\widetilde{x}_{s-}'^n,z)\widetilde{q}^n(ds,dz) \to \int_0^t \int_Z c(s,\widetilde{x}_{s-}'^0,z)\widetilde{q}^0(ds,dz)$,
in probability $\widetilde{P}$. Write
$x_t'^n = x_0 + A_t^n + \int_0^t \sigma(s,\widetilde{x}_s'^n)d\widetilde{w}_s'^n + \int_0^t \int_Z c(s,\widetilde{x}_{s-}'^n,z)\widetilde{q}^n(ds,dz)$,
where $A_t^n = \int_0^t b^n(s,\widetilde{x}_s'^n)ds$. Hence there exists a limit (in probability)
$A_t = \lim_{n\to\infty} A_t^n$,
and we have

$$x_t = x_0 + A_t + \int_0^t \sigma(s,\widetilde{x}_s'^0)d\widetilde{w}_s'^0 + \int_0^t \int_Z c(s,\widetilde{x}_{s-}'^0,z)\widetilde{q}^0(ds,dz). \tag{10.12}$$

Write
$\tau_N = \inf\left\{t \geq 0 : \left|\widetilde{x}_t'^0\right| > N\right\}$.
So it only remains for us to prove that
$\left|\int_0^{t\wedge\tau_N}(b^n(s,\widetilde{x}_s'^n) - b(s,\widetilde{x}_s'^0))ds\right|$
$= \left|\int_0^{t\wedge\tau_N -}(b^n(s,\widetilde{x}_s'^n) - b(s,\widetilde{x}_s'^0))ds\right| \to 0$, in probability $\widetilde{P}$.
If this can be done, then letting $n \to \infty$ in (10.11) we can show that $\widetilde{x}_t'^0$ satisfies: $\widetilde{P} - a.s.$

$$\begin{aligned}\widetilde{x}_{t\wedge\tau_N}'^0 &= x_0 + \int_0^{t\wedge\tau_N} b(s,\widetilde{x}_s'^0)ds + \int_0^{t\wedge\tau_N} \sigma(s,\widetilde{x}_s'^0)d\widetilde{w}_s'^0 \\ &+ \int_0^{t\wedge\tau_N} \int_Z c(s,\widetilde{x}_{s-}'^0,z)\widetilde{q}^n(ds,dz), \forall t \geq 0.\end{aligned}$$

Notice that $\tau_N \uparrow \infty$, as $N \uparrow \infty$. We obtain that $\widetilde{x}_t'^0$ is a weak solution of (10.7), as $t \geq 0$. Now observe that for any $\varepsilon > 0$
$\widetilde{P}(\left|\int_0^{t\wedge\tau_N -}(b^n(s,\widetilde{x}_s'^n) - b(s,\widetilde{x}_s'^0))ds\right| > \varepsilon)$
$\leq \sup_n \widetilde{P}(\sup_{s\leq T}|\widetilde{x}_s'^n| + \sup_{s\leq T}|\widetilde{x}_s'^0| > \widetilde{N})$
$+\frac{2k_{\widetilde{N}}}{\varepsilon^2}E^{\widetilde{P}}\int_0^{t\wedge\tau_N -} I_{\widetilde{x}_s'^0 \in G_2} I_{\sup_{s\leq T}|\widetilde{x}_s'^0| \leq \widetilde{N}} ds$
$+\frac{1}{\varepsilon^2}E^{\widetilde{P}}\int_0^{t\wedge\tau_N -} I_{\widetilde{x}_s'^0 \notin G_2}\left|b^n(s,\widetilde{x}_s'^n) - b(s,\widetilde{x}_s'^0)\right| I_{\sup_{s\leq T}|\widetilde{x}_s'^n| + \sup_{s\leq T}|\widetilde{x}_s'^0| \leq \widetilde{N}} ds]$
$= I_1^{n,\widetilde{N}} + I_2^{\widetilde{N}} + I_3^{n,\widetilde{N}}$,
where $k_{\widetilde{N}} = 2(1 + \widetilde{N}\prod_{k=1}^m g_k(\widetilde{N}))$, and $F \subset G_2 \subset G_1$, G_2 is an bounded open set. Obviously, for arbitrary $\overline{\varepsilon} > 0$ one can take a large enough $\widetilde{N}$

such that $I_1^{n,\widetilde{N}} < \frac{\varepsilon}{3}$. (See the proof in Thoerem 176). Notice that by the occupation density formula of local time (Lemma 159)

$I_2^{\widetilde{N}} \leq \frac{2k_{\widetilde{N}}}{\varepsilon^2} E^{\widetilde{P}} \int_0^{t\wedge\tau_N -} I_{\widetilde{x}_s'^0 \in G_2} \frac{\left|\sigma(s,\widetilde{x}_s'^0)\right|^2}{\delta_0} ds = \frac{2k_{\widetilde{N}}}{\delta_0 \varepsilon^2} E^{\widetilde{P}} \int_{G_2} L^a_{t\wedge\tau_N -}(\widetilde{x}'^0) da.$

However,

$E\left|A_{t\wedge\tau_N -}\right| \leq E\left|\widetilde{x}'^0_{t\wedge\tau_N -}\right| + E\left|x_0\right| + E\left|\int_0^{t\wedge\tau_N -} \sigma(s,\widetilde{x}_s'^0) d\widetilde{w}_s'^0\right|$

$+E\left|\int_0^{t\wedge\tau_N -} \int_Z c(s,\widetilde{x}_{s-}'^0,z)\widetilde{q}^0(ds,dz)\right| \leq \widetilde{k}_N.$

Hence by Lemma 158 as $a \in \overline{G}_2$ (the closure of G_2)

$EL^a_{t\wedge\tau_N -}(\widetilde{x}'^0) \leq 4E[\left|(x_{t\wedge\tau_N -} - a)^+ - (x_0 - a)^+\right| + \left|A_{t\wedge\tau_N -}\right|$

$+\left|\int_0^{t\wedge\tau_N -} I_{(x_s > a)}\sigma(s,\widetilde{x}_s'^0) dw_s\right| + \left|\int_0^{t\wedge\tau_N -} \int_Z I_{(x_{s-} > a)} c(s,z,\omega)\widetilde{N}_k(ds,dz)\right|]$

$\leq \widetilde{k}'_N,$

since $\overline{G}_2$ is a bounded set. Thus, we can choose a small enough $\delta_{N,\widetilde{N}} > 0$ and an bounded open set G_2 such that $F \subset G_2 \subset G_1$, and $m_1 G_2 < \delta_{N,\widetilde{N}} < \frac{\delta_0 \varepsilon^2}{2k_{\widetilde{N}} k_N k_{\widetilde{N}}} \frac{\varepsilon}{3}$. Hence

$I_2^{\widetilde{N}} \leq \frac{\varepsilon}{3}.$

Since F is a compact set, we can also choose an open set G_3 such that $F \subset G_3 \subset \overline{G}_3 \subset G_2 \subset G_1$.

Write

$\delta_1 = \inf\left\{d(x,y) : x \in \overline{G}_3, y \in G_2\right\}.$

Then $\delta_1 > 0$. Noice that

$F_N^3 = G_3^c \cap \left\{x \in R^1 : |x| \leq \widetilde{N}\right\}$

is a compact set, where G_3^c is the complement of G_3; and for each s, $b(s,x)$ is uniformly continuous on $x \in F_N^3$. So for any $\varepsilon > 0$ there exists a $\delta_1 > \eta > 0$ (η may depend on the given s) such that as $|x' - x''| \leq \eta$ and $x', x'' \in F_N^3$,

$|b(s,x') - b(s,x'')| < \varepsilon.$

Now take $N \geq \widetilde{N}$, one finds that as $n \geq N$,

$I_3^{n,\widetilde{N}} = \frac{1}{\varepsilon^2} E^{\widetilde{P}} \int_0^{t\wedge\tau_N -} I_{\widetilde{x}_s'^0 \notin G_2} \left|b(s,\widetilde{x}_s'^n) - b(s,\widetilde{x}_s'^0)\right|$

$\cdot I_{\sup_{s\leq T}|\widetilde{x}_s'^n| + \sup_{s\leq T}|\widetilde{x}_s'^0| \leq \widetilde{N}} ds.$

However, for each given s

$\widetilde{P}(\left|b(s,\widetilde{x}_s'^n) - b(s,\widetilde{x}_s'^0)\right| I_{\widetilde{x}_s'^0 \in F_N^2} > \varepsilon) \leq \widetilde{P}(\left|\widetilde{x}_s'^n - \widetilde{x}_s'^0\right| > \eta)$

$+\widetilde{P}(I_{|\widetilde{x}_s'^n - \widetilde{x}_s'^0| \leq \eta} \left|b(s,\widetilde{x}_s'^n) - b(s,\widetilde{x}_s'^0)\right| I_{\widetilde{x}_s'^0 \in F_N^2} > \varepsilon)$

$\leq \widetilde{P}(\left|\widetilde{x}_s'^n - \widetilde{x}_s'^0\right| > \eta)..$

where

$F_N^2 = G_2^c \cap \left\{x \in R^d : |x| \leq \widetilde{N}\right\}.$

Hence, for each given s as $n \to \infty$,

$\widetilde{P}(\left|b(s,\widetilde{x}_s'^n) - b(s,\widetilde{x}_s'^0)\right| I_{\widetilde{x}_s'^0 \in F_N^2} > \varepsilon) \leq \widetilde{P}(\left|\widetilde{x}_s'^n - \widetilde{x}_s'^0\right| > \eta) \to 0.$

So applying Lebesgue's dominated convergence theorem, one finds that as $n \to \infty$, $I_3^{n,\widetilde{N}} \to 0$. Therefore, we have proved that as $n \to \infty$,

$\left|\int_0^{t\wedge\tau_N} (b^n(s,\widetilde{x}_s'^n) - b(s,\widetilde{x}_s'^0))ds\right| \to 0$, in probability $\widetilde{P}$..

So that $\widetilde{x}_s'^0$ is a weak solution of (10.7). Finally, if 7° holds, then by the Tanaka type formula the pathwise uniqueness of solutions to (10.7) also holds. Hence by the Y-W theorem (Theorem 137) (10.7) has a pathwise unique strong solution. ■

10.4 Stochastic Pathwise Bang-Bang Control for a Non-linear System

10.4.1 Non-Degenerate Case

Now let us formulate an existence theorem of the optimal Bang-Bang control for a very non-linear stochastic system with jumps. Let

$J(u) = E\int_0^T |x_t^u|^2 \, dt,$

where $x_t^u$is the pathwise unique strong solution of the following 1–dimensional SDE with jumps: $\forall t \in [0,T]$,

$$\begin{aligned} dx_t &= (A_t^1 x_t g_1(t, |x_t|^2) - A_t^2 x_t g_2(t, |x_t|) - A_t^3 (x_t)^{2k_3+1} \\ &\quad - A_t^4 (x_t)^{(2k_1+1)/(2k_2+1)} + A_t^0 x_t + B_t u_t)dt + (C_t^0 + C_t^1 |x_t|)dw_t \\ &\quad + \int_Z D_t(z)(x_{t-} I_{|x_{t-}| \leq \sqrt{N_0}} + \sqrt{N_0} \frac{x_{t-}}{|x_{t-}|} I_{|x_{t-}| > \sqrt{N_0}}) \widetilde{N}_k(dt,dz), \quad x_0 = x \in R^1, \end{aligned} \tag{10.13}$$

and the admissible control $u = u(t, x_t)$ is such that $u(t,x)$ is jointly measurable, $|u(t,x)| \leq 1, \forall t \in [0,T], x \in R^1$, and it makes that (10.13) has a pathwise unique strong solution. Denote the admissible control set by $\mathfrak{U}$. Our object is to find out an optimal control $u^0 \in \mathfrak{U}$ such that $J(u^0) = \min_{u \in \mathfrak{U}} J(u)$. We have the following theorem.

Theorem 313 *Assume that*

1° $A_t^i, i = 0,1,2,3,4, C_t^0$, and B_t all are non-random real continuous functions of t; $N_0 \geq 0$ is a constant, k_1, k_2 and k_3 are any even natural numbers; and $D_t(z) \geq 0$ is jointly measurable, non-random such that

$\int_Z |D_t(z)|^2 \, \pi(dz) < \infty,$

$\int_Z |D_{t+h}(z) - D_t(z)|^2 \, \pi(dz) \to 0,$ *as* $h \to 0$,

moreover, $Z = R - \{0\}$, $\int_Z \frac{|z|^2}{1+|z|^2} \pi(dz) < \infty;$

2° $A_t^i \geq 0, i = 2,3,4; C_t^1 \geq 0; C_t^0 \geq \delta_0 > 0$, where δ_0 is a constant,

$B_t > C_t^0 C_t^1;$

3° $g_2(t,u)$ is jointly continuous in $(t,u) \in [0,\infty) \times R$ such that

$g_2(t,0) = 0, g_2(t,u) \geq 0,$ *as* $u \geq 0;$

$x \geq y \geq 0 \implies g_2(t,x) \geq g_2(t,y),$

4° $g_1(t,|x|^2)$ is jointly continuous in (t,x) such that

$g_1(t,|x|^2) \leq k_0(1 + \prod_{k=1}^m \widetilde{g}_k(x)),$

where
$\widetilde{g}_k(x) = 1 + \underbrace{\ln(1+\ln(1+\cdots\ln}_{k-times}(1+|x|^{2n_0}))),$
(n_0 is some natural number), and $\frac{\partial g_1}{\partial x}$ exists such that it is uniformly locally bounded in x, that is, for each $r<\infty$, $\left|\frac{\partial g_1}{\partial x}\right| \le k_r$, as $|x|\le r$, where $k_r \ge 0$ is a constant only depending on r.
Then an optimal stochastic control exists, which is admissible and Bang-Bang, that is, there exists an admissible control $u^0 \in \mathfrak{U}$ such that
$J(u^0) = \min_{u\in\mathfrak{U}} J(u)$, $u_t^0 = -sgn\ x_t^0$,
where the optimal trajectory x_t^0 is the pathwise unique strong solution of (10.13) with $u_t = -sgn\ x$. That is, the SDE for the otimal trajectory is: $\forall t\in[0,T]$,

$$dx_t = (A_t^0 x_t + A_t^1 x_t g_1(t,|x_t|^2) - A_t^2 x_t g_2(t,|x_t|) - A_t^3 (x_t)^{2k_3+1}$$
$$-A_t^4(x_t)^{(2k_1+1)/(2k_2+1)} - B_t sgn(x_t))dt + (C_t^0 + C_t^1|x_t|)dw_t$$
$$+ \int_Z D_t(z)(x_{t-} I_{|x_{t-}|\le\sqrt{N_0}} + \sqrt{N_0}\frac{x_{t-}}{|x_{t-}|} I_{|x_{t-}|>\sqrt{N_0}})\widetilde{N}_k(dt,dz),\ x_0 = x\in R^1. \tag{10.14}$$

Let us explain the meaning of Theorem 313. If we consider $J(u) = E\int_0^T |x_t^u|^2\,dt$, to be the target functional as an energy functional, where x_t^u is a trajectory subject to a stochastic system like (10.13), controlled by some bounded process u_t (its bound is assumed to be 1 for simplicity), which we call an admissible control, ("admissible" means that it can be applied to the system and makes the system have a solution), then Theorem 313 tells us that an optimal control exists that makes the energy expended by this system the smallest. The idea is, if the energy reaches zero, thas is, of course, its smallest value. If so, then $x_t \equiv 0, \forall t\in[0,T]$. This is impossible in general, because 0 is not neccessary a solution of the given system. However, if the trajectory is closer to the original central point, then the energy expended is smaller. Thus to make the enrgy expended as small as possible, we should apply an optimal control as follows: At each time when the trajectory x_t, the solution of the system, departs from the original central point, we should apply a control u_t such that it immediately fully pulls back the tracjectory x_t directed towards the original central point such that it makes the energy $|x_t^u|^2$ smaller. Obviously, such a control should be $u_t = -sgn(x_t)$, which is called a Bang-Bang control. Theorem 313 tells us that the Bang-Bang control actually is admissible and optimal.

To prove this theorem we first need to introduce a lemma.

Lemma 314 *Let $(x_t^u, w_t^u, \zeta_t^u)$ be a triple of $d-$dimensional $\mathfrak{F}_t-$adapted random processes defined on a probability space $(\Omega, \mathfrak{F}, (\mathfrak{F})_{t\ge 0}, P)$ such that w_t^u is a BM, and if we write*
$\widetilde{p}^u((0,t],U) = \sum_{0<s\le t} I_{0\ne\triangle\widetilde{\zeta^u}_s\in U}$, *for $t\ge 0, U\in\mathfrak{B}(Z)$,*

$\widetilde{q}^u(dt, dz) = \widetilde{p}^u(dt, dz) - \pi(dz)dt,$
and we assume that
$Z = R^1 - \{0\}$, *and* $\int_Z \frac{|z|^2}{1+|z|^2}\pi(dz) < \infty;$
then $\widetilde{q}^u(dt, dz)$ *is a Poisson martingale measure with the compensator* $\pi(dz)dt$.
Let $(x_t^0, w_t^0, \zeta_t^0)$ *be a similar triple defined on another space* $(\Omega', \mathfrak{F}', (\mathfrak{F}')_{t\geq 0}, P')$.
Then there exist a probability space $(\widetilde{\Omega}, \widetilde{\mathfrak{F}}, (\widetilde{\mathfrak{F}}_t)_{t\geq 0}, \widetilde{P})$ *and four d–dimensional* $\widetilde{\mathfrak{F}}_t$*–adapted random processes* $(\widetilde{x}_t^u, \widetilde{x}_t^0, \widetilde{w}_t, \widetilde{\zeta}_t)$ *such that the finite probabilty distributions of* $(\widetilde{x}_t^u, \widetilde{w}_t, \widetilde{\zeta}_t)$ *and* $(\widetilde{x}_t^0, \widetilde{w}_t, \widetilde{\zeta}_t)$ *coincide with those of* $(x_t^u, w_t^u, \zeta_t^u)$ *and* $(x_t^0, w_t^0, \zeta_t^0)$*, respectively. Moreover,* $\widetilde{w}_t$ *is a d–dimensional* $\widetilde{\mathfrak{F}}_t$*–BM under the probability* $\widetilde{P}$*, and if we write*
$\widetilde{p}((0,t], U) = \sum_{0<s\leq t} I_{0\neq\Delta\widetilde{\zeta}_s\in U}$, *for* $t \geq 0, U \in \mathfrak{B}(Z)$,
$\widetilde{q}(dt, dz) = \widetilde{p}(dt, dz) - \pi(dz)dt,$
then $\widetilde{q}(dt, dz)$ *is a Poisson martingale measure with the same compensator* $\pi(dz)dt$ *under the probability* $\widetilde{P}$.

Lemma 314 is easily seen from the proof of Theorem 137. Now let us prove Theorem 313.

Proof. First let us prove that (10.14) has a pathwise unique strong solution, so $u_t^0 = -sgn\ x_t^0 \in \mathfrak{U}$. In fact, if we write
$b(t, x) = A_t^0 x + A_t^1 x g_1(t, |x|^2) - A_t^2 x g_2(t, |x|) - A_t^3(x)^{2k_3+1}$
$-A_t^4(x)^{(2k_1+1)/(2k_2+1)} - B_t sgn(x),$
$\sigma(t, x) = C_t^0 + C_t^1 |x|,$
$c(t, x, z) = D_t(z)(xI_{|x|\leq\sqrt{N_0}} + \sqrt{N_0}\frac{x}{|x|}I_{|x|>\sqrt{N_0}}),$
then
$xb(t, x) \leq A_t^1 x^2 g_1(t, |x|^2) + A_t^0 x^2 - B_t x \cdot sgn(x)$
$\leq [k_0 |A_t^1| + |A_t^0|]x^2(1 + \prod_{k=1}^m \widetilde{g}_k(x)),$
$\sigma(t, x) \geq \delta_0 > 0,$
$\int_Z |c(t, x, z)|^2 \pi(dz) \leq \widetilde{k}_0 < \infty,$
$x \geq y \implies x + c(t, x, z) \geq y + c(t, y, z).$
Hence Theorem 305 applies, and (10.14) has a pathwise unique strong solution. Secondly, let us show that
$J(u^0) \leq J(u), \forall u \in \mathfrak{U}.$
In fact, for any $u \in \mathfrak{U}$ let x_t^u be the pathwise unique strong solution of (10.13). Write
$w_t^u = \int_0^t [I_{x_s^u\neq 0}\frac{x_s^u}{|x_s^u|} + I_{x_s^u=0}]dw_s = \int_0^t p(x_s^u)dw_s.$
Then
$E[(w_t^u)^2 - (w_s^u)^2|\mathfrak{F}_s] = E[(w_t^u - w_s^u)^2|\mathfrak{F}_s] = t - s.$
Hence $\{w_t^u\}_{t\geq 0}$ is still a BM by Theorem 97. Moreover, one easily sees that $p(x)^{-1}$ exists, and $p(x)^{-1} = 1$, as $x = 0$; $p(x)^{-1} = \frac{|x|}{x}$, as $x \neq 0$. Hence $xp(x)^{-1} = |x|$, and we can write
$dw_t = p(x_t^u)^{-1}dw_t^u$. So x_t^u will satisfy (10.13) with $p(x_t^u)^{-1}dw_t^u$ substituting dw_t. Similarly, x_t^0 will satisfy (10.14) with $p(x_t^0)^{-1}dw_t^0$ substituting dw_t. Applying Lemma 314 we find a probability space $(\widetilde{\Omega}, \widetilde{\mathfrak{F}}, (\widetilde{\mathfrak{F}}_t)_{t\geq 0}, \widetilde{P})$

and four 1−dimensional $\widetilde{\mathfrak{F}}_t$−adapted random processes $(\widetilde{x}_t^u, \widetilde{x}_t^0, \widetilde{w}_t, \widetilde{\zeta}_t)$ such that the finite probabilty distributions of $(\widetilde{x}_t^u, \widetilde{w}_t, \widetilde{\zeta}_t)$ and $(\widetilde{x}_t^0, \widetilde{w}_t, \widetilde{\zeta}_t)$ coincide with those of $(x_t^u, w_t^u, \zeta_t^u)$ and $(x_t^0, w_t^0, \zeta_t^0)$, respectively. Moreover, $\widetilde{w}_t$ is a 1−dimensional $\widetilde{\mathfrak{F}}_t$−BM under the probability $\widetilde{P}$, and if we write

$\widetilde{p}((0,t], U) = \sum_{0<s\leq t} I_{0\neq\Delta\widetilde{\zeta}_s\in U}$, for $t \geq 0, U \in \mathfrak{B}(Z)$,

$\widetilde{q}(dt, dz) = \widetilde{p}(dt, dz) - \pi(dz)dt$,

then $\widetilde{q}(dt, dz)$ is a Poisson martingale measure with the same compensator $\pi(dz)dt$ under the probability $\widetilde{P}$. So we have that $\widetilde{x}_t^u$ sastisfies the following SDE: $\widetilde{P} - a.s.$ $\forall t \in [0,T]$,

$$\begin{aligned}
&dx_t = (A_t^1 x_t g_1(t, |x_t|^2) - A_t^2 x_t g_2(t, |x_t|) - A_t^3 (x_t)^{2k_3+1} \\
&-A_t^4 (x_t)^{(2k_1+1)/(2k_2+1)} + A_t^0 x_t + B_t u_t)dt + (C_t^0 + C_t^1 x_t) p(x_t)^{-1} d\widetilde{w}_t \\
&+ \int_Z D_t(z)(x_{t-} I_{|x_{t-}|\leq N_0} + \sqrt{N_0} \tfrac{x_{t-}}{|x_{t-}|} I_{|x_{t-}|>N_0}) \widetilde{q}(dt, dz), \; x_0 = x \in R^1.
\end{aligned}$$

Applying Ito's formula to $|\widetilde{x}_t^u|^2$, one finds that (for simplicity we still write x_t for $\widetilde{x}_t^u$, and u_t for $u(t, \widetilde{x}_t^u)$) $\widetilde{y}_t = \widetilde{y}_t^u = |\widetilde{x}_t^u|^2$ satisfies the following SDE: $\widetilde{P} - a.s.$ $\forall t \in [0,T]$,

$$\begin{aligned}
&dy_t = [2(A_t^0 y_t + A_t^1 y_t g_1(t, y_t) - A_t^2 y_t g_2(t, \sqrt{y_t \vee 0}) - A_t^3 (y_t)^{k_3+1} \\
&-A_t^4 (y_t)^{(k_1+k_2+1)/(2k_2+1)} + B_t x_t u_t) + (C_t^0)^2 + (C_t^1)^2 y_t + 2C_t^0 C_t^1 \sqrt{y_t \vee 0} \\
&+ \int_Z D_t(z)^2 (y_t I_{\sqrt{y_t \vee 0}\leq\sqrt{N_0}} + N_0 I_{\sqrt{y_t \vee 0}>\sqrt{N_0}}) \pi(dz)]dt + (C_t^0 \sqrt{y_t \vee 0} \\
&+C_t^1 y_t) d\widetilde{w}_t + \int_Z [2D_t(z)(y_{t-} I_{\sqrt{y_{t-}\vee 0}\leq\sqrt{N_0}} + \sqrt{N_0}\sqrt{y_{t-} \vee 0} I_{\sqrt{y_{t-}\vee 0}>\sqrt{N_0}}) \\
&+D_t(z)^2 (y_{t-} I_{\sqrt{y_{t-}\vee 0}\leq\sqrt{N_0}} + N_0 I_{\sqrt{y_{t-}\vee 0}>\sqrt{N_0}})] \widetilde{q}(dt, dz), \; y_0 = |x_0|^2.
\end{aligned}$$

Similarly, by Ito's formula to to $|\widetilde{x}_t^0|^2$, where $\widetilde{x}_t^0$ is the pathwise unique strong solution for $u_t = u_t^0 = -sgn\ x_t^0$, one sees that $\widetilde{y}_t^0 = |\widetilde{x}_t^0|^2$ satisfies the following SDE: $\widetilde{P} - a.s.$ $\forall t \in [0,T]$,

$$\begin{aligned}
&dy_t^0 = [2(A_t^0 y_t^0 + A_t^1 y_t^0 g_1(t, y_t^0) - A_t^2 y_t^0 g_2(t, \sqrt{y_t^0 \vee 0}) - A_t^3 (y_t^0)^{k_3+1} \\
&-A_t^4 (y_t^0)^{(k_1+k_2+1)/(2k_2+1)} - (B_t - 2C_t^0 C_t^1)\sqrt{y_t^0 \vee 0}) + (C_t^0)^2 + (C_t^1)^2 y_t^0 \\
&+ \int_Z D_t(z)^2 (y_t^0 I_{\sqrt{y_t^0 \vee 0}\leq\sqrt{N_0}} + N_0 I_{\sqrt{y_t^0 \vee 0}>\sqrt{N_0}}) \pi(dz)]dt + (C_t^0 \sqrt{y_t^0 \vee 0} \\
&+C_t^1 y_t^0) d\widetilde{w}_t + \int_Z [2D_t(z)(y_{t-}^0 I_{\sqrt{y_{t-}^0 \vee 0}\leq\sqrt{N_0}} + \sqrt{N_0}\sqrt{y_{t-}^0 \vee 0} I_{\sqrt{y_{t-}^0 \vee 0}>\sqrt{N_0}}) \\
&+D_t(z)^2 (y_{t-}^0 I_{\sqrt{y_{t-}^0 \vee 0}\leq\sqrt{N_0}} + N_0 I_{\sqrt{y_{t-}^0 \vee 0}>\sqrt{N_0}})] \widetilde{q}(dt, dz), \; y_0^0 = |x_0|^2.
\end{aligned}$$

Now write

$$\begin{aligned}
&b(t,y) = 2(A_t^0 y + A_t^1 y g_1(t,y) - A_t^2 y g_2(t, \sqrt{y \vee 0}) - A_t^3 (y)^{k_3+1} \\
&-A_t^4 (y)^{(k_1+k_2+1)/(2k_2+1)} - (B_t - C_t^0 C_t^1)\sqrt{y \vee 0}) + (C_t^0)^2 + (C_t^1)^2 y \\
&+ \int_Z D_t(z)^2 (y I_{\sqrt{y \vee 0}\leq\sqrt{N_0}} + N_0 I_{\sqrt{y \vee 0}>\sqrt{N_0}}) \pi(dz), \\
&\sigma(t,y) = C_t^0 \sqrt{y \vee 0} + C_t^1 y, \\
&c(t,y,z) = 2D_t(z)(y I_{\sqrt{y \vee 0}\leq\sqrt{N_0}} + \sqrt{N_0}\sqrt{y_{t-}^0 \vee 0} I_{\sqrt{y \vee 0}>\sqrt{N_0}}) \\
&+D_t(z)^2 (y I_{\sqrt{y \vee 0}\leq\sqrt{N_0}} + N_0 I_{\sqrt{y \vee 0}>\sqrt{N_0}}).
\end{aligned}$$

Then $\widetilde{y}_t^0$ is a strong solution of the SDE with coefficients (b, σ, c). Moreover, since

$$\begin{aligned}
&(x-y)(b(t,x) - b(t,y)) \leq 2\left|A_t^0\right| |x-y|^2 \\
&+k_N \left|A_t^1\right| |x-y|^2 + (C_t^1)^2 |x-y|^2 + |x-y|^2 \int_Z D_t(z)^2 \pi(dz)
\end{aligned}$$

$\leq k_{T,N}|x-y|^2$, as $|x|,|y|\leq N$, $t\in[0,T]$,
where $k_{T,N}$ is a non-negative constant depending only on T and N. Dividing both sides by $I_{x\neq y}|x-y|$, one easily finds that
$sgn(x-y)\cdot(b(t,x)-b(t,y))\leq k_{T,N}|x-y|$,
as $|x|,|y|\leq N$, $t\in[0,T]$.
On the other hand, one also sees that as $|x|,|y|\leq N$, $t\in[0,T]$,
$|\sigma(t,x)-\sigma(t,y)|^2\leq k'_{T,N}\sqrt{|x-y|}$.
Moreover, as $x\geq y$,
$x+c(t,x,z)\geq y+c(t,y,z)$.
Let

$$\beta^1(t,\omega)=$$
$$2(A_t^0\widetilde{y}_t^u+A_t^1\widetilde{y}_t^u g_1(t,\widetilde{y}_t^u)-A_t^2 y_t g_2(t,\sqrt{\widetilde{y}_t^u\vee 0})-A_t^3(\widetilde{y}_t^u)^{k_3+1}$$
$$-A_t^4(\widetilde{y}_t^u)^{(k_1+k_2+1)/(2k_2+1)}+B_t\widetilde{x}_t^u u_t)+(C_t^0)^2+(C_t^1)^2\widetilde{y}_t^u$$
$$+2C_t^0C_t^1\sqrt{\widetilde{y}_t^u\vee 0}+\int_Z D_t(z)^2(\widetilde{y}_t^u I_{\sqrt{\widetilde{y}_t^u\vee 0}\leq\sqrt{N_0}}+N_0 I_{\sqrt{\widetilde{y}_t^u\vee 0}>\sqrt{N_0}})\pi(dz).$$

Then one finds that
$\beta^1(t,\omega)\geq b(t,\widetilde{y}_t^u)$,
because that $\widetilde{x}_t^u u_t\geq-\widetilde{x}_t^u sgn(\widetilde{x}_t^u)$ for $u(\cdot)\in\mathfrak{U}$. Therefore, the comparison theorem for solutions of SDE with jumps (Theorem 295) applies, and we obtain that $\widetilde{P}-a.s.$ $|\widetilde{x}_t^u|^2=\widetilde{y}_t^u\geq\widetilde{y}_t^0=|\widetilde{x}_t^0|^2$, $\forall t\in[0,T]$. That is, $P-a.s.$
$|x_t^u|^2\geq|x_t^0|^2$, $\forall t\in[0,T]$,
because $\{\widetilde{x}_t^u\}_{t\in[0,T]}$ and $\{x_t^u\}_{t\in[0,T]}$ have the same finite probability distributions. Therefore,
$J(u)=E\int_0^T|x_t^u|^2dt\geq J(u^0)=E\int_0^T|x_t^0|^2dt$, $\forall u(\cdot)\in\mathfrak{U}$. ■

10.4.2 Partially-Degenerate Case

Consider the following 1−dimensional stochastic system: $\forall t\in[0,T]$,

$$dx_t=(A_t^0x_t-A_t^1x_tg_1(t,|x_t|)-A_t^2\frac{x_t}{|x_t|^\alpha}I_{x_t\neq 0}+B_tu_t)dt$$
$$+(C_t^0+C_t^1|x_t|)dw_t+\int_Z D_t(z)x_{t-}\widetilde{N}_k(dt,dz),\ x_0=x\in R^1. \quad (10.15)$$

The admissible control set is:
$\mathfrak{U}=\{u=u(t,\omega):u(t,\omega)$ is $\mathfrak{F}_t-$adapted, and $|u(t,\omega)|\leq 1\}$.
In general, if $u\in\mathfrak{U}$, u is called an open-loop control. However, if $u=u(t,x_t^u(\omega))$, where $x_t^u(\omega)$ is the pathwise unique solution of (10.15) with this u_t, and $|u(t,x_t^u(\omega))|\leq 1$, then $u(t,x_t^u(\omega))$ is called a feedback control, or a closed-loop control. We still want to minimize the energy functional $J(u)=E\int_0^T|x_t^u|^2dt$ among all $u\in\mathfrak{U}$. We can have the following theorem.

Theorem 315 *Assume that*

1° $A_t^i, C_t^i, i = 0,1; A_t^2$ *and* B_t *all are non-random real continuous functions of* t*; and* $D_t(z) \geq 0$ *is jointly measurable, non-random such that* $\int_Z |D_t(z)|^2 \pi(dz) < \infty$,

$\int_Z |D_{t+h}(z) - D_t(z)|^2 \pi(dz) \to 0$, *as* $h \to 0$,

moreover, $Z = R - \{0\}$, $\int_Z \frac{|z|^2}{1+|z|^2}\pi(dz) < \infty$*;*

2° $A_t^1, A_t^2 \geq 0, |C_t^0| > 0,\ B_t > C_t^0 C_t^1$, *and* $0 \leq \alpha < 1$ *is a constant;*

3° $g_1(t,u)$ *is jointly continuous in* $(t,u) \in [0,\infty) \times R$ *such that*

$|g_1(t,u)| \leq k_0, g_1(t,0) = 0;$

$g_1(t,u) \geq 0$, *as* $u \geq 0$;

$x \geq y \geq 0 \implies g_1(t,x) \geq g_1(t,y)$.

Then an optimal stochastic control exists, which is admissible, feedback and Bang-Bang, that is, there exists an admissible control $u^0 \in \mathfrak{U}$ *such that*

$J(u^0) = \min_{u \in \mathfrak{U}} J(u),\ u_t^0 = -sgn\ x_t^0$,

where x_t^0 *is the pathwise unique strong solution of (10.15) with* $u_t = -sgn\ x$.

Proof. First let us prove that (10.15) with $u_t = -sgn\ x$ has a pathwise unique strong solution, so $u_t^0 = -sgn\ x_t^0 \in \mathfrak{U}$. In fact, if we write

$b(t,x) = A_t^0 x - A_t^1 x^\alpha g_1(t,|x|) - A_t^2 \frac{x}{|x|^\alpha} I_{x_t \neq 0} - B_t sgn(x)$,

$\sigma(t,x) = C_t^0 + C_t^1 |x|$,

$c(t,x,z) = D_t(z)x$,

then

$xb(t,x) \leq A_t^0 x^2 - B_t x \cdot sgn(x) \leq |A_t^0| x^2$,

$|\sigma(t,0)|^2 \geq \delta_0 > 0$,

$\int_Z |c(t,x,z)|^2 \pi(dz) \leq x^2 \int_Z |D_t(z)|^2 \pi(dz) < \infty$,

$x \geq y \implies x + c(t,x,z) \geq y + c(t,y,z)$,

and $b(t,x)$ is discontinuous at $x = 0$. However,

$|\sigma(t,x)|^2 \geq (C_t^0)^2 > 0$, as $x = 0$.

Moreover, $2(x_1 - x_2)(b(t,x_1) - b(t,x_2)) \leq |A_t^0| |x-y|^2$,

$|\sigma(t,x_1) - \sigma(t,x_2)|^2 \leq |C_t^1| |x-y|^2$,

$x \geq y \implies x + c(t,x,z) \geq y + c(t,y,z)$.

Hence Theorem 310 applies, and (10.15) with $u_t = -sgn\ x$ has a pathwise unique strong solution. Now the proof of the remaining part is similar to that of Theorem 313. ∎

Notice that in the above theorem the coefficient $b(t,x)$ has a discontinuous point $x = 0$, and at this point $|\sigma(t,0)|^2 > 0$. That is, $\sigma(t,x)$ is non-degenerate at the discontinuous points of $b(t,x)$, and the Lebesgue's measure of the set of all discontinuous points is zero.

Now let us consider another partially degenerate stochastic system, where its coefficient $b(t,x)$ can be greater than linear growth: $\forall t \in [0,T]$,

$$dx_t = (A_t^0 x_t - A_t^1 x_t g_1(t, |x_t|) - A_t^2 \frac{x_t}{|x_t|^\alpha} I_{x_t \neq 0}$$
$$+ A_t^3 x_t g_2(t, |x_t|^2) + B_t u_t)dt + (C_t^0 + C_t^1 |x_t|)dw_t$$
$$+ \int_Z D_t(z)(x_{t-} I_{|x_{t-}| \leq \sqrt{N_0}} + \sqrt{N_0} \frac{x_{t-}}{|x_{t-}|} I_{|x_{t-}| > \sqrt{N_0}}) \widetilde{N}_k(dt, dz), \; x_0 = x \in R^1. \tag{10.16}$$

Theorem 316 *Assume that*
1° $A_t^i, C_t^i, i = 0, 1; A_t^2, A_t^3$ *and* B_t *all are non-random real continuous functions of* t; *and* $D_t(z) \geq 0$ *is jointly measurable, non-random such that* $\int_Z |D_t(z)|^2 \pi(dz) < \infty$,
$\int_Z |D_{t+h}(z) - D_t(z)|^2 \pi(dz) \to 0$, *as* $h \to 0$,
moreover, $Z = R - \{0\}$, $\int_Z \frac{|z|^2}{1+|z|^2} \pi(dz) < \infty$;
2° $A_t^1, A_t^2 \geq 0$, $C_t^0 (C_t^1)^{-1} < 0$, $B_t > C_t^0 C_t^1$, $B_t > 0$ *and* $0 \leq \alpha < 1$ *is a constant;*
3° $g_1(t, u)$ *is jointly continuous in* $(t, u) \in [0, \infty) \times R$ *such that*
$|g_1(t, u)| \leq k_0, g_1(t, 0) = 0$;
$g_1(t, u) \geq 0$, *as* $u \geq 0$;
$x \geq y \geq 0 \implies g_1(t, x) \geq g_1(t, y)$;
4° $g_2(t, |x|^2)$ *is jointly continuous in* (t, x) *such that*
$\left| g_2(t, |x|^2) \right| \leq k_0 (1 + \prod_{k=1}^m \widetilde{g}_k(x))$,
where
$\widetilde{g}_k(x) = 1 + \underbrace{\ln(1 + \ln(1 + \cdots \ln}_{k-times}(1 + |x|^{2n_0})))$,
(n_0 *is some natural number), and* $\frac{\partial g_2}{\partial x}$ *exists such that it is uniformly locally bounded in* x, *that is, for each* $r < \infty$, $\left| \frac{\partial g_2}{\partial x} \right| \leq k_r$, *as* $|x| \leq r$, *where* $k_r \geq 0$ *is a constant only depending on* r.
Then the optimal stochastic control exists, which is admissible, feedback and Bang-Bang, that is, there exists an admissible control $u^0 \in \mathfrak{U}$ *such that*
$J(u^0) = \min_{u \in \mathfrak{U}} J(u)$, $u_t^0 = -sgn\ x_t^0$,
where x_t^0 *is the pathwise unique strong solution of (10.16) with* $u_t = -sgn\ x$.

Proof. First let us prove that (10.16) with $u_t = -sgn\ x$ has a pathwise unique strong solution, so $u_t^0 = -sgn\ x_t^0 \in \mathfrak{U}$. In fact, if we write
$b(t, x) = A_t^0 x - A_t^1 x g_1(t, |x|) - A_t^2 \frac{x}{|x|^\alpha} I_{x \neq 0} + A_t^3 x g_2(t, |x|^2) - B_t sgn(x)$,
$\sigma(t, x) = C_t^0 + C_t^1 |x|$,
$c(t, x, z) = D_t(z)(x I_{|x_t| \leq \sqrt{N_0}} + \sqrt{N_0} \frac{x}{|x|} I_{|x-| > \sqrt{N_0}})$,
then
$|b(t, x)| \leq |B_t| + |A_t^2| |x|^{1-\alpha} + (|A_t^0| + k_0 |A_t^1|) |x| + |A_t^3| |x| k_0 (1 + \prod_{k=1}^m \widetilde{g}_k(x))$,
$|\sigma(t, 0)|^2 \geq \delta_0 > 0$,

$\int_Z |c(t,x,z)|^2 \pi(dz) \leq N_0 \int_Z |D_t(z)|^2 \pi(dz) < \infty,$
$x \geq y \Longrightarrow x + c(t,x,z) \geq y + c(t,y,z),$
and $b(t,x)$ is discontinuous at $x = 0$. Obviously, $F = \{0\}$ is a compact set in R^1. Moreover, the degenerate points of $\sigma(t,x)$ should satisfy that $|x| = -C_t^0(C_t^1)^{-1} = f_t$. Since by assumption $-C_t^0(C_t^1)^{-1}$ is positive and continuous, then one should have $f_t \geq \delta_1 > 0$, as $t \in [0,T]$, where δ_1 is a constant, which exists. Write
$G_{\delta_1} = (-\delta_1, \delta_1)$.
Then $m_1 G_{\delta_1} = 2\delta_1 > 0$, and $\sigma(t,x)$ is non-degenerate in $x \in G_{\delta_1}, \forall t \in [0,T]$. Since $\sigma(t,x)$ is jointly continuous, there exists a constant $\delta_0 > 0$ such that $|\sigma(t,x)|^2 \geq \delta_0$, as $x \in G_{\delta_1/2}, \forall t \in [0,T]$. Write $G_1 = G_{\delta_1/2}$. Then G_1 satisfies the condition 2° in Thoerem 312. Moreover, as $|x|, |y| \leq N$,
$(x-y)(b(t,x) - b(t,y)) \leq |A_t^0| |x-y|^2 + k_N |A_t^3| |x-y|^2;$
so, as $|x|, |y| \leq N$,
$sgn(x_1 - x_2)(b(t,x_1) - b(t,x_2)) \leq (|A_t^0| + k_N |A_t^3|) |x-y|.$
Furthermore,
$|\sigma(t,x) - \sigma(t,y)|^2 \leq |C_t^1| |x-y|^2,$
$x \geq y \Longrightarrow x + c(t,x,z) \geq y + c(t,y,z).$
Hence Theorem 312 applies, and (10.16) with $u_t = -sgn\ x$ has a pathwise unique strong solution. Now the proof of the remaining part is similar to that of Theorem 313. ■

10.5 Bang-Bang Control for $d-$Dimensional Non-linear Systems

10.5.1 Non-Degenerate Case

Consider the following $d-$dimensional stochastic system: $\forall t \in [0,T]$,

$$\begin{aligned} dx_t &= (A_t^0 x_t + A_t^1 x_t g_1(t,|x_t|^2) - A_t^2 x_t g_2(t,|x_t|) - A_t^3 x_t |x_t|^{2k_3} \\ &\quad - A_t^4 x_t |x_t|^{2(k_1-k_2)/(2k_2+1)} + B_t u_t)dt + (C_t^0 + C_t^1 |x_t|)dw_t \\ &+ \int_Z D_t(z)(x_{t-} I_{|x_{t-}| \leq \sqrt{N_0}} + \sqrt{N_0} \frac{x_{t-}}{|x_{t-}|} I_{|x_{t-}| > \sqrt{N_0}}) \widetilde{N}_k(dt,dz),\ x_0 = x \in R^d. \end{aligned} \tag{10.17}$$

The admissible control set is:
$\mathfrak{U} = \{u = u(t,\omega) : u(t,\omega)$ is $\mathfrak{F}_t-$adapted, and $|u(t,\omega)| \leq 1\}$.
We still want to minimize the energy functional $J(u) = E\int_0^T |x_t^u|^2 dt$ among all $u \in \mathfrak{U}$. We have the following theorem.

Theorem 317 *Assume that*
1° $A_t^i, i = 0,1,2,3,4,$ and B_t all are non-random continuous $d \times d$ matrices of t; $N_0 \geq 0$ is a constant, k_1, k_2 and k_3 are any even natural numbers;

and $d \times d$ matrix $D_t(z) \geq 0$ (non-negative definite) is jointly measurable, non-random and such that $\int_Z |D_t(z)|^2 \pi(dz) < \infty$,

$\int_Z |D_{t+h}(z) - D_t(z)|^2 \pi(dz) \to 0$, *as* $h \to 0$,

moreover, $Z = R^d - \{0\}$, $\int_Z \frac{|z|^2}{1+|z|^2} \pi(dz) < \infty$;

2° $d \times d$ matrices $A_t^i \geq 0, i = 2,3,4$; and $d \times d$ matrix $C_t^1 \geq 0$, and $C_t^0 \geq \delta_0 I_{d\times d}$, that is, $C_t^0 - \delta_0 I_{d\times d} \geq 0$, where $\delta_0 > 0$ is a constant, and $I_{d\times d}$ is a $d \times d$ unit matrix; moreover,

$B_t > C_t^0 C_t^1$,

that is, $d \times d$ matrix $B_t - C_t^0 C_t^1$ is positive definite;

3° $g_2(t,u)$ is real, jointly continuous in $(t,u) \in [0,\infty) \times R^d$ such that $\frac{\partial g_2}{\partial u}$ exists such that it is uniformly locally bounded in u, that is, for each $r < \infty$, $\left|\frac{\partial g_2}{\partial u}\right| \leq k_r$, as $|u| \leq r$, where $k_r \geq 0$ is a constant depending only on r; moreover,

$g_2(t,0) = 0, g_2(t,u) \geq 0$, *as* $u \geq 0$;

$x \geq y \geq 0 \implies g_2(t,x) \geq g_2(t,y)$;

4° $g_1(t,|x|^2)$ is real, jointly continuous in (t,x) such that

$g_1(t,|x|^2) \leq k_0(1 + \prod_{k=1}^m \widetilde{g}_k(x))$,

where $\widetilde{g}_k(x)$ is defined in 4° of Theorem 316.

Then an optimal stochastic control exists which is admissible and Bang-Bang, that is, there exists an admissible control $u^0 \in \mathfrak{U}$ such that

$J(u^0) = \min_{u\in\mathfrak{U}} J(u)$, $u_t^0 = -\frac{x_t^0}{|x_t^0|} I_{|x_t^0|\neq 0}$,

where the optimal trajectory x_t^0 is the pathwise unique strong solution of the SDE (10.17) with $u_t = -\frac{x}{|x|} I_{|x|\neq 0}$.

Proof. First let us prove that (10.17) with $u_t = -\frac{x}{|x|} I_{|x|\neq 0}$ has a pathwise unique strong solution, so $u_t^0 = -\frac{x_t^0}{|x_t^0|} I_{|x_t^0|\neq 0} \in \mathfrak{U}$. In fact, if we write

$b(t,x) = A_t^0 x + A_t^1 x g_1(t,|x|^2) - A_t^2 x g_2(t,|x|) - A_t^3 x |x|^{2k_3}$
$-A_t^4 x |x|^{2(k_1-k_2)/(2k_2+1)} - B_t \frac{x}{|x|} I_{|x|\neq 0}$

$\sigma(t,x) = C_t^0 + C_t^1 |x|$,

$c(t,x,z) = D_t(z)(x I_{|x|\leq\sqrt{N_0}} + \sqrt{N_0}\frac{x}{|x|} I_{|x|>\sqrt{N_0}})$,

then

$\langle x, b(t,x)\rangle \leq A_t^0 |x|^2 - B_t |x| + k_0 |A_t^1| |x|^2 (1 + \prod_{k=1}^m \widetilde{g}_k(x))$,

$|\sigma(t,x)|^2 \geq \delta_0^2 > 0$,

$\int_Z |c(t,x,z)|^2 \pi(dz) \leq N_0 \int_Z |D_t(z)|^2 \pi(dz) < \infty$,

and $b(t,x)$ is discontinuous at $x = 0$. Moreover,

$2\langle x-y, b(t,x) - b(t,y)\rangle \leq |A_t^0| |x-y|^2 + k_{N,T} |x-y|^2$,

as $|x|, |y| \leq N$, $t \in [0,T]$, where $k_{N,T} \geq 0$ is a constant depending only on N and T; and we have applied the facts that

$|g_2(t,|x|) - g_2(t,|y|)| \leq \left|\frac{\partial}{\partial u} g_2(t, |x| + \theta(|y|-|x|)\right| ||x|-|y|| \leq k_N |x-y|$,

as $|x|, |y| \leq N$; etc. Obviously,

$|\sigma(t,x) - \sigma(t,y)|^2 \leq |C_t^1| |x-y|^2$,

$\int_Z |c(t,x,z)-\sigma(t,y,z)|^2 \pi(dz) \leq \int_Z |D_t(z)|^2 \pi(dz) |x-y|^2$.
Hence Theorem 185 applies, and (10.15) with $u_t = -\frac{x}{|x|} I_{|x|\neq 0}$ has a pathwise unique strong solution. Now for any $u \in \mathfrak{U}$ write the solution of (10.17) corresponding to this u as x_t^u. Let

$$dw_t^u = p(x_t^u)dw_t,$$

where $p(x)$ is a $d \times d$ orthogonal matrix, that is, $p(x) \cdot p^*(x) = I$, ($p^*(x)$ is the transpose of $p(x)$), such that the first row vector is $(\frac{x^1}{|x|}, \frac{x^2}{|x|}, \cdots, \frac{x^d}{|x|})$, as $x \neq 0$; and is 1, as $x = 0$. Then $w_t^u, t \geq 0$, is a BM in the same probability space, which satisfies the conditions that

$$dw_t = p^{-1}(x_t^u)dw_t^u = p^*(x_t^u)dw_t^u$$

and

$$x_t^u \cdot p^{-1}(x_t^u)dw_t^u = \sum_{j=1}^d (x_t^u)^j \frac{(x_t^u)^j}{|x_t^u|} d(w_t^u)^1 + \sum_{j=1}^d (x_t^u)^j p_{2j}(x_t^u)d(w_t^u)^2$$
$$+ \cdots + \sum_{j=1}^d (x_t^u)^j p_{dj}(x_t^u)d(w_t^u)^d = |x_t^u| \, d(w_t^u)^1,$$

where $w_t^u = ((w_t^u)^1, \cdots, (w_t^u)^d)$. In fact, $\forall 0 \leq s \leq t$,

$$E[(w_t^u - w_s^u)^i (w_t^u - w_s^u)^j | \mathfrak{F}_s] = E\int_s^t \delta_{ij} dr = (t-s)\delta_{ij}.$$

Hence by Theorem 97 $w_t^u, t \geq 0$, is a $d-$dimensional BM. The other conditions for $w_t^u, t \geq 0$ stated above are obviously true.

So x_t^u satisfies (10.17) with $p(x_t^u)^{-1}dw_t^u$ substituting dw_t. Similarly, x_t^0 satisfies (10.17) with $p(x_t^0)^{-1}dw_t^0$ substituting dw_t. By Lemma 314 we find a probability space $(\widetilde{\Omega}, \widetilde{\mathfrak{F}}, (\widetilde{\mathfrak{F}}_t)_{t\geq 0}, \widetilde{P})$ and four $\widetilde{\mathfrak{F}}_t-$adapted random processes $(\widetilde{x}_t^u, \widetilde{x}_t^0, \widetilde{w}_t, \widetilde{\zeta}_t)$, where the first three componenets are $d-$dimensional and the fourth is $1-$dimensional, and such that the finite probabilty distributions of $(\widetilde{x}_t^u, \widetilde{w}_t, \widetilde{\zeta}_t)$ and $(\widetilde{x}_t^0, \widetilde{w}_t, \widetilde{\zeta}_t)$ coincide with those of $(x_t^u, w_t^u, \zeta_t^u)$ and $(x_t^0, w_t^0, \zeta_t^0)$, respectively. Moreover, $\widetilde{w}_t$ is a $d-$dimensional $\widetilde{\mathfrak{F}}_t-$BM under the probability $\widetilde{P}$, and if we write

$$\widetilde{p}((0,t],U) = \sum_{0<s\leq t} I_{0\neq \Delta\widetilde{\zeta}_s \in U}, \text{ for } t \geq 0, U \in \mathfrak{B}(Z),$$
$$\widetilde{q}(dt,dz) = \widetilde{p}(dt,dz) - \pi(dz)dt,$$

then $\widetilde{q}(dt,dz)$ is a Poisson martingale measure with the same compensator $\pi(dz)dt$ under the probability $\widetilde{P}$. So we have that $\widetilde{x}_t^u$ sastisfies the following SDE: $\widetilde{P} - a.s.$ $\forall t \in [0,T]$,

$$dx_t = (A_t^1 x_t g_1(t,|x_t|^2) - A_t^2 x_t g_2(t,|x_t|) - A_t^3 (x_t)^{2k_3+1}$$
$$-A_t^4 (x_t)^{(2k_1+1)/(2k_2+1)} + A_t^0 x_t + B_t u_t)dt + (C_t^0 + C_t^1 x_t)p(x_t)^{-1} d\widetilde{w}_t$$
$$+ \int_Z D_t(z)(x_{t-} I_{|x_{t-}|\leq N_0} + \sqrt{N_0}\frac{x_{t-}}{|x_{t-}|} I_{|x_{t-}|>N_0})\widetilde{q}(dt,dz), \; x_0 = x \in R^1.$$

Applying Ito's formula to $|\widetilde{x}_t^u|^2$, one finds that (for simplicity we still write x_t for $\widetilde{x}_t^u$, and u_t for $u(t,\widetilde{x}_t^u)$) $\widetilde{y}_t = \widetilde{y}_t^u = |\widetilde{x}_t^u|^2$ satisfies the following SDE: $\widetilde{P} - a.s.$ $\forall t \in [0,T]$,

$$dy_t = [2(A_t^0 y_t + A_t^1 y_t g_1(t,y_t) - A_t^2 y_t g_2(t,\sqrt{y_t \vee 0}) - A_t^3 (y_t)^{k_3+1}$$
$$-A_t^4 (y_t)^{(k_1+k_2+1)/(2k_2+1)} + B_t \langle x_t, u_t \rangle) + (C_t^0)^2 + (C_t^1)^2 y_t + 2C_t^0 C_t^1 \sqrt{y_t \vee 0}$$
$$+ \int_Z D_t(z)^2 (y_t I_{\sqrt{y_t \vee 0} \leq \sqrt{N_0}} + N_0 I_{\sqrt{y_t \vee 0} > \sqrt{N_0}})\pi(dz)]dt + (C_t^0 \sqrt{y_t \vee 0}$$
$$+ C_t^1 y_t)d\widetilde{w}_t^1 + \int_Z [2D_t(z)(y_{t-} I_{\sqrt{y_{t-} \vee 0} \leq \sqrt{N_0}} + \sqrt{N_0}\sqrt{y_{t-} \vee 0} I_{\sqrt{y_{t-} \vee 0} > \sqrt{N_0}})$$
$$+ D_t(z)^2 (y_{t-} I_{\sqrt{y_{t-} \vee 0} \leq \sqrt{N_0}} + N_0 I_{\sqrt{y_{t-} \vee 0} > \sqrt{N_0}})]\widetilde{q}(dt,dz), \; y_0 = |x_0|^2 \in R^1;$$

where $\widetilde{w}_t = (\widetilde{w}_t^1, \cdots, \widetilde{w}_t^d)$, and $\widetilde{w}_t^1$ is a 1−dimensional BM, which is the first component of $\widetilde{w}_t$. Notice that the SDE for y_t now is a 1−dimensional SDE, and

$\langle x, u \rangle \geq -\left\langle x, \frac{x}{|x|} I_{|x|\neq 0} \right\rangle = -|x|$, for all $u \in R^d$ such that $|u| \leq 1$.

So the proof of the remaining part follows as in Theorem 313. ∎

10.5.2 Partially-Degenerate Case

Before we discuss the stochastic Bang-Bang control for a system which can be partially degenerate, we first need to establish a theorem on the existence of a strong solution for a SDE with jumps and with discontinuous coefficients and with partially degenerate σ in d−dimensional space. Consider the following d−dimensional SDE with jumps as (5.15): $\forall t \geq 0$,

$$x_t = x_0 + \int_0^t b(s, x_s)ds + \int_0^t \sigma(s, x_s)dw_s + \int_0^t \int_Z c(s, x_{s-}, z)\widetilde{N}_k(ds, dz). \tag{10.18}$$

Theorem 318 *Assume that*
1° $Z = R^d - \{0\}$, and $\int_Z \frac{|z|^2}{1+|z|^2}\pi(dz) < \infty$
2° $b = b(t, x) : [0, \infty) \times R^d \to R^d$,
$\sigma = \sigma(t, x) : [0, \infty) \times R^d \to R^{d\otimes d}$,
$c = c(t, x, z) : [0, \infty) \times R^d \times Z \to R^d$,
are jointly Borel measurable such that
$|b(t, x)| \leq c_1(t)(1 + |x| \prod_{k=1}^m g_k(x))$,
$|\sigma(t, x)|^2 \leq k_0(1 + |x|^2 \prod_{k=1}^m g_k(x))$,
$\int_Z |c(t, x, z)|^2 \pi(dz) \leqslant c_1(t)$,
where
$$g_k(x) = 1 + \underbrace{\ln(1 + \ln(1 + \cdots \ln}_{k-times}(1 + |x|^{2n_0}))),$$
(n_0 is some natural number), and $c_1(t) \geq 0$ is non-random such that for each $T < \infty$, $\int_0^T c_1(t)dt < \infty$; furthermore, $b(t, x)$ is locally bounded for x, that is, for each $r > 0$, as $|x| \leq r$,
$|b(t, x)| \leq k_r$,
where $k_r > 0$ is a constant depending only on r;
3° $\sigma(t, x)$ is jointly continuous in (t, x); and
$\lim_{h,h'\to 0} \int_Z |c(t + h', x + h, z) - c(t, x, z)|^2 \pi(dz) = 0$;
4° $b(t, x)$ is continuous in $x \in R^d \backslash F$, where $F \subset R^d$ is a compact set such that $m_d F = 0$, where m_d is the Lebesgue measure in R^d; moreover, there exist a $\delta_0 > 0$ and an open set $G_1 \supset F$ such that $x \in G_1 \Longrightarrow |\sigma(t, x)| \geq \delta_0 > 0, \forall t \geq 0$.
Then for any given constant $x_0 \in R^d$ (10.18) has a weak solution on $t \geq 0$. Furthermore, if, in addition, the following condition (for the pathwise uniqueness) holds:

(PWU1) for each $N = 1, 2, \cdots$, *and each* $T < \infty$,
$2\langle(x_1 - x_2), (b(t, x_1) - b(t, x_2))\rangle$
$+ |\sigma(t, x_1) - \sigma(t, x_2)|^2 + \int_Z |c(t, x_1, z) - c(t, x_2, z)|^2 \pi(dz)$
$\leqslant c_T^N(t)\rho_T^N(|x_1 - x_2|^2)$,
as $|x_i| \leqslant N, i = 1, 2,\ t \in [0, T]$; *where* $c_T^N(t) \geq 0$ *such that* $\int_0^T c_T^N(t)dt < \infty$; *and* $\rho_T^N(u) \geq 0$, *as* $u \geq 0$, *is strictly increasing, continuous and concave such that*
$\int_{0+} du/\rho_T^N(u) = \infty$;
then (10.18) has a pathwise unique strong solution.

Proof. Let us smooth out $b(t, x)$ and $\sigma(t, x)$ only with respect to x to get $b^n(t, x)$ and $\sigma^n(t, x)$, respectively. (See Lemma 172 and its proof). Then by Theorem 175 for each n there exists a weak solution x_t^n with a BM w_t^n and a Poisson martingale measure $\widetilde{N}_{k^n}(dt, dz)$, which has the same compensator $\pi(dz)dt$, defined on some probability space $(\Omega^n, \mathfrak{F}^n, \{\mathfrak{F}_t^n\}, P^n)$ such that $P^n - a.s.\ \forall t \geq 0$,
$x_t^n = x_0 + \int_0^t b^n(s, x_s^n)ds + \int_0^t \sigma^n(s, x_s^n)dw_s^n + \int_0^t \int_Z c(s, x_{s-}^n, z)\widetilde{N}_{k^n}(ds, dz)$.
As in the proof of Theorem 176, one finds that "the result os SDE from the Skorohod weak convergence technique" holds. In particular, we have that $(\widetilde{x}_t'^n)$ satisfies the following SDE with $\widetilde{w}_t'^n$ and $\widetilde{q}^n(dt, dz)$ on $(\widetilde{\Omega}, \widetilde{\mathfrak{F}}, \widetilde{P})$, where $(\widetilde{x}_t'^n, \widetilde{w}_t'^n, \widetilde{q}^n(dt, dz))$ come from "the result of SDE from the Skorohod weak convergence technique", etc:

$$\begin{aligned}\widetilde{x}_t'^n &= x_0 + \int_0^t b^n(s, \widetilde{x}_s'^n)ds + \int_0^t \sigma^n(s, \widetilde{x}_s'^n)d\widetilde{w}_s'^n \\ &\quad + \int_0^t \int_Z c(s, \widetilde{x}_{s-}'^n, z)\widetilde{q}^n(ds, dz). \qquad (10.19)\end{aligned}$$

Write
$\tau_N = \inf\{t \geq 0 : |\widetilde{x}_t'^0| > N\}$.
Let us show that for each N as $n \to \infty$
1) $\left|\int_0^{t\wedge\tau_N}(b^n(s, \widetilde{x}_s'^n) - b(s, \widetilde{x}_s'^0))ds\right| \to 0$, in probability $\widetilde{P}$.
For this write
$A_{t\wedge\tau_N}^n = \int_0^{t\wedge\tau_N} b^n(s, \widetilde{x}_s'^n)ds$.
Then there exists a limit (in probability)
$A_{t\wedge\tau_N} = \lim_{n\to\infty} A_{t\wedge\tau_N}^n$,
and we find that

$$\widetilde{x}_t'^0 = x_0 + A_t + \int_0^t \sigma(s, \widetilde{x}_s'^0)d\widetilde{w}_s'^0 + \int_0^t \int_Z c(s, \widetilde{x}_{s-}'^0, z)\widetilde{q}^0(ds, dz). \qquad (10.20)$$

Observe that for any $\varepsilon > 0$
$\widetilde{P}(\left|\int_0^{t\wedge\tau_N -}(b^n(s, \widetilde{x}_s'^n) - b(s, \widetilde{x}_s'^0))ds\right| > \varepsilon)$
$\leq \sup_n \widetilde{P}(\sup_{s\leq T}|\widetilde{x}_s'^n| + \sup_{s\leq T}|\widetilde{x}_s'^0| > \widetilde{N})$

$+\frac{1}{\varepsilon^2}E^{\widetilde{P}}\int_0^{t\wedge\tau_N-} I_{\widetilde{x}_s'^0\notin G_2}\left|b^n(s,\widetilde{x}_s'^n)-b(s,\widetilde{x}_s'^0)\right| I_{\sup_{s\leq T}|\widetilde{x}_s'^n|+\sup_{s\leq T}|\widetilde{x}_s'^0|\leq\widetilde{N}}ds]$
$= I_1^{\widetilde{N}} + I_2^{n,\widetilde{N}}.$

Obviously, for any $\frac{\overline{\varepsilon}}{2}>0$ one can take an $\widetilde{N}$ large enough such that
$I_1^{\widetilde{N}} < \frac{\overline{\varepsilon}}{2}.$

However, the term $I_2^{n,\widetilde{N}}$ requires more discussion. Notice that
$I_2^{n,\widetilde{N}} \leq \frac{2k_{\widetilde{N}}}{\varepsilon^2}E^{\widetilde{P}}\int_0^{t\wedge\tau_N-} I_{\widetilde{x}_s'^0\in G_2} I_{\sup_{s\leq T}|\widetilde{x}_s'^0|\leq\widetilde{N}}ds$
$+\frac{1}{\varepsilon^2}E^{\widetilde{P}}\int_0^{t\wedge\tau_N-} I_{\widetilde{x}_s'^0\notin G_2}\left|b^n(s,\widetilde{x}_s'^n)-b(s,\widetilde{x}_s'^0)\right| I_{\sup_{s\leq T}|\widetilde{x}_s'^n|+\sup_{s\leq T}|\widetilde{x}_s'^0|\leq\widetilde{N}}ds]$
$= I_{21}^{\widetilde{N}} + I_{22}^{n,\widetilde{N}},$

where $k_{\widetilde{N}} = 2(1+\widetilde{N}\prod_{k=1}^m g_k(\widetilde{N}))$, and
$F\subset G_3\subset \overline{G}_3\subset G_2\subset G_1,$
G_2 and G_3 are open sets, $\overline{G}_3$ is the closure of G_3, all of which will be determined below. Notice that $y_t = \widetilde{x}_{t\wedge\tau_N}'^0$ satisfies the following SDE: $\forall t\geq 0$,
$y_t = x_0 + A_{t\wedge\tau_N} + \int_0^t \sigma(s,y_s)I_{|y_s|\leq N}d\widetilde{w}_s'^0 + \int_0^t\int_Z c(s,y_{s-},z)I_{|y_{s-}|\leq N}\widetilde{q}^0(ds,dz).$

Moreover,
$\left|\sigma(s,y)I_{|y|\leq N}\right|^2 + \int_Z |c(s,y,z)|^2 I_{|y_{s-}|\leq N}\pi(dz) \leq \widetilde{k}_N + c_1(s),$
where $\widetilde{k}_N = k_0(1+|N|^2\prod_{k=1}^m g_k(N))$, and $\int_0^T c_1(t)dt<\infty$, for each $T<\infty$. Hence Lemma 165 applies.
$I_{21}^{\widetilde{N}} \leq \frac{2k_{\widetilde{N}}}{\varepsilon^2}E^{\widetilde{P}}\int_0^{t\wedge\tau_N-} I_{\widetilde{x}_s'^0\in G_2}\frac{(\det A(s,x_s))^{1/(d+1)}}{\delta_0^{2/(d+1)}}ds$
$\leq k(p,k_T,d,T)\frac{2k_{\widetilde{N}}}{\delta_0^{2/(d+1)}\varepsilon^2}(\int_0^T\int_{G_2}dads)^{1/p}$
$\leq k'(p,k_T,d,T)\frac{2k_{\widetilde{N}}}{\delta_0^{2/(d+1)}\varepsilon^2}(m_d(G_2))^{1/p}.$

Thus, we can choose a small enough $\delta_{N,\widetilde{N}}>0$ and an open set G_2 such that $F\subset G_2\subset G_1$, and
$(m_d(G_2))^{1/p} < \delta_{N,\widetilde{N}} < \frac{\delta_0^{2/(d+1)}\varepsilon^2}{2k_{\widetilde{N}}k'(p,k_T,d,T)}\frac{\overline{\varepsilon}}{3}.$

Since F is a compact set, we can also choose an open set G_3 such that
$F\subset G_3\subset\overline{G}_3\subset G_2\subset G_1.$

Write
$\delta_1 = \inf\left\{d(x,y): x\in\overline{G}_3, y\in G_2\right\}.$

Then $\delta_1>0$. Noice that
$F_N^3 = G_3^c\cap\left\{x\in R^d: |x|\leq\widetilde{N}\right\}$
is a compact set, where G_3^c is the complement of G_3; and for each s, $b(s,x)$ is uniformly continuous on $x\in F_N^3$. So for any $\varepsilon>0$ there exists a $\delta_1>\eta>0$ (η may depend on the given s) such that as $|x'-x''|\leq\eta$ and $x',x''\in F_N^3$,
$|b(s,x')-b(s,x'')|<\varepsilon.$

Now one finds that
$I_{22}^{n,N} \leq \frac{1}{\varepsilon^2}E^{\widetilde{P}}\int_0^{t\wedge\tau_N-}\int_{|\overline{x}|\leq 1} I_{\widetilde{x}_s'^0\notin G_2}\left|b(s,\widetilde{x}_s'^n-\frac{\overline{x}}{n})-b(s,\widetilde{x}_s'^0)\right|$
$\cdot I_{\sup_{s\leq T}|\widetilde{x}_s'^n|+\sup_{s\leq T}|\widetilde{x}_s'^0|\leq\widetilde{N}}J(\overline{x})d\overline{x}ds.$

However, for each given s as $n>\frac{1}{\eta}$

$\widetilde{P}(\left|b(s,\widetilde{x}_s^{\prime n})-b(s,\widetilde{x}_s^{\prime 0})\right| I_{\widetilde{x}_s^{\prime 0}\in F_N^2}>\varepsilon)\leq \widetilde{P}(\left|\widetilde{x}_s^{\prime n}-\widetilde{x}_s^{\prime 0}\right|>\eta)$
$+\widetilde{P}(I_{\left|\widetilde{x}_s^{\prime n}-\widetilde{x}_s^{\prime 0}\right|\leq\eta}\left|b(s,\widetilde{x}_s^{\prime n})-b(s,\widetilde{x}_s^{\prime 0})\right| I_{\widetilde{x}_s^{\prime 0}\in F_N^2}>\varepsilon)$
$\leq \widetilde{P}(\left|\widetilde{x}_s^{\prime n}-\widetilde{x}_s^{\prime 0}\right|>\eta)..$
where
$F_N^2=G_2^c\cap\left\{x\in R^d:|x|\leq\widetilde{N}\right\}.$
Therefore, for each given s as $n\to\infty$,
$\widetilde{P}(\left|b(s,\widetilde{x}_s^{\prime n})-b(s,\widetilde{x}_s^{\prime 0})\right| I_{\widetilde{x}_s^{\prime 0}\in F_N^2}>\varepsilon)\leq \widetilde{P}(\left|\widetilde{x}_s^{\prime n}-\widetilde{x}_s^{\prime 0}\right|>\eta)\to 0.$
So applying Lebesgue's dominated convergence theorem, one finds that as $n\to\infty$, $I_{22}^{n,\widetilde{N}}\to 0$. Therefore, we have proved 1). However, since the conditions on σ and c are the same as those in Theorem 176, so the other terms in (10.18) also have similar limits to those in Theorem 176. So that $\widetilde{x}_s^{\prime 0}$ is a weak solution of (10.18). Finally, if 7° holds, then by the Tanaka type formula the pathwise uniqueness of solutions to (10.18) also holds. Hence by the Y-W theorem (Theorem 137) (10.18) has a pathwise unique strong solution. ∎

Now let us discuss the stochastic Bang-Bang control for $d-$dimensional SDE with jumps and with partially degenerate coefficients. Consider the following partially degenerate stochastic system in $d-$dimensional space, where its coefficient $b(t,x)$ can have a greater than linear growth: $\forall t\in[0,T]$,

$$\begin{aligned} dx_t &= (A_t^0 x_t - A_t^1 x_t g_1(t,|x_t|) - A_t^2\frac{x_t}{|x_t|^\alpha} I_{x_t\neq 0} \\ &\quad + A_t^3 x_t g_2(t,|x_t|^2) + B_t u_t)dt + (C_t^0 + C_t^1|x_t|)dw_t \\ &+ \int_Z D_t(z)(x_{t-} I_{|x_{t-}|\leq\sqrt{N_0}} + \sqrt{N_0}\frac{x_{t-}}{|x_{t-}|} I_{|x_{t-}|>\sqrt{N_0}})\widetilde{N}_k(dt,dz),\ x_0 = x\in R^d. \end{aligned} \tag{10.21}$$

Theorem 319 *Assume that*
1° $A_t^i, C_t^i, i=0,1; A_t^2, A_t^3$ *and* B_t *all are non-random* $d\times d$ *continuous matrices of* t; *and* $D_t(z)\geq 0$ *is a jointly measurable, non-random non-negative definite* $d\times d$ *matrix such that* $\int_Z |D_t(z)|^2\pi(dz)<\infty$,
$\int_Z |D_{t+h}(z)-D_t(z)|^2\pi(dz)\to 0$, *as* $h\to 0$,
and, moreover, $Z=R^d-\{0\}$, $\int_Z \frac{|z|^2}{1+|z|^2}\pi(dz)<\infty$;
2° $A_t^1, A_t^2\geq 0$, $C_t^0(C_t^1)^{-1}<0$, $B_t>C_t^0C_t^1$, $B_t>0$ *and* $0\leq\alpha<1$ *is a constant;*
3° $g_1(t,u)$ *is real, jointly continuous in* $(t,u)\in[0,\infty)\times R^d$ *such that* $\frac{\partial g_1}{\partial u}$ *exists and is uniformly locally bounded in* u, *that is, for each* $r<\infty$, $\left|\frac{\partial g_1}{\partial u}\right|\leq k_r$, *as* $|u|\leq r$, *where* $k_r\geq 0$ *is a constant depending only on* r; *and, moreover,* $g_1(t,u)$ *is jointly continuous in* $(t,u)\in[0,\infty)\times R$ *such that*
$|g_1(t,u)|\leq k_0, g_1(t,0)=0;$
$g_1(t,u)\geq 0,$ *as* $u\geq 0$;

$x \geq y \geq 0 \implies g_1(t,x) \geq g_1(t,y);$
4° $g_2(t,|x|^2)$ *is jointly continuous in* (t,x) *such that*
$\left|g_2(t,|x|^2)\right| \leq k_0(1+\prod_{k=1}^m \widetilde{g}_k(x)),$
where $\widetilde{g}_k(x)$ *is defined in* 4° *of Theorem 316.*
Then an optimal stochastic control exists, which is admissible, feedback and Bang-Bang; that is, there exists an admissible control $u^0 \in \mathfrak{U}$ *such that*
$J(u^0) = \min_{u\in\mathfrak{U}} J(u),\ u_t^0 = -\frac{x_t^0}{|x_t^0|} I_{x_t^0\neq 0},$
where x_t^0 *is the pathwise unique strong solution of the SDE (10.21) with* $u_t = -\frac{x}{|x|} I_{x\neq 0}$.

Proof. First we prove that (10.21) with $u_t = -\frac{x}{|x|} I_{x\neq 0}$ has a pathwise unique strong solution x_t^0, so $u_t^0 = -\frac{x_t^0}{|x_t^0|} I_{x_t^0\neq 0}$ is well defined. Since $|u_t^0| \leq 1$, we have $u_t^0 \in \mathfrak{U}$. In fact, if we write

$b(t,x) = A_t^0 x - A_t^1 x g_1(t,|x|) - A_t^2 \frac{x}{|x|^\alpha} I_{x\neq 0} + A_t^3 x g_2(t,|x|^2) - B_t \frac{x_t^0}{|x_t^0|} I_{x_t^0\neq 0},$
$\sigma(t,x) = C_t^0 + C_t^1 |x|,$
$c(t,x,z) = D_t(z)(x I_{|x_t|\leq\sqrt{N_0}} + \sqrt{N_0}\frac{x}{|x|} I_{|x-|>\sqrt{N_0}}),$

then

$|b(t,x)| \leq |B_t| + |A_t^2| |x|^{1-\alpha} + (|A_t^0| + k_0 |A_t^1|)|x| + |A_t^3| |x| k_0(1+\prod_{k=1}^m \widetilde{g}_k(x)),$
$|\sigma(t,0)|^2 \geq \delta_0 > 0,$
$\int_Z |c(t,x,z)|^2 \pi(dz) \leq N_0 \int_Z |D_t(z)|^2 \pi(dz) < \infty,$

and $b(t,x)$ is discontinuous at $x=0$. Obviously, $F=\{0\}$ is a compact set in R^d. Moreover, the degenerate points of $\sigma(t,x)$ should satisfy that
$|x| I_{d\times d} = -C_t^0 (C_t^1)^{-1}$. Let us discuss all situations.
1) If there is a solution $|x| = f_t$, since by assumption $-C_t^0(C_t^1)^{-1}$ is positive definite and continuous, then one should have $f_t \geq \delta_1 > 0$, as $t\in[0,T]$, where δ_1 is a constant, which exists. Write
$G_{\delta_1} = \{x \in R^d : |x| < \delta_1\}.$
Then $m_d G_{\delta_1} = \pi\delta_1^2 > 0$, and $\sigma(t,x)$ is non-degenerate in $x \in G_{\delta_1}, \forall t \in [0,T]$. Since $\sigma(t,x)$ is jointly continuous, there exists a constant $\delta_0 > 0$ such that $\langle \sigma(t,x)\sigma^*(t,x)\lambda, \lambda\rangle \geq \delta_0 |\lambda|^2, \forall\lambda \in R^d$, as $x \in G_{\delta_1/2}, \forall t \in [0,T]$. Write $G_1 = G_{\delta_1/2}$. Then G_1 satisfies the condition 4° in Thoerem 318. Furthermore, one finds that as $|x|, |y| \leq N$,

$\langle (x-y), (b(t,x) - b(t,y)\rangle \leq |A_t^0| |x-y|^2 + k_N |A_t^3| |x-y|^2,$

where we have applied the facts that

$\left\langle x-y, -\frac{x}{|x|^\alpha} + \frac{y}{|y|^\alpha}\right\rangle = -|x|^{2-\alpha} - |y|^{2-\alpha} + \frac{\langle x,y\rangle}{|x|^\alpha} + \frac{\langle x,y\rangle}{|y|^\alpha}$
$\leq -|x|^{2-\alpha} - |y|^{2-\alpha} + |x|^{1-\alpha}|y| + |y|^{1-\alpha}|x|$
$= -(|x|^{1-\alpha} - |y|^{1-\alpha})(|x| - |y|) \leq 0,$

etc. Furthermore,

$|\sigma(t,x) - \sigma(t,y)|^2 \leq |C_t^1| |x-y|^2,$
$\int_Z |c(t,x,z) - \sigma(t,y,z)|^2 \pi(dz) \leq \int_Z |D_t(z)|^2 \pi(dz) |x-y|^2.$

Hence Theorem 318 applies, and (10.21) with $u_t = -\frac{x}{|x|}I_{x\neq 0}$ has a pathwise unique strong solution. Now the proof of the remaining part is similar to that of Theorem 313 and Theorem 317.

2) If there is no solution for $|x|\, I_{d\times d} = -C_t^0(C_t^1)^{-1}$. Then $\sigma(t,x)$ is non-degenerate. In particular, since $\sigma(t,x)$ is jointly continuous, there exists a constant $\delta_0 > 0$ such that $\langle \sigma(t,x)\sigma^*(t,x)\lambda, \lambda\rangle \geq \delta_0 |\lambda|^2, \forall \lambda \in R^d$, as $|x| \leq N, \forall t \in [0,T]$. Write $G_1 = \left\{x \in R^d : |x| < N\right\}$. Then G_1 also satisfies the condition 4° in Thoerem 318. So the proof still can be completed. ■

11

Stochastic Population Control and Reflecting SDE

The components of many important physical quantities can only take non-negative values, and usually the quantities can be considered to satisfy some dynamic evolution systems. Therefore these situations can be examined by studying the dynamic systems. In the deterministic case one can put some conditions on the coefficients to make all components of the solutions to the dynamic systems non-negative. However, if the system is disturbed by a random process, e.g. a Wiener process, even in the simplest case the components of solutions to stochastic differential equations (SDE) can still change between arbitrary large positive and negative values. This prevents research proceeding any further. To overcome this difficulty we are motivated to examine the reflecting stochastic differential equation (RSDE) with coordinate planes as its boundaries, or with a more general boundary. In this chapter the general theory and applications of RSDE with jumps, which can be used as a stochastic population control model, etc., are systematically studied. In particular, for the case when the coefficients of RSDE are discontinuous, with greater than linear growth, the questions of the existence, uniqueness, comparison, convergence and stability of strong solutions to such RSDE or population RSDE are examined. Applications to the stochastic population control problem is also developed.

11.1 Introduction

Yu, Guo and Zhu (1987) has considered the following deterministic population control system

$$\begin{cases} dx_t = (A(t)x_t + B(t)x_t\beta_t)dt \\ x_0 = x, \ t \in [0,T], \end{cases}$$

where x_t^i is the size of the population with age between $[i, i+1)$, and $x_t = (x_t^1, \cdots, x_t^{r_m})$; β_t is the specific fertility rate of females,

$$A(t) = \begin{pmatrix} -(1+\eta_1(t)) & 0 & \cdots & \cdots & 0 \\ 1 & -(1+\eta_2(t)) & 0 & \cdots & 0 \\ \cdots & \cdots & \cdots & \cdots & \cdots \\ 0 & \cdots & 0 & 1 & -(1+\eta_{r_m}(t)) \end{pmatrix},$$

$$B(t) = \begin{pmatrix} 0 & \cdots & 0 & b_{r_1}(t) & \cdots & b_{r_2}(t) & 0 & \cdots & 0 \\ 0 & \cdots & \cdots & \cdots & \cdots & \cdots & \cdots & \cdots & 0 \end{pmatrix},$$

$b_i(t) = (1 - \mu_{00}(t))k_i(t)h_i(t) \neq 0$, $i = r_1, \cdots, r_2; 1 < r_1 < r_2 < r_m$, $\mu_{00}(t)$ is the death rate of babies, $\eta_i(t)$ is the forward death rate by ages, $k_i(t)$ and $h_i(t)$ are the corresponding sex rate and fertility model, respectively, x_t^i is the size of the population with ages between $[i, i+1)$,

$x_t = (x_t^1, \cdots, x_t^{r_m})$,

r_m is the largest age of the people, and β_t is the specific fertility rate of females. The size of the population should be non-negative, i.e. $x_t^i \geq 0, 1 \leq i \leq r_m$. Under appropriate consideration Yu, Guo and Zhu showed that this is true. Furthermore, they and others, developed the theory of this model, used β_t as a control, and established a population control theory. By means of their theory, after some practical computations; they were able to make population control suggestions to the Chinese goverment and, finally, such suggestions were considered by the goverment. Hence the population control problem is interesting and useful in both its theory and in its applications.

In the practical case, the population control system is disturbed by some stochastic perturbation. So we will consider that the stochastic population control system is a more realistic way. However, in the stochastic case the situation is very different. As the system is disturbed by a random process, e.g. a Wiener process, even in the simplest case the components of solutions to stochastic differential equations (SDE) can still change between arbitrary large positive and negative values. This means that if we consider the following SDE system:

$$\begin{cases} dx_t = (A(t)x_t + B(t)x_t\beta_t)dt + \sigma_t dw_t \\ x_0 = x, \ t \in [0,T]. \end{cases}$$

Then, in general, we cannot have the components $x_t^i, i = 1, \cdots, r_m$ of the solution non-negative.

Now, for simplicity, let us explain first how could we get a non-negative solution for a SDE similar to the above in 1−dimensional space. To make $x_t \geq 0$, it is natural to consider the following SDE:

$$\begin{cases} dx_t = dy_t + d\phi_t \\ dy_t = (A(t)x_t + B(t)x_t\beta_t)dt + \sigma_t dw_t, \\ x_0 = y_0 = x, \ \phi_0 = 0, \\ x_t \geq 0, \ t \in [0,T], \end{cases}$$

where, intuitively, we introduce a process ϕ_t to make $x_t \geq 0, \ t \in [0,T]$. Notice that

$$x_{t+\triangle t} = x_t + \triangle x_t = x_t + \triangle y_t + \triangle \phi_t. \tag{11.1}$$

So, intuitively, we require that the process ϕ_t has the following properties: 1) ϕ_t is continuous in t such that $\phi_0 = 0, \phi_t \uparrow$, as $t \uparrow$. 2) If at time t, $x_t > 0$, then at time t, we can require that $d\phi_t = 0$; because $\triangle\phi_t = d\phi_t + o(\triangle t)$ and y_t is right continuous in t, so there exists a $\delta > 0$ such that by (11.1) as $0 < \triangle t < \delta$, $x_{t+\triangle t} > 0$. 3) If at time t, $x_t = 0$, and there does not exist a $\delta > 0$ such that as $0 < \triangle t < \delta$, $\triangle y_t \geq 0$, then at time t, we should require that $d\phi_t > 0$, so there still exists a $\delta > 0$ such that as $0 < \triangle t < \delta$, $x_{t+\triangle t} > 0$. Or, roughly speaking, ϕ_t is an instantly reflecting process such that when x_t hits the boundary 0 and wants to become negative, ϕ_t immediatly has a positive increment to make x_t remain in the region $[0,\infty)$. A SDE that has an instantly reflecting process in it, is naturally called a reflecting SDE (RSDE). Let us write it in a more precise form:

$$\begin{cases} dx_t = dy_t + d\phi_t \\ dy_t = (A(t)x_t + B(t)x_t\beta_t)dt + \sigma_t dw_t, \\ \phi_t \text{ s a } \mathfrak{F}_t - \text{adapted, continuous and increasing process such that} \\ \phi_0 = 0, \ \int_0^t x_s d\phi_s = 0, \forall t \geq 0, \\ x_0 = y_0 = x, \ \phi_0 = 0, \\ x_t \in [0,\infty), \ \forall t \geq 0. \end{cases} \tag{11.2}$$

Here $\int_0^t x_s d\phi_s = 0, \forall t \geq 0$ means that $x_s > 0 \implies d\phi_s = 0$, and $d\phi_s > 0 \implies x_s = 0$. This is properties 2) and 3) of ϕ_t, which we required above. In (11.2) if we write a domain $\Theta = (0,\infty)$, its closure $\overline{\Theta} = [0,\infty)$, its boundary $\partial\Theta = \{0\}$, and its inner normal vector at the boundary $n = 1$, then $\int_0^t x_s d\phi_s = 0, \forall t \geq 0$ can be rewritten as

$\int_0^t x_s n_s d\phi_s = 0, \forall t \geq 0,$
where $n_s = n = 1$.

For a solution with non-negative components to a SDE in a 2−dimensional space the situation will be more complicated. First, in this case we want the solution value in a domain $\Theta = (0,\infty) \times (0,\infty)$. So its closure is $\overline{\Theta} = [0,\infty) \times [0,\infty)$, its boundary is $\{\{0\} \times [0,\infty)\} \cup \{[0,\infty) \times \{0\}\}$. For the inner normal vectors at the boundary there are three kinds: $n_x = (1,0)$, as $x = (x_1, x_2) \in \{0\} \times (0, \infty)$; $n_x = (0,1)$, as $x \in (0,\infty) \times \{0\}$; and $n_x = (\cos\theta, \sin\theta), \forall \theta \in [0, \frac{\pi}{2}]$, as $x = (0,0)$. So one sees that at some special boundary point the inner normal vector may be non-unique. Secondly, the reflection process ϕ_t is now also two dimensional, so the properties that we require should become 1) ϕ_t is continuous in t such that $\phi_0 = 0, \phi_t$ is finite variational on any finite interval; that is, $|\phi|_T < \infty$, for each $T < \infty$, where $|\phi|_T$ is the total variation of ϕ on $[0,T]$; 2) if at time t, $x_t \in \Theta$, then at time t, we can require that $d\,|\phi|_t = 0$; 3) if at time t, $d\,|\phi|_t > 0$, then we should require that $x_t \in \partial\Theta$.

For a solution value in a general domain Θ to a SDE with jumps in a $d-$dimensional space similar requirements for the reflection process can also be made. See the next section.

11.2 Notation

Now let us consider a $d-$dimensional reflecting stochastic differential equation (RSDE) with Poisson jumps as follows: (As before we write "RCLL" for "right continuous with left limit".)

$$\begin{cases} dx_t = b(t,x_t,\omega)dt + \sigma(t,x_t,\omega)dw_t + \int_Z c(t,x_{t-},z,\omega)\widetilde{N}_k(dt,dz) + d\phi_t, \\ x_0 = x \in \overline{\Theta}, \ t \geq 0, \\ x_t \in \overline{\Theta}, \ t \geq 0, \\ \phi_t \text{ is a } R^d - \text{valued } \mathfrak{F}_t - \text{adapted RCLL process with finite variation} \\ |\phi|_t \text{ on each finite interval } [0,t] \text{ such that } \phi_0 = 0, \text{and} \\ |\phi|_t = \int_0^t I_{\partial\Theta}(x_s)d\,|\phi|_s\,, \\ \phi_t = \int_0^t n(s)d\,|\phi|_s\,, \\ n(t) \in \mathcal{N}_{x_t}, \text{ as } x_t \in \partial\Theta, \end{cases} \tag{11.3}$$

where w_t is a $d-$dimensional standard Brownian motion process (BM), $\widetilde{N}_k(dt,dz)$ is a Poisson martingale measure generated by a $\mathfrak{F}_t-$Poisson point process $k(\cdot)$ with a compensator $\pi(dz)dt$, $\pi(.)$ is a $\sigma-$finite measure on a measurable space $(Z, \Re(Z))$ such that

$\widetilde{N}_k(dt,dz) = N_k(dt,dz) - \pi(dz)dt,$

and

$\mathcal{N}_x = \cup_{r\neq 0}\mathcal{N}_{x,r}, \mathcal{N}_{x,r} = \left\{n \in R^d : |n| = 1, B(x - nr, r) \cap \Theta = \emptyset\right\},$

$B(x_0,\varepsilon) = \left\{y \in R^d : |y - x_0| < \varepsilon\right\},$

and Θ is a given convex domain in R^d, $\partial\Theta$ is its boundary, $\overline{\Theta}$ is its closure, where we assume that Θ satisfies the so-called
uniform inner normal vector positive projection condition: if there exists a unit vector $e \in R^d$ and a constant $c > 0$ such that $e \cdot n = \langle e, n\rangle \geq c, \forall n \in \cup_{x\in\partial\Theta}\mathcal{N}_x$.

It is obvious that $\Theta = R^d_+ = \left\{x = (x^1, \cdots, x^d) \in R^d : x^i > 0, 1 \leq i \leq d\right\}$

obviously satisfies this condition with $e = (\frac{1}{\sqrt{d}}, \cdots, \frac{1}{\sqrt{d}})$, and $c_0 = \frac{1}{\sqrt{d}}$.

Condition $|\phi|_t = \int_0^t I_{\partial\Theta}(x_s) d|\phi|_s$ in (11.3) obviously means that $|\phi|_s$ increases only when $x_s \in \partial\Theta$, and the condition $\phi_t = \int_0^t n(s) d|\phi|_s$ means that the reflection $d\phi_t$ happens along the direction of the inner normal vector $n(t)$, so we may call such a reflection a normal reflection.

The geometrical meaning of $\mathcal{N}_x$ is that it is the set of all inner normal vectors at x, when $x \in \partial\Theta$. Obviously, when $x \in \Theta$, by definition $\mathcal{N}_x = \Phi$ (the empty set). Actually, in the case that $x_t \in \Theta$ no refelction is needed, so we do not need an inner normal vector.

Remark 320 *1° In the case $\Theta = R^d$ one can take $\phi_t \equiv 0$, then (11.3) reduces to the usual SDE (without reflection).*
2° By the convexity of Θ and the normal reflection of ϕ_t one easily sees that in (11.3)
conditions 3): $|\phi|_t = \int_0^t I_{\partial\Theta}(x_s) d|\phi|_s$,
and 4): $\phi_t = \int_0^t n(s) d|\phi|_s$,
are equivalent to
3'): for all $f : \overline{\Theta} \to R^d$, bounded and continuous such that $f|_{\partial\Theta} = 0$, and for all $t > 0$
$\int_0^t f(x_s) \cdot d\phi_s = 0$;
and 4') for all $y(\cdot) \in D^d([0,\infty), \overline{\Theta})$, where $D^d([0,\infty), \overline{\Theta})$ is the totality of RCLL functions $f : [0,\infty) \to \overline{\Theta}$,
$\int_0^t (y_s - x_s) \cdot d\phi_s$ is increasing, as t is increasing.

Let us show 2° in Remark 320. In fact, 3) and 4) can be rewritten as
$d|\phi|_t = I_{\partial\Theta}(x_t) d|\phi|_t$, or $0 = I_\Theta(x_t) d|\phi|_t$, and $d\phi_t = n(t) d|\phi|_t$.
So if they are true, then $\forall f \in C_b(\overline{\Theta}; R^d)$, with $f|_{\partial\Theta} = 0$, where $C_b(\overline{\Theta}; R^d)$ is the totality of bounded continuous functions $f : \overline{\Theta} \to R^d$,
$f(x_t) d|\phi|_t = 0, \forall t > 0$;
hence $f_1(x_t) n_1(t) d|\phi|_t = 0, \cdots, f_d(x_t) n_d(t) d|\phi|_t = 0, \forall t > 0$; that is,
$f(x_t) \cdot d\phi_t = 0$.
So 3') is true. Furthermore, since Θ is a convex domain, and $d\phi_t = n(t) d|\phi|_t$,
$(y_t - x_t) \cdot d\phi_t = (y_t - x_t) \cdot n(t) d|\phi|_t \geq 0, \forall y(\cdot) \in D^d([0,\infty), \overline{\Theta})$.
Hence 4') is true.

Conversely, suppose that 3') and 4') are true. Since Θ is a convex domain, and $(y_t - x_t) \cdot d\phi_t \geq 0, \forall y(\cdot) \in D^d([0,\infty), \overline{\Theta})$, $d\phi_t = n(t) d|\phi|_t$.
That is, 4) is true. Furthermore, $\forall f \in C_b(\overline{\Theta}; R^d)$, with $f|_{\partial\Theta} = 0$, by $f(x_t) \cdot d\phi_t = 0$, $f(x_t) \cdot n(t) d|\phi|_t = f(x_t) \cdot d\phi_t = 0$. From this one has that
$I_{\partial\Theta}(x_t) d|\phi|_t = 0$.
So 3) holds true. The proof of 2° in Remark 320 is complete.

Obviously, according to (11.3), the population SDE discussed in the introduction should be presented as follows:

$$\begin{cases} dx_t = (A(t)x_t + B(t)x_t\beta_t)dt + \sigma(t,x_t)dw_t + \int_Z c(t,x_{t-},z)\widetilde{N}_k(dt,dz) \\ +d\phi_t, t \geq 0, x_0 = x,\ x_t \in \overline{R}^d_+, \text{for all } t \geq 0, \\ \text{where } R^d_+ = \left\{x = (x^1,\cdots,x^d) \in R^d : x^i > 0, 1 \leq i \leq d\right\}, \\ \phi_t \text{ is a } R^d - \text{ valued } \mathfrak{F}_t - \text{adapted continuous process with finite} \\ \text{variation on any finite interval such that } \phi_0 = 0, \text{ and} \\ |\phi|_t = \int_0^t I_{\partial R^d_+}(x_s)d|\phi|_s,\ \phi_t = \int_0^t n(s)d|\phi|_s, \\ n(t) \in \mathcal{N}_{x_t},\ \text{as } x_t \in \partial R^d_+. \end{cases} \tag{11.4}$$

Remark 321 *For (11.4), by definition as $x \in R^d_+$,*
$\mathcal{N}_x = \Phi$ *(empty set).*
If $x \in \partial R^d_+$, then $n = (n_1, n_2, \cdots, n_d) \in \mathcal{N}_x$, iff, as $x_i = 0,\ x_j \neq 0, j \neq i$,
$n_i = 1,\ n_j = 0, j \neq i;$
and as $x_i = x_j = 0,\ x_k \neq 0, k \neq i, j$,
$n_i = \cos\theta, n_j = \sin\theta,\ \theta \in [0, \frac{\pi}{2}]$,
$n_k = 0, \forall k \neq i, j;$
and as $x_{i_1} = x_{i_2} = x_{i_k} = 0,\ x_j \neq 0, j \neq i_1, i_2, \cdots, i_k$,
$n_{i_1} = \sin\theta\cos\phi_1\cos\phi_2 \cdots\cdots\cdots \cos\phi_{k-2}$,
$n_{i_2} = \sin\theta\cos\phi_1\cos\phi_2\cdots\cos\phi_{k-3}\sin\phi_{k-2}$,
$\cdots\cdots\cdots$
$n_{i_{k-1}} = \sin\theta\sin\phi_1$,
$n_{i_k} = \cos\theta,\ \theta \in [0, \frac{\pi}{2}], \phi_1, \phi_2 \cdots,, \phi_k \in [0, \frac{\pi}{2}]$,
$n_j = 0$, *as* $j \neq i_1, i_2, \cdots, i_k,\ k \geq 3$.
Since all $n_i \geq 0, i = 1, 2, \cdots, d$. So all $\phi_i(t), i = 1, 2, \cdots, d$, in (11.4) are increasing, where
$\phi(t) = (\phi_1(t), \phi_2(t), \cdots, \phi_d(t))$.

For the solution of RSDE (11.3) we introduce the following definition.

Definition 322 *1) We say that (x_t, ϕ_t) is a strong solution of (11.3), if it is RCLL and $\mathfrak{F}_t^{w,\widetilde{N}_k}$−adapted, where*
$\mathfrak{F}_t^{w,\widetilde{N}_k} = \sigma(w_s, \widetilde{N}_k((0,s\}, U); s \leq t, U \in \mathfrak{B}(Z))$,
and it satisfies (11.3).

2) We say that (11.3) has a weak solution, if there exist a probability space $(\Omega, \mathfrak{F}, (\mathfrak{F}_t), P)$ and two $\mathfrak{F}_t$−adapted RCLL processes (x_t, ϕ_t) with a BM w_t and a Poisson martingale measure $\widetilde{N}_{k'}(dt, dz)$ which has the same compensator $\pi(dz)dt$ as the given one in (11.3) defined on it such that $(x_t, \phi_t, w_t, \widetilde{N}_{k'}(dt, dz))$ satisfy (11.3) on $(\Omega, \mathfrak{F}, (\mathfrak{F}_t), P)$.
3) We say that the pathwise uniqueness for solutions of (11.3) holds, if for any two solutions $(x_t^i, \phi_t^i), i = 1, 2$, of (11.3), which are defined on the same probability space $(\Omega, \mathfrak{F}, (\mathfrak{F}_t), P)$ with the same BM w_t and Poisson martingale measure $\widetilde{N}_k(dt, dz)$,

$P(\sup_{t\geq 0}|x_t^1 - x_t^2| = 0) = 1,\ P(\sup_{t\geq 0}|\phi_t^1 - \phi_t^2| = 0) = 1.$
4) We say that the uniqueness in (probability) law for solutions of (11.3) holds, if for any two solutions $(x_t^i, \phi_t^i), i = 1, 2,$ *of (11.3), which may be defined on different probability spaces* $(\Omega^i, \mathfrak{F}^i, (\mathfrak{F}_t^i), P^i)$, *with different BM* w_t^i *and Poisson martingale measures* $\widetilde{N}_{k^i}(dt, dz)$, $i = 1, 2$, *respectively, but with Poisson martingale measures* $\widetilde{N}_{k^i}(dt, dz)$, $i = 1, 2$, *which have the same compensator* $\pi(dz)dt$ *as the given one, the probability law of* $\{(x_t^1, \phi_t^1), t \geq 0\}$ *and* $\{(x_t^1, \phi_t^1), t \geq 0\}$ *are the same.*

11.3 Skorohod's Problem and its Solutions

To find the solution of (11.3) we may discuss the simplest deterministic case first. That is the so-called the Skorohod problem. Now let us introduce the Skorohod's problem. Suppose we are given a convex domain Θ in R^d; denote its boundary by $\partial\Theta$, and its closure by $\overline{\Theta}$. We introduce the following

Definition 323 *For a given* $Y \in D$ *and* $Y_0 \in \overline{\Theta}$, *if there exists an* $X \in D(\Theta) = \{X \in D : X_t \in \overline{\Theta}, t \geq 0\}$ *and a* $\phi \in D$ *such that* (X, ϕ) *satisfies*

$$\begin{cases} X_t = Y_t + \phi_t, t \geq 0, X_0 = Y_0 \in \overline{\Theta}, \\ X_t \in \overline{\Theta},\ t \geq 0, \\ \phi_t \text{ is a } R^d - \text{valued } \mathfrak{F}_t - \text{adapted RCLL function with finite} \\ \text{variation } |\phi|_t \text{ on each finite interval } [0,t] \text{ such that } \phi_0 = 0, \text{ and} \\ |\phi|_t = \int_0^t I_{\partial\Theta}(X_s) d|\phi|_s,\ \phi_t = \int_0^t n(s) d|\phi|_s, \\ n(t) \in \mathcal{N}_{X_t}, \text{ as } X_t \in \partial\Theta, \end{cases} \tag{11.5}$$

where $\mathfrak{F}_t = \bigcap_{n=1}^{\infty} \sigma(X_s; s \leq t + n^{-1})$; then (X, ϕ) is called a solution of Skorohod's problem (11.5) with given (Y, Y_0).(For space D see Appendix B).

Obviously, the last three conditions in (11.5) indicate that $|\phi|_t$ increases only when x_t takes values on the boundary, and $d\phi_t$ has the same direction as that of n_t. Besides, in the case that $\mathcal{N}_{x(t)}$ has only one element, then n_t is the inner normal vector at $x_t \in \partial\Theta$; in the case that $\mathcal{N}_{x(t)}$ has more than one element, then n_t is the one belonging to $\mathcal{N}_{x(t)}$, and it exists. Furthermore, it should be noticed that the reflection ϕ_t here can have jumps, but if its jump increment is not equal to zero, then it only happens at such times t, that x_t takes values on the boundary.

For the existence of solutions of the above Skorohod's problem we need to introoduce conditions on the domain in which the solutions take their values. We have already introduced **the uniform inner normal vector positive projection condition** in the "Notatation" section. Let us restate it here.

Definition 324 *We will say that the domain* Θ *in* R^d *satisfies* ***the uniform inner normal vector positive projection condition,*** *if there exists an* $e \in R^d$ *with* $|e| = 1$ *and a constant* $c_0 > 0$ *such that* $e \cdot n = \langle e, n \rangle \geq c_0, \forall n \in \cup_{x \in \partial\Theta} \mathcal{N}_x$.

The geometric meaning of the uniform inner normal vector positive projection condition actually means that all inner normal vectors on the boundary have a uniform positive projection: There exists a vector $e \in R^d$ with $|e| = 1$, amd there exists a positive constant $c_0 > 0$ such that all inner normal vectors n at the boundary have a uniform positive projection on this vector e; that is $n \cdot e \geq c_0 > 0$.

Let us explain why we need the condition that Θ is convex and satisfies the uniform inner normal vector positive projection condition for the solution to Skorohod's problem. Notice that Remark 320 still holds true for (11.5). In particular, by means of the convexity of Θ one has that if (x_t, ϕ_t) is a solution of (11.5), then for all $y(\cdot) \in D^d([0,\infty), \overline{\Theta})$,

$\int_0^t (y_s - x_s) \cdot d\phi_s$ is increasing, as t is increasing;
or equivalently, for all $y(\cdot) \in D^d([0,\infty), \overline{\Theta})$,

$$(y_t - x_t) \cdot d\phi_t \geq 0. \tag{11.6}$$

Such a property will be used many times in the discussion of the solutions. For example, we can use it to prove the uniqueness of the solutions to (11.5):

Suppose that (x_t, ϕ_t) and (x_t', ϕ_t') are two solutions of (11.5) with a given $Y_0, Y_t, \forall t \geq 0$ and $Y_0', Y_t', \forall t \geq 0$, respectively. We want to prove that $(x_t, \phi_t) = (x_t', \phi_t')$, $\forall t \geq 0$, when $Y_0 = Y_0'$, and $Y_t = Y_t', \forall t \geq 0$. Naturally, we discuss the difference $|X_t - X_t'|^2$. By means of the elementary inequality

$|a+b|^2 \leq (|a| + |b|)^2 \leq 2[|a|^2 + |b|^2]$, $\forall a, b \in R^d$,
we have

$|X_t - X_t'|^2 \leq 2(\ |Y_t - Y_t'|^2 + |\phi_t - \phi_t'|^2)$.
Notice that as ϕ_t and ϕ_t' are finite variational, we can use Ito's formula to find

$$\begin{aligned}
&|\phi_t - \phi_t'|^2 = 2\int_0^t (\phi_{s-} - \phi_{s-}') \cdot d(\phi_s - \phi_s') \\
&+ \textstyle\sum_{0<s\leq t} [|\phi_{s-} - \phi_{s-}' + \Delta\phi_s - \Delta\phi_s'|^2 - |\phi_{s-} - \phi_{s-}'|^2 \\
&-2(\phi_{s-} - \phi_{s-}') \cdot (\Delta\phi_s - \Delta\phi_s')] \\
&= 2\int_0^t (\phi_{s-} - \phi_{s-}') \cdot d(\phi_s - \phi_s') + \textstyle\sum_{0<s\leq t} |\Delta\phi_s - \Delta\phi_s'|^2 \\
&= 2\int_0^t (\phi_s - \phi_s') \cdot d(\phi_s - \phi_s') - 2\int_0^t (\phi_s - \phi_{s-} - (\phi_s' - \phi_{s-}')) \cdot d(\phi_s^c - \phi_s'^c) \\
&- \textstyle\sum_{0<s\leq t} |\Delta\phi_s - \Delta\phi_s'|^2 = \sum_{i=1}^3 I_i
\end{aligned}$$

where

$\phi_t^c = \phi_t - \sum_{0<s\leq t} \Delta\phi_s$
is the continuous part of ϕ_t. Now we estimate $\sum_{i=1}^3 I_i$. Obviously, $I_3 \leq 0$. Observe that ϕ_t and ϕ_t' are RCLL, so they have only countable many discontinuous points in t. However, ϕ_t^c and $\phi_t'^c$ are continuous, so

$I_2 = -2\sum_{i=1}^{\infty} \int_{\{t_i\}} (\phi_s - \phi_{s-} - (\phi'_s - \phi'_{s-})) \cdot d(\phi_s^c - \phi_s'^c) = 0,$
where $\{t_i\}_{i=1}^{\infty}$ is the set of all discontinuous points of ϕ_s^c and/or $\phi_s'^c$ such that $0 \leq t_i \leq t$. Finally, we apply the fact that Θ is a convex domain, so by (??)
$I_1 = 2\int_0^t (X_s - X'_s) \cdot d\phi_s + 2\int_0^t (X'_s - X_s) \cdot d\phi'_s$
$-2\int_0^t (Y_s - Y'_s) \cdot d(\phi_s - \phi'_s) \leq 0 - 2\int_0^t (Y_s - Y'_s) \cdot d(\phi_s - \phi'_s).$
Hence we have proved that

$$\begin{cases} |\phi_t - \phi'_t|^2 \leq -2\int_0^t (Y_s - Y'_s) \cdot d(\phi_s - \phi'_s), \\ |X_t - X'_t|^2 \leq 2|Y_t - Y'_t|^2 - 4\int_0^t (Y_s - Y'_s) \cdot d(\phi_s - \phi'_s). \end{cases} \tag{11.7}$$

Therefore, when $(Y_0, Y_t) = (Y'_0, Y'_t), \forall t \geq 0$; we have that $X_t = X'_t, \phi_t = \phi'_t, \forall t \geq 0$. So the uniqueness of the solution to (11.5) is proved under assumption that Θ is a convex domain.

Similarly, if (X_t, ϕ_t) is a solution of (11.5) with a given $Y_0, Y_t, \forall t \geq 0$, one has that
$|X_t - X_s|^2 \leq 2(\,|Y_t - Y_s|^2 + |\phi_t - \phi_s|^2).$
By Ito's formula and again by (??) as $s \leq t$
$|\phi_t - \phi_s|^2 = 2\int_s^t (\phi_{u-} - \phi_s) \cdot d\phi_u + \sum_{s<u\leq t} |\triangle\phi_u|^2$
$\leq 2\int_s^t (\phi_u - \phi_s) \cdot d\phi_u - \sum_{s<u\leq t} |\triangle\phi_u|^2 \leq 2\int_s^t (\phi_u - \phi_s) \cdot d\phi_u$
$= -2\int_s^t (Y_u - Y_s) \cdot d\phi_u + 2\int_s^t (X_u - X_s) \cdot d\phi_u$
$\leq -2\int_s^t (Y_u - Y_s) \cdot d\phi_u$
Hence under the assumption that Θ is a convex domain one also has that for $s \leq t$

$$\begin{cases} |\phi_t - \phi_s|^2 \leq -2\int_s^t (Y_u - Y_s) \cdot d\phi_u, \\ |X_t - X_s|^2 \leq 2(|Y_t - Y_s|^2 - 2\int_s^t (Y_u - Y_s) \cdot d\phi_u). \end{cases} \tag{11.8}$$

Now let us explain why we need the uniform inner normal vector positve projection condition. Notice that this condition says that there exists a vector $e \in R^d$ with $|e| = 1$ and a positive constant $c_0 > 0$ such that all inner normal vector n at the boundary $\partial\Theta$ of the domain have a positive projection greater than c_0 at vector e; that is,

$$n \cdot e \geq c_0 > 0, \forall n \in \cup_{x\in\partial\Theta} \mathcal{N}_x. \tag{11.9}$$

Recall that in the discussion of the existence of the solution to an SDE when the coefficients satisfy the Lipschitzian condition and the less than linear growth condition, we can always use an iteration technique to find the approximate solutions. After that we can take the limit to find the solution. However, now for the RSDE or for the special case - the Skorohod's problem - even if we can get approximate solutions, they are a sequence of paired functions (X_t, ϕ_t). So to discuss their convergenc we need some estimate which derived from the uniform inner normal vector positve projection condition. More precisely, we have the following lemma.

Lemma 325 *Assume that* Θ *is convex and satisfies the uniform inner normal vector positive projection condition, and* (X, ϕ) *is a solution of (2.1) with* (Y, Y_0). *Then as* $s \leq t$

$|\phi|_t - |\phi|_s \leq c_0^{-1} |\phi_t - \phi_s|,$

$|X_t - X_s| + (|\phi|_t - |\phi|_s) \leq k_0 \sup_{s \leq r \leq t} |Y_r - Y_s|,$

where $k_0 \geq 0$ *is a constant depending on* c_0 *only. In particular,*

$|\phi|_t \leq 2k_0 \sup_{s \leq t} |Y_s|.$

Proof. For $s < t$ let

$\triangle_t^{Y_s} = \sup_{s \leq r \leq t} |Y_r - Y_s|.$

By (11.8)

$$|X_t - X_s|^2 \leq 2(\left|\triangle_t^{Y_s}\right|^2 + 2\triangle_t^{Y_s}(|\phi|_t - |\phi|_s)).$$

Now by using the uniform inner normal positive projection condition we can estimate the second term in the right side of the above inequality. In fact, by (11.9) and (11.5)

$$|\phi|_t - |\phi|_s \leq c_0^{-1} e \cdot \int_s^t n_r d|\phi|_r = c_0^{-1} e \cdot (\phi_t - \phi_s).$$

One finds that

$$|\phi|_t - |\phi|_s \leq c_0^{-1} |\phi_t - \phi_s|,$$

and

$$|\phi|_t - |\phi|_s \leq c_0^{-1}(|X_t - X_s| + |Y_t - Y_s|).$$

So applying the elementary inequality

$4ab \leq \frac{1}{2}a^2 + 8b^2, \forall a, b \geq 0,$

one has that

$$|X_t - X_s|^2 \leq 2\left|\triangle_t^{Y_s}\right|^2 + 4c_0^{-1}\triangle_t^{Y_s}(|X_t - X_s| + |Y_t - Y_s|)$$

$$\leq 2\left|\triangle_t^{Y_s}\right|^2 + 2^{-1}|X_t - X_s|^2 + 16c_0^{-2}\left|\triangle_t^{Y_s}\right|^2 + 2^{-1}|Y_t - Y_s|^2.$$

Hence

$$\frac{1}{2}|X_t - X_s|^2 \leq (\tfrac{5}{2} + 16c_0^{-2})\left|\triangle_t^{Y_s}\right|^2. \blacksquare$$

From now on let us always assume that

Θ *is convex and satisfies the uniform inner normal vector positive projection condition.*

In the following let us show that for any given step function Y (11.5) has a unique solution.

Lemma 326 *Let* $(Y, Y_0) \in D \times \overline{\Theta}$, *and* Y *is a step function, then (2.1) has a unique solution.*

Proof. Write $[x]_{\partial} \in \overline{\Theta}$ for the unique vector satisfying the following equality

$$|x - [x]_{\partial}| = Min\left\{|x - y| : y \in \overline{\Theta}\right\},$$

i.e. $[x]_{\partial}$ is the point which attains at the shortest distance from $x \in R^d$ to $\overline{\Theta}$. We will use a step by step constructive procedure to find the solution

(X_t, ϕ_t). Notice that Y is a step function. So for any given $T < \infty$, as $t \in [0, T]$, Y_t only takes finite many different vector values. Moreover, by assumption $Y_0 \in \overline{\Theta}$, and Y is is RCLL. So, there should be a smallest time T_1 such that $Y_{T_1} \notin \overline{\Theta}$. (If $T_1 = \infty$, then we immediately get a solution $X_t = Y_t, \phi_t = 0$, as $t \in [0, \infty)$. The construction of a solution is completed at this first step. The same remark holds true for the 2nd step and so on. We do not repeat it anymore). To construct an X_t always lives in $\overline{\Theta}$ we may define

$X_t = Y_t, \phi_t = 0$, as $t \in [0, T_1)$,

and

$X_{T_1} = [Y_{T_1}]_\partial \in \partial\Theta$,

$\phi_{T_1} = X_{T_1} - Y_{T_1} = [Y_{T_1}]_\partial - Y_{T_1}$.

By this definition one sees that

$X_t = Y_t + \phi_t \in \overline{\Theta}$, as $t \in [0, T_1]$,

and $d\phi_t > 0$ only when $X_t \in \partial\Theta$, (here it happens at $t = T_1$); moreover, $d\phi_t = n_t d\,|\phi|_t$ (by the definition of ϕ_{T_1}). Now starting from T_1 and noticing that $Y_t + \phi_{T_1}, t \geq T_1$ is still a step function, we can do the same thing to $Y_t + \phi_{T_1}, t \geq T_1$ to find a T_2 and then to define X_t and ϕ_t on $t \in [T_{n-1}, T_n]$ such that (X_t, ϕ_t) is a solution of (11.5) for $t \in [T_{n-1}, T_n]$. Such solution can be constructed by induction. More precisely, let

$T_0 = 0, \phi_0 = 0,$

$T_n = \inf\left\{t \geq T_{n-1} : Y_t + \phi_{T_{n-1}} \notin \overline{\Theta}\right\}, n \geq 1,$

$X_t = Y_t + \phi_{T_{n-1}}$, as $t \in [T_{n-1}, T_n)$;

$X_t = [Y_{T_n} + \phi_{T_{n-1}}]_\partial$, as $t = T_n$;

$\phi_t = \phi_{T_{n-1}}$, as $t \in [T_{n-1}, T_n)$;

$\phi_t = [Y_{T_n} + \phi_{T_{n-1}}]_\partial - Y_{T_n}$, as $t = T_n$.

Then one sees that (X, ϕ) is a solution of (11.5) with (Y, Y_0). Moreover, it is already known that if the solution of (11.5) with (Y, Y_0) exists, then it must be unique under assumption that Θ is convex. ∎

Lemma 327 *Assume that $(Y^n, Y_0^n) \in D \times \overline{\Theta}$, and (X^n, ϕ^n) is the solution of (11.5) with (Y^n, Y_0^n). If there exists a $Y \in D$ and a $Y_0 \in \overline{\Theta}$ such that*

$\lim_{n\to\infty} \sup_{t\leq T} |Y_t^n - Y_t| = 0$, *for any* $T > 0$;

then (11.5) has a unique solution (X, ϕ) with (Y, Y_0), and for any $T > 0$ it satisfies

$\lim_{n\to\infty} \sup_{t\leq T} |X_t^n - X_t| = 0,$

$\lim_{n\to\infty} \sup_{t\leq T} |\phi_t^n - \phi_t| = 0.$

Proof. Notice that $Y \in D$, so $Y_t, t \in [0, T]$ is bounded for any given $T < \infty$. By assumption we also have that

$\sup_n \sup_{t\leq T} |Y_t^n| \leq k_T < \infty,$

where k_T is a constant depending only on T. So by (11.7)

$\sup_{t\leq T} |X_t^n - X_t^m|^2 \leq 2\sup_{t\leq T} |Y_t^n - Y_t^m|^2 + 4\sup_{t\leq T} |Y_t^n - Y_t^m| \cdot$

$\cdot(|\phi^n|_T + |\phi^m|_T) \leq 2\sup_{t\leq T} |Y_t^n - Y_t^m|^2 + 8k_T \sup_{t\leq T} |Y_t^n - Y_t^m| \to 0.$

By this argument one also sees that

$\sup_{t\leq T}|\phi_t^n - \phi_t^m|^2 \to 0.$
Hence there exists a $(X,\phi) \in D(\Theta)\times D$, such that for any $T>0$
$\lim_{n\to\infty}(\sup_{t\leq T}|X_t^n - X_t| + \sup_{t\leq T}|\phi_t^n - \phi_t|) = 0.$
Now let us show that (X,ϕ) is a solution of (11.5). By
$X_t^n = Y_t^n + \phi_t^n,$
obviously, letting $n\to\infty$, one has that
$X_t = Y_t + \phi_t.$
Now, let us show that ϕ_t is a finite variational function. For this, and for any $t>0$, let us make the subdivision in $[0,t]$:
$0 = t_0 < t_1 < ... < t_k = t.$
Then one has that $\forall\varepsilon>0, \exists N$, as $n\geq N$,
$\sum_{j=1}^k \left|\phi_{t_j} - \phi_{t_{j-1}}\right| \leq \sum_{j=1}^k \left|\phi_{t_j}^n - \phi_{t_{j-1}}^n\right| + 2\sum_{j=1}^k \left|\phi_{t_j} - \phi_{t_j}^n\right|$
$\leq \sup_n |\phi^n|_t + 2\sum_{j=1}^k \left|\phi_{t_j} - \phi_{t_j}^n\right| \leq \sup_n |\phi^n|_t + \varepsilon.$
Hence
$|\phi|_t \leq \sup_n |\phi^n|_t \leq k_t$, for all $t\geq 0$.
So, ϕ_t is finite variational. Furthermore, let us show that 3') and 4') in Remark 320 hold. If this is done, then (X,ϕ) is a solution of (11.5). We will use the step function approximation to complete the proof.

Let us show 4'). For any given $\overline{X}\in D(\Theta)$ we know that for any $s\leq t$
$\int_s^t(\overline{X_r} - X_r^n)\cdot d\phi_s^n \geq 0$, as $n\geq 1$,
because (X^n,ϕ^n) is a solution of (11.5) with (Y^n, Y_0^n). Now take step functions $Z^N \in D$ such that
$\sup_{s\leq t}\left|\overline{X_s} - X_s - Z_s^N\right| \to 0$, as $N\to\infty$, for any $0\leq t<\infty$.
Then
$\left|\int_s^t(\overline{X_r} - X_r)\cdot d\phi_r - \int_s^t(\overline{X_r} - X_r^n)\cdot d\phi_r^n\right| \leq \left|\int_s^t(\overline{X_r} - X_r - Z_r^N)\cdot d\phi_r\right|$
$+\left|-\int_s^t(\overline{X_r} - X_r - Z_r^N)\cdot d\phi_r^n\right| + \left|\int_s^t Z_r^N\cdot d(\phi_r - \phi_r^n)\right|$
$+\left|-\int_s^t(X_r - X_r^n)\cdot d\phi_r^n\right| = \sum_{I=1}^4 I_t^i.$
Obviously, as $n\to\infty, N\to\infty$,
$\sum_{I=1}^2 I_t^i + I_t^4 \leq \sup_{s\leq t}\left|\overline{X_s} - X_s - Z_s^N\right|\left[|\phi|_t + |\phi^n|_t\right]$
$+\sup_{s\leq t}|X_s - X_s^n|\,|\phi^n|_t \leq 2k_t(\sup_{s\leq t}\left|\overline{X_s} - X_s - Z_s^N\right|$
$+\sup_{s\leq t}|X_s - X_s^n|) \to 0.$
Notice that Z^N is a step function and $\lim_{n\to\infty}\sup_{t\leq T}|\phi_t^n - \phi_t| = 0$, so for each N as $n\to\infty$,
$I_t^3 = \left|\int_s^t Z_r^N\cdot d(\phi_r - \phi_r^n)\right| \to 0.$
Thus, for arbitrary $\varepsilon>0$, one can choose a large enough N such that $I_t^1 + I_t^2 < \frac{\varepsilon}{2}$. Then for this N there exists an $\widetilde{N}$ such that when $n\geq\widetilde{N}$, $I_t^3 + I_t^4 < \frac{\varepsilon}{2}$. So as $n\to\infty$, $\sum_{I=1}^4 I_t^i \to 0$, and we have that for any $s\leq t$,
$\int_s^t(\overline{X_r} - X_r)\cdot d\phi_s \geq 0.$
4') is proved.

3') can be shown in a similar way. In fact, we can choose step functions $Z^N\in D$ such that

$\sup_{s\leq t}\left|X_s - Z_s^N\right| \to 0$, as $N \to \infty$, for any $0 \leq t < \infty$.
For any $f : \Theta^{cl} \to R^d$, which is bounded and continuous and such that $f\mid_{\partial\Theta} = 0$, one has that

$$\left|\int_0^t f(X_s)\cdot d\phi_s\right| = \left|\int_0^t f(X_s)\cdot d\phi_s - \int_0^t f(X_s^n)\cdot d\phi_s^n\right|$$
$$\leq \left|\int_0^t (f(X_s) - f(Z_s^N))\cdot d\phi_s\right| + \left|\int_0^t (f(X_s) - f(Z_s^N))\cdot d\phi_s^n\right|$$
$$+ \left|\int_0^t f(Z_s^N)\cdot d(\phi_s - \phi_s^n)\right| + \left|\int_0^t (f(X_s) - f(X_s^n))\cdot d\phi_s^n\right|.$$

So we can similarly show that $\left|\int_0^t f(X_s)\cdot d\phi_s\right| = 0$. Therefore 3') is derived. The proof is complete. ∎

Now we are in a position to establish an existence and uniqueness theorem for the solution of (11.5). (For Skorohod's topology see Appendix B).

Theorem 328 *Assume that Θ is convex and satisfies the uniform inner normal vector positive projection condition, and $Y \in D$ with $Y_0 \in \overline{\Theta}$, then (11.5) has a unique solution (X, ϕ) with (Y, Y_0). If we write this solution corresponding to the given $Y \in D$ as $(X(Y), \phi(Y)) \in D \times D$, then this is a continuous map from D to $D \times D$, that is, if $Y_n \to Y$ (in D under Skorohod topology), then*
$X(Y_n) \to X(Y) \in D$ (in D under Skorohod topology), and
$\phi(Y_n) \to \phi(Y) \in D$ (in D under Skorohod topology).
Moreover,
$|\phi(Y)|_t \leq k_0' \sup_{s\leq t} |Y_s|$,
where the constant $k_0' \geq 0$ depends only on c_0. Therefore
$\sup_{s\leq t} |X_s(Y)| \leq (k_0' + 1) \sup_{s\leq t} |Y_s|$.

Proof. First, for any given $(Y, Y_0) \in D \times \overline{\Theta}$, by Lemma 381 one can take step functions $Y^n \in D$ such that
$\sup_{s\leq T} |Y_s - Y_s^n| \to 0$, as $n \to \infty$, for any $0 \leq T < \infty$.
Now by Lemma 326 (11.5) has a unique solution (X^n, ϕ^n) with (Y^n, Y_0^n) for each n. By Lemma 325
$|\phi^n|_T \leq 2k_0 \sup_{s\leq T} |Y_s^n|$.
Hence
$\sup_n |\phi^n|_T \leq 2k_0(\sup_{s\leq T} |Y_s| + \sup_n \sup_{s\leq T} |Y_s^n - Y_s|) \leq k_T$,
where $k_T \geq 0$ depends on c_0 (introduced in the uniform inner normal vector positive projection condition), Y and T only. Therefore applying Lemma 327, one finds that (2.1) has a unique solution (X, ϕ) with (Y, Y_0) such that
$\lim_{n\to\infty} \sup_{s\leq T} |X_s^n - X_s| = 0$,
$\lim_{n\to\infty} \sup_{s\leq T} |\phi_s^n - \phi_s| = 0$.
Finally, if there exists a sequence $\lambda_n \in \Lambda, n = 1, 2, \cdots$, such that for each $T < \infty$
$\lim_{n\to\infty} \sup_{s\leq T} |\lambda_s - \lambda_s^n| = 0$, and
$\lim_{n\to\infty} \sup_{s\leq T} \left|Y_{\lambda_s} - Y_{\lambda_s^n}^n\right| = 0$,
(that is, $Y_n \to Y$ in D under the Skorohod topology), then we can write

$\overline{Y}_s^n = Y_{\lambda_s^n}^n$, $\overline{Y}_s = Y_{\lambda_s}$,

etc. and from that $(\overline{X}_s^n, \overline{\phi}_s^n)$ is the solution of the Skorohod's problem

$$\begin{cases} \overline{X}_s^n = \overline{Y}_s^n + \overline{\phi}_s^n, \\ \text{and the other statements of (11.5) hold true for } (\overline{X}_s^n, \overline{\phi}_s^n), \end{cases}$$

and also from that $\lim_{n\to\infty}\sup_{s\leq T}\left|\overline{Y}_s - \overline{Y}_s^n\right| = 0$, so we find that

$\lim_{n\to\infty}\sup_{s\leq T}\left|\overline{X}_s - \overline{X}_s^n\right| = 0$, and

$\lim_{n\to\infty}\sup_{s\leq T}\left|\overline{\phi}_s - \overline{\phi}_s^n\right| = 0$,

by means of Lemma 327. These mean that exactly

$X(Y_n) \to X(Y) \in D$ and

$\phi(Y_n) \to \phi(Y) \in D$ (both in D under Skorohod topology).

11.4 Moment Estimates and Uniqueness of Solutions to RSDE

Now let us return to discuss the reflecting stochastic differential equations (RSDE) presented as (11.3) in the "Notation" section. Because we will use the results from the previous section to discuss the existence of solutions and so on, so we will always assume that the domain Θ satisfies the follwing condition:

(H$_1$) *Θ is a convex domain and it satisfies the uniform inner normal vector positive projection condition.*

An important property of the solution to (11.3) should be mentioned first. See the following lemma.

Lemma 329 *If (x_t, ϕ_t) is a solution of (11.3), then for each $t > 0$*

$\triangle\phi_t = \phi_t - \phi_{t-} = 0 \Longleftrightarrow x_{t-} + \int_Z c(t, x_{t-}, z, \omega)N_k(\{t\}, dz) \in \overline{\Theta}.$

Proof. Indeed,

$\triangle x_t = \int_Z c(t, x_{t-}, z, \omega)N_k(\{t\}, dz) + \triangle\phi_t.$

Hence, by $x_t \in \overline{\Theta}$, if $\triangle\phi_t = 0$, then

$x_{t-} + \int_Z c(t, x_{t-}, z, \omega)N_k(\{t\}, dz) = x_{t-} + \triangle x_t = x_t \in \overline{\Theta}.$

The implication "$\Longrightarrow$" is proved.

Conversely, notice that the RSDE in (11.3) can be rewritten as

$dx_t = b(t, x_t, \omega)dt + \sigma(t, x_t, \omega)dw_t + \int_Z c(t, x_{t-}, z, \omega)\widetilde{N}_k(dt, dz) + d\phi_t$

$= dy_t + d\phi_t,$

or

$x_t = y_t + \phi_t.$

Moreover, from $x_t = x_{t-} + \triangle y_t + \triangle\phi_t$ one finds that

$x_t - (x_{t-} + \triangle y_t) = \triangle\phi_t.$

So, if

$x_{t-} + \triangle y_t = x_{t-} + \int_Z c(t, x_{t-}, z, \omega)\widetilde{N}_k(\{t\}, dz) \in \overline{\Theta},$

then, by 4') in Remark 320 one finds that

$|\triangle\phi_t|^2 = \triangle\phi_t \cdot [x_t - (x_{t-} + \triangle y_t)] \leq 0.$

Thus $\triangle\phi_t = 0$. So the inverse implication "$\Longleftarrow$" is true. ■

From now on, for simplicity, let us also always assume that

(H$_2$) $x + c(t,x,z,\omega) \in \overline{\Theta}$, for all $t \geq 0, z \in Z, \omega \in \Omega$ and $x \in \overline{\Theta}$.

Obviously, by assumption (H$_2$) it is always true that

$x_{t-} + \int_Z c(t, x_{t-}, z, \omega) N_k(\{t\}, dz) \in \overline{\Theta}$.

Now, under assumptions (H$_1$) and (H$_2$) the RSDE (11.3) can be rewritten as follows:

$$\begin{cases} dx_t = b(t,x_t,\omega)dt + \sigma(t,x_t,\omega)dw_t + \int_Z c(t,x_{t-},z,\omega)\widetilde{N}_k(dt,dz) \\ +d\phi_t,\ t \geq 0, x_0 = x \in \overline{\Theta},\ x_t \in \overline{\Theta},\ t \geq 0, \\ \phi_t \text{ is a continuous } R^d - \text{valued } \mathfrak{F}_t - \text{adapted RCLL process with} \\ \text{finite variation } |\phi|_t \text{ on each finite interval } [0,t] \text{ such that } \phi_0 = 0, \text{ and} \\ |\phi|_t = \int_0^t I_{\partial\Theta}(x_s)d|\phi|_s,\ \phi_t = \int_0^t n(s)d|\phi|_s, \\ n(t) \in \mathcal{N}_{x_t}, \text{ as } x_t \in \partial\Theta. \end{cases} \tag{11.10}$$

The difference between (11.3) and (11.10) is that in (11.3) ϕ_t is RCLL, but here in (11.10) we can require that ϕ_t is continuous by means of assumption (H$_2$).

For the solutions of (11.10) we have the following a priori estimate.

Theorem 330 *If (x_t, ϕ_t) is a solution of (11.10) with $E|x_0|^2 < \infty$, and*

$|b(t,x,\omega)|^2 + \|\sigma(t,x,\omega)\|^2 + \int_Z |c(t,x,z,\omega)|^2 \pi(dz) \leq c_1(t)(1+|x|^2),$

where $c_1(t) \geq 0$ is non-random such that for each $0 \leq T < \infty$ $\int_0^T c_1(t)dt < \infty$, then for any $0 \leq T < \infty$

$E\sup_{0 \leq t \leq T} |x_t|^2 + E|\phi|_T^2 \leq k_T,$

$E\sup_{s \leq r \leq t} |x_r - x_s|^2 + E\sup_{s \leq r \leq t}(|\phi|_r - |\phi|_s)^2 \leq k_T \int_s^t c_1(u)du,$

as $0 \leq s \leq t \leq T$,

where $0 \leq k_T$ is a constant depending only on $\int_0^T c_1(t)dt, c_0$ (appeared in Assumption (H$_1$)), $E|x_0|^2$ and T.

Proof. By Ito's formula

$$\begin{aligned} |x_t - x_s|^2 &= 2\int_s^t (x_r - x_s)\cdot b(r,x_r,\omega)dr + 2\int_s^t (x_r - x_s)\cdot\sigma(r,x_r,\omega)dw_r \\ &+ \int_s^t \|\sigma(r,x_r,\omega)\|^2 dr + 2\int_s^t (x_{r-} - x_s)\cdot\int_Z c(r,x_{r-},z,\omega)\widetilde{N}_k(dr,dz) \\ &+2\int_s^t (x_{r-} - x_s)\cdot d\phi_r + \sum_{s<r\leq t} |\triangle x_r|^2 = \sum_{i=1}^6 I_{s,t}^i. \end{aligned} \tag{11.11}$$

Since ϕ_t is continuous in t, and x_t is RCLL in t, it has at most countable many discontinuous points in t. So by 4') in Remark 320 one finds that

$I_{s,t}^5 = 2\int_s^t (x_r - x_s)\cdot d\phi_r + 2\int_s^t (x_{r-} - x_r)\cdot d\phi_r \leq 0.$
Again by the continuity of ϕ_t one finds that as $0 \leq s \leq t \leq T$,
$EI_{s,t}^6 = E\sum_{s<r\leq t} |\Delta x_r|^2 = E\int_s^t \int_Z |c(r, x_{r-}, z, \omega)|^2 N_k(dr, dz)$
$= E\int_s^t \int_Z |c(r, x_r, z, \omega)|^2 \pi(dz)dr \leq k_0[2\int_s^t c_1(u)du$
$+4\int_s^t c_1(r)E|x_r - x_0|^2 dr + 4E|x_0|^2 \int_s^t c_1(u)du].$
Let $s = 0$. By the martingale inequality, Gronwall's inequality, and Fatou's lemma (introduce stopping times $\tau_N = \inf\{t \in [0,T] : |x_t| > N\}, N = 1, 2, ...,$ if necessary), from (11.11) it is easily seen that
$E\sup_{0\leq t\leq T} |x_t - x_0|^2 \leq k_T'.$
Hence
$E\sup_{0\leq t\leq T} |x_t|^2 \leq 2E\sup_{0\leq t\leq T} |x_t - x_0|^2 + 2E|x_0|^2 \leq k_T.$
Now by Lemma 325

$$(|\phi|_t - |\phi|_s) \leq k_0' \sup_{s<r\leq t} |y_r - y_s|, \tag{11.12}$$

where
$y_t = x_0 + \int_0^t b(r, x_r, \omega)dr + \int_0^t \sigma(r, x_r, \omega)dw_r + 2\int_0^t \int_Z c(r, x_{r-}, z, \omega)\widetilde{N}_k(dr, dz).$
Hence
$E|\phi|_T^2 \leq k_T'.$
Again by (11.11) one finds that as $0 \leq s \leq t \leq T$
$E\sup_{s\leq r\leq t} |x_r - x_s|^2 \leq k_T'' \int_s^t c_1(u)du.$
Applying (11.12) again one also has that
$E(|\phi|_t - |\phi|_s)^2 \leq k_T''' \int_s^t c_1(u)du$, as $0 \leq s \leq t \leq T.$
The proof is complete. ■

For the uniqueness of solutions to (11.3) we have the following theorem.

Theorem 331 *Assume that $(x_t^i, \phi_t^i), i = 1, 2,$ satisfy (11.3) with the same initial value and with the same BM and the same Poisson martingale measure on the same probability space, and assume that the conditions in Theorem 330 hold. Moreover, assume that for each $T < \infty$ and $N = 1, 2, \cdots$*
$2(x^1 - x^2)\cdot(b(t, x^1, \omega) - b(t, x^2, \omega)) + \|\sigma(t, x^1, \omega) - \sigma(t, x^2, \omega)\|^2$
$+\int_Z |c(t, x^1, z, \omega) - c(t, x^2, z, \omega)|^2 \pi(dz) \leq k_{N,T}(t)\rho_{N,T}(|x^1 - x^2|^2),$
as $t \in [0,T]$, $x^1, x^2 \in R^d$ and $|x^1|, |x^2| \leq N$; where $k_{N,T}(t) \geq 0$ is non-random such that
$\int_0^T k_{N,T}(s)ds < \infty;$
and $\rho_{N,T}(u)$ is non-random, concave and strictly increasing in $u \geq 0$ such that $\rho_{N,T}(0) = 0$, and $\int_{0+} du/\rho_N(u) = \infty$.
Then
$P(\sup_{t\geq 0} |x_t^1 - x_t^2| + \sup_{t\geq 0} |\phi_t^1 - \phi_t^2| = 0) = 1.$

Proof. For arbitrary given $T < \infty$ write
$\tau_N = \inf\{t \in [0,T] : |x_t^1| + |x_t^2| > N\}.$
By Ito's formula one finds that as $t \leq T$
$E|x_{t\wedge\tau_N}^1 - x_{t\wedge\tau_N}^2|^2 \leq K_0 E\int_0^{t\wedge\tau_N} k_{N,T}(s)\rho_{N,T}(|x_s^1 - x_s^2|^2)ds.$

Hence one easily sees that $E\left|x_t^1 - x_t^2\right|^2 = 0, \forall t \geq 0$ and $P - a.s.\ x_t^1 = x_t^2, \forall t \geq 0$. Now again by Ito's formula and after an application of the martingale inequality one finds that

$P(\sup_{t\in[0,T]} \left|x_t^1 - x_t^2\right| = 0) = 1.$

From this and (11.3), and again by means of the martingale inequality it is also seen that

$P(\sup_{t\in[0,T]} \left|\phi_t^1 - \phi_t^2\right| = 0) = 1.$ ∎

11.5 Solutions for RSDE with Jumps and with Continuous Coefficients

For RSDE (11.10) we discuss the case for non-random coefficients. Recall that we always make the assumptions (H_1) and (H_2) from the previous section; that is, we always assume that

(H_1): Θ is a convex domain, and there exists a constant $c_0 > 0$ and a vector $e \in R^d, |e| = 1$ such that

$e \cdot n \geq c_0 > 0$, for all $n \in \cup_{x\in\partial\Theta}\aleph_x$.

(H_2): $x + c(t,x,z,\omega) \in \overline{\Theta}$, for all $t \geq 0, z \in Z, \omega \in \Omega$ and $x \in \overline{\Theta}$.

We have the following theorem.

Theorem 332 *Assume that for $t \geq 0, x \in \overline{\Theta}, z \in Z$*

1° $b(t,x), \sigma(t,x), c(t,x,z)$ are jointly measurable such that there exists a constant $k_0 \geq 0$

$|b|^2 + \|\sigma\|^2 + \int_Z |c|^2 \pi(dz) \leq k_0,$

2° $b(t,x), \sigma(t,x)$ are jointly continuous, and as $|x-y| \to 0, |t-s| \to 0$

$\int_Z |c(t,x,z) - c(s,y,z)|^2 \pi(dz) \to 0,$

3° $\pi(dz) = dz/|z|^{d+1}$, $(Z = R^d - \{0\})$.

Then (11.10) has a weak solution.

Proof. We will use the step coefficient approximation technique to show this theorem. Let

$h_n(0) = 0,$

$h_n(t) = (k-1)2^{-n}$, as $(k-1)2^{-n} < t \leq k2^{-n}$.

Then by Theorem 328 there exists a unique solution (x_t^n, ϕ_t^n) satisfying

$$\begin{cases} x_t^n = x_0 + \int_0^t b^n(s,x_s^n)ds + \int_0^t \sigma^n(s,x_s^n)dw_s \\ \quad + \int_0^t \int_Z c^n(s,x_{s-}^n,z)q(ds,dz) + \phi_t^n, \\ \text{and the other statements in (11.3) hold for } (x_t^n, \phi_t^n), \end{cases} \tag{11.13}$$

where for simplification of the notation in this proof we write $q(ds,dz) = \widetilde{N}_k(ds,dz)$, $p(ds,dz) = N_k(ds,dz)$, and

$b^n(s,x_s) = b(h_n(s), x_{h_n(s)})$, etc.

In fact, once x_t^n is obtained for $0 \leq t \leq k2^{-n}$, then (x_t^n, ϕ_t^n) is uniquely determined as the solution of the Skorohod problem:

$$\left\{ \begin{array}{c} x_t^n = x_{k2^{-n}}^n + b(k2^{-n}, x_{k2^{-n}}^n)(t - k2^{-n}) + \sigma(k2^{-n}, x_{k2^{-n}}^n) \\ \cdot(w(t) - w(k2^{-n})) + \int_Z c(k2^{-n}, x_{k2^{-n}}^n, z)q((k2^{-n}, t], dz) + \phi_t^n, \\ \text{and the other statements in (11.3) hold for } (x_t^n, \phi_t^n) \\ \text{on } k2^{-n} \leq t \leq (k+1)2^{-n}. \end{array} \right. \tag{11.14}$$

Notice that in (11.13) the reflection process ϕ_t^n is RCLL, but not necessary continuous. Now we show that for arbitrary $\varepsilon > 0$ and $T > 0$

$$\lim_{c \to \infty} \sup_n \sup_{t \leq T} P(|\eta_t^n| > c) = 0, \tag{11.15}$$

$$\lim_{c \to \infty} \sup_n \sup_{t,s \leq T, |t-s| \leq h} P(|\eta_t^n - \eta_s^n| > \varepsilon) = 0, \tag{11.16}$$

hold for $\eta_t^n = x_t^n, \phi_t^n, w_t, \zeta_t$, where
$\zeta_t = \int_0^t \int_{|z| \leq 1} zq(ds, dz) + \int_0^t \int_{|z|>1} zp(ds, dz)$.
However, by the proof of Theorem 180, it can be shown in the same way that (11.15) and (11.16) are true for $\eta_t^n = w_t$ and ζ_t. Notice that by (11.8) one has that

$$\left\{ \begin{array}{c} |\phi_t^n - \phi_s^n|^2 \leq -2 \int_s^t (y_u^n - y_s^n) \cdot d\phi_u^n, \\ |x_t^n - x_s^n|^2 \leq 2(|y_t^n - y_s^n|^2 - 2 \int_s^t (y_u^n - y_s^n) \cdot d\phi_u^n), \end{array} \right. \tag{11.17}$$

where
$y_t^n = x_0 + \int_0^t b^n(s, x_s^n)ds + \int_0^t \sigma^n(s, x_s^n)dw_s + \int_0^t \int_Z c^n(s, x_{s-}^n, z)q(ds, dz)$.
Moreover, by Theorem 328

$$\sup_{s \leq t} |x_s^n| + |\phi^n|_t \leq k_0' \sup_{s \leq t} |y_s^n|, \tag{11.18}$$

where the constant $k_0' \geq 0$ depends only on c_0 which come from assumption (H_1). Now by the condition 1°, and the martingale inequality,
$E \sup_{s \leq t} |y_s^n|^2 \leq k_t$,
where $k_t \geq 0$ is a constant depending only on t. Hence, by (11.18)

$$E \sup_{s \leq T} |x_s^n| + |\phi^n|_T \leq k_T'. \tag{11.19}$$

On the other hand, by Lemma 325 as $s \leq t$
$|\phi^n|_t - |\phi^n|_s \leq c_0^{-1} |\phi_t^n - \phi_s^n|$.
So by (11.17) one easily sees that
$E(|\phi^n|_t - |\phi^n|_s)^2 \leq k_T E \sup_{s \leq u \leq t} |y_u^n - y_s^n|^2$.
Therefore, again by (11.17), as $s \leq t$

$$E |x_t^n - x_s^n|^2 + E(|\phi^n|_t - |\phi^n|_s)^2 \leq k_T' |t - s|. \tag{11.20}$$

From (11.19) and (11.20) one easily sees that (11.15) and (11.16) still hold true for $\eta_t^n = x_t^n$ and ϕ_t^n. Therefore Skorohod's theorem (Theorem 398) applies: There exist a probability space $(\widetilde{\Omega}, \widetilde{\mathfrak{F}}, \widetilde{P})$ and four RCLL processes $(\widetilde{x}_t^n, \widetilde{\phi}_t^n, \widetilde{w}_t^n, \widetilde{\zeta}_t^n)$ defined on it with the same finite probability distributions as that of $(x_t^n, \phi_t^n, w_t, \zeta_t)$, and there exist $(\widetilde{x}_t^0, \widetilde{\phi}_t^0, \widetilde{w}_t^0, \widetilde{\zeta}_t^0)$, where $\widetilde{w}_t^0$ is a BM on $(\widetilde{\Omega}, \widetilde{\mathfrak{F}}, \widetilde{P})$, and there also exists a subsequence $\{n_k\}$ of $\{n\}$ denoted by $\{n\}$ again, such that as $n \to \infty$

$\widetilde{\eta}_t^n \to \widetilde{\eta}_t^0$, in probability, as $\widetilde{\eta}_t^n = \widetilde{x}_t^n, \widetilde{\phi}_t^n, \widetilde{w}_t^n, \widetilde{\zeta}_t^n, n = 0, 1, 2, ...$

Now set

$\widetilde{p}^n(dt, dz) = \sum_{s \in dt} I_{(0 \neq \triangle \widetilde{\zeta}_s^n \in dz)}(s)$, $\widetilde{q}^n(dt, dz) = \widetilde{p}^n(dt, dz) - \pi(dz)dt$, $n = 0, 1, 2, ...$

Then $\widetilde{p}^n(dt, dz)$ is a Poisson random counting measure with the compensator $\pi(dz)dt$ for each $n = 0, 1, 2, ...$, and it satisfies the condition

$\widetilde{\zeta}_t^n = \int_0^t \int_{|z| \leq 1} z\widetilde{q}^n(ds, dz) + \int_0^t \int_{|z|>1} z\widetilde{p}^n(ds, dz), n = 0, 1, 2, ...$

(See the proof in Theorem 180). By the coincidence of finite probability distributions one easily sees that $(\widetilde{x}_t^n, \widetilde{\phi}_t^n)$ satisfies (11.13) with $\widetilde{w}_t^n$ and $\widetilde{q}^n(dt, dz)$ on $(\widetilde{\Omega}, \widetilde{\mathfrak{F}}, \widetilde{P})$; that is,

$\widetilde{x}_t^n = \widetilde{y}_t^n + \widetilde{\phi}_t^n,$

where

$\widetilde{y}_t^n = x_0 + \int_0^t b^n(s, \widetilde{x}_s^n)ds + \int_0^t \sigma^n(s, \widetilde{x}_s^n)d\widetilde{w}_s^n + \int_0^t \int_Z c^n(s, \widetilde{x}_{s-}^n, z)\widetilde{q}^n(ds, dz),$

and $(\widetilde{x}_t^n, \widetilde{\phi}_t^n)$ also satisfy the other statements in (11.3).

Now notice that as $n \to \infty$, $\forall t \geq 0, x \in R^d$,

$b^n(s, x) \to b(s, x), \ |b^n(s, x)| \leq k_0,$

$\sigma^n(s, x) \to \sigma(s, x), \ |\sigma^n(s, x)| \leq k_0,$

$\int_Z |c^n(s, x, z) - c(s, x, z)|^2 \pi(dz) \to 0, \int_Z |c^n(s, x, z)|^2 \pi(dz) \leq k_0.$

So, similar to the proof of Theorem 180, one finds that as $n \to \infty$

$\left| \int_0^t (b^n(s, \widetilde{x}_s^n) - b(s, \widetilde{x}_s^0))ds \right| \to 0$, in probability.

$\int_0^t \sigma^n(s, \widetilde{x}_s^n)d\widetilde{w}_s^n \to \int_0^t \sigma(s, \widetilde{x}_s^0)d\widetilde{w}_s^0$, in probability,

$\int_0^t \int_Z c^n(s, \widetilde{x}_{s-}^n, z)\widetilde{q}^n(ds, dz) \to \int_0^t \int_Z c(s, \widetilde{x}_{s-}^0, z)\widetilde{q}^0(ds, dz)$, in probability,

$\int_0^t \int_Z \left| c^n(s, \widetilde{x}_s^n, z) - c(s, \widetilde{x}_s^0, z) \right|^2 \pi(dz)ds \to 0$, in probability.

Therefore, letting $n \to \infty$ one finds that

$\widetilde{x}_t^0 = \widetilde{y}_t^0 + \widetilde{\phi}_t^0,$

where

$\widetilde{y}_t^0 = x_0 + \int_0^t b(s, \widetilde{x}_s^0)ds + \int_0^t \sigma(s, \widetilde{x}_s^0)d\widetilde{w}_s^0 + \int_0^t \int_Z c(s, \widetilde{x}_{s-}^0, z)\widetilde{q}^0(ds, dz).$

In the rest of the proof one only needs to show that the other statements in (11.3) hold true for $(\widetilde{x}_t^0, \widetilde{\phi}_t^0)$. However, by the above, one also finds that as $n \to \infty$

$E \sup_{t \in [0,T]} \left| \widetilde{y}_t^n - \widetilde{y}_t^0 \right|^2 \to 0.$

Hence one can take a subsequence of $\{n\}$ denoted again by $\{n\}$ such that $P - a.s.$, as $n \to \infty$,

$\sup_{t \in [0,T]} \left| \widetilde{y}_t^n - \widetilde{y}_t^0 \right|^2 \to 0.$

Thus, applying Lemma 327 $P-a.s.$, as $n \to \infty$,
$\sup_{t\in[0,T]} \left|\widetilde{x}_t^n - \widetilde{x}_t^0\right|^2 \to 0$,
$\sup_{t\in[0,T]} \left|\widetilde{\phi}_t^n - \widetilde{\phi}_t^0\right|^2 \to 0$,
and the other statements in (11.3) hold true for $(\widetilde{x}_t^0, \widetilde{\phi}_t^0)$. Furthermore, by assumption (H$_2$) $\widetilde{\phi}_t^0$ is also continuous. So $(\widetilde{x}_t^0, \widetilde{\phi}_t^0)$ is a weak solution of (11.10). The proof is complete. ∎

To obtain a strong solution of (11.10) we need a Yamada-Watanabe theorem as in the SDE case. Since the proof is almost the same as that of Theorem 137, so we will omit the proof.

Theorem 333 *(Yamada-Watanabe type theorem). Assume that the coefficients $b(t,x), \sigma(t,x)$, and $c(t,x,z)$ in (11.10) do not depend on ω, and make the assumption:*

$$Z = R^d - \{0\}, \text{ and } \int_{R^d-\{0\}} \frac{|z|^2}{1+|z|^2}\pi(dz) < \infty. \tag{11.21}$$

Then we have the following two assertions.
1) If (11.10) has a weak solution and the pathwise uniqueness holds for (11.10), then (11.10) has a pathwise unique strong solution (x_t, ϕ_t), $t \geq 0$.
2) If (11.10) has two weak solutions $(x_t^i, \phi_t^i, w_t^i, q^i(dt,dz)), i = 1,2$, defined on two different probability spaces $(\Omega^i, \mathfrak{F}^i, (\mathfrak{F}_t^i), P^i), i = 1,2$, respectively, where w_t^i is a P^i-BM, $q^i(dt,dz)$ is a P^i-Poisson martingale measure with the same compensator $\pi(dz)dt, i = 1,2$, then there exist a probability space $(\Omega, \mathfrak{F}, (\mathfrak{F}_t), P)$ and four $\mathfrak{F}_t-$adapted RCLL processes $(\widetilde{x}_t^i, \widetilde{\phi}_t^i), i = 1,2$, with a BM $\widetilde{w}_t$ and a Poisson martingale measure $\widetilde{q}(dt,dz)$ having the same compensator $\pi(dz)dt$ as the given one defined on it such that $(\widetilde{x}_t^i, \widetilde{\phi}_t^i), i = 1,2$, are adapted to $\mathfrak{F}_t$, and the probability law of $(\widetilde{x}_t^i, \widetilde{\phi}_t^i, \widetilde{w}_t, \widetilde{q}(dt,dz))$ coincides with that of $(x_t^i, \phi_t^i, w_t^i, q^i(dt,dz)), i = 1,2$. Moreover, $(\widetilde{x}_t^i, \widetilde{\phi}_t^i, \widetilde{w}_t, \widetilde{q}(dt,dz))$ satisfy (11.10) on the same probability space $(\Omega, \mathfrak{F}, (\mathfrak{F}_t), P), i = 1,2$.

Applying the previous two theorems and the uniqueness theorem (Theorem 331) we immediately obtain the following theorem. From now on let us always make the following assumption in this chapter:
(H$_3$) $\pi(dz) = dz/|z|^{d+1}$, $(Z = R^d - \{0\})$.

Theorem 334 *Under the assumption of Theorem 332 and the assumption (H$_3$), if for $t \geq 0, x \in \overline{\Theta}, z \in Z$*
$2(x-y)\cdot(b(t,x)-b(t,y)) + \|\sigma(t,x)-\sigma(t,y)\|^2$
$+\int_Z |c(t,x,z)-c(t,y,z)|^2\pi(dz) \leq k_{N,T}(t)\rho_{N,T}(|x-y|^2)$,
as $|x|, |y| \leq N, t \in [0,T]$,
where for each $T < \infty$, $0 \leq k_{N,T}(t), \int_0^T k_{N,T}(t)(t)dt < \infty$ and $\rho_{N,T}(u)$ is concave and strictly increasing in $u \geq 0$ such that $\rho_{N,T}(0) = 0$, and

$\int_{0+} du/\ \rho_{N,T}(u) = \infty$, *for each* $N = 1, 2, ...,$
then (11.10) has a pathwise unique strong solution.

Now let us relax the bounded condition for coefficients b and σ to the less than linear growth condition.

Theorem 335 *Assume that for* $t \geq 0, x \in \overline{\Theta}, z \in Z$
1° $b(t,x), \sigma(t,x), c(t,x,z)$ *are jointly measurable, and there exists a constant* $k_0 > 0$ *such that*
$|b|^2 + \|\sigma\|^2 \leq k_0(1 + |x|^2), \int_Z |c|^2 \pi(dz) \leq k_0,$
2° $b(t,x), \sigma(t,x)$ *are jointly continuous, and as* $|x - y| \to 0, |t - s| \to 0$
$\int_Z |c(t,x,z) - c(s,y,z)|^2 \pi(dz) \to 0,$
3° assumption (H_1)*–*(H_3) *hold,*
4° $2(x-y) \cdot (b(t,x) - b(t,y)) + \|\sigma(t,x) - \sigma(t,y)\|^2$
$+ \int_Z |c(t,x,z) - c(t,y,z)|^2 \pi(dz) \leq k_{N,T}(t)\rho_{N,T}(|x-y|^2),$
as $|x|, |y| \leq N, t \in [0,T]$,
where for each $T < \infty$, $0 \leq k_{N,T}(t), \int_0^T k_{N,T}(t)dt < \infty$ *and* $\rho_{N,T}(u)$ *is concave and strictly increasing in* $u \geq 0$ *such that* $\rho_{N,T}(0) = 0$, *and*
$\int_{0+} du/\ \rho_{N,T}(u) = \infty$, *for each* $N = 1, 2, ...$
Then (11.10) has a pathwise unique strong solution.

Proof. Let
$b^N(t,x) = b(t,x)$, as $|x| \leq N$; and $b^N(t,x) = b(t, Nx/|x|)$, as $|x| > N$; σ^N is similarly defined. Then by the previous thoerem there exists a pathwise unique strong solution (x_t^N, ϕ_t^N), where ϕ_t^N is continuous, satisfying (11.10) with the coefficients b^N, σ^N and c. Now set
$\tau_N = \inf\{t \geq 0 : |x_t^N| > N\},$
$x_t = x_t^N,\ \phi_t = \phi_t^N$, as $0 \leq t \leq \tau_N$.
By pathwise uniqueness the above definition is well posed, and

$$x_{t\wedge\tau_N} = x_0 + \int_0^{t\wedge\tau_N} b(s,x_s)ds + \int_0^{t\wedge\tau_N} \sigma(s,x_s)dw_s$$
$$+ \int_0^{t\wedge\tau_N} \int_Z c(s,x_{s-},z)\widetilde{N}_k(ds,dz) + \phi_{t\wedge\tau_N}, \text{ for all } t \geq 0.$$

Notice that by the a priori estimate (Theorem 330) for any $T < \infty$
$N^2 P(\tau_N \leq T) \leq E|x_{T\wedge\tau_N}|^2 \leq k_T$, for all $N = 1, 2, ...$
Hence it follows that
$P(\lim_{N\to\infty} \tau_N = \infty) = 1.$
The proof is complete. ∎

11.6 Solutions for RSDE with Jumps and with Discontinuous Coefficients

By means of the Girsanov type theorems we can establish the existence of solutions for RSDE with jumps and with discontinuous coefficients. One key assumption here is that the diffusion coefficients $\sigma(t,x)$ cannot be degenerate. So the results obtained in this section cannot refer back to the RODE (reflecting ordinary differential equation) case.

Theorem 336 *Assume that $(H_1) - (H_3)$ hold true and for $t \geq 0, x \in \overline{\Theta}, z \in Z$*
1° $b(t,x), \sigma(t,x), c(t,x,z)$ are jointly measurable, and there exists a constant $k_0 > 0$ such that
$|b|^2 + \|\sigma\|^2 + \int_Z |c|^2 \pi(dz) \leq k_0,$
2° $\sigma(t,x)$ is jointly continuous, and as $|x-y| \to 0, |t-s| \to 0$
$\int_0^t |c(t,x,z) - c(s,y,z)|^2 \pi(dz) \to 0,$
3° there exists a constant $\delta > 0$ such that as $(t,x) \in [0,T] \times \overline{\Theta}$
$(A\lambda, \lambda) \geq \delta |\lambda|^2,$
where $A = 2^{-1}\sigma\sigma^*$.
Then (11.10) has a weak solution.
Furthermore, if the following conditions also hold:
4° $2(x-y) \cdot (b(t,x,\omega) - b(t,y,\omega)) + \|\sigma(t,x,\omega) - \sigma(t,y,\omega)\|^2$
$+ \int_0^t |c(t,x,z,\omega) - c(t,y,z,\omega)|^2 \pi(dz) \leq k_N(t)\rho_N(|x-y|^2),$
as $|x|, |y| \leq N$, where $0 \leq k_N(t)$, $\int_0^T k_N(t)dt < \infty$, for each $T < \infty$, and $\rho_N(u)$ is concave and strictly increasing in $u \geq 0$ such that $\rho_N(0) = 0$, and
$\int_{0+} du/\ \rho_N(u) = \infty$, *for each $N = 1, 2, ...$;*
but if condition 1° is weaken to
1°′ $\|\sigma\|^2 \leq k_0(1 + |x|^2)$, $\int_Z |c|^2 \pi(dz) \leq k_0$,
and $\langle x, b(t,x)\rangle \leq c_1(t)(1 + |x|^2 \prod_{k=1}^m g_k(x))$,
where
$$g_k(x) = 1 + \underbrace{\ln(1 + \ln(1 + \cdots \ln}_{k-times}(1 + |x|^{2n_0}))),$$
(n_0 is some natural number), and $c_1(t) \geq 0$ is non-random such that for each $T < \infty$, $\int_0^T c_1(t)dt < \infty$; furthermore, $b(t,x)$ is locally bounded for x, that is, for each $r > 0$, as $|x| \leq r$,
$|b(t,x)| \leq k_r,$
where $k_r > 0$ is a constant depending only on r;
then (11.10) has a pathwise unique strong solution.

Proof. Case 1. Assume that
$|b| + \|\sigma\|^2 + \int_Z |c|^2 \pi(dz) \leq k_0',$
and assume that conditions $2° - 3°$ hold. By Theorem 332 there exists a weak solution (x_t, ϕ_t) defined on some probability space $(\Omega, \mathfrak{F})$ (without loss of any generality we may assume that $(\Omega, \mathfrak{F})$ is a standard measurable space) satisfying the following RSDE
$$\begin{cases} x_t = x_0 + \int_0^t \sigma(s,x_s)dw_s + \int_0^t \int_Z c(s,x_{s-},z)\widetilde{N}_k(ds,dz) + \phi_t, \\ \quad \text{and the other statements in (11.3) hold for } (x_t, \phi_t), \end{cases} \quad P-a.s.$$
Let $d\widetilde{P}_t = z_t dP$, where
$z_t = \exp[\int_0^t \theta_s \cdot dw_s - \frac{1}{2}\int_0^t |\theta_s|^2 ds]$, $\theta_t = \sigma^{-1}(s,x_s)b(t,x_s)$.

Then by Theorem 124 there exists a probability measure $\widetilde{P}$ defined on $(\Omega, \mathfrak{F})$ such that $\widetilde{P}\mid_{\mathfrak{F}_T} = \hat{P}_T$, for each $0 \leq T < \infty$; moreover,
1) $w'_t = w_t - \int_0^t \theta_s ds, 0 \leq t$,
is a BM under probability $\hat{P}$;
2) $\widetilde{N}_k(dt, dz) = N_k(dt, dz) - \pi(dz)dt$
is still a Poisson random martingale measure with the same compensator $\pi(dz)dt$ under the probability $\hat{P}$. Therefore we find that (x_t, ϕ_t) satisfies the following RSDE

$$\begin{cases} x_t = x_0 + \int_0^t b(s, x_s)ds + \int_0^t \sigma(s, x_s)dw'_s + \int_0^t \int_Z c(s, x_{s-}, z)\widetilde{N}_k(ds, dz) \\ \quad +\phi_t, \text{ and the other statements in (11.3) hold for } (x_t, \phi_t). \end{cases}$$

$\tilde{P} - a.s.$
Now assume that conditions 4° and 5° also hold. Then by the same reasoning as in Theorem 334 we find that (11.10) has a pathwise unique strong solution.
Case 2. (General case). In the case that $1^{\circ\prime}$ is satisfied, let
$b^N(t, x) = b(t, x)$, as $|x| \leq N$; and $b^N(t, x) = b(t, Nx/|x|)$, as $|x| > N$; σ^N is similarly defined. Then by the case 1 there exists a pathwise unique strong solution (x_t^N, ϕ_t^N), where ϕ_t^N is continuous, satisfying (11.10) with the coefficients b^N, σ^N and c. Set now

$$\tau_N = \inf\left\{t \geq 0 : \left|x_t^N\right| > N\right\},$$

$$x_t = x_t^N,\ \phi_t = \phi_t^N, \text{ as } 0 \leq t \leq \tau_N.$$

By the pathwise uniqueness the above definition is well posed, and

$$x_{t\wedge\tau_N} = x_0 + \int_0^{t\wedge\tau_N} b(s, x_s)ds + \int_0^{t\wedge\tau_N} \sigma(s, x_s)dw_s$$
$$+\int_0^{t\wedge\tau_N} \int_Z c(s, x_{s-}, z)\widetilde{N}_k(ds, dz) + \phi_{t\wedge\tau_N}, \text{ for all } t \geq 0.$$

Applying Ito's formula to $g(x_t) = g_{m+1}(x_t)$ we have that as $0 \leq t \leq T$

$$0 \leq g(x_{t\wedge\tau_N}) = g(x_0) + \int_0^{t\wedge\tau_N} g'(x_s)b(s, x_s)ds$$
$$+\int_0^{t\wedge\tau_N} g'(x_s)\sigma(s, x_s)dw_s + \int_0^{t\wedge\tau_N} \int_Z g'(x_{s-})c(s, x_{s-}, z)\widetilde{N}_k(ds, dz)$$
$$+\int_0^{t\wedge\tau_N} \int_Z [g(x_{s-} + c(s, x_{s-}, z)) - g(x_{s-}) - g'(x_{s-})c(s, x_{s-}, z)]N_k(ds, dz)$$
$$+2\int_0^{t\wedge\tau_N} g'(x_s) \cdot d\phi_s = \sum_{i=1}^6 I^i,\ \hat{P}_T^N - a.s.$$

where

$$\frac{\partial}{\partial x_i} g_{m+1}(x) = \prod_{k=1}^m g_k^{-1}(x) \frac{2n_0 x_i |x|^{2n_0-2}}{1+|x|^{2n_0}},$$
$$g'(x) = grad\ g(x) = (\tfrac{\partial}{\partial x_i} g(x), \cdots, \tfrac{\partial}{\partial x_i} g(x)),$$

and we write $g_0(x) = 1$. Notice that

$$\int_0^{t\wedge\tau_N} g'(x_s) \cdot d\phi_s$$
$$= \sum_{i=1}^d \int_0^{t\wedge\tau_N} (\prod_{k=1}^m g_k^{-1}(x(s))\frac{2n_0|x(s)|^{2n_0-2}}{1+|x(s)|^{2n_0}}(x_i(s) - x_i(0))d\phi_i(s)$$
$$+\sum_{i=1}^d \int_0^{t\wedge\tau_N} (\prod_{k=1}^m g_k^{-1}(x(s))\frac{2n_0|x(s)|^{2n_0-2}}{1+|x(s)|^{2n_0}} x_i(0)d\phi_i(s)$$
$$= I^{61} + I^{62} \leq I^{62}.$$

So as $t \leq T$,

$$E\int_0^{t\wedge\tau_N} g'(x_s) \cdot d\phi_s \leq EI^{62} \leq k_T.$$

Now discussing the other terms I^i, as in the proof of Theorem 133, one finds that for any $T < \infty$

$g(N)P(\tau_N < T) \leq E(g(x_{T\wedge\tau_N})I_{\tau_N<T}) \leq k_T,$
Therefore, as $N \to \infty$,
$P(\tau_N < T) \to 0.$
This means that $P - a.s.\lim_{N\to\infty} \tau_N = \infty$. So the unique strong solution $(x_t, \phi_t), t \geq 0$ is obtained. ■

11.7 Solutions to Population SDE and Their Properties

Recall that the population SDE can be given as follows (See (11.4) in the section "Notation"):

$$\begin{cases} dx_t = (A(t)x_t + B(t)x_t\beta_t)dt + \sigma(t,x_t)dw_t + \int_Z c(t,x_{t-},z)\widetilde{N}_k(dt,dz) \\ +d\phi_t, t \geq 0, x_0 = x;\ x_t \in \overline{R}^d_+, \text{for all } t \geq 0, \\ \text{where } R^d_+ = \left\{x = (x^1,\cdots,x^d) \in R^d : x^i > 0, 1 \leq i \leq d\right\}, \\ \phi_t \text{ is a } R^d - \text{ valued } \mathfrak{F}_t - \text{adapted continuous process with finite} \\ \text{variation on any finite interval such that } \phi_0 = 0, \text{ and} \\ |\phi|_t = \int_0^t I_{\partial R^d_+}(x_s)d\,|\phi|_s,\ \phi_t = \int_0^t n(s)d\,|\phi|_s, \\ n(t) \in \mathcal{N}_{x_t}, \text{ as } x_t \in \partial R^d_+. \end{cases} \tag{11.22}$$

Now we easily see that if $\beta_t-$the specific fertility rate of females is bounded, and coefficients b and σ satisfy the conditions in Theorem 335, then (11.22) has a unique solution. That is, if in (11.22) we denote by x^i_t the size of the population with ages between $[i, i+1)$, and write
$x_t = (x^1_t, \cdots, x^d_t),$
where d is the largest age of the people, then for any given $\beta_t-$the specific fertility rate of females, we can find the population size vector x_t with $x^i_t \geq 0, \forall i = 1, 2, \cdots, d$. More precisely, we have the following theorem.

Theorem 337 *Assume that for* $t \geq 0, x \in \overline{R}^d_+, z \in Z$
1° $A(t)$ and $B(t)$ are bounded and non-random, $\beta_t(x)$, $\sigma(t,x)$, and $c(t,x,z)$ are also non-random and jointly measurable, and there exists a constant $k_0 > 0$ such that
$|\beta_t(x)|^2 + \int_Z |c|^2 \pi(dz) \leq k_0, \|\sigma\|^2 \leq k_0(1 + |x|^2);$
2° $\beta_t(x)$ and $\sigma(t,x)$ are jointly continuous, and as $|x - y| \to 0, |t - s| \to 0$
$\int_Z |c(t,x,z) - c(s,y,z)|^2 \pi(dz) \to 0;$
3° $x + c(t,x,z) \in \overline{R}^d_+, \forall t \geq 0, x \in \overline{R}^d_+, z \in Z;$
4° $|\beta_t(x) - \beta_t(y)|^2 + \|\sigma(t,x) - \sigma(t,y)\|^2$
$+ \int_Z |c(t,x,z) - c(t,y,z)|^2 \pi(dz) \leq k_{N,T}(t)\rho_N(|x-y|^2),$
as $|x|, |y| \leq N, t \in [0,T]$,

where $0 \leq k_{N,T}(t)$ *is non-random,* $\int_0^T k_{N,T}(t)dt < \infty$ *for each* $T < \infty$*, and* $\rho_N(u)$ *is non-random, concave and strictly increasing in* $u \geq 0$ *such that* $\rho_N(0) = 0$*, and*

$\int_{0+} du/\ \rho_N(u) = \infty$*, for each* $N = 1, 2, ...$

Then (11.22) has a pathwise unique strong solution $(x_t, \phi_t), t \geq 0$.

Theorem 337 is obviously a direct conclusion from Theorem 335. A special case for the above theorem is when

$\beta_t = \beta_t(x)$ does not depend on x, and it is bounded and, moreover,

$$\begin{cases} \sigma(t, x) = \left(c_0^{(1)} + c_1^{(1)}x, \cdots, c_0^{(d)} + c_1^{(d)}x\right), \\ c(t, x, z) = \overline{c}_0 + \overline{c}_1(xI_{|x| \leq k_0} + k_0 \frac{x}{|x|} I_{|x| > k_0}) f(z), \end{cases} \tag{11.23}$$

where $k_0 \geq 0$ is a constant, $c_0^{(i)}, i = 1, \cdots, d$, and $\overline{c}_0$ are constant R^d-vectors, (i.e. each component in a vector is a constant), $c_1^{(i)}, i = 1, \cdots, d$, and $\overline{c}_1$ are constant $d \times d$ matrices, (i.e. each element in a matrix is a constant) and, in addition, $\overline{c}_0^{(i)}, \overline{c}_{1,ij} \geq 0, \forall 1 \leq i, j \leq d$, where $\overline{c}_1 = (\overline{c}_{1,ij})_{i,j=1}^d$, and $f(z) \geq 0$ is such that $\int_{R^d - \{0\}} \frac{f(z)^2}{|z|^{d+1}} dz < \infty$. Then all conditions of Theorem 337 are satisfied, and hence the population SDE has a unique solution.

Now let us discuss the properties of solutions to the population SDE.

1. The Convergence Property.

Let us consider a sequence of solutions (x_t^n, ϕ_t^n) to the population SDEs:

$$\begin{cases} dx_t^n = (A^n(t)x_t^n + B^n(t)x_t^n \beta_t^n)dt + \sigma^n(t, x_t^n)dw_t \\ \quad + \int_Z c^n(t, x_{t-}^n, z)\widetilde{N}_k(dt, dz) + d\phi_t^n, \\ x_0^n = x_0^n \in \overline{R}_+^d, t \geq 0, \\ x_t \in \overline{R}_+^d, \text{for all } t \geq 0, \\ \text{and } (x_t^n, \phi_t^n) \text{ satisfies the other statements as in (11.22).} \end{cases}$$

$n = 0, 1, 2, ...$

We have the following convergence theorem.

Theorem 338 *Assume that*

1° *there exists a constant* $k_0 \geq 0$*, for all* $n = 0, 1, 2, ...$

$|A^n(t)| + |B^n(t)| + \beta_t^n \leq k_0,$

where $\beta_t^n \geq 0$ *are non-random such that they do not depend on* x*, and*

$\|\sigma^n(t, x)\|^2 + \int_Z |c^n(t, x, z)|^2 \pi(dz) \leq k_0(1 + |x|^2),$

2° $\|\sigma^0(t, x) - \sigma^0(t, y)\|^2 + \int_Z |c^0(t, x, z) - c^0(t, y, z)|^2 \pi(dz)$

$\leq k(t)\rho(|x - y|^2),$

where $k(t) \geq 0, \rho(u)$ *is continuous, strictly increasing, concave,* $\rho(0) = 0$*, and they are non-random such that for any* $0 < T < \infty$

$\int_0^T k(t)dt < \infty$, $\int_{0+} du/\ \rho(u) = \infty;$

3°(a) $lim_{n \to \infty} \int_0^T [|A^n(t) - A^0(t)| + |B^n(t) - B^0(t)|]dt = 0,$

$lim_{n \to \infty} \int_0^T [\sup_{x \in \overline{R}_+^d} \|\sigma^n(s, x) - \sigma^0(s, x)\|^2$

$+\sup_{x\in\overline{R}^d_+}\int_Z \left|c^n(s,x,z)-c^0(s,x,z)\right|^2\pi(dz)]ds=0,$ *for all* $T<\infty$;

(b) $\lim_{n\to\infty}\int_0^T \left|\beta^n(t)-\beta^0(t)\right|]dt=0$;

4° $E\left|x_0^n-x_0^0\right|^2\to 0$, *as* $n\to\infty$, *and* $E\left|x_0^0\right|^2\le k_0$.

Then for all $t\ge 0$, *as* $n\to\infty$

$E\sup_{s\le t}\left|x_s^n-x_s^0\right|^2\to 0,$

$E\sup_{s\le t}\left|\phi_s^n-\phi_s^n\right|^2\to 0.$

Proof. By Ito's formula

$\left|x_t^n-x_t^0\right|^2=\left|x_0^n-x_0^0\right|^2+2\int_0^t(x_s^n-x_s^0)\cdot(b^n(s,x_s^n)-b^0(s,x_s^0))ds$
$+\int_0^t\left\|\sigma^n(s,x_s^n)-\sigma^0(s,x_s^0)\right\|^2ds+2\int_0^t(x_s^n-x_s^0)\cdot(\sigma^n(s,x_s^n)-\sigma^0(s,x_s^0))dw_s$
$+2\int_0^t(x_s^n-x_s^0)\cdot\int_Z(c^n(s,x_{s-}^n,z)-c^0(s,x_{s-}^0,z))\widetilde{N}_k(ds,dz)$
$+2\int_0^t(x_s^n-x_s^0)\cdot d(\phi_s^n-\phi_s^0)+\int_0^t\int_Z\left|c^n(s,x_{s-}^n,z)-c^0(s,x_{s-}^0,z)\right|^2N_k(ds,dz)$
$=\sum_{i=1}^7 I_i^n(t),$

where we have written

$b^n(t,x_t^n)=A^n(t)x_t^n+B^n(t)x_t^n\beta_t^n,\ n=0,1,2,\cdots.$

Notice that

$I_6^n(t)=2\int_0^t(x_s^n-x_s^0)\cdot d(\phi_s^n-\phi_s^0)=2\int_0^t(x_s^n-x_s^0)\cdot d\phi_s^n$
$+2\int_0^t(x_s^0-x_s^n)\cdot d\phi_s^0\le 0,$

and by the condition 2° that

$E[I_3^n(t)+I_7^n(t)]\le 2\int_0^t k(s)\rho(E\left|x_s^n-x_s^0\right|^2)ds+J_1^n(t),$

where

$J_1^n(t)=2\int_0^t[\sup_{x\in\overline{R}^d_+}\left\|\sigma^n(s,x)-\sigma^0(s,x)\right\|^2$
$+\sup_{x\in\overline{R}^d_+}\int_Z\left|c^n(s,x,z)-c^0(s,x,z)\right|^2\pi(dz)]ds\to 0,$ as $n\to\infty$.

Moreover, by Theorem 330

$E(\sup_{0\le t\le T}\left|x_t^n\right|^2+\sup_{0\le t\le T}\left|\phi_t^n\right|^2)\le k_T,$ for all $n=0,1,2,...$

On the other hand,

$\frac{1}{2}EI_2^n(t)=E\int_0^t(x_s^n-x_s^0)\cdot(A^n(s)x_s^n+B^n(s)x_s^n\beta_s^n-A^0(s)x_s^0-B^0(s)x_s^0\beta_s^0)ds$
$\le k_0E\int_0^t\left|x_s^n-x_s^0\right|^2ds+E\int_0^t\left|A_s^n-A_s^0\right|\left|x_s^0\right|\left|x_s^n-x_s^0\right|ds$
$+k_0^2E\int_0^t\left|x_s^n-x_s^0\right|^2ds+k_0E\int_0^t\left|B_s^n-B_s^0\right|\left|x_s^0\right|\left|x_s^n-x_s^0\right|ds$
$+k_0E\int_0^t\left|\beta_s^n-\beta_s^0\right|\left|x_s^0\right|\left|x_s^n-x_s^0\right|ds.$

Notice that by the condition 3°, as $n\to\infty$

$E\int_0^t\left|A_s^n-A_s^0\right|\left|x_s^0\right|\left|x_s^n-x_s^0\right|ds$
$\le\frac{1}{2}\int_0^t\left|A_s^n-A_s^0\right|ds[3E\sup_{s\le t}\left|x_s^0\right|^2+2E\sup_{s\le t}\left|x_s^n\right|^2]$
$\le\frac{5}{2}k_t\int_0^t\left|A_s^n-A_s^0\right|ds\to 0,$

where by Theorem 330

$E\sup_{s\le t}\left|x_s^n\right|^2\le k_t,\ \forall n=0,1,2,\cdots.$

Similar results hold for the last two terms on the right hand side of the above inequality for $\frac{1}{2}EI_2^n(t)$. Hence by Fatou's lemma one finds that $\forall t\in[0,T]$

$\overline{\lim}_{n\to\infty}E\left|x_t^n-x_t^0\right|^2\le\widetilde{k}_T\int_0^t[\overline{\lim}_{n\to\infty}E\left|x_s^n-x_s^0\right|^2$
$+k(s)\rho(\overline{\lim}_{n\to\infty}E\left|x_s^n-x_s^0\right|^2)ds.$

So, $\forall t \in [0,T]$,

$\overline{\lim}_{n\to\infty} E\left|x_t^n - x_t^0\right|^2 = 0.$

Making similar observation, one also finds that

$\overline{\lim}_{n\to\infty} \sup_{s\leq t} E\left|x_s^n - x_s^0\right|^2 = 0.$

Therefore by the martingale inequality and by Ito's formula above we find that

$\overline{\lim}_{n\to\infty} E\sup_{s\leq t}\left|x_s^n - x_s^0\right|^2 \leq k_0'[\int_0^t k(r)\rho(\overline{\lim}_{n\to\infty} E\sup_{s\leq r}\left|x_s^n - x_s^0\right|^2)dr$
$+ \int_0^t \overline{\lim}_{n\to\infty} E\sup_{s\leq r}\left|x_s^n - x_s^0\right|^2 dr].$

Hence $\forall t \in [0,T]$

$\overline{\lim}_{n\to\infty} E\sup_{s\leq t}\left|x_s^n - x_s^0\right|^2 = 0.$

So, $\forall t \geq 0$,

$\lim_{n\to\infty} E\sup_{s\leq t}\left|x_s^n - x_s^0\right|^2 = 0.$

Furthermore, observe that

$|\phi_t^n - \phi_t^n|^2 \leq 5(\left|x_t^n - x_t^0\right|^2 + \left|x_0^n - x_0^0\right|^2$
$+ \left|\int_0^t (b^n(s,x_s^n) - b^0(s,x_s^n))ds\right|^2 + \left|\int_0^t (\sigma^n(s,x_s^n) - \sigma^0(s,x_s^0))dw_s\right|^2$
$+ \left|\int_0^t \int_Z (c^n(s,x_{s-}^n,z) - c^0(s,x_{s-}^0,z))\widetilde{N}_k(ds,dz)\right|^2).$

We also notice that as $n \to \infty$

$E[\int_0^t \left|A_s^n - A_s^0\right| \left|x_s^0\right| ds]^2 \leq [\int_0^t \left|A_s^n - A_s^0\right| ds]^2 E\sup_{s\leq t}\left|x_s^0\right|^2$
$\leq k_t[\int_0^t \left|A_s^n - A_s^0\right| ds]^2 \to 0,$

etc. Hence by the martingale inequality, as in the above, one also easily finds that as $n \to \infty$,

$E\sup_{s\leq t}\left|\phi_s^n - \phi_s^0\right|^2 \to 0.$ ■

Theorem 338 actually tells us the following facts:

1) For a given stochastic population dynamics and a given specific fertility rate of females, if an initial population size vector x_0^n "in closes to" a known initial population size vector x_0^0 "in the mean square", then the population size vector x_t^n corresponding to x_0^n will "uniformly close to" the population size vector x_t^0 corresponding to x^0 on any given interval $[0,T]$ "in the mean square"; that is, for a given $A(t), B(t), \sigma(t,x), c(t,x,z)$ and β_t,

$E\left|x_0^n - x_0^0\right|^2 \to 0 \implies E\sup_{s\leq T}\left|x_t^n - x_t^0\right|^2 \to 0.$

2) For given stochastic population dynamics, and a given initial population size vector, if a specific fertility rate of females β_t^n "is close to" a known specific fertility rate of females β_t^0 "in $L^1([0,T])$", then the population size vector x_t^n corresponding to β_t^n will be "uniformly close to" the population size vector x_t^0 corresponding to β_t^0 on any given interval $[0,T]$ "in the mean square"; that is, for a given $A(t), B(t), \sigma(t,x), c(t,x,z)$ and x_0,

$\int_0^T \left|\beta^n(t) - \beta^0(t)\right| dt \to 0 \implies E\sup_{s\leq T}\left|x_t^n - x_t^0\right|^2 \to 0.$

3) If the population dynamics can only be "approximately calculated in some sense", then for a given initial population size vector, and a given specific fertility rate of females, the corresponding population size vector will

also be "obtained approximatey uniformly on $[0,T]$ in the mean square"; that is, for a given x_0 and β_t,

$$\int_0^T [\left|A^n(t) - A^0(t)\right| + \left|B^n(t) - B^0(t)\right|]dt$$
$$+ \int_0^T [\sup_{x\in \overline{R}^d_+} \left\|\sigma^n(s,x) - \sigma^0(s,x)\right\|^2$$
$$+ \sup_{x\in \overline{R}^d_+} \int_Z \left|c^n(s,x,z) - c^0(s,x,z)\right|^2 \pi(dz)]ds \to 0$$
$$\Longrightarrow E\sup_{s\leq T} \left|x^n_t - x^0_t\right|^2 \to 0.$$

Furthermore, we have the following approximate error estimate calculation:

Corollary 339 *Assume that all conditions in Theorem 338 hold, and* $\forall n = 0,1,2,\cdots$

$$\sigma^n(t,x) = \sigma(t,x),$$
$$c^n(t,x,z) = c(t,x,z),\ f(z) = I_U(z),\ \pi(U) = 1,$$

where $\sigma(t,x)$ and $c(t,x,z)$ are defined by (11.23), together with the properties there.

Then

$$E\sup_{t\leq T} \left|x^n_t - x^0_t\right|^2 \leq 2[E\left|x^n_0 - x^0_0\right|^2 + 5k_T \int_0^T [\left|A^n_s - A^0_s\right|$$
$$+k_0\left|B^n_s - B^0_s\right| + k_0\left|\beta^n_s - \beta^0_s\right|]ds]e^{2[\widetilde{c}_1 + 18\overline{c}_1^2 + 2(k_0+k_0^2)]T},$$

where

$$\widetilde{c}_1 = \sum_{i=1}^d \left|c_1^{(i)}\right|^2 + \left|\overline{c}_1\right|^2 (k_0^2+1),$$

and the constant k_0 comes from the condition 1° in Theorem 338, k_T comes from Theorem 330. Actually, here we can get that

$$E\sup_{s\leq t} \left|x^n_s\right|^2 \leq k_T,$$

where

$$k_T = 2[\widetilde{k}_0 + 74(\sum_{i=1}^d \left|c_0^{(i)}\right|^2 + \left|\overline{c}_0\right|^2 + \left|\overline{c}_1\right|^2 k_0^2)T]e^{4[k_0+k_0^2+37\sum_{i=1}^d \left|c_1^{(i)}\right|^2]T},$$

and we assume that $E\left|x^n_0\right|^2 \leq \widetilde{k}_0, \forall n$. (Recall that $c_0^{(i)}, c_1^{(i)}, 1\leq i\leq d$ come from $\sigma(t,x)$, and $\overline{c}_0, \overline{c}_1$ come from $c(t,x,z)$. See (11.23). Moreover, $c_0^{(i)}, \overline{c}_0, 1\leq i\leq d$ are R^d-vectors, and $c_1^{(i)}, \overline{c}_1, 1\leq i\leq d$ are $R^{d\otimes d}-$matrices).

Proof. By Ito's formula, as in the proof of Theorem 338, one finds that

$$\left|x^n_t - x^0_t\right|^2 = \left|x^n_0 - x^0_0\right|^2 + 2\int_0^t (x^n_s - x^0_s)\cdot(b^n(s,x^n_s) - b^0(s,x^0_s))ds$$
$$+ \int_0^t \left\|\sigma^n(s,x^n_s) - \sigma^0(s,x^0_s)\right\|^2 ds + 2\int_0^t (x^n_s - x^0_s)\cdot(\sigma^n(s,x^n_s) - \sigma^0(s,x^0_s))dw_s$$
$$+2\int_0^t\int_Z (x^n_s - x^0_s)\cdot(c^n(s,x^n_{s-},z) - c^0(s,x^0_{s-},z))\widetilde{N}_k(ds,dz)$$
$$+ \int_0^t\int_Z \left|c^n(s,x^n_{s-},z) - c^0(s,x^0_{s-},z)\right|^2 N_k(ds,dz)$$
$$+2\int_0^t (x^n_s - x^0_s)\cdot d(\phi^n_s - \phi^0_s) = \sum_{i=1}^7 I^n_i(t) \leq \sum_{i=1}^6 I^n_i(t),$$

where

$$b^n(t,x^n_t) = A^n(t)x^n_t + B^n(t)x^n_t\beta^n_t,\ n = 0,1,2,\cdots.$$

Hence

$$EI^n_3(t) \leq E\int_0^t \sum_{i=1}^d \left|c_1^{(i)}\right|^2 \left|x^n_s - x^0_s\right|^2 ds,$$
$$EI^n_6(t) \leq E\int_0^t \left|\overline{c}_1\right|^2 (k_0^2+1)\left|x^n_s - x^0_s\right|^2 ds,$$

$E\left|x_t^n - x_t^0\right|^2 \leq E\left|x_0^n - x_0^0\right|^2 + \widetilde{c}_1 \int_0^t E\left|x_s^n - x_s^0\right|^2 ds + EI_2^n(t),$

where

$\widetilde{c}_1 = \sum_{i=1}^d \left|c_1^{(i)}\right|^2 + \left|\overline{c}_1\right|^2 (k_0^2 + 1).$

However, by the proof in Theorem 338 one sees that as $t \leq T$

$EI_2^n(t) \leq 2(k_0 + k_0^2) \int_0^t E\left|x_s^n - x_s^0\right|^2 ds$

$+5k_t \int_0^t [\left|A_s^n - A_s^0\right| + k_0 \left|B_s^n - B_s^0\right| + k_0 \left|\beta_s^n - \beta_s^0\right|]ds$

$\leq 2(k_0 + k_0^2) \int_0^t E\left|x_s^n - x_s^0\right|^2 ds$

$+5k_T \int_0^T [\left|A_s^n - A_s^0\right| + k_0 \left|B_s^n - B_s^0\right| + k_0 \left|\beta_s^n - \beta_s^0\right|]ds.$

Notice that from Theorem 330 one finds that

$E\sup_{s\leq T} \left|x_s^n\right|^2 \leq k_T, \ \forall n = 0, 1, 2, \cdots.$

Hence, by the above Ito formula

$E\sup_{s\leq t} \left|x_s^n - x_s^0\right|^2 \leq E\left|x_0^n - x_0^0\right|^2 + \widetilde{c}_1 \int_0^t E\left|x_s^n - x_s^0\right|^2 ds$

$+EI_2^n(t) + 6E\sup_{s\leq t} \left|x_s^n - x_s^0\right| (\int_0^t \sum_{i=1}^d \left|c_1^{(i)}\right|^2 \left|x_s^n - x_s^0\right|^2 ds)^{1/2},$

where we have used the following fact:[75], pp69

If $M \in \mathcal{M}^{2,loc}$ (i.e. M_t is a locally square integrable martingale) with $M_0 = 0$, then for any $T < \infty$ as $p \in (0, 2)$

$E\sup_{t\leq T} \left|M_t\right|^p \leq \frac{4-p}{2-p} E \left\langle M \right\rangle_T^{p/2}.$

Hence

$\frac{1}{2}E\sup_{s\leq t} \left|x_s^n - x_s^0\right|^2 \leq E\left|x_0^n - x_0^0\right|^2 + \widetilde{c}_1 \int_0^t E\left|x_s^n - x_s^0\right|^2 ds$

$+18\widetilde{c}_1^2 \int_0^t E\left|x_s^n - x_s^0\right|^2 ds + 2(k_0 + k_0^2) \int_0^t E\left|x_s^n - x_s^0\right|^2 ds$

$+5k_T \int_0^T [\left|A_s^n - A_s^0\right| + k_0 \left|B_s^n - B_s^0\right| + k_0 \left|\beta_s^n - \beta_s^0\right|]ds.$

By Gronwall's inequality one finds that

$E\sup_{s\leq t} \left|x_s^n - x_s^0\right|^2 \leq 2[E\left|x_0^n - x_0^0\right|^2$

$+5k_T \int_0^T [\left|A_s^n - A_s^0\right| + k_0 \left|B_s^n - B_s^0\right| + k_0 \left|\beta_s^n - \beta_s^0\right|]ds] e^{2[\widetilde{c}_1 + 18\widetilde{c}_1^2 + 2(k_0 + k_0^2)]T}.$

Now let us calculate

$E\sup_{s\leq T} \left|x_s^n\right|^2 \leq k_T, \ \forall n = 0, 1, 2, \cdots.$

More precisely, by Ito's formula,

$\left|x_t^n\right|^2 \leq \left|x_0^n\right|^2 + 2\int_0^t x_s^n \cdot b^n(s, x_s^n)ds$

$+2\int_0^t \sum_{i=1}^d (\left|c_0^{(i)}\right|^2 + \left|c_1^{(i)}\right|^2 \left|x_s^n\right|^2)ds + 2\int_0^t x_s^n \cdot \sum_{i=1}^d (c_0^{(i)} + c_1^{(i)} x_s^n)dw_s^i$

$+2\int_0^t \int_Z x_s^n \cdot (\overline{c}_0 + \overline{c}_1(x_s^n I_{\left|x_s^n\right| \leq k_0} + k_0 \frac{x_s^n}{\left|x_s^n\right|} I_{\left|x_s^n\right| > k_0}) I_U(z) \widetilde{N}_k(ds, dz)$

$+\int_0^t \int_Z \left|\overline{c}_0 + \overline{c}_1(x_s^n I_{\left|x_s^n\right| \leq k_0} + k_0 \frac{x_s^n}{\left|x_s^n\right|} I_{\left|x_s^n\right| > k_0}\right|^2 I_U(z) N_k(ds, dz)$

$+2\int_0^t x_s^n \cdot d\phi_s^n = \sum_{i=1}^7 I_i^n(t) = \sum_{i=1}^6 I_i^n(t),$

where

$b^n(t, x_t^n) = A^n(t)x_t^n + B^n(t)x_t^n \beta_t^n, \ n = 0, 1, 2, \cdots.$

Hence as $t \leq T$

$E(I_2^n(t) + I_3^n(t) + I_6^n(t)) \leq 2(k_0 + k_0^2 + \sum_{i=1}^d \left|c_1^{(i)}\right|^2) \int_0^t E\left|x_s^n\right|^2 ds$

$+2(\sum_{i=1}^d \left|c_0^{(i)}\right|^2 + \left|\overline{c}_0\right|^2 + \left|\overline{c}_1\right|^2 k_0^2)T.$

Since $E\left|x_0^0\right|^2 \leq k_0$ and $E\left|x_0^n - x_0^0\right|^2 \to 0$, we may estimate (assume) that

$$E\left|x_0^n\right|^2 \leq \widetilde{k}_0.$$

Now by the above Ito formula, and by applying the martingale inequality, one easily finds that

$$E\sup_{s\leq t}\left|x_s^n\right|^2 \leq E\left|x_0^n\right|^2 + 2(k_0 + k_0^2 + \sum_{i=1}^d \left|c_1^{(i)}\right|^2)\int_0^t E\left|x_s^n\right|^2 ds$$
$$+2(\sum_{i=1}^d \left|c_0^{(i)}\right|^2 + \left|\overline{c}_0\right|^2 + \left|\overline{c}_1\right|^2 k_0^2)T$$
$$+6E\sup_{s\leq t}\left|x_s^n\right|(\int_0^t \sum_{i=1}^d \left|c_0^{(i)} + c_1^{(i)}x_s^n\right|^2 ds)^{1/2}$$
$$+6E\sup_{s\leq t}\left|x_s^n\right|(\int_0^t \left|\overline{c}_0 + \overline{c}_1(x_s^n I_{\left|x_s^n\right|\leq k_0} + k_0\tfrac{x_s^n}{\left|x_s^n\right|}I_{\left|x_s^n\right|>k_0})\right|^2 ds)^{1/2}.$$

Hence as $t \leq T$

$$\tfrac{1}{2}E\sup_{s\leq t}\left|x_s^n\right|^2 \leq \widetilde{k}_0 + 2(k_0 + k_0^2 + \sum_{i=1}^d \left|c_1^{(i)}\right|^2)\int_0^t E\left|x_s^n\right|^2 ds$$
$$+2(\sum_{i=1}^d \left|c_0^{(i)}\right|^2 + \left|\overline{c}_0\right|^2 + \left|\overline{c}_1\right|^2 k_0^2)T + 36E\int_0^t \sum_{i=1}^d \left|c_0^{(i)} + c_1^{(i)}x_s^n\right|^2 ds$$
$$+36\int_0^t \left|\overline{c}_0 + \overline{c}_1(x_s^n I_{\left|x_s^n\right|\leq k_0} + k_0\tfrac{x_s^n}{\left|x_s^n\right|}I_{\left|x_s^n\right|>k_0})\right|^2 ds$$
$$\leq \widetilde{k}_0 + 2(k_0 + k_0^2 + \sum_{i=1}^d \left|c_1^{(i)}\right|^2)\int_0^t E\left|x_s^n\right|^2 ds$$
$$+2(\sum_{i=1}^d \left|c_0^{(i)}\right|^2 + \left|\overline{c}_0\right|^2 + \left|\overline{c}_1\right|^2 k_0^2)T + 72\sum_{i=1}^d \left|c_0^{(i)}\right|^2 T$$
$$+72\sum_{i=1}^d \left|c_1^{(i)}\right|^2 \int_0^t E\left|x_s^n\right|^2 ds + 72[\left|\overline{c}_0\right|^2 + \left|\overline{c}_1\right|^2 k_0^2]T.$$

By Gronwall's inequality, as $t \leq T$

$$E\sup_{s\leq t}\left|x_s^n\right|^2 \leq e^{4[k_0 + k_0^2 + 37\sum_{i=1}^d \left|c_1^{(i)}\right|^2]T}.$$
$$\cdot 2[\widetilde{k}_0 + 74(\sum_{i=1}^d \left|c_0^{(i)}\right|^2 + \left|\overline{c}_0\right|^2 + \left|\overline{c}_1\right|^2 k_0^2)T] = k_T. \blacksquare$$

In the previous convergence results we have assumed that β_t—the specific fertility rate of females, does not depend on x_t—the population size vector. However, when we consider some population control problems and use β_t as a control, then in some cases we need to use β_t as a feedback control, that is, $\beta_t = \beta_t(x_t)$ must depend on the instant x_t. Because, from the state (x_t) situation at the presemt time, we want to adjust it instantly by using a control (β_t). In such a case we must use could we obtain a convergence theorem for x_t^n? The following theorem answers this question.

Theorem 340 *Under assumptions 1°, 3°(a), and 4° of Theorem 338, but now with the assumption 2° is weakened to*

2°' $\left\|\sigma^0(t,x) - \sigma^0(t,y)\right\|^2 + \int_Z \left|c^0(t,x,z) - c^0(t,y,z)\right|^2 \pi(dz)$
$\leq k_{N,T}(t)\rho_N(\left|x-y\right|^2)$, *as* $t \in [0,T], \left|x\right|, \left|y\right| \leq N$,
where $\rho_N(u)$ *is continuous, strictly increasing, concave,* $\rho_N(0) = 0$, *and* $k_{N,T}(t) \geq 0$, *and they are non-random such that*
$\int_0^T k_{N,T}(t)dt < \infty$, $\int_{0+} du/\rho_N(u) = \infty, \forall T < \infty, N = 1,2,\cdots;$
and we also suppose that

3°(b)' $\beta^n(t,x), n = 0,1,2,\cdots$ *all depend on* x *in such a way that*
$lim_{n\to\infty}\int_0^T \sup_{x\in\overline{R}_d^+}\left|\beta^n(t,x) - \beta^0(t,x)\right| dt = 0,$
$\left|\beta^0(t,x) - \beta^0(t,y)\right| \le k_{N,T}(t)\rho_N(|x-y|),$ *as* $t\in[0,T], |x|, |y| \le N,$
where $k_{N,T}(t)$ *and* $\rho_N(u)$ *have the same properties as that in 2°';*
then for any $\varepsilon > 0$ *and* $t \ge 0$
$P(\sup_{s\le t}\left|x_s^n - x_s^0\right| > \varepsilon) \to 0,$ *and*
$P(\sup_{s\le t}\left|\phi_s^n - \phi_s^0\right| > \varepsilon) \to 0.$

Notice that $\lim_{n\to\infty} P(\sup_{s\le t}\left|x_s^n - x_s^0\right| > \varepsilon) = 0$ is equivalent to
$\lim_{n\to\infty} P(\sup_{s\le t}\left|x_s^n - x_s^0\right| \le \varepsilon) = 1.$
So, roughly speaking, the physical meaning of the conclusion is that with a very good chance (the possibility P is close to 100%), x_s^n can be sufficiently and uniformly close to x_s on $[0,t]$.

Remark 341 *1) If we take* $\rho_N(u) = u$, $k_{N,T}(t) = k_0$, *where* $k_0 \ge 0$ *is a constant, then the condition 2°' actually means that* $\sigma^0(t,x)$ *satisfies a local Lipschitz condition in* x*, and* $c^0(t,x,z)$ *also satisfies a local Lipschitz-type condition in* x*.*
2) If we let $\beta^n(t,x) = \beta^0(t,x) = xI_{|x|\le k_0} + k_0\frac{x}{|x|}I_{|x|\le k_0}$, *where* $k_0 \ge 0$ *is a constant, then the condition 3°(b)' is satisfied.*
3) The convergence in probability: $P(\sup_{s\le t}\left|x_s^n - x_s^0\right| > \varepsilon) \to 0$ *is weaker than the convergence in mean square:* $E\sup_{s\le t}\left|x_s^n - x_s^0\right|^2 \to 0$*. In fact,*
$P(\sup_{s\le t}\left|x_s^n - x_s^0\right| > \varepsilon) \le \frac{1}{\varepsilon^2}E\sup_{s\le t}\left|x_s^n - x_s^0\right|^2.$

Proof. Notice that now
$b^n(t,x_t^n) = A^n(t)x_t^n + B^n(t)x_t^n\beta_t^n(x_t^n).$
Set
$\tau^N = \inf\left\{t : \left|x_t^0\right| > N\right\},$
$\tau^{n,\varepsilon} = \inf\left\{t : \left|x_t^n - x_t^0\right| > \varepsilon\right\},$
$\tau_\varepsilon^N(n) = \tau^N \wedge \tau^{n,\varepsilon}.$
Then by Ito's formula, as in the proof of Theorem 338, we have
$$\lim_{n\to\infty} E\sup_{s\le t}\left|x^n_{s\wedge\tau_\varepsilon^N(n)} - x^0_{s\wedge\tau_\varepsilon^N(n)}\right|^2 = 0.$$
However, for all $t \ge 0$
$$E\left|x^n_{t\wedge\tau_\varepsilon^N(n)} - x^0_{t\wedge\tau_\varepsilon^N(n)}\right|^2 \ge \varepsilon^2 P(\tau^{n,\varepsilon} \le \tau^N \wedge t).$$
Hence
$$\lim_{n\to\infty} P(\tau^{n,\varepsilon} \le \tau^N \wedge t) = 0.$$
We have that
$$P(\sup_{s\le\tau^N\wedge t}\left|x_s^n - x_s^0\right| > 2\varepsilon) \le P(\tau^{n,\varepsilon} \le \tau^N \wedge t) \to 0, \text{ as } n\to\infty.$$
Notice that
$$P(\sup_{s\le T}\left|x_s^n - x_s^0\right| > \varepsilon) \le P(\sup_{s\le\tau^N\wedge T}\left|x_s^n - x_s^0\right| > \varepsilon)$$
$$+P(\sup_{\tau^N\wedge T<s\le T}\left|x_s^n - x_s^0\right| > \varepsilon) \le P(\sup_{s\le\tau^N\wedge T}\left|x_s^n - x_s^0\right| > \varepsilon)$$
$$+P(\tau^N\wedge T < T).$$
Since

$\lim_{N\to\infty} \tau^N = \infty$.

Hence

$\lim_{N\to\infty} P(\tau^N \wedge T < T) = 0$.

Therefore the first conclusion is established. Now, by (1.1)

$$P(\sup_{s\leq t} |\phi_s^n - \phi_s^0| > 4\varepsilon) \leq P(\sup_{s\leq t} |x_s^n - x_s^0| > \varepsilon)$$
$$+P(\sup_{r\leq t} |\int_0^r b^n(s, x_s^n)ds - \int_0^r b^0(s, x_s^0)ds| > \varepsilon)$$
$$+P(\sup_{r\leq t} |\int_0^r \sigma^n(s, x_s^n)dw_s - \int_0^r \sigma^0(s, x_s^0)dw_s| > \varepsilon)$$
$$+P(\sup_{r\leq t} | \int_0^r \int_Z c^n(s, x_{s-}^n, z)\widetilde{N}_k(ds, dz)$$
$$- \int_0^r \int_Z c^0(s, x_{s-}^0, z)\widetilde{N}_k(ds, dz)| > \varepsilon) = \sum_{i=1}^4 I_i.$$

However, it is already known that as $n \to \infty$

$I_1 \to 0$.

Let us show that as $n \to \infty$

$I_4 \to 0$.

In fact, for any $\overline{\delta} > 0$, by the martingale inequality,

$$I_4 \leq P(\sup_{0\leq s\leq t} |x_s^n - x_s^0| > \overline{\delta})$$
$$+k_t\varepsilon^{-2}E(\int_0^t k(s)\rho(\sup_{r\leq s} |x_r^n - x_r^0|^2)ds)I_{\sup_{0\leq s\leq t}|x_s^n - x_s^0|\leq\overline{\delta}}$$
$$+\sup_{x\in \overline{R}_+^d} \int_Z |c^n(s, x, z) - c^0(s, x, z)|^2 \pi(dz)]ds = \sum_{i=1}^3 I_{4i}^n.$$

Obviously, by the result just obtained, and by the assumption as $n \to \infty$,

$I_{41}^n + I_{43}^n \to 0$.

However, for $\overline{\delta}$ is small enough,

$I_{42}^n \leq k_t\varepsilon^{-2}\rho(\overline{\delta}^2) \int_0^t k(s)ds < \varepsilon$.

So, it is seen that as $n \to \infty$

$I_4 \to 0$.

The proof that $I_3 \to 0$,as $n \to \infty$ is obtained in a similar fashion. Let us show that as $n \to \infty$, $I_2 \to 0$. In fact, notice that

$$|B^n(t)x_t^n\beta_t^n(x_t^n) - B^0(t)x_t^0\beta_t^0(x_t^0)|$$
$$= |B^n(t)x_t^n\beta_t^n(x_t^n) \mp B^0(t)x_t^n\beta_t^n(x_t^n)$$
$$\mp B^0(t)x_t^n\beta_t^0(x_t^n) \mp B^0(t)x_t^n\beta_t^0(x_t^0) - B^0(t)x_t^0\beta_t^0(x_t^0)|$$
$$\leq k_0 |B^n(t) - B^0(t)| |x_t^n| + k_0 |\beta_t^n(x_t^n) - \beta_t^0(x_t^n)| |x_t^n|$$
$$+k_0 |x_t^0| |\beta_t^0(x_t^n) - \beta_t^0(x_t^0)| + k_0^2 |x_t^n - x_t^0|,$$

and

$E\sup_{s\leq T} |x_s^n|^2 \leq k_T, \forall n = 0, 1, 2, \cdots$.

The last inequality implies that $\forall\varepsilon > 0$ as $N \to \infty$,

$\sup_n P(\sup_{s\leq T} |x_s^n|^2 > N) \leq k_T/N \to 0$.

So, in similar fashion, one can similarly prove that $\forall\varepsilon > 0$, as $n \to \infty$,

$P(|B^n(t)x_t^n\beta_t^n(x_t^n) - B^0(t)x_t^0\beta_t^0(x_t^0)| > \varepsilon) \to 0$.

From this, one easily finds that as $n \to \infty$, $I_2 \to 0$. Therefore, we arrive at the second conclusion. ∎

2. The Stability Property

Now we are going to discuss the stability of solutions to (11.22). We have the following stability property for the solutions of the population dynamical system.

Theorem 342 *Assume that coefficients* b, σ *and* c *in (11.22), where* $b(t.x) = A(t)x + B(t)x\beta_t$*, satisfy all conditions in Theorem 337, and that*
$\sigma(t, 0) = 0, c(t, 0, z) = 0.$
Then
1) $(0, 0)$ *is a pathwise unique strong solution of (11.22) with initial condition* $x_0 = 0$.
2) For any given $x_0 \in \bar{R}^d_+$ *(11.22) has a pathwise unique strong solution* $(x_t, \phi_t), t \geq 0$.
3) Furthermore, if there exists a positive constant $k_1 \geq 0$ *such that*
$2x \cdot b(t, x) + \|\sigma(t, x)\|^2 + \int_Z |c(t, x, z)|^2 \pi(dz) \leq -k_1 |x|^2,$
then
$E|x_t|^2 \leq E|x_0|^2 e^{-k_1 t}$*, for all* $t \geq 0$.
and there exists a constant k_4 *such that*
$E|\phi_t|^2 \leq k_4 E|x_0|^2 (1 + e^{-k_1 t})$*, for all* $t \geq 0$.
Furthermore, write $\phi_\infty = \lim_{n\to\infty} \phi_t$*, then*
$E(|\phi_\infty|^2 - |\phi_t|^2) \leq k''' E|x_0|^2 e^{-k_1 t}.$
4) In the case that there exists a positive constant $k_5 > 0$ *such that*
$2x \cdot b(t, x) + \|\sigma(t, x)\|^2 + \int_Z |c(t, x, z)|^2 \pi(dz) \geq k_5 |x|^2,$
and
$|b(t, x)|^2 + \|\sigma(t, x)\|^2 + \int_Z |c(t, x, z)|^2 \pi(dz) \leq k_0(1 + |x|^2),$
then
$E|x_t|^2 \geq E|x_0|^2 e^{k_5 t}$*, for all* $t \geq 0$.
Hence
$lim_{t\to\infty} E|x_t|^2 = \infty.$

This theorem tells us that under appropriate conditions the solutions of the population dynamical system are exponentially stable. Roughly speaking, this means that, if 0 is a solution of the dynamical system with the initial condition 0, then under appropriate conditions any solution x_t even if it has a initial condition $x_0 \neq 0$, will "be close to" the solution 0 as time tends to infinity. Moreover, the approximation speed is a negative exponencial. (See the conclusion 3)).

Proof. 1) and 2) are true by Theorem 337. Let us show 3): By Ito's formula

$$|x_t|^2 = |x_0|^2 + \int_0^t (2x_s \cdot b(s, x_s) + \|\sigma(s, x_s)\|^2) ds + 2\int_0^t x_s \cdot \sigma(s, x_s) dw_s$$
$$+ 2\int_0^t \int_Z x_s \cdot c(s, x_{s-}, z) \widetilde{N}_k(ds, dz) + \int_0^t \int_Z |c(s, x_{s-}, z|^2 N_k(ds, dz),$$

where we have used the fact that

$$2\int_0^t x_s \cdot d\phi_s = 0.$$

Set $\tau_N = \inf\{t \geq 0 : |x|_t > N\}$. Then

$$E|x_{t\wedge\tau_N}|^2 = E|x_0|^2 + E\int_0^{t\wedge\tau_N} (2x_s \cdot b(s, x_s) + \|\sigma(s, x_s)\|^2$$
$$+ \int_Z |c(s, x_s, z|^2 \pi(dz)) ds.$$

Hence

$$\tfrac{d}{dt} E|x_{t\wedge\tau_N}|^2 = E(2x_{t\wedge\tau_N} \cdot b(t\wedge\tau_N, x_{t\wedge\tau_N}) + \|\sigma(t\wedge\tau_N, x_{t\wedge\tau_N})\|^2$$
$$+ \int_Z |c(t\wedge\tau_N, x_{t\wedge\tau_N}, z|^2 \pi(dz)) \leq -k_1 E|x_{t\wedge\tau_N}|^2, \text{ a.e. } t \geq 0.$$

Therefore

$E|x_{t\wedge\tau_N}|^2 \leq E|x_0|^2 e^{-k_1 t}$, for all $t \geq 0$.

Letting $N \uparrow \infty$, by Fatou's lemma one finds that

$E|x_t|^2 \leq E|x_0|^2 e^{-k_1 t}$, for all $t \geq 0$.

On the other hand, by Ito's formula

$|\phi_T|^2 - |\phi_t|^2 = 2\int_t^T \phi_s \cdot d\phi_s = 2\int_t^T \phi_s \cdot (dx_s - b(s,x_s)ds - \sigma(s,x_s)dw_s$
$- \int_Z c(s,x_{s-},z)\widetilde{N}_k(ds,dz)) = 2(\phi_T \cdot x_T - \phi_t \cdot x_t)$
$-2\int_t^T \phi_s \cdot (b(s,x_s)ds + \sigma(s,x_s)dw_s + \int_Z c(s,x_{s-},z)\widetilde{N}_k(ds,dz))$,

where we have again used the result that $\int_t^T x_s \cdot d\phi_s = 0$. Hence, by assumption we get that $|b(t,x)| \leq (k_0 + k_0^2)|x|$, and

$E(|\phi_T|^2 - |\phi_t|^2) \leq k''(E|x_t|^2 + \int_t^T E|x_s|^2 ds + E|x_T|^2)$
$\leq k'E|x_0|^2 (e^{-k_1 t} + e^{-k_1 T})$.

Letting $t = 0$, the second conclusion of 3) then follows. Now, since ϕ_t^i is increasing in t for each i, (see Remark 321), ϕ_∞ exists, and

$E(|\phi_\infty|^2 - |\phi_t|^2) \leq k'''E|x_0|^2 e^{-k_1 t}$.

3) is proved.

Finally, let us discuss the unstable case 4). In the above discussion, and by assumption, we have

$\frac{d}{dt}E|x_t|^2 = E(2x_t \cdot b(t,x_t) + \|\sigma(t,x_t)\|^2 + \int_Z |c(t,x_t,z|^2 \pi(dz))ds \geq k_5 E|x_t|^2$.

Hence

$E|x_t|^2 \geq E|x_0|^2 e^{k_5 t}$, for all $t \geq 0$. ∎

Corollary 343 *Let $\sigma(t,x)$ and $c(t,x,z)$ be defined by (11.23) with $\bar{c}_0 = 0, c_0^{(i)} = 0, 1 \leq i \leq d$, and with the properties required there. Moreover, let $f(z) = I_U(z), \pi(U) = 1$. Suppose that $A(t)$ and $B(t)$ are as given in the "Introduction" to this Chapter, and that $0 \leq \beta \leq \beta_0$. Let*

$\delta_0 = \min_{t\geq 0}(\frac{1}{2} + \eta_1(t), \eta_2(t), \cdots, \eta_{d-1}(t), \frac{1}{2} + \eta_d(t))$,
$b_M = \max_{t\geq 0}(b_{r_1}(t), \cdots, b_{r_2}(t))$,

where $d = r_m -$the largest age in the population.
Then as

$\delta_1 = 2\delta_0 - \sum_{i=1}^d \left|c_1^{(i)}\right|^2 - |\bar{c}_1|^2 - \beta_0 b_M \max(r_2 - r_1, 1) > 0$,
$E|x_t|^2 \leq E|x_0|^2 e^{-\delta_1 t}$, *for all* $t \geq 0$.

That is, the population dynamics of (11.22) is exponentially stable in the mean square under the above assumptions. In particular, for a given constant $a > 0$ we have that

$t \geq \frac{1}{\delta_1}\ln\frac{E|x_0|^2}{a} \implies E|x_t|^2 \leq a$.

This corollary actually tells us that if the stochastic perturbation is not too large (that is, $\sum_{i=1}^d \left|c_1^{(i)}\right|^2 + |\bar{c}_1|^2$ is very small), and the forward death rate is greater than a positive constant (that is, $\delta_0 > 0$), then the population dynamics (11.22) can be exponentially stable "in the mean square" if the fertility rate of females is small enough. Furthermore, if we have a target

$a > 0$, we can find out when the population size vector "in the mean square" can be less than this target a.

Proof. By calculation

$2x \cdot b(t,x) = -2\sum_{i=1}^{d}(1+\eta_i)x_i^2 + 2\sum_{i=1}^{d-1} x_i x_{i+1} + 2(\sum_{i=r_1}^{r_2} b_i x_i x_1)\beta_t,$

$\|\sigma(t,x)\|^2 = \sum_{i=1}^{d} \left|c_1^{(i)}\right|^2 |x|^2,$

$\int_Z |c(t,x,z)|^2 \pi(dz) = |\bar{c}_1|^2 \left|x I_{|x|\leq k_0} + k_0 \frac{x}{|x|} I_{|x|>k_0}\right|^2 \leq |\bar{c}_1|^2 |x|^2.$

Notice that $x_i \geq 0, \forall i = 1,2,\cdots,d$. Hence

$\sum_{i=1}^{d-1} x_i x_{i+1} \leq \frac{1}{2}(x_1^2 + x_d^2) + \sum_{i=2}^{d-1} x_i^2,$

and

$0 \leq (\sum_{i=r_1}^{r_2} b_i x_i x_1)\beta_t \leq b_M \beta_0 [\frac{r_2 - r_1}{2} x_1^2 + \frac{1}{2}\sum_{i=r_1}^{r_2} x_i^2]$

$\leq \frac{1}{2} b_M \beta_0 \max(r_2 - r_1, 1) |x|^2,$

So

$2x \cdot b(t,x) + \|\sigma(t,x)\|^2 + \int_Z |c(t,x,z)|^2 \pi(dz)$

$\leq -(2\delta_0 - \sum_{i=1}^{d} \left|c_1^{(i)}\right|^2 - |\bar{c}_1|^2 - b_M \beta_0 \max(r_2 - r_1, 1)) |x|^2,$

where $\delta_0 = \min(\frac{1}{2} + \eta_1, \eta_2, \cdots, \eta_{d-1}, \frac{1}{2} + \eta_d)$. Thus as

$\delta_1 = 2\delta_0 - \sum_{i=1}^{d} \left|c_1^{(i)}\right|^2 - |\bar{c}_1|^2 - \beta_0 b_M \max(r_2 - r_1, 1) > 0,$

the population dynamics (11.22) is exponentially stable in the mean square, i.e. by 3) of Theorem 342

$E|x_t|^2 \leq E|x_0|^2 e^{-\delta_1 t}$, for all $t \geq 0$.

Finally, noticing that

$E|x_0|^2 e^{-\delta_1 t} \leq a \iff t \geq \frac{1}{\delta_1} \ln \frac{E|x_0|^2}{a}.$

We arrive at the final conclusion. ■

11.8 Comparison of Solutions and Stochastic Population Control

1. The Tanaka type Formula and Comparison Theorems

In this section we will discuss the comparison of solutions to the stochastic population dynamics equations and optimal stochastic population control. However, to show a comparison theorem for the solutions of a $d-$dimensional RSDE we need the help of a Tanaka type formula for such a $d-$dimensional RSDE. So we first present the following Tanaka type formula. Consider two $d-$dimensional RSDEs, which is a little more general than (11.4): $i = 1, 2,$

$$\begin{cases} dx_t^i = b^i(t,x_t^i)dt + \sigma(t,x_t^i)dw_t + \int_Z c(t,x_{t-}^i,z)\widetilde{N}_k(dt,dz) + d\phi_t^i, \\ x_0^i = x^i \in \overline{R}_+^d, \ t \geq 0, \\ x_t^i \in \overline{R}_+^d, \ t \geq 0, \text{ and the other stantements in (11.4)} \\ \text{also holds for } (x_t^i, \phi_t^i). \end{cases} \tag{11.24}$$

Theorem 344 *(Tanaka type formula). Assume that*

1° b^1, b^2, σ and $\int_Z |c|^2 \pi(dz)$ are locally bounded, i.e. for each $r = 1, 2, ...$
$|h(t,x)| \leq k_r$, as $|x| \leq r$, $h = b^1, b^2, \sigma$ and $\int_Z |c|^2 \pi(dz)$;
where $0 \leq k_r$ is a constant depending on r only;
2° $\sigma(t,x) = (\sigma_{ik}(t,x))_{i,k=1}^d$ satisfies the condition that
$|\sigma_{ik}(t,x) - \sigma_{ik}(t,y)|^2 \leq k_{N,T}(t)\rho_{N,T}(|x_i - y_i|^2)$,
as $|x|, |y| \leq N, t \in [0,T]$, $\forall i, k = 1, 2, \cdots, d$;
where $0 \leq k_{N,T}(t)$,
$\int_0^T k_{N,T}(t)dt < \infty$, for each $T < \infty$,
and $\rho_{N,T}(u) > 0$,as $u > 0$; $\rho_{N,T}(0) = 0$; and $\rho_{N,T}(u)$ is strictly increasing in u and such that
$\int_{0+} du/\rho_{N,T}(u) = \infty$, for $N = 1, 2, ...$ and $T < \infty$;
3° $c(t,x,z) = (c_i(t,x,z))_{i=1}^d$ satisfies conditions that
$x + c(t,x,z) \in \overline{R}_+^d$, $\forall t \geq 0, x \in \overline{R}_+^d, z \in Z$;
$x_i \geq y_i \implies x_i + c_i(t,x,z) \geq y_i + c_i(t,y,z)$,
$\forall t \geq 0,\ x, y \in \overline{R}_+^d, z \in Z$,
where $x = (x_1, \cdots, x_d), y = (y_1, \cdots, y_d)$.
If (x_t^j, ϕ_t^j) satisfies (11.24), $j = 1, 2$, then $\forall i = 1, 2, ..., d$,
$(x_i^1(t) - x_i^2(t))^+ = (x_i^1(0) - x_i^2(0))^+ + \int_0^t I_{(x_i^1(s-) > x_i^2(s-)} d(x_i^1(s) - x_i^2(s))$.
Furthermore, in detail, $\forall i = 1, 2, ..., d$,

$$\begin{aligned} &(x_i^1(t) - x_i^2(t))^+ = (x_i^1(0) - x_i^2(0))^+ \\ &+ \int_0^t I_{(x_i^1(s) > x_i^2(s)}(b_i^1(s,x^1(s)) - b_i^2(s,x^2(s)))ds \\ &+ \int_0^t I_{(x_i^1(s) > x_i^2(s)} \textstyle\sum_{k=1}^d (\sigma_{ik}(s,x^1(s)) - \sigma_{ik}(s,x^2(s)))dw_k(s) \\ &+ \int_0^t \int_Z I_{(x_i^1(s-) > x_i^2(s-)}(c_i(s,x^1(s-),z) - c_i(s,x^2(s-),z))\widetilde{N}_k(ds,dz) \\ &+ \int_0^t I_{(x_i^1(s) > x_i^2(s))} d(\phi_i^1(s) - \phi_i^2(s)). \end{aligned}$$

Remark 345 *An example in which the above assumptions 2° and 3° are satisfied is*

$\sigma_{ik}(t,x) = c_0^i + c_1^{ik} x_i$,
$c^i(t,x,z) = \overline{c}_0^i + \overline{c}_1^i x_i$,
where all $c_0^i, c_1^{ik}, \overline{c}_0^i, \overline{c}_1^i, i, k = 1, 2, \cdots, d$ are constants; and $\overline{c}_0^i, \overline{c}_1^i \geq 0, i = 1, 2, \cdots, d$.

Now let us establish the theorem.

Proof. By Theorem 153 one finds that the first and second formulas are true. However, by the definition of the solution one finds that
$\int_0^t I_{(x_i^1(s)>x_i^2(s))} d\phi_i^1(s) = 0.$
Because $d\phi_i^1(s) > 0$ only when $x_i^1(s) = 0$, and now $x_i^2(s) \geq 0$. Hence the last formula follows. ■

Now let us derive some comparison theorems for solutions to the stochastic population dynamics (11.22). For conveneince let us write out it again as follows:

$$\begin{cases} dx_t = (A(t)x_t + B(t)x_t\beta_t)dt + \sigma(t,x_t)dw_t + \int_Z c(t,x_{t-},z)\widetilde{N}_k(dt,dz) \\ +d\phi_t,\ x_0 = x,\ t \geq 0, \\ x_t \in \overline{R}_+^d, \text{for all } t \geq 0, \\ \text{and the other statements hold for } (x_t,\phi_t), \end{cases} \tag{11.25}$$

where x_t^i is the size of the population with age between $[i, i+1)$, and $x_t = (x_t^1, \cdots, x_t^d)$; while β_t is the specific fertility rate of females,

$$A(t) = \begin{pmatrix} -(1+\eta_1(t)) & 0 & \cdots & \cdots & 0 \\ 1 & -(1+\eta_2(t)) & 0 & \cdots & 0 \\ \cdots & \cdots & \cdots & \cdots & \cdots \\ 0 & \cdots & 0 & 1 & -(1+\eta_d(t)) \end{pmatrix}, \tag{11.26}$$

$$B(t) = \begin{pmatrix} 0 & \cdots & 0 & b_{r_1}(t) & \cdots & b_{r_2}(t) & 0 & \cdots & 0 \\ 0 & \cdots & \cdots & \cdots & \cdots & \cdots & \cdots & \cdots & 0 \end{pmatrix}, \tag{11.27}$$

$b_i(t) = (1-\mu_{00}(t))k_i(t)h_i(t) \neq 0,\ i = r_1, \cdots, r_2; 1 < r_1 < r_2 < d,$
$\mu_{00}(t)$ is the death rate of babies, $\eta_i(t)$ is the forward death rate by ages, $k_i(t)$ and $h_i(t)$ are the corresponding sex rate and fertility models, respectively, x_t^i is the size of the population with ages between $[i, i+1)$,
$x_t = (x_t^1, \cdots, x_t^d),$
$d = r_m$ is the largest age of people. Because of the physical meaning, we may assume that all $\eta_i(t), b_j(t), \beta_t, i = 1,2,\cdots,d; j = r_1,\cdots,r_2$ are non-negative and bounded.

To consider the comparison of solution to (11.25) first we write out all its component equations as follows:

$$\begin{cases} dx_t^1 = [-(1+\eta_t^1)x_t^1 + \sum_{i=r_1}^{r_2} b_t^i x_t^i \beta_t]dt + \sum_{k=1}^d \sigma^{1k}(t,x_t)dw_t^k \\ \qquad + \int_Z c^1(t,x_{t-},z)\widetilde{N}_k(dt,dz) + d\phi_t^1, \\ dx_t^i = [-(1+\eta^i)x_t^i + x_t^{i-1}]dt + \sum_{k=1}^d \sigma^{ik}(t,x_t)dw_t^k \\ \qquad + \int_Z c^i(t,x_{t-},z)\widetilde{N}_k(dt,dz) + d\phi_t^i, i = 2,\cdots,d. \end{cases} \tag{11.28}$$

Now assume that conditions 2° and 3° in Theorem 344 hold, and all coefficients η_t^i, b_t^i, β_t in the above population dynamics are bounded, non-negative and that $|\sigma(t,x)|^2$, $\int_Z |c(t,x,z)|^2 \pi(dz)$ have less than linear growth in $|x|^2$; that is,

$0 \leq \eta_t^i + b_t^j + \beta_t \leq k_0,\ \forall i = 1, 2, \cdots, d; j = r_1, \cdots, r_2;$

$|\sigma(t,x)|^2 + \int_Z |c(t,x,z)|^2 \pi(dz) \leq k_0(1 + |x|^2),$

where $k_0 \geq 0$ is a constant. We also assume that there is another solution $(\overline{x}_t, \overline{\phi}_t)$ satisfying the same stochastic population dynamics (11.25) but with another initial value $\overline{x}_0$ and another fertility rate for females $\overline{\beta}_t$. Then by the Tanaka type formula (Theorem 344) one finds that

$$(x_t^1 - \overline{x}_t^1)^+ = (x_0^1 - \overline{x}_0^1)^+ + \int_0^t I_{x_s^1 > \overline{x}_s^1}[-(1+\eta_s^1)(x_s^1 - \overline{x}_s^1)$$
$$+ \sum_{i=r_1}^{r_2} b_s^i (x_s^i \beta_s - \overline{x}_s^i \overline{\beta}_s)]ds + M_t^1 - \int_0^t I_{x_s^1 > \overline{x}_s^1} d\overline{\phi}_s^1$$
$$\leq (x_0^1 - \overline{x}_0^1)^+ + \int_0^t [-(1+\eta_s^1)(x_s^1 - \overline{x}_s^1)^+$$
$$+ \sum_{i=r_1}^{r_2} b_s^i ((x_s^i - \overline{x}_s^i)^+ \beta_s + \overline{x}_s^i(\beta_s - \overline{\beta}_s))]ds + M_t^1,$$

where M_t^1 is a martingale. Similarly,

$$(x_t^i - \overline{x}_t^i)^+ \leq (x_0^i - \overline{x}_0^i)^+ + \int_0^t [-(1+\eta_s^i)(x_s^i - \overline{x}_s^i)^+$$
$$+ (x_s^{i-1} - \overline{x}_s^{i-1})^+]ds + M_t^i,\ i = 2, \cdots, d,$$

where M_t^i, $i = 2, \cdots, d$ are martingales. Furthermore, now assume that

$x_0^i \leq \overline{x}_0^i, \forall i = 1, \cdots, d;$

$\beta_t \leq \overline{\beta}_t, \forall t \geq 0.$

Then

$$0 \leq y_t = E \sum_{i=1}^d (x_t^i - \overline{x}_t^i)^+ \leq \int_0^t [k_0^2 y_s + y_s] ds.$$

Hence $y_t = 0, \forall t \geq 0$. This implies that $P - a.s.$

$x_t^i \leq \overline{x}_t^i,\ \forall t \geq 0,\ \forall i = 1, \cdots, d.$

Thus we arrive at the following theorem.

Theorem 346 *Assume that*

1° $0 \leq \eta_t^i, b_t^j, \beta_t \leq k_0,\ \forall i = 1, 2, \cdots, d; j = r_1, \cdots, r_2;$

$|\sigma(t,x)|^2 + \int_Z |c(t,x,z)|^2 \pi(dz) \leq k_0(1 + |x|^2);$

2° the same as 2° in Theorem 344;

3° the same as 3° in Theorem 344.

If (x_t, ϕ_t) and $(\overline{x}_t, \overline{\phi}_t)$ are solutions of the stochastic population dynamics (11.25) with the initial value x_0, the fertility rate of females β_t and $\overline{x}_0, \overline{\beta}_t$, respectively; then

$x_0^i \geq \overline{x}_0^i, \forall i = 1, \cdots, d;$ and

$\beta_t \geq \overline{\beta}_t, \forall t \geq 0$

implies that $P - a.s.$

$x_t^i \geq \overline{x}_t^i,\ \forall t \geq 0,\ \forall i = 1, \cdots, d.$

The comparison theorem (Theorem 346) actually tells us the following facts:

1) In a stochastic population dynamic if the initial size of the population takes a larger value, then the size of the population will also take larger values forever, as time evolves; that is, with the same η_t^i, b_t^j, β_t

$$x_0^i \geq \overline{x}_0^i, \forall i = 1, \cdots, d \Longrightarrow x_t^i \geq \overline{x}_t^i,\ \forall t \geq 0,\ \forall i = 1, \cdots, d.$$

2) In a stochastic population dynamic if the fertility rate of females always takes larger values, then the size of the population will also take larger values forever, as time evolves; that is, with the same $\eta_t^i, b_t^j, x_0^i = \overline{x}_0^i$, $\beta_t \geq \overline{\beta}_t, \forall t \geq 0 \implies x_t^i \geq \overline{x}_t^i, \forall t \geq 0, \forall i = 1, \cdots, d$.

Furthermore, the proof of Theorem 346 also motivates the following more general theorem.

Consider two more general RSDEs as (11.24). More precisely, consider the following two RSDEs: $i = 1, 2$,

$$\begin{cases} dx_t^{(i)} = b^{(i)}(t, x_t^{(i)})dt + \sigma(t, x_t^{(i)})dw_t + \int_Z c(t, x_{t-}^{(i)}, z)\widetilde{N}_k(dt, dz) + d\phi_t^{(i)}, \\ x_0^{(i)} = x^{(i)} \in \overline{R}_+^d, \ t \geq 0, \\ x_t^{(i)} \in \overline{R}_+^d, \ t \geq 0, \\ \text{and the other stantements in (11.4) also holds for } (x_t^{(i)}, \phi_t^{(i)}). \end{cases} \tag{11.29}$$

We have the following comparison theorem.

Theorem 347 *Assume that all conditions 1° − 3° in Theorem 344 hold, and assume that one of the $b^{(i)}(t, x), i = 1, 2$; say, $b^{(1)}(t, x)$ satisfies the following condition:*

4° $I_{x^i > y^i}(b^{(1)i}(t, x) - b^{(1)i}(t, y)) \leq \sum_{k=1}^d c_{N,T}^{i,k}(t)\rho_{N,T}^{i,k}((x^k - y^k)^+)$,
as $t \in [0, T]$; $|x|, |y| \leq N$, *for each* $T < \infty$, *and* $N = 1, 2, \cdots$,
$\forall i = 1, 2, \cdots, d$; $\forall x = (x^1, \cdots, x^d), y = (y^1, \cdots, y^d) \in \overline{R}_+^d$;
where
$b^{(1)}(t, x) = (b^{(1)1}(t, x), \cdots, b^{(1)d}(t, x))$,
and $c_{N,T}^{i,k}(t) \geq 0, \forall i, k$ *are such that* $\int_0^T c_{N,T}^{i,k}(t)dt < \infty$; *and* $\rho_{N,T}^{i,k}(u) \geq 0, \forall i, k, \forall u \geq 0$ *are strictly increasing, continuous and concave such that* $\rho_{N,T}^{i,k}(0) = 0$ *and* $\int_{0+} \frac{du}{\rho_{N,T}^{i,k}(u)} = \infty$.

If $(x_t^{(1)}, \phi_t^{(1)})$ *and* $(x_t^{(2)}, \phi_t^{(2)})$ *are solutions of (11.29) with the initial values* $x_0^{(1)}$ *and* $x_0^{(2)}$, *respectively; then*
$x_0^{(1)i} \leq x_0^{(2)i}, \forall i = 1, \cdots, d$; *and*
$b^{(1)i}(t, x) \leq b^{(2)i}(t, x), \forall i = 1, \cdots, d; \forall t \geq 0, \forall x \in \overline{R}_+^d$
implies that $P - a.s.$
$x_t^{(1)i} \leq x_t^{(2)i}, \ \forall t \geq 0, \ \forall i = 1, \cdots, d.$

Proof. For each $T < \infty$ and $N = 1, 2, \cdots$ Write
$\widetilde{c}_{N,T}(t) = \sum_{i,k=1}^d c_{N,T}^{i,k}(t)(t), \ \forall t \geq 0,$
$\widetilde{\rho}_{N,T}(u) = \sum_{i,k=1}^d \rho_{N,T}^{i,k}(u), \ \forall u \geq 0.$
Then $\int_0^T \widetilde{c}_{N,T}(t)dt < \infty$ and $\widetilde{\rho}_{N,T}(u) \geq 0$ is still strictly increasing, continuous and concave such that $\widetilde{\rho}_{N,T}(0) = 0$ and $\int_{0+} \frac{du}{\widetilde{\rho}_{N,T}(u)} = \infty$. (See Lemma 144). By Theorem 344 one sees that $\forall i = 1, 2, \cdots, d$,
$(x_t^{(1)i} - x_t^{(2)i})^+ = (x_0^{(1)i} - x_0^{(2)i})^+$

$+\int_0^t I_{(x_s^{(1)i}>x_s^{(2)i})}(b_i^{(1)}(s,x_s^{(1)})-b_i^{(2)}(s,x_s^{(2)}))ds$
$+\int_0^t I_{(x_s^{(1)i}>x_s^{(2)i})}\sum_{k=1}^d(\sigma_{ik}(s,x_s^{(1)})-\sigma_{ik}(s,x_s^{(2)}))dw_k(s)$
$+\int_0^t\int_Z I_{(x_{s-}^{(1)i}>x_{s-}^{(2)i})}(c_i(s,x_{s-}^{(1)},z)-c_i(s,x_{s-}^{(2)},z))\widetilde{N}_k(ds,dz)$
$-\int_0^t I_{(x_s^{(1)i}>x_s^{(2)i})}d\phi_s^{2i}=\sum_{k=1}^5 I_t^k\le\sum_{k=1}^4 I_t^k.$

However,

$I_t^2\le\int_0^t I_{(x_s^{(1)i}>x_s^{(2)i})}(b_i^{(1)}(s,x_s^{(1)})-b_i^{(1)}(s,x_s^{(2)}))ds$
$\le\int_0^t\sum_{k=1}^d c_{N,T}^{i,k}(s)\rho_{N,T}^{i,k}((x_s^{(1)k}-x_s^{(2)k})^+)ds.$

Now let

$\tau_N=\inf\{t\ge 0:\left|x_t^{(1)}\right|+\left|x_t^{(2)}\right|>N\}.$

Then one easily finds that $\forall t\ge 0$,

$y_t=E\sum_{i=1}^d(x_{t\wedge\tau_N}^{(1)i}-x_{t\wedge\tau_N}^{(2)i})^+\le\int_0^t\widetilde{c}(s)\widetilde{\rho}(y_s)ds.$

Hence $y_t=0,\forall t\ge 0$. Since $\lim_{N\to\infty}\tau_N=\infty$. These imply that $P-a.s.$

$x_t^{(2)i}\ge x_t^{(1)i},\ \forall t\ge 0,\ \forall i=1,\cdots,d.$ ■

Analyzing the component equations of the stochastic population dynamics (11.28) one finds that the fertility rate of females β_t only influnces the popuation aged in $[r_1,r_2]$. (See the first term in (11.28)). This is reasonable, because only females with age in some interval can have babies. However, one also may think that for a better control of the population, β_t itself may also depend on the population size x_t. Is it still true that in such a case the comparison result still holds? The following theorem gives a partial answer to this question.

Theorem 348 *Assume that all η_t^i, b_t^j, σ and c satisfy all conditions $1°-3°$ in Theorem 346, moreover, assume that*

4° $\beta_t=\beta_t(x)$ satisfies the following condition: $\forall x=(x^1,\cdots,x^d)\in\overline{R}_+^d$,

$\beta_t(x)=\beta_t^0+\sum_{i=1}^d c_t^i(x^i I_{0\le x^i\le k_0^i}+k_0^i I_{x^i>k_0^i}),$

where $0\le\beta_t^0,c_t^1,\cdots,c_t^d\le k_0$, they do not depend on x, and k_0 is a constnat; besides, $k_0^1,\cdots,k_0^d\ge 0$ are also constants.

Now let $\forall x=(x^1,\cdots,x^d)\in\overline{R}_+^d$,

$\overline{\beta}_t(x)=\overline{\beta}_t^0+\sum_{i=1}^d c_t^i(x^i I_{0\le x^i\le k_0^i}+k_0^i I_{x^i>k_0^i}).$

Then the conclusion of Theorem 346 still holds, that is, if (x_t,ϕ_t) and $(\overline{x}_t,\overline{\phi}_t)$ are solutions of (11.25) with x_0, β_t and $\overline{x}_0,\overline{\beta}_t$, respectively; then

$x_0^i\le\overline{x}_0^i,\forall i=1,\cdots,d$; *and*

$\beta_t(x)\le\overline{\beta}_t(x),\forall t\ge 0,\forall x\in\overline{R}_+^d$

implies that $P-a.s.$

$x_t^i\le\overline{x}_t^i,\ \forall t\ge 0,\ \forall i=1,\cdots,d.$

Remark 349 *Theorem 348 implies Theorem 346. In fact, let $c_t^i=0,i=1,2,\cdots,d$. Then from Theorem 348 we obtain Theorem 346.*

Now let us establish the above thoerem.

Proof. Let

$b^{(1)}(t,x) = A(t)x + B(t)x\beta_t(x),$
$b^{(2)}(t,x) = A(t)x + B(t)x\overline{\beta}_t(x),$
where $A(t)$ and $B(t)$ are defined by (11.26) and (11.27), respectively. Then by assumption

$b^{(1)i}(t,x) \leq b^{(2)i}(t,x), \forall i = 1, \cdots, d; \forall t \geq 0, \forall x \in \overline{R}^d_+.$

So if we want to apply Theorem 347, we only need to check that condition 4° in Theorem 347 holds. In fact,

$$I_{x^1>y^1}(b^{(1)1}(t,x) - b^{(1)1}(t,y)) = I_{x^1>y^1}[-(1+\eta^1_t)(x^1 - y^1)$$
$$+ \sum_{k=r_1}^{r_2} b^k_t(x^k\beta_t(x) - y^k\beta_t(y))] \leq I_{x^1>y^1}\sum_{k=r_1}^{r_2} b^k_t(x^k\beta_t(x) - y^k\beta_t(y))]$$
$$= I_{x^1>y^1}\sum_{k=r_1}^{r_2}[b^k_t(x^k - y^k)\beta_t(x) + b^k_t y^k(\beta_t(x) - \beta_t(y))]$$
$$\leq k_0^2\sum_{k=r_1}^{r_2}(x^k - y^k)^+ + I_{x^1>y^1}\sum_{k=r_1}^{r_2} b^k_t y^k \cdot$$
$$\cdot \sum_{i=1}^d c^i_t[x^i I_{0\leq x^i\leq k^i_0} + k^i_0 I_{x^i>k^i_0} - y^i I_{0\leq x^i\leq k^i_0} - k^i_0 I_{y^i>k^i_0}]$$
$$\leq k_0^2\sum_{k=r_1}^{r_2}(x^k - y^k)^+ + I_{x^1>y^1}\sum_{k=r_1}^{r_2} b^k_t y^k \sum_{i=1}^d c^i_t[(x^i - y^i)\vee 0]$$
$$\leq k_0^2\sum_{k=r_1}^{r_2}(x^k - y^k)^+ + (r_2 - r_1)k_0^2 N\sum_{i=1}^d (x^i - y^i)^+$$
$$\leq k'_0 N\sum_{i=1}^d (x^i - y^i)^+,$$
as $|x|, |y| \leq N$; $x = (x^1, \cdots, x^d), y = (y^1, \cdots, y^d) \in \overline{R}^d_+$.

More easily, $\forall i = 2, \cdots, d,$

$$I_{x^i>y^i}(b^{(1)i}(t,x) - b^{(1)i}(t,y)) = I_{x^i>y^i}[-(1+\eta^i_t)(x^i - y^i)$$
$$+(x^{i-1} - y^{i-1})] \leq I_{x^i>y^i}(x^{i-1} - y^{i-1}) \leq (x^{i-1} - y^{i-1})^+.$$

Therefore, condition 4° in Theorem 347 holds, and Theorem 347 applies.

■

2. <u>The Optimal Stochastic Population Control</u>

Now let us discuss an optimal stochastic population control problem for RSDE (11.25). Denote the admissible control set by

$$\widetilde{U} = \left\{ \begin{array}{l} \beta : \beta = \beta_t(x) \text{ is jointly continuous, } \overline{\beta}_0 \leq \beta \leq \beta_0, \\ \text{and } \beta \text{ satisfies the condition 4° in Theorem 348} \end{array} \right\},$$

where $0 < \overline{\beta}_0$ and β_0 are constants. Suppose we want to minimize the following functional

$$J(\beta) = E[\int_0^T F(t, x^\beta_t)dt + G(x^\beta_T)],$$

among all $\beta \in \widetilde{U}$, where $0 \leq T < \infty$ is an arbitrarily given constant, and $(x^\beta_t, \phi^\beta_t)$ is the pathwise unique strong solution of (11.25) corresponding to the given $\beta \in \widetilde{U}$; $F(t,x)$ and $G(x)$ are jointly Borel measurable and Borel measurable, repectively. Then we have the following theorem.

Theorem 350 *Assume that conditions 1° − 4° in Theorem 348 hold, and assume that*

5° $\int_Z |c(t,x,z)|^2 \pi(dz) \leq k_0$, *and as* $|t-s| \to 0, |x-y| \to 0,$

$\int_Z |c(t,x,z) - c(s,y,z)|^2 \pi(dz) \to 0;$

6° $\sigma(t,x)$ *is jointly continuous.*

Furthermore, assume that $F(t,x)$ *and* $G(x)$ *are jointly Borel measurable functions defined on* $(t,x) \in [0,T] \times \overline{R}^d_+$ *and* $x \in \overline{R}^d_+$, *respectively, such that as* $x_i \leq y_i$, $i = 1,2,..,d$; $x,y \in \overline{R}^d_+, \forall t \in [0,T],$

$G(x) \leq G(y)$,
$F(t,x) \leq F(t,y)$.
Then the smallest constant $\overline{\beta}_0$ is an admissible optimal control for the minimization of the functional $J(\beta)$; that is,
1) $\overline{\beta}_0 \in \widetilde{U}$,
2) $J(\overline{\beta}_0) = \inf_{\beta \in \widetilde{U}} J(\beta)$.
Furthermore, Assume that

$$\begin{cases} \sigma(t,x) = \left(c_1^{(1)}x, \cdots, c_1^{(d)}x\right), \\ c(t,x,z) = \overline{c}_1(xI_{|x| \leq k_0} + k_0 \frac{x}{|x|} I_{|x|>k_0}) I_U(z)), \end{cases}$$

where $\pi(U) = 1, k_0 \geq 0$ is a constant, $c_1^{(i)}, i = 1, \cdots, d$, and $\overline{c}_1$ are constant $d \times d$ matrices, moreover, $\overline{c}_{1,ij} \geq 0, \forall 1 \leq i,j \leq d$, where $\overline{c}_1 = (\overline{c}_{1,ij})_{i,j=1}^d$, and let

$$\delta_0 = \min_{t \geq 0}(\tfrac{1}{2} + \eta_1(t), \eta_2(t), \cdots, \eta_{d-1}(t), \tfrac{1}{2} + \eta_d(t)),$$
$$b_M = \max_{t \geq 0}(b_{r_1}(t), \cdots, b_{r_2}(t)),$$

where $d = r_m$—the largest age of the population.
Then as

$$\delta_1 = 2\delta_0 - \sum_{i=1}^d \left|c_1^{(i)}\right|^2 - |\overline{c}_1|^2 - \overline{\beta}_0 b_M (r_2 - r_1) > 0,$$
$$E\left|x_t^{\overline{\beta}_0}\right|^2 \leq E|x_0|^2 e^{-\delta_1 t}, \text{ for all } t \geq 0;$$

where $x_t^{\overline{\beta}_0}$ is the so-called optimal trajectory corresponding to the optimal control $\overline{\beta}_0$.

This theorem actually tells us the following facts:

1) If a target functional monotonically depends on the population size, for example, the energy, the consumption, and so on, spent by the population, then to minimize this target functional we should control the fertility rate of females to take a value as small as possible.

2) It is also possible to make the optimal population size to be exponentially stable in mean square, if we can take the fertility rate of females small enough and when the stochastic perturbation is not too large.

To show the truth of this theorem one only needs to apply the existence theorem, the comparison theorem, and the stability theorem of solutions to the stochatic population dynamics. (See Theorem 337, Theorem 348 and Corollary 343). We leave this to the reader.

3. Some Explanations and Conclusions on Population RSDE

Finally, by using the results obtained here, the following conclusions can be drawn: The RSDE (11.22) is a suitable model for the stochastic population control system. In fact, by this model one sees that

$$x_{t+1}^1 - x_t^1 \approx -(1+\eta_t^1)x_t^1 + \sum_{i=r_1}^{r_2}(1-\mu_{00}(t))k_i(t)h_i(t)x_t^i\beta_t$$
$$+(\text{a stochastic perturbation between } [t,t+1]) + \phi_{t+1}^1 - \phi_t^1,$$
$$x_{t+1}^i - x_t^i \approx -(1+\eta_t^i)x_t^i + x_t^{i-1}$$
$$+(\text{a stochastic perturbation between } [t,t+1]) + \phi_{t+1}^1 - \phi_t^1,$$
$$\forall i = 2, \cdots, r_m.$$

The second expression tells us intuitively that when $i \geq 2$, x_t^i–the size of the population with ages between $[i, i+1)$, will have an increment $x_{t+1}^i - x_t^i$ when the time t increases from t to $t+1$. However, this increment is contributed by two terms when we do not consider the stochastic perturbation. (In the deterministic case $\phi_t = 0$). A positive term due to people with ages in $[i-1, i)$ at time t, who will be in ages $[i, i+1)$, when time goes to $t+1$. That is, the term "$+x_t^{i-1}$". A negative term is due to the fact that people may die with a death rate η_t^i caused by a disease or some other cause when the time evolves from t to $t+1$; moreover, the people with ages in $[i, i+1)$ at time t, will also arrive in ages $[i+1, i+2)$, when time goes to $t+1$. So from time t to $t+1$ the size of population with ages in $[i, i+1)$ will lose in all the amount $(1+\eta_t^i)x_t^i$. That is the term "$-(1+\eta_t^i)x_t^i$".

However, for the first expression above, the only difference is that x_t^1 is the size of the population with ages between $[1, 2)$. So when time goes to $t+1$ the positive contribution term can only be the number of babies born during this time interval $[t, t+1]$, since women, who can have babies, can only be in some age interval $[r_1, r_2]$, and a baby if born, may die with a death rate $\mu_{00}(t)$. Furthermore, the number of babies born and living, also depends on the size of the population x_t^i, the fertility rate of females β_t, the sex model $k_i(t)$, and the fertility model $h_i(t)$ at time t and aged $i \in [r_1, r_2]$. That is, the positive contribution term from time t to time $t+1$ is

"$\sum_{i=r_1}^{r_2}(1-\mu_{00}(t))k_i(t)h_i(t)x_t^i\beta_t$".

Now if the system is disturbed by a continuous and a jump type stochastic perturbation, then as we said in the "Introduction" section, we need a reflecting SDE to discribe the population dynamics to keep all population sizes non-negative, that is, $x_t^i \geq 0, \forall i = 1, 2, \cdots, r_m$. So we need the stochastic population control system (11.22).

In this model we have shown the following facts:

1) The size of the population depends continuously in some sense (Theorem 338 and 340) on the initial size of the population, the fertility rate of females, and the coefficients in the stochastic population dynamics system. Moreover, the resultant error can be calculated in some cases (Corollary 339).

2) If the initial size of the population or the fertility rate of females takes larger value, then so does the size of the population forever as time evolves (Theorem 348).

3) If the stochastic perturbation is not large, and the forward death rate is greater than zero, then it is possible to take a fertility rate of females that makes the system exponentially stable in some sense, and the time when the population size can in some sense be less than a given level, can also be calculated. (Corollary 343).

4) If a payoff value functional (or say, a target functional) depends monotonically on the size of the population, then the payoff value will take the smallest value, as does the fertility rate of females (Theorem 350).

11.9 Caculation of Solutions to Population RSDE

Now let us discuss how to calculate, or say, to construct a practical solution of the stochastic population dynamics. For simplicity we will discuss the population RSDE without jumps. Suppose we have already got stochastic population dynamics as in (11.25) with $c(t,x,z)=0$. To calculate its solution practically, first let us see what will happen if (x_t,ϕ_t) is a solution of (11.25) with $c(t,x,z)=0$. For this purpose we now give a lemma for a more general RSDE.

Lemma 351 *If (x_t,ϕ_t) satisfies*

$$\begin{cases} dx_t = b(t,x_t)dt + \sigma(t,x_t)dw_t + d\phi_t, \\ x_0 = x \in \overline{R}_+^d, \ t \geq 0, \\ x_t \in \overline{R}_+^d, \ t \geq 0, \\ \textit{and the other stantements in (11.4) also hold for } (x_t,\phi_t), \end{cases} \tag{11.30}$$

and the pathwise uniqueness of solutions holds for this RSDE, then (x_t,ϕ_t) is such that

$\phi_i(t) = \sup_{0\leq s\leq t}((-y_i(s))\vee 0),$
$x_i(t) = y_i(t) + \phi_i(t),\ y_i(0)\geq 0,$

for all $t\geq 0$, where

$y_i(t) = x_i(0) + \int_0^t b_i(s,x(s))ds + \sum_{k=1}^d \int_0^t \sigma_{ik}(s,x(s))dw_k(s),$
$i = 1,2,...,d.$

Proof. If (x_t,ϕ_t) solves (11.30), set

$\overline{\phi}_i(t) = \sup_{0\leq s\leq t}((-y_i(s))\vee 0),$
$\overline{x}_i(t) = y_i(t) + \overline{\phi}_i(t),$

then it is obvious that $0\leq \overline{\phi}_i(t)$ is increasing, continuous, $\overline{\phi}_i(0)=0$, and $\overline{x}_i(t)\geq 0$, for all $i=1,2,...,d$. Moreover,

$\overline{\phi}_i(t) = \int_0^t I_{(\overline{x}_i(s)=0)}d\overline{\phi}_i(s),\ \overline{\phi}_i(t) = \int_0^t n_i(s)d\left|\overline{\phi}\right|(s),$

where $n(t) = (n_1(t),\cdots,n_d(t)) \in \mathcal{N}_{\overline{x}(t)}$, as $\overline{x}(t)\in\partial\overline{R}_+^d$, and $\overline{\phi}(t) = (\overline{\phi}_1(t),\cdots,\overline{\phi}_d(t)), \overline{x}(t) = (\overline{x}_1(t),\cdots,\overline{x}_d(t))$. So $(\overline{x}(t),\overline{\phi}(t))$ solves

$$\begin{cases} d\overline{x}(t) = dy(t) + d\overline{\phi}(t), \\ x_0 = x \in \overline{R}_+^d, \ t\geq 0, \\ x_t \in \overline{R}_+^d, \ t\geq 0, \\ \text{and the other stantements in (11.4) hold for } (\overline{x}(t),\overline{\phi}(t)). \end{cases}$$

In particular, $(\overline{x}_i(t),\overline{\phi}_i(t))$ satisfies

$$\begin{cases} \overline{x}_i(t) = y_i(t) + \overline{\phi}_i(t), \\ \overline{x}_i(t)\geq 0, \ \text{for all } t\geq 0, \\ 0\leq \overline{\phi}_i(t) \text{ is increasing, continuous}, \overline{\phi}_i(0)=0, \text{and} \\ \overline{\phi}_i(t) = \int_0^t I_{(\overline{x}_i(s)=0)}d\overline{\phi}_i(s), \end{cases} \tag{11.31}$$

i.e. it is a solution of the Skorohod's problem (11.31) in 1-dimensional space. Since $(x_i(t), \phi_i(t))$ also satisfies (11.31); by the uniqueness of solution to the Skorohod problem we have that $\forall i = 1, 2, \cdots, d$,

$(x_i(t), \phi_i(t)) = (\overline{x}_i(t), \overline{\phi}_i(t))$. ■

This lemma motivates us to use the Picard's iteration technique to calculate the unknown solution $(x_i(t), \phi_i(t)), i = 1, 2, \cdots, d$ of the RSDE (11.30). In fact, we may let

$y_i^1(t) = x_i(0) + \int_0^t b_i(s, x(0))ds + \sum_{k=1}^d \int_0^t \sigma_{ik}(s, x(0))dw_k(s),$

$i = 1, 2, ...d.$

By this $y_i^1(t)$ we can get

$\phi_i^1(t) = \sup_{0 \leq s \leq t}((-y_i^1(s)) \vee 0),$

$x_i^1(t) = y_i^1(t) + \phi_i^1(t).$

By the proof of the previous lemma $(x_i^1(t), \phi_i^1(t))$ satisfies a Skorohod problem similar to (11.31) but with the given $y_i^1(t)$, and the same initial value $x_i(0)$. By using the induction we may construct

$$\begin{cases} y_i^n(t) = x_i(0) + \int_0^t b_i(s, x^{n-1}(s))ds + \sum_{k=1}^d \int_0^t \sigma_{ik}(s, x^{n-1}(s))dw_k(s), \\ \phi_i^n(t) = \sup_{0 \leq s \leq t}((-y_i^n(s)) \vee 0), \\ x_i^n(t) = y_i^n(t) + \phi_i^n(t), \end{cases} \tag{11.32}$$

and we see that $(x_i^n(t), \phi_i^n(t))$ still satisfies a Skorohod problem similar to (11.31) but with the given $y_i^n(t)$, and the same initial value $x_i(0)$. Thus, after some calculation, we have a sequence of $(x^n(t), \phi^n(t)), n = 1, 2, \cdots$, which we may call the approximated solutions of (11.30). Can we show that they actually "converge" to the solution of (11.30)? The following theorem answers this question.

Theorem 352 *Assume that for all* $t \geq 0$ $\omega \in \Omega, z \in Z, x \in \overline{\Theta}$

1° $|b(t,x)|^2 + \|\sigma(t,x)\|^2 \leq k_0(1 + |x|^2)$,

where $k_0 \geq 0$ *is a constant,*

2° $|b(t,x) - b(t,y)|^2 + \|\sigma(t,x) - \sigma(t,y)\|^2 \leq k_0 |x-y|^2$.

Then we have the following conclusions:

1) (11.30) has a pathwise unique strong solution (x_t, ϕ_t).

2) For $i = 1, 2, \cdots, d$ *the sequence* $(x_i^n(t), \phi_i^n(t)), n = 1, 2, \cdots$ *constructed by (11.32) satisfies that*

$\lim_{n\to\infty}[E \sup_{t \leq T} |x^n(t) - x(t)|^2 + E \sup_{t \leq T} |\phi^n(t) - \phi(t)|^2] = 0,$

where $x^n(t) = (x_1^n(t), \cdots, x_d^n(t))$, *etc.*

3) $\forall \eta > 0$, *let* N_η *be such that*

$c_T \sum_{n=N_\eta}^\infty n^4 (122k_0T)^{n-1}/(n-1)! < \eta,$

where $c_T = 8E|x_0|^2 + (30k_0 + 30k_0 E|x_0|^2)T$,

$\forall \varepsilon > 0$, *let* N_ε *be such that*

$\sum_{n=N_\varepsilon}^\infty 1/n^2 < \varepsilon,$

then as $n \geq \max(N_\eta, N_\varepsilon)$,

$P(\sup_{s \leq T} |x_s^{n-1} - x_s| < \varepsilon) \geq 1 - \eta.$

In this theorem the conclusion 1) and 2) mean that a uniform limit x_t exists, which is just the unique solution of the RSDE (11.30) and conclusion 3) tells us that with a greater than $(1-\eta)\%$ possibility, the error of the uniform approximation is less than a given number $\varepsilon > 0$, if n is large enough. So the approximae solutions and the approximate error can all be calculated.

Proof. 1) is true by Theorem 335. Now let us show 2).

By assumption $n = 1, 2, ...$

$x_t^n = x_0 + \int_0^t b(s, x_s^{n-1}, \omega)ds + \int_0^t \sigma(s, x_s^{n-1}, \omega)dw_s + \phi_t^n = y_t^n + \phi_t^n,$

$x_0^n = x_0,$

Applying Ito's formula, one finds that

$|x_t^n - x_t|^2 \leq 2\int_0^t (x_s^n - x_s) \cdot (b(s, x_s^{n-1}) - b(s, x_s))ds$

$+2\int_0^t (x_s^n - x_s^{n-1}) \cdot (\sigma(s, x_s^{n-1}) - \sigma(s, x_s^{n-2}))dw_s$

$+\int_0^t \left\|\sigma(s, x_s^{n-1}) - \sigma(s, x_s^{n-2})\right\| ds.$

It follows by the martingale inequality that as $0 \leq t \leq T$

$I_t^n = E\sup_{s\leq t}\left|x_t^n - x_t^{n-1}\right|^2 \leq \widetilde{k}_0 \int_0^t I_s^{n-1}ds \leq c_t(\widetilde{k}_0 t)^{n-1}/(n-1)!,$

$I_t'^n = E\sup_{s\leq t}\left|\phi_t^n - \phi_t^{n-1}\right|^2 \leq c_t'(\widetilde{k}_0' t)^{n-1}/(n-1)!,$

where c_t and c_t' are constants depending on t only.

Therefore for any $T < \infty$

$$P(\sup_{s\leq T}\left|x_s^n - x_s^{n-1}\right| > 1/n^2) \leq n^4 c_t(\widetilde{k}_0 t)^{n-1}/(n-1)!, \quad (11.33)$$

$$P(\sup_{s\leq T}\left|\phi_s^n - \phi_s^{n-1}\right| > 1/n^2) \leq n^4 c_t'(\widetilde{k}_0' t)^{n-1}/(n-1)!. \quad (11.34)$$

Now, by the Borel-Cantelli type lemma below, there exist a $(\overline{x}_t, \overline{\phi}_t)$ such that $P-a.s.$

$\overline{x}_t = \lim_{n\to\infty} x_t^n,\ \overline{\phi}_t = \lim_{n\to\infty}\phi_t^n$, uniformly in $t \in [0, T]$,

for any given $T < \infty$.

Since by Theorem 330

$E\sup_{s\leq t}|x_t^n|^2 + E\sup_{s\leq t}|\phi_t^n|^2 \leq k_T, \forall n.$

Hence by Fatou's lemma

$E\sup_{s\leq t}|\overline{x}_t|^2 + E\sup_{s\leq t}\left|\overline{\phi}_t\right|^2 \leq k_T.$

On the other hand, by Theorem 328 there exists a unique solution $(\widetilde{x}_t, \widetilde{\phi}_t)$ solving

$\widetilde{x}_t = x_0 + \int_0^t b(s, \overline{x}_s, \omega)ds + \int_0^t \sigma(s, \overline{x}_s, \omega)dw_s + \widetilde{\phi}_t = y_t + \widetilde{\phi}_t,$

and the other statements in (11.30) for $(\widetilde{x}_t, \widetilde{\phi}_t)$ hold. Now by the Lipschitz condition one easily shows that

$E\sup_{t\leq T}|y_t - y_t^n|^2 = 0.$

Hence one can take a subsequence of $\{n\}$, denoted again by $\{n\}$, such that $P-a.s.$, as $n \to \infty$,

$\sup_{t\in[0,T]}|y_t - y_t^n|^2 \to 0.$

Thus, applying Lemma 327 $P-a.s.$, as $n \to \infty$,

$\sup_{t\in[0,T]} |x_t^n - \widetilde{x}_t|^2 \to 0,$

$\sup_{t\in[0,T]} \left|\phi_t^n - \widetilde{\phi}_t\right|^2 \to 0,$

By the uniqueness of a limit $P-a.s.\forall t \in [0,T]$

$\widetilde{x}_t = \overline{x}_t,\ \widetilde{\phi}_t = \overline{\phi}_t.$

So $(\overline{x}_t, \overline{\phi}_t)$ is a solution of (11.30). Since we already know that (x_t, ϕ_t) is the pathwise unique solution of (11.30), we have $P-a.s.\forall t \in [0,T]$

$x_t = \overline{x}_t,\ \phi_t = \overline{\phi}_t.$

Now let us show 3). For this we only need to calculate the $\widetilde{k}_0$ and c_T in (11.33). Observe that by Ito's formula

$I_t^n = E\sup_{s\le t} \left|x_s^n - x_s^{n-1}\right|^2$

$\le E[2\int_0^t \left|x_s^n - x_s^{n-1}\right| \left|b(s,x_s^{n-1}) - b(s,x_s^{n-2})\right| ds$

$+2\sup_{r\le t} \left|\int_0^r (x_s^n - x_s^{n-1})(\sigma(s,x_s^{n-1}) - \sigma(s,x_s^{n-2}))dw_s\right|$

$+\int_0^r \left\|\sigma(s,x_s^{n-1}) - \sigma(s,x_s^{n-2})\right\|^2 ds].$

Notice that by the martingale inequality

$E\sup_{r\le t} \left|\int_0^r (x_s^n - x_s^{n-1})(\sigma(s,x_s^{n-1}) - \sigma(s,x_s^{n-2}))dw_s\right|$

$\le 3E[\int_0^t \left|x_s^n - x_s^{n-1}\right|^2 \left|\sigma(s,x_s^{n-1}) - \sigma(s,x_s^{n-2})\right|^2 ds]^{1/2}.$

So applying the elementary inequality

$ab \le \frac{a^2}{2} + \frac{b^2}{2}, \forall a,b \ge 0,$

and by virtue of the Lipschitz assumption one finds that

$I_t^n \le 2k_0[6+54+1]\int_0^t I_s^{n-1}ds = 122k_0\int_0^t I_s^{n-1}ds$

$\le (122k_0)^2 \int_0^t \int_0^s I_r^{n-2}drds \le (122k_0)^2 \int_0^t (t-r)I_r^{n-2}dr$

$\le \cdots \le (122k_0)^{n-1}\int_0^t \frac{(t-r)^{n-2}}{(n-2)!} I_r^1 dr \le (122k_0)^{n-1}\frac{t^{n-1}}{(n-1)!} I_t^1$

$\le c_t(122k_0)^{n-1}\frac{t^{n-1}}{(n-1)!}.$

However,

$I_t^1 = E\sup_{s\le t} \left|x_s^1 - x_s^0\right|^2 \le 2E\left|x_0\right|^2 + 2E\sup_{s\le t}\left|x_s^1\right|^2$

$\le 2E\left|x_0\right|^2 + 6E[\left|x_0\right|^2 + k_0\int_0^t (1+\left|x_0\right|^2)ds + 4k_0\int_0^t (1+\left|x_0\right|^2)ds]$

$\le 8E\left|x_0\right|^2 + (30k_0 + 30k_0E\left|x_0\right|^2)t = c_t.$

Now the conclusion 3) follows from the lemma below. ∎

Finally, let us give the following Borel-Cantelli type lemma.

Lemma 353 *(Borel-Cantelli type lemma). Assume that* $\{x_t^n\}_{n=1}^\infty$ *is a sequence of random processes such that*

$P(\sup_{s\le T}\left|x_s^n - x_s^{n-1}\right| > 1/n^2) \le n^4 c_T(\widetilde{k}_0 T)^{n-1}/(n-1)!, \forall n = 1,2,\cdots.$

where $k_T \ge 0$ *is a constant.*

Then 1) there exists a random process x_t *such that* $P-a.s.$

$\lim_{n\to\infty}\sup_{s\le T}\left|x_s^n - x_s\right| = 0;$

2) $\forall \eta > 0$, *let* N_η *be such that*

$c_T\sum_{n=N_\eta}^\infty n^4(122k_0T)^{n-1}/(n-1)! < \eta,$

$\forall \varepsilon > 0$, *let* N_ε *be such that*

$\sum_{n=N_\varepsilon}^\infty 1/n^2 < \varepsilon,$

then as $n \ge \max(N_\eta, N_\varepsilon)$,

$P(\sup_{s \leq T} \left| x_s^{n-1} - x_s \right| < \varepsilon) \geq 1 - \eta.$

Proof. 1): If we can show $P - a.s.$ that $\forall \varepsilon > 0$ there exists a N such that as $n \geq N$

$\sup_{s \leq T} |x_s^{n+p} - x_s^n| < \varepsilon$, $\forall p = 1, 2, \cdots$;

then $P - a.s.$ $\{x_t^n\}_{n=1}^{\infty}$ is a uniform Cauchy sequence on $t \in [0, T]$, and 1) is proved. However, by assumption

$P(\cap_{N=1}^{\infty} \cup_{n=N}^{\infty} \{\sup_{s \leq T} \left| x_s^n - x_s^{n-1} \right| > 1/n^2\})$

$\leq \sum_{n=N}^{\infty} P(\sup_{s \leq T} \left| x_s^n - x_s^{n-1} \right| > 1/n^2)$

$\leq \sum_{n=N}^{\infty} n^4 c_T (\widetilde{k}_0 T)^{n-1}/(n-1)! \to 0$, as $N \to \infty$.

So $P(\cap_{N=1}^{\infty} \cup_{n=N}^{\infty} \{\sup_{s \leq T} \left| x_s^n - x_s^{n-1} \right| > 1/n^2\}) = 0$, or

$P(\cup_{N=1}^{\infty} \cap_{n=N}^{\infty} \{\sup_{s \leq T} \left| x_s^n - x_s^{n-1} \right| \leq 1/n^2\}) = 1.$

This means that there exists a $\Lambda \in \mathfrak{F}$ with $P(\Lambda) = 0$ such that

$\Omega = (\cup_{N=1}^{\infty} \cap_{n=N}^{\infty} \{\sup_{s \leq T} \left| x_s^n - x_s^{n-1} \right| \leq 1/n^2\}) \cup \Lambda.$

Now fix any $\omega \in \cup_{N=1}^{\infty} \cap_{n=N}^{\infty} \{\sup_{s \leq T} \left| x_s^n - x_s^{n-1} \right| \leq 1/n^2\}$. $\forall \varepsilon > 0$ take a N_ε such that $\sum_{n=N_\varepsilon}^{\infty} 1/n^2 < \varepsilon$. By assumption there also exists a N_2 such that $\omega \in \cap_{n=N_2}^{\infty} \{\sup_{s \leq T} \left| x_s^n - x_s^{n-1} \right| \leq 1/n^2\}$. Let $N = \max\{N_\varepsilon, N_2\}$. Then as $k \geq N$,

$\sup_{s \leq T} \left| x_s^{k+p} - x_s^k \right| \leq \sum_{n=k+1}^{k+p} \sup_{s \leq T} \left| x_s^n - x_s^{n-1} \right|$

$\leq \sum_{n=k+1}^{\infty} \sup_{s \leq T} \left| x_s^n - x_s^{n-1} \right| \leq \sum_{n=k+1}^{\infty} 1/n^2 < \varepsilon$, $\forall p = 1, 2, \cdots$.

So 1) follows.

Now let us show 2). Let N_η and N_ε be taken as in 2). Then by assumption

$P(\cup_{n=N_\eta}^{\infty} \{\sup_{s \leq T} \left| x_s^n - x_s^{n-1} \right| > 1/n^2\})$

$\leq c_T \sum_{n=N_\eta}^{\infty} n^4 (\widetilde{k}_0 T)^{n-1}/(n-1)! < \eta$, or

$P(\cap_{n=N_\eta}^{\infty} \{\sup_{s \leq T} \left| x_s^n - x_s^{n-1} \right| \leq 1/n^2\}) \geq 1 - \eta.$

Hence as $k \geq \max(N_\eta, N_\varepsilon)$, and

$\omega \in \cap_{n=k}^{\infty} \{\sup_{s \leq T} \left| x_s^n - x_s^{n-1} \right| \leq 1/n^2\}$,

$\sup_{s \leq T} \left| x_s(\omega) - x_s^{k-1}(\omega) \right| \leq \sum_{n=k}^{\infty} \sup_{s \leq T} \left| x_s^n(\omega) - x^{n-1}(\omega) \right|$

$\leq \sum_{n=k}^{\infty} 1/n^2 < \varepsilon.$

So as $k \geq \max(N_\eta, N_\varepsilon)$,

$P(\sup_{s \leq T} \left| x_s(\omega) - x_s^{k-1}(\omega) \right| < \varepsilon)$

$\geq P(\cap_{n=k}^{\infty} \{\sup_{s \leq T} \left| x_s^n - x_s^{n-1} \right| \leq 1/n^2\}) \geq 1 - \eta.$

Now 2) follows. ∎

12
Maximum Principle for Stochastic Systems with Jumps

In this chapter we will establish the maximum principle for stochastic systems with jumps (that is, the necessary conditions for a control to be optimal for controlling the system).

12.1 Introduction

In the optimal control problem for a dynamical system one important subject concerns finding an optimal control when we know one exists. It is well known in the Calculus that if a function $f(x)$ attains at its maximum at the point $x = x_0$, and $f(x)$ is differentiable, then $f'(x_0) = 0$. So one can find the maximum point x_0 by solving the equation $f'(x_0) = 0$ (the necessary condition for the maximum point). However, to find out a necessary condition for an optimal control to be satisfied is more complicated. This is because now the target we treat is a functional of the control (for example, $J(v) = E\int_0^T |x_t^v|^2 \, dt$), subject to a SDE ($x_t^v$ satisfies a SDE depending on the given control $v(\cdot)$). The statement of necessary conditions for an optimal control is usually called the maximum principle. To establish such maximum principle we need to work analysis.

Consider a $d-$dimensional stochastic systems with Poisson jumps as follows:

$$
\begin{aligned}
y_t = x_0 &+ \int_0^t b(s, y_s, v_s)ds + \int_0^t \sigma(s, y_s, v_s)dw_s \\
&+ \int_0^t \int_Z c(s, y_{s-}, z)\widetilde{N}_k(ds, dz), \ \forall t \in [0, T]
\end{aligned} \tag{12.1}
$$

where w_t is a $d-$dimensional BM, $\widetilde{N}_k(ds, dz)$ is a centralized Poisson martingale measure with compensator $\pi(dz)ds$ such that
$\widetilde{N}_k(ds, dz) = N_k(ds, dz) - \pi(dz)ds$,
and $\pi(\cdot)$ is a $\sigma-$finite measure on the measurable space $(Z, \mathfrak{B}(Z))$. Let $U \subset R^r$ be a non-empty subset. Set

$$
\mathfrak{U}_{ad} = \left\{ \begin{array}{c} v(.) : v(t) \text{ is } \mathfrak{F}_t - \text{adapted and } U \ - \text{valued such that} \\ E\left(\sup_{t\in[0,T]} |v(t)|^{2k}\right) < \infty, k = 1, 2, 3, 4 \end{array} \right\}. \tag{12.2}
$$

$\mathfrak{U}_{ad}$ is called an admissible control set. Under appropriate conditions (12.1) will have a unique solution y_t for each $v(.) \in \mathfrak{U}_{ad}$. Let
$J(v(\cdot)) = E[\int_0^T l(t, y_t, v_t)dt + h(y_T)]$.
If $u(\cdot) \in \mathfrak{U}_{ad}$ is such that

$$
J(u(\cdot)) = \inf_{v(\cdot)\in\mathfrak{U}_{ad}} J(v(\cdot)) \tag{12.3}
$$

then $u(\cdot)$ is called an optimal control, and $J(u(\cdot))$ an optimal value functional. Under much stronger conditions the optimal value functional for the optimal control problem (12.1) - (12.3) will be a Sobolev solution of a Hamilton-Jacobi-Bellman equation. (See [65]). This book will not discuss Sobolev solutions.

12.2 Basic Assumption and Notation

In this section we always make the following assumption
(A): b, σ, c, l, h are all continuously twice differentiable for x with all bounded derivatives such that
$|b(t, x, v)|^2 + |\sigma(t, x, v)|^2 + \int_Z |c(t, x, z)|^2 \pi(dz) + |l(t, x, v)|^2 + |h(x)|^2$
$\leqslant k_0(1 + |x|^2 + |v|^2)$;
$\int_Z |c(t, x, z)|^{2p} \pi(dz) \leqslant k_0(1 + |x|^{2p}), \forall p = 1, 2, 3, 4$;
$\int_Z |c_x(t, x, z)|^{2\overline{p}} \pi(dz) + \int_Z |c_{xx}(t, x, z)|^{2p} \pi(dz) \leqslant k_0, \forall \overline{p} = 1, 2, 3, 4; p = 1, 2$;
$\int_Z |c(t, x, z) - c(t, y, z)|^2 \pi(dz) \leqslant k_0 |x - y|^2$,

$\int_0^T \sup_x \int_Z |c_x(s,x,z)|^2 N_k(ds,dz) \leqslant k_0,$
$\int_0^T \sup_x \int_Z |c_{xx}(s,x,z)|^2 N_k(ds,dz) \leqslant k_0.$

Under assumption (A), by Theorem 117 (12.1) has a unique strong solution. For simplicity we use the following notation:

$\sigma_x =$ the derivative of the matrix $\sigma \in R^{d\otimes d}$ with respect to x;

$$\sigma_x = \left(\sigma_x^{(1)}, \cdots, \sigma_x^{(d)}\right) \in \underbrace{R^{d\otimes d} \times \cdots \times R^{d\otimes d}}_{d-\text{times}} = (R^{d\otimes d})^d,$$

$$\sigma_x y dw = \left(\sigma_x^{(1)} y, \cdots, \sigma_x^{(d)} y\right) dw = \begin{pmatrix} \sum_{jk} \sigma_{x_j}^{1k} y_j dw^k \\ \cdots \\ \sum_{jk} \sigma_{x_j}^{dk} y_j dw^k \end{pmatrix} \in R^d,$$

where $y^* = (y_1, \cdots, y_d)$, $w_t^* = (w_t^1, \cdots, w_t^d)$, and y^* is the transpose of y. Since we will discuss the $d-$dimensional SDE, it is natural to assume that

$b(t,x,v) : [0,T] \times R^d \times U \longmapsto R^d,$
$\sigma(t,x,v) : [0,T] \times R^d \times U \longmapsto R^{d\otimes d},$
$c(t,x,z) : [0,T] \times R^d \times Z \longmapsto R^d,$

are all jointly Borel measurable

12.3 Maximum Principle and Adjoint Equation as BSDE with Jumps

In the following we always assume that $u(\cdot) \in \mathfrak{U}_{ad}$ is an optimal control for (12.1) - (12.3), and $x(\cdot)$ is the unique solution of (12.1) corresponding to $u(\cdot)$. Introduce a Hamiltonian function as follows:

$H : [0,T] \times R^d \times U \times R^d \times R^{d\otimes d_1} \mapsto R^1,$
$H(t,x,v,P,Q) = l(t,x,v) + \langle P, b(t,x,v)\rangle + trace Q\sigma^*(t,x,v).$

We have the following Maximum Principle.

Theorem 354 *There exists a unique*
$(P(\cdot), Q(\cdot), R(\cdot)) \in L^2_{\mathfrak{F}}(R^d) \times L^2_{\mathfrak{F}}(R^{d\otimes d}) \times \mathcal{F}^2_{\mathfrak{F}}(R^d),$
and there exits a unique
$(\overline{P}(\cdot), \overline{Q}(\cdot), \overline{R}(\cdot)) \in L^2_{\mathfrak{F}}(R^{d\otimes d}) \times (L^2_{\mathfrak{F}}(R^{d\otimes d}))^{d_1} \times \mathcal{F}^2_{\mathfrak{F}}(R^{d\otimes d})$
satisfying the following respective BSDE

$$\begin{cases} -dP_t = [b_x^*(t,x_t,u_t)P_t + \sum_{j=1}^d \sigma_x^{j*}(t,x_t,u_t)Q_t^j + \int_Z c_x^*(t,x_t,z)R_t(z)\pi(dz) \\ +l_x(t,x_t,u_t)]dt - Q_t dw_t - \int_Z R_t(z)\widetilde{N}_k(ds,dz), \\ P_T = h_x(x_T), \forall t \in [0,T]; \end{cases} \tag{12.4}$$

$$\left\{\begin{array}{l} -d\overline{P}_t = [b_x^*(t,x_t,u_t)\overline{P}_t + \overline{P}_t b_x(t,x_t,u_t) \\ +\sigma_x^*(t,x_t,u_t)\overline{P}_t\sigma_x(t,x_t,u_t) + \int_Z c_x^*(t,x_t,z)\overline{P}_t(z)c_x(t,x_t,z)\pi(dz) \\ +\sigma_x^*(t,x_t,u_t)\overline{Q}_t + \overline{Q}_t\sigma_x(t,x_t,u_t) + \int_Z (c_x^*(t,x_t,z)\overline{R}_t(z) \\ +\overline{R}_t(z)c_x(t,x_t,z) + c_x^*(t,x_t,z)\overline{R}_t(z)c_x(t,x_t,z))\pi(dz) \\ +H_{xx}(t,x_t,u_t,P_t,Q_t) + \int_Z \overline{H}_{xx}(t,x_t,z,R_t(z))\pi(dz)]dt \\ -\overline{Q}_t dw_t - \int_Z \overline{R}_t(z)\widetilde{N}_k(ds,dz), \\ \overline{P}_T = h_{xx}(x_T), \forall t\in[0,T]; \end{array}\right. \tag{12.5}$$

where $\sigma_x^*\overline{Q}_t = \sum_{j=1}^{d_1}\sigma_x^{j*}\overline{Q}_t^j \in R^{d\otimes d}$, $\overline{H}(t,x,z,\widehat{R}) = \left\langle \widehat{R}, c(t,x_t,z)\right\rangle$ *for* $\widehat{R}\in R^d$ *such that* $\forall v\in U$, *a.e., a.s.*

$$\begin{aligned} &H(t,x_t,v,P_t,Q_t-\overline{P}_t\sigma(t,x_t,u_t)) + \frac{1}{2}trace\left(\sigma\sigma^*(t,x_t,v)\overline{P}_t\right) \qquad (12.6) \\ &\geqslant H(t,x_t,u_t,P_t,Q_t-\overline{P}_t\sigma(t,x_t,u_t)) + \frac{1}{2}trace\left(\sigma\sigma^*(t,x_t,u_t)\overline{P}_t\right). \end{aligned}$$

Equations (12.4) and (12.5) are called adjoint equations. Obviously, they are BSDEs with jumps. However, they are linear. So by the theory of BSDEs with jumps (Theorem 234) they have a unique solution. Actually, we can solve the first one, then the second one.

12.4 A Simple Example

To illustrate how to use the above Maximum Principle let us give a simple example as follows: For simplicity let us assume that all processes and coefficients appearing below are in 1−dimensional space. Consider the following linear stochastic system:

$$\left\{\begin{array}{l} dy_s = (Ay_s + Bv_s)ds + Cy_s dw_s + \int_Z F(z)y_s\widetilde{N}_k(ds,dz), \\ y_0 = x,\ 0\leqslant s\leqslant T, \end{array}\right. \tag{12.7}$$

where $v\in\mathfrak{U}_{ad}$ and $\mathfrak{U}_{ad}$ is defined by (12.2) such that $E\left(\sup_{t\in[0,T]}|v(t)|^2\right)\leq k_0<\infty, \forall v\in\mathfrak{U}_{ad}$ and the target functional is

$$J(v(\cdot)) = E\left\{\int_0^T[\widetilde{R}y_s^v + \widetilde{N}|v_s|^2]ds\right\},$$

and y_s^v with $y_0^v = x$ is the unique solution of (12.7) for all $0\leqslant s\leqslant T$ and we assume that $\widetilde{N}>0$, and all $A,B,C,F,\widetilde{R},\widetilde{N}$ are real valued. Suppose that $u(\cdot)\in\mathfrak{U}_{ad}$ is an optimal control, and x_s is the optimal trajectory satisfying (12.7) corresponding to this control $u(\cdot)$, and we also assume that in the definition of $\mathfrak{U}_{ad}$, U is a closed and convex set. Notice that now we have

$$H(t,y,v,P,Q) = \widetilde{R}y + \widetilde{N}|v|^2 + P(Ay+Bv) + QCy.$$

Applying Theorem 354, by (12.6) we have that $u(\cdot)$ and x_s satisfy the following variational inequality: $\forall v \in U$

$$\widetilde{N}\,|v|^2 + P_t B v \geq \widetilde{N}\,|u_t|^2 + P_t B u_t,\ a.e.a.s., \tag{12.8}$$

where by (12.4) and by the uniqueness of solutions $P(\cdot) \in L^2_{\mathfrak{F}}(R^d)$ satisfy

$$-dP_t = [AP_t + \widetilde{R}]dt, P_T = 0, \forall t \in [0,T]. \tag{12.9}$$

Let $f(t,v) = \widetilde{N}\,|v|^2 + P_t B v$. Then $f'_v(t,v) = 2\widetilde{N}v + P_t B$, and one sees that
$f'_v(t,u_t) = 0$, as $u_t = -\frac{P_t B}{2\widetilde{N}}$;
$f'_v(t,v) < 0$, as $v < u_t$; $f'_v(t,v) > 0$, as $v > u_t$;
where P_t satisfies (12.9). So we have that such u_t satisfies (12.8). However, solving (12.9) one finds that
$P_t = \int_t^T e^{-A(t-s)}\widetilde{R}ds$.
From these results one finds that for such a u_t, (12.7) has a unique solution. Since by the approximation to the infimum, (see Lemma below), one easily finds that there should exist an optimal control $u(\cdot) \in \mathfrak{U}_{ad}$ such that $J(u(\cdot)) = \inf_{v(\cdot)\in\mathfrak{U}_{ad}} J(v(\cdot))$. So the u_t obtained above must be an optimal control.

Lemma 355 *There exists an optimal control* $u(\cdot) \in \mathfrak{U}_{ad}$ *such that* $J(u(\cdot)) = \inf_{v(\cdot)\in\mathfrak{U}_{ad}} J(v(\cdot))$.

Proof. Take $v_n \in \mathfrak{U}_{ad}$ such that $J(v_n(\cdot)) < \inf_{v(\cdot)\in\mathfrak{U}_{ad}} J(v(\cdot)) + \frac{1}{2^n}$, for $n = 1, 2, \cdots$ Since $E\left(\sup_{t\in[0,T]} |v_n(t)|^2\right) \leq k_0 < \infty$, there exists a subsequence of $\{v_n\}$ denoted again by $\{v_n\}$ and there also exists a $u \in \mathfrak{U}_{ad}$ such that $v_n \xrightarrow{w} u$ in $L^2_{\mathfrak{F}}(R^1)$. Hence there also exists a subsequence denoted again by $\{v_n\}$ such that its arithmetic means from this sequence converge strongly to u in $L^2_{\mathfrak{F}}(R^1)$, that is, $\frac{v_1+\cdots+v_n}{n} \to u$, in $L^2_{\mathfrak{F}}(R^1)$. However, one easily sees that $\frac{v_1+\cdots+v_n}{n} \in \mathfrak{U}_{ad}$ and $J\left(\frac{v_1+\cdots+v_n}{n}\right) \leq \frac{J(v_1)+\cdots+J(v_n)}{n} \leq \frac{1}{n}$. Hence by letting $n \to \infty$, one easily obtains that $J(u(\cdot)) = \inf_{v(\cdot)\in\mathfrak{U}_{ad}} J(v(\cdot))$. ∎

12.5 Intuitive thinking on the Maximum Principle

In Calculus if a real-valued function with d variables $f(x), x \in R^d$, attains its minimum at the point x_0, then we will have $f(x) \geq f(x_0), \forall x \in R^d$; or equivalently, $f(x_0 + y) \geq f(x_0), \forall y \in R^d$. To explain clearly the intuitive derivation of a Maximum Principle for a functional, let us simplify the notation. Consider a non-random functional $J(v) = \int_0^T l(x_t^v, v_t)dt + h(x_T^v)$, where all functions are real-valued, x_t is a solution of a real ODE

$$dx_t^v = b(x_t^v, v_t)dt, x_0^v = x, \tag{12.10}$$

for $v(\cdot) \in \mathfrak{U}$, and where v_t takes values in a non-empty set $U \subset R^1$. If $J(v)$ attains at its minimu at $v = u \in \mathfrak{U}$, then obviously we will have $J(v) \geq J(u), \forall v \in \mathfrak{U}$. Or, equivalently, $\int_0^T (l(x_t^v, v_t) - l(x_t^u, u_t))dt + h(x_T^v) - h(x_T^u) \geq 0, \forall v \in \mathfrak{U}$. By Taylor's formula we will have, approximately, that $\forall v \in \mathfrak{U}$

$$\int_0^T (l(x_t^u, v_t) - l(x_t^u, u_t) + l_x(x_t^u, u_t)(x_t^v - x_t^u))dt + h_x(x_T^v)(x_T^v - x_T^u) \\ +o(\sup_{t\in[0,T]} |x_t^v - x_t^u|) \geq 0, \tag{12.11}$$

where l_x is the derivative of l with respect to x, provided that l is smooth enough. Actually, all kinds of maximum principle are only trying to transform this inequality into a more applicable statement. The way here by which to derive a necessary condition for the inequality (12.11) is by using the following two techniques: 1) We use a "spike variation" $u^\varepsilon(t)$ of the optimal control to obtain a first order variational equation $dx_t^{1,\varepsilon}$ such that $d(x_t^{u^\varepsilon} - x_t^u) = dx_t^{1,\varepsilon} + o(\varepsilon^2)dt$, where $u^\varepsilon(t) = v$, as $t \in [t_0, t_0 + \varepsilon)$; $u^\varepsilon(t) = u(t)$, otherwise; and $x_t^{u^\varepsilon}$ is the solution of (12.10) corresponding to the control $u^\varepsilon(\cdot)$ provided that $u^\varepsilon(\cdot) \in \mathfrak{U}$. In fact by Taylor's formula, and under appropriate conditions, one has that
$d(x_t^{u^\varepsilon} - x_t^u) = (b(x_t^{u^\varepsilon}, v_t) - b(x_t^u, u_t))dt$
$= b_x(x_t^u, u_t) \cdot (x_t^{u^\varepsilon} - x_t^u)dt + (b(x_t^u, v_t) - b(x_t^u, u_t))dt + o(\varepsilon)$.
So we may introduce a first order variational equation:

$$dx_t^{1,\varepsilon} = b_x(x_t^u, v_t) \cdot x_t^{1,\varepsilon}dt + (b(x_t^u, v_t) - b(x_t^u, u_t))dt, \; x_0^{1,\varepsilon} = 0. \tag{12.12}$$

2) By using integration by parts we have
$-\int_0^T (dp_t)x_t^{1,\varepsilon} = \int_0^T p_t dx_t^{1,\varepsilon} - \int_0^T d(p_t x_t^{1,\varepsilon})$,
and we can introduce a function P_t such that
$\int_0^T l_x(x_t^u, u_t)x_t^{1,\varepsilon}dt + h_x(x^u(T))(x_T^{1,\varepsilon}) = -\int_0^T (x_t^{1,\varepsilon})dP_t + h_x(x^u(T))(x_T^{1,\varepsilon})$
$-\int_0^T \left(x_t^{1,\varepsilon}\right) b_x(x_t^u, u_t)P_t dt = \int_0^T P_t(b(x_t^u, u_t^\varepsilon) - b(x_t^u, u_t))dt$,
if

$$-dP_t = (b_x(x_t^u, u_t)P_t + l_x(x_t^u, u_t))dt, P_T = h_x(x^u(T)). \tag{12.13}$$

This means that (12.11) can be rewritten as that
$0 \leq \int_0^T (l(x_s^u, u^\varepsilon) - l(x_s^u, u_s) + P_s(b(x_s^u, u^\varepsilon) - b(x_s^u, u_s))ds + o(\varepsilon)$.
By the definition of u^ε one sees that
$0 \leq \frac{1}{\varepsilon}\int_{t_0}^{t_0+\varepsilon} (l(x_s^u, v) - l(x_s^u, u_s) + P_s(b(x_s^u, v) - b(x_s^u, u_s))ds + \frac{o(\varepsilon)}{\varepsilon}$.
Letting $\varepsilon \to 0$, one obtains the following variational inequality for the optimal control $u(\cdot)$:

$$l(x_t^u, v) + P_t b(x_t^u, v) \geq l(x_t^u, u_t) + P_t b(x_t^u, u_t), a.e.t, \forall v \in U. \tag{12.14}$$

This is the Maximum Principle. In what follows, by using such idea we will make rigorous the proof of the Maximum Principle for a stochastic system with jumps. That is, we will arrive at Theorem 354.

12.6 Some Lemmas

For any $t_0 \in [0,T], \varepsilon > 0$, set

$$u^\varepsilon(t) = \begin{cases} v, & \text{as } t \in [t_0, t_0+\varepsilon), \\ u(t), & \text{otherwise,} \end{cases} \tag{12.15}$$

where $v(\cdot) \in \mathfrak{U}_{ad}$ is arbitrary. It is easily seen that $u^\varepsilon(\cdot) \in \mathfrak{U}_{ad}$. Hence there corresponds a unique trajectory $x^\varepsilon(\cdot)$ (the solution of (12.1)). Let us introduce the first and second order R^d-valued variational equations, respectively, as follows:

$$\begin{aligned} x_t^{1,\varepsilon} &= \int_0^t b_x(s,x_s,u_s)x_s^{1,\varepsilon}ds + \int_0^t (b(s,x_s,u_s^\varepsilon) - b(s,x_s,u_s))ds \\ &\quad + \int_0^t \sigma_x(s,x_s,u_s)x_s^{1,\varepsilon}dw_s + \int_0^t (\sigma(s,x_s,u_s^\varepsilon) - \sigma(s,x_s,u_s))dw_s \\ &\quad + \int_0^t \int_Z c_x(s,x_s,z)x_s^{1,\varepsilon}\widetilde{N}_k(ds,dz), \end{aligned} \tag{12.16}$$

and

$$\begin{aligned} x_t^{2,\varepsilon} &= \int_0^t [b_x(s,x_s,u_s)x_s^{2,\varepsilon} + \varphi_s^\varepsilon]ds + \int_0^t [\sigma_x(s,x_s,u_s)x_s^{2,\varepsilon} + \psi_s^\varepsilon]dw_s \\ &\quad + \int_0^t \int_Z [c_x(s,x_s,z)x_s^{2,\varepsilon} + \chi_s^\varepsilon(z)]\widetilde{N}_k(ds,dz), \end{aligned} \tag{12.17}$$

where
$\varphi_s^\varepsilon = \frac{1}{2}b_{xx}(s,x_s,u_s)x_s^{1,\varepsilon}x_s^{1,\varepsilon} + (b_x(s,x_s,u_s^\varepsilon) - b_x(s,x_s,u_s))\,x_s^{1,\varepsilon}$,
$\psi_s^\varepsilon = \frac{1}{2}\sigma_{xx}(s,x_s,u_s)x_s^{1,\varepsilon}x_s^{1,\varepsilon} + (\sigma_x(s,x_s,u_s^\varepsilon) - \sigma_x(s,x_s,u_s))\,x_s^{1,\varepsilon}$,
$\chi_s^\varepsilon(z) = \frac{1}{2}c_{xx}(s,x_s,z)x_s^{1,\varepsilon}x_s^{1,\varepsilon}$,
where we will write
$f_{xx}yy = \sum_{i,j=1}^d f_{x^ix^j}y^iy^j$, for $f = b,\sigma,c,l,h$.
By Theorem 117 it is known that (12.16) has a unique strong solution and hence again, by Theorem 117, (12.17) also has a unique strong solution.

We have the following lemmas.

Lemma 356 *1° For* $p = 1,2,3,4$
$E\sup_{t\in[0,T]}|x_t|^{2p} \leqslant k_T$,

where $k_T = k_0 T^p(1 + \sup_{t\in[0,T]} |u_t|^{2p})$, *and* $k_0 \geqslant 0$ *is a constant;*

2° *for* $p = 1, 2, 3, 4$

$E\sup_{t\in[0,T]} \left|x_t^{1,\varepsilon}\right|^{2p} \leqslant k_T\varepsilon^p, E\sup_{t\in[0,T]} \left|x_t^{2,\varepsilon}\right|^{2p} \leqslant k_1\varepsilon^{2p},$

where $k_1 = k_0(1 + E\sup_{t\in[0,T]} |u_t|^{2p} + E\sup_{t\in[0,T]} |v_t|^{2p})$;

3° $E\int_0^T |\varphi_t^\varepsilon| dt \leqslant \frac{1}{2}E\int_0^T \left|b_{xx}(t,x_t,u_t)x_t^{1,\varepsilon}x_t^{1,\varepsilon}\right| dt + o(\varepsilon) \leqslant k_T'\varepsilon,$

$E\int_0^T |\psi_t^\varepsilon| dt \leqslant k_T'\varepsilon, \ E\int_0^T \int_Z |\chi_s^\varepsilon(z)|^2 \pi(dz)dt \leqslant k_T'\varepsilon^2.$

Remark 357 *In* [188] *there are some errors: In (2.10) there is the estimate for* $p > 1$

$$E\left|\int_0^\infty \int_{Z\times I_\rho} g_0(s,z)\widetilde{N}_k(ds,dz)\right|^{2p}$$
$$\leqslant c_p |I\rho|^{p-1} E\int_{I_\rho} \left|\int_Z |g_0(s,z)|^2 \pi(dz)\right|^p ds.$$

This estimate is used in [188] *to get an estimate similar to* 2°. *However, the above inequality is not correct. To see this, if we take* $I_\rho = [s,t]$, *and* $p = 3$, *then by the Kolmogorov continuous trajectory version theorem* $\int_0^t \int_Z g_0(s,z)\widetilde{N}_k(ds,dz)$ *will have a continuous (in t) version. This is obviously incorrect. Hence the proof that the general Maximum Principle still holds for stochastic systems with jumps such that the jump coefficients also depend on controls in* [188] *is at the very least incomplete. Actually, by the Burkholder-Gundy-Davis inequality*[75] *one can only get that for* $p > 1$

$$E\left|\int_0^\infty \int_{Z\times I_\rho} g_0(s,z)\widetilde{N}_k(ds,dz)\right|^{2p} \leqslant k_p E\left|\int_0^\infty \int_{Z\times I_\rho} |g_0(s,z)|^2 N_k(ds,dz)\right|^p,$$

but not that

$$E\left|\int_0^\infty \int_{Z\times I_\rho} g_0(s,z)\widetilde{N}_k(ds,dz)\right|^{2p} \leqslant k_p E\left|\int_0^\infty \int_{Z\times I_\rho} |g_0(s,z)|^2 \pi(dz)ds\right|^p.$$

Now let us show Lemma 356.

Proof. 1° can be proved by Ito's formula and Gronwall's inequality. Let us use Gronwall's inequality to show 2° by 1°. In fact, by assumption and by (12.16)

$$y_t^1 \hat{=} E\sup_{s\leqslant t} \left|x_s^{1,\varepsilon}\right|^{2p} \leqslant 5^{2p}[k_0' \int_0^t y_s^1 ds$$
$$+E(\int_0^t |b(s,x_s,u_s^\varepsilon) - b(s,x_s,u_s)| ds)^{2p}$$
$$+E(\left|\int_0^t (\sigma(s,x_s,u_s^\varepsilon) - \sigma(s,x_s,u_s))dw_s)\right|^{2p} = 5^{2p}[I_t^1 + I_t^2 + I_t^3].$$

However,

$$I_t^3 \leqslant k_0'' E(\int_{t_0}^{t_0+\varepsilon} |\sigma(s,x_s,u_s^\varepsilon) - \sigma(s,x_s,u_s)|^2 ds)^p$$
$$\leqslant k_0'' E\int_{t_0}^{t_0+\varepsilon} |\sigma(s,x_s,u_s^\varepsilon) - \sigma(s,x_s,u_s)|^{2p} ds \cdot \varepsilon^{p-1} \leqslant 2k_0'' k_1 \varepsilon^p.$$

Similar estimate holds for I_t^2. Hence by Gronwall's inequality we obtain the first estimate in 2°. 3° is proved by the first estimate of 2°. Finally by (12.17) and by Gronwall's inequality one easily derives the second estimate of 2°. The proof is complete. ■

Lemma 358 $E\sup_{t\in[0,T]}\left|x_t^\varepsilon - x_t - x_t^{1,\varepsilon} - x_t^{2,\varepsilon}\right|^2 = o\left(\varepsilon^2\right).$

Proof. Set
$y_t^\varepsilon = x_t^\varepsilon - x_t - x_t^{1,\varepsilon} - x_t^{2,\varepsilon}.$
Then
$$\begin{aligned}
y_t^\varepsilon &= \textstyle\int_0^t (b(s,x_s^\varepsilon,u_s^\varepsilon) - b(s,x_s + x_s^{1,\varepsilon} + x_s^{2,\varepsilon},u_s^\varepsilon)\\
&+b(s,x_s + x_s^{1,\varepsilon} + x_s^{2,\varepsilon},u_s^\varepsilon) - b(s,x_s,u_s^\varepsilon) + b(s,x_s,u_s^\varepsilon) - b(s,x_s,u_s))ds\\
&-x_s^{1,\varepsilon} - x_s^{2,\varepsilon} + \textstyle\int_0^t (\sigma(s,x_s^\varepsilon,u_s^\varepsilon) - \sigma(s,x_s + x_s^{1,\varepsilon} + x_s^{2,\varepsilon},u_s^\varepsilon)\\
&+\sigma(s,x_s + x_s^{1,\varepsilon} + x_s^{2,\varepsilon},u_s^\varepsilon) - \sigma(s,x_s,u_s^\varepsilon) + \sigma(s,x_s,u_s^\varepsilon) - \sigma(s,x_s,u_s))dw_s\\
&+ \textstyle\int_0^t \int_Z (c(s,x_s^\varepsilon,z) - c(s,x_s + x_s^{1,\varepsilon} + x_s^{2,\varepsilon},z)\\
&+c(s,x_s + x_s^{1,\varepsilon} + x_s^{2,\varepsilon},z) - c(s,x_s,z))\widetilde{N}_k(ds,dz)\\
&= \textstyle\int_0^t \int_0^1 b_x(s,x_s + x_s^{1,\varepsilon} + x_s^{2,\varepsilon} + \lambda y_s^\varepsilon,u_s^\varepsilon)d\lambda y_s^\varepsilon ds\\
&+ \textstyle\int_0^t \int_0^1 \sigma_x(s,x_s + x_s^{1,\varepsilon} + x_s^{2,\varepsilon} + \lambda y_s^\varepsilon,u_s^\varepsilon)d\lambda y_s^\varepsilon dw_s\\
&+ \textstyle\int_0^t \int_Z \int_0^1 c_x(s,x_s + x_s^{1,\varepsilon} + x_s^{2,\varepsilon} + \lambda y_s^\varepsilon,z)d\lambda y_s^\varepsilon \widetilde{N}_k(ds,dz)\\
&+ \textstyle\int_0^t \displaystyle\int_Z [\tfrac{1}{2}c_{xx}(s,x_s,z)(2x_s^{1,\varepsilon}x_s^{2,\varepsilon} + x_s^{2,\varepsilon}x_s^{2,\varepsilon})\\
&+ \textstyle\int_0^1 (1-\lambda)(c_{xx}(s,x_s + \lambda(x_s^{1,\varepsilon} + x_s^{2,\varepsilon}),z) - c_{xx}(s,x_s z))d\lambda\\
&\cdot(x_s^{1,\varepsilon} + x_s^{2,\varepsilon})(x_s^{1,\varepsilon} + x_s^{2,\varepsilon})]\widetilde{N}_k(ds,dz)\\
&+ \textstyle\int_0^t R^\varepsilon(b)(s)ds + \int_0^t R^\varepsilon(\sigma)(s)dw_s,
\end{aligned}$$
where
$$\begin{aligned}
R^\varepsilon(b)(s) &= (b_x(s,x_s,u_s^\varepsilon) - b_x(s,x_s,u_s))x_s^{2,\varepsilon}\\
&+\tfrac{1}{2}b_{xx}(s,x_s,u_s)\cdot(2x_s^{1,\varepsilon}x_s^{2,\varepsilon} + x_s^{2,\varepsilon}x_s^{2,\varepsilon})\\
&+\tfrac{1}{2}(b_{xx}(s,x_s,u_s^\varepsilon) - b_{xx}(s,x_s,u_s))(x_s^{1,\varepsilon} + x_s^{2,\varepsilon})(x_s^{1,\varepsilon} + x_s^{2,\varepsilon})\\
&+ \textstyle\int_0^1 (1-\lambda)(b_{xx}(s,x_s + \lambda(x_s^{1,\varepsilon} + x_s^{2,\varepsilon}),u_s^\varepsilon)\\
&-b_{xx}(s,x_s u_s^\varepsilon))d\lambda(x_s^{1,\varepsilon} + x_s^{2,\varepsilon})(x_s^{1,\varepsilon} + x_s^{2,\varepsilon}),
\end{aligned}$$
$R^\varepsilon(\sigma)(s)$ is similarly defined. Now by applying Ito's formula, Gronwall's inequality and Lemma 356, one obtains the stated conclusion. ∎

Lemma 359

$$\begin{aligned}
&E\int_0^T l_x(t,x_t,u_t)(x_t^{1,\varepsilon} + x_t^{2,\varepsilon})dt + Eh_x(x(T))(x_T^{1,\varepsilon} + x_T^{2,\varepsilon})\\
&+\frac{1}{2}E\int_0^T (x_t^{1,\varepsilon})^* l_{xx}(t,x_t,u_t)x_t^{1,\varepsilon}dt + \frac{1}{2}E(x_t^{1,\varepsilon})^* h_{xx}(x(T))x_T^{1,\varepsilon}\\
&+E\int_0^T (l(t,x_t,u_t^\varepsilon) - l(t,x_t,u_t))dt \geqslant o(\varepsilon).
\end{aligned} \tag{12.18}$$

Proof. Since (x_t,u_t) is optimal, we have
$$\begin{aligned}
0 &\leqslant E[\textstyle\int_0^T l(t,x_t^\varepsilon,u_t^\varepsilon)dt + h(x_T^\varepsilon)] - E[\int_0^T l(t,x_t,u_t)dt + h(x_T)]\\
&= E[\textstyle\int_0^T (l(t,x_t + x_t^{1,\varepsilon} + x_t^{2,\varepsilon},u_t^\varepsilon) - l(t,x_t,u_t))dt\\
&+E[h(x_T + x_T^{1,\varepsilon} + x_T^{2,\varepsilon}) - h(x_T)] + o(\varepsilon)\\
&= E[\textstyle\int_0^T (l(t,x_t + x_t^{1,\varepsilon} + x_t^{2,\varepsilon},u_t) - l(t,x_t,u_t))dt
\end{aligned}$$

$+E[h(x_T + x_T^{1,\varepsilon} + x_T^{2,\varepsilon}) - h(x_T)] + E[\int_0^T (l(t, x_t + x_t^{1,\varepsilon} + x_t^{2,\varepsilon}, u_t^\varepsilon)$
$-l(t, x_t + x_t^{1,\varepsilon} + x_t^{2,\varepsilon}, u_t))dt] + o(\varepsilon) = E[\int_0^T [l_x(t, x_t, u_t) \left(x_t^{1,\varepsilon} + x_t^{2,\varepsilon}\right)$
$+\frac{1}{2} l_{xx}(t, x_t, u_t) \left(x_t^{1,\varepsilon} + x_t^{2,\varepsilon}\right) \left(x_t^{1,\varepsilon} + x_t^{2,\varepsilon}\right)]dt$
$+E[\int_0^T (l(t, x_t, u_t^\varepsilon) - l(t, x_t, u_t))dt$
$+E \int_0^T (l_x(t, x_t, u_t^\varepsilon) - l_x(t, x_t, u_t))dt \left(x_t^{1,\varepsilon} + x_t^{2,\varepsilon}\right)$
$+\frac{1}{2} E \int_0^T (l_{xx}(t, x_t, u_t^\varepsilon) - l_{xx}(t, x_t, u_t)) x_t^{1,\varepsilon} x_t^{1,\varepsilon} dt$
$+E[h_x(x_T)(x_T^{1,\varepsilon} + x_T^{2,\varepsilon})] + \frac{1}{2} E[h_{xx}(x_T) x_T^{1,\varepsilon} x_T^{1,\varepsilon}] + o(\varepsilon)$.
Thus the conclusion now follows from Lemma 356 ■

12.7 Proof of Theorem 354

To derive the Maximum Principle from Lemma 359 we need to introduce an adjoint process p_t such that it makes
$E \int_0^T l_x(t, x_t, u_t) x_t^{1,\varepsilon} dt + E h_x(x(T))(x_T^{1,\varepsilon}) = E \int_0^T f(t, u_t, v_t, x_t, p_t) dt.$
For this it is necessary to notice that by Ito's formula (or say, using integration by parts)
$-E \int_0^T (dp_t) x_t^{1,\varepsilon} = E[\int_0^T p_t dx_t^{1,\varepsilon} + \int_0^T dp_t dx_t^{1,\varepsilon} - \int_0^T d(p_t x_t^{1,\varepsilon})].$
So if we introduce (P_t, Q_t, R_t) as the solution of (12.4), then by (12.16) we have
$E \int_0^T l_x(t, x_t, u_t) x_t^{1,\varepsilon} dt + E h_x(x(T))(x_T^{1,\varepsilon}) = -E \int_0^T (x_t^{1,\varepsilon})^* dP_t$
$+E h_x(x(T))(x_T^{1,\varepsilon}) - E \int_0^T \left(x_t^{1,\varepsilon}\right)^* [b_x^*(t, x_t, u_t) P_t + \sum_{j=1}^d \sigma_x^{j*}(t, x_t, u_t) Q_t^j$
$+ \int_Z c_x^*(t, x_t, z) R_t(z) \pi(dz)] dt = E \int_0^T [P_s(b(s, x_s, u_s^\varepsilon) - b(s, x_s, u_s))$
$+Q_s(\sigma(s, x_s, u_s^\varepsilon) - \sigma(s, x_s, u_s))] ds.$
Similarly, by (12.4) and (12.17)
$E \int_0^T l_x(t, x_t, u_t) x_t^{2,\varepsilon} dt + E h_x(x(T))(x_T^{2,\varepsilon})$
$= E \int_0^T [\langle P_s, \ (b_x(s, x_s, u_s^\varepsilon) - b_x(s, x_s, u_s)) x_s^{1,\varepsilon} \rangle$
$+Q_s(\sigma_x(s, x_s, u_s^\varepsilon) - \sigma_x(s, x_s, u_s)) x_s^{1,\varepsilon} + \frac{1}{2} P_s b_{xx}(s, x_s, u_s) x_s^{1,\varepsilon} x_s^{1,\varepsilon}$
$+\frac{1}{2} Q_s \sigma_{xx}(s, x_s, u_s) x_s^{1,\varepsilon} x_s^{1,\varepsilon} + \frac{1}{2} \int_Z R_s c_{xx}(s, x_s, z) x_s^{1,\varepsilon} x_s^{1,\varepsilon} \pi(dz))] ds.$
Hence we have
L.H.S. of (12.18) $= E \int_0^T [H(t, x_t, u_t^\varepsilon, P_t, Q_t) - H(t, x_t, u_t, P_t, Q_t)] dt$
$+\frac{1}{2} E \int_0^T (x_s^{1,\varepsilon})^* H_{xx}(t, x_t, u_t, P_t, Q_t) x_t^{1,\varepsilon} dt + \frac{1}{2} E (x_T^{1,\varepsilon})^* h_{xx}(x(T)) x_T^{1,\varepsilon}$
$+\frac{1}{2} E \int_0^T \int_Z (x_s^{1,\varepsilon})^* R_s c_{xx}(s, x_s, z) x_s^{1,\varepsilon} \pi(dz) ds + o(\varepsilon)$,
where we have used the result that
$E \int_0^T [P_s(b_x(s, x_s, u_s^\varepsilon) - b_x(s, x_s, u_s)) x_s^{1,\varepsilon} ds$
$+E \int_0^T [Q_s(\sigma_x(s, x_s, u_s^\varepsilon) - \sigma_x(s, x_s, u_s)) x_s^{1,\varepsilon} ds = o(\varepsilon)$.
Now let

$$X_s^\varepsilon = x_s^{1,\varepsilon}(x_s^{1,\varepsilon})^* = \begin{bmatrix} x_{1s}^{1,\varepsilon}(x_{1s}^{1,\varepsilon}) \cdots x_{1s}^{1,\varepsilon}(x_{ds}^{1,\varepsilon}) \\ \cdots\cdots\cdots\cdots \\ x_{ds}^{1,\varepsilon}(x_{1s}^{1,\varepsilon}) \cdots x_{ds}^{1,\varepsilon}(x_{ds}^{1,\varepsilon}) \end{bmatrix}$$

Then we have

$$dX_s^\varepsilon = [X_s^\varepsilon b_x^*(s,x_s,u_s) + b_x(s,x_s,u_s)X_s^\varepsilon + \Phi^\varepsilon(s)$$
$$+\sum_{i=1}^d \sigma_x^i(s,x_s,u_s)X_s^\varepsilon \sigma_x^{*i}(s,x_s,u_s) + \int_Z c_x(s,x_s,z)X_s^\varepsilon c_x^*(s,x_s,z)\pi(dz)]ds$$
$$+[X_s^\varepsilon \sigma_x^*(s,x_s,u_s) + \sigma_x(s,x_s,u_s)X_s^\varepsilon + \Psi^\varepsilon(s)]dw_s + \int_Z [X_s^\varepsilon c_x^*(s,x_s,.z)$$
$$+c_x(s,x_s,z)X_s^\varepsilon + c_x(s,x_s,z)X_s^\varepsilon c_x^*(s,x_s,z)]\widetilde{N}_k(ds,dz), \qquad (12.19)$$

where
$\Phi^\varepsilon(s) = x_s^{1,\varepsilon}(b(s,x_s,u_s^\varepsilon) - b(s,x_s,u_s))^*$
$+(b(s,x_s,u_s^\varepsilon) - b(s,x_s,u_s))(x_s^{1,\varepsilon})^* + \sigma_x(s,x_s,u_s)x_s^{1,\varepsilon}(\sigma(s,x_s,u_s^\varepsilon)$
$-\sigma(s,x_s,u_s))^* + (\sigma(s,x_s,u_s^\varepsilon) - \sigma(s,x_s,u_s))(x_s^{1,\varepsilon})^*\sigma_x(s,x_s,u_s)^*$
$+(\sigma(s,x_s,u_s^\varepsilon) - \sigma(s,x_s,u_s))(\sigma(s,x_s,u_s^\varepsilon) - \sigma(s,x_s,u_s))^*,$
$\Psi^\varepsilon(s) = x_s^{1,\varepsilon}(\sigma(s,x_s,u_s^\varepsilon) - \sigma(s,x_s,u_s))^* + (\sigma(s,x_s,u_s^\varepsilon) - \sigma(s,x_s,u_s))(x_s^{1,\varepsilon})^*.$
Notice that for all $x \in R^d$, $A \in R^{d\times d}$ we define the inner product of matrices xx^* and A by $(xx^*, A)_* = trace((xx^*)A) = x^*Ax$. Then by (12.19) and (12.5)
$E\int_0^T (x_s^{1,\varepsilon})^* H_{xx}(t,x_t,u_t,P_t,Q_t)x_t^{1,\varepsilon}dt + E(x_T^{1,\varepsilon})^* h_{xx}(x(T))x_T^{1,\varepsilon}$
$+E\int_0^T \int_Z \left\langle R_s, c_{xx}(s,x_s,z)x_s^{1,\varepsilon}x_s^{1,\varepsilon}\right\rangle \pi(dz)ds = -E\int_0^T \left(d\overline{P}_s, X_s^\varepsilon\right)_*$
$-\{(\overline{P}_s, [X_s^\varepsilon b_x^*(s,x_s,u_s) + b_x(s,x_s,u_s)X_s^\varepsilon + \sum_{i=1}^d \sigma_x^i(s,x_s,u_s)X_s^\varepsilon \sigma_x^{*i}(s,x_s,u_s)$
$+\int_Z c_x(s,x_s,z)X_s^\varepsilon c_x^*(s,x_s,z)\pi(dz)])_*$
$-(\overline{Q}_s, [X_s^\varepsilon \sigma_x^*(s,x_s,u_s) + \sigma_x(s,x_s,u_s)X_s^\varepsilon])_*$
$-\int_Z (\overline{R}_s(z), [X_s^\varepsilon c_x^*(s,x_s,.z) + c_x(s,x_s,z)X_s^\varepsilon + c_x(s,x_s,z)X_s^\varepsilon c_x^*(s,x_s,z)])_*$
$+E(x_T^{1,\varepsilon})^* h_{xx}(x(T))x_T^{1,\varepsilon} + E\int_0^T \int_Z \left(x_s^{1,\varepsilon}(x_s^{1,\varepsilon})^*, R_s c_{xx}(s,x_s,z)\right)_* \pi(dz)ds$
$= E\int_0^T (\overline{P}_s, (\sigma(s,x_s,u_s^\varepsilon) - \sigma(s,x_s,u_s))(\sigma(s,x_s,u_s^\varepsilon) - \sigma(s,x_s,u_s))^*)_* ds + o(\varepsilon)$.
Hence we have
L.H.S. of (12.18)
$= E\int_0^T [H(t,x_t,u_t^\varepsilon,P_t,Q_t) - H(t,x_t,u_t,P_t,Q_t)]dt$
$+\frac{1}{2}E\int_0^T tr.(\sigma(s,x_s,u_s^\varepsilon) - \sigma(s,x_s,u_s))\overline{P}_s$
$\cdot(\sigma(s,x_s,u_s^\varepsilon) - \sigma(s,x_s,u_s))^* ds \geqslant o(\varepsilon)$.
Dividing the above inequality by ε, and letting $\varepsilon \to 0$, by the arbitrariness of $v(\cdot) \in L^\infty(\Omega, (\mathfrak{F}_t); U)$ one finds that Theorem 354 holds.

Appendix A

A Short Review on Basic Probability Theory

A.1 Probability Space, Random Variable and Mathematical Expectation

In real analysis a real-valued function $f(x)$ defined on R^1 is called a Borel - measurable function, if for any $c \in R^1$ $\{x \in R^1 : f(x) \leq c\}$ is a Borel-measurable set. If we denote the totality of Borel - measurable sets in R^1 by $\mathfrak{B}(R^1)$, and call $(R^1, \mathfrak{B}(R^1))$ a measurable space, then we may write $\{x \in R^1 : f(x) \leq c\} \in \mathfrak{B}(R^1), \forall c \in R^1$, for a Borel - measurable function f, and say that f is a measurable function defined on the measurable space $(R^1, \mathfrak{B}(R^1))$. The definition for a Lebesgue - measurable function is given in similar fashion.. Furthermore, in real analysis, the Lebesgue measure m is also introduced and used to construct a measure space $(R^1, \widetilde{\mathfrak{B}}(R^1), m)$, so that one can define the Lebesgue - integral for a Lebesgue - measurable function under very general conditions and obtain its many important properties and applications.

Such concepts and ideas can be applied to probability theory. For this we first need to introduce a probability space $(\Omega, \mathfrak{F}, P)$.

Definition 360 $(\Omega, \mathfrak{F}, P)$ *is called a probability space, if* Ω *is a set,* $\mathfrak{F}$ *is a family of subsets from* Ω*, which is a* $\sigma-$*field, and* P *is a probability measure.*

Definition 361 $\mathfrak{F}$ *is called a* $\sigma-$*field (*$\sigma-$*algebra), if it satisfies the following conditions:*

1) $\Omega \in \mathfrak{F}$,

2) $A \in \mathfrak{F} \implies A^c = \Omega - A \in \mathfrak{F}$,

3) $A_i \in \mathfrak{F}, i = 1, 2, \cdots \Longrightarrow \cup_{i=1}^{\infty} A_i \in \mathfrak{F}$.

It is easily seen that a $\sigma-$field is closed under the operations of $\cup, \cap, -$ in a countable number of times.

Definition 362 *A set function P defined on a $\sigma-$field $\mathfrak{F}$ is called a probability measure, or simply a probability, if*
1) $P(A) \geq 0, \forall A \in \mathfrak{F}$,
2) $P(\phi) = 0$, *where ϕ is the empty set,*
3) $A_i \in \mathfrak{F}, A_i \cap A_j = \phi, \forall i \neq j; i, j = 1, 2, \cdots$
$\Longrightarrow P(\cup_{i=1}^{\infty} A_i) = \sum_{i=1}^{\infty} P(A_i)$,
4) $P(\Omega) = 1$.

Usually, a set function P with properties 1) - 3) is called a measure. Now we can introduce the concept of a random variable.

Definition 363 *A real - valued function ξ defined on a probability space $(\Omega, \mathfrak{F}, P)$ is called a random variable, if for any $c \in R^1$,*
$\{\omega \in \Omega : \xi(\omega) \leq c\} \in \mathfrak{F}$.

So, actually, a random variable is a measurable function defined on a probability space. Recall that in real analysis if two Lebesgue - measurable functions f and g are equal except at a set with a zero Lebesgue measure, then we will consider them to be the same function and show this by writing
$f(x) = g(x)$, *a.e.*
Such a concept is also introduced into probability theory.

Definition 364 *If ξ and η are two random variables on the same probability space $(\Omega, \mathfrak{F}, P)$, such that they equal except at a set with a zero probability, then we will consider them to be the same random variable, and show this by writing*
$\xi(\omega) = \eta(\omega)$, $P -$ *a.s. Or, simply,* $\xi = \eta$, *a.s.*

Here "a.s." means "almost sure". Similar to real analysis we may also define the integral of a random variable as follows: (Recall that a random variable is a measurable function defined on a probability measure space). If $\xi \geq 0$ is a random variable defined on a probability space $(\Omega, \mathfrak{F}, P)$, we will write
$E\xi = \int_{\Omega} \xi(\omega) P(d\omega) = \int_{\Omega} \xi(\omega) dP(\omega) = \int_{\Omega} \xi dP$
$= \lim_{n \to \infty} \sum_{i=1}^{n2^n} \frac{i-1}{2^n} P(\{\omega : \xi(\omega) \in [\frac{i-1}{2^n}, \frac{i}{2^n})\})$.
For a general random variable $\xi = \xi^+ - \xi^-$ we will write
$E\xi = E\xi^+ - E\xi^-$,
if the right side makes sense.

Definition 365 *$E\xi$ defined above is called the mathematical expectation (or, simply, the expectation) of ξ and $D(\xi) = E|\xi - E\xi|^2$ is called the variance of ξ.*

By definition $E\xi$ may not exist for a general random variable ξ. However, if $E\xi^+ < \infty$ or $E\xi^- < \infty$, then $E\xi$ exists. By definition one also sees that $E\xi$ is simply the mean value of the random variable ξ. So to call $E\xi$ the expectation of ξ is quite natural. (Sometimes we also call $E\xi$ the mean of ξ). Obviously, $D(\xi)$ describes the mean square error or difference between ξ and its expectation $E\xi$.

In real analysis the Fourier transform is a powerful tool when simplifying and treating many kinds of mathematical problem. Such idea can also be introduced into Probability Theory, leading to the so-called a characteristic function of a random variable ξ.

Definition 366 *If ξ is a random variable, $\varphi(t), t \in R^1$ defined by $\varphi(t) = Ee^{it\xi}$, where i is the imaginary unit number, is called the characteristic function of ξ.*

The charateristic function is a powerful tool when simplifying and treating many probability problems. Here, we can use it to distinguish different specific useful random variables.

Example 367 *ξ is called a Normal (distributed) random variable, or a Guassian random variable, if and only if its characteristic function $\varphi(t) = e^{imt-\frac{1}{2}\sigma^2 t}$. In this case*
$E\xi = m, E\left|\xi - E\xi\right|^2 = \sigma^2.$
That is, m and σ^2 are the expectation and variance of ξ, respectively.

One can show that if ξ is such a random variable that $\forall x \in R^1$,
$P(\xi \leq x) = \frac{1}{\sqrt{2\pi}\sigma}\int_{-\infty}^{x} e^{-\frac{|x-m|^2}{2\sigma^2}}dx,$
then $\varphi(t) = Ee^{it\xi} = e^{imt-\frac{1}{2}\sigma^2 t}$. So ξ is a Normal random variable. Moreover, by the definition of expectation and by the definition of derivative one can show that
$E\xi = \int_\Omega \xi dP = \int_{-\infty}^{\infty} xf(x)dx$, and
$\frac{d}{dx}P(\xi \leq x) = f(x),$
where $f(x) = \frac{1}{\sqrt{2\pi}\sigma}e^{-\frac{|x-m|^2}{2\sigma^2}}$. Naturally, we call $f(x) = \frac{1}{\sqrt{2\pi}\sigma}e^{-\frac{|x-m|^2}{2\sigma^2}}$ the probability density function of the Normal random variable ξ. So this (probability) density function also completely describes a Normal random variable. Furthermore, we also call the function denoted by
$F(x) = P(\xi \leq x), \forall x \in R^1,$
the (probability) distribution function of a random variable ξ. Obviously, any (probability) distribution function $F(x)$ has the following properties: 1) $F(-\infty) = \lim_{x\to-\infty} F(x) = 0, F(\infty) = \lim_{x\to\infty} F(x) = 1$; 2) $F(x)$ is an increasing and right continuous function on $x \in R^1$. Conversely, one can show[40],[78] that if $F(x)$ is such a function satisfying the above two properties, then there exists a probability space $(\Omega, \mathfrak{F}, P)$ and a random variable ξ defined on it such that $F(x) = P(\xi \leq x), \forall x \in R^1$. That is,

$F(x)$ must be a distribution function of some random variable ξ. So the distribution function completely describes a random variable.

Now let us discuss the $n-$dimensional case.

For a random vector $\xi = (\xi_1, \cdots, \xi_n)$, where each component $\xi_i, i = 1, \cdots, n$ is a real-valued random variable, we may also introduce a $n-$variable characteristic function $\varphi(t)$ as follows:

Definition 368 $\varphi(t) = \varphi(t_1, \cdots, t_n) = Ee^{i\langle t,\xi\rangle} = Ee^{i\sum_{k=1}^{n} t_k\xi_k}$ *is called the characteristic function of* ξ.

Now, we can use the characteristic function to define the independence of two random variables.

Definition 369 *We say that two real random variables* ξ *and* η *are independent, if they satisfy that*

$\varphi_{\xi,\eta}(t) = \varphi_{\xi,\eta}(t_1, t_2) = \varphi_\xi(t_1)\varphi_\eta(t_2), \forall t = (t_1, t_2) \in R^2,$

where $\varphi_{\xi,\eta}(t_1, t_2) = Ee^{i(t_1\xi + t_2\eta)}, \varphi_\xi(t_1) = Ee^{it_1\xi}$, *etc.*

or, equivalently, $\forall A, B \in \mathfrak{B}(R^1)$,

$P(\xi \in A, \eta \in B) = P(\xi \in A)P(\eta \in B).$

In similar fashion we can define the independence of the three random variables ξ, ζ and η. If n random variables $\xi_1, \cdots, \xi_n$ are independent, then we will say that $\{\xi_k\}_{k=1}^n$ is an independent system. The intuitive meaning of independence for two random variables is that they can take values independently in probability. One easily sees that if $\{\xi_k\}_{k=1}^n$ is an independent system, then each pair of random variables is independent, i.e. all (ξ_i, ξ_j) form an independent system, $i \neq j$; $i, j = 1, \cdots, n$. One should pay attention to the fact that, the inverse statement is not necessary true.[126] However, if $\{\xi_k\}_{k=1}^n$ is a Gaussian system, then it will certainly hold true. See the next section. This is why the Gaussian system is quite useful and easy to treat in practical cases, because it is much easier to check whether each pair of random variables forms an independent system, than to check the whole $n-$dimensional system. The amount of calculation work will be considerably saved. (Calculations on involving $2 \times 2-$ matrices are much easier than calculation on involving an $n \times n-$matrix).

A.2 Gaussian Vectors and Poisson Random Variables

Naturally, we generalize the concept of a real normal random variable to an $n-$dimensional normal random variable (also called a Gaussian vector) as follows:

Definition 370 *An* $n-$*dimensional random variable* $\xi^* = (\xi_1, \cdots, \xi_n)$ *is called a Gaunssian vector if it its characteristic function is*

$\varphi(t_1, \cdots, t_n) = E[e^{it\cdot\xi}] = e^{im\cdot t - \frac{1}{2}t^* \sum_n t}$,
where $t^* = (t_1, \cdots, t_n), m^* = (m_1, \cdots, m_n), \sum_n = [\sigma_{ij}]_{i,j=1}^n$, *and*
$m_i = E\xi_i, \sigma_{ij} = E[(\xi_i - m_i) \cdot (\xi_j - m_j)]$.
(Here $t^* = t^T$ *means the transpose of* t*, etc.).*

Usually, m and $\sum_n$ are called the mean vector and covariance matrix of the Gaussian vector ξ, respectively. As in the 1−dimensional normal random variable case one can show that if $\xi^* = (\xi_1, \cdots, \xi_n)$ is such a random vector that $\forall(x_1, \cdots, x_n) \in R^n$,
$P(\xi_1 \leq x_1, \cdots, \xi_n \leq x_n)$
$= \frac{1}{(\sqrt{2\pi})^n \left|\det \sum_n\right|^{1/2}} \int_{-\infty}^{x_1} \cdots \int_{-\infty}^{x_n} e^{-\frac{1}{2}(x-m)^*(\sum_n)^{-1}(x-m)} dx_n \cdots dx_1$,
then $\varphi(t) = Ee^{i\langle t,\xi\rangle} = e^{im\cdot t - \frac{1}{2}t^* \sum_n t}$. So ξ is a normal random vector. Moreover, by the definition of the expectation one can also show that $\forall i = 1, \cdots, n$,
$E\xi_i = \int_\Omega \xi_i dP = \int_{-\infty}^{\infty} \cdots \int_{-\infty}^{\infty} x_i f(x_1, \cdots, x_n) dx_n \cdots dx_1 = m_i$,
where $f(x_1, \cdots, x_n) = \frac{1}{(\sqrt{2\pi})^n \left|\det \sum_n\right|^{1/2}} e^{-\frac{1}{2}(x-m)^*(\sum_n)^{-1}(x-m)}$ is the so-called (n−dimensional) probability density function of the normal random vector ξ. So this (probability) density function also completely describes a normal random vector. Now as in the 1−dimensional case let us call the function denoted by
$F(x_1, \cdots, x_n) = P(\xi_1 \leq x_1, \cdots, \xi_n \leq x_n), \forall x = (x_1, \cdots, x_n) \in R^n$,
the n−dimensional (probability) distribution function of a random vector ξ. Obviously, any n−dimensional (probability) distribution function $F(x_1, \cdots, x_n)$ has the following properties:
1) $F(-\infty, x_2, \cdots, x_n) = \lim_{x_1 \to -\infty} F(x_1, \cdots, x_n) = 0$,
$\cdots\cdots\cdots, F(x_1, x_2, \cdots, -\infty) = \lim_{x_n \to -\infty} F(x_1, \cdots, x_n) = 0$;
$F(\infty, \cdots, \infty) = \lim_{x_1 \to \infty, \cdots, x_n \to \infty} F(x_1, \cdots, x_n) = 1$;
2) $F(x_1, \cdots, x_n)$ is increasing and right continuous in each $x_i \in R^1, \forall i = 1, \cdots, n$;
3) for $\forall(x_1, \cdots, x_n), (y_1, \cdots, y_n) \in R^n$
$x_1 \leq y_1, \cdots, x_n \leq y_n \Longrightarrow \triangle(x_1, \cdots, x_n; y_1, \cdots, y_n) =$
$= F(y_1, \cdots, y_n) - \sum_{i=1}^n F(y_1, \cdots, y_{i-1}, x_i, y_i, \cdots, y_n)$
$+ \sum_{i_1,i_2=1, i_1<i_2}^n F(y_1, \cdots, y_{i_1-1}, x_{i_1}, y_{i_1}, \cdots, y_{i_2-1}, x_{i_2}, y_{i_2}, \cdots, y_n)$
$+ \cdots\cdots\cdots + (-1)^n F(x_1, \cdots, x_n) \geq 0$.
(For the property 3) one sees that
$\triangle(x_1, \cdots, x_n; y_1, \cdots, y_n) = P(x_i \leq \xi_i \leq y_i, 1 \leq i \leq n) \geq 0$).

Conversely, one can show[40],[78] that if $F(x_1, \cdots, x_n)$ is such a function satisfying the above three properties, then there exists a probability space $(\Omega, \mathfrak{F}, P)$ and a random vector $\xi^* = (\xi_1, \cdots, \xi_n)$ defined on it such that
$F(x_1, \cdots, x_n) = P(\xi_1 \leq x_1, \cdots, \xi_n \leq x_n), \forall x = (x_1, \cdots, x_n) \in R^n$.
That is, $F(x_1, \cdots, x_n)$ must be a distribution function of some random vector ξ. So the distribution function completely describes a random vector.

Now let us discuss some nice properties of the Gaussian vector.

Proposition 371 *If*
$\xi^* = (\xi_1, \cdots, \xi_{n_1}, \xi_{n_1+1}, \cdots, \xi_{n_2}, \cdots, \xi_{n_{k-1}+1}, \cdots, \xi_n)$ *is a* $n-$*dimensional Gaussian vector, let*
$\eta_i = \sum_{j=n_{i-1}+1}^{n_i} c_j \xi_j,\ j = 1, \cdots, k,$
where we set $n_0 = 0$, *and let all of the* $c_j, i = 1, \cdots, n$ *be constants, then* $\eta^* = (\eta_1, \cdots, \eta_k)$ *is a* $k-$*dimensional Gaussian vector.*

Proof. Obviously, the characteristic function of η is
$\varphi(t_1, \cdots, t_k) = E[e^{t \cdot \eta}] = e^{i\overline{m} \cdot t - \frac{1}{2} t^* \overline{\sum}_k t},$
where $t^* = (t_1, \cdots, t_k), \overline{m}^* = (\overline{m}_1, \cdots, \overline{m}_k), \sum_k = [\overline{\sigma}_{ij}]_{i,j=1}^k$, and
$\overline{m}_i = E\eta_i = \sum_{j=n_{i-1}+1}^{n_i} c_j m_j,$
$\overline{\sigma}_{ij} = E[(\eta_i - \overline{m}_i) \cdot (\eta_j - \overline{m}_j)] = \sum_{j=n_{i-1}+1}^{n_i} \sum_{\overline{j}=n_{j-1}+1}^{n_j} c_j c_{\overline{j}} \sigma_{i\overline{j}}.$
The conclusion is now follows. ∎

Proposition 372 *If* $\xi^* = (\xi_1, \cdots, \xi_n)$ *is a Gaussian vector, then the following statements are equivalent:*
1) $\{\xi_i\}_{i=1}^n$ *is an independent system;*
2) each two random variables ξ_i *and* ξ_j $(i \neq j)$ *are independent,* $i \neq j$; $i, j = 1, \cdots, n$;
3) each two random variables ξ_i *and* ξ_j $(i \neq j)$ *are uncorrelated, that is,*
$E[(\xi_i - m_i) \cdot (\xi_j - m_j)] = 0,\ i \neq j;\ i, j = 1, \cdots, n.$

Proof. Since $\xi^* = (\xi_1, \cdots, \xi_n)$ is a Gaunssian vector,
$\{\xi_i\}_{i=1}^n$ is an independent system
$\iff \varphi(t_1, \cdots, t_k) = e^{im \cdot t - \frac{1}{2} t^* \sum_n t} = \prod_{i=1}^n \varphi_i(t_i) = \prod_{i=1}^n e^{im_i t_i - \frac{1}{2}\sigma_{ii} t_i^2}$
$\iff \sigma_{ij} = 0,\ i \neq j;\ i, j = 1, \cdots, n.$ ∎

Proposition 373 *If* ξ *is a* $1-$*dimensional Gaunssian variable with the mean* $m = E\xi$, *and the variance* $\sigma^2 = E[|\xi - m|^2]$, *then*
$E[|\xi - m|^{2k+1}] = 0,$
$E[|\xi - m|^{2k}] = \sigma^2 (2k-1)(2k-3) \cdots 3 \cdot 1.$

Proof. Notice that the characteristic function of ξ is
$\varphi(t) = Ee^{it\xi} = e^{imt - \frac{1}{2}\sigma^2 t^2},$
and the probability density function of ξ is $f(x) = \frac{1}{\sqrt{2\pi}\sigma} e^{-\frac{|x-m|^2}{2\sigma^2}}$. Hence
$E[|\xi - m|^{2k+1}] = \frac{1}{\sqrt{2\pi}\sigma} \int_{-\infty}^{\infty} |x - m|^{2k+1} e^{-\frac{|x-m|^2}{2\sigma^2}} dx = 0,$
and from the integration by parts
$I_{2k} = E[|\xi - m|^{2k}] = \frac{1}{\sqrt{2\pi}} \int_{-\infty}^{\infty} |x|^{2k} e^{-\frac{|x|^2}{2}} dx = \frac{1}{\sqrt{2\pi}} \int_{-\infty}^{\infty} x^{2k-1} e^{-\frac{|x|^2}{2}} d(\frac{x^2}{2})$
$= (2k-1) I_{2k-2} = \cdots = \sigma^2 (2k-1)(2k-3) \cdots 3 \cdot 1.$ ∎
From the above discussion one sees how powerful and useful the characteristic function technique can be, when proving some probabilistic statements! This also happens as in real analysis by using the Fourier transform to treat some problems. Now let us look at the Poisson random variables.

Definition 374 *A* $1-$*dimensional random variable* ξ *only taking values on the set* $\{0, 1, 2, 3, \cdots\}$ *is called a Poisson variable if its characteristic function is*
$\varphi(t) = E[e^{it\xi}] = e^{\lambda(e^{it}-1)}$,
where $\lambda > 0$ *is its intensity such that* $E\xi = \lambda$.

One easily sees that for the above Poisson random variable its probability distribution is
$P(\xi = n) = e^{-\lambda}\frac{\lambda^n}{n!}$,
and $E\xi = \lambda, D(\xi) = E(|\xi - E\xi|^2) = \lambda$.

A.3 Conditional Mathematical Expectation and its Properties

Up to now we have reviewed the elementary theory of Probability needed in this book. Now we will enter a higher level of Probability theory. The first important concept is the conditional mathematical expectation of ξ under a given $\sigma-$algebra $\mathcal{G}$, where $\mathcal{G} \subset \mathfrak{F}$.

Definition 375 *Suppose that* $\mathcal{G} \subset \mathfrak{F}$ *is a* $\sigma-$*algebra,* ξ *is a random variable. If there exists a* $\mathcal{G}-$*measurable function denoted by* $E[\xi|\mathcal{G}]$ *such that* $\forall A \in \mathcal{G}$
$\int_A E[\xi|\mathcal{G}]dP = \int_A \xi dP$,
then $E[\xi|\mathcal{G}]$ *is called the conditional expectation of* ξ *given* $\mathcal{G}$.

If one takes $\mathcal{G} = \{\Omega, A, A^c, \phi\}$, where $A \in \mathfrak{F}$ is any fixed set with $P(A) > 0$, then one easily sees that $E[\xi|\mathcal{G}](\omega) = \frac{E\xi I_A}{P(A)}$, as $\omega \in A$; and $E[\xi|\mathcal{G}](\omega) = \frac{E\xi I_{A^c}}{P(A^c)}$, as $\omega \in A^c$. (To sees notice that $E[\xi|\mathcal{G}]$ is $\mathcal{G}-$measurable. So one should have that $E[\xi|\mathcal{G}](\omega) = c_1 I_A(\omega) + c_2 I_{A^c}(\omega)$, where c_1 and c_1 are constants only depending on the given ξ. On the other hand, by $c_1 P(A) = \int_A E[\xi|\mathcal{G}]dP = \int_A \xi dP$ one sees that $c_1 = \frac{E\xi I_A}{P(A)}$. Similarly, one has that $c_2 = \frac{E\xi I_{A^c}}{P(A^c)}$. From which the conclusion follows). This tells us that if one takes the conditional expectation of ξ given the condition that A has happened (that is, $\omega \in A$), then one should consider $(A, A \cap \mathfrak{F}, P(\cdot)/P(A))$ as a new probability space, and take the expectation of ξ in this new probability space, that equals $\int_A \xi dP/P(A) = \frac{E\xi I_A}{P(A)}$.

The intuitive meaning of the conditional expectation $E[\xi|\mathcal{G}]$ is the expectation of ξ taken under the given information $\mathcal{G}$. Notice that by definition the conditional expectation is usually a random variable, unlike the expectation which is a deterministic value. This also can be explained by intuition. For example, a person at the present time s has wealth x_s for a game, and at the future time t he will have the wealth x_t for this game. In this game the expected money for this person at the future time t is naturally expressed as $E[x_t|\mathfrak{F}_s]$, where $\mathfrak{F}_s$ means the information up to time s, which is known by the gambler. Naturally, $E[x_t|\mathfrak{F}_s]$ is a random variable for any $t > s$.

The important thing is that the conditional expectation has many useful properties which we will use a lot in this book, so we must discuss it.

Existence and Properties of the Conditional Expectation

Now let us use the Radon-Nikodyn theorem to show the existence of the conditional expectation. Assume that ξ is a random variable such that $E|\xi| < \infty$, and $\mathcal{G} \subset \mathfrak{F}$ is a $\sigma-$algebra. Write $\mu(A) = \int_A \xi dP, \forall A \in \mathcal{G}$. Then obviously, $\mu << P$ on $\mathcal{G}$. (μ is absolutely continuous with respect to P on $\mathcal{G}$, or, equivalently, $P(A) = 0 \implies \mu(A) = 0, \forall A \in \mathcal{G}$). So, by the Radon-Nikodyn theorem[28], there exists a $\mathcal{G}-$measurable function denoted by $\frac{d\mu}{dP} = E[\xi|\mathcal{G}]$ such that $\mu(A) = \int_A \frac{d\mu}{dP} dP = \int_A E[\xi|\mathcal{G}]dP, \forall A \in \mathcal{G}$.

The following important properties of the conditional expectation are frequently used:

Assume that $\xi, \eta, \xi_n, n = 1, 2, \cdots$ are random variables such that $E|\xi| < \infty, E|\eta| < \infty, E|\xi_n| < \infty, n = 1, 2, \cdots$ and $\mathcal{G}_1 \subset \mathfrak{F}, \mathcal{G} \subset \mathfrak{F}$ are all $\sigma-$algebras. (All properties that hold below are $P - a.s.$ true).

1) (Linear property). For any constants c_1 and c_2
$E[c_1\xi + c_2\eta|\mathcal{G}] = c_1 E[\xi|\mathcal{G}] + c_2 E[\eta|\mathcal{G}]$.

2) (Monotone property). If $\xi \leq \eta$, then $E[\xi|\mathcal{G}] \leq E[\eta|\mathcal{G}]$.

3) (Measurable property). If ξ is $\mathcal{G}-$measurable, then $E[\xi\eta|\mathcal{G}] = \xi E[\eta|\mathcal{G}]$.

4) (Independent property). If ξ is independent of $\mathcal{G}$, (that is, $\forall A \in \mathfrak{B}(R^1), \forall B \in \mathcal{G}, P(\xi \in A, B) = P(\xi \in A)P(B))$), then $E[\xi|\mathcal{G}] = E[\xi]$.

5) (Expectation property). $E(E[\xi|\mathcal{G}]) = E[\xi]$.

6) ("Snake swallows elephant" property). If $\mathcal{G}_1 \subset \mathcal{G}$, then $E(E[\xi|\mathcal{G}]|\mathcal{G}_1) = E(E[\xi|\mathcal{G}_1]|\mathcal{G}) = E[\xi|\mathcal{G}_1]$.

7) (Lebesgue's dominated convergence property). If $\xi_n \to \xi$, $P - a.s.$, (that is, except on a $P-$zero set Λ, one has that $\forall \omega \notin \Lambda, \xi_n(\omega) \to \xi(\omega)$), and there exists a random variable $g \geq 0$ with $Eg < \infty$ such that $|\xi_n| \leq g$, then $E[\xi_n|\mathcal{G}] \to E[\xi|\mathcal{G}]$, $P - a.s.$

8) (Fatou's lemma property). If $0 \leq \xi_n \uparrow \xi$, $P - a.s.$, then $0 \leq E[\xi_n|\mathcal{G}] \uparrow E[\xi|\mathcal{G}]$, $P - a.s.$

If we take $\mathcal{G} = \{\Omega, \phi\}$, then by 4) $E[\xi_n] = E[\xi_n|\mathcal{G}]$ and $E[\xi] = E[\xi|\mathcal{G}]$. So from 7) and 8) we obtain the usual Lebesgue's dominated convergence theorem and Fatou's lemma for the expectation. Be careful to take note that for an expectation (i.e. for an integral) the following statement is true:[28]

If $\xi_n \to \xi$, $P - a.s.$, and $\{\xi_n\}_{n=1}^{\infty}$ is uniformly integrable with respect to the probability measure P, (that is, $\lim_{N\to\infty} \sup_n E|\xi_n| I_{|\xi_n|>N} = 0$), then $E\xi_n \to E\xi$.

However, under the same condition it is not true[198] that $E[\xi_n|\mathcal{G}] \to E[\xi|\mathcal{G}], P - a.s.$!

As examples of what technique to use when proving the properties of conditional expectations let us show 6) and 7). For 6) we have that $\forall A \in \mathcal{G}_1$,

$\int_A E[\xi|\mathcal{G}_1]dP = \int_A \xi dP = \int_A E[\xi|\mathcal{G}]dP = \int_A E(E[\xi|\mathcal{G}]|\mathcal{G}_1)dP.$
So $E[\xi|\mathcal{G}_1] = E(E[\xi|\mathcal{G}]|\mathcal{G}_1), P - a.s.$ Simlarly, one also sees that $E[\xi|\mathcal{G}_1] = E(E[\xi|\mathcal{G}_1]|\mathcal{G}), P - a.s.$ The property 6) holds true. Now for 7) we have that $\forall A \in \mathcal{G}$,

$\int_A E[\xi|\mathcal{G}]dP = \int_A E[\lim_{n\to\infty} \xi_n|\mathcal{G}]dP = \int_A \lim_{n\to\infty} \xi_n dP$
$= \lim_{n\to\infty} \int_A \xi_n dP = \lim_{n\to\infty} \int_A E[\xi_n|\mathcal{G}]dP = \int_A \lim_{n\to\infty} E[\xi_n|\mathcal{G}]dP,$
where we have applied the Lebesgue's dominated convergence theorem to the expectation, because that $|\xi_n| \leq g, Eg < \infty$, and $|E[\xi_n|\mathcal{G}]| \leq E[g|\mathcal{G}], EE[g|\mathcal{G}] < \infty$. Thus $E[\xi|\mathcal{G}]dP = \lim_{n\to\infty} E[\xi_n|\mathcal{G}], P - a.s.$ The property 7) holds true.

The other properties can be proved in completely the same way. Let us give one more property of the conditional expectation which is used a lot in the discussion of the filtering problem.

9) (Integral property). If $\alpha_t(\omega) = \alpha(t,\omega)$ is jointly measurable in $([0,T] \times \Omega, \mathfrak{B}([0,T] \times \mathfrak{F})$, with $\int_0^T E|\alpha_t| dt < \infty$ and $\mathcal{G} \subset \mathfrak{F}$ is a $\sigma-$algebra, then

$E(\int_0^t \alpha_s ds\, |\mathcal{G}) = \int_0^t E(\alpha_s\, |\mathcal{G})\, ds,\ P - a.s.$ for all $t \in [0,T]$.

This conclusion is easily derived from the definition of the conditional expectation and Fubini's theorem. Indeed, $\forall A \in \mathcal{G}$

$E[I_A E(\int_0^t \alpha_s ds\, |\mathcal{G})] = E(I_A \int_0^t \alpha_s ds) = \int_0^t E(I_A \alpha_s)ds$
$= \int_0^t E(I_A E(\alpha_s\, |\mathcal{G}))ds = E[I_A \int_0^t E(\alpha_s\, |\mathcal{G}))ds].$
So the conclusion follows.

10) (Jensen's inequality).[25],[95]Suppose that $\mathcal{G} \subset \mathfrak{F}$ is a $\sigma-$algebra, ξ is a random variable such that $E|\xi| < \infty$. If $f(x), x \in R^1$ is a convex function such that $E|f(\xi)| < \infty$, then

$f(E(\xi|\mathcal{G})) \leq E(f(\xi)|\mathcal{G}).$

A.4 Random Processes and the Kolmogorov Theorem

In nature we meet a lot of "random processes" $x_t(\omega)$: like the temporature in some area, the wind speed over the sea, the population of fish in some large pond, etc. All of them have much the same features: at each time t_0, $x_{t_0}(\omega)$ is a random variable; after each observation, that is for each ω_0, $x_t(\omega_0)$ can be recorded as a function or a trajectory of changing time t. These are exactly the concepts of random processes.

Definition 376 *We call a family of random variables $\{x_t(\omega)\}_{t\geq 0}$ a random process, i.e. for each $t \geq 0$, x_t is a random variable. If the $x_t(\omega)$ are jointly measurable with respect to $\mathfrak{B}([0,\infty)) \times \mathfrak{F}$, then $\{x_t(\omega)\}_{t\geq 0}$ is called a measurable random process.*

As in the case of Gaussian vector, we can have a Gaussian process as follows:

Definition 377 *A random process* $\{x_t\}_{t\geq 0}$ *is called a Gaussian process, if for each* $0 \leq t_1 < \cdots < t_n$, $(x_{t_1}, \cdots, x_{t_n})$ *is a Gaussian vector.*

We can also have a Poisson process as follows:

Definition 378 *A random process* $\{N(t)\}_{t\geq 0}$ *is called a Poisson process, if for each* t, $N(t)$ *is a Poisson random variable (r.v.) with intensity* λt, *and it has independent increments; that is, for any*
$0 \leq t_1 < t_2 < \cdots < t_n$
$\{N(t_2) - N(t_1), \cdots, N(t_n) - N(t_{n-1})\}$ *is an independent system.*

Obviously, the trajectory of a Poisson r.v. $N(t, \omega)$, as ω is fixed, is RCLL (right continuous with left limit), and it is a step function with the jump equal to 1 at each jump time.

Similar to the random vector we want to introduce some distribution functions to describe a random process $\{x_t\}_{t\geq 0}$. Naturally, for arbitrary $(t_1, t_2, \cdots, t_n) \in R^n_+$, where R^n_+ is the set of all points in R^n with all non-negative components, we will have a finite - dimensional (probability) distribution function $F_{t_1,\cdots,t_n}(x_1, \cdots, x_n)$ from the random vector $(x_{t_1}, \cdots, x_{t_n})$. Thus we will have a family of all finite - dimensional distribution functions $\{F_{t_1,\cdots,t_n}(x_1, \cdots, x_n), \forall (t_1, t_2, \cdots, t_n) \in R^n_+, \forall n\}$ from a random process $\{x_t\}_{t\geq 0}$. Obviously, this family has the following properties:

1) (Symmetric property). Exchanging any positions of the indexes, the values of the distribution functions do not change. For example, if we exchange the positions of the indexes i and k, then
$F_{t_1,\cdots,t_{i-1},t_i,t_{i+1},\cdots,t_{k-1},t_k,t_{k+12},\cdots t_n}(x_1, \cdots, x_{i-1}, x_i, x_{i+1}, \cdots,$
$x_{k-1}, x_k, x_{k+1}, \cdots, x_n) = F_{t_1,\cdots,t_{i-1},t_k,t_{i+1},\cdots,t_{k-1},t_i,t_{k+12},\cdots t_n}(x_1, \cdots,$
$x_{i-1}, x_k, x_{i+1}, \cdots, x_{k-1}, x_i, x_{k+1}, \cdots, x_n).$

2) (Consistent property). Restricting the higher dimensional distribution function to a lower dimensional space will cause it to coincide with the lower dimensional distribution function. For example,
$F_{t_1,\cdots,t_k,\cdots,t_n}(x_1, \cdots, x_k, \infty, \cdots, \infty) = F_{t_1,\cdots,t_k}(x_1, \cdots, x_k).$

Conversely, one can show that the following Kolmogorov's theorem holds.

Theorem 379 [40],[78] *Suppose that*
$\{F_{t_1,\cdots,t_n}(x_1, \cdots, x_n), \forall (t_1, t_2, \cdots, t_n) \in R^n_+, \forall n\}$
is a family of finite - dimensional distribution functions. If it satisfies the Symmetric property and Consistent property stated above, then there exists a probability sapce $(\Omega, \mathfrak{F}, P)$ *and a random process* $\{x_t\}_{t\geq 0}$ *defined on it such that* $\{F_{t_1,\cdots,t_n}(x_1, \cdots, x_n), \forall (t_1, t_2, \cdots, t_n) \in R^n_+, \forall n\}$ *is just the family of finite - dimensional distribution functions from* $\{x_t\}_{t\geq 0}$.

So the Symmetric and Consistent family of finite - dimensional distribution functions completely describes a random process.

An important special kind of random processes is useful for us in this book.

Definition 380 *A real random process* $\{x_t\}_{t\geq 0}$ *is called a stationary Markov process, if for any* $A \in \mathfrak{B}(R^1)$*, that is* A *is a Borel set in* R^1*,* $\forall s \leq t$ *(*$\forall \sigma \leq \tau$*)*
$P(x_t \in A|x_u, u \leq s) = P(x_{t-s} \in A|\mathfrak{x}_s)$,
where $P(\widetilde{A}|x_u, u \leq s) = E[I_{\widetilde{A}}|\sigma(x_u, u \leq s)]$ *is the so-called conditional probability.*

The physical meaning of a stationary Markov process $\{x_t\}_{t\geq 0}$ is as follows: 1) It is stationary; that is, starting from any time, its probability distributions remain the same; or say, its probability distributions do not depend on the starting time. More precisely, $\forall s \geq 0, t_1, , \cdots, t_n \geq 0, \forall A_1, A_2, \cdots, A_n \in \mathfrak{B}(R^1)$,
$P(x_{t_1} \in A_1, \cdots, x_{t_n} \in A_n) = P(x_{t_1+s} \in A_1, \cdots, x_{t_n+s} \in A_n)$.
2) If its "present" is known, then its "future" behavior does not depend on its "past" behavior.

A stationary strong Markov process can be similarly defined, if we substitute the $\forall s \leq t$ by $\forall \sigma \leq \tau$, where σ and τ are two $\mathfrak{F}_t^x$–stopping times. Here $\mathfrak{F}_t^x = \sigma(x_u, u \leq t)$. (For the definition of stopping times see Definition 5).

Appendix B

Space D and Skorohod's Metric

Write

$D = D([0,\infty), R^d)$,

where $D([0,\infty), R^d)$ is the space of RCLL functions $X = (X_t)_{t\geq 0}$,with values in R^d. Let us introduce a metric in this space. We will first define a metric d in space $D([0,1]) = D([0,1]), R)$. Recall that for space $C([0,1]) = C([0,1]), R)$–the totality of continuous functions $X = (X_t)_{t\in[0,1]}$,with values in R– it is well known that the metric between two element $X, Y \in C([0,1])$ is defined by $\max_{t\in[0,1]} |x_t - Y_t|$, under this metric the space $C([0,1])$ becomes a complete metric space, and the convergence in this space is uniform convergence.

However, for the space $D([0,1])$ such metric is not suitable, because now elements have discontinuous points. In particular if we want to discuss how $X^n \in D([0,1])$ converges to $X \in D([0,1])$, where the discontinuous points of X^n also approximate the discontinuous points of X, then such convergence usually cannot be uniform. For example, $X^n(t) = I_{[\frac{1}{2}-\frac{1}{n},1]}(t) \downarrow I_{[\frac{1}{2},1]}(t) = X(t)$, for each $t \in [0,1]$. But

$\sup_{t\in[0,1]} |X^n(t) - X(t)| = 1 \nrightarrow 0$.

This is because of the discontinuous point t of $X^n(t)$ also moves. (Notice that by Dini's theorem, if $X^n, X \in C([0,1])$, and $X^n(t) \downarrow X(t)$, $\forall\, t \in [0,1]$, then $\sup_{t\in[0,1]} |X^n(t) - X(t)| \to 0$). So we need to introduce a metric that allows t to change a little between two elements from $D([0,1])$.

Skorohod introduce the following metric: for $X, Y \in D([0,1])$

$$d(X,Y) = \inf_{\lambda\in\Lambda} \left\{ \sup_{0\leq t\leq 1} \left| X_t - Y_{\lambda(t)} \right| + \sup_{0\leq t\leq 1} |\lambda(t) - t| \right\}, \tag{B.1}$$

where

$$\Lambda = \left\{ \begin{array}{l} \lambda = \lambda(t) : \lambda \text{ is strictly increasing, continuous} \\ \quad \text{on } t \in [0,1], \text{ such that } \lambda(0) = 0, \lambda(1) = 1 \end{array} \right\}.$$

One can verify that $d(X, Y)$ is a metric in $D([0, 1])$. To show this let us first establish a lemma which indicates that, similar to the continuous function case, the elements in $D([0, 1])$ can also be approximated by step functions uniformly. Such a result is very useful. Moreover, the proof of this lemma also gives us with a nice property of the RCLL functions.

Lemma 381 [13] *For each $X \in D([0,1])$ there exist step functions $X^n \in D$ such that*

$sup_{s\in[0,1]} |X_s - X_s^n| \to 0,$ *as* $n \to \infty$.

Proof. In fact, since X is RCLL, there exists a point $\widetilde{t}_1 > 0$ such that $\sup_{s\in[0,\widetilde{t}_1)} |X_s - X_0| < \varepsilon/2$. So $\sup_{s,t\in[0,\widetilde{t}_1)} |X_s - X_t| < \varepsilon$. Write $t_1 = \sup_{\widetilde{t}_1 \leq 1} \{\widetilde{t}_1\}$. Then

$\sup_{s,t\in[0,t_1)} |X_s - X_t| \leq \varepsilon$.

Again starting from t_1 there exists a maximum point $t_2 > t_1$ such that $\sup_{s,t\in[t_1,t_2)} |X_s - X_t| \leq \varepsilon$. By induction there exists a maximum point $t_n > t_{n-1}$ such that $\sup_{s,t\in[t_{n-1},t_n)} |X_s - X_t| \leq \varepsilon$. Now such a procedure should stop at finitely many steps. Otherwise, let $T = \sup_n t_n$. As X_{T-} exists, one easily finds a $t_n' > t_n$ such that $\sup_{s,t\in[t_{n-1},t_n')} |X_s - X_t| \leq \varepsilon$. This is a contradiction with the definition of t_n. Obviously, we should have $T = 1$. So we have proved that there exist finite many points $0 = t_0 < t_1 < \cdots < t_{n_\varepsilon} = 1$ such that

$$sup_{s,t\in[t_{n-1},t_n)} |X_s - X_t| \leq \varepsilon, n = 1, 2, \cdots, n_\varepsilon. \tag{B.2}$$

Now let $X^\varepsilon(t) = \sum_{n=1}^{n_\varepsilon} I_{[t_{n-1},t_n)}(t) X_{t_{n-1}} + I_{\{1\}} X_1$. Then $X^\varepsilon \in D([0,1])$ is a step function and $\sup_{s\in[0,1]} |X_s - X_s^\varepsilon| \leq \varepsilon$. ■

By (B.2) we immediately obtain the following corollary.

Corollary 382 *If $X \in D([0,1])$, then $\forall \varepsilon > 0$ there exist finitely many points $0 = t_0 < t_1 < \cdots < t_{n_\varepsilon} = 1$ such that*

$sup_{s,t\in[t_{n-1},t_n)} |X_s - X_t| \leq \varepsilon, n = 1, 2, \cdots, n_\varepsilon.$

By means of this corollary one immediately finds the following facts: If $X \in D([0,1])$, then

1) $sup_{t\in[0,1]} |X_t| < \infty$;

2) $\forall \varepsilon > 0$, there exist only finitely many points $0 = t_0 < t_1 < \cdots < t_{n_\varepsilon} = 1$ such that $\triangle X_{t_i} > \varepsilon$, $i = 1, 2, \cdots, n_\varepsilon$. Thus, X only has at most countably many discontinuous points.

In fact, by Corollary 382 taking $\varepsilon = 1$, one sees that

$\sup_{t\in[0,1]} |X_t| \leq \sum_{n=1}^{n_1} (|X_{t_{n-1}}| + |X_{t_{n-1}} - X_t|) I_{t\in[t_{n-1},t_n)}$
$\leq n_1 + \sum_{n=1}^{n_1} |X_{t_{n-1}}| < \infty.$

Furthermore, by Corollary 382 for any given $\varepsilon > 0$ it only can be true that at the points $0 = t_0 < t_1 < \cdots < t_{n_\varepsilon} = 1$, $|\triangle X_{t_i}| > \varepsilon$, $i = 1, 2, \cdots, n_\varepsilon$. Finally,

$$\{|\triangle X_t| > 0\} = \{t \in [0,1] : |\triangle X_t| > 0\} = \cup_{m=1}^\infty \{|\triangle X_t| > \tfrac{1}{m}\}.$$

So X only has at most countably many discontinuous points. Thus, facts 1) and 2) hold true. These two facts mean that any RCLL function defined on [0,1] is bounded, and for any given $\varepsilon > 0$ it has only finitely many jump points where its jumps can exceed ε, and so it can only have at most countably many discontinuous points.

Now let us show the following lemma.

Lemma 383 *$d(X,Y)$ defined by (B.1) is a metric in $D([0,1])$.*

Proof. First, take $\lambda(t) = t, \forall t \in [0,1]$. then

$$0 \leq d(X,Y) \leq \sup_{t\in[0,1]} |X_t - Y_t| < \infty,$$

because $X - Y \in D([0,1])$, and any function in $D([0,1])$ is bounded. Secondly, let us show that $d(X,Y)$ satisfies the three properties for metric. 1) Notice that $\lambda \in \Lambda \iff \lambda^{-1} \in \Lambda$. For each $s_0 \in [0,1]$ write $t_0 = \lambda(s_0)$. Then

$$s_0 - \lambda(s_0) = \lambda^{-1}(t_0) - t_0, \text{ and } X_{s_0} - Y_{\lambda(s_0)} = X_{\lambda^{-1}(t_0)} - Y_{t_0}.$$

So

$$\sup_{0\leq t\leq 1} |\lambda(t) - t| = \sup_{0\leq t\leq 1} \left|t - \lambda^{-1}(t)\right|, \text{ and}$$
$$\sup_{0\leq t\leq 1} \left|X_{\lambda^{-1}(t)} - Y_t\right| = \sup_{0\leq t\leq 1} \left|X_t - Y_{\lambda(t)}\right|.$$

This shows that $d(Y,X) = d(Y,X)$. 2) If $X = Y$, obviously, $d(X,Y) = 0$. Conversely, if $d(X,Y) = 0$, then for each n there exists a $\lambda^n \in \Lambda$ such that

$$\sup_{0\leq t\leq 1} \left|X_t - Y_{\lambda^n(t)}\right| + \sup_{0\leq t\leq 1} |\lambda^n(t) - t| < \tfrac{1}{n}.$$

So as $n \to \infty, \lambda^n(t) \to t$, uniformly for $0 \leq t \leq 1$. Now for any given $t \in [0,1]$ if there exists a subsequence $\{\lambda^{n_k}(t)\}_{k=1}^\infty$ of $\{\lambda^n(t)\}_{n=1}^\infty$ such that $\lambda^{n_k}(t) \downarrow t$ as $k \uparrow \infty$, then $X_t = Y_t$. (In this case t may or may not be the discontinuous point of X). Otherwise, $X_t = Y_{t-}$. From this one sees that if t is a continuous point of Y, then we always have $X_t = Y_{t-} = Y_t$. However, we can show that if $X_t = Y_{t-}$, then t cannot be a discontinuous point of Y. In fact, for any $\varepsilon > 0$ there exists a sequence of $\{t_n\}$ such that all $\{t_n\}$ are continuous points of Y and $|Y_{t_n} - Y_{t-}| \leq \varepsilon, t_n \uparrow t,$ because there are only finitely many points s such that $|\triangle Y_s| > \varepsilon$, and Y can only have countably many discontinuous points. So

$$|X_{t_n} - Y_{t-}| \leq \varepsilon, t_n \uparrow t, \text{ and } |X_{t-} - Y_{t-}| \leq \varepsilon.$$

This shows that $X_{t-} = Y_{t-} = X_t$. By symmetry one also has that $Y_{t-} = X_{t-} = Y_t$. So t is a continuous point of Y, and we complete the proof that $d(X,Y) = 0 \implies X_t = Y_t, \forall t \in [0,1]$. 3) Notice that if $\lambda_1, \lambda_2 \in \Lambda$, then their composition $\lambda_1 \circ \lambda_2 \in \Lambda$. By this one sees that for any $\lambda_1, \lambda_2 \in \Lambda$; $X, Y, Z \in D([0,1])$,

$$\sup_{t\in[0,1]} |\lambda_1 \circ \lambda_2(t) - t| \leq \sup_{t\in[0,1]} |\lambda_1 \circ \lambda_2(t) - \lambda_2(t)|$$
$$+ \sup_{t\in[0,1]} |\lambda_2(t) - t| = \sup_{t\in[0,1]} |\lambda_1(t) - t| + \sup_{t\in[0,1]} |\lambda_2(t) - t|,$$

and

$\sup_{t\in[0,1]} \left|X_{\lambda_1\circ\lambda_2(t)} - Y_t\right| \leq \sup_{t\in[0,1]} \left|X_{\lambda_1\circ\lambda_2(t)} - Z_{\lambda_2(t)}\right|$
$+\sup_{t\in[0,1]} \left|Z_{\lambda_2(t)} - Y_t\right| = \sup_{t\in[0,1]} \left|X_{\lambda_1(t)} - Z_t\right| + \sup_{t\in[0,1]} \left|Z_{\lambda_2(t)} - Y_t\right|.$
So, if we take the infimum among $\lambda_1 \in \Lambda$ first, then take the infimum among $\lambda_2 \in \Lambda$ secondly, we arrive at the result

$d(X,Y) \leq d(X,Z) + d(Z,Y).$

Thus $d(X,Y)$ is a metric. ■

Now one sees that under metric d

$X^n = I_{[\frac{1}{2}-\frac{1}{n},1]} \to I_{[\frac{1}{2},1]} = X,$ in $D([0,1])$.

In fact, one easily finds that $d(X^n, X) = \frac{1}{n} \to 0$.

However, one should notice that this convergence is not equivalent to pointwise convergence; that is, for $X^n, X \in D([0,1])$,

$d(X^n, X) \to 0 \not\Leftrightarrow X^n(t) \to X(t), \forall\, t \in [0,1].$

For example, write $X^n(t) = t^n, \forall t \in [0,1]$; and $X(t) = 0$, as $t \in [0,1)$; $X(t) = 1$, as $t = 1$. Then $X^n(t) \to X(t)$, $\forall\, t \in [0,1]$. However, $d(X^n, X) = 1 \not\to 0$. So

$X^n(t) \to X(t), \forall\, t \in [0,1] \not\Rightarrow d(X^n, X) \to 0.$

Such a situation happens here because $X^n \in C([0,1])$ is continuous for each n, but its limit $X \in D([0,1])$ has a discontinuous point.

Another example is $X^n = I_{[\frac{1}{2}+\frac{1}{n},1]}, X = I_{[\frac{1}{2},1]}$. Then $X^n, X \in D([0,1])$, and $d(X^n, X) = 0 \to 0$. However, $X^n(\frac{1}{2}) = 0 \not\to X(\frac{1}{2}) = 1$. So

$d(X^n, X) \to 0 \not\Rightarrow X^n(t) \to X(t), \forall\, t \in [0,1].$

The situation happens here because the jump points move, and at the limit of jump points we change the value of the limit function to fit that $d(X^n, X) = 0$.

Anyway, we have the following lemma, which indicates that $d(X^n, X) \to 0$ is equivalent to some kind of uniformly convergence.

Lemma 384 *For $X^n, X \in D([0,1])$, $d(X^n, X) \to 0$ is equivalent to there existing a sequence $\lambda^n \in \Lambda$ such that as $n \to \infty$,*

$\sup_{t\in[0,1]} |\lambda^n(t) - t| + \sup_{t\in[0,1]} |X^n(\lambda^n(t)) - X(t)| \to 0.$

Proof. In fact, if $d(X^n, X) \to 0$, then for each n there exists a $\lambda^n \in \Lambda$ such that

$\sup_{t\in[0,1]} |\lambda^n(t) - t| + \sup_{t\in[0,1]} |X^n(\lambda^n(t)) - X(t)| < \frac{1}{n}.$

Conversely, if as $n \to \infty$,

$\sup_{t\in[0,1]} |\lambda^n(t) - t| + \sup_{t\in[0,1]} |X^n(\lambda^n(t)) - X(t)| \to 0,$

then $\forall \varepsilon > 0$ there exists a N such that as $n \geq N$

$d(X^n, X) \leq \sup_{t\in[0,1]} |\lambda^n(t) - t| + \sup_{t\in[0,1]} |X^n(\lambda^n(t)) - X(t)| < \varepsilon.$

So as $n \to \infty$, $d(X^n, X) \to 0$. ■

Furthermore, we have that for $C([0,1])$ the convergence in d here is equivalent to uniform convergence. Shown in the following lemma.

Lemma 385 *If $X^n, X \in C([0,1])$, then $d(X^n, X) \to 0$ is equivalent to that* $\sup_{t\in[0,1]} |X^n(t) - X(t)| \to 0.$

Proof. For $X^n, X \in D([0,1])$, taking $\lambda(t) = t$, $t \in [0,1]$, one sees that

$d(X^n, X) \leq \sup_{t\in[0,1]} |X^n(t) - X(t)|,$
so,
$\sup_{t\in[0,1]} |X^n(t) - X(t)| \to 0 \Longrightarrow d(X^n, X) \to 0.$
Conversely, notice that
$|X^n(t) - X(t)| \leq \left|X^n(t) - X((\lambda^n)^{-1}(t))\right| + \left|X((\lambda^n)^{-1}(t)) - X(t)\right|,$
where $\lambda^n \in \Lambda$ are taken from Lemma 384, so if $X^n, X \in C([0,1])$, and $d(X^n, X) \to 0$, then
$\sup_{t\in[0,1]} |X^n(t) - X(t)| \leq \frac{1}{n} + \sup_{t\in[0,1]} \left|X((\lambda^n)^{-1}(t)) - X(t)\right|.$
As $X(t)$ is uniformly continuous on $t \in [0,1]$ and
$\lim_{n\to\infty} \left|(\lambda^n)^{-1}(t) - t\right| = 0,$
so
$\lim_{n\to\infty} \sup_{t\in[0,1]} \left|X((\lambda^n)^{-1}(t)) - X(t)\right| = 0.$
Hence as $n \to \infty$, $\sup_{t\in[0,1]} |X^n(t) - X(t)| \to 0.$ ∎

So, one sees that the Skorohod's metric is a nice generalization of the uniform metric from the space $C([0,1])$ to the space $D([0,1])$. However, this Skorohod's metric is not a complete metric, that is, the space $D([0,1])$ is not a complete metric space under this metric. So it is not so convenient, and we need to introduce an equivalent complete metric

$$d_0(X,Y) = \inf_{\lambda\in\Lambda} \left\{\sup_{0\leq t\leq 1} \left|X_t - Y_{\lambda(t)}\right| + \sup_{0\leq s\leq t\leq 1} \left|\log \frac{\lambda(t)-\lambda(s)}{t-s}\right|\right\}.$$

Lemma 386 [13] *The space $(D([0,1]), d_0)$ is a complete separable metric space. The metric d_0 is equivalent to the above Skorohod's metric.*

Remark 387 *We say that two metrices d_0 and d are equivalent, if the topologies generated by d_0 and d, respectively, are equivalent. In other words, write $S_{d_0}(x,\varepsilon) = \{y \in D([0,1]) : d_0(x,y) < \varepsilon\}$, (that is, the ball with center x and radius ε under the metric d_0), we have that for each given $S_{d_0}(x,\varepsilon)$, there exists a $S_d(x,\varepsilon')$ such that $S_d(x,\varepsilon') \subset S_{d_0}(x,\varepsilon)$. Vice versa. Actually, for the Skorohod's metric one can show that*
$d(x,y) \leq 2d_0(x,y).$
However, the inverse is not true. So the completeness in metric d_0 cannot imply the completenes in metric d.

Now let us introduce a metric to the space $D = D([0,\infty), R^d)$. Let
$D^0([0,1]) = \{X \in D([0,1]) : X_1 = X_{1-}\}.$
Obviously, $(D^0([0,1]), d_0)$ is a closed subspace of $(D([0,1]), d_0)$. Similarly, define
$D^0 = \{X \in D : \lim_{t\to\infty} X_t \text{ exists and finite}\}.$
Let
$\rho^0(X,Y) = d_0(X',Y'),$
where $X'_t = X_{\psi^{-1}(t)}$, $Y'_t = Y_{\psi^{-1}(t)}$,
$\psi(t) = -\log(1-t)$, $0 \leq t < 1$; and $\psi(t) = \infty$, as $t = 1$.
Then obviously, (D^0, ρ^0) is a complete separable metric space. Now let
$X^k_t = X_t \cdot g_k(t),$
where

$g_k(t) = I_{[0,k]}(t) + (k+1-t)I_{(k,k+1]}(t).$
Set

$$\rho(X,Y) = \sum_{k=1}^{\infty} 2^{-k} \frac{\rho^0(X^k,Y^k)}{1+\rho^0(X^k,Y^k)}.$$

Then we have the following

Lemma 388 [74] *(Lindvall, 1973). (D,ρ) is a complete separable metric space. Moreover, the convergence in the metric ρ is equivalent to the existence of functions $\lambda_n \in \Lambda_\infty, n \geq 1$, such that for each $L > 0$*

$$\sup_{0\leq t\leq L} \left| X_t - X^n_{\lambda_n(t)} \right| \to 0, \ \sup_{0\leq t\leq L} |\lambda_n(t) - t| \to 0, \ \textit{as } n \to \infty,$$

where Λ_∞ is the set of continuous strictly increasing functions $\lambda(t)$ with $\lambda(0) = 0$.

Appendix C

Monotone Class Theorems. Convergence of Random Processes

C.1 Monotone Class Theorems

In this Appendix we will present some useful tools used quite frequently in this book. The first one is the so-called: monotone class theorems.

Definition 389 *Let $\widetilde{\Omega}$ be a set, $\mathcal{C}$ be a family of subsets of $\widetilde{\Omega}$.*
1) $\mathcal{C}$ is called a $\pi-$system, if it is closed under finite intersection;
2) $\mathcal{C}$ is called a $\lambda-$system, if
(i) $\widetilde{\Omega} \in \mathcal{C}$,
(ii) $A, B \in \mathcal{C},\ A \subset B \Longrightarrow B \setminus A \in \mathcal{C}$
(iii) $A_n \in \mathcal{C},\ A_n \uparrow A \Longrightarrow A \in \mathcal{C}$.
3) $\mathcal{C}$ is called a monotone system, if it is closed under all monotone limits.

Obviously, a $\lambda-$system is a monotone system. If a $\lambda-$system is also a $\pi-$system, then it is a $\sigma-$field. If a field is also a monotone system, then it is a $\sigma-$field. The following monotone class theorem on a family of sets is useful.

Theorem 390 *1) If $\mathcal{C}$ is a $\pi-$system, denote by $\lambda(\mathcal{C})$ the smallest $\lambda-$system containing $\mathcal{C}$, then $\lambda(\mathcal{C}) = \sigma(\mathcal{C})$.*

2) If $\mathcal{C}$ is a field, denote by $\mathfrak{M}(\mathcal{C})$ the smallest monotone system containing $\mathcal{C}$, then $\mathfrak{M}(\mathcal{C}) = \sigma(\mathcal{C})$.

Proof. 1): Let us show this by using a method we will call the technique of "collecting right sets". First, we note that the intersection of all $\lambda-$systems containing $\mathcal{C}$ is $\lambda(\mathcal{C})$. Now set

$\mathcal{F}_1 = \{B \in \lambda(\mathcal{C}) : \forall A \in \mathcal{C}, B \cap A \in \lambda(\mathcal{C})\}$.
Then $\mathcal{F}_1 \supset \mathcal{C}$, and $\mathcal{F}_1$ is a $\lambda-$system. So $\mathcal{F}_1 = \lambda(\mathcal{C})$. Set
$\mathcal{F}_2 = \{B \in \lambda(\mathcal{C}) : \forall A \in \mathcal{F}_1, B \cap A \in \lambda(\mathcal{C})\}$.
Again $\mathcal{F}_2 \supset \mathcal{C}$, and $\mathcal{F}_2$ is a $\lambda-$system. So $\mathcal{F}_2 = \lambda(\mathcal{C})$. This means that $\lambda(\mathcal{C})$ is a also $\pi-$system. So it is a $\sigma-$field, i.e. $\lambda(\mathcal{C}) \supset \sigma(\mathcal{C})$. However, a $\sigma-$field is also a $\lambda-$system. Therefore $\lambda(\mathcal{C}) \subset \sigma(\mathcal{C})$.

2): Notice that the intersection of all monotone systems containing $\mathcal{C}$ is $\mathfrak{M}(\mathcal{C})$. Similarly as in 1) we find that it is a $\pi-$system. Now set
$\mathcal{F}_3 = \{A \in \mathfrak{M}(\mathcal{C}) : A^c \in \mathfrak{M}(\mathcal{C})\}$.
Since $\mathcal{C}$ is a field, then $\mathcal{C} \subset \mathcal{F}_3$. Moreover, $\mathcal{F}_3$ is obviously a monotone system. So $\mathcal{F}_3 = \mathfrak{M}(\mathcal{C})$. This means that $\mathfrak{M}(\mathcal{C})$ is also a field, since it is a $\pi-$system, and $\mathcal{F}_3$ is closed under the complement operation. So it is a $\sigma-$field, i.e. $\mathfrak{M}(\mathcal{C}) \supset \sigma(\mathcal{C})$. However, a $\sigma-$field is also a monotone system. Therefore $\mathfrak{M}(\mathcal{C}) \subset \sigma(\mathcal{C})$. ∎

The following theorem is the monotone class theorem on a function family.

Theorem 391 *Assume that $\mathcal{H}$ is a linear space of some real functions defined on $\widetilde{\Omega}$, and $\mathcal{C}$ is a $\pi-$system of some subsets in $\widetilde{\Omega}$. If*
(i) $1 \in \mathcal{H}$,
(ii) $f_n \in \mathcal{H}, 0 \leq f_n \uparrow f$, *$f$ is finite (resp. bounded)*$\Longrightarrow f \in \mathcal{H}$;
(iii) $A \in \mathcal{C} \Longrightarrow I_A \in \mathcal{H}$;
then $\mathcal{H}$ contains all $\sigma(\mathcal{C})-$measurable real (resp. bounded) functions defined on $\widetilde{\Omega}$.

Proof. Let $\mathcal{F} = \left\{A \subset \widetilde{\Omega} : I_A \in \mathcal{H}\right\}$. Then by assumption one sees that $\mathcal{F}$ is a $\lambda-$system, and $\mathcal{F} \supset \mathcal{C}$. Hence $\sigma(\mathcal{C}) = \lambda(\mathcal{C}) \subset \mathcal{F}$. Now for any $\sigma(\mathcal{C})-$measurable real (resp. bounded) function ξ let
$\xi_n = \sum_{i=0}^{n2^n} \frac{i}{2^n} I_{\{\xi \in [\frac{i}{2^n}, \frac{i+1}{2^n})\}}$.

Then $\xi_n \in \mathcal{H}$, since $\{\xi \in [\frac{i}{2^n}, \frac{i+1}{2^n})\} \in \sigma(\mathcal{C}) \subset \mathcal{F}$, and $\mathcal{H}$ is a linear space. Obviously, $0 \leq \xi_n \uparrow \xi^+ = \xi \vee 0$. Hence $\xi^+ \in \mathcal{H}$. Similarly, we argue that $\xi^- \in \mathcal{H}$. Therefore $\xi = \xi^+ - \xi^- \in \mathcal{H}$. ∎

Now we formulate the following frequently used a monotone class theorem:

Theorem 392 *If $\mathcal{H}$ is a linear space of real finite (bounded) measurable processes satisfying*
(i) $0 \leq f_n \in \mathcal{H}$, $f_n \uparrow f$, *f is finite (resp. bounded)* $\Longrightarrow f \in \mathcal{H}$;
(ii) $\mathcal{H}$ contains all $\mathfrak{F}_t-$ left continuous processes;
then $\mathcal{H}$ contains all real (resp. bounded) $\mathfrak{F}_t-$ predictable processes.

(For the definition of a predictable process, etc. see section 2 of Chapter 1 in this book).

Remark 393 *If in the condition (ii) we substitute the left - continuous processes by the right - continuous processes, then in the conclusion the* *predictable processes will also be substituted by the optional processes. The proof for this new case is the same.*

Now let us prove Theorem 392.

Proof. Notice that f is $\mathfrak{F}_t-$ predictable, means that, $f \in \mathcal{P}$, where
$\mathcal{P} = \sigma(\mathcal{C})$,
and
$\mathcal{C} = \{\cap_{i=1}^k f_i^{-1}\{(a_i, b_i)\} \subset [0,\infty) \times \Omega : \forall k, \forall a_i < b_i,$
$\forall f_i - \mathfrak{F}_t-$ left continuous process$\}$

Obviously, $\mathcal{C}$ is a $\pi-$system. By Theorem 391 one only needs to show that $\forall A \in \mathcal{C} \Longrightarrow I_A \in \mathcal{H}$. In fact, there exists a sequence of bounded continuous functions $0 \leq \varphi_n^i$ defined on R such that $\varphi_n^i(x) \uparrow I_{(a_i,b_i)}(x)$, as $n \uparrow \infty$. Hence as $n \uparrow \infty$,
$\prod_{i=1}^k \varphi_n^i(f_i(t,\omega)) \uparrow \prod_{i=1}^k I_{(a_i,b_i)}(f_i(t,\omega)) = I_{\cap_{i=1}^k f_i^{-1}\{(a_i,b_i)\}}(t,\omega)$.
Since the left hand side is $\mathfrak{F}_t-$left continuous, so it belongs to $\mathcal{H}$. Now by assumption (i) $I_{\cap_{i=1}^k f_i^{-1}\{(a_i,b_i)\}}(t,\omega) \in \mathcal{H}$. The proof for predictable processes is complete. ■

C.2 Convergence of Random Variables

Now let us discuss the convergence of random variables. Recall that in real analysis we have several different concepts for the convergence of measurable functions: a.e. convergence (almost everywhere convergence); convergence in measure; and weak convergence. Since random variavles are measurable functions defined on a probaility space, all such convergent concepts can be introduced into probability theory.

Definition 394 *A sequence of real random variables ξ_n defined on a probability space is said to be convergent to a real random variable ξ almost surely (a.s.), and we will write:*

$$\xi_n \to \xi, \ a.s., \ as\ n \to \infty, \tag{C.1}$$

if
$P(\omega : \lim_{n\to\infty} |\xi_n(\omega) - \xi(\omega)| = 0) = 1;$
and ξ_n is said to be convergent to ξ in probability P, and we will write:

$$\xi_n \to \xi, \ in\ P, \ as\ n \to \infty, \tag{C.2}$$

if for every $\varepsilon > 0$
$\lim_{n\to\infty} P(\omega : |\xi_n(\omega) - \xi(\omega)| > \varepsilon) = 0;$
and ξ_n is said to be weakly convergent to ξ, and we will write:

$$\xi_n \xrightarrow{w} \xi, a.s., \quad as\ n \to \infty, \tag{C.3}$$

if for arbitrary $f \in C_b(R)$, where $C_b(R)$ is the totality of all real bounded continuous functions defined on R,
$\int f(\xi_n(\omega))dP \to \int f(\xi_n(\omega))dP$, *as $n \to \infty$.*

Naturally, we will have the following relations between these types of convergence.

Proposition 395 *Assume that $\xi_n, n = 1, 2, \cdots$, and ξ are real random variables. Then $(C.1) \implies (C.2) \implies (C.3)$.*

Proof. C.1) $\implies$ C.2): Let
$A = \cap_{m=1}^{\infty} \cup_{N=1}^{\infty} \cap_{n \geq N} \left\{ |\xi_n - \xi| < \frac{1}{m} \right\}$.
If C.1) holds, then $P(A^c) = 0$, where A^c is the complement of A. So
$\lim_{N\to\infty} P(\{|\xi_N - \xi| \geq \frac{1}{m}\}) \leq \lim_{N\to\infty} P(\cup_{n\geq N} \{|\xi_n - \xi| \geq \frac{1}{m}\})$
$= P(\cap_{N=1}^{\infty} \cup_{n\geq N} \{|\xi_n - \xi| \geq \frac{1}{m}\}) \leq P(A^c) = 0$, for each m.
C.2) $\implies$ $C.3$): For any $f \in C_b(R)$, and f is uniformly continuous,
$\left|\int f(\xi_n(\omega))dP \to \int f(\xi(\omega))dP\right| \leq E\left|f(\xi_n) - f(\xi)\right|$
$\leq E\left|f(\xi_n) - f(\xi)\right| I_{|\xi_n(\omega) - \xi(\omega)| < \delta} + k_0 P(|\xi_n(\omega) - \xi(\omega)| \geq \delta)$.
So in this case if C.2) is true, then $C.3$) holds true. Now for any given $f \in C_b(R)$ since f is bounded, we may assume that $0 < f < 1$. Write
$F_i = \left\{x : \frac{i}{k} \leq f(x)\right\}, i = 0, 1, \cdots, k$. Then $F_i, i = 0, 1, \cdots, k$, are closed sets. Write

$$\phi(t) = \begin{cases} 1, & \text{as } t \leq 0, \\ 1 - t, & \text{as } 0 \leq t \leq 1, \\ 0, & \text{as } t \geq 1. \end{cases}$$

We have that for each i, $\rho(x, F_i)$ is uniformly continuous in x, where $\rho(x, F_i) = \inf\{|x - y| : y \in F_i\}$, and as $m \to \infty$, $\phi(m\rho(x, F_i)) \to I_{F_i}(x)$, $\phi(m\rho(x, F_i)) \geq I_{F_i}(x)$. So by the result just proved $\forall m$

$$P(\xi \in F_i) = \int_{\Omega} \phi(m\rho(\xi(\omega), F_i))dP$$

$$= \lim_{n\to\infty} \int \phi(m\rho(\xi_n(\omega), F_i))dP \geq \overline{\lim}_{n\to\infty} P(\xi_n \in F_i). \tag{C.4}$$

Now write
$A_i = \left\{x : \frac{i-1}{k} \leq f(x) < \frac{i}{k}\right\} = F_{i-1} - F_i, , i = 1.2. \cdots, k$.
Then
$\sum_{i=0}^{k-1} \frac{1}{k} P(\xi \in F_i) - \frac{1}{k} = \sum_{i=1}^{k} \frac{i-1}{k} P(\xi \in A_i) \leq \int f(\xi(\omega))dP$
$\leq \sum_{i=1}^{k} \frac{i}{k} P(\xi \in A_i) = \sum_{i=0}^{k-1} \frac{1}{k} P(\xi \in F_i)$.
The same inequality holds for ξ_n. Therefore, by this and by (C.4)
$\overline{\lim}_{n\to\infty} \int f(\xi_n(\omega))dP \leq \overline{\lim}_{n\to\infty} \sum_{i=0}^{k-1} \frac{1}{k} P(\xi_n \in F_i)$
$\leq \sum_{i=0}^{k-1} \frac{1}{k} P(\xi \in F_i) \leq \frac{1}{k} + \int f(\xi(\omega))dP$.
Letting $k \to \infty$, one obtains that $\overline{\lim}_{n\to\infty} \int f(\xi_n(\omega))dP \leq \int f(\xi(\omega))dP$.
Write $\overline{f} = 1 - f$. One easily sees that $\underline{\lim}_{n\to\infty} \int \overline{f}(\xi_n(\omega))dP \geq \int \overline{f}(\xi(\omega))dP$.
Thus $C.3$) is proved. ∎

Using real analysis one easily construct a counter example that C.2) $\nRightarrow$C.1). Let us give an example to show that $C.3) \nRightarrow$ C.1).

Example 396 *Write* $\xi(t) = w(t), \xi_n(t) = -w(t), \forall n = 1, 2, \cdots$; *where* $w(t), t \geq 0$, *is a BM. Then, obviously, 3) holds, and 2) fails.*

The convergence in probability has a nice physical meaning. Since
$\xi_n \to \xi$, in P
is equivalent to the case that $\forall \varepsilon, \delta > 0, \exists N$ such that as $n \geq N$
$P(\omega : |\xi_n(\omega) - \xi(\omega)| > \varepsilon) < \delta.$
So this obviously means that with any given probability $1 - \delta$ (or say possibility $(1 - \delta)\%$) the error between ξ_n and ξ can be less than or equal to any given positive number ε, when n is large enough (that is, $n \geq N$).

In this book we also see that convergence in probability is quite helpful when discussing whether approximation solutions from the approximation SDE converge to the true solution of the true SDE, because such convergence needs weaker conditions. Moreover, it has a nice property that simplifies the discussions. See the following remark.

Remark 397 *Suppose that a sequence of random process* $x_t^0; x_t^1, \cdots, x_t^n, \cdots$, $t \in [0, T]$ *is given, which is uniformly bounded in probability, that is, as* $N \to \infty$, $\sup_{n=0,1,\cdots} P(\sup_{t\in[0,T]} |x_t^n| > N) \to 0$. *Then*

$$P(\left|F_n(x_t^n, t \leq T) - F_0(x_t^0, t \leq T)\right| > \varepsilon)$$
$$\leq \sup_{n=0,1,\cdots} P(\sup_{t\in[0,T]} |x_t^n| > N) + P(\sup_{t\in[0,T]} |x_t^0| > N)$$
$$+P(|F_n - F_0| > \varepsilon, \sup_{t\in[0,T]} |x_t^n| \leq N, \sup_{t\in[0,T]} |x_t^0| \leq N).$$

This obviously means that proving

$$P(\left|F_n(x_t^n, t \leq T) - F_0(x_t^0, t \leq T)\right| > \varepsilon) \to 0$$

is equivalent to proving that

$$P(\left|F_n(x_t^n, t \leq T) - F_0(x_t^0, t \leq T)\right| > \varepsilon) \to 0$$

under the condition that all $\{x_t^n, t \in [0, T]\}_{n=0}^{\infty}$ *are uniformly bounded, that is,* $|x_t^n| \leq k_0, \forall t \in [0, T], \forall n = 0, 1, 2, \cdots$.

Finally, let us point out that $E \sup_{t\leq T} |x_t^n|^2 \leq k_T < \infty, \forall n = 0, 1, 2, \cdots$ *imples that as* $N \to \infty$, $\sup_{n=0,1,\cdots} P(\sup_{t\in[0,T]} |x_t^n| > N) \to 0$.

(In fact, $\sup_{n=0,1,\cdots} P(\sup_{t\in[0,T]} |x_t^n| > N) \leq \frac{k_T}{N}$*).*

C.3 Convergence of Random Processes and Stochastic Integrals

Finally, let us discuss the convergence of random processes and stochastic integrals. First we will quote a Skorohod theorem, which is very useful in the discussion of the convergence of random processes.

Theorem 398 *(Skorohod, 1961)*[184]. *Let* $\{\xi_t^n\}_{t\geq 0}, n = 1,2,\cdots$, *be a sequence of* $d-$*dimensional RCLL random processes satisfying the following two conditions: for each* $T \geq 0, \varepsilon > 0$

$\lim_{N\to\infty} \sup_n \sup_{t\leq T} P\{|\xi_t^n| > N\} = 0,$

$\lim_{h\downarrow 0} \sup_n \sup_{t_1,t_2\leq T,|t_1-t_2|\leq h} P\{|\xi_{t_1}^n - \xi_{t_2}^n| > \varepsilon\} = 0.$

Then there exists a subsequence $\{n_k\}$ *of* $\{n\}$, *a probability space* $(\widetilde{\Omega}, \widetilde{\mathfrak{F}}, \widetilde{P})$, *(actually,* $\widetilde{\Omega} = [0,1], \widetilde{\mathfrak{F}} = \mathfrak{B}([0,1])$*), and* $d-$*dimensional RCLL random processes* $\left\{\widetilde{\xi}_t^{n_k}\right\}_{t\geq 0}, k = 1,2,\cdots$, *and* $\left\{\widetilde{\xi}_t\right\}_{t\geq 0}$ *defined on it such that*

1) all finite-dimensional distributions of $\left\{\widetilde{\xi}_t^{n_k}\right\}_{t\geq 0}$ *coincide with the finite-dimensional distributions of* $\{\xi_t^{n_k}\}_{t\geq 0}, k = 1,2,\cdots$;

2) $\widetilde{\xi}_t^{n_k} \to \widetilde{\xi}_t$, *in probability* $\widetilde{P}$, *as* $k \to \infty$, $\forall t \geq 0$.

Here we give the following lemma, and quote two more lemmas that will be needed for the discussion in the book. To save space we will not give the proofs of the last two lemmas. (For their proofs see [184]).

Lemma 399 *Let* $\{x_t^n, w_t^n, \zeta_t^n\}_{t\geq 0}, n = 1,2,\cdots$, *be a sequence of* $d \times r \times \bar{d}-$*dimensional RCLL random processes satisfying the two conditions stated in Theorem 398.*

1) If, in addition, $\{w_t^n\}_{t\geq 0}$ *are BMs,* $n = 1,2,\cdots$, *then the* $\{\widetilde{w}_t^{n_k}\}_{t\geq 0}$ *obtained from Theorem 398 also are BMs on* $(\widetilde{\Omega}, \widetilde{\mathfrak{F}}, \widetilde{P}), k = 1,2,\cdots$ *and so also is the limit* $\{\widetilde{w}_t\}_{t\geq 0}$.

2) If, in addition, $\{\zeta_t^n\}_{t\geq 0}, n = 1,2,\cdots$, *are random processes with independent increments, then* $\left\{\widetilde{\zeta}_t^{n_k}\right\}_{t\geq 0}$ *obtained from Theorem 398 also are random processes with independent increments on* $(\widetilde{\Omega}, \widetilde{\mathfrak{F}}, \widetilde{P}), k = 1,2,\cdots$ *and so also is the limit* $\left\{\widetilde{\zeta}_t\right\}_{t\geq 0}$. *Write*

$p^n(dt,dz) = \sum_{s\in dt} I_{\Delta\zeta_s^n\neq 0} I_{\Delta\zeta_s^n\in dz},$

$\widetilde{p}^{n_k}(dt,dz) = \sum_{s\in dt} I_{\Delta\widetilde{\zeta}_s^{n_k}\neq 0} I_{\Delta\widetilde{\zeta}_s^{n_k}\in dz},$

$\widetilde{p}(dt,dz) = \sum_{s\in dt} I_{\Delta\widetilde{\zeta}_s\neq 0} I_{\Delta\widetilde{\zeta}_s\in dz}$

If, besides, all $p^n(dt,dz)$ *are Poisson random measures with the same compensator* $\frac{dz}{|z|^{\bar{d}+1}}$ *such that* $\forall n = 1,2,\cdots$

$q^n(dt,dz) = p^n(dt,dz) - \frac{dz}{|z|^{\bar{d}+1}},$

where $q^n(dt,dz), n = 1,2,\cdots$, *are Poisson martingale measures, then*

$\widetilde{p}^{n_k}(dt,dz), k = 1,2,\cdots,$

also are Poisson random measures with the same compensator $\frac{dz}{|z|^{\bar{d}+1}}$ *on* $(\widetilde{\Omega}, \widetilde{\mathfrak{F}}, \widetilde{P})$ *such that* $\forall k = 1,2,\cdots$

$\widetilde{q}^{n_k}(dt,dz) = \widetilde{p}^{n_k}(dt,dz) - \frac{dz}{|z|^{\bar{d}+1}},$

where $\widetilde{q}^{n_k}(dt,dz)$ *are Poisson martingale measures on* $(\widetilde{\Omega}, \widetilde{\mathfrak{F}}, \widetilde{P})$*; and the same conclusions hold true for the limits* $\widetilde{p}(dt,dz)$ *and* $\widetilde{q}(dt,dz)$.

Proof. 1). Write $\mathfrak{F}_t^n = \sigma(x_s^n, w_s^n, \zeta_s^n; s \leq t)$, $\widetilde{\mathfrak{F}}_t^n = \sigma(\widetilde{x}_s^n, \widetilde{w}_s^n, \widetilde{\zeta}_s^n; s \leq t)$, and $\widetilde{\mathfrak{F}}_t = \sigma(\widetilde{x}_s, \widetilde{w}_s, \widetilde{\zeta}_s; s \leq t)$. Since by assumption for each n, $\{w_t^n, \mathfrak{F}_t^n\}_{t \geq 0}$ is a BM, so $w_{t_2}^n - w_{t_1}^n$, $t \leq t_1 < t_2$, is independent of $\mathfrak{F}_t^n$, and $w_t^n - w_s^n$, $0 \leq s < t$, is Guassian $N(0, t-s)$. However, $\{\widetilde{w}_t^{n_k}\}_{t \geq 0}$ has the same finite-distributions as that of $\left\{w_{tt \geq 0}^{n_k}\right\}_{t \geq 0}$. So $\widetilde{w}_{t_2}^{n_k} - \widetilde{w}_{t_1}^{n_k}$, $t \leq t_1 < t_2$, is also independent of $\widetilde{\mathfrak{F}}_t^{nK}$, and $\widetilde{w}_t^{n_k} - \widetilde{w}_s^{n_k}$, $0 \leq s < t$, is also Guassian $N(0, t-s)$; that is, $\left\{\widetilde{w}_t^{n_k}, \widetilde{\mathfrak{F}}_t^{nK}\right\}_{t \geq 0}$ also is a BM. Notice that $\widetilde{w}_t^{n_k} \to \widetilde{w}_t$, in probability $\widetilde{P}$, as $k \to \infty$. Hence the same conclusion holds true for $\left\{\widetilde{w}_t, \widetilde{\mathfrak{F}}_t\right\}_{t \geq 0}$.

2) is proved in a similar fashion. ∎

Lemma 400 [184] *Assume that* $p^n(dt, dz)$ $n = 1, 2, \ldots$,*are the Poisson counting measures with the same compensator* $\pi(dz)dt = \frac{dz}{|z|^{d+1}} dt$ *such that*

$q^n(dt, dz) = p^n(dt, dz) - \pi(dz)dt,\ n = 1, 2, \ldots.$

Set

$\zeta_t^n = \int_0^t \int_{|z| \leq 1} z q^n(ds, dz) + \int_0^t \int_{|z| > 1} z p^n(ds, dz), n = 1, 2, \ldots.$

If as $n \to \infty$

$\zeta_t^n \to \zeta_t^0$, *in probability, for all* $t \geq 0$,

then

1) for any finite Borel measurable function $f(z)$ *and for any* $\varepsilon > 0$ *as* $n \to \infty$

$\int_0^t \int_{|z| > \varepsilon} f(z) p^n(ds, dz) \to \int_0^t \int_{|z| > \varepsilon} f(z) p^0(ds, dz)$, *in probability,*

for all $t \geq 0$, *where*

$\widetilde{p}^0(dt, dz) = \sum_{s \in dt} I_{\triangle \widetilde{\zeta}_s^0 \neq 0} I_{\triangle \widetilde{\zeta}_s^0 \in dz}$

is the Poisson counting measure with the same compensator $\frac{dz}{|z|^{d+1}} dt$;

2) $\zeta_t^0 = \int_0^t \int_{|z| \leq 1} z q^0(ds, dz) + \int_0^t \int_{|z| > 1} z p^0(ds, dz)$,

where

$q^0(dt, dz) = p^0(dt, dz) - \frac{dz}{|z|^{d+1}} dt.$;

3) if $0 \leq t \leq t + h$, *and* $x_t^n \to x_t$, *in probability, as* $n \to \infty$, *such that* x_t^n, x_t *are finite, then for any bounded jointly continuous function* $f(s, x, z)$ *as* $n \to \infty$,

$\int_t^{t+h} \int_{|z| > \varepsilon} f(t, x_t^n, z) p^n(ds, dz) \to \int_t^{t+h} \int_{|z| > \varepsilon} f(t, x_t, z) p^0(ds, dz)$,

in probability.

Lemma 401 [184] *Assume that* $f_n(t)$ *satisfies the following conditions: for any* $0 \leq T < \infty$

1° $lim_{N \to \infty} \sup_n P(\sup_{t \leq T} |f_n(t)| > N) = 0$

2° *for any* $\varepsilon > 0$

$lim_{\delta \to 0} \sup_n \sup_{|t-s| < \delta} P(|f_n(t) - f_n(s)| > \varepsilon) = 0.$

Moreover, if as $n \to \infty$

$f_n(t) \to f(t)$, *in probability,*

where the stochastic integrals $\int_0^T f_n(t)dw_n(t)$ *and* $\int_0^T f(t)dw(t)$ *exist,* $w_n(t)$ *and* $w(t)$ *are* $r-$*dimensional BM's,* $0 \leq T < \infty$ *is arbitrary given, and as* $n \to \infty$, $w_n(t) \to w(t)$, *in probability, then as* $n \to \infty$
$\int_0^T f_n(t)dw_n(t) \to \int_0^T f(t)dw(t)$, *in probability.*

References

[1] Adams, R. A. (1975). *Sobolev spaces.* Pure and Applied Mathematics, Vol. 65. Academic Press, New York-London, 1975.

[2] Arnold, L. (1974). *Stochastic Differential Equations: Theory and Applications.* Wiley, New York.

[3] Anulova, S.V. (1978). On process with Levy generating operator in a half-space. *Izv. AH. SSSR ser. Math.* 47:4, 708-750. (In Russian).

[4] Anulova, S.V. (1982). On stochastic differential equations with boundary conditions in a half-space. *Math. USSR-Izv.* 18, 423-437.

[5] Anulova, S.S. and Pragarauskas, H. (1977). On strong Markov weak solutions of stochastic equations. *Lit. Math. Sb.* XVII:2, 5-26. (In Russian).

[6] Anulova, A., Veretennikov, A., Krylov, N.. Liptser, R. and Shiryaev, A. (1998). *Stochastic Calculus.* Encyclopedia of Mathemetical Sciences. Vol 45, Springer-Verlag.

[7] Aase K.K. (1988). Contingent claims valuation when the security price is a combination of an Ito process and a random point process. *Stochastic Process. Appl.,* 28: 185-220.

[8] Bardhan I. and Chao X. (1993). Pricing options on securities with discontinuous returns. *Stochastic Process. Appl.,* 48: 123-137.

[9] Barles G., Buckdahn R. and Pardoux E. (1997). Backward stochastic differential equations and integral-partial differential equations. *Stochastics and Stochastics Reports, 60: 57-83*

[10] Bensoussan A. (1982). Lectures on stochastic control. In: Mittler S.K. & Moro A. eds. *Nonlinear Filtering and Stochastic Control. LNM, 972*: 1-62.

[11] Bensoussan A. and Lions J.L. (1984). *Impulse Control and Quasi-Variational Inequalities.* Gautheir-Villars.

[12] Bismut J. M. (1973). Theorie probabiiste du controle des diffusions. *Mem. Amer. Math. Soc., 176: 1-30.*

[13] Billingsleys, P. (1968). *Convergence of Probability Measures.* John Wiley and Sons, New York.

[14] Black, F. and Scholes, M. (1971). The pricing of options and corporate liabilities. *J. Polit. Econom.* 8, 637-869.

[15] Chen S. and Yu X. (1999). Linear quadratic optimal control: from deterministic to stochastic cases. In: Chen S., Li X., Yong J., Zhou X.Y. eds. *Control of Distributed Parameter and Stochastic Systems.* Kluwer Acad. Pub., Boston, 181-188.

[16] Chen, Z. J. and Peng, S. G. (2000). A general downcrossing inequality for $g-$martingales. *Statistics & Prob. Letters, 46, 169-175.*

[17] Chen, Z.J. and Wang, B. (2000). Infinite time interval BSDEs and the convergence of $g-$martingales. *J. Austral. Math. Soc. (Ser A) 69, 187-211.*

[18] Chen Z. J. (1997). Existence of solution to backward stochastic differential equation with stopping times. *Chin. Bulletin*, 42(2): 2379-2382.

[19] Chow G. C. (1997). *Dynamic Economics: Optimization by The Lagrange Method*, NewYork: Oxford Univ. Press.

[20] Cox J. C., Ingersoll Jr. J. E., and Ross S. A. (1985). An International general equilibrium model of asset prices. *Econometrica*, 53(2): 363-384.

[21] Curtain R. F. and Pritchard A. J. (1977). *Functional Analysis in Modern Applied Mathematics.* New York: Academic Press.

[22] Cvitanic J. and Karatzas I. (1993). Hedging contingent claims with constrained portfolios. *Annals of Applied Probab.*, 3: 652-681.

[23] Cvitanic J. and Karatzas I. (1996). Backward SDE's with reflection and Dynkin games. Ann. Probab. 24(4): 2021-2056.

[24] Darling R. & Pardoux E. (1997). Backward SDE with random time and applications to semi-linear elliptic PDE. *Annal. Prob.*, 25: 1135-1159.

[25] Dellacherie, C. and Meyer,, P.A. (1980). *Probabilités et Potentiel. Théorie des Martingales.* Hermann, Paris.

[26] Doob, J.L. (1953). *Stochastic Processes.* John Willey & Sons, Inc.

[27] Doob, J.L. (1983). *Classical Potential Theory and Its Probabilistic Counterpart.* Springer-Verlag.

[28] Dunford, N. and Schwartz, J. T. (1958). *Linear Operators - Part I: General Theory.* Interscience Publishers, INC., New York.

[29] Durrett, R. (1996). *Stochastic Calculus : A Practical Introduction.* CRC Press.

[30] El Karoui N., Kapoudjian C., Pardoux E., Peng S. and Quenez M.C. (1997). Reflected solutions of backward SDE's and related obstacle problems for PDE's. *The Annals of Probab.*, 25(2): 702-737.

[31] El Karoui N. and Mazliak L. (eds.) (1997). *Backward Stochastic Differential Equations.* Addison Wesley Longman Inc. USA.

[32] El Karoui N., Pardoux E., and Quenez M.C. (1997). Reflected backward SDEs and American options. In: Rogers L.C.G. & Talay D. eds. *Numerical Methods in Finance.* Pub. Newton Inst., Cambridge Univ. Press, 215-231.

[33] El Karoui N., Peng S., and Quenez M.C. (1997). Backward stochastic differential equations in finance. *Math. Finance,* 7, 1-71.

[34] Elliott, R.J. (1982). *Stochastic Calculus and Application.* Springer-Verlag. N.Y.

[35] Friedman, A. (1975). *Stochastic Differential Equations and Applications. I.* Academic Press.

[36] Friedman, A. (1976). *Stochastic Differential Equations and Applications. II.* Academic Press.

[37] Elliott, R.J. (1982). *Stochastic Calculus and Application.* Springer-Verlag. N.Y.

[38] Gihman, I.I. and Skorohod, A.V. (1979). *The Theory of Stochastic Processes III*, Springer-Verlag, Berlin.

[39] Gikhman, I. I .; Skorokhod, A. V. (1982). *Stochastic differential equations and their applications.* "Naukova Dumka", Kiev, 612 pp. (In Russian).

[40] Halmos, P.R. (1950). *Measure Theory.* Van Nostrand, New York.

[41] Hamadène S. (1998). Reflected BSDEs and mixed game Problem. *Stoch. Proc. Appl., 85(2): 177-188.*

[42] Hamadène S. and Lepeltier J.P. (2000). Backward - forward SDE's and stochastic differential games. *Stoch. Proc. Appl., 77: 1-15.*

[43] He S.W., Wang J.G. and Yan J.A. (1992). *Semimartingale Theory and Stochastic Calculus.* Science Press & CRC Press Inc.

[44] Hida, T. (1980). *Brownian motion.* Springer-Verlag.

[45] Hu, Y. and Peng, S. (1990). Maximum principle for semilinear stochastic evolution control systems. *Stochastics and Stochastic Reports, 33: 159-180.*

[46] Hu, Y. and Peng, S. (1991). Adapted solution of a backward semilinear stochastic evolution equation. *Stochastic Analysis and Applications,* 9(4): 445-459.

[47] Hu Y. and Peng S. (1995). Solution of forward-backward stochastic differential equations. *Probab. Theory Relat. Fields,* 103: 273-283.

[48] Hu Y. and Yong J. (2000). Forward-backward stochastic differential equations with nonsmooth coefficients. *Stoch. Proc. Appl. 87: 93-106.*

[49] Ikeda N. and Watanabe S. (1989). *Stochastic Differential Equations and Diffusion Processes.* North-Holland.

[50] Ito, K. (1942). Differential equations determining Markov processes. *Zenkoku Shijo Sugako Danwakai,* 244:1077, 1352-1400. (In Japanese).

[51] Ito, K. (1944). Stochastic integral. *Proc. Imp. Acad. Tokyo,* 20, 519-524.

[52] Ito, K. (1951). *On Stochastic differential Equations.* Mem. Amer. Math. Soc., 4.

[53] Ito, K. (1951). On a formula concerning stochastic differentials. *Nagoya Math. J. 3,* 55-65.

[54] Ito, K. (1972). Poisson point process attached to Markov processes. *Proc. Sixth Berkeley Symp. Math. Statist. Prob. III,* 225-239. Univ. California Press, Berkeley.

[55] Jacobson D.H., Martin D.H., Pachter M. and Geveci T. (1980). *Extensions of Linear-Quadratic Control Theory.* LNCIS 27, Springer-Verlag.

[56] Jacod, J. (1979). Calcul Stochastique et Problemes de Martingales. *L.N.Math. 714*.

[57] Kallianpur, G. (1980). *Stochastic Filtering Theory.* Springer-Verlag.

[58] Karatzas I., Lehoczky J. P., Shreve S. E. and Xu G. L. (1991). Martingale and duality methods for utility maximization in incomplete market. *SIAM J. Control Optim.,* 29(3): 702-730.

[59] Karatzas I. and Shreve S.E. (1987). *Brownian Motion and Stochastic Calculus.* Springer-Verlag.

[60] Karatzas I. and Shreve S.E. (1998). *Methods of Mathematical Finance.* Springer-Verlag.

[61] Klebaner, F. C. (1998). *Introduction to stochastic calculus with applications.* Imperial College Press, London.

[62] Kohlmann M. (1999). Reflected forward backward stochastic differential equations and contingent claims. In Chen S., Li X., Yong J., Zhou X.Y. eds. *Control of Distributed Parameter and Stochastic Systems.* Kluwer Acad. Pub., Boston, 223-230.

[63] Krylov, N. V. (1974). Some estimates on the distribution densities of stochastic integrals. *Izv. A.N. USSR, Math. Ser., 38:1,* 228-248. (In Russian).

[64] Krylov N.V. (1980). *Controlled Diffusion Processes.* Springer-Verlag.

[65] Krylov N. V. and Paragarauskas, H. (1980). On the Bellman equations for uniformly non-degenerate general stochastic processes. *Liet. Matem. Rink.*, XX: 85-98. (In Russian).

[66] Kunita, H. and Watanabe, S. (1969). On square integrable martingales. *Nagoya Math. J.* 30, 209-245.

[67] Kunita H. (1990). *Stochastic Flows and Stochastic Differential Equations.* Cambridge Univ. Press, World Publ. Corp.

[68] Ladyzenskaja O. A., Solonnikov V. A. and Uralceva N. N. (1968). *Linear and Quasilinear Equations of Parabolic Type.* Translation of Monographs 23, AMS Providence, Rode Island.

[69] Lamberton, D. and Lapeyre, B. (1996). *Introduction to stochastic calculus applied to finance.* Chapman & Hall, New York.

[70] Le Gall, J.F. (1983). Applications du temps local aux equations differentielles stochastiques unidimensionnelles. *Lect. Notes Math. 986,* Springer-Verlag, 15-31.

[71] Le Gall, J.F. (1984). One- dimensional stochastic differential equations involving the local time of the unknown processes. *Lect. Notes Math. 1095*, Springer-Verlag.

[72] Lepeltier J.P. and San Martin J. (1997). Backward stochastic differential equations with continuous coefficient. *Statistics & Probab. Letters,* 32: 425-430.

[73] Li, C. W., Dong, Z. and Situ, R. (Situ Rong). (2002). Almost sure stability of linear stochastic differential equations with jumps. *Probab. Theory Related Fields* 123(1), 121-155.

[74] Lindvall, T. (1973). Weak convergence of probability measures and random functions in the function space $D[0,\infty)$. *Appl.Probab.,* 10, 109-121.

[75] Liptser, R.S. and Shiryaev. (1989). *Theory of Martingales.* Kluwer Academic Publishers. Boston.

[76] Liptser, R.S. and Shiryaev. (2001). *Statistics of Random Processes, I, II.* 2nd ed. Springer-Verlag. N.Y.

[77] Li Xiaojun. (1993). Maximum principle of stochastic evolution equation. *Chin. Ann. Math.* 14A(6): 647-653.

[78] Loeve, M. (1977). *Probability Theory.* Springer-Verlag, New York.

[79] Ma J., Protter P. and Yong J. (1994). Solving forward-backward stochastic differential equations explicitly - a four step scheme. *Probab. Theory Relat. Fields,* 98: 339-359.

[80] Ma J. and Yong J. (1995). Solvability of forward-backward SDEs and the nodal set of Hamilton-Jacobi-Bellman equations. *Chinese Ann. of Math. 16*B(3): 279-298.

[81] Ma J. and Yong J. (1997). Adapted solution of a degenerate backward SPDE with applications. *Stoch. Proc. Appl., 70: 59-84.*

[82] Ma J. and Yong J. (1999). On linear degenerate backward stochastic partial differential equations. *Prob. Th. & Relat. Fields, 113(2): 135-170.*

[83] Ma J. and Yong J. (1999). *Forward - Backward Stochastic Differential Equations and their Applications.* LNM 1702, Springer-Verlag.

[84] Malliavin, P. and Thalmaier, A. (2003). *Stochastic calculus of variations in mathematical finance. Springer-Verlag.*

[85] Mao X. (1995). Adapted solutions of backward stochastic differential equations with non-Lipschitz coefficients. *Stoch. Proc. Appl., 58: 281-292.*

[86] Marhno S. (1976). The first boundary problem for integral-differential equations. *Theory of Stochastic Processes* 4, Kiev, 66-72. (In Russian).

[87] Mckean, H. (1969). *Stochastic Integrals.* Academic, New York.

[88] McShane, E. J. (1974). *Stochastic calculus and stochastic models.* Academic Press, New York.

[89] Menaldi, J.L. and Robins, M. (1985). Reflected diffusion processes with jumps. *The Annals of Probab.* 13:2, 319-342.

[90] Merton R. (1971). Optimal consumption and portfolio rules in a continuous time model, *Journal of Economic Theory,* 3*:* 373-413.

[91] Merton, R.C. (1976). Option pricing when underlying stock returns are discontinuous. *J. Finan. Econom. 3, 125-144.*

[92] Métivier, M. and Pellaumail, J. (1980). *Stochastic Integration.* Academic, New York.

[93] Meyer, M. (2001). *Continuous stochastic calculus with applications to finance.* Chapman & Hall/CRC.

[94] Meyer P.A. (1966). *Probabilités et Potentiel,* Hermann, Paris.

[95] Meyer P.A. (1976). Un Cours sur les integrales stochasticques. *Lect. Notes Math., 511,* Springer.

[96] Meyer P.A. (1989). A Short presentation of stochastic calculus. In: *Stochastic Calculus in Manifolds,* M. Emery (ed). Springer-Verlag.

[97] Modeste N'zi. (1997). Multivalued backward stochastic differential equations with local Lipschitz drift. *Stochastics and Stochastics Reports, 60: 205-218.*

[98] Musiela, M. and Rutkowski. (1997). *Martingale Methods in Financial Modelling.* Springer-Verlag.

[99] Novikov A.A. (1973). On moment inequalities and identities for stochastic integrals. *Proc. Second Japan-USSR Symp. Prob. Theor., Lect. Notes in Math., 330,* Springer-Verlag, 333-339.

[100] Øksendal, B. (2003). *Stochastic Differential Equations: An Introduction with Applications.* 6th ed. Springer-Verlag, N.Y.

[101] Pardoux E. (1979). Stochastic partial differential equations and filtering of diffusion processes. *Stochastic, 3: 127-167.*

[102] Pardoux E. and Peng S. (1990). Adapted solution of a backward stochastic differential equation, *Systems & Contr. Lett.*, 14: 55-61.

[103] Pardoux E. and Peng S. (1992). Backward stochastic differential equations and quasi-linear parabolic partial differential equations. *Lect. Notes in CIS, 176,* Springer, *200-217.*

[104] Pardoux E. and Rǎscanu A. (1988). Backward stochastic differential equations with subdifferential operator and related variational inequalities. *Stoch.Proc. Appl., 76(2): 191-215.*

[105] Pardoux E. and Veretennikov A. Yu. (1997). Averaging of backward stochastic differential equations, with application to semi-linear PDE's. *Stochastics and Stochastics Reports, 60: 255-270.*

[106] Pardoux E. and Zhang S. (1998). Generalized BSDE's and nonlinear Neumann boundary value problems. *Probab. Th. Relat. Fields,* 110: 535-558.

[107] Peng S. (1990). A general stochastic maximum principle for optimal control problems. *SIAM J. Control and Optimization, 28(4): 966-979.*

[108] Peng S. (1991). Probabilistic interpretation for systems of quasilinear parabolic partial differential equations. *Stochastics and Stochastics Reports,* 32: 61-74.

[109] Peng S. (1992). A generalized dynamic programming principle and Hamilton-Jacobi-Bellman equation. *Stochastics and Stochastics Reports,* 38: 119-134.

[110] Peng S. (1992). Stochastic Hamilton-Jacobi-Bellman Equations. *SIAM J. Control & Optim.*, 30(2): 284-304.

[111] Peng S. (1993). Backward stochastic differential equations and applications to optimal control. *Appl. Math.Optim.*, 27: 125-144.

[112] Peng S. (1994). Backward stochastic differential equations. *CIMPA School: Lect. Notes on Stochastic Calculus and Applications to Mathematical Finance*, Beijing.

[113] Peng S. (1994). Backward stochastic differential equations and exact controllability of stochastic control systems. *Progress in Natural Science 4(3): 274-284.*

[114] Peng S. (1997). BSDE and related g-expectation. *Pitman Research Notes in Mathematics Series 364: 141-159.*

[115] Peng S. (1999). Open problems on backward stochastic differential equations. In Chen S., Li X., Yong J., Zhou X.Y. eds. *Control of Distributed Parameter and Stochastic Systems.* Kluwer Acad. Pub., Boston, 265-273.

[116] Peng S. (1999). Monotonic limit theorem of BSDE and its application to Doob-Meyer decomposition theorem. *Probab. Theory Relat. Fields,* 113: 473-499.

[117] Peng S. (2000). Problem of eigenvalues of stochastic Hamiltonian systems with boundary conditions. *Stoch.Proc. Appl. 88(2): 259-290.*

[118] Peng S. and Shi Y. 2000. Infinite horizon forward-backward stochastic differential equations. *Stoch.Proc. Appl. 85(1): 75-92.*

[119] Peng S. and Wu Z. (1999). Fully coupled forward-backward stochastic differential equations and applications to optimal control. *SIAM J. Control. Optim., 37(3): 825-843.*

[120] Protter, P. (1990). *Stochastic Integral and Differential Equations - A New Approach.* Springer-Verlag.

[121] Revuz, D. and Yor, M. (1991). *Continuous Martingales and Brownian Motion.* Springer-Verlag.

[122] Mikosch, T. (1998). *Elementary stochastic calculus with finance in view.* World Scientific Publ. Singapore.

[123] Rogers, L.C. and Williams, D. (1987). *Diffusions, Markov Processes and Martingales. V.2: Ito Calculus.* John Wiley & Sons, New York.

[124] Roman, S. (2004). *Introduction to the Mathematics of Finance.* Springer-Verlag.

[125] Saisho, Y.S. (1987). Stochastic differential equation for multi-dimensional domain with reflecting boundary. *Probability Th.& Rel. Fields* 74, 455-477.

[126] Shiryayev, A. N. (1996). *Probabilty.* 2nd edition. Springer-Verlag. New York.

[127] Shiryayev, A. N. (1999). *Essentials of Stochastic Finance.* World Scientific, Singapore.

[128] Shreve, S. E. (2004). *Stochastic Calculus for Finance II: Continuopus-Time Models.* Springer-Verlag.

[129] Situ Rong. (1983). Non-convex stochastic optimal control and maximum principle. *IFAC 3rd Symposium, Control of Distributed Parameter System 1982*, Eds. J.P. Babary and L.L. Letty, Pergamon Press, 401-407.

[130] Situ Rong. (1984). An application of local time to stochastic differential equations in m-dimensional space. *ACTA Scientiarum Natur. Univ. Sunyatseni*, 3, 1-12.

[131] Situ Rong. (1984). On strong solutions and stochastic pathwise Bang-Bang control for a non-linear system. *IFAC 9th Triennial Word Congress, Hungary, 1984, Proceedings vol. 3, Large-Scale Systems., Decision Making, Mathematics of Control*, Ed. A. Titili, etc.1433-1438.(1985).

[132] Situ Rong. (1985). On strong solution, uniqueness, stability and comparison theorem for a stochastic system with Poisson jumps. In: Kappel F., Kunisch K., & Schappacher W. eds. *Distributed Parameter System. Lect. Notes in Contr. Inform. Scienc. 75,* Springer, 352-381.

[133] Situ Rong. (1985). Strong solutions and pathwise Bang-Bang control for stochastic system with Poisson jumps in 1-dimensional space. *Proceedings of 5th National Conference on Control Theory and its Applications*, 319-321.

[134] Situ Rong. (1986). Strong solutions and pathwise Bang-Bang control for multidimensional non-linear stochastic system with jumps. *4th IFAC Symposium of Control of Distributed Parameter System,* LA. Calif. USA, June, 1986.

[135] Situ Rong. (1987). On weak, strong solutions and pathwise Bang-Bang controls for non-linear degenerate stochastic systems. In: *Shinha N.K. and Telksnys L. eds. IFAC Stochastic Control, USSR, 1986. Proceedings Series (1987), No.2,* Pergamon Press, 145-150.

[136] Situ Rong. (1987). A non-linear filtering problem and its application. *Chin. Ann. Math.* 8B(3), 296-310.

[137] Situ Rong. (1987). Strong solutions, pathwise control for stochastic nonlinear system with Poisson jumps in n-dimensional space and its application to parameter estimation. *Proc. of Sino-American Statistical Meeting*, Beijing, 389-392.

[138] Situ Rong. (1987). Local time, Krylov estimation, Tanaka formula, strong comparison theorem and limit theorem for 1-dimensional semimartingales. *Acta Sci. Naturali Univ. Sunyatseni* 10, 14-26.

[139] Situ Rong. (1987). On non-linear Voterra Ito stochastic system and pathwise control. *Proc. of National Annual Conf. on Control Theory and its Applications*, China, 341-344.

[140] Situ Rong (1990). On existence and stability of solutions for reflecting stochastic differential equations with rectangular boundary. *Acta Scientiarum Naturalium Universitatis Sunyatseni* 29:3, 1-11.

[141] Situ Rong (1990). Stochastic differential equations with jump reflection on half space and pathwise optimal control. *Proc. of National Annual Conference on Control Theory and its Applications*, 679-682.

[142] Situ Rong (1990). Strong solutions and pathwise control for nonlinear stochastic system with Poisson jumps in n-dimensional space. *Chinese Ann. Math.*, Ser. B, 11:4, 513-524.

[143] Situ Rong. (Rong Situ). (1991). Theory and Application of stochastic differential equations in China. *Contemporary Math. 118*, Amer. Math. Soci., 263-280.

[144] Situ Rong (1991). Strong solutions and optimal control for stochastic differential equations in duals of nuclear spaces. *Lect. Notes in Control and Inform. Sci.* 159, Control Theory of Distributed Parameter Systems and Applications, Springer-Verlag, 144-153.

[145] Situ Rong (1991). Reflecting stochastic differential equations with jumps and stochastic population control. In *Control Theory, Stochastic Analysis and Applications*, eds. S. Chen and J. Yong, World Scientific, Singapore, 193-202.

[146] Situ Rong (1991). Stochastic population control problem on white noise measure space. *International Conf. on White Noise Analysis and Applications* , May Nagoya, Japan.

[147] Situ Rong (1993). Non-linear filtering for reflecting stochastic differential systems. *Proc. of National Annual Conf. on Control Theory and its Appl.*, Ocean Press, 210-213.

[148] Situ Rong (1994). Anticipating stochastic differential equations and optimal control. *Proc. of CSIAM Conf. on Systems and Control*, Shangdong, China, 64-69.

[149] Situ Rong (1995). On existence of strong solutions and optimal controls for anticipating stochastic differential equations with non-Lipschitzian coefficients. *Proc. of 1995 Chinese Control Conference,* Chinese Sci.& Tech. Press, 537-540.

[150] Situ Rong. (1995). On backward stochastic differential equations with jumps and with non-Lipschitzian coefficients and applications. *The 23rd Conference on Stochastic Processes and their Applications*, 19-23 June, Singapore.

[151] Situ Rong (1996). On strong solutions and pathwise optimal control for stochastic differential equations in Hilbert space. *Proc. of 1996 Chinese Control Conference*, Shangdong, 396-399.

[152] Situ Rong (1996). On comparison theorem of solutions to backward stochastic differential equations with jumps and its applications. *Proc. of the 3rd CSIAM Conf. on Systems and Control*, LiaoNing, China, 46-50.

[153] Situ Rong. (1997). On solutions of backward stochastic differential equations with jumps and applications. *Stochastic Process. Appl.*, 66: 209 -236.

[154] Situ Rong (1997). One kind of solution for optimal consumption to financial market with jumps by Lagrange multipliers in stochastic control. *Proc. of 1997 Chinese Control Conference,* Wuhan Press, 1205-1209.

[155] Situ Rong (1997). Optimization for financial market with jumps by Lagrange's method. *The 2nd International Symposium of Econometrics on Financial Topics*, Guangzhou.

[156] Situ Rong (1998). FBSDE and financial market with applications to stochastic control. *Proceedings of 1998 Chinese Control Conference*, National Defense University Press, Beijing, 301-305.

[157] Situ Rong (1998). On solutions of reflecting backward stochastic differential equations with jumps. *Conference on Mathematical Finance*, Oct. Shanghai, China.

[158] Situ Rong (1998). Comparison theorem of solutions to BSDE with jumps and viscosity solution to a generalized Hamilton-Jacobi-Bellman equation. *Conference on Control of Distributed Parameter and Stochastic Systems*, June 19-22, Hangzhou, China.

[159] SITU Rong. (R. Situ) (1999). Comparison theorem of solutions to BSDE with jumps and viscosity solution to a generalized Hamilton-Jacobi-Bellman equation. In Chen S., Li X., Yong J., Zhou X.Y. eds. *Control of Distributed Parameter and Stochastic Systems.* Kluwer Acad. Pub., Boston, 275-282.

[160] Situ Rong (1999). On solutions of FBSDE with jumps and with non-Lipschitzian coefficients. *Workshop on Infinite Dimensional Analysis and Dirichlet Forms*, March 28 - April 1, WuHan, China.

[161] Situ Rong. (1999). On comparison theorems and existence of solutions to backward stochastic differential equations with jumps and with discontinuous coefficients. *The 26th Conference on Stochastic Processes and their Applications*, 14-18, June, Beijing.

[162] Situ Rong. (1999). Optimization for financial market with jumps by Lagrange's method. *Pacific Economic Review*, 4(3): 261-275.

[163] Situ Rong. (1999). On solutions of backward stochastic differential equations with jumps and stochastic control. *Intern. Workshop on Markov Processes & Controlled Markov Chains*, Changsha, Hunan, China, 22-28 Aug.

[164] Situ Rong. (1999). Option Pricing in Mathematical Financial Market with Jumps and Related Problems. *Colloquium* held in Math. Institute of Vietnam Academy, Dec. 1999, Hanoi, Vietnam.

[165] Situ Rong. (2000). *Reflecting Stochastic Differential Equations with Jumps and Applications.* Chapman & Hall/CRC Research Notes in Mathematics 408, Chapman & Hall/CRC.

[166] Situ Rong. (2000). *Backward Stochastic Differential Equations with Jumps and Applications.* Guangadong Science and Technology Press, Guangzhou.

[167] Situ Rong. (2002). Option Pricing in Mathematical Financial Market with Jumps and Related Problems. *Vietnam Journal of Mathematics* 30(2), 103-112.

[168] Situ Rong. (2002). On solutions of backward stochastic differential equations with jumps and with non-Lipschitzian coefficients in Hilbert spaces and stochastic control. *Statistics and Probability Letters* 60(3), 279-288.

[169] Situ Rong. (2002). On solutions of backward stochastic differential equations with jumps and stochastic control. In Hou Z.T. and Jerzy Filar et al eds. *Markov Processes and Controlled Markov Chains.* 331-340, Kluwer Acad. Pub., Boston.

[170] Situ Rong (2002). On solutions of reflecting backward stochastic differential equations with jumps in finite and infinite horizon. *The 3rd Conference on "Backward Stochastic Differential Equations and Applications" - Satellite Conference of International Congress Mathematics 2002*, Aug 29 - Sept. 2, 2002, Weihai, China.

[171] Situ Rong. (2004). Hedging contingent claims in financial market with jump perturbation and Backward Stochastic Differential Equations. *Proceedings of the International Conference "Abstract and Applied Analysis"*, eds. N.M. Chuong, L. Nirenberg, and W. Tutschke, Hanoi, Vietnam, 13-17 August 2002, Word Scientific, pp 515-532.

[172] Situ Rong & W.L.Chen. (1992). Existence of solutions and optimal control for reflecting stochastic differential equations with applications to population control theory. *Stochastic Analysis and Applications* 10:1, 45-106.

[173] Situ Rong and Hwang Min. (2001). On solutions of backward stochastic differential equations with jumps in Hilbert space (I). *Acta Sci. Naturali Univ. Sunyatseni.* 40(3), 1-5.

[174] Situ Rong and Hwang Min. (2001). On solutions of backward stochastic differential equations with jumps in Hilbert space (II). *Acta Sci. Naturali Univ. Sunyatseni.* 40(4), 20-23.

[175] SITU Rong and HUANG Wei. (2004). On solutions of BSDEs with jumps and with coefficients having a quadractic growth (I). *Acta Sci. Naturali Univ. Sunyatseni.* 43(6), 48-51.

[176] SITU Rong and HUANG Wei. (2005). On solutions of BSDEs with jumps and with coefficients having a quadractic growth (II). *Acta Sci. Naturali Univ. Sunyatseni.* 44(1), 1-5.

[177] SITU Rong and WANG Yueping. (2000). On solutions of BSDE with jumps, with unbounded stopping times as terminals and with non-Lipschitzian coefficients, and probabilistic interpretation of solutions to quasi-linear elliptic integro-differential equations. *Applied Math. & Mechanics, 21(6): 659-672.*

[178] Situ Rong and Xu Huanyao. (2001). Adapted solutions of backward stochastic evolution equations with jumps on Hilbert space (I). *Acta Sci. Naturali Univ. Sunyatseni.* 40(1), 1-5.

[179] Situ Rong and Xu Huanyao. (2001). Adapted solutions of backward stochastic evolution equations with jumps on Hilbert space (II). *Acta Sci. Naturali Univ. Sunyatseni.* 40(2), 1-5.

[180] Situ Rong and Yang Yan. (2003). Upcrossing inequality and Strong $g-$supermartingales (I). *Acta Sci. Naturali Univ. Sunyatseni* 42(5), 1-5.

[181] Situ Rong and Yang Yan. (2003). Upcrossing inequality and Strong $g-$supermartingales (II). *Acta Sci. Naturali Univ. Sunyatseni.* 42(6), 1-3.

[182] Situ Rong and Zeng A.T. (1999). On solutions of BSDE with jumps and with unbounded stopping terminals I. *Acta Sci. Naturali Univ. Sunyatseni*, 38(2), 1-6.

[183] Situ Rong and Zeng, A.T. (1999). On solutions of BSDE with jumps and with unbounded stopping terminals II. *Acta Sci. Naturali Univ. Sunyatseni.* 38(4), 1-5.

[184] Skorohod, A.V. (1965). *Studies in the Theory of Random Processes.* Addison-Wesley, Reading, Massachusetts.

[185] Skorohod, A. V. (1991). *Random processes with independent increments.* Translated from the second Russian edition by P. V. Malyshev. Mathematics and its Applications (Soviet Series), 47. Kluwer Academic Publishers Group, Dordrecht, 1991.

[186] Steele, J. M. (2001). *Stochastic calculus and financial applications.* Springer-Verlag.

[187] Stroock, D.W. and Varadhan, S.R.S. (1979). *Multidimensional Diffusion Processes.* Springer-Verlag. N.Y.

[188] Tang S. and Li X. (1994). Necessary conditions for optimal control of stochastic systems with random jumps. *SIAM J. Control and Optimization,* 32(5): 1447-1475.

[189] Teo, K. L. and Wu, Z. S. (1984). *Computational methods for optimizing distributed systems.* Mathematics in Science and Engineering, 173. Academic Press, Inc., Orlando, FL, 1984.

[190] Veretennikov, A.Ju. (1981). On the strong and weak solutions of one-dimensional stochastic equations with boundary conditions. *Theor. Prob. Appl.,* 26:4, 685-700. (In Russian).

[191] Watanabe, S. and Nair, M. G. (1984). *Lectures on Stochastic differential Equations and Malliavin Calculus.* Springer-Verlag.

[192] Yeh, J. (1973). *Stochastic Processes and Wiener Integral.* Marcel Dekker Inc. N.Y.

[193] Yin Juliang and Situ Rong. (2003). On solutions of forward-backward stochastic differential equations with Poission jumps. *Stochastic Analysis and Applications* 21:6, 1419-1448.

[194] Yin Juliang and Situ Rong. (2004). Existence of solutions for forward-backward stochastic differential equations with jumps and non-Lipschitzian coefficients. *J. Math. Res. Exposition* 24:4, 577–588.

[195] Yong J. (1997). Finding adapted solution of forward-backward stochastic differential equations - method of continuation. *Prob. Th. & Relat. Fields, 107: 537-572.*

[196] Yong J. and Zhou X.Y. (1999). *Stochastic Controls. Hamiltonian Systems and HJB Equations.* Appl. of Math. 43, Springer-Verlag.

[197] Yu, J.Y., Guo, B.Z. and Zhu, G.T. (1987). The control of the semi-discrete population evolution system. *J. Syst. & Math. Scien.* 7:3, 214-219. (In Chinese).

[198] Zheng, W. (1980). A note on the convergence of sequence of conditional expectations of random variables. *Z. Wahrsch. Verw. Gebiete,* 53:3, 291-2

Index

a.s., 390
arbitrage-free market, 253

backward stochastic differential equation (BSDE), 206
    solution, 212
Black-Scholes formula
    for continuous market, 225
    for market with jumps, 233
Brownian Motion
    nowhere differentiable, 42
    stationary strong Markov property, 68
Brownian Motion (BM)
    definition, 39
    standard, 42

comparison theorem
    for solutions of BSDE, 243
    for solutions of population SDE
        in components, 366
    for solutions of RSDE
        in components, 367
    for solutions of SDE
        in 1- dimensional space, 292
        in components, 293
contingent claim, 206
    hedging, 222
convergence
    almost surely (a.s.), 409
    in probability P, 409
    weakly, 409

Doob's stopping theorem, 8, 19
Doob-Meyer decomposition
    class (D), 20
    class (DL), 20
    theorem, 22

Fatou's lemma
    for conditional expectation, 396
    for expectation, 396
Feynman-Kac formula
    by solution of BSDE, 273
    by solution of SDE, 274
field
    $\sigma-$, 389
filtering
    linear
        continuous, 194

more general continuous, 200
optimal, 193
non-linear, 184
filtration
σ−field, 4

Girsanov type theorem, 89, 96, 97
for SDE with functional coefficients, 173
Gronwall's inequality, 82, 217
stochastic, 115

Hamilton-Jacobi-Bellman equation, 278

independence
of BM and Poisson point process, 64
independent random variables, 392
independent system, 392
integration by parts, 63
Ito's formula
for continuous martingale, 57
for continuous semi-martingale, 58
for semi-martingale with jumps, 58
d-dimensional case, 62
Ito's integral
w.r.t Brownian Motion, 43

Jensen's inequality, 397

Kalman-Bucy filtering equation
in 1-dimensional case, 194
in multi-dimensional case, 196
Kolmogolov's theorem, 398
Krylov type estimate
in 1-dimensional space, 129
in d-dimensional space, 131
Kunita-Watanabe's inequality, 52

Lagrange
functional, 282
method, 279
multiplier, 280, 282

Lebesgue's dominated convergence theorem
for conditional expectation, 396
for expectation, 396
Levi's theorem, 15
local time, 124
occupation density formula, 126

martingale, 4
inequality, 8, 17
limit theorem, 11, 17
locally square integral, 35
representation theorem, 69
for SDE with functional coefficients, 177
square integrable, 35
with continuous times, 17
with discrete time, 8
maximum principle
adjoint equations, 380
for SDE, 379
measure
characteristic, 29
counting, 28
compensator (of), 33
random, 26
Poisson, 27, 28
Weiner, 41
monotone class thoerems, 407

Novikov's theorem, 174

optimal
consumption, 262, 279
control, 264
option, 207
pricing, 208
by partial differential equation approach, 227
by probability approach, 225

P-a.s., 390
Pathwise Bang-Bang control
in 1-dimensional space

non-degenerate case, 312
partial- degenerate case, 316
in d-dimensional space
non-degenerate case, 319
partial-degenerate case, 325
population SDE
optimal control, 369
solution
calculation, 372
convergence, 353
stability, 360
strong, 352
portfolio, 206
self-financed, 206
price equation
bond, 205
stock, 206
probability
measure, 389
space, 389
process
$\mathfrak{F}_t-$adapted, 4
characteristic
cross predictable, 49
predictable, 49
innovation, 181
integrable and increasing, 21
measurable, 6
progressive, 6
natural, 21
optional, 7
point, 28
$\mathfrak{F}_t-$adapted, 32
$\sigma-$finite, 32
class (QL), 32
Poisson, 28
$\mathfrak{F}_t-$Poisson, 33
stationary Poisson, 29
Poisson, 29, 398
with intensity, 29
Poisson point
stationary strong Markov property, 68
predictable, 7, 34
stationary Markov, 399
stationary strong Markov, 399
variational, 49
predictable quadratic, 49, 184
Weiner, 39

Radon-Nikodym derivative, 184, 187, 188
random process, 397
finite-dimensional distribution functions, 398
Gaussian, 397
random variable, 390
characteristic function, 391
distribution function, 391
Gaussian, 391
mathematical expectation, 390
conditional, 395
Normal, 391
Poisson, 395
variance, 390
random vector, 392
characteristic function, 392
distribution function, 393
Gaussian, 392
reflecting stochastic differential equation (RSDE), 332
solution
pathwise uniqueness, 334
strong, 334
uniqueness in law, 335
weak, 334
right continuous with left limit (RCLL), 17

semi-martingale
continuous, 58
with jumps, 58
Skorohod
metric, 401
problem, 335
theorem, 411
topology, 341
weak convergence technique, 145

result of SDE from, 145
standard measurable space, 89
stochastic differential equation (SDE), 75
linear
solution, 86
solution, 76
minimal, 306
pathwise uniqueness, 76
strong, 76
weak, 93
weak uniqueness, 94
with functional coefficient, 169
strong solution, 171
stochastic Fubini theorem, 127
stochastic integral
w.r.t. Brownian Motion, 46, 48
w.r.t. counting measure, 32, 34
w.r.t. martingale, 49
w.r.t. martingale measure, 34
stochastic population control, 330
stopping time, 5
submartingale, 4
regular, 25
right-closed, 14
uniformly integrable, 14
with continuous time, 17
with discrete time, 8
supermartingale, 4
with continuous times, 17
with discrete time, 8

Tanaka type formula
for RSDE
in components, 364
for SDE
d-dimensional case, 110
in 1-dimensional space, 116
in components, 121

uniform inner normal vector positive projection condition, 332, 335
uniformly integrable, 12

variational equation
first order, 383
second order, 383

wealth process, 206
more general, 234

Yamada-Watanabe type theorem
for RSDE, 348
for SDE, 104

Zakai's equation, 202